T0184321

Lecture Notes in Computer Science　　11646

Commenced Publication in 1973
Founding and Former Series Editors:
Gerhard Goos, Juris Hartmanis, and Jan van Leeuwen

Editorial Board Members

David Hutchison
　Lancaster University, Lancaster, UK
Takeo Kanade
　Carnegie Mellon University, Pittsburgh, PA, USA
Josef Kittler
　University of Surrey, Guildford, UK
Jon M. Kleinberg
　Cornell University, Ithaca, NY, USA
Friedemann Mattern
　ETH Zurich, Zurich, Switzerland
John C. Mitchell
　Stanford University, Stanford, CA, USA
Moni Naor
　Weizmann Institute of Science, Rehovot, Israel
C. Pandu Rangan
　Indian Institute of Technology Madras, Chennai, India
Bernhard Steffen
　TU Dortmund University, Dortmund, Germany
Demetri Terzopoulos
　University of California, Los Angeles, CA, USA
Doug Tygar
　University of California, Berkeley, CA, USA

More information about this series at http://www.springer.com/series/7407

Zachary Friggstad · Jörg-Rüdiger Sack ·
Mohammad R. Salavatipour (Eds.)

Algorithms
and Data Structures

16th International Symposium, WADS 2019
Edmonton, AB, Canada, August 5–7, 2019
Proceedings

 Springer

Editors
Zachary Friggstad
Department of Computing Science
University of Alberta
Edmonton, AB, Canada

Jörg-Rüdiger Sack
School of Computer Science
Carleton University
Manotick, ON, Canada

Mohammad R. Salavatipour
Department of Computing Science
University of Alberta
Edmonton, AB, Canada

ISSN 0302-9743   ISSN 1611-3349   (electronic)
Lecture Notes in Computer Science
ISBN 978-3-030-24765-2   ISBN 978-3-030-24766-9   (eBook)
https://doi.org/10.1007/978-3-030-24766-9

LNCS Sublibrary: SL1 – Theoretical Computer Science and General Issues

© Springer Nature Switzerland AG 2019
This work is subject to copyright. All rights are reserved by the Publisher, whether the whole or part of the material is concerned, specifically the rights of translation, reprinting, reuse of illustrations, recitation, broadcasting, reproduction on microfilms or in any other physical way, and transmission or information storage and retrieval, electronic adaptation, computer software, or by similar or dissimilar methodology now known or hereafter developed.
The use of general descriptive names, registered names, trademarks, service marks, etc. in this publication does not imply, even in the absence of a specific statement, that such names are exempt from the relevant protective laws and regulations and therefore free for general use.
The publisher, the authors and the editors are safe to assume that the advice and information in this book are believed to be true and accurate at the date of publication. Neither the publisher nor the authors or the editors give a warranty, expressed or implied, with respect to the material contained herein or for any errors or omissions that may have been made. The publisher remains neutral with regard to jurisdictional claims in published maps and institutional affiliations.

This Springer imprint is published by the registered company Springer Nature Switzerland AG
The registered company address is: Gewerbestrasse 11, 6330 Cham, Switzerland

Preface

This proceeding volume contains the papers presented at the 16th International Algorithms and Data Structures Symposium (WADS 2019), which was held during August 5–7, 2019, in Edmonton, Alberta, Canada. WADS, which alternates with the Scandinavian Symposium and Workshops on Algorithm Theory, SWAT, is a veune for researchers in the area of design and analysis of algorithms and data structures to present their works. In response to the call for papers, 89 papers were submitted to WADS this year. From these, the Program Committee selected 42 papers for presentation.

In addition, three invited lectures were given by David Eppstein (University of California, Irvine), Rasmus Pagh (IT University of Copenhagen), and Robert E. Tarjan (Princeton University). Special issues of papers selected from WADS 2019 are planned for two journals, *Algorithmica* and *Computational Geometry: Theory and Applications*.

The 2017 Alejandro López-Ortiz Best Paper Award for WADS 2017 was awarded to the paper "Universal Hinge Patterns for Folding Strips Efficiently into Any Grid Polyhedron" by Nadia Benbernou, Erik D. Demaine, Martin L. Demaine, and Anna Lubiw, and the best student paper award went to Sebastian Brandt for the paper "Approximating Small Balanced Vertex Separators in Almost Linear Time," which was coauthored with Roger Wattenhofer.

We appreciate all the work done by the Program Committee and also gratefully acknowledge the support of the WADS 2019 sponsors: University of Alberta, PIMS Institute, Elsevier, and Springer.

May 2019

Zachary Friggstad
Jörg-Rüdiger Sack
Mohammad Salavatipour

Organization

Program Committee

Jaroslaw Byrka	University of Wroclaw, Poland
Amit Chakrabarti	Dartmouth College, USA
Arnaud De Mesmay	IST Austria
Feodor Dragan	Kent State University, USA
Khaled Elbassioni	Masdar Institute of Technology, UAE
Andreas Emil Feldmann	Charles University in Prague, Czech Republic
Dimitris Fotakis	National Technical University of Athens, Greece
Zachary Friggstad	University of Alberta, Canada
Martin Gross	TU Berlin, Germany
Martin Hoefer	Goethe University Frankfurt/Main, Germany
Takehiro Ito	Tohoku University, Japan
Michael Kerber	Graz University of Technology, Austria
Christian Knauer	Universität Bayreuth, Germany
Euiwoong Lee	Carnegie Mellon University, USA
Dániel Marx	Computer and Automation Research Institute, Hungarian Academy of Sciences, Hungary
Bojan Mohar	SFU and IMFM, Canada
Wolfgang Mulzer	Freie Universität Berlin, Germany
Amir Nayyeri	Oregon State University, USA
Alantha Newman	CNRS, Laboratoire G-SCOP, France
Sang-Il Oum	Institute for Basic Science (IBS) and KAIST, South Korea
Richard Peng	Georgia Institute of Technology, USA
Marcin Pilipczuk	Institute of Informatics, University of Warsaw, Poland
Dömötör Pálvölgyi	ELTE, Hungary
Günter Rote	Freie Universität Berlin, Germany
Jorg Sack	Carleton University, Canada
Kunihiko Sadakane	The University of Tokyo, Japan
Mohammad Salavatipour	University of Alberta, Canada
Anke Vanzuylen	College of William & Mary, USA
Kasturi Varadarajan	University of Iowa, USA
Laszlo Vegh	London School of Economics and Political Science, UK
Justin Ward	Queen Mary University of London, UK

Additional Reviewers

Abam, Mohammad Ali
Aboulker, Pierre
Alman, Josh
An, Hyung-Chan
Antoniadis, Antonios
Bandyapadhyay, Sayan
Becker, Amariah
Belmonte, Rémy
Bilò, Davide
Bodlaender, Hans
Bodlaender, Hans L.
Bonnet, Edouard
Bressan, Alberto
Buchin, Kevin
Buchin, Maike
Bui, Van Vuong
Bunde, David
Böhm, Martin
Cabello, Sergio
Cao, Yixin
Catusse, Nicolas
Chakraborty, Sankar Deep
Chambers, Erin
Chang, Hsien-Chih
Charbit, Pierre
Chepoi, Victor
Chitnis, Rajesh
Chiu, Man Kwun
Choudhary, Aruni
Christman, Ananya
Chung, Christine
Cleve, Jonas
Croft, Lee
Cseh, Agnes
Damásdi, Gábor
Devroye, Luc
Driemel, Anne
Dvořák, Pavel
Escolar, Emerson G.
Fox, Kyle
Ganian, Robert
Ghosal, Pratik
Giannopoulos, Panos

Golin, Mordecai J.
Gourves, Laurent
Habib, Michel
Hanaka, Tesshu
Har-Peled, Sariel
Haverkort, Herman
He, Meng
Holm, Jacob
Hočevar, Tomaž
Husic, Edin
Iacono, John
Ibrahimpur, Sharat
Issac, Davis
Kaufmann, Michael
Keszegh, Balázs
Klemz, Boris
Klimošová, Tereza
Klost, Katharina
Kobayashi, Yasuaki
Kozma, Laszlo
Kozma, László
Kulkarni, Janardhan
Kumar, Nirman
Kwon, O-Joung
Köhler, Ekkehard
Lampis, Michael
Le, Hung
Leitert, Arne
Lenger, Dániel
Li, Shi
Lianeas, Thanasis
Lintzmayer, Carla Negri
Liotta, Giuseppe
Liu, Hsiang-Hsuan
Locatelli, Marco
Lubiw, Anna
MacGillivray, Gary
Magnard, Thomas
Mandric, Igor
McConnell, Ross
Mihalák, Matúš
Miltzow, Till
Mizuta, Haruka

Mondal, Debajyoti
Muller, Haiko
Munro, Ian
Nadara, Wojciech
Nagarajan, Viswanath
Naves, Guyslain
Nicholson, Patrick K.
Ophelders, Tim
Ponomarenko, Ilia
Quanrud, Kent
Raghvendra, Sharath
Rai, Ashutosh
Raichel, Benjamin
Raman, Venkatesh
Ray, Saurabh
Roussillon, Tristan
Sandlund, Bryce
Sau, Ignasi
Schlosser, Benjamin
Schmand, Daniel
Schröder, Marc
Schweitzer, Pascal
Sheehy, Don
Sidiropoulos, Anastasios
Sitters, Rene
Skoulakis, Stratis
Skums, Pavel

Snoeyink, Jack
Soma, Tasuku
Soto, José A.
Staals, Frank
Stehn, Fabian
Stephen, Tamon
Stout, Quentin
Tsourakakis, Charalampos
Uehara, Ryuhei
Uznański, Przemysław
Van Cleemput, Nico
van Goethem, Arthur
van Koert, Otto
van Stee, Rob
Vaxès, Yann
Verbeek, Kevin
Veselý, Pavel
Vondrak, Jan
Wagner, Hubert
Wang, Junxing
Warode, Philipp
Willert, Max
Wulff-Nilsen, Christian
Xu, Chao
Yu, Huacheng
Zamfirescu, Carol
Zhang, Jingru

Abstracts

Graphs in Nature

David Eppstein

Computer Science Department, University of California, Irvine, CA 92697, USA
eppstein@uci.edu.

Many natural processes produce planar structures that can be modeled mathematically as graphs. These include cracking of sheets of glass or mud [1], the growth of needle-like crystals [2], foams of soap bubbles [3, 4], and the folding patterns of crumpled paper [5, 6]. We survey graph-theoretic models for these phenomena, the properties of the graphs arising from them, and algorithms for recognizing these graphs and reconstructing their geometry.

References

1. Gray, N.H., Anderson, J.B., Devine, J.D., Kwasnik, J.M.: Topological properties of random crack networks. Math. Geol. **8**(6), 617–626 (1976)
2. Gilbert, E.N.: Random plane networks and needle-shaped crystals. In: Noble, B. (ed.) Applications of Undergraduate Mathematics in Engineering. Macmillan, New York (1967)
3. Eppstein, D.: A Möbius-invariant power diagram and its applications to soap bubbles and planar Lombardi drawing. Discrete Comput. Geom. **52**(3), 515–550 (2014)
4. Rivier, N.: Statistical thermodynamics of foam. In: Sadoc, J.F., Rivier, N. (eds.) NATO ASI Series, pp. 105–126, vol. 354. Springer, Dordrecht (1999). https://doi.org/10.1007/978-94-015-9157-7_7
5. Andresen, C.A., Hansen, A., Schmittbuhl, J.: Ridge network in crumpled paper. Phys. Rev. E **76**(2), August 2007
6. Eppstein, D.: Realization and connectivity of the graphs of origami flat foldings. In: Biedl, T., Kerren, A. (eds.) GD 2018. LNCS, vol. 11282, pp. 541–554. Springer, Cham (2018). https://doi.org/10.1007/978-3-030-04414-5_38

D. Eppstein—Supported in part by NSF grants CCF-1618301 and CCF-1616248.

Set Similarity – A Survey

Rasmus Pagh[1,2]

[1]IT University of Copenhagen and BARC
https://itu.dk/people/pagh/
pagh@itu.dk
[2]Google Research

Abstract. Many types of data can be represented as a sets of elements from some universe such that the size of the intersection of two sets has semantic meaning. More generally, notions of *set similarity* are important in fields such as databases, information retrieval, and machine learning. Algorithmic problems involving set similarity tend to be hard in the sense that brute-force algorithms are essentially the best known. This has given rise to the study of *approximate* versions of these problems that are more tractable. The talk will survey algorithmic techniques that have been proposed in recent years addressing: (1) the problem of estimating set similarity, and (2) the problem of searching a collection of sets for members similar to a given query set.

Towards the end of the talk we discuss approaches to understanding algorithmic limitations by showing fine-grained, conditional lower bounds. Finally, we present some open problems on closing the gap between upper and lower bounds.

Keywords: Sets · Similarity · Approximation · Search

This work has received funding from the European Research Council under the European Union's 7th Framework Programme (FP7/2007-2013)/ERC grant agreement no. 614331. It is also supported by Villum Foundation grant 16582 to Basic Algorithms Research Copenhagen (BARC). Part of this work was done as a visiting faculty researcher at Google Research.

Concurrent Connected Components Algorithms: Recent Results and Open Problems

Robert E. Tarjan[1,2]

[1]Princeton University, Princeton, NJ, USA
[2]Intertrust Technologies, Sunnyvale, CA, USA
ret@cs.princeton.edu

Abstract. The problem of finding the connected components of an undirected graph is one of the most basic in graph algorithms. It can be solved sequentially in linear time using graph search or in almost-linear time using a disjoint-set data structure. The latter solves the incremental version of the problem, in which edges are added singly or in batches on-line.

With the growth of the internet, computing connected components on huge graphs has become important, and both experimentalists and theoreticians have explored the use of concurrency in speeding up the computation. We shall survey recent work. Even simple concurrent algorithms are hard to analyze, as we discuss. This work is joint with Sixue Liu of Princeton.

Research at Princeton University partially supported by an innovation research grant from Princeton and a gift from Microsoft.

Contents

Succinct Data Structures for Families of Interval Graphs

Hüseyin Acan[1] ⓘ, Sankardeep Chakraborty[2], Seungbum Jo[3](✉) ⓘ,
and Srinivasa Rao Satti[4] ⓘ

[1] Drexel University, Philadelphia, USA
huseyin.acan@drexel.edu
[2] RIKEN Center for Advanced Intelligence Project, Chūō, Japan
sankar.chakraborty@riken.jp
[3] University of Siegen, Siegen, Germany
Seungbum.Jo@uni-siegen.de
[4] Seoul National University, Seoul, South Korea
ssrao@cse.snu.ac.kr

Abstract. We consider the problem of designing succinct data structures for *interval graphs* with n vertices while supporting degree, adjacency, neighborhood and shortest path queries in optimal time. Towards showing succinctness, we first show that at least $n \log_2 n - 2n \log_2 \log_2 n - O(n)$ bits. are necessary to represent any unlabeled interval graph G with n vertices, answering an open problem of Yang and Pippenger [Proc. Amer. Math. Soc. 2017]. This is augmented by a data structure of size $n \log_2 n + O(n)$ bits while supporting not only the above queries optimally but also capable of executing various combinatorial algorithms (like proper coloring, maximum independent set etc.) on interval graphs efficiently. Finally, we extend our ideas to other variants of interval graphs, for example, *proper/unit, k-improper interval graphs, and circular-arc graphs*, and design succinct data structures for these graph classes as well along with supporting queries on them efficiently.

Keywords: Space efficient data structures · Succinct encoding · Interval graphs

1 Introduction

A simple undirected graph G is called an *interval graph* if its vertices can be assigned to intervals on the real line so that two vertices are adjacent in G if and only if their assigned intervals intersect. The set of intervals assigned to the vertices of G is called a *realization* of G. These graphs were first introduced by Hajós [20] who also asked for the characterization of them. The same problem was also asked, independently, by [3] while studying the structure of genes. Interval graphs naturally appear in a variety of contexts, for example, operations research and scheduling theory [2], biology especially in physical mapping of DNA [28],

© Springer Nature Switzerland AG 2019
Z. Friggstad et al. (Eds.): WADS 2019, LNCS 11646, pp. 1–13, 2019.
https://doi.org/10.1007/978-3-030-24766-9_1

temporal reasoning [17] and many more. We refer the reader to [15, 16] for a thorough treatment of interval graphs and its applications. Eventually answering the question of Hajós [20], several researchers came up with different characterizations of interval graphs, including linear time algorithms for recognizing them; see, for example, [16, Chap. 8] for characterizations, and [4] and [19] for linear time algorithms. Moreover, by exploiting the special structure of interval graphs, many otherwise NP-hard problems in general graphs are also shown to have polynomial time algorithms for interval graphs [15]. These include computing maximum independent set, reporting a proper coloring, returning a maximum clique etc. In spite of having many applications in practically motivated problems, we are not aware of any study of interval graphs from the point of view of *succinct data structures* where the goal is to store a set Z of objects using the information theoretic minimum $\log(|Z|) + o(\log(|Z|))$ bits of space[1] while still being able to support the relevant set of queries efficiently, which we focus on in this paper. We also assume the usual model of computation, namely a $\Theta(\log n)$-bit word RAM model where n is the size of the input.

1.1   Related Work

There already exists a large body of work on representing various classes of graphs succinctly. This is partly motivated by theoretical curiosity and partly by the practical needs as these combinatorial structures do arise quite often in various applications. A partial list of such special graph classes would be trees [23], planar graphs [1], chordal graphs [24], partial k-tree [11] among others, while succinct encoding for arbitrary graphs is also considered [12] in the literature. For interval graphs, other than the algorithmic works mentioned earlier, there are plenty of attempts in exactly counting the number of unlabeled interval graphs [21, 22], and the state-of-the-art result is due to [27], which is what we improve in this work. For the variants of the interval graphs that we study in this paper, there exists also a fairly large number of algorithmic results on them as well as structural results. See [15, 16] for details.

1.2   Our Results and Paper Organization

Given an unlabeled interval graph G with n vertices, in Sect. 3 we first show that at least $n \log n - 2n \log \log n - O(n)$ bits are necessary to represent G, answering an open problem of Yang and Pippenger [27]. More specifically, Yang and Pippenger [27] showed a lower bound of $(n \log n)/3 + O(n)$-bit for representing any unlabeled interval graph and asked whether this lower bound can be further improved. Augmenting this lower bound, in Sect. 4 we also propose a succinct representation of G using $n \log n + O(n)$ bits while still being able to support the relevant queries optimally, where the queries are defined as follows. For any two vertices $u, v \in G$,

---

[1] Throughout the paper, we use log to denote the logarithm to the base 2.

- degree(v): returns the number of vertices that are adjacent to v in G,
- adjacent(u, v): returns true if u and v are adjacent in G, and false otherwise,
- neighborhood(v): returns all the vertices that are adjacent to v in G, and
- spath(u, v): returns the shortest path between u and v in G.

We show that all these queries can be supported optimally using our succinct data structure for interval graphs. More precisely, for any two vertices $v, u \in G$, we can answer degree(v) and adjacent(u, v) queries in $O(1)$ time, neighborhood(v) queries in $O(\text{degree}(v))$ time, and spath(u, v) queries in $O(|\text{spath}(u, v)|)$ time. Furthermore, we also show how one can implement various fundamental graph algorithms in interval graphs, for example depth-first search (DFS), breadth-first search (BFS), computing a maximum independent set, determining a maximum clique, both time and space efficiently using our succinct representation for interval graphs. In Sect. 5, we extend our ideas to other variants of interval graphs, for example, *proper/unit interval graphs, k-proper and k-improper interval graphs, and circular-arc graphs*, and design succinct data structures for these graph classes as well along with supporting queries on them efficiently. For definitions of these graphs, see Sect. 5. Finally we conclude in Sect. 6 with some remarks on possible future directions for exploring. We list all the preliminary data structures and graph theoretic terminologies that will be used throughout this paper, in Sect. 2.

2   Preliminaries

We will use the following data structures in the rest of this paper.

Rank and Select Queries: Let $S = s_1, \ldots, s_n$ be a sequence of size n over an alphabet $\Sigma = \{0, 1, \ldots, \sigma - 1\}$. Then for $1 \le i \le n$, and $\alpha \in \Sigma$, one can define rank and select queries as follows.

- $\text{rank}_\alpha(S, i)$ = the number of occurrences of α in $s_1 \ldots s_i$.
- $\text{select}_\alpha(S, i)$ = the position j where s_j is the i-th α in S.

The following lemma shows that these operations can be supported efficiently using optimal space.

Lemma 1 ([8,18]). *Given a sequence $S = s_1, \ldots, s_n$ of size n over an alphabet $\Sigma = \{0, 1, \ldots, \sigma - 1\}$ for any $\sigma > 1$, for any $\alpha \in \Sigma$, there exists an $n \log \sigma + o(n \log \sigma)$-bit data structure which answers rank_α queries on S in $O(\log(1 + \log(\sigma)))$ time and select_α queries on S in $O(1)$ time.*

Note that one can access any element of the input sequence (at a given index) in $O(1)$ (resp. $O(\log \log \sigma)$) time with the $n + o(n)$ (resp. $n \log \sigma + o(n \log \sigma)$)-bit data structure of Lemma 1.

Range Maximum Queries: Given a sequence $S = s_1, \ldots, s_n$ of size n, for $1 \le i, j \le n$, the *Range Maximum Query* on range $[i, j]$ (denoted by $\text{RMax}_S(i, j)$)

returns the position $i \leq k \leq j$ such that s_k is a maximum value in $s_i \ldots s_j$ (if there is a tie, we return the leftmost such position). One can define the *Range Minimum Queries* on range $[i, j]$ ($\mathsf{RMin}_S(i, j)$) analogously. The following lemma shows that there exist data structures which can answer these queries efficiently using optimal space.

Lemma 2 ([6,14]). *Given a sequence S of size n and for any $1 \leq c \leq n$,*

1. *there exists a data structure of size $O(n/c)$ bits, in addition to storing the sequence S, which supports RMax_S and RMin_S queries in $O(c)$ time while supporting access on S in $O(1)$ time.*
2. *there exists a data structure of size $2n + o(n)$ bits (that does not store the sequence S) which supports RMax_S or RMin_S queries in $O(1)$ time.*

Graph Terminology and Input Representation: We will assume the knowledge of basic graph theoretic terminology as given in [10] and basic graph algorithms as given in [9]. Throughout this paper, $G = (V, E)$ will denote a simple undirected graph with the vertex set V of cardinality n and the edge set E having cardinality m. We call G an *interval graph* if (a) with every vertex we can associate a closed interval on the real line, and (b) two vertices share an edge if and only if the corresponding intervals are not disjoint (see Fig. 1 for an example). It is well known that given an interval graph with n vertices, one can assign intervals to vertices such that every end point is a distinct integer from 1 to $2n$ using $O(n \log n)$ time [21], and in the rest of this paper, we deal exclusively with such representations. Moreover, for vertex $v \in V$, we refer to I_v as the interval corresponding to v.

3   Counting the Number of Unlabeled Interval Graphs

This section deals with counting unlabeled interval graphs. Let $\mathcal{I}_n$ denote the number of unlabeled graphs on n vertices. This is the sequence with id A005975 in the On–Line Encyclopedia of Integer Sequences [26]. Initial values of this sequence are given by Hanlon [21] but he did not prove an asymptotic form for enumerating the sequence. Answering a question posed by Hanlon [21], Yang and Pippenger [27] proved that the generating function $\mathcal{I}(x) = \sum_{n \geq 1} \mathcal{I}_n x^n$ diverges for any $x \neq 0$ and they established the bounds

$$\frac{n \log n}{3} + O(n) \leq \log \mathcal{I}_n \leq n \log n + O(n). \tag{1}$$

The upper bound in (1) follows from $\mathcal{I}_n \leq (2n - 1)!! = \prod_{j=1}^{n}(2j - 1)$, where the right hand side is the number of matchings on $2n$ points on a line. For the lower bound, the authors showed $\mathcal{I}_{3k} \geq k!/3^{3k}$ by finding an injection from S_k, the set of permutations of length k, to three-colored interval graphs of size $3k$. Furthermore, they left it open whether the leading terms of the lower and upper bounds in (1) can be matched, which is what show in affirmative by improving the lower bound. In other words, we find the asymptotic value of $\log \mathcal{I}_n$. In what follows, for a set S, we denote by $\binom{S}{k}$ the set of k-subsets of S.

Theorem 1. *Let $\mathcal{I}_n$ be the number of unlabeled interval graphs with n vertices. As $n \to \infty$, we have*

$$\log \mathcal{I}_n \geq n \log n - 2n \log \log n - O(n). \tag{2}$$

Proof. We consider certain interval graphs on n vertices with colored vertices. Let k be a positive integer smaller than $n/2$ such that $k^2 \geq n - 2k$, and ε a positive constant smaller than $1/2$. For $1 \leq j \leq k$, let B_j and R_j denote the intervals $[-j-\varepsilon, -j+\varepsilon]$ and $[j-\varepsilon, j+\varepsilon]$, respectively. These $2k$ pairwise-disjoint intervals will make up $2k$ vertices in the graphs we consider. Now let $\mathcal{W}$ denote the set of k^2 closed intervals with one endpoint in $\{-k, \ldots, -1\}$ and the other in $\{1, \ldots, k\}$. We color $B_1, \ldots, B_k$ with blue, $R_1, \ldots, R_k$ with red, and the k^2 intervals in $\mathcal{W}$ with white.

Together with $\mathcal{S} := \{B_1, \ldots, B_k, R_1, \ldots, R_k\}$, each $\{J_1, \ldots, J_{n-2k}\} \in \binom{\mathcal{W}}{n-2k}$ gives an n-vertex, three-colored interval graph. For a given $\mathcal{J} = \{J_1, \ldots, J_{n-2k}\}$, let $G_{\mathcal{J}}$ denote the colored interval graph whose vertices correspond to n intervals in $\mathcal{S} \cup \mathcal{J}$, and let $\mathcal{G}$ denote the set of all $G_{\mathcal{J}}$.

Now let $G \in \mathcal{G}$. For a white vertex $w \in G$, the pair $(d_B(w), d_R(w))$, which represents the numbers of blue and red neighbors of w, uniquely determine the interval corresponding to w; this is the interval $[-d_B(w), d_R(w)]$. In other words, $\mathcal{J}$ can be recovered from $G_{\mathcal{J}}$ uniquely. Thus $|\mathcal{G}| = \binom{k^2}{n-2k}$. Since there are at most 3^n ways to color the vertices of an interval graph with blue, red, and white, we have

$$\mathcal{I}_n \cdot 3^n \geq |\mathcal{G}| = \binom{k^2}{n-2k} \geq \left(\frac{k^2}{n-2k}\right)^{n-2k} \geq \left(\frac{k^2}{n}\right)^{n-2k}$$

for any $k < n/2$. Setting $k = \lfloor n/\log n \rfloor$ and taking the logarithms, we get

$$\log \mathcal{I}_n \geq (n - 2k) \log(k^2/n) - O(n) = n \log n - 2n \log \log n - O(n).$$

Remark 1. Yang and Pippenger [27] also posed the question whether $\log \mathcal{I}_n = Cn \log n + O(n)$ for some C or not. According to Theorem 1, this boils down to getting rid of the $2n \log \log n$ term in (2). Such a result would imply that the exponential generating function $J(x) = \sum_{n \geq 1} \mathcal{I}_n x^n / n!$ has a finite radius of convergence. (As noted in [27], the bound $\mathcal{I}_n \leq (2n-1)!!$ implies that the radius of convergence of $J(x)$ is at least $1/2$.)

4   Succinct Representation of Interval Graphs

In this section, we introduce a succinct $n \log n + (3 + \epsilon)n + o(n)$-bit representation of unlabeled interval graph G on n vertices with constant $\epsilon > 0$, and show that the navigational queries (**degree**, **adjacent**, **neighborhood**, and **spath** queries) and some basic graph algorithms (BFS, DFS, PEO traversals, proper coloring, computing the size of a maximum clique and maximum independent set) on G can be answered/executed efficiently using our representation of G.

4.1   Succinct Representation of G

We first label the vertices of G using the integers from 1 to n, as described in the following. By the assumption in Sect. 2, the vertices in G can be represented by n intervals $I = \{I_1 = [l_1, r_1], I_2 = [l_2, r_2], \ldots, I_n = [l_n, r_n]\}$ where all the endpoints in I are distinct integers in the range $[1, 2n]$. Since there are $2n$ distinct endpoints for the n intervals in I, every integer in $[1, 2n]$ corresponds to a unique l_i or r_i for some $1 \leq i \leq n$. We assign the labels to the vertices in G based on the sorted order of left endpoints of their corresponding intervals, i.e., for any two vertices $a, b \in G$, $a < b$ if and only if $l_a < l_b$.

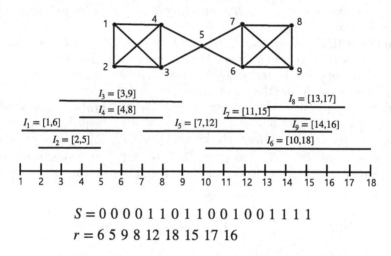

$$S = 0\,0\,0\,0\,1\,1\,0\,1\,1\,0\,0\,1\,0\,0\,1\,1\,1\,1$$
$$r = 6\ 5\ 9\ 8\ 12\ 18\ 15\ 17\ 16$$

Fig. 1. Example of the interval graph and its representation.

Now we describe the representation of G. Let $S = s_1 \ldots s_{2n}$ be the binary sequence of length $2n$ such that for $1 \leq i \leq 2n$, $s_i = 0$ if $i \in \{l_1, l_2, \ldots, l_n\}$ (i.e., if i corresponds to the left end point of an interval in I), and $s_i = 1$ otherwise. If $i = l_k$ or $i = r_k$, we say that s_i corresponds to the interval I_k. We represent the sequence S using the data structure of Lemma 1, using a $2n + o(n)$ bits to support rank and select queries on S in $O(1)$ time. Next, we store the sequence $r = r_1 \ldots r_n$ using $n \log 2n = n \log n + n$ bits, and for some fixed constant $\epsilon > 0$, we also store an ϵn-bit data structure of Lemma 2(1) (with $c = 1/\epsilon$) to support RMax and RMin queries on r in $O(1)$ time. Using the representations of S and r, it is easy to show that for any vertex $v \in G$, we can return its corresponding interval $I_v = [l_v, r_v]$ in $O(1)$ time by computing $l_v = \text{select}_0(S, v)$, and r_v can be accessed from the sequence r. Thus, the total space usage of our representation is $n \log n + (3 + \epsilon)n + o(n)$ bits. See Fig. 1 for an example.

4.2   Supporting Navigational Queries

In this section, we show that degree, adjacent, neighborhood, and spath queries on G can be answered in asymptotically optimal time using the representation described in the Sect. 4.1.

degree(v) Query: We count the number of vertices in G which are not adjacent to v, which is a disjoint union of the two sets: (i) the set of intervals that end before the starting point l_v, and (ii) the set of intervals that start after the end point r_v. Using our representation the cardinalities of these two sets can be computed as follows. The number of intervals u with $r_u < l_v$ is given by $\mathsf{rank}_1(S, l_v)$. Similarly, the number of intervals u with $r_v < l_u$ is given by $n - \mathsf{rank}_0(S, r_v)$. Therefore, we can answer degree(v) query in $O(1)$ time by returning $n - \mathsf{rank}_1(S, l_v) - (n - \mathsf{rank}_0(S, r_v)) = \mathsf{rank}_0(S, r_v) - \mathsf{rank}_1(S, l_v)$.

adjacent(u, v) Query: Since we can compute the intervals I_u and I_v in $O(1)$ time, adjacent(u, v) query can be answered in $O(1)$ by checking $r_u < l_v$ or $r_v < l_u$ (u and v are not adjacent if and only if one of these conditions is satisfied).

neighborhood(v) Query: The set of all neighbors of a vertex v can be reported by considering all the intervals I_u whose left end points are within the range $[1, \ldots, r_v]$ and returning all such u's with $r_u > l_v$ (i.e., which start to the left of r_v and end after l_v). With our data structure, this query can be supported by returning the set $\{u \mid 1 \leq u \leq rank_0(S, r_v) \text{ and } r_u > l_v\}$. Using the RMax structure stored on r, this can be supported in $O(\mathsf{degree}(v))$ time. Note that given a threshold value l_v and a query range $[a, b]$ of the sequence r, the range max data structure can be used to report all the elements r_u within the range $[a, b]$ such that $r_u > t$, in $O(1)$ time per element, using the following recursive procedure. Compute the position $c = \mathsf{RMax}_r(a, b)$. If $r_c > l_v$, then return r_c, and recurse on the subintervals $[a, c - 1]$ and $[c + 1, b]$; else stop.

spath(u, v) Query: We first define the SUCC query as described in [7]. For an interval I_u, SUCC(I_u) returns the interval $I_{u'}$ such that $I_u \cap I_{u'} \neq \emptyset$ and there is no $I_{u''}$ with $I_u \cap I_{u''} \neq \emptyset$ and $r_{u'} < r_{u''}$. (For example in Fig. 1, SUCC(I_2) = I_3 and SUCC(I_5) = I_6). To answer the spath(u, v) query, let P_{uv} be the shortest path from u to v initialized with $\emptyset$ (without loss of generality, we assume that $u \leq v$). If u and v are identical, we simply add u to P_{uv} and return P_{uv}. If not, we first add u to P_{uv} and consider two cases as follows [7].

- If u is adjacent to v, add v to P_{uv} and return P_{uv}.
- If I_u is not adjacent to I_v, we perform spath(SUCC(u), v) query recursively.

Since we can answer adjacent queries in $O(1)$ time, it is enough to show how to answer the SUCC queries in $O(1)$ time. Let k be the number of vertices v which satisfies $l_v < r_u$, which can be answered in $O(1)$ time by $k = rank_0(S, r_u)$. Then by the definition of SUCC query, I_i with $i = \mathsf{RMax}_r(1, k)$ gives an answer of SUCC(I_u) if $r_i > l_u$ (if not, there is no vertex in G adjacent to u). Therefore we can answer the SUCC query in $O(1)$ time, which implies spath(u, v) query can be answered in $O(|spath(u, v)|)$ time. Thus, we obtain a following theorem.

Theorem 2. *Given an interval graph G with n vertices, there exists an $n \log n + (3 + \epsilon)n + o(n)$-bit representation of G which answers degree(v) and adjacent(u, v) queries in $O(1)$ time, neighborhood(v) queries in $O(\text{degree}(v))$ time, and spath(u, v) queries in $O(|\text{spath}(u, v)|)$ time, for any vertices $u, v \in G$.*

In the extended version, we discuss how to support some basic graph algorithms (BFS, DFS, PEO traversals, proper coloring, computing the size of maximum clique, maximum independent set and minimum vertex cover) efficiently on G with the above set of operations along with the representation of Sect. 4.1.

5   Representation of Some Related Families of Interval Graphs

In this section, we propose space-efficient representations for proper interval graphs, k-proper and k-improper interval graphs, and circular arc graphs. Since these graphs are restrictions or extensions (i.e., sub/super-classes) of interval graphs, we can represent them by modifying the representation in Sect. 4.1 (to make the representation asymptotically optimal in terms of space). We also show that navigation queries on these graph classes can be answered efficiently with the modified representation.

5.1   Proper Interval Graphs

An interval graph G is *proper* if there exists an interval representation of G such that for any two vertices $u, v \in G$, $I_u \not\subset I_v$ and $I_v \not\subset I_u$ (let such interval representation of G be *proper representation* of G). Also it is known that proper interval graphs are equivalent to the *unit interval graphs*, which have an interval representation such that every interval has the same length [25].

Now we consider how to represent a proper interval graph G with n vertices while supporting navigational queries efficiently on G. We first obtain an interval representation of the graph G where the intervals satisfy the property of proper interval graph. We then assign labels to vertices of G based on the sorted order left end points of their corresponding intervals, as described in Sect. 4.1. Let S be the bit sequence obtained from this representation, as defined in Sect. 4.1. Then by the definition of G, there are no two vertices $u, v \in G$ with $l_u < l_v$ and $r_u > r_v$ (if so, $I_v \subset I_u$). Thus by the Lemma 1, for any vertex $i \in G$ we can compute l_i and r_i in $O(1)$ time by $\text{select}_0(S, i)$ and $\text{select}_1(S, i)$ respectively using $2n + o(n)$ bits. Also note that r is strictly increasing sequence when G is a proper interval graph, and hence one can support the RMax queries on $r = r_1 \ldots r_n$ in $O(1)$ time without maintaining any data structure, by simply returning the rightmost position of the query range. Thus, we obtain the following theorem.

Theorem 3. *Given a proper interval graph or unit interval graph G with n vertices, there exists a $2n + o(n)$-bit representation of G which answers degree(v) and adjacent(u, v) queries in $O(1)$ time, neighborhood(v) queries in $O(\text{degree}(v))$ time, and spath(u, v) queries in $O(|\text{spath}(u, v)|)$ time, for any vertices $u, v \in G$.*

It is known that there are asymptotically $\frac{1}{8\kappa\sqrt{\pi}}n^{-3/2}4^n$ non-isomorphic unlabeled unit interval graphs with n vertices, for some constant $\kappa > 0$ [13], and hence $2n - O(\log n)$ bits is an information-theoretic lower bound on representing an arbitrary proper interval graph. Thus our representation in Theorem 3 gives a succinct representation for proper interval graphs.

5.2 k-proper and k-improper Interval Graphs

One can generalize the proper interval graph to the following two sub-classes of interval graphs. An interval graph G with n vertices, G is a *k-proper interval graph* (resp. *k-improper interval graph*) if there exists an interval representation of G such that for any vertex $v \in G$, I_v is contained by (resp., contains) at most $k \leq n$ intervals in G other than I_v. We call such an interval representation of G as the *k-proper representation* (resp. *k-improper representation*) of G. Note that every proper interval graph is both a 0-proper and a 0-improper graph. The graph in Fig. 1 is a 2-proper, and a 3-improper graph. Now we consider how to represent a k-proper interval graph G with n vertices and support navigation queries efficiently on G. We first represent G k-properly into n intervals, and assign the labels to vertices of G based on the sorted order of their left end points, as described in Sect. 4.1. Same as the representation in Sect. 4.1, we first maintain the data structure for supporting rank and select queries on S in $O(1)$ time, using $2n + o(n)$ bits in total. Also we maintain the $2n + o(n)$-bit data structure of Lemma 2 on $r = r_1, \ldots, r_n$ for supporting RMax queries on r in $O(1)$ time. Next, to access r without using $n \log n$ bits, we define the sequence $T = t_1 \ldots t_{2n}$ of size $2n$ over the alphabet $\{0, \ldots, 2k+1\}$ such that $t_i = 2k'$ (resp. $t_i = 2k'+1$) if $s_i = 0$ (resp. $s_i = 1$) and its corresponding interval is contained by $k' \leq k$ intervals in $I = \{I_1 \ldots I_n\}$. Now for any $0 \leq i \leq k$, let $R_i \subset I$ be the set of all intervals such that for any $[a, b] \in R_i$, $t_a = 2i$ and $t_b = 2i + 1$. It is easy to show that each R_i corresponds to a proper interval graph. For example the graph in Fig. 1 is a 2-proper interval graph, and $T = 0\ 2\ 0\ 2\ 3\ 1\ 0\ 3\ 1\ 0\ 2\ 1\ 2\ 4\ 3\ 5\ 3\ 1$, $R_0 = \{I_1, I_3, I_5, I_6\}$, $R_1 = \{I_2, I_4, I_7, I_8\}$, and $R_2 = \{I_9\}$. By Lemma 1, we can maintain T using $2n \log(2k + 2) + o(n \log k) = 2n \log k + 2n + o(n \log k)$ bits with supporting rank and select queries in $O(\log \log k)$ and $O(1)$ time respectively. Then for any vertex $v \in G$, we can answer its corresponding interval $I_v = [l_v, r_v]$ in $O(\log \log k)$ time by $l_v = \mathsf{select}_0(S, v)$ and $r_v = \mathsf{select}_{(t_{l_v}+1)}(T, \mathsf{rank}_{t_{l_v}}(T, l_v))$. Thus, we obtain the following theorem.

Theorem 4. *Given a k-proper interval graph G with n vertices, there exists a $(2n \log k + 6n + o(n \log k))$-bit representation of G which answers degree(v) and adjacent(u, v) queries in $O(\log \log k)$ time, neighborhood(v) queries in $O(\log \log k \cdot degree(v))$ time, and spath(u, v) queries in $O(\log \log k \cdot |spath(u, v)|)$ time, for any vertices $u, v \in G$.*

Note that we can represent k-improper interval graphs in same space with same query time as in Theorem 4 by changing the definition of T to be $t_i = 2k'$ (resp. $t_i = 2k'+1$) if $s_i = 0$ (resp. $s_i = 1$) and its corresponding interval contains $k' \leq k$ intervals in $\{I_1 \ldots I_n\}$.

5.3   Circular-Arc Graphs

In this section, we propose a succinct representation for circular-arc graphs, and show how to support navigation queries efficiently on the representation. A *circular-arc graph* G is a graph whose vertices can be assigned to arcs on a circle so that two vertices are adjacent in G if and only if their assigned arcs intersect. It is easy to see that every interval graph is a circular-arc graph. Thus, by the Lemma 3, we need at least $n \log n - 2n \log \log n - O(n)$ bits to represent an arbitrary circular-arc graph G.

Suppose that G is represented by the circle C together with n arcs of C. For an arc, we define its start point to be the unique point on it such that the arc continues from that point in the clock-wise direction but stops in the anti-clockwise direction; and similarly define its end point to be the unique point on it such that the arc stops in the clockwise direction but continues in the anti-clockwise direction. As in the case of interval graphs, we assume, without loss of generality, that all the start and end points of all the arcs are distinct. We label the vertices of G with the integers form 1 to n as described below. We first select an arbitrary arc, and label the vertex (and the arc) corresponding to this arc by 1. We then traverse the circle from the starting point of that arc in the clockwise direction, and label the remaining vertices and arcs in the order in which their starting points are encountered during the traversal, and finish the traversal when we return to the starting point of the first arc. We also map all the start and end points of all arcs, in the order in which they are encountered in the above traversal, into the range $[1, \ldots, 2n]$ (since the start and end points of all the n arcs are distinct). With the above defined labeling of the arcs, and the numbering of their start and end points, let l_i and r_i start and end points of the arc labeled i, for $1 \le i \le n$. Now the arcs can be thought of as two types of intervals in the range $[1, \ldots, 2n]$; we call an interval i as *normal* if $l_i < r_i$ (i.e., we traverse l_i prior to r_i), and *reversed* otherwise. A normal interval i corresponds to the interval $[l_i, r_i]$, while a reversed interval i actually corresponds to the union of the two intervals $[1, \ldots, r_i]$ and $[l_i, \ldots, 2n]$. See Fig. 2 for an example; intervals numbered 4 and 7 are reversed, while the others are normal. Our representation of G consists of the following substructures.

1. Define a binary sequence $S = s_1, \ldots, s_{2n}$ of length $2n$ such that for $1 \le i \le 2n$, $s_i = 0$ (resp. $s_i = 1$) if i-th end point encountered during the traversal of C is in $\{l_1, \ldots, l_n\}$ (resp. $\{r_1, \ldots, r_n\}$). Now, construct a sequence $S' = s'_1, \ldots, s'_{2n}$ of size $2n$ over an alphabet $\{0, 1, 2, 3\}$ such that for all $1 \le i \le 2n$, $s'_i = s_i + 2$ if the position s_i corresponds to the start or end point of a reversed interval, and $s'_i = s_i$ otherwise (i.e., if s_i corresponds to a normal interval). We represent S' using the structure of Lemma 1, using $4n + o(n)$ bits, so that we can answer rank and select queries on S' in $O(1)$ time. In addition, we also store auxiliary structures (of $o(n)$ bits) on top of S' to support rank and select queries on S (without explicitly storing S – note that one can efficiently reconstruct any subsequence of S from S').

2. To store the interval end points efficiently, we introduce two 2-dimensional grids of points, R_1 and R_2, defined as follows. Suppose there are $q \le n$ vertices

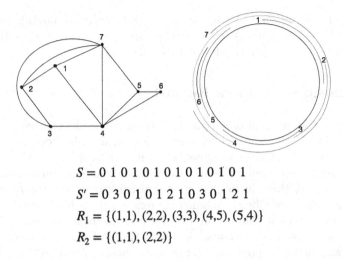

$$S = 0\ 1\ 0\ 1\ 0\ 1\ 0\ 1\ 0\ 1\ 0\ 1\ 0\ 1$$

$$S' = 0\ 3\ 0\ 1\ 0\ 1\ 2\ 1\ 0\ 3\ 0\ 1\ 2\ 1$$

$$R_1 = \{(1,1), (2,2), (3,3), (4,5), (5,4)\}$$

$$R_2 = \{(1,1), (2,2)\}$$

Fig. 2. Example of the circular graph and its representation. (Color figure online)

in G which correspond to normal intervals (and $n - q$ vertices correspond to reversed intervals). Then let R_1 be a set of q points on the 2-dimensional grid $[1, q] \times [1, q]$ which consist of $(\mathsf{rank}_0(S', l_i), \mathsf{rank}_1(S', r_i))$, for all $1 \leq i \leq n$ with $l_i < r_i$. Similarly let R_2 be a set of $n - q$ points on the 2-dimensional grid $[1, n - q] \times [1, n - q]$ which consist of $(\mathsf{rank}_2(S', l_i), \mathsf{rank}_3(S', r_i))$, for all $1 \leq i \leq n$ with $r_i < l_i$. Given a set of points R on 2-dimensional grid, we define the following queries (for any rectangular range A):

- $Y(R, x)$: returns y with $(x, y) \in R$.
- $count(R, A)$: returns the number of points in R within the range A.

We represent R_1 and R_2 using $n \log n + o(n \log n)$ bits in total, such that Y and $count$ queries can be supported in $O(\log n / \log \log n)$ time [5].

Using these data structures, when the vertex $1 \leq i \leq n$ is given, we can answer l_i and r_i in $O(\log n / \log \log n)$ time by $l_i = \mathsf{select}_0(S, i)$, and $r_i = \mathsf{select}_1(S', Y(R_1, \mathsf{rank}_0(S', l_i)))$ if $S'_{l_i} = 0$ (i.e., if l_i is the left end point of a normal interval), and $r'_i = \mathsf{select}_3(Y(R_2, \mathsf{rank}_2(S', l'_i)))$ otherwise (i.e., if if l_i is the left end point of a reversed interval). Finally, let $r' = r'_1, \ldots, r'_q$ be a sequence such that for $1 \leq i \leq q$, $r'_i = r_{j_i}$ with $j_i = \mathsf{select}_0(S', i)$. Similarly, let $r'' = r''_1, \ldots, r''_{n-q}$ be a sequence such that for $1 \leq i \leq n - q$, $r''_i = r_{j_i}$ with $j_i = \mathsf{select}_2(S', i)$. Then we maintain the data structure of Lemma 2 on r' and r'', using a total of $2n + o(n)$ bits, to support RMax queries on each of them. Thus, the overall representation takes $n \log n + o(n \log n)$ bits in total. One can show that this representation supports the degree, adjacent, neighborhood and spath queries efficiently, to prove the following theorem (proof omitted).

Theorem 5. *Given a circular arc graph G with n vertices, there exists a $(n \log n + o(n \log n))$-bit representation of G which answers degree(v) and* *adjacent*(u,v) *queries in* $O(\log n/\log\log n)$ *time,* **neighborhood**(v) *queries in* $O(\log n/\log\log n \cdot$ **degree**$(v))$ *time, and* **spath**(u,v) *queries in* $O(|spath(u,$ $v)|\log n/\log\log n)$ *time for any two vertices* $u,v \in G$.

6   Conclusion and Final Remarks

We considered the problem of succinctly encoding an unlabeled interval graph with n vertices so as to support adjacency, degree, neighborhood and shortest path queries. To this end, we designed a succinct data structure that can support these queries optimally. We also showed how one can implement various combinatorial algorithms in interval graphs using our succinct data structure in both time and space efficient manner. Extending these ideas, finally, we also showed succinct/compact data structures for multiple other variants of interval graphs. For some of these variants, the query times of our data structures are super constant, hence non-optimal and we leave them as open problems whether we can design data structures for supporting these queries in constant time.

References

1. Aleardi, L.C., Devillers, O., Schaeffer, G.: Succinct representations of planar maps. Theor. Comput. Sci. **408**(2–3), 174–187 (2008)
2. Bar-Noy, A., Bar-Yehuda, R., Freund, A., Naor, J., Schieber, B.: A unified approach to approximating resource allocation and scheduling. J. ACM **48**(5), 1069–1090 (2001)
3. Benser, S.: On the topology of the genetic fine structure. Proc. Nat. Acad. Sci. **45**, 1607–1620 (1959)
4. Booth, K.S., Lueker, G.S.: Testing for the consecutive ones property, interval graphs, and graph planarity using pq-tree algorithms. J. Comput. Syst. Sci. **13**(3), 335–379 (1976)
5. Bose, P., He, M., Maheshwari, A., Morin, P.: Succinct orthogonal range search structures on a grid with applications to text indexing. In: Dehne, F., Gavrilova, M., Sack, J.-R., Tóth, C.D. (eds.) WADS 2009. LNCS, vol. 5664, pp. 98–109. Springer, Heidelberg (2009). https://doi.org/10.1007/978-3-642-03367-4_9
6. Brodal, G.S., Davoodi, P., Rao, S.S.: On space efficient two dimensional range minimum data structures. Algorithmica **63**(4), 815–830 (2012)
7. Chen, D.Z., Lee, D.T., Sridhar, R., Sekharan, C.N.: Solving the all-pair shortest path query problem on interval and circular-arc graphs. Networks **31**(4), 249–258 (1998)
8. Clark, D.R., Munro, J.I.: Efficient suffix trees on secondary storage. In: Proceedings of the Seventh Annual ACM-SIAM Symposium on Discrete Algorithms, pp. 383–391 (1996)
9. Cormen, T.H., Leiserson, C.E., Rivest, R.L., Stein, C.: Introduction to Algorithms, 3rd edn. MIT Press, Cambridge (2009)
10. Diestel, R.: Graph Theory. Graduate Texts in Mathematics, vol. 173, 4th edn. Springer, Heidelberg (2012)
11. Farzan, A., Kamali, S.: Compact navigation and distance oracles for graphs with small treewidth. Algorithmica **69**(1), 92–116 (2014)

12. Farzan, A., Munro, J.I.: Succinct encoding of arbitrary graphs. Theor. Comput. Sci. **513**, 38–52 (2013)
13. Finch, S.R.: Mathematical Constants. Cambridge University Press, Cambridge (2003)
14. Fischer, J., Heun, V.: Space-efficient preprocessing schemes for range minimum queries on static arrays. SIAM J. Comput. **40**(2), 465–492 (2011)
15. Golumbic, M.C.: Interval graphs and related topics. Discrete Math. **55**(2), 113–121 (1985)
16. Golumbic, M.C.: Algorithmic Graph Theory and Perfect Graphs (2004)
17. Golumbic, M.C., Shamir, R.: Complexity and algorithms for reasoning about time: a graph-theoretic approach. J. ACM **40**(5), 1108–1133 (1993)
18. Golynski, A., Munro, J.I., Rao, S.S.: Rank/select operations on large alphabets: a tool for text indexing. In: Proceedings of the Seventeenth Annual ACM-SIAM Symposium on Discrete Algorithms, SODA 2006, pp. 368–373 (2006)
19. Habib, M., McConnell, R.M., Paul, C., Viennot, L.: Lex-BFS and partition refinement, with applications to transitive orientation, interval graph recognition and consecutive ones testing. Theor. Comput. Sci. **234**(1–2), 59–84 (2000)
20. Hajós, G.: Über eine art von graphen. Int. Math. Nachr. **11**, 1607–1620
21. Hanlon, P.: Counting interval graphs. Trans. Am. Math. Soc. **272**(2), 383–426 (1982)
22. Klavzar, S., Petkovsek, M.: Intersection graphs of halflines and halfplanes. Discrete Math. **66**(1–2), 133–137 (1987)
23. Munro, J.I., Raman, V.: Succinct representation of balanced parentheses and static trees. SIAM J. Comput. **31**(3), 762–776 (2001)
24. Munro, J.I., Wu, K.: Succinct data structures for chordal graphs. In: 29th International Symposium on Algorithms and Computation, ISAAC 2018, pp. 67:1–67:12 (2018). https://doi.org/10.4230/LIPIcs.ISAAC.2018.67
25. Roberts, F.S.: Indifference graphs. In: Harary, F. (ed.) Proof Techniques in Graph Theory (1969)
26. Sloane, N.J.E.: The on-line encyclopedia of integer sequences. http://oeis.org
27. Yang, J.C., Pippenger, N.: On the enumeration of interval graphs. Proc. Am. Math. Soc. Ser. B **4**(1), 1–3 (2017)
28. Zhang, P., et al.: An algorithm based on graph theory for the assembly of contigs in physical mapping of DNA. Bioinformatics **10**(3), 309–317 (1994)

On Polynomial-Time Combinatorial Algorithms for Maximum L-Bounded Flow

Kateřina Altmanová, Petr Kolman$^{(\boxtimes)}$, and Jan Voborník

Department of Applied Mathematics, Faculty of Mathematics and Physics,
Charles University, Prague, Czech Republic
kolman@kam.mff.cuni.cz

Abstract. Given a graph $G = (V, E)$ with two distinguished vertices $s, t \in V$ and an integer L, an *L-bounded flow* is a flow between s and t that can be decomposed into paths of length at most L. In the *maximum L-bounded flow problem* the task is to find a maximum L-bounded flow between a given pair of vertices in the input graph.

The problem can be solved in polynomial time using linear programming. However, as far as we know, no polynomial-time combinatorial algorithm for the L-bounded flow is known. The only attempt, that we are aware of, to describe a combinatorial algorithm for the maximum L-bounded flow problem was done by Koubek and Říha in 1981. Unfortunately, their paper contains substantional flaws and the algorithm does not work; in the first part of this paper, we describe these problems.

In the second part of this paper we describe a combinatorial algorithm based on the exponential length method that finds a $(1+\varepsilon)$-approximation of the maximum L-bounded flow in time $\mathcal{O}(\varepsilon^{-2} m^2 L \log L)$ where m is the number of edges in the graph. Moreover, we show that this approach works even for the NP-hard generalization of the maximum L-bounded flow problem in which each edge has a length.

1 Introduction

Given a graph $G = (V, E)$ with two distinguished vertices $s, t \in V$ and an integer L, an *L-bounded flow* is a flow between s and t that can be decomposed into paths of length at most L. In the *maximum L-bounded flow problem* the task is to find a maximum L-bounded flow between a given pair of vertices in the input graph. The L-bounded flow was first studied, as far as we know, in 1971 by Adámek and Koubek [1]. In connection with telecommunication networks, L-bounded flows in networks with unit edge lengths have been widely studied and are known as *hop-constrained* flows [7].

For networks with unit edge lengths (or, more generally, with polynomially bounded edge lengths, with respect to the number of vertices), the problem can be solved in polynomial time using linear programming. Linear programming is

This research was partially supported by project GA17-09142S of GA ČR.

© Springer Nature Switzerland AG 2019
Z. Friggstad et al. (Eds.): WADS 2019, LNCS 11646, pp. 14–27, 2019.
https://doi.org/10.1007/978-3-030-24766-9_2

a very general tool that does not make use of special properties of the problem at hand. This often leaves space for superior combinatorial algorithms that do exploit the structure of the problem. For example, maximum flow, matching, minimum spanning tree or shortest path problems can all be described as linear programs but there are many algorithms that outperform general linear programming approaches. However, as far as we know, no polynomial-time combinatorial algorithm[1] for the L-bounded flow is known.

1.1   Related Results

For clarity we review the definitions of a few more terms that are used in this paper. A *network* is a quintuple $G = (X, R, c, s, t)$, where $G = (X, R)$ is a directed graph, X denotes the set of vertices, R the set of edges, c is the edge capacity function $c : R \to \mathbb{R}^+$, s and t are two distinguished vertices called the source and the sink. We use m and n to denote the number of edges and the number of vertices, respectively, in the network G, that is, $m = |R|$ and $n = |X|$. Given an L-bounded flow f, we denote by $|f|$ the size of the flow, and for an edge $e \in R$, we denote by $f(e)$ the total amount of flow f through the edge e.

An *L-bounded flow problem with edge lengths* is a generalization of the L-bounded flow problem: each edge has also an integer length and the length of a path is computed not with respect to the number of edges on it but with respect the sum of lengths of edges on it.

Given a network G and an integer parameter L, an *L-bounded cut* is a subset C of edges R in G such that there is no path from s to t of length at most L in the network $G = (X, R \setminus C, c, s, t)$. The objective is to find an L-bounded cut of minimum size. We sometimes abbreviate the phrase L-bounded cut to L-cut and, similarly, we abbreviate the phrase L-bounded flow to L-flow.

Although the problems of finding an L-flow and an L-cut are easy to define and they have been studied since the 1970's, still some fundamental open problems remain unsolved. Here we briefly survey the main known results.

L-Bounded Flows. As far as we know, the L-bounded flow was first considered in 1971 by Adámek and Koubek [1]. They published a paper introducing the L-bounded flows and cuts and describing some interesting properties of them. Among other results, they show that, in contrast to the ordinary flows and cuts, the duality between the maximum L-flow and the minimum L-cut does not hold.

The maximum L-flow can be computed in polynomial time using linear programming [4,17,21]. The only attempt, that we are aware of, to describe a combinatorial algorithm for the maximum L-bounded flow problem was done by Koubek and Říha in 1981 [18]. The authors say the algorithm finds a maximum L-flow in time $O(m \cdot |I|^2 \cdot S/\psi(G))$, where I denotes the set of paths in the constructed L-flow, S is the size of the maximum L-flow, and $\psi(G) = \min(|c(e) - c(g)| : c(e) \neq c(g), e, g \in R \cup \{e'\})$, where $c(e') = 0$. Unfortunately, their paper contains substantional flaws and the algorithm does not work as we show in the first part of

---

[1] Combinatorial in the sense that it does not explicitly use linear programming methods or methods from linear algebra or convex geometry.

this paper. Thus, it is a challenging problem to find a polynomial time combinatorial algorithm for the maximum L-bounded flow.

Surprisingly, the maximum L-bounded flow problem with edge lengths is NP-hard [4] even in outer-planar graphs. Baier [3] describes a FPTAS for the maximum L-bounded flow with edge lengths that is based on the ellipsoid algorithm. He also shows that the problem of finding a decomposition of a given L-bounded flow into paths of length at most L is NP-hard, again even if the graph is outer-planar.

A related problem is that of L-bounded disjoint paths: the task is to find the maximum number of vertex or edge disjoint paths, between a given pair of vertices, each of length at most L. The vertex version of the problem is known to be solvable in polynomial time for $L \leq 4$ and NP-hard for $L \geq 5$ [15], and the edge version is solvable in polynomial time for $L \leq 5$ and NP-hard for $L \geq 6$ [6].

L-Bounded Cuts. The L-bounded cut problem is NP-hard [22]. Baier et al. [4] show that it is NP-hard to approximate it by a factor of 1.377 for $L \geq 5$ in the case of the vertex L-cut, and for $L \geq 4$ in the case of the edge L-cut. Assuming the Unique Games Conjecture, Lee et al. [19] proved that the minimum L-bounded cut problem is NP-hard to approximate within any constant factor. For planar graphs, the problem is known to be NP-hard [10,24], too.

The best approximations that we are aware of are by Baier et al. [4]: they describe an algorithm with an $\mathcal{O}(\min\{L, n/L\}) \subseteq \mathcal{O}(\sqrt{n})$-approximation for the vertex L-cut, and $\mathcal{O}(\min\{L, n^2/L^2, \sqrt{m}\}) \subseteq \mathcal{O}(n^{2/3})$-approximation for the edge L-cut. The approximation factors are closely related with the cut-flow gaps: there are instances where the minimum edge L-cut (vertex L-cut) is $\Theta(n^{2/3})$-times ($\Theta(\sqrt{n})$-times) bigger than the maximum L-flow [4]. For the vertex version of the problem, there is a τ-approximation algorithm for graphs of treewidth τ [16].

The L-bounded cut was also studied from the perspective of parameterized complexity. It is fixed parameter tractable (FPT) with respect to the treewidth of the underlying graph [8,16]. Golovach and Thilikos [12] consider several parameterizations and show FPT-algorithms for many variants of the problem (directed/undirected graphs, edge/vertex cuts). On planar graphs, it is FPT with respect to the length bound L [16].

The L-bounded cut appears in the literature also as the short paths interdiction problem [5,16,19] or as the most vital edges for shortest paths [5].

1.2   Our Contributions

In the first part of the paper, we show that the combinatorial algorithm by Koubek and Říha [18] for the maximum L-bounded flow is not correct.

In the second part of the paper we describe an iterative combinatorial algorithm, based on the exponential length method, that finds a $(1+\varepsilon)$-approximation of the maximum L-bounded flow in time $\mathcal{O}(\varepsilon^{-2}m^2L \log L)$; that is, we describe a fully polynomial approximation scheme (FPTAS) for the problem.

Moreover, we show that this approach works even for the NP-hard generalization of the maximum L-bounded flow problem in which each edge has a length. This approach is more efficient than the FPTAS based on the ellipsoid method [3].

Our result is not surprising (e.g., Baier [3] mentions the possibility, without giving the details, to use the exponential length method to obtain a FPTAS for the problem); however, considering the absence of other polynomial time algorithms for the problem that are not based on the general LP algorithms, despite of the effort to find some, we regard it as a meaningful contribution. The paper is based on the results in the bachelor's thesis of Kateřina Altmanová [2] and in the master's thesis of Jan Voborník [23].

2 The Algorithm of Koubek And Říha

2.1 Increasing an L-bounded Flow

Before describing the problem with the algorithm by Koubek and Říha [18], we informally describe the purpose and the main attributes of *an increasing L-system*, a key structure used in the algorithm.

Consider a network $G = (X, R, c, s, t)$ and an arbitrary L-bounded flow f from s to t in G, together with its decomposition into paths of length at most L (say paths $p_1, p_2, \ldots$ carrying $r_1, r_2, \ldots$ units of flow, resp.) that is not a maximum L-bounded flow. Given G and f, Koubek and Říha [18] build a labeled oriented tree $T = (V, E, v_0, LABV, LABE)$ where V is the set of nodes, E is the set of edges, v_0 is the root, $LABV$ is a vertex labelling and $LABE$ is an edge labeling. The tree is called *an increasing L system with respect to f*.

There are four types of the nodes of the tree T; to explain the error in the paper, it is sufficient to deal with three of them: 1-son, 3-son, 4-son. With (almost) each node u in T, are associated two consecutive paths in G: the first one, denoted by $q(u)$, contains only edges that are not used by the current L-flow f, and the second one, denoted by $\bar{q}(u)$, coincides with a subpath of some path from the current L-flow f. The tree T encodes a combination of these paths with paths in f and this combination is supposed to yield a larger L-flow than the L-flow f.

The label of a vertex v in the tree T, denoted by LABV in the original paper, and the label of the edge e connecting v to its immediate ancestor, if there is one, denoted by LABE, are of the following form:

	LABV	LABE
1-son	$(q(v), i(v), a(v), b(v))$	none
3-son or 4-son	$(q(v), i(v), a(v), b(v))$	$(h(e), j(e), d(e), o(e))$

where

- $q(v)$ is a path in G that is edge disjoint with every path in the L-flow f,
- $i(v), j(v)$ are indices of paths in the L-flow f,
- $a(v), b(v), d(e)$ are positive integers (distances),
- $o(e)$ is a positive integer, if v is a 3-son, and $o(v)$ is a pointer to a 3-son, if v is a 4-son,
- $h(e)$ is a subset of edges in G.

As for every node v in the tree (except for the root) there is a unique edge e connecting it to its parent, Koubek and Říha often refer to the label of the edge e, and to its attributes, by the name of the vertex v, e.g., they write $h(v)$ instead of $h(e)$; we shall use the same convention.

The tree T is supposed to describe an L-flow f' derived from f. In particular, each path $q(v)$ and $\bar{q}(v)$ is a subpath of a new path between s and t of length at most L. Very roughly speaking, the attributes $a(v)$ and $d(v)$ store information about the distance of the path segments $q(v)$ and $\bar{q}(v)$ from s along the paths used in the new L-flow f', the attribute $i(v)$ specifies the index of a path from f s.t. $\bar{q}(v)$ is a subpath of $p_{i(v)}$, and the attributes $b(v)$ and $o(v)$, resp., specify the number of edges along which the paths $p_{i(v)}$ and $p_{j(v)}$ are being followed by some of the new paths.

Consider a node w in the tree T such that at least one edge in $\bar{q}(w)$, say an edge e, is saturated in the L-flow f (i.e., $f(e) = c(e)$). In this case, the properties of the tree T enforce that the node w has at least one 3-son u whose responsibility is to desaturate the edge e by diverting one of the paths that use e in f along a new route; the attribute $j(u)$ specifies the index of the path from f that is being diverted by the 3-son u of w (Fig. 1), and $h(w)$ specifies which saturated edges from $\bar{q}(v)$ are desaturated by the son u of w.

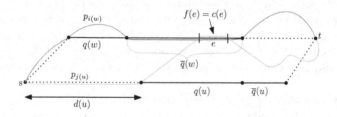

Fig. 1. Desaturation of a saturated edge e in a $\bar{q}(w)$ by a 3-son u.

As the definition of the tree T does not pose any requirements on the disjointness of the $\bar{q}$-paths corresponding to different nodes of T, it may happen that the paths $\bar{q}(w)$ and $\bar{q}(w')$ for two different nodes w and w' of the tree T overlap in a saturated edge e. In this case, Koubek and Říha allow an *exception* (our terminology) to the rule described in previous paragraph: if one of the nodes w and w', say the node w, has a 3-son u that desaturates e, the other node, the node w', need not have a 3-son but it may have a 4-son instead. The purpose of this 4-son is just to provide a pointer to the 3-son u of w that takes care about the desaturation of the edge e.

2.2   The Main Error

We start by recalling a few definitions and lemmas from the original paper [18]; for space reasons, for the definition of the increasing system we refer to [18]. In this section we view an *L*-bounded flow f as a collection of paths $p_1, p_2, \ldots$ together with positive numbers $r_1, r_2, \ldots$ specifying the amount of flow carried by the respective paths.

Definition 1 (Definition 4.2 in [18]). *Let T be an increasing L-system with respect to an L-flow $f = \{(p_i, r_i) : i \in I\}$ in a network $G = (X, R, c, s, t)$. Given an edge $u \in R$, we define:*

- *$T_1(u)$ is the number of vertices x in the tree T such that $u \in \bar{q}(x)$ and if there is a saturated edge $v \in \bar{q}(x)$ then there is a 3-son y of x with $v \in h(y)$, $u \notin p_{j(y)}$.*
- *$T_2(u)$ is the number of vertices x in the tree T such that $u \in q(x)$.*
- *$T_3(u)$ is the number of vertices x which are 3-sons or 4-sons with $u \in h(x)$.*

For $i \in I$ we denote $m_i = \sup\{T_3(u) : u \in p_i\}$, $|T| = \min\{\frac{c(u)}{T_2(u)} : u \in R, f(u) = 0\} \cup \{\frac{c(u)-f(u)}{T_1(u)} : u \in R\} \cup \{\frac{r_i}{m_i} : i \in I\}$, where the expressions that are not defined are omitted.

Lemma 1 (Lemma 4.2 in [18]). *If there is an increasing L-system with respect to an L-flow f, then there is an L-flow g with $|g| = |f| + |T|$.*

Definition 2 (Definition 4.3 in [18]). *Let $\overline{R} = R \cup \{u'\}$, where $u' \notin R$ and $c(u') = 0$. We put $\psi(G) = \min(|c(u) - c(v)| : c(u) \neq c(v), u, v \in \overline{R})$.*

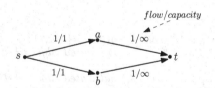

Fig. 2. A network G with a 2-bounded flow f.

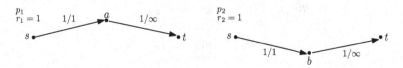

Fig. 3. A decomposition of the 2-bounded flow f into paths p_1, p_2.

Lemma 2 (Lemma 4.4 in [18]). *For each increasing L-system T (with respect to an L-flow $f = \{(p_i, r_i) : i \in I\}$) constructed by the above procedure it holds $|T| \geq \psi(G)/|I|$.*

The *above procedure* in Lemma 2 refers to a construction of an increasing L-system that is outlined in the original paper. As Definition 2 implies $\psi(G) > 0$, we also know by Lemma 2 that for every increasing L-system T, $|T| > 0$.

Now we are ready to describe the counter example. Take $k = 2$ and consider the following network G with a 2-bounded flow f of size 2 (Figs. 2 and 3); clearly, this is a maximum 2-bounded flow.

We are going to show that there exists an increasing system T for f. According to Lemmas 1 and 2 this implies the existence of a 2-bounded flow g of size $|f| + |T| > |f|$. As the flow f is a maximum 2-bounded flow in G, this is a contradiction.

The increasing system T is depicted in Fig. 4; for the sake of simplicity, we list only the most relevant attributes. It is just a matter of a mechanical effort to check that it meets Definition 4.1 of the increasing system from the original paper.

In words, the essence of the counter example is the following. The purpose of the root of the tree, the node u_0, is to increase the flow from s to t along the path $q(u_0)\bar{q}(u_0)$ which is (accidently) the path p_1. As there is a saturated edge on this path, namely the edge sa, there is a 3-son of the node u_0, the node u_1, whose purpose is to desaturate the edge sa by diverting one of the paths that use the edge sa along an alternative route; in particular, the node u_1 is diverting the path p_1 and it is diverting it from the very beginning, from s, along the path $q(u_1)\bar{q}(u_1)$ which is (accidently) the path p_2.

As there is a saturated edge on this path, namely the edge sb, there is a 3-son of the node u_1, the node u_2, whose purpose is to desaturate the edge sb by diverting one of the paths that use the edge sb along an alternative route; in particular, the node u_2 is diverting the path p_2 and it is diverting it from the very beginning, from s, along the path $q(u_2)\bar{q}(u_2)$ which is (accidently) again the path p_1.

$u_0 : 1 - son$ $\bar{q}(u_0) = \{s, a, t\}$
$q(u_0) = \emptyset$ $saturated\ edge = \{sa\}$

$u_1 : 3 - son$ $saturated\ edge = \{sb\}$
$q(u_1) = \emptyset$ $h(u_1) = \{sa\}$
$\bar{q}(u_1) = \{s, b, t\}$ $j(u_1) = 1$

$u_2 : 3 - son$ $saturated\ edge = \{sa\}$
$q(u_2) = \emptyset$ $h(u_2) = \{sb\}$
$\bar{q}(u_0) = \{s, a, t\}$ $j(u_2) = 2$

$u_3 : 4 - son$ $h(u_3) = \{sa\}$
   $o(u_3) = u_1$

Fig. 4. Increasing 2-system T.

As there is a saturated edge on this path, namely the edge sa, and as there is already another node in the tree that is desaturating sa, namely the node u_1, the node u_2 does not have a 3-son but it has a 4-son u_3 instead, which is a pointer to the 3-son u_1. This way, there is a kind of a deadlock cycle in the increasing system: u_1 is desaturating the edge sa for the node u_0 but it itself needs u_2 to desaturate the edge sb in it and u_2 in turn needs u_3 to desaturate the edge sa, but u_3 delegates this task back to u_1.

At this point, we know that Lemma 1 or Lemma 2 is not correct. By Definition 1, one can check that $|T| = 1/2$ which implies, as we started with a maximum flow, that it is Lemma 1 that does not hold.

3   FPTAS for Maximum L-bounded Flow

We first describe a fully polynomial approximation scheme for maximum L-bounded flow on networks with unit edge length. The algorithm is based on the algorithm for the maximum multicommodity flow by Garg and Könemann [11].

Then we describe a FPTAS for the L-bounded flow problem with general edge lengths. Our approximation schemas for the maximum L-bounded flow on unit edge lengths and the maximum L-bounded flow with edge lengths are almost identical, the only difference is in using an approximate subroutine for resource constrained shortest path in the general case which slightly complicates the analysis.

3.1   FPTAS for Unit Edge Lengths

Let us consider the path based linear programming (LP) formulation of the maximum L-bounded flow, $\mathbf{P}_{\text{path}}$, and its dual, $\mathbf{D}_{\text{path}}$. We assume that $G = (V, E, c, s, t)$ is a given network and L is a given length bound. Let $\mathcal{P}_L$ denote the set of all s-t paths of length at most L in G. There is a primal variable $x(p)$ for each path $p \in \mathcal{P}_L$, and a dual variable $y(e)$ for each edge $e \in E$. Note that the dual LP is a relaxation of an integer LP formulation of the minimum L-bounded cut problem.

$$\max \sum_{P \in \mathcal{P}_L} x(P) \qquad\qquad \min \sum_{e \in E} c(e) y(e)$$

$$\text{s.t.} \sum_{\substack{P \in \mathcal{P}_L: \\ e \in P}} x(P) \le c(e) \quad \forall e \in E \qquad\qquad \text{s.t.} \sum_{e \in P} y(e) \ge 1 \quad \forall P \in \mathcal{P}_L$$

$$x \ge 0 \qquad\qquad\qquad\qquad y \ge 0$$

The algorithm simultaneously constructs solutions for the maximum L-bounded flow and the minimum fractional L-bounded cut. It iteratively routes flow over shortest paths with respect to properly chosen dual edge lengths and at the same time increases these dual lengths; dual edge length of the edge e after i iterations will be denoted by $y_i(e)$. The progress of the algorithm depends on two positive parameters, $\varepsilon < 1$, $\delta < 1$. During the runtime of the algorithm, the constructed flow need not respect the edge capacities, however, with the right choice of the parameters ε, δ the resulting flow can be scaled down to a feasible (i.e., respecting the edge capacities) flow (Lemma 3) that is a $(1 + \varepsilon)$-approximation of the maximum L-bounded flow (Theorem 1).

For a vector y of dual variables, let $d_y^L(s, t)$ denote the length of the y-shortest $s-t$ path from the set of paths $\mathcal{P}_L$ and let $\alpha^L(i) = d_{y_i}^L(s, t)$. Note that a shortest $s - t$ path with respect to edge lengths y that uses at most a given number of edges can be computed in polynomial time by a modification of the Dijkstra's shortest path algorithm.

Algorithm 1. APPROX(ε, δ)

1: $i \leftarrow 0$, $y_0(e) \leftarrow \delta$ $\forall e \in E$, $x_0(P) \leftarrow 0$ $\forall P \in \mathcal{P}_L$
2: **while** $\alpha^L(i) < 1$ **do**
3:     $i \leftarrow i + 1$
4:     $x_i \leftarrow x_{i-1}, y_i \leftarrow y_{i-1}$
5:     $P \leftarrow y_i$-shortest s-t path with at most L edges
6:     $c \leftarrow \min_{e \in P} c(e)$
7:     $x_i(P) \leftarrow x_i(P) + c$
8:     $y_i(e) \leftarrow y_i(e)(1 + \varepsilon c / c(e))$ $\forall e \in P$
9: **end while**
10: **return** x_i

Let f_i denote the size of the flow after i iterations, $f_i = \sum_{P \in \mathcal{P}_L} x_i(P)$, and let τ denote the total number of iterations performed by APPROX; then x_τ is the output of the algorithm and f_τ its size.

Lemma 3. *The flow x_τ scaled down by a factor of $\log_{1+\varepsilon} \frac{1+\varepsilon}{\delta}$ is a feasible L-bounded flow.*

Proof. By construction, for every i, x_i is an L-bounded flow. Thus, we only have to care about the feasibility of the flow

$$\frac{x_\tau}{\log_{1+\varepsilon} \frac{1+\varepsilon}{\delta}}. \tag{1}$$

For every iteration i and every edge $e \in E$, as $\alpha^L(i - 1) < 1$, we also have $y_{i-1}(e) < 1$ and so $y_i(e) < 1 + \varepsilon$. It follows that

$$y_\tau(e) < 1 + \varepsilon. \tag{2}$$

Consider an arbitrary edge $e \in E$ and suppose that the flow $f_\tau(e)$ along e has been routed in iterations $i_1, i_2, \ldots, i_r$ and the amount of flow routed in iteration i_j is c_j. Then $f_\tau(e) = \sum_{j=1}^r c_j$ and $y_\tau(e) = \delta \prod_{j=1}^r (1 + \varepsilon c_j / c(e))$. Because each c_j was chosen such that $c_j \leq c(e)$, we have by Bernoulli's inequality that $1 + \varepsilon c_j / c(e) \geq (1 + \varepsilon)^{c_j / c(e)}$ and

$$y_\tau(e) \geq \delta \prod_{j=1}^{r}(1 + \varepsilon)^{c_j/c(e)} = \delta(1 + \varepsilon)^{f_\tau(e)/c(e)}. \tag{3}$$

Combining inequalities (2) and (3) gives

$$\frac{f_\tau(e)}{c(e)} \leq \log_{1+\varepsilon}\frac{1+\varepsilon}{\delta}$$

which completes the proof.

Claim. For $i = 1, \ldots, \tau$,

$$\alpha^L(i) \leq \delta L e^{\varepsilon f_i/\beta}. \tag{4}$$

Proof. For a vector y of dual variables, let $D(y) = \sum_e c(e)y(e)$ and let $\beta = \min_y D(y)/d_y^L(s,t)$. Note that β is equal to the optimal value of the dual linear program. For notational simplicity we abbreviate $D(y_i)$ as $D(i)$.

Let P_i be the path chosen in iteration i and c_i be the value of c in iteration i. For every $i \geq 1$ we have

$$D(i) = \sum_{e \in E} y_i(e)c(e)$$
$$= \sum_{e \in E} y_{i-1}(e)c(e) + \varepsilon \sum_{e \in P_i} y_{i-1}(e)c_i$$
$$= D(i-1) + \varepsilon(f_i - f_{i-1})\alpha^L(i-1)$$

which implies that

$$D(i) = D(0) + \varepsilon \sum_{j=1}^{i}(f_j - f_{j-1})\alpha^L(j-1). \tag{5}$$

Now consider the length function $y_i - y_0$. Note that $D(y_i - y_0) = D(i) - D(0)$ and $d_{y_i-y_0}^L(s,t) \geq \alpha^L(i) - \delta L$. Hence,

$$\beta \leq \frac{D(y_i - y_0)}{d_{y_i-y_0}^L(s,t)} \leq \frac{D(i) - D(0)}{\alpha^L(i) - \delta L}. \tag{6}$$

By combining relations (5) and (6) we get

$$\alpha^L(i) \leq \delta L + \frac{\varepsilon}{\beta}\sum_{j=1}^{i}(f_j - f_{j-1})\alpha^L(j-1).$$

Now we define $z(0) = \alpha^L(0)$ and for $i = 1, \ldots, \tau$, $z(i) = \delta L + \frac{\varepsilon}{\beta}\sum_{j=1}^{i}(f_j - f_{j-1})z(j-1)$. Note that for each i, $\alpha^L(i) \leq z(i)$. Furthermore,

$$z(i) = \delta L + \frac{\varepsilon}{\beta}\sum_{j=1}^{i}(f_j - f_{j-1})z(j-1)$$
$$= \left(\delta L + \frac{\varepsilon}{\beta}\sum_{j=1}^{i-1}(f_j - f_{j-1})z(j-1)\right) + \frac{\varepsilon}{\beta}(f_i - f_{i-1})z(i-1)$$

$$= z(i-1)(1 + \varepsilon(f_i - f_{i-1})/\beta)$$
$$\leq z(i-1)e^{\varepsilon(f_i - f_{i-1})/\beta}.$$

Since $z(0) \leq \delta L$, we have $z(i) \leq \delta L e^{\varepsilon f_i/\beta}$, and thus also, for $i = 1, \ldots, \tau$, $\alpha^L(i) \leq \delta L e^{\varepsilon f_i/\beta}$.

Theorem 1. *For every $0 < \varepsilon < 1$ there is an algorithm that computes an $(1 + \varepsilon)$-approximation to the maximum L-bounded flow in a network with unit edge lengths in time $\mathcal{O}(\varepsilon^{-2}m^2 L \log L)$.*

Proof. We start by showing that for every $\varepsilon < \frac{1}{3}$ there is a constant $\delta = \delta(\varepsilon)$ such that x_τ, the output of APPROX(ε, δ), scaled down by $\log_{1+\varepsilon} \frac{1+\varepsilon}{\delta}$ as in Lemma 3, is a $(1 + 3\varepsilon)$-approximation.

Let γ denote the approximation ratio of such an algorithm, that is, let γ denote the ratio of the optimal dual solution (β) to the appropriately scaled output of APPROX(ε, δ),

$$\gamma = \frac{\beta \log_{1+\varepsilon} \frac{1+\varepsilon}{\delta}}{f_\tau}, \tag{7}$$

where the constant δ will be specified later.

By Claim 3.1 and the stopping condition of the while cycle we have

$$1 \leq \alpha^L(\tau) \leq \delta L e^{\varepsilon f_\tau/\beta}$$

and hence

$$\frac{\beta}{f_\tau} \leq \frac{\varepsilon}{\log \frac{1}{\delta L}}.$$

Plugging this bound in the equality for the approximation ratio γ, we obtain

$$\gamma \leq \frac{\varepsilon \log_{1+\varepsilon} \frac{1+\varepsilon}{\delta}}{\log \frac{1}{\delta L}} = \frac{\varepsilon}{\log(1+\varepsilon)} \frac{\log \frac{1+\varepsilon}{\delta}}{\log \frac{1}{\delta L}}.$$

Setting $\delta = \frac{1+\varepsilon}{((1+\varepsilon)L)^{1/\varepsilon}}$ yields

$$\frac{\log \frac{1+\varepsilon}{\delta}}{\log \frac{1}{\delta L}} = \frac{\frac{1}{\varepsilon} \log((1+\varepsilon)L)}{(\frac{1}{\varepsilon} - 1) \log((1+\varepsilon)L)} = \frac{1}{1-\varepsilon}.$$

Taylor expansion of $\log(1+\varepsilon)$ gives a bound $\log(1+\varepsilon) \geq \varepsilon - \frac{\varepsilon^2}{2}$ for $\varepsilon < 1$ and it follows for $\varepsilon < \frac{1}{3}$ that

$$\gamma \leq \frac{\varepsilon}{(1-\varepsilon)\log(1+\varepsilon)} \leq \frac{\varepsilon}{(1-\varepsilon)(\varepsilon - \varepsilon^2/2)} \leq \frac{1}{1 - \frac{3}{2}\varepsilon} \leq 1 + 3\varepsilon.$$

To complete the proof, we just put $\varepsilon' = \varepsilon/3$ and run APPROX$(\varepsilon', \delta(\varepsilon'))$. It remains to prove the time complexity of the algorithm. In every iteration i of APPROX, the length $y_i(e)$ of an edge e with the smallest capacity on the chosen path P is increased by a factor of $1 + \varepsilon'$. Because P was chosen such that $y_i(P) < 1$ also $y_i(e) < 1$ for every edge $e \in P$. Lengths of other edges get increased by a factor of at most $1 + \varepsilon'$, therefore $y_\tau(e) < 1 + \varepsilon'$ for every edge $e \in E$. Every edge has the minimum capacity on the chosen path in at most $\left\lceil \log_{1+\varepsilon'} \frac{1+\varepsilon'}{\delta} \right\rceil = \mathcal{O}(\frac{1}{\varepsilon} \log_{1+\varepsilon} L)$ iterations, so APPROX makes at most $\mathcal{O}(\frac{m}{\varepsilon} \log_{1+\varepsilon} L) = \mathcal{O}(\frac{m}{\varepsilon^2} \log L)$ iterations.

Each iteration takes time $\mathcal{O}(Lm)$ so the total time taken by APPROX is $\mathcal{O}(\varepsilon^{-2} m^2 L \log L)$.

3.2 FPTAS for General Edge Lengths

Now we extend the approximation algorithm to networks with general edge lengths that are given by a length function $\ell : E \to \mathbb{N}$. The dynamic programming algorithm for computing shortest paths that have a restricted length with respect to another length function, does not work in this case. In fact, the problem of finding shortest path with respect to a given edge length function while restricting to paths of bounded length with respect to another length function is NP-hard in general [13]. On the other hand, there exists a FPTAS for it [14,20].

We assume that we are given as a black-box an algorithm that for a given graph G, two edge length functions y and ℓ, two distinguished vertices s and t from G, a length bound L and an error parameter $w > 0$, computes a $(1 + w)$-approximation of the y-shortest path of ℓ-length at most L; we denote by $d_{y,\ell}^L(s,t;w)$ the length of such a path and we also introduce an abbreviation $\bar{\alpha}^L(i) = d_{y_i,\ell}^L(s,t;w)$. Note that for every i, $\bar{\alpha}^L(i) \le (1+w)\alpha^L(i)$. We can use the FPTAS of Lorenz and Raz [20] for this task.

The algorithm of Garg and Könemann [11] for approximating maximal multicommodity flow has been improved by Fleischer [9]. The original algorithm computes the shortest path between every terminal pairs in every iteration. Fleischer divided the algorithm to phases where she worked with commodities one by one. This way her algorithm effectively works with approximations of shortest paths while eliminates the dependency on the number of commodities and still gets a good approximation ratio. Using a similar analysis we show that we can work with an approximation shortest path algorithm to get an FPTAS to otherwise intractable maximum L-bounded flow problem with general edge lengths.

The structure of the L-bounded flow algorithm with general edge lengths stays the same as in the unit edge lengths case. The only difference is that instead of y-shortest L-bounded paths, approximations of y-shortest L-bounded paths are used: in step 2, the new condition is **while** $\bar{\alpha}^L(i) < 1 + w$, and in step 5, we set $P \leftarrow (1 + w)$-approximation of the y_i-shortest L-bounded path. The analysis of the algorithm follows the same steps as the analysis of Algorithm 1 but one has to be more careful when dealing with the lengths. For the lack of space we omit the proofs.

Lemma 4. *The flow x_τ scaled down by a factor of* $\log_{1+\varepsilon} \frac{(1+\varepsilon)(1+w)}{\delta}$ *is a feasible L-bounded flow.*

Theorem 2. *There is an algorithm that computes an* $(1 + \varepsilon)$*-approximation to the maximum L-bounded flow in a graph with general edge lengths in time* $\mathcal{O}(\frac{m^2 n}{\varepsilon^2} \log L (\log \log n + \frac{1}{\varepsilon}))$.

We note that the exponential length method can be used for many fractional packing problems and using the same technique we could get an approximation algorithm for maximum multicommodity L-bounded flow.

References

1. Adámek, J., Koubek, V.: Remarks on flows in network with short paths. Comment. Math. Univ. Carolin. **12**(4), 661–667 (1971)
2. Altmanová, K.: Toky cestami omezené délky. Technical report, Bachelor's thesis, Charles University, Faculty of Mathematics and Physics, Department of Applied Mathematics (2018). (in Czech)
3. Baier, G.: Flows with path restrictions. Ph.D. thesis, TU Berlin (2003)
4. Baier, G., et al.: Length-bounded cuts and flows. ACM Trans. Algorithms **7**(1), 4:1–4:27 (2010)
5. Bazgan, C., Fluschnik, T., Nichterlein, A., Niedermeier, R., Stahlberg, M.: A more fine-grained complexity analysis of finding the most vital edges for undirected shortest paths. CoRR, abs/1804.09155 (2018)
6. Bley, A.: On the complexity of vertex-disjoint length-restricted path problems. Comput. Complex. **12**(3–4), 131–149 (2003)
7. Bley, A., Neto, J.: Approximability of 3- and 4-Hop bounded disjoint paths problems. In: Eisenbrand, F., Shepherd, F.B. (eds.) IPCO 2010. LNCS, vol. 6080, pp. 205–218. Springer, Heidelberg (2010). https://doi.org/10.1007/978-3-642-13036-6_16
8. Dvořák, P., Knop, D.: Parametrized complexity of length-bounded cuts and multicuts. In: Jain, R., Jain, S., Stephan, F. (eds.) TAMC 2015. LNCS, vol. 9076, pp. 441–452. Springer, Cham (2015). https://doi.org/10.1007/978-3-319-17142-5_37
9. Fleischer, L.K.: Approximating fractional multicommodity flow independent of the number of commodities. SIAM J. Discret. Math. **13**(4), 505–520 (2000)
10. Fluschnik, T., Hermelin, D., Nichterlein, A., Niedermeier, R.: Fractals for kernelization lower bounds, with an application to length-bounded cut problems. CoRR, abs/1512.00333 (2015)
11. Garg, N., Könemann, J.: Faster and simpler algorithms for multicommodity flow and other fractional packing problems. SIAM J. Comput. **37**(2), 630–652 (2007)
12. Golovach, P.A., Thilikos, D.M.: Paths of bounded length and their cuts: parameterized complexity and algorithms. In: International Symposium on Parameterized and Exact Computation (2009)
13. Handler, G.Y., Zang, I.: A dual algorithm for the constrained shortest path problem. Networks **10**, 293–310 (1980)
14. Hassin, R.: Approximation schemes for the restricted shortest path problem. Math. Oper. Res. **17**(1), 36–42 (1992)
15. Itai, A., Perl, Y., Shiloach, Y.: The complexity of finding maximum disjoint paths with length constraints. Networks **12**(3), 277–286 (1982)
16. Kolman, P.: On algorithms employing treewidth for L-bounded cut problems. J. Graph Algorithms Appl. **22**, 177–191 (2018)

17. Kolman, P., Scheideler, C.: Improved bounds for the unsplittable flow problem. J. Algorithms **61**(1), 20–44 (2006)
18. Koubek, V., Říha, A.: The maximum k-flow in a network. In: Gruska, J., Chytil, M. (eds.) MFCS 1981. LNCS, vol. 118, pp. 389–397. Springer, Heidelberg (1981). https://doi.org/10.1007/3-540-10856-4_106
19. Lee, E.: Improved hardness for cut, interdiction, and firefighter problems. In: International Colloquium on Automata Languages, and Programming (2017)
20. Lorenz, D.H., Raz, D.: A simple efficient approximation scheme for the restricted shortest path problem. Oper. Res. Lett. **28**(5), 213–219 (2001)
21. Mahjoub, R.A., McCormick, T.S.: Max flow and min cut with bounded-length paths: complexity, algorithms, and approximation. Math. Program. **124**(1–2), 271–284 (2010)
22. Schieber, B., Bar-Noy, A., Khuller, S.: The complexity of finding most vital arcs and nodes. Technical report, College Park, MD, USA (1995)
23. Vorборník, J.: Algorithms for *L*-bounded flows. Master's thesis, Charles University, Faculty of Mathematics and Physics, Department of Applied Mathematics (2016)
24. Zschoche, P., Fluschnik, T., Molter, H., Niedermeier, R.: The computational complexity of finding separators in temporal graphs. ArXiv e-prints, November 2017

Efficient Nearest-Neighbor Query and Clustering of Planar Curves

Boris Aronov[1] , Omrit Filtser[2], Michael Horton[3] , Matthew J. Katz[2(✉)], and Khadijeh Sheikhan[1]

[1] Tandon School of Engineering, New York University, Brooklyn, NY 11201, USA
boris.aronov@nyu.edu, khadije.sheikhan@gmail.com
[2] Ben-Gurion University of the Negev, 84105 Beer-Sheva, Israel
{omritna,matya}@cs.bgu.ac.il
[3] SPORTLOGiQ Inc., Montreal, Quebec H2T 3B3, Canada

Abstract. We study two fundamental problems dealing with curves in the plane, namely, the nearest-neighbor problem and the center problem. Let $\mathcal{C}$ be a set of n polygonal curves, each of size m. In the nearest-neighbor problem, the goal is to construct a compact data structure over $\mathcal{C}$, such that, given a query curve Q, one can efficiently find the curve in $\mathcal{C}$ closest to Q. In the center problem, the goal is to find a curve Q, such that the maximum distance between Q and the curves in $\mathcal{C}$ is minimized. We use the well-known discrete Fréchet distance function, both under L_∞ and under L_2, to measure the distance between two curves.

For the nearest-neighbor problem, despite discouraging previous results, we identify two important cases for which it is possible to obtain practical bounds, even when m and n are large. In these cases, either Q is a line segment or $\mathcal{C}$ consists of line segments, and the bounds on the size of the data structure and query time are nearly linear in the size of the input and query curve, respectively. The returned answer is either exact under L_∞, or approximated to within a factor of $1 + \varepsilon$ under L_2. We also consider the variants in which the location of the input curves is only fixed up to translation, and obtain similar bounds, under L_∞.

As for the center problem, we study the case where the center is a line segment, i.e., we seek the line segment that represents the given set as well as possible. We present near-linear time exact algorithms under L_∞, even when the location of the input curves is only fixed up to translation. Under L_2, we present a roughly $O(n^2 m^3)$-time exact algorithm.

Keywords: Polygonal curves · Nearest-neighbor queries · Clustering · Fréchet distance · Data structures · (Approximation) algorithms

1 Introduction

We consider efficient algorithms for two fundamental data-mining problems for sets of polygonal curves in the plane: nearest-neighbor query and clustering. Both

---

A more complete version of this paper is available on arXiv [7].

© Springer Nature Switzerland AG 2019
Z. Friggstad et al. (Eds.): WADS 2019, LNCS 11646, pp. 28–42, 2019.
https://doi.org/10.1007/978-3-030-24766-9_3

of these problems have been studied extensively and bounds on the running time and storage consumption have been obtained. In general, these bounds suggest that the existence of algorithms that can efficiently process large datasets of curves of high complexity is unlikely. Therefore we study special cases of the problems where some curves are assumed to be directed line segments (henceforth referred to as segments), and the distance metric is the discrete Fréchet distance.

Such analysis of curves has many practical applications, where the position of an object as it changes over time is recorded as a sequence of readings from a sensor to generate a *trajectory*. For example, the location readings from GPS devices attached to migrating animals [5], the traces of players during a football match captured by a computer vision system [22], or stock market prices [27]. In each case, the output is an ordered sequence C of m vertices (i.e., the sensor readings), and by interpolating the location between each pair of vertices as a segment, a polygonal chain is obtained.

Given a collection $\mathcal{C}$ of n curves, a natural question to ask is whether it is possible to preprocess $\mathcal{C}$ into a data structure so that the nearest curve in the collection to a query curve Q can be determined efficiently. This is the *nearest-neighbor* problem for curves (NNC).

Indyk [25] gave a near-neighbor data structure for polygonal curves under the discrete Fréchet distance. The data structure achieves an approximation factor of $O(\log m + \log \log n)$, where n is the number of curves and m is the maximum size of a curve. Its space consumption is very high, $O(|X|^{\sqrt{m}}(m^{\sqrt{m}}n)^2)$, where $|X|$ is the size of the domain on which the curves are defined, and the query time is $O(m^{O(1)} \log n)$.

Later, Driemel and Silvestri [17] presented a locality-sensitive-hashing scheme for curves under the discrete Fréchet distance, improving the result of Indyk for short curves. They also provide a trade-off between approximation quality and computational performance: for a parameter $k \in [m]$, a data structure using $O(2^{2k}m^{k-1}n \log n + mn)$ space is constructed that answers queries in $O(2^{2k}m^k \log n)$ time with an approximation factor of $O(m/k)$.

Recently, Emiris and Psarros [19] presented near-neighbor data structures for curves under both discrete Fréchet and dynamic time warping distance. Their algorithm achieves approximation factor of $1 + \varepsilon$, at the expense of increasing space usage and preprocessing time. For curves in the plane, the space used by their data structure is $\tilde{O}(n) \cdot (2 + \frac{1}{\log m})^{O(m^{1/\varepsilon} \cdot \log(1/\varepsilon))}$ for discrete Fréchet distance and $\tilde{O}(n) \cdot O(\frac{1}{\varepsilon})^m$ for dynamic time warping distance, while the query time in both cases is $O(2^{2m} \log n)$.

De Berg et al. [9] described a dynamic data structure for approximate nearest neighbor for curves (which can also be used for other types of queries such as approximate range searching), under the (continuous) Fréchet distance. Their data structure uses $n \cdot O\left(\frac{1}{\varepsilon}\right)^{2m}$ space and has $O(1)$ query time (for a segment query), but with an *additive* error of $\varepsilon \cdot reach(Q)$, where $reach(Q)$ is the maximum distance between the start vertex of the query curve Q and any other vertex of Q. Furthermore, when the distance from Q to its nearest neighbor is relatively large, the query procedure might fail.

Afshani and Driemel [2] studied range searching under both the discrete and continuous Fréchet distance. In this problem, the goal is to preprocess $\mathcal{C}$ such that, given a query curve Q of length m_q and a radius r, all curves in $\mathcal{C}$ that are within distance r of Q can be found efficiently. For the discrete Fréchet distance in the plane, their data structure uses space in $O(n(\log\log n)^{m-1})$ and has query time in $O(\sqrt{n}\cdot\log^{O(m)} n \cdot m_q^{O(1)})$, assuming $m_q = \log^{O(1)} n$. They also show that any data structure in the pointer model that achieves $Q(n) + O(k)$ query time, where k is the output size, has to use roughly $\Omega(n/Q(n))^2)$ space in the worst case, even if queries are just points, for discrete Fréchet distance!

De Berg, Cook, and Gudmundsson [8] considered range counting queries for curves under the continuous Fréchet distance. Given a single polygonal curve C with m vertices, they show how to preprocess it into a data structure in $O(k\,\mathrm{polylog}\,m)$ time and space, so that, given a query segment s, one can return a constant approximation of the number of subcurves of C that lie within distance r of s in $O(\frac{n}{\sqrt{k}}\,\mathrm{polylog}\,m)$ time, where k is a parameter between m and m^2.

Driemel and Har-Peled [15] preprocess a curve C into a data structure of linear size, which, given a query segment s and a subcurve of C, returns a $(1+\varepsilon)$-approximation of the distance between s and the subcurve in logarithmic time.

Clustering is another fundamental problem in data analysis that aims to partition an input collection of curves into clusters where the curves within each cluster are similar in some sense, and a variety of formulations have been proposed [1,14,16]. The k-CENTER problem [3,21,24] is a classical problem in which a point set in a metric space is clustered. The problem is defined as follows: given a set $\mathcal{P}$ of n points, find a set $\mathcal{G}$ of k center points, such that the maximum distance from a point in $\mathcal{P}$ to a nearest point in $\mathcal{G}$ is minimized.

Given an appropriate metric for curves, such as the discrete Fréchet distance, one can define a metric space on the space of curves and then use a known algorithm for point clustering. The clustering obtained by the k-CENTER problem is useful in that it groups similar curves together, thus uncovering a structure in the collection, and furthermore the center curves are of value as each can be viewed as a representative or exemplar of its cluster, and so the center curves are a compact summary of the collection. However, an issue with this formulation, when applied to curves, is that the optimal center curves may be *noisy*, i.e., the size of such a curve may be linear in the total number of vertices in its cluster, see [16] for a detailed description. This can significantly reduce the utility of the centers as a method of summarizing the collection, as the centers should ideally be of low complexity. To address this issue, Driemel et al. [16] introduced the (k, ℓ)-CENTER problem, where the k desired center curves are limited to at most ℓ vertices each.

Inherent in both problems is a notion of *similarity* between pairs of curves, which is expressed as a distance function. Several such functions have been proposed to compare curves, including the continuous [6,20] and discrete [18] Fréchet distance, the Hausdorff distance [23], and dynamic time warping [10]. We consider the problems under the discrete Fréchet distance, which is often informally described by two frogs, each hopping from vertex to vertex along a polygonal curve. At each step, one or both of the frogs may advance to the next vertex on its curve, and then the distance between them is measured using some point metric. The discrete Fréchet distance is defined as the smallest maximum distance between the frogs that can be achieved in such a joint sequence of hops of the frogs. The point metrics that we consider are the L_∞ and L_2 metrics. The problem of computing the Fréchet distance has been widely investigated [4,11,12], and in particular Bringmann and Mulzer [11] showed that strongly subquadratic algorithms for the discrete Fréchet distance are unlikely to exist.

Several hardness of approximation results for both the NNC and (k, ℓ)-CENTER problems are known. For the NNC problem under the discrete Fréchet distance, no data structure exists requiring $O(n^{2-\varepsilon} \operatorname{polylog} m)$ preprocessing and $O(n^{1-\varepsilon} \operatorname{polylog} m)$ query time for $\varepsilon > 0$, and achieving an approximation factor of $c < 3$, unless the strong exponential time hypothesis fails [16,26]. In the case of the (k, ℓ)-CENTER problem under the discrete Fréchet distance, Driemel et al. showed that the problem is NP-hard to approximate within a factor of $2 - \varepsilon$ when k is part of the input, even if $\ell = 2$ and $d = 1$. Furthermore, the problem is NP-hard to approximate within a factor $2 - \varepsilon$ when ℓ is part of the input, even if $k = 2$ and $d = 1$, and when $d = 2$ the inapproximability bound is $3 \sin \pi/3 \approx 2.598$ [13].

However, we are interested in algorithms that can process large inputs, i.e., where n and/or m are large, which suggests that the processing time ought to be near-linear in nm and the query time for NNC queries should be near-linear in m only. The above results imply that algorithms for the NNC and (k, ℓ)-CENTER problems that achieve such running times are not realistic. Moreover, given that strongly subquadratic algorithms for computing the discrete Fréchet distance are unlikely to exist, an algorithm that must compute pairwise distances explicitly will incur a roughly $O(m^2)$ running time. To circumvent these constraints, we focus on specific important settings: for the NNC problem, either the query curve is assumed to be a segment or the input curves are segments; and for the (k, ℓ)-CENTER problem the center is a segment and $k = 1$, i.e., we focus on the $(1, 2)$-CENTER problem.

While these restricted settings are of theoretical interest, they also have a practical motivation when the inputs are trajectories of objects moving through space, such as migrating birds. A segment ab can be considered a trip from a starting point a to a destination b. Given a set of trajectories that travel from point to point in a noisy manner, we may wish to find the trajectory that most closely follows a direct path from a to b, which is the NNC problem with a segment query. Conversely, given an input of (directed) segments and a query trajectory, the NNC problem would identify the segment (the simplest possible trajectory, in a sense) that the query trajectory most closely resembles. In the case of the $(1, 2)$-CENTER problem, the obtained segment center for an input of trajectories would similarly represent the summary direction of the input, and the radius r^* of the solution would be a measure of the maximum deviation from that direction for the collection.

Our Results. We present algorithms for a variety of settings (summarized in the table below) that achieve the desired running time and storage bounds. Under the L_∞ metric, we give exact algorithms for the NNC and (1,2)-CENTER problems, including under translation, that achieve the roughly linear bounds. For the L_2 metric, $(1 + \varepsilon)$-approximation algorithms with near-linear running times are given for the NNC problem, and for the (1,2)-CENTER problem, an exact algorithm is given whose running time is roughly $O(n^2 m^3)$ and whose space requirement is quadratic. (An asterisk marks results under translation, presented in [7].)

	Input/query: m-curves/segment	Input/query: segments/m-curve	Input: (1,2)-center
L_∞	Sect. 3.1 (∗)	Sect. 3.2 (∗)	Sect. 5.1 (∗)
L_2	See [7]	See [7]	See [7]

2   Preliminaries

The discrete Fréchet distance is a measure of similarity between two curves, defined as follows. Consider the curves $C = (p_1, \ldots, p_m)$ and $C' = (q_1, \ldots, q_{m'})$, viewed as sequences of vertices. A (monotone) *alignment* of the two curves is a sequence $\tau := \langle (p_{i_1}, q_{j_1}), \ldots, (p_{i_v}, q_{j_v}) \rangle$ of pairs of vertices, one from each curve, with $(i_1, j_1) = (1, 1)$ and $(i_v, j_v) = (m, m')$. Moreover, for each pair (i_u, j_u), $1 < u \leq v$, one of the following holds: (i) $i_u = i_{u-1}$ and $j_u = j_{u-1} + 1$, (ii) $i_u = i_{u-1} + 1$ and $j_u = j_{u-1}$, or (iii) $i_u = i_{u-1} + 1$ and $j_u = j_{u-1} + 1$. The discrete Fréchet distance is defined as

$$d_{dF}^d(C, C') = \min_{\tau \in \mathcal{T}} \max_{(i,j) \in \tau} d(p_i, q_j),$$

with the minimum taken over the set $\mathcal{T}$ of all such alignments τ, and where d denotes the metric used for measuring interpoint distances.

We now give two alternative, equivalent definitions of the discrete Fréchet distance between a segment $s = ab$ and a polygonal curve $C = (p_1, \ldots, p_m)$ (we will drop the point metric d from the notation, where it is clear from the context). Let $C[i, j] := \{p_i, \ldots, p_j\}$. Denote by $B(p, r)$ the ball of radius r centered at p, in metric d. The discrete Fréchet distance between s and C is at most r, if and only if there exists a partition of C into a prefix $C[1, i]$ and a suffix $C[i + 1, m]$, such that $B(a, r)$ contains $C[1, i]$ and $B(b, r)$ contains $C[i + 1, m]$.

A second equivalent definition is as follows. Consider the intersections of balls around the points of C. Set $I_i(r) = B(p_1, r) \cap \cdots \cap B(p_i, r)$ and $\bar{I}_i(r) = B(p_{i+1}, r) \cap \cdots \cap B(p_m, r)$, for $i = 1, \ldots, m-1$. Then, the discrete Fréchet distance between s and C is at most r, if and only if there exists an index $1 \leq i \leq m - 1$ such that $a \in I_i(r)$ and $b \in \bar{I}_i(r)$.

Given a set $\mathcal{C} = \{C_1, \ldots, C_n\}$ of n polygonal curves in the plane, the nearest-neighbor problem for curves is formulated as follows:

Problem 1 (NNC). Preprocess $\mathcal{C}$ into a data structure, which, given a query curve Q, returns a curve $C \in \mathcal{C}$ with $d_{dF}(Q, C) = \min_{C_i \in \mathcal{C}} d_{dF}(Q, C_i)$.

We consider two variants of Problem 1: (i) when the query curve Q is a segment, and (ii) when the input $\mathcal{C}$ is a set of segments.

Secondly, we consider a particular case of the (k, ℓ)-CENTER problem for curves [16].

Problem 2 ((1,2)-CENTER). Find a segment s^* that minimizes $\max_{C_i \in \mathcal{C}}$ $d_{dF}(s, C_i)$, over all segments s.

3   NNC and L_∞ Metric

When d is the L_∞ metric, each ball $B(p_i, r)$ is a square. Denote by $S(p, d)$ the axis-parallel square of radius d centered at p.

Given a curve $C = (p_1, \ldots, p_m)$, let d_i, for $i = 1, \ldots, m - 1$, be the smallest radius such that $S(p_1, d_i) \cap \cdots \cap S(p_i, d_i) \neq \emptyset$. In other words, d_i is the radius of the smallest enclosing square of $C[1, i]$. Similarly, let $\overline{d}_i$, for $i = 1, \ldots, m - 1$, be the smallest radius such that $S(p_{i+1}, \overline{d}_i) \cap \cdots \cap S(p_m, \overline{d}_i) \neq \emptyset$.

For any $d > d_i$, $S(p_1, d) \cap \cdots \cap S(p_i, d)$ is a rectangle, $R_i = R_i(d)$, defined by four sides of the squares $S(p_1, d), \ldots, S(p_i, d)$, see Fig. 1. These sides are fixed and do not depend on the specific value of d. Furthermore, the left, right, bottom and top sides of $R_i(d)$ are provided by the sides corresponding to the right-, left-, top- and bottom-most vertices in $C[1, i]$, respectively, i.e., the sides corresponding to the vertices defining the bounding box of $C[1, i]$.

Denote by p_ℓ^i the vertex in the ith prefix of C that contributes the left side to $R_i(d)$, i.e., the left side of $S(p_\ell^i, d)$ defines the left side of $R_i(d)$. Furthermore, denote by p_r^i, p_b^i, and p_t^i the vertices of the ith prefix of C that contribute the right, bottom, and top sides to $R_i(d)$, respectively. Similarly, for any $d > \overline{d}_i$, we denote the four vertices of the ith suffix of C that contribute the four sides of the rectangle $\overline{R}_i(d) = S(p_{i+1}, d) \cap \cdots \cap S(p_m, d)$ by $\overline{p}_\ell^i$, $\overline{p}_r^i$, $\overline{p}_b^i$, and $\overline{p}_t^i$, respectively.

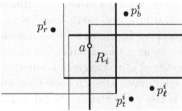

Fig. 1. The rectangle $R_i = R_i(d)$ and the vertices of the ith prefix of C that define it.

Finally, we use the notation $R_i^j = R_i^j(d)$ ($\overline{R}_i^j = \overline{R}_i^j(d)$) to refer to the rectangle $R_i = R_i(d)$ ($\overline{R}_i = \overline{R}_i(d)$) of curve C_j.

Observation 1. Let $s = ab$ be a segment, C be a curve, and let $d > 0$. Then, $d_{dF}(s, C) \leq d$ if and only if there exists i, $1 \leq i \leq m - 1$, such that $a \in R_i(d)$ and $b \in \overline{R}_i(d)$.

3.1 Query Is a Segment

Let $\mathcal{C} = \{C_1, \ldots, C_n\}$ be the input curves, each of size m. Given a query segment $s = ab$, the task is to find a curve $C \in \mathcal{C}$ such that $d_{dF}(s, C) = \min_{C' \in \mathcal{C}} d_{dF}(s, C')$.

The Data Structure. The data structure is an eight-level search tree. The first level of the data structure is a search tree for the x-coordinates of the vertices p_ℓ^i, over all curves $C \in \mathcal{C}$, corresponding to the nm left sides of the nm rectangles $R_i(d)$. The second level corresponds to the nm right sides of the rectangles $R_i(d)$, over all curves $C \in \mathcal{C}$. That is, for each node u in the first level, we construct a search tree for the subset of x-coordinates of vertices p_r^i which corresponds to the canonical set of u. Levels three and four of the data structure correspond to the bottom and top sides, respectively, of the rectangles $R_i(d)$, over all curves $C \in \mathcal{C}$, and they are constructed using the y-coordinates of the vertices p_b^i and the y-coordinates of the vertices p_t^i, respectively. The fifth level is constructed as follows. For each node u in the fourth level, we construct a search tree for the subset of x-coordinates of vertices $\overline{p}_\ell^i$ which corresponds to the canonical set of u; that is, if the y-coordinate of p_t^j is in u's canonical subset, then the x-coordinate of $\overline{p}_\ell^j$ is in the subset corresponding to u's canonical set. The bottom four levels correspond to the four sides of the rectangles $\overline{R}_i(d)$ and are built using the x-coordinates of the vertices $\overline{p}_\ell^i$, the x-coordinates of the vertices $\overline{p}_r^i$, the y-coordinates of the vertices $\overline{p}_b^i$, and the y-coordinates of the vertices $\overline{p}_t^i$, respectively.

The Query Algorithm. Given a segment $s = ab$ and a distance $d > 0$, we can use our data structure to determine whether there exists a curve $C \in \mathcal{C}$, such that $d_{dF}(s, C) \leq d$. The search in the first and second levels of the data structure is done with $a.x$, the x-coordinate of a, in the third and fourth levels with $a.y$, in the fifth and sixth levels with $b.x$ and in the last two levels with $b.y$. When searching in the first level, instead of performing a comparison between $a.x$ and the value v that is stored in the current node (which is an x-coordinate of some vertex p_ℓ^i), we determine whether $a.x \geq v - d$. Similarly, when searching in the second level, at each node that we visit we determine whether $a.x \leq v + d$, where v is the value that is stored in the node, etc.

Notice that if we store the list of curves that are represented in the canonical subset of each node in the bottom (i.e., eighth) level of the structure, then curves whose distance from s is at most d may also be reported in additional time roughly linear in their number.

Finding the Closest Curve. Let $s = ab$ be a segment, let C be the curve in $\mathcal{C}$ that is closest to s and set $d^* = d_{dF}(s, C)$. Then, there exists $1 \leq i \leq m - 1$, such that $a \in R_i(d^*)$ and $b \in \overline{R}_i(d^*)$. Moreover, one of the endpoints a or b lies on the boundary of its rectangle, since, otherwise, we could shrink the rectangles without 'losing' the endpoints. Assume without loss of generality that a lies on the left side of $R_i(d^*)$. Then, the difference between the x-coordinate of the vertex p_ℓ^i and $a.x$ is exactly d^*. This implies that we can find d^* by performing a binary search in the set of all x-coordinates of vertices of curves in $\mathcal{C}$. In each step of the binary search, we need to determine whether $d \geq d^*$, where $d = v - a.x$ and v is the current x-coordinate, and our goal is to find the smallest such d for which the answer is still yes. We resolve a comparison by calling our data structure with the appropriate distance d. Since we do not know which of the two endpoints, a or b, lies on the boundary of its rectangle and on which of its sides, we perform 8 binary searches, where each search returns a candidate distance. Finally, the smallest among these 8 candidate distances is the desired d^*.

In other words, we perform 4 binary searches in the set of all x-coordinates of vertices of curves in $\mathcal{C}$. In the first we search for the smallest distance among the distances $d_\ell = v - a.x$ for which there exists a curve at distance at most d_ℓ from s; in the second we search for the smallest distance $d_r = a.x - v$ for which there exists a curve at distance at most d_r from s; in the third we search for the smallest distance $\overline{d}_\ell = v - b.x$ for which there exists a curve at distance at most $\overline{d}_\ell$ from s; and in the fourth we search for the smallest distance $\overline{d}_r = b.x - v$ for which there exists a curve at distance at most $\overline{d}_r$ from s. We also perform 4 binary searches in the set of all y-coordinates of vertices of curves in $\mathcal{C}$, obtaining the candidates d_b, d_t, $\overline{d}_b$, and $\overline{d}_t$. We then return the distance $d^* = \min\{d_\ell, d_r, \overline{d}_\ell, \overline{d}_r, d_b, d_u, \overline{d}_b, \overline{d}_u\}$.

Theorem 2. *Given a set $\mathcal{C}$ of n curves, each of size m, one can construct a search structure of size $O(nm \log^7(nm))$ for segment nearest-curve queries. Given a query segment s, one can find in $O(\log^8(nm))$ time the curve $C \in \mathcal{C}$ and distance d^* such that $d_{dF}(s, C) = d^*$ and $d^* \leq d_{dF}(s, C')$ for all $C' \in \mathcal{C}$, under the L_∞ metric.*

3.2 Input Is a Set of Segments

Let $\mathcal{S} = \{s_1, \ldots, s_n\}$ be the input set of segments. Given a query curve $Q = (p_1, \ldots, p_m)$, the task is to find a segment $s = ab \in \mathcal{S}$ such that $d_{dF}(Q, s) = \min_{s' \in \mathcal{S}} d_{dF}(Q, s')$, after suitably preprocessing $\mathcal{S}$. We use an overall approach similar to that used in Sect. 3.1, however the details of the implementation of the data structure and algorithm differ.

The Data Structure. Preprocess the input $\mathcal{S}$ into a four-level search structure $\mathcal{T}$ consisting of a two-dimensional range tree containing the endpoints a, and where the associated structure for each node in the second level of the tree is another two-dimensional range tree containing the endpoints b corresponding to the points in the canonical subset of the node.

This structure answers queries consisting of a pair of two-dimensional ranges (i.e., rectangles) $(R, \overline{R})$ and returns all segments $s = ab$ such that $a \in R$ and $b \in \overline{R}$. The preprocessing time for the structure is $O(n \log^4 n)$, and the storage is $O(n \log^3 n)$. Querying the structure with two rectangles requires $O(\log^3 n)$ time, by applying fractional cascading [28].

The Query Algorithm. Consider the decision version of the problem where, given a query curve Q and a distance d, the objective is to determine if there exists a segment $s \in S$ with $d_{dF}(s, Q) \leq d$. Observation 1 implies that it is sufficient to query the search structure $\mathcal{T}$ with the pair of rectangles $(R_i(d), \overline{R}_i(d))$ of the curve Q, for all $1 \leq i \leq m - 1$. If $\mathcal{T}$ returns at least one segment for any of the partitions, then this segment is within distance d of Q.

As we traverse the curve Q left-to-right, the bounding box of $Q[1, i]$ can be computed at constant incremental cost. For a fixed $d > 0$, each rectangle $R_i(d)$ can be constructed from the corresponding bounding box in constant time. Rectangle $\overline{R}_i(d)$ can be handled similarly by a reverse traversal. Hence all the rectangles can be computed in time $O(m)$, for a fixed d. Each pair of rectangles requires a query in $\mathcal{T}$, and thus the time required to answer the decision problem is $O(m \log^3 n)$.

Finding the Closest Segment. In order to determine the nearest segment s to Q, we claim, using an argument similar to that in Sect. 3.1, for a segment $s = ab$ of distance d^* from Q that either a lies on the boundary of $R_i(d^*)$ or b lies on the boundary of $\overline{R}_i(d^*)$ for some $1 \leq i < m$. Thus, in order to determine the value of d^* it suffices to search over all $8m$ pairs of rectangles where either a or b lies on one of the eight sides of the obtained query rectangles. The sorted list of candidate values of d for each side can be computed in $O(n)$ time from a sorted list of the corresponding x- or y-coordinates of a or b. The smallest value of d for each side is then obtained by a binary search of the sorted list of candidate values. For each of the $O(\log n)$ evaluated values d, a call to $\mathcal{T}$ decides on the existence of a segment within d of Q.

Theorem 3. *Given an input S of n segments, a search structure can be pre-processed in $O(n \log^4 n)$ time and requiring $O(n \log^3 n)$ storage that can answer the following. For a query curve Q of m vertices, find the segment $s^* \in S$ and distance d^* such that $d_{dF}(Q, s^*) = d^*$ and $d_{dF}(Q, s) \geq d^*$ for all $s \in S$ under the L_∞ metric. The time to answer the query is $O(m \log^4 n)$.*

3.3 NNC Under Translation and L_∞ Metric

An analogous approach yields algorithms with similar running times for the problems under translation. The algorithms are presented in [7], and are summarized in the following two theorems. Let s_t and C_t be the images of segment s and curve C, respectively, under the translation t. When the query is a segment, we have:

Theorem 4. *Given a set C of n curves, each of size m, one can construct a search structure of size $O(nm \log^4(nm))$, such that, given a query segment s, one can find in $O(\log^6(nm))$ time the curve $C \in C$ nearest to s under translation, that is the curve minimizing $\min_t d_{dF}(s_t, C')$, where the discrete Fréchet distance is computed using the L_∞ metric.*

When the input is a set of segments, we have the following theorem.

Theorem 5. *Given a set S of n segments, one can construct a search structure of size $O(n \log^2 n)$, so that, given a query curve Q of size m, one can find in $O(m \log^2 n)$ time the segment $s \in S$ nearest to Q under translation, that is the segment minimizing $\min_t d_{dF}(Q, s_t')$, where the discrete Fréchet distance is computed using the L_∞ metric.*

4   NNC and L_2 Metric

In this section, we present algorithms for approximate nearest-neighbor search under the discrete Fréchet distance using L_2. Notice that the algorithms from Sect. 3 for the L_∞ version of the problem, already give $\sqrt{2}$-approximation algorithms for the L_2 version. Next, we provide $(1 + \varepsilon)$-approximation algorithms. The details can be found in [7].

Theorem 6. *Given a set C of n curves, each of size m, and $0 < \varepsilon \leq 1$, one can construct a search structure of size $O(\frac{n}{\varepsilon^4} \log^4(\frac{n}{\varepsilon}))$ for approximate segment nearest-neighbor queries. Given a query segment s, one can find in $O(\log^5(\frac{n}{\varepsilon}))$ time a curve $C' \in C$ such that $d_{dF}(s, C') \leq (1+\varepsilon)d_{dF}(s, C)$, under the L_2 metric, where C is the curve in C closest to s.*

Theorem 7. *Given an input S of n segments, and $0 < \varepsilon \leq 1$, one can construct a search structure of size $O(n \log^{O(\frac{1}{\sqrt{\varepsilon}})} n)$ for approximate segment nearest-neighbor queries. Given a query curve Q of size m, one can find in $O(m \log^{O(\frac{1}{\sqrt{\varepsilon}})} n)$ time a segment $s' \in S$ such that $d_{dF}(s', Q) \leq (1 + \varepsilon)d_{dF}(s, Q)$, under the L_2 metric, where s is the segment in S closest to Q.*

5 $(1, 2)$-CENTER

The objective of the $(1, 2)$-CENTER problem is to find a segment s such that $\max_{C_i \in C} d_{dF}(s, C_i)$ is minimized. This can be reformulated equivalenly as: Find a pair of balls $(B, \overline{B})$, such that (i) for each curve $C \in C$, there exists a partition at $1 \leq i < m$ of C into prefix $C[1, i]$ and suffix $C[i + 1, m]$, with $C[1, i] \subseteq B$ and $C[i + 1, m] \subseteq \overline{B}$, and (ii) the radius of the larger ball is minimized.

5.1 $(1, 2)$-CENTER and L_∞ Metric

An optimal solution to the $(1, 2)$-CENTER problem under the L_∞ metric is a pair of squares $(S, \overline{S})$, where S contains all the prefix vertices and $\overline{S}$ contains all the suffix vertices. Assume that the optimal radius is r^*, and that it is determined by S, i.e., the radius of S is r^* and the radius of $\overline{S}$ is at most r^*. Then, there must exist two *determining vertices* p, p', belonging to the prefixes of their respective curves, such that p and p' lie on opposite sides of the boundary of S. Clearly, $||p - p'||_\infty = 2r^*$. Let the positive normal direction of the sides be the *determining direction* of the solution.

The proofs for Lemmas 1 and 2, below, are given in [7]. Let R be the axis-aligned bounding rectangle of $C_1 \cup \cdots \cup C_n$, and denote by e_ℓ, e_r, e_t, and e_b the left, right, top, and bottom edges of R, respectively.

Lemma 1. *At least one of p, p' must lie on the boundary of R.*

We say that a corner of S (or $\overline{S}$) *coincides* with a corner of R when the corner points are incident, and they are both of the same type, i.e., top-left, bottom-right, etc.

Lemma 2. *There exists an optimal solution $(S, \overline{S})$ where at least one corner of S or $\overline{S}$ coincides with a corner of R.*

Lemma 2 implies that for a given input $\mathcal{C}$ where the determining vertices are in S, there must exist an optimal solution where S is positioned so that one of its corners coincides with a corner of the bounding rectangle, and that one of the determining vertices is on the boundary of R. The optimal solution can thus be found by testing all possible candidate squares that satisfy these properties and returning the valid solution that yields the smallest radius. The algorithm presented in the sequel will compute the radius r^* of an optimal solution $(S^*, \overline{S^*})$ such that r^* is determined by the prefix square S^*, see Fig. 2. The solution where r^* is determined by $\overline{S^*}$ can be computed in a symmetric manner.

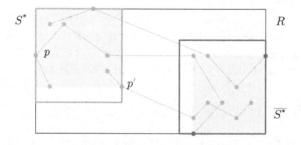

Fig. 2. The optimal solution is characterized by a pair of points p, p' lying on the boundary of S^*, and a corner of S^* coincides with a corner of R.

For each corner v of the bounding rectangle R, we sort the $(m-2)n$ vertices in $C_1 \cup \cdots \cup C_n$ that are not endpoints—the initial vertex of each curve must always be contained in the prefix, and the final vertex in the suffix—by their L_∞ distance from v. Each vertex p in this ordering is associated with a square S of radius $\|v - p\|_\infty / 2$, coinciding with R at corner v.

A sequential pass is made over the vertices, and their respective squares S, and for each S we compute the radius of S and $\overline{S}$ using the following data structures. We maintain a balanced binary tree T_C for each curve $C \in \mathcal{C}$, where the leaves of T_C correspond to the vertices of C, in order. Each node of the tree contains a single bit: The bit at a leaf node corresponding to vertex p_j indicates whether $p_j \in S$, where S is the current square. The value of the bit at a leaf of T_C can be updated in $O(\log m)$ time. The bit of an internal node is 1 if and only if all the bits in the leaves of its subtree are 1, and thus the longest prefix of C can be determined in $O(\log m)$ time. At each step in the pass, the radius of $\overline{S}$ must also be computed, and this is obtained by determining the bounding box of the suffix vertices. Thus, two balanced binary trees are maintained: $\overline{T}_x$ contains a leaf for each of the suffix vertices ordered by their x-coordinate; and $\overline{T}_y$ where the leaves are ordered by the y-coordinate. The extremal vertices that determine the bounding box can be determined in $O(\log mn)$ time. Finally, the current optimal squares S^* and $\overline{S^*}$, and the radius r^* of S^* are persisted.

The trees $T_{C_1}, \ldots, T_{C_n}$ are constructed with all bits initialized to 0, except for the bit corresponding to the initial vertex in each tree which is set to 1, taking $O(nm)$ time in total. $\overline{T}_x$ and $\overline{T}_y$ are initialized to contain all non-initial vertices in $O(mn \log mn)$ time. The optimal square S^* containing all the initial vertices is computed, and $\overline{S^*}$ is set to contain the remaining vertices. The optimal radius r^* is the larger of the radii induced by S^* and $\overline{S^*}$.

At the step in the pass for vertex p of curve C_j whose associated square is S, the leaf of T_C corresponding to p is updated from 0 to 1 in $O(\log m)$ time. The index i of the longest prefix covered by S can then be determined, also in $O(\log m)$ time. The vertices from C_j that are now in the prefix must be deleted from $\overline{T}_x$ and $\overline{T}_y$, and although there may be $O(m)$ of them in any iteration, each will be deleted exactly once, and so the total update time over the entire sequential pass is $O(mn \log mn)$. The radius of the square S is $\|v - p\|_\infty / 2$, and the radius of $\overline{S}$ can be computed in $O(\log mn)$ time as half the larger of x- and y-extent of the suffix bounding box. The optimal squares S^*, $\overline{S^*}$, and the cost r^* are updated if the radius of S determines the cost, and the radius of S is less than the existing value of r^*.

Finally, we return the optimal pair of squares $(S^*, \overline{S^*})$ with the minimal cost r^*.

Theorem 8. *Given a set of curves C as input, an optimal solution to the $(1,2)$-* CENTER *problem using the discrete Fréchet distance under the L_∞ metric can be computed in time $O(mn \log mn)$ using $O(mn)$ storage.*

5.2 $(1,2)$-CENTER Under Translation and L_∞ Metric

The $(1,2)$-CENTER problem under translation and the L_∞ metric can be solved using a similar approach. The solution is presented in [7], leading to the following theorem.

Theorem 9. *Given a set of curves C as input, an optimal solution to the $(1,2)$-* CENTER *problem under translation using the discrete Fréchet distance under the L_∞ metric can be computed in $O(nm)$ time and $O(nm)$ space.*

5.3　$(1, 2)$-CENTER and L_2 metric

For the $(1, 2)$-CENTER problem and L_2 we need some more sophisticated arguments, but again we use a similar basic approach. The solution is presented in [7], yielding the following theorem.

Theorem 10. *Given a set of curves C as input, an optimal solution to the $(1, 2)$-CENTER problem using the discrete Fréchet distance under the L_2 metric can be computed in $O(n^2 m^3 \log^3(nm))$ time and $O(n^2 m^2)$ space.*

Acknowledgements. B. Aronov was supported by NSF grants CCF-12-18791 and CCF-15-40656, and by grant 2014/170 from the US-Israel Binational Science Foundation. O. Filtser was supported by the Israeli Ministry of Science, Technology & Space, and by grant 2014/170 from the US-Israel Binational Science Foundation. Most of the work on this project by M. Horton was performed while visiting the Department of Computer Science and Engineering at the Tandon School of Engineering, New York University in the spring/summer of 2018, partially supported by NSF grant CCF-12-18791. M. Katz was supported by grant 1884/16 from the Israel Science Foundation and by grant 2014/170 from the US-Israel Binational Science Foundation. Part of the work on this project by M. Katz was performed while visiting the Department of Computer Science and Engineering at the Tandon School of Engineering, New York University in the spring of 2018, partially supported by NSF grants CCF-12-18791 and CCF-15-40656. Work of K. Sheikhan on this paper was performed while at the Tandon School of Engineering, New York University, supported by NSF grant CCF-12-18791.

References

1. Abraham, C., Cornillon, P.A., Matzner-Lober, E., Molinari, N.: Unsupervised curve clustering using b-splines. Scand. J. Stat. **30**(3), 581–595 (2003). https://doi.org/10.1111/1467-9469.00350
2. Afshani, P., Driemel, A.: On the complexity of range searching among curves. In: Proceedings of the 29th Annual ACM-SIAM Symposium on Discrete Algorithms, pp. 898–917. SIAM (2018)
3. Agarwal, P.K., Procopiuc, C.M.: Exact and approximation algorithms for clustering. Algorithmica **33**(2), 201–226 (2002). https://doi.org/10.1007/s00453-001-0110-y
4. Agarwal, P.K., Avraham, R.B., Kaplan, H., Sharir, M.: Computing the discrete Fréchet distance in subquadratic time. SIAM J. Comput. **43**(2), 429–449 (2014). https://doi.org/10.1137/130920526
5. Alewijnse, S.P.A., Buchin, K., Buchin, M., Kölzsch, A., Kruckenberg, H., Westenberg, M.A.: A framework for trajectory segmentation by stable criteria. In: Proceedings of the 22nd ACM SIGSPATIAL International Conference on Advances in Geographic Information Systems. ACM Press, Dallas, November 2014. https://doi.org/10.1145/2666310.2666415
6. Alt, H., Godau, M.: Computing the Fréchet distance between two polygonal curves. Intern. J. Comput. Geom. Appl. **05**(01n02), 75–91 (1995). https://doi.org/10.1142/S0218195995000064
7. Aronov, B., Filtser, O., Horton, M., Katz, M.J., Sheikhan, K.: Efficient nearest-neighbor query and clustering of planar curves. arXiv preprint arXiv:1904.11026 (2019)

8. de Berg, M., Cook, A.F., Gudmundsson, J.: Fast Fréchet queries. Comput. Geom. **46**(6), 747–755 (2013). https://doi.org/10.1016/j.comgeo.2012.11.006
9. de Berg, M., Gudmundsson, J., Mehrabi, A.D.: A dynamic data structure for approximate proximity queries in trajectory data. In: Proceedings of the 25th ACM SIGSPATIAL International Conference on Advances in Geographic Information Systems, p. 48. ACM (2017)
10. Berndt, D.J., Clifford, J.: Using dynamic time warping to find patterns in time-series. In: Papers from the AAAI Knowledge Discovery in Databases Workshop: Technical report WS-94-03, pp. 359–370. AAAI Press, Seattle, July 1994
11. Bringmann, K.: Why walking the dog takes time: Fréchet distance has no strongly subquadratic algorithms unless SETH fails. In: Proceedings of the 55th IEEE Symposium Foundations of Computer Science. IEEE, Philadelphia, October 2014. https://doi.org/10.1109/focs.2014.76
12. Bringmann, K., Mulzer, W.: Approximability of the discrete Fréchet distance. J. Comput. Geom. **7**(2), 46–76 (2016). http://jocg.org/index.php/jocg/article/view/261
13. Buchin, K., et al. Approximating (k, l)-center clustering for curves. In: Proceedings of the 30th Annual ACM-SIAM Symposium on Discrete Algorithms, San Diego, California, USA, 6–9 January 2019, pp. 2922–2938 (2019). https://doi.org/10.1137/1.9781611975482.181
14. Chiou, J.M., Li, P.L.: Functional clustering and identifying substructures of longitudinal data. J. Roy. Stat. Soc.: Ser. B (Stat. Methodol.) **69**(4), 679–699 (2007). https://doi.org/10.1111/j.1467-9868.2007.00605.x
15. Driemel, A., Har-Peled, S.: Jaywalking your dog—computing the Fréchet distance with shortcuts. In: Proceedings of the 23rd ACM-SIAM Symposium on Discrete Algorithms, pp. 318–355. Society for Industrial and Applied Mathematics, Kyoto, January 2012. https://doi.org/10.1137/1.9781611973099.30
16. Driemel, A., Krivošija, A., Sohler, C.: Clustering time series under the Fréchet distance. In: Proceedings of the 27th ACM-SIAM Symposium on Discrete Algorithms, pp. 766–785. SIAM, January 2016. https://doi.org/10.1137/1.9781611974331.ch55
17. Driemel, A., Silvestri, F.: Locality-sensitive hashing of curves. In: Proceedings of the 33rd International Symposium on Computational Geometry, SoCG 2017, Brisbane, Australia, pp. 37:1–37:16 (2017). http://drops.dagstuhl.de/opus/volltexte/2017/7203
18. Eiter, T., Mannila, H.: Computing discrete Fréchet distance. Technical report CD-TR 94/64, Christian Doppler Labor. für Expertensysteme, Technische Uni. Wien (1994)
19. Emiris, I.Z., Psarros, I.: Products of Euclidean metrics and applications to proximity questions among curves. In: Proceedings of the 34th International Symposium on Computational Geometry, SoCG 2018, 11–14 June 2018, Budapest, Hungary, pp. 37:1–37:13 (2018). https://doi.org/10.4230/LIPIcs.SoCG.2018.37. arXiv:1712.06471
20. Fréchet, M.M.: Sur quelques points du calcul fonctionnel. Rendiconti del Circolo Matematico di Palermo **22**(1), 1–72 (1906). https://doi.org/10.1007/BF03018603
21. Gonzalez, T.F.: Clustering to minimize the maximum intercluster distance. Theor. Comput. Sci. **38**, 293–306 (1985). https://doi.org/10.1016/0304-3975(85)90224-5
22. Gudmundsson, J., Horton, M.: Spatio-temporal analysis of team sports. ACM Comput. Surv. **50**(2), 1–34 (2017). https://doi.org/10.1145/3054132
23. Hausdorff, F.: Mengenlehre. Walter de Gruyter, Berlin (1927)
24. Hsu, W.L., Nemhauser, G.L.: Easy and hard bottleneck location problems. Discr. Appl. Math. **1**(3), 209–215 (1979). https://doi.org/10.1016/0166-218x(79)90044-1

25. Indyk, P.: Approximate nearest neighbor algorithms for Fréchet distance via product metrics. In: Proceedings of the 8th Symposium on Computational Geometry, pp. 102–106. ACM Press, Barcelona, June 2002. https://doi.org/10.1145/513400.513414

26. Indyk, P., Matoušek, J.: Low-distortion embeddings of finite metric spaces. In: Handbook of Discrete and Computational Geometry, 2 edn. Chapman and Hall/CRC, April 2004. https://doi.org/10.1201/9781420035315.ch8

27. Niu, H., Wang, J.: Volatility clustering and long memory of financial time series and financial price model. Digit. Signal Process. **23**(2), 489–498 (2013). https://doi.org/10.1016/j.dsp.2012.11.004

28. Willard, D.E., Lueker, G.S.: Adding range restriction capability to dynamic data structures. J. ACM **32**(3), 597–617 (1985). https://doi.org/10.1145/3828.3839

Positive-Instance Driven Dynamic Programming for Graph Searching

Max Bannach[1]([⊠])[iD] and Sebastian Berndt[2][iD]

[1] Institute for Theoretical Computer Science,
Universität zu Lübeck, Lübeck, Germany
bannach@tcs.uni-luebeck.de
[2] Department of Computer Science, Kiel University, Kiel, Germany
seb@informatik.uni-kiel.de

Abstract. Research on the similarity of a graph to being a tree – called the *treewidth* of the graph – has seen an enormous rise within the last decade, but a practically fast algorithm for this task has been discovered only recently by Tamaki (ESA 2017). It is based on dynamic programming and makes use of the fact that the number of positive subinstances is typically substantially smaller than the number of all subinstances. Algorithms producing only such subinstances are called *positive-instance driven* (PID). We give an alternative and intuitive view on this algorithm from the perspective of the corresponding configuration graphs in certain two-player games. This allows us to develop PID-algorithms for a wide range of important graph parameters such as treewidth, pathwidth, and treedepth. We analyse the worst case behaviour of the approach on some well-known graph classes and perform an experimental evaluation on real world and random graphs.

Keywords: Treewidth · Pathwidth · Treedepth · Graph searching

1 Introduction

Treewidth, a concept to measure the similarity of a graph to being a tree, is arguably one of the most used tools in modern combinatorial optimization. It is a cornerstone of parameterized algorithms [14] and its success has led to its integration into many different fields: For instance, treewidth and its close relatives treedepth and pathwidth have been theoretically studied in the context of machine learning [5,15,20], model-checking [3,32], SAT-solving [7,21,27], QBF-solving [12,18], CSP-solving [31,33], or ILPs [19,24,25,34,40]. Some of these results (e. g. [3,7,12,21,27,31–33]) show quite promising experimental results giving hope that the theoretical results lead to actual practical improvements.

To utilize the treewidth for this task, we have to be able to compute it quickly. More crucially, most algorithms also need a witness for this fact in form of a tree-decomposition. In theory we have a beautiful algorithm for this task [8], which is unfortunately known to *not* work in practice due to huge constants [38].

© Springer Nature Switzerland AG 2019
Z. Friggstad et al. (Eds.): WADS 2019, LNCS 11646, pp. 43–56, 2019.
https://doi.org/10.1007/978-3-030-24766-9_4

We may argue that, instead, a heuristic is sufficient, as the attached solver will work correctly independently of the actual treewidth – and the heuristic may produce a decomposition of "small enough" width. However, even a small error, something as "off by 5", may put the parameter to a computationally intractable range, as the dependency on the treewidth is usually at least exponential. It is therefore a very natural and important task to build practical fast algorithms to determine parameters as the treewidth or treedepth exactly.

To tackle this problem, the fpt-community came up with an implementation challenge: the PACE [16,17]. Besides many, one very important result of the challenge was a new combinatorial algorithm due to Hisao Tamaki, which computes the treewidth of an input graph exactly and astonishingly fast on a wide range of instances. An implementation of this algorithm by Tamaki himself [41] won the corresponding track in the PACE challenge in 2016 [16] and an alternative implementation due to Larisch and Salfelder [36] won in 2017 [17]. The algorithm is based on a dynamic program by Arnborg et al. [1] for computing tree decompositions. This algorithm has a game theoretic characterisation that we will utilities in order to apply Tamaki's approach to a broader range of problems. It should be noted, however, that Tamaki has improved his algorithm for the second iteration of the PACE by applying his framework to the algorithm by Bouchitté and Todinca [11,42]. This algorithm has a game theoretic characterisation as well [23], but it is unclear how this algorithm can be generalized to other parameters. Therefore, we focus on Tamaki's first algorithm and analyze it both, from a theoretical and a practical perspective. Furthermore, we will extend the algorithm to further graph parameters, which is surprisingly easy due to the new game-theoretic representation. In detail, our contributions are the following, but due to space constraints some of the proofs are only included in the technical report version.

Contribution I: A simple description of Tamaki's first algorithm. We describe Tamaki's algorithm based on a well-known graph searching game for treewidth. This provides a nice link to known theory and allows us to analyze the algorithm in depth.

Contribution II: Extending Tamaki's algorithm to other parameters. The game theoretic point-of-view allows us to extend the algorithm to various other parameters that can be defined in terms of similar games – including pathwidth, treedepth, and more.

Contribution III: Experimental and theoretical analysis. We provide, for the first time, theoretical bounds on the runtime of the algorithm on certain graph classes. Furthermore, we count the number of subinstances generated by the algorithm on various random and named graphs.

2   Graph Searching

A *tree decomposition* of a graph $G = (V, E)$ is a tuple (T, ι) consisting of a rooted tree T and a mapping ι from nodes of T to sets of vertices of G (called *bags*) such that (1) for all $v \in V$ the set $\{ x \mid v \in \iota(x) \}$ is nonempty and connected in

T, and (2) for every edge $\{v, w\} \in E$ there is a node m in T with $\{v, w\} \subseteq \iota(m)$. The *width* of a tree decomposition is the maximum size of one of its bags minus one, its *depth* is the maximum of the width and the depth of T. The *treewidth* of G, denoted by $tw(G)$, is the minimum width any tree decomposition of G must have. If T is a path we call (T, ι) a *path decomposition*; if for all nodes x, y of T we have $\iota(x) \subsetneq \iota(y)$ whenever y is a descendent of x we call (T, ι) a *treedepth decomposition*; and if on any path from the root to a leaf there are at most q nodes with more then one children we call (T, ι) a *q-branched tree decomposition*. Analogous to the treewidth, we define the *pathwidth* and *q-branched-treewidth* of G, denoted by $pw(G)$ and $tw_q(G)$, respectively. The *treedepth* $td(G)$ is the minimum depth any treedepth decomposition must have. Another important variant of this parameter is *dependency-treewidth*, which is used primarily in the context of quantified Boolean formulas [18]. For a graph $G = (V, E)$ and a partial order $\prec$ of V the dependency-treewidth $dtw(G)$ is the minimum width any tree-decomposition (T, ι) with the following property must have: Consider the natural partial order $\leq_T$ that T induces on its nodes, where the root is the smallest elements and the leaves form the maximal elements; define for any $v \in V$ the node $F_v(T)$ that is the $\leq_T$-minimal node t with $v \in \iota(t)$ (which is well defined); then define a partial order $<_T$ on V such that $u <_T v \iff F_u(T) \leq_T F_v(T)$; finally for all $u, v \in V$ it must hold that $F_u(T) <_T F_v(T)$ implies that that $u \prec v$ does not hold.

We study classical graph searching in a general setting proposed by Fomin, Fraigniaud, and Nisse [22]. The input is an undirected graph $G = (V, E)$ and a number $k \in \mathbb{N}$, and the question is whether a team of k searchers can catch an *invisible* fugitive on G by the following set of rules: At the beginning, the fugitive is placed at a vertex of her choice and at any time, she knows the position of the searchers. In every turn she may move with *unlimited speed* along edges of the graph, but may never cross a vertex occupied by a searcher. This implies that the fugitive does not occupy a single vertex but rather a subgraph, which is separated from the rest of the graph by the searchers. The vertices of this subgraph are called *contaminated* and at the start of the game all vertices are contaminated. The searchers, trying to catch the fugitive, can perform one of the following operations during their turn:

1. *place* a searcher on a contaminated vertex;
2. *remove* a searcher from a vertex;
3. *reveal* the current position of the fugitive.

When a searcher is placed on a contaminated vertex it becomes *clean*. When a searcher is removed from a vertex v, the vertex may become *recontaminated* if there is a contaminated vertex adjacent to v. The searchers win the game if they manage to clean all vertices, i. e., if they catch the fugitive; the fugitive wins if, at any point, a recontamination occurs, or if she can escape infinitely long. Note that this implies that the searchers have to catch the fugitive in a *monotone* way. A priori one could assume that the later condition gives the fugitive an advantage (recontamination could be necessary for the cleaning strategy), however, a crucial result in graph searching is that "recontamination does not help" in all variants of the game that we consider [6, 26, 35, 37, 39].

2.1   Entering the Arena and the Colosseum

Our primary goal is to determine whether the searchers have a winning strategy. A folklore algorithm for this task is to construct an alternating graph $\text{arena}(G, k) = ((V_s \cup V_f), E_{\text{ar}})$ that contains for each position of the searchers ($S \subseteq V$ with $|S| \le k$) and each position of the fugitive ($f \in V$) two copies of the vertex (S, f), one in V_s and one in V_f (see e. g. Sect. 7.4 in [14]). Vertices in V_s correspond to a configuration in which the searchers do the next move (they are existential) and vertices in V_f correspond to fugitive moves (they are universal). The edges E_{ar} are constructed according to the possible moves. Clearly, our task is now reduced to the question whether there is an alternating path from a start configuration to some configuration in which the fugitive is caught. Since alternating paths can be computed in linear time (see e. g., Sect. 3.4 in [28]), we immediately obtain an $O(n^{k+1})$ algorithm.

Modeling a configuration of the game as tuple (S, f) comes, however, with a major drawback: The size of the arena does directly depend on n and k and does *not* depend on some further structure of the input. For instance, the arena of a path of length n and any other graph on n vertices will have the same size for any fixed value k. As the major goal of parameterized complexity is the understanding of structural parameters beyond the input size n, such a fixed-size approach is usually not practically feasible. In contrast, we will define the configuration graph $\text{colosseum}(G, k)$, which might be larger then $\text{arena}(G, k)$ in general, but is also "prettier" in the sense that it adapts to the input structure of the graph. Moreover, the resulting algorithms are *self-adapting* in the sense that it needs no knowledge about this special structure to make use of it (in constrast to other parameterized algorithms, where the parameter describing this structure needs to be given explicitly).

2.2   Simplifying the Game

Our definition is based upon a similar formulation by Fomin et al. [22], but we simplify the game to make it more accessible to our techniques. First of all, we restrict the fugitive in the following sense. Since she is invisible to the searchers and travels with unlimited speed, there is no need for her to take regular actions. Instead, the only moment when she is actually active is when the searchers perform a reveal. If C is the set of contaminated vertices, consisting of the induced components $C_1, \ldots, C_\ell$, a reveal will uncover the component in which the fugitive hides and, as a result, reduce C to C_i for some $1 \le i \le \ell$. The only task of the fugitive is, thus, to answer a reveal with such a number i. We call the whole process of the searcher performing a reveal, the fugitive answering it, and finally of reducing C to C_i a *reveal-move*.

We will also restrict the searchers by the concept of *implicit searcher removal*. Let $S \subseteq V(G)$ be the vertices currently occupied by the searchers, and let $C \subseteq V(G)$ be the set of contaminated vertices. We call a vertex $v \in S$ *covered* if every path between v and C contains a vertex $w \in S$ with $w \ne v$.

Lemma 1. *A covered searcher can be removed safely.*

Proof. As we have $N(v) \cap C = \emptyset$, the removal of v will not increase the contaminated area. Furthermore, at no later point of the game v can be recontaminated, unless a neighbor of v gets recontaminated as well (in which case the game would already be lost for the searchers). □

Lemma 2. *Only covered searchers can be removed safely.*

Proof. Since for any other vertex $w \in S$ we have $N(w) \cap C \neq \emptyset$, the removal of w would recontaminate w and, hence, would result in a defeat of the searchers. □

Both lemmas together imply that the searchers never have to decide to remove a searcher, but rather do it *implicitly*. We thus restrict the possible moves of the searchers to a combined move of placing a searcher and *immediately* removing the searchers from all covered vertices. We call this a *fly-move*. Observe that the sequence of original moves mimicked by a fly-move does not contain a reveal and, thus, may be performed independently of any action of the fugitive.

We are now ready to define the colosseum. We could, as for the arena, define it as an alternating graph. However, as the searcher is the only player that performs actions in our simplified game, we find it more natural to express this game as *edge-alternating graph* – a generalization of alternating graphs. An edge-alternating graph is a triple $H = (V, E, A)$ consisting of a *vertex set* V, an existential edge relation $E \subseteq V \times V$, and an universal edge relation $A \subseteq V \times V$. We define the neighborhood of a vertex v as $N_\exists(v) := \{ w \mid (v,w) \in E \}$, $N_\forall(v) = \{ w \mid (v,w) \in A \}$, and $N_H(v) = N_\exists(v) \cup N_\forall(v)$. An *edge-alternating s-t-path* is a set $P \subseteq V$ such that (1) $s, t \in P$ and (2) for all $v \in P$ with $v \neq t$ we have either $N_\exists(v) \cap P \neq \emptyset$ or $\emptyset \neq N_\forall(v) \subseteq P$ or both. We write $s \prec t$ if such a path exists and define $\mathcal{B}(Q) = \{ v \mid v \in Q \lor (\exists w \in Q : v \prec w) \}$ for $Q \subseteq V$ as the set of vertices on edge-alternating paths leading to Q. We say that an edge-alternating s-t-path P is *q-branched*, if (i) H is acyclic and (ii) every (classical) directed path π from s to t in H with $\pi \subseteq P$ uses at most q universal edges.

For an undirected graph $G = (V, E)$ and a number $k \in \mathbb{N}$ we now define the colosseum(G, k) to be the edge-alternating graph H with vertex set $V(H) = \{ C \mid \emptyset \neq C \subseteq V \text{ and } |N_G(C)| \leq k \}$ and the following edge sets: for all pairs $C, C' \in V(H)$ there is an edge $e = (C, C') \in E(H)$ if, and only if, $C \setminus \{v\} = C'$ for some $v \in C$ and $|N_G(C)| < k$; furthermore, for all $C \in V(H)$ with at least two components $C_1, \ldots, C_\ell$ we have edges $(C, C_i) \in A(H)$. The *start configuration* of the game is the vertex $C = V$, that is, all vertices are contaminated. We define $Q = \{ \{v\} \subseteq V : |N_G(\{v\})| < k \}$ to be the set of *winning configurations*, as at least one searcher is available to catch the fugitive. Therefore, the searchers have a winning strategy if, and only if, $V \in \mathcal{B}(Q)$ and we will therefore refer to $\mathcal{B}(Q)$ as the *winning region*. Observe that the colosseum is acyclic (that is, the digraph $(V, E \cup A)$ is acyclic) as we have for every edge (C, C') that $|C| > |C'|$, and observe further that Q is a subset of the sinks of H. Hence, we can test if $V \in \mathcal{B}(Q)$ in time $O(|\text{colosseum}(G, k))|)$. Finally, note that the size of colosseum(G, k) may be of order 2^n rather than n^{k+1}, giving us a slightly worse overall runtime.

The reader that is familiar with graph searching or with exact algorithms for treewidth will probably notice the similarity of the colosseum and an exact "Robertson–Seymour fashioned" algorithm for that task. In fact, the colosseum is essentially the configuration graph of such a procedure if it is used with memoization.

2.3   Fighting in the Pit

Both algorithms introduced in the previous section run asymptotically in the size of the generated configuration graph $|\text{arena}(G, k)|$ or $|\text{colosseum}(G, k)|$. Both of these graphs might be very large, as the arena has fixed size of order $O(n^{k+1})$, while the colosseum may even have size $O(2^n)$. Additionally, both graphs contain many unnecessary configurations, that is, configurations that are not contained in the winning region of the searchers. In the light of dynamic programming this is the same as listing all possible configurations; and in the light of positive-instance driven dynamic programming we would like to list only the positive instances – which is exactly the winning region in this context.

To realize this idea, we consider the *pit* inside the colosseum, which is the area where only true champions can survive – formally we define $\text{pit}(G, k)$ as the subgraph of $\text{colosseum}(G, k)$ induced by $\mathcal{B}(Q)$, that is, as the induced subgraph on the winning region. The key-insight is that $|\text{pit}(G, k)|$ may be smaller than $|\text{colosseum}(G, k)|$ or even $|\text{arena}(G, k)|$ on various graph classes. Our primary goal for the next section will therefore be the development of an algorithm that computes the pit in time $O(|\text{pit}(G, k)|^2)$.

3   Computing the Pit

Our aim for this section is to develop an algorithm that computes $\text{pit}(G, k)$. Of course, a simple way to do this is to compute the whole colosseum and to extract the pit afterwards. However, this will cost time $O(2^n)$ and is surely not what we aim for. Our algorithm traverses the colosseum "backwards" by starting at the set Q of winning configurations and by uncovering $\mathcal{B}(Q)$ layer by layer. In order to achieve this, we need to compute the predecessors of a configuration C. This is easy if C was reached by a fly-move as we can simply enumerate the n possible predecessors. Reversing a reveal-move, that is, finding the universal predecessors, is significantly more involved. A simple approach is to test for every subset of already explored configurations if we can "glue" them together – but this would result in an even worse runtime of $2^{|\text{pit}(G,k)|}$. Fortunately, we can avoid this exponential blow-up as the colosseum has the following useful property:

Definition 1 (Universal Consistent). *We say that an edge-alternating graph $H = (V, E, A)$ is* universal consistent *with respect to a set $Q \subseteq V$ if for all $v \in V \setminus Q$ with $v \in \mathcal{B}(Q)$ and $N_\forall(v) = \{w_1, \ldots, w_r\}$ we have (1) $N_\forall(v) \subseteq \mathcal{B}(Q)$ and (2) for every $I \subseteq \{w_1, \ldots, w_r\}$ with $|I| \geq 2$ there is a vertex $v' \in V$ with $N_\forall(v') = I$ and $v' \in \mathcal{B}(Q)$.*

Intuitively, this definition implies that for every vertex with high universal-degree there is a set of vertices that we can arrange in a tree-like fashion to realize the same adjacency relation. This allows us to glue only two configurations at a time and, thus, removes the exponential dependency.

Lemma 3. *For every graph $G = (V, E)$ and number $k \in \mathbb{N}$, the edge-alternating graph* $\mathrm{colosseum}(G, k)$ *is universal consistent.*

Proof. For the first property just observe that "reveals do not harm" in the sense that if the searchers can catch the fugitive without knowing where she hides, they certainly can do if they do know.

For the second property consider any configuration $C \in V(H)$ that has universal edges to $C_1, \ldots, C_\ell$. By definition we have $|N(C)| \leq k$ and $N(C_i) \subseteq N(C)$ for all $1 \leq i \leq \ell$. Therefore we have for every $I \subseteq \{1, \ldots, \ell\}$ and $C' = \cup_{i \in I} C_i$ that $N(C') \subseteq N(C)$ and $|N(C')| \leq k$ and, thus, $C' \in V(H)$. □

We are now ready to formulate the algorithm for computing the pit shown in Listing 1.1. In essence, the algorithm runs in three phases: first it computes the set Q of winning configurations; then the winning region $\mathcal{B}(Q)$ (that is, the vertices of $\mathrm{pit}(G, k)$); and finally, it computes the edges of $\mathrm{pit}(G, k)$.

Theorem 1. *The algorithm* $\mathsf{Discover}(G, k)$ *finishes in at most* $O\big(|\mathcal{B}(Q)|^2 \cdot |V|^2\big)$ *steps and correctly outputs* $\mathrm{pit}(G, k)$.

Proof. The algorithm is supposed to compute Q in phase I, $\mathcal{B}(Q)$ in phase II, and the edges of $\mathrm{colosseum}(G, k)[\mathcal{B}(Q)]$ in phase III. First observe that Q is correctly computed in phase I by the definition of Q.

To show the correctness of the second phase we argue that the computed set $V(\mathrm{pit}(G, k))$ equals $\mathcal{B}(Q)$. Let us refer to the set $V(\mathrm{pit}(G, k))$ during the computation as K and observe that this is exactly the set of vertices inserted into the queue. We first show $K \subseteq \mathcal{B}(Q)$ by induction over the ith inserted vertex. The first vertex C_1 is in $\mathcal{B}(Q)$ as $C_1 \in Q$. Now consider C_i. As $C_i \in K$, it was either added in Line 14 or Line 18. In the first case there was a vertex $\tilde{C}_i \in K$ such that $C_i = \tilde{C}_i \cup \{v\}$ for some $v \in N(\tilde{C}_i)$. By the induction hypothesis we have $\tilde{C}_i \in \mathcal{B}(Q)$ and by the definition of the colosseum we have $(C_i, \tilde{C}_i) \in E(H)$ and, thus, $C_i \in \mathcal{B}(Q)$. In the second case there where vertices $\tilde{C}_i$ and $\hat{C}_i$ with $\tilde{C}_i, \hat{C}_i \in K$ and $C_i = \tilde{C}_i \cup \hat{C}_i$. By the induction hypothesis we have again $\tilde{C}_i, \hat{C}_i \in \mathcal{B}(Q)$. Let $t_1, \ldots, t_\ell$ be the connected components of $\tilde{C}_i$ and $\hat{C}_i$. Since the colosseum H is universal consistent with respect to Q by Lemma 3, we have $t_1, \ldots, t_\ell \in \mathcal{B}(Q)$. By the definition of the colosseum we have $N_\forall(C_i) = t_1, \ldots, t_\ell$ and, thus, $C_i \in \mathcal{B}(Q)$.

To see $\mathcal{B}(Q) \subseteq K$ consider for a contradiction the vertices of $\mathcal{B}(Q)$ in reversed topological order (recall that H is acyclic) and let C be the first vertex in this order with $C \in \mathcal{B}(Q)$ and $C \notin K$. If $C \in Q$ we have $C \in K$ by phase I and are done, so assume otherwise. Since $C \in \mathcal{B}(Q)$ we have either $N_\exists(C) \cap \mathcal{B}(Q) \neq \emptyset$ or $\emptyset \neq N_\forall(C) \subseteq \mathcal{B}(Q)$. In the first case there is a $\tilde{C} \in \mathcal{B}(Q)$ with $(C, \tilde{C}) \in E(H)$. Therefore, $\tilde{C}$ precedes C in the reversed topological order and, by the choice of

C, we have $\tilde{C} \in K$. Therefore, at some point of the algorithm $\tilde{C}$ gets extracted from the queue and, in Line 14, would add C to K, a contradiction.

In the second case there are vertices $t_1, \ldots, t_\ell$ with $N_\forall(C) = \{t_1, \ldots, t_\ell\}$ and $t_1, \ldots, t_\ell \in \mathcal{B}(Q)$. By the choice of C, we have again $t_1, \ldots, t_\ell \in K$. Since H is universal consistent with respect to Q, we have for every $I \subseteq \{1, \ldots, \ell\}$ that $\bigcup_{i \in I} t_i$ is contained in $\mathcal{B}(Q)$. In particular, the vertices $t_1 \cup t_2$, $t_3 \cup t_4$, $\ldots$, $t_{\ell-1} \cup t_\ell$ are contained in $\mathcal{B}(Q)$, and these elements are added to K whenever the t_i are processed (for simplicity assume here that ℓ is a power of 2). Once these elements are processed, Line 18 will also add their union, that is, vertices of the form $(t_1 \cup t_2) \cup (t_3 \cup t_4)$. In this way, the process will add vertices that correspond to increasing subgraphs of G to K, resulting ultimately in adding $\bigcup_{i=1}^{\ell} t_i = C$ into K, which is the contradiction we have been looking for.

Finally, once the set $\mathcal{B}(Q)$ is known, it is easy to compute the subgraph $\text{colosseum}(G, k)[\mathcal{B}(Q)]$, that is, to compute the edges of the subgraph induced by $\mathcal{B}(Q)$. Phase III essentially iterates over all vertices and adds edges according to the definition of the colosseum.

For the runtime, observe that the queue will contain exactly the set $\mathcal{B}(Q)$ and, for every element extracted, we search through the current $K' \subseteq \mathcal{B}(Q)$, which leads to the quadratic timebound of $|\mathcal{B}(Q)|^2$. Furthermore, we have to compute the neighborhood of every extracted element, and we have to test whether two such configurations intersect – both can easily be achieved in time $O(|V|^2)$. Finally, in phase III we have to compute connected components of the elements in $\mathcal{B}(Q)$, but since this is possible in time $O(|V| + |E|)$ per element, it is clearly possible in time $|\mathcal{B}(Q)| \cdot |V|^2$ for the whole graph. $\square$

Listing 1.1. Discover(G, k)

```
1   V(pit(G, k)) := ∅
2   E(pit(G, k)) := ∅
3   A(pit(G, k)) := ∅
4   initialize empty queue
5   // Phase I: compute Q
6   for v ∈ V(G) do
7       insert({v}, k − 1)
8   end
9   // Phase II: compute B(Q) = V(pit(G, k))
10  while queue not empty do
11      extract C from queue
12      // reverse fly−moves
13      for v ∈ N(C) do
14          insert(C ∪ {v}, k − 1)
15      end
16      // reverse reveal−moves
17      for C' ∈ V(pit(G, k)) with C ∩ C' = ∅ do
18          insert(C ∪ C', k)
19      end
20  end
21  // Phase III: compute E and A
22  discoverEdges()
23  return (V(pit(G, k)), E(pit(G, k)), A(pit(G, k)))
```

Listing 1.2. insert(C, t)

```
1   if C ∉ V(pit(G, k)) and |N_G(C)| ≤ t then
2       add C to V(pit(G, k))
3       insert C into queue
4   end
```

Listing 1.3. discoverEdges$()$

```
1   for C ∈ V(pit(G, k)) do
2       // add fly−move edges
3       for v ∈ C do
4           if C \ {v} ∈ V(pit(G, k)) then
5               add (C, C \ {v}) to E(pit(G, k))
6           end
7       end
8       // add reveal−move edges
9       let C_1, ..., C_ℓ be
10          the connected components of G[C]
11      if C_1, ..., C_ℓ ∈ K then
12          for i = 1 to ℓ do
13              add (C, C_i) to A(pit(G, k))
14          end
15      end
16  end
```

4   Distance Queries in Edge-Alternating Graphs

In the previous section we have discussed how to compute the pit for a given graph and a given value k. The computation of treewidth now boils down to a reachability problem within this pit. But, intuitively, the pit should be able to give us much more information. In the present section we formalize this claim: We will show that we can compute shortest edge-alternating paths. To get an intuition of "distance" in edge-alternating graphs think about such a graph as in our game and consider some vertex v. There is always one active player that may decide to take *one* existential edge (a fly-move in our game), or the player may decide to ask the opponent to make a move and, thus, has to handle *all* universal edges (a reveal-move in our game). From the point of view of the active player, the distance is thus the *minimum* over the minimum of the distances of the existential edges and the maximum of the universal edges.

Definition 2 (Edge-Alternating Distance). *Let $H = (V, E, A)$ be an edge-alternating graph with $v \in V$ and $Q \subseteq V$, let further $c_0 \in \mathbb{N}$ be a constant and $\omega_E : E \to \mathbb{N}$ and $\omega_A : A \to \mathbb{N}$ be weight functions. The distance $d(v, Q)$ from v to Q is inductively defined as $d(v, Q) = c_0$ for $v \in Q$ and otherwise:*

$$d(v, Q) = \min \left(\min_{w \in N_\exists(v)} (d(w, Q) + \omega_E(v, w)), \max_{w \in N_\forall(v)} (d(w, Q) + \omega_A(v, w)) \right).$$

Lemma 4. *Given an acyclic edge-alternating graph $H = (V, E, A)$, weight functions $\omega_E : E \to \mathbb{N}$ and $\omega_A : A \to \mathbb{N}$, a source vertex $s \in V$, a subset of the sinks Q, and a constant $c_0 \in \mathbb{N}$. The value $d(s, Q)$ can be computed in time $O(|V| + |E| + |A|)$ and a corresponding edge-alternating path can be computed in the same time.*

Proof. Since H is acyclic we can compute a topological order of V using the algorithm from [30]. We iterate over the vertices v in reversed order and compute the distance as follows: if v is a sink we either set $d(v, Q) = c_0$ or $d(v, Q) = \infty$, depending on whether we have $v \in Q$. If v is not a sink we have already computed $d(w, Q)$ for all $w \in N(v)$ and, hence, can compute $d(v, Q)$ by the formula of the definition. Since this algorithm has to consider every edge once, the whole algorithm runs in time $O(|V| + |E| + |A|)$. A path from s to Q of length $d(s, Q)$ can be found by backtracking the labels starting at s. □

Theorem 2. *Given a graph $G = (V, E)$ and a number $k \in \mathbb{N}$, we can decide in time $O(|\mathrm{pit}(G, k + 1)|^2 \cdot |V|^2)$ whether G has { treewidth, pathwidth, treedepth, q-branched-treewidth, dependency-treewidth } at most k.*

Sketch of Proof. All five problems have game theoretic characterizations in terms of the same search game with the same configuration set [6,22,26]. More precisely, they condense to various distance questions within the colosseum by assigning appropriate weights to the edges.

treewidth: To solve treewidth, it is sufficient to find *any* edge-alternating path from the vertex $C_s = V(G)$ to a vertex in Q. We can find a path by choosing ω_E and ω_A as $(x, y) \mapsto 0$, and by setting $c_0 = 0$.

pathwidth: In the pathwidth game, the searchers are not allowed to perform any reveal [6]. Hence, universal edges cannot be used and we set ω_A to $(x, y) \mapsto \infty$. By setting ω_E to $(x, y) \mapsto 0$ and $c_0 = 0$, we again only need to find some path from $V(G)$ to Q with weight less than ∞.

treedepth: In the game for treedepth, the searchers are not allowed to remove a placed searcher again [26]. Hence, the searchers can only use k existential edges. Choosing ω_E as $(x, y) \mapsto 1$, ω_A as $(x, y) \mapsto 0$, and $c_0 = 1$ is sufficient. We have to search a path of weight at most k.

q-branched-treewidth: For q-branched-treewidth we wish to use at most q reveals [22]. By choosing ω_E as $(x, y) \mapsto 0$, ω_A as $(x, y) \mapsto 1$, and $c_0 = 0$, we have to search for a path of weight at most q.

dependency-treewidth: This parameter is in essence defined via graph searching game that is equal to the game we study with some fly- and reveal-moves forbidden. Forbidding a move can be archived by setting the weight of the corresponding edge to ∞ and by searching for an edge-alternating path of weight less then ∞. □

5   Theoretical Bounds for Certain Graph Classes

In general, it is hard to compare the size of the arena, the colosseum, and the pit. For instance, already simple graph classes as paths (P_n) and stars (S_n) reveal that the colosseum may be smaller or larger than the arena (the arena has size $O(n^3)$ on both, but the colosseum has size $O(n)$ on P_n and $O(2^n)$ on S_n, both with regard to their optimal treewidth 1). However, experimental data of the PACE challenge [16,17] shows that the pit is very small in practice. In the following, we are thus interested in graph classes where we can give theoretical guarantees on the size of the pit. We will first show that the colosseum is indeed often smaller than the arena (Lemma 5) and furthermore, that the pit might be much smaller than the colosseum (Lemma 7).

Lemma 5. *For every connected claw-free graph $G = (V, E)$ and integer $k \in \mathbb{N}$, it holds that $|\text{colosseum}(G, k)| \leq \sum_{i=1}^{k} \binom{n}{i} \cdot 2^{2i} \in O(\binom{n}{k} \cdot 4^k)$.*

Proof. Observe that in a claw-free graph every $X \subseteq V$ separates G in at most $2 \cdot |X|$ components, as every component is connected to a vertex in X (since G is connected), but every vertex in X may be connected to at most two components (otherwise it forms a claw). In the colosseum, every configuration C corresponds to a separator $N(C)$ of size at most k, and there are at most $\sum_{i=1}^{k} \binom{n}{i}$ such separators. For each separator we may combine its associated components in an arbitrary fashion to build configurations of the colosseum, but since there are at most $2 \cdot i$ components, we can build at most $2^{2 \cdot i}$ configurations. □

We remark that the result of Lemma 5 can easily be extended to $K_{1,t}$-free graphs for every fixed t, and that this result is rather tight:

Lemma 6. *Let $G = (V, E)$ be a graph and $k \in \mathbb{N}$ be an integer. It holds that $|\text{colosseum}(G, k)| \geq \sum_{i=1}^{k} \binom{|V_i|}{i}$, where $V_i = \{v \in V : |N(v)| \geq i\}$.*

Proof. Let X be any subset of at most i vertices from V_i with $i \leq k$. As $|X| \leq i$, every vertex in X has a neighbour in $V \setminus X$. Hence, $N(V \setminus X) = X$ and thus $|N(V \setminus X)| \leq k$ and $V \setminus X \in V(\text{colosseum}(G, k))$.   □

We now show that the pit, on the other hand, can be substantially smaller than the colosseum even for graphs with many high-degree vertices. For $n, k \in \mathbb{N}$ with $n \geq 2k$, we define the graph $P_{n,k}$ on vertices $V(P_{n,k}) = \{v_0, v_1, \ldots, v_{n \cdot k}, v_{n \cdot k+1}\}$. For $i = 1, \ldots, n$, let $X_i = \{v_{(i-1) \cdot k+1}, v_{(i-1) \cdot k+2}, \ldots, v_{(i-1) \cdot k+k}\}$, $X_0 = \{v_0\}$, and $X_{n+1} = \{v_{n \cdot k+1}\}$. The edges $E(P_{n,k})$ are defined as

$$E(P_{n,k}) = \bigcup_{i=1}^{n} \{\{u, v\} \mid u, v \in X_i\} \cup \bigcup_{i=0}^{n} \{\{u, v\} \mid u \in X_i, v \in X_{i+1}\}.$$

Informally, $P_{n,k}$ is constructed by taking a path of length $n+2$ and replacing the inner vertices by cliques of size k that are completely connected to each other.

Lemma 7. *It holds:*

(i) $\text{tw}(P_{n,k}) = \text{pw}(P_{n,k}) = 2k - 1$;   *(ii)* $|\text{arena}(P_{n,k}, 2k)| = 2 \cdot \binom{n \cdot k+2}{2k+1}$;

(iii) $|\text{colosseum}(P_{n,k}, 2k)| \geq \sum_{i=1}^{2k} \binom{n \cdot k}{i}$;   *(iv)* $|\text{pit}(P_{n,k}, 2k)| \in O(n^2 + n \cdot 2^{6k})$.

6   Experimental Estimation of the Pit Size

A heavily optimized version of the treewidth algorithm described above has been implemented in the Java library Jdrasil [2, 4]. To show the usefulness of our general approach, we experimentally compared the size of the pit, the arena, and the colosseum for various named graphs known from the DIMACS Coloring Challenge [29] or the PACE [16, 17]. For each graph the values are taken for the minimal k such that k searchers can win. Note that $|\text{arena}(G, k)| \leq |\text{pit}(G, k)|$ holds only in 6 of 24 cases, emphasized by underlining.

Graph	$|V|$	$|E|$	k	Pit	Arena	Col.	Graph	$|V|$	$|E|$	k	Pit	Arena	Col.
Grotzsch	11	20	6	1,235	<u>660</u>	1,853	Hoffman	16	32	7	5,851	25,740	30,270
Heawood	14	21	6	5,601	6,864	9,984	Friendship 10	21	30	3	57,554	11,970	58,695
Chvatal	12	24	7	3,170	<u>990</u>	3,895	Poussin	15	39	7	3,745	12,870	17,358
Goldner Harary	11	27	4	103	924	639	Markstroem	24	36	5	13,846	269,192	71,604
Sierpinski Gasket	15	27	4	488	6,006	2,494	McGee	24	36	8	487,883	2,615,008	1,905,241
Blanusa 2. Snark	18	27	5	861	37,128	15,413	Naru	24	36	7	41,623	1,470,942	708,044
Icosahedral	12	30	7	2,380	990	3,575	Clebsch	16	40	9	20,035	<u>16,016</u>	55,040
Pappus	18	27	7	54,004	87,516	97,970	Folkman	20	40	7	21,661	251,940	151,791
Desargues	20	30	7	85,146	251,940	202,661	Errera	17	45	7	3,527	48,620	42,418
Dodecahedral	20	30	7	112,924	251,940	207,165	Shrikhande	16	48	10	50,627	8,736	61,456
Flower Snark	20	30	7	79,842	251,940	203,473	Paley	17	68	12	114,479	<u>4,760</u>	129,474
Gen. Petersen	20	30	7	78,384	251,940	202,685	Goethals Seidel	16	72	12	54,833	<u>1,120</u>	65296

We have performed the same experiment on various random graph models. For each model we picked 25 graphs at random and build the mean over all instances, where each instance contributed values for its minimal k. We used all 3 models with $N = 25$ and, for the first two with $p = 0.33$; and for the later two with $K = 5$. For a detailed description of the models see for instance [10].

Finally, we observe the growth of the pit, the arena, and the colosseum for a fixed graph if we raise k from 2 to the optimal value. While the arena shows its

Model	$|\text{pit}(G, \text{OPT})|$	$|\text{arena}(G, \text{OPT})|$	$|\text{colosseum}(G, \text{OPT})|$
Erdős–Rényi	66,320	342,918	503,767
Watts Strogats	15,323	192,185	108,074
Barabási Albert	61,147	352,716	551,661

binomial behavior, the colosseum is in many early stages actually smaller then the arena. This effect is even more extreme for the pit, which is *very* small for k that are smaller then the optimum. This makes the technique especially well suited to establish lower bounds, an observation also made by Tamaki [42].

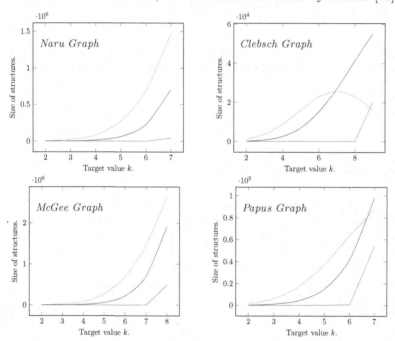

7   Conclusion and Outlook

Treewidth is one of the most useful graph parameters that is successfully used in many different areas. The Positive-Instance Driven algorithm of Tamaki has led to the first practically relevant algorithm for this parameter. We have formalized Tamaki's algorithm in the more general setting of graph searching, which has allowed us to (i) provide a clean and simple formulation; and (ii) extend the algorithm to many natural graph parameters. With a few further modification of the colosseum, our approach can also be used for the notion of *special-treewidth* [13]. We assume that a similar modification may also be possible for other parameters such as *spaghetti-treewidth* [9].

Acknowledgements. The authors would like to thank Jan Arne Telle and Fedor Fomin for helpful discussions about the topic and its presentation.

References

1. Arnborg, S., Corneil, D.G., Proskurowski, A.: Complexity of finding embeddings in a k-tree. SIAM J. Algebraic Discrete Methods **8**(2), 277–284 (1987)
2. Bannach, M., Berndt, S., , T.E.: Jdrasil (2017). https://github.com/maxbannach/Jdrasil. Accessed 09 Feb 2019
3. Bannach, M., Berndt, S.: Practical access to dynamic programming on tree decompositions. In: ESA, pp. 6:1–6:13 (2018)
4. Bannach, M., Berndt, S., Ehlers, T.: Jdrasil: a modular library for computing tree decompositions. In: SEA, pp. 28:1–28:21 (2017)
5. Berg, J., Järvisalo, M., Malone, B.: Learning optimal bounded treewidth bayesian networks via maximum satisfiability. In: Artificial Intelligence and Statistics, pp. 86–95 (2014)
6. Bienstock, D., Seymour, P.D.: Monotonicity in graph searching. J. Algorithms **12**(2), 239–245 (1991)
7. Bjesse, P., Kukula, J., Damiano, R., Stanion, T., Zhu, Y.: Guiding SAT diagnosis with tree decompositions. In: Giunchiglia, E., Tacchella, A. (eds.) SAT 2003. LNCS, vol. 2919, pp. 315–329. Springer, Heidelberg (2004). https://doi.org/10.1007/978-3-540-24605-3_24
8. Bodlaender, H.L.: A linear-time algorithm for finding tree-decompositions of small treewidth. SIAM J. Comput. **25**(6), 1305–1317 (1996)
9. Bodlaender, H.L., Kratsch, S., Kreuzen, V.J.C., Kwon, O., Ok, S.: Characterizing width two for variants of treewidth. Discrete Appl. Math. **216**, 29–46 (2017)
10. Bollobás, B.: Random graphs. In: Bollobás, B. (ed.) Modern Graph Theory, pp. 215–252. Springer, New York (1998). https://doi.org/10.1007/978-1-4612-0619-4_7
11. Bouchitté, V., Todinca, I.: Listing all potential maximal cliques of a graph. Theor. Comput. Sci. **276**(1–2), 17–32 (2002)
12. Charwat, G., Woltran, S.: Dynamic programming-based QBF solving. In: QBF, pp. 27–40 (2016)
13. Courcelle, B.: On the model-checking of monadic second-order formulas with edge set quantifications. Discrete Appl. Math. **160**(6), 866–887 (2012)
14. Cygan, M., et al.: Parameterized Algorithms. Springer, Berlin Heidelberg (2015). https://doi.org/10.1007/978-3-319-21275-3
15. Darwiche, A.: A differential approach to inference in bayesian networks. J. ACM (JACM) **50**(3), 280–305 (2003)
16. Dell, H., Husfeldt, T., Jansen, B., Kaski, P., Komusiewicz, C., Rosamond, F.: The first parameterized algorithms and computational experiments challenge. In: IPEC 2016, pp. 30:1–30:9 (2017)
17. Dell, H., Komusiewicz, C., Talmon, N., Weller, M.: The PACE 2017 parameterized algorithms and computational experiments challenge: the second iteration. In: IPEC 2017 (2018)
18. Eiben, E., Ganian, R., Ordyniak, S.: Small resolution proofs for QBF using dependency treewidth. In: STACS. LIPIcs, vol. 96, pp. 28:1–28:15. Schloss Dagstuhl - Leibniz-Zentrum fuer Informatik (2018)
19. Eisenbrand, F., Hunkenschröder, C., Klein, K.: Faster algorithms for integer programs with block structure. In: ICALP. LIPIcs, vol. 107, pp. 49:1–49:13. Schloss Dagstuhl - Leibniz-Zentrum fuer Informatik (2018)
20. Elidan, G., Gould, S.: Learning bounded treewidth bayesian networks. J. Mach. Learn. Res. **9**(Dec), 2699–2731 (2008)

21. Fichte, J.K., Hecher, M., Woltran, S., Zisser, M.: Weighted model counting on the GPU by exploiting small treewidth. In: ESA, pp. 28:1–28:16 (2018)
22. Fomin, F., Fraigniaud, P., Nisse, N.: Nondeterministic graph searching: from pathwidth to treewidth. Algorithmica **53**(3), 358–373 (2009)
23. Fomin, F.V., Kratsch, D.: Exact Exponential Algorithms. Texts in Theoretical Computer Science. An EATCS Series. Springer, Heidelberg (2010). https://doi.org/10.1007/978-3-642-16533-7
24. Ganian, R., Ordyniak, S.: The complexity landscape of decompositional parameters for ILP. Artif. Intell. **257**, 61–71 (2018)
25. Ganian, R., Ordyniak, S., Ramanujan, M.S.: Going beyond primal treewidth for (M)ILP. In: AAAI, pp. 815–821. AAAI Press (2017)
26. Giannopoulou, A., Hunter, P., Thilikos, D.: LIFO-search: a min-max theorem and a searching game for cycle-rank and tree-depth. Discrete Appl. Math. **160**(15), 2089–2097 (2012)
27. Habet, D., Paris, L., Terrioux, C.: A tree decomposition based approach to solve structured SAT instances. In: ICTAI, pp. 115–122 (2009)
28. Immerman, N.: Descriptive Complexity. Springer, Heidelberg (1999). https://doi.org/10.1007/978-1-4612-0539-5
29. Johnson, D.S., Trick, M.A.: Cliques, coloring, and satisfiability: second DIMACS implementation challenge, 11–13 October 1993, vol. 26. American Mathematical Soc. (1996)
30. Kahn, A.B.: Topological sorting of large networks. Commun. ACM **5**(11), 558–562 (1962)
31. Karakashian, S., Woodward, R.J., Choueiry, B.Y.: Improving the performance of consistency algorithms by localizing and bolstering propagation in a tree decomposition. In: AAAI (2013)
32. Kneis, J., Langer, A., Rossmanith, P.: Courcelle's theorem - a game-theoretic approach. Discrete Optim. **8**(4), 568–594 (2011)
33. Koster, A.M.C.A., van Hoesel, S.P.M., Kolen, A.W.J.: Solving partial constraint satisfaction problems with tree decomposition. Networks **40**(3), 170–180 (2002)
34. Koutecký, M., Levin, A., Onn, S.: A parameterized strongly polynomial algorithm for block structured integer programs. In: ICALP. LIPIcs, vol. 107, pp. 85:1–85:14. Schloss Dagstuhl - Leibniz-Zentrum fuer Informatik (2018)
35. LaPaugh, A.: Recontamination does not help to search a graph. J. ACM **40**, 224–245 (1993)
36. Larisch, L., Salfelder, F.: p17 (2017). https://github.com/freetdi/p17. Accessed 02 Aug 2017
37. Mazoit, F., Nisse, N.: Monotonicity of non-deterministic graph searching. TCS **399**(3), 169–178 (2008)
38. Röhrig, H.: Tree decomposition: a feasibility Study. Diploma thesis, Max-Planck-Institut für Informatik in Saarbrücken (1998)
39. Seymour, P., Thomas, R.: Graph searching and a min-max theorem for tree-width. JCT **58**, 22–33 (1993)
40. Szeider, S.: On fixed-parameter tractable parameterizations of SAT. In: Giunchiglia, E., Tacchella, A. (eds.) SAT 2003. LNCS, vol. 2919, pp. 188–202. Springer, Heidelberg (2004). https://doi.org/10.1007/978-3-540-24605-3_15
41. Tamaki, H.: treewidth-exact (2016). https://github.com/TCS-Meiji/treewidth-exact. Accessed 02 Aug 2017
42. Tamaki, H.: Positive-instance driven dynamic programming for treewidth. In: ESA, pp. 68:1–68:13 (2017)

How to Morph a Tree on a Small Grid

Fidel Barrera-Cruz[1], Manuel Borrazzo[2], Giordano Da Lozzo[2(✉)],
Giuseppe Di Battista[2], Fabrizio Frati[2], Maurizio Patrignani[2],
and Vincenzo Roselli[2]

[1] Sunnyvale, CA, USA
fidel.barrera@gmail.com
[2] Roma Tre University, Rome, Italy
{manuel.borrazzo,giordano.dalozzo,giuseppe.dibattista,
fabrizio.frati,maurizio.patrignani,vincenzo.roselli}@uniroma3.it

Abstract. In this paper we study planar morphs between straight-line
planar grid drawings of trees. A morph consists of a sequence of morphing steps, where in a morphing step vertices move along straight-line
trajectories at constant speed. We show how to construct planar morphs
that simultaneously achieve a reduced number of morphing steps and a
polynomially-bounded resolution. We assume that both the initial and
final drawings lie on the grid and we ensure that each morphing step
produces a grid drawing; further, we consider both upward drawings of
rooted trees and drawings of arbitrary trees.

1 Introduction

The problem of morphing combinatorial structures is a consolidated research
topic with important applications in several areas of Computer Science such as
Computational Geometry, Computer Graphics, Modeling, and Animation. The
structures of interest typically are drawings of graphs; a *morph* between two
drawings Γ_0 and Γ_1 of the same graph G is defined as a continuously changing
family of drawings $\{\Gamma_t\}$ of G indexed by time $t \in [0, 1]$, such that the drawing
at time $t = 0$ is Γ_0 and the drawing at time $t = 1$ is Γ_1. A morph is usually
required to preserve a certain drawing standard and pursues certain qualities.

The *drawing standard* is the set of the geometric properties that are maintained at any time during the morph. For example, if both Γ_0 and Γ_1 are planar
drawings, then the drawing standard might require that all the drawings of
the morph are planar. Other properties that might be required to be preserved
are the convexity of the faces, or the fact that the edges are straight-line segments, etc.

Regarding the *qualities* of the morph, the research up to now mainly focused
on limiting the number of *morphing steps*, where in a morphing step vertices

This research was supported in part by MIUR Project "MODE" under PRIN
20157EFM5C, by MIUR Project "AHeAD" under PRIN 20174LF3T8, by H2020-
MSCA-RISE project 734922 – "CONNECT", and by MIUR-DAAD JMP N° 34120.

© Springer Nature Switzerland AG 2019
Z. Friggstad et al. (Eds.): WADS 2019, LNCS 11646, pp. 57–70, 2019.
https://doi.org/10.1007/978-3-030-24766-9_5

move along straight-line trajectories at constant speed. A morph $\mathcal{M}$ can then be described as a sequence of drawings $\mathcal{M} = \langle \Gamma_0 = \Delta_0, \Delta_1, \ldots, \Delta_k = \Gamma_1 \rangle$ where the morph $\langle \Delta_{i-1}, \Delta_i \rangle$, for $i = 1, \ldots, k$, is a morphing step. Following the pioneeristic works of Cairns and Thomassen [8,13], most of the literature focused on the straight-line planar drawing standard. A sequence of recent results in [1–5] proved that a linear number of morphing steps suffices, and is sometimes necessary, to construct a morph between any two straight-line planar drawings of a graph.

Although the results mentioned in the previous paragraph establish strong theoretical foundations for the topic of morphing graph drawings, they produce morphs that are not appealing from a visualization perspective. Namely, such algorithms produce drawings that have poor *resolution*, i.e., they may have an exponential ratio of the distances between the farthest and closest pairs of geometric objects (points representing vertices or segments representing edges), even if the same ratio is polynomially bounded in the initial and final drawings. Indeed, most of the above cited papers mention the problem of constructing morphs with bounded resolution as the main challenge in this research area.

The only paper we are aware of where the resolution problem has been successfully addressed is the one by Barrera-Cruz et al. [6], who showed how to construct a morph with polynomially-bounded resolution between two *Schnyder drawings* Γ_0 and Γ_1 of the same planar triangulation. The model they use in order to ensure a bound on the resolution requires that $\Gamma_0 = \Delta_0, \Delta_1, \ldots, \Delta_k = \Gamma_1$ are *grid drawings*, i.e., vertices have integer coordinates, and the resolution is measured by comparing the area of Γ_0 and Γ_1 with the area of the Δ_i's. We remark that morphs between planar orthogonal drawings of maximum-degree-4 planar graphs, like those in [7,12], inherently have polynomial resolution.

In this paper we show how to construct morphs of tree drawings that simultaneously achieve a reduced number of morphing steps and a polynomially-bounded resolution. Adopting the setting of [6], we assume that Γ_0 and Γ_1 are grid drawings and we ensure that each morphing step produces a grid drawing.

We present three algorithms. The first two algorithms construct morphs between any two strictly-upward straight-line planar grid drawings Γ_0 and Γ_1 of n-node rooted trees; *strictly-upward* drawings are such that each node lies above its children. Both algorithms construct morphs in which each intermediate grid drawing has linear width and height, where the input size is measured by n and by the width and the height of Γ_0 and Γ_1. The first algorithm employs $\Theta(n)$ morphing steps. The second algorithm employs $\Theta(1)$ morphing steps, however it only applies to binary trees. The third algorithm allows us to achieve our main result, namely that for any two straight-line planar grid drawings Γ_0 and Γ_1 of an n-node tree, there is a planar morph with $\Theta(n)$ morphing steps between Γ_0 and Γ_1 such that each intermediate grid drawing has polynomial area, where the input size is again measured by n and by the width and the height of Γ_0 and Γ_1.

The first algorithm uses recursion; namely, it eliminates a leaf in the tree, it recursively morphs the drawings of the remaining tree and it then reintroduces the removed leaf in suitable positions during the morph. The second algorithm morphs the given drawings by independently changing their x- and y-coordinates; this technique is reminiscent of a recent paper by Da Lozzo et al. [10]. Finally, the third algorithm scales the given drawings up in order to make room for a bottom-up modification of each drawing into a "canonical" drawing of the tree.

Missing proofs can be found in the full version of the paper.

2   Preliminaries

In this section we introduce some definitions and preliminaries; see also [11].

Trees. The node and edge sets of a tree T are denoted by $V(T)$ and $E(T)$, respectively. The *degree* $\deg(v)$ of a node v of T is the number of its neighbors. In an *ordered* tree, a counter-clockwise order of the edges incident to each node is specified.

A *rooted tree* T is a tree with one distinguished node, which is called *root* and is denoted by $r(T)$. For any node $u \in V(T)$ with $u \neq r(T)$, the *parent* $p(u)$ of u is the neighbor of u in the unique path from u to $r(T)$. For any node $u \in V(T)$ with $u \neq r(T)$, the *children* of u are the neighbors of u different from $p(u)$; the *children* of $r(T)$ are all its neighbors. The nodes that have children are called *internal*; a non-internal node is a *leaf*. For any node $u \in V(T)$ with $u \neq r(T)$, the *subtree* T_u of T rooted at u is defined as follows: remove from T the edge $(u, p(u))$, thus separating T in two trees; the one containing u is the subtree of T rooted at u. If each node of T has at most two children, then T is a *binary tree*.

An *ordered rooted tree* is a tree that is rooted and ordered. In an ordered rooted tree T, for each node $u \in V(T)$, a *left-to-right* (linear) order $u_1, \ldots, u_k$ of the children of u is specified. If T is binary then the first (second) child in the left-to-right order of the children of any node u is the *left* (*right*) *child* of u, and the subtree rooted at the left (right) child of u is the *left* (*right*) *subtree* of u.

Tree Drawings. In a *straight-line drawing* Γ of a tree T each node u is represented by a point of the plane (whose coordinates are denoted by $x_\Gamma(u)$ and $y_\Gamma(u)$) and each edge is represented by a straight-line segment between its endpoints. All the drawings considered in this paper are straight-line, even when not specified. In a *planar* drawing no two edges intersect except, possibly, at common end-points. For a rooted tree T, a *strictly-upward* drawing Γ is such that each edge $(u, p(u)) \in E(T)$ is represented by a curve monotonically increasing in the y-direction from u to $p(u)$; if Γ is a straight-line drawing, this is equivalent to requiring that $y_\Gamma(u) < y_\Gamma(p(u))$. For an ordered tree T, an *order-preserving* drawing Γ is such that, for each node $u \in V(T)$, the counter-clockwise order of the edges incident to u in Γ is the same as the order associated with u in T.

The *bounding box* of a drawing Γ is the smallest axis-parallel rectangle enclosing Γ. In a *grid* drawing Γ each node has integer coordinates; then the *width* and the *height* of Γ, denoted by $w(\Gamma)$ and $h(\Gamma)$, respectively, are the number of grid columns and rows intersecting the bounding box of Γ, while the *area* of Γ is its width times its height. For a node v in a drawing Γ, an *ℓ-box centered at* v is the convex hull of the square whose corners are $(x_\Gamma(v) \pm \frac{\ell}{2}, y_\Gamma(v) \pm \frac{\ell}{2})$.

Morphs. A morph is *planar* if all its intermediate drawings are planar. A morph between two strictly-upward drawings of a rooted tree is *upward* if all its intermediate drawings are strictly-upward. A morph is *linear* if each node moves along a straight-line trajectory at constant speed. Whenever the linear morph between two straight-line planar drawings Γ_0 and Γ_1 of a graph G is not planar, one is usually interested in the construction of a piecewise-linear morph with small complexity between Γ_0 and Γ_1. This is formalized by defining a *morph* between Γ_0 and Γ_1 as a sequence $\langle \Gamma_0 = \Delta_0, \Delta_1, \ldots, \Delta_k = \Gamma_1 \rangle$ of drawings of G such that the linear morph $\langle \Delta_{i-1}, \Delta_i \rangle$ is planar, for $i = 1, \ldots, k$; each linear morph $\langle \Delta_{i-1}, \Delta_i \rangle$ is called a *morphing step* or simply a *step*.

The *width* $w(\mathcal{M})$ of a morph $\mathcal{M} = \langle \Delta_0, \Delta_1, \ldots, \Delta_k \rangle$, where Δ_i is a grid drawing, for $i = 0, 1, \ldots, k$, is equal to $\max\{w(\Delta_0), w(\Delta_1), \ldots, w(\Delta_k)\}$. The *height* $h(\mathcal{M})$ of $\mathcal{M}$ is defined analogously. The *area* of a morph $\mathcal{M}$ is defined as $w(\mathcal{M}) \times h(\mathcal{M})$.

The algorithms we design in this paper receive in input two order-preserving straight-line planar grid drawings Γ_0 and Γ_1 of an ordered tree and construct morphs $\langle \Gamma_0 = \Delta_0, \Delta_1, \ldots, \Delta_k = \Gamma_1 \rangle$ with few steps and small area.

Remark 1. A necessary and sufficient condition for the existence of a planar morph between two straight-line planar drawings Γ_0 and Γ_1 of a tree T is that they are "topologically-equivalent", i.e., the counter-clockwise order of the edges incident to each node $u \in V(T)$ is the same in Γ_0 and Γ_1. In order to better exploit standard terminology about tree drawings, we ensure that Γ_0 and Γ_1 are topologically-equivalent by assuming that T is ordered and that Γ_0 and Γ_1 are order-preserving drawings; hence, dealing with ordered trees and with order-preserving drawings is not a loss of generality.

Remark 2. The width and height of the morphs we construct are expressed not only in terms of the number of nodes of the input tree T, but also in terms of the width and height of the input drawings Γ_0 and Γ_1 of T; this is necessary, given that $\max\{w(\Gamma_0), w(\Gamma_1)\}$ and $\max\{h(\Gamma_0), h(\Gamma_1)\}$ are obvious lower bounds for the width and height of any morph between Γ_0 and Γ_1, respectively.

Remark 3. The morphs $\langle \Delta_0, \Delta_1, \ldots, \Delta_k \rangle$ we construct in this paper are such that $\Delta_0, \Delta_1, \ldots, \Delta_k$ are *grid* drawings, even when not explicitly specified.

3   Upward Planar Morphs of Rooted-Tree Drawings

In this section we study small-area morphs between order-preserving strictly-upward straight-line planar grid drawings of rooted ordered trees.

Our first result shows that such morphs can always be constructed consisting of a linear number of steps. This is obtained via an inductive algorithm which is described in the following. Let T be an n-node rooted ordered tree. The *rightmost path* of T is the maximal path $(s_0, \ldots, s_m)$ such that $s_0 = r(T)$ and s_i is the rightmost child of s_{i-1}, for $i = 1, \ldots, m$. Note that s_m is a leaf, which

Γ_0 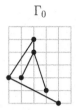 Γ_0' 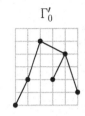 Γ_1' 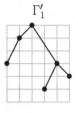 Γ_1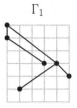

Fig. 1. The 3-step morph $\langle \Gamma_0, \Gamma_0', \Gamma_1', \Gamma_1 \rangle$.

is called the *rightmost leaf* $l_{\overrightarrow{T}}$ of T. For a straight-line grid drawing Γ, denote by ℓ_Γ the rightmost vertical line intersecting Γ; note that ℓ_Γ is a grid column.

Let Γ_0 and Γ_1 be two order-preserving strictly-upward straight-line planar grid drawings of T. We inductively construct a morph $\mathcal{M}$ from Γ_0 to Γ_1 as follows.

In the base case $n = 1$; then $\mathcal{M}$ is the linear morph $\langle \Gamma_0, \Gamma_1 \rangle$.

In the inductive case $n > 1$. Let $l = l_{\overrightarrow{T}}$ be the rightmost leaf of T. Let $\pi = p(l)$ be the parent of l. Let T' be the $(n-1)$-node tree obtained from T by removing the node l and the edge (π, l). Let Γ_0' and Γ_1' be the drawings of T' obtained from Γ_0 and Γ_1, respectively, by removing the node l and the edge (π, l). Inductively compute a k-step upward planar morph $\mathcal{M}' = \langle \Gamma_0' = \Delta_1', \Delta_2', \ldots, \Delta_k' = \Gamma_1' \rangle$.

We now construct a morph $\mathcal{M} = \langle \Gamma_0, \Delta_1, \Delta_2, \ldots, \Delta_k, \Gamma_1 \rangle$. For each $i = 2, 3, \ldots, k-1$, we define Δ_i as the drawing obtained from Δ_i' by placing l one unit below π and one unit to the right of $\ell_{\Delta_i'}$. Further, we define Δ_1 (Δ_k) as the drawing obtained from Δ_1' (resp. from Δ_k') by placing l one unit below π and one unit to the right of ℓ_{Γ_0} (resp. ℓ_{Γ_1}). Note that the point at which l is placed in Δ_1 (in Δ_k) is one unit to the right of $\ell_{\Delta_1'}$ (resp. $\ell_{\Delta_k'}$), similarly as in $\Delta_2, \Delta_3, \ldots, \Delta_{k-1}$, except if l is to the right of every other node of Γ_0 (of Γ_1); in that case l might be several units to the right of $\ell_{\Delta_1'}$ (resp. $\ell_{\Delta_k'}$). This completes the construction of $\mathcal{M}$. We get the following.

Theorem 1. *Let T be an n-node rooted ordered tree, and let Γ_0 and Γ_1 be two order-preserving strictly-upward straight-line planar grid drawings of T. There exists a $(2n-1)$-step upward planar morph $\mathcal{M}$ from Γ_0 to Γ_1 with $h(\mathcal{M}) = \max\{h(\Gamma_0), h(\Gamma_1)\}$ and $w(\mathcal{M}) = \max\{w(\Gamma_0), w(\Gamma_1)\} + n - 1$.*

In view of Theorem 1, it is natural to ask whether a sub-linear number of steps suffices to construct a small-area morph between any two order-preserving strictly-upward straight-line planar grid drawings of a rooted ordered tree. In the following we prove that this is indeed the case for binary trees, for which just three morphing steps are sufficient.

Our algorithm borrows ideas from a recent paper by Da Lozzo et al. [10], which deals with upward planar morphs of *upward plane graphs*.

Consider any two order-preserving strictly-upward straight-line planar grid drawings Γ_0 and Γ_1 of an n-node rooted ordered binary tree T. We define two order-preserving strictly-upward straight-line planar grid drawings Γ_0' and Γ_1' of T such that the 3-step morph $\langle \Gamma_0, \Gamma_0', \Gamma_1', \Gamma_1 \rangle$ is upward and planar.

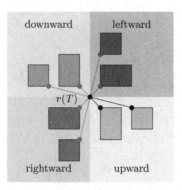

Fig. 2. Four canonical drawings of a tree T (each shown in a differently colored quadrant).

For $i = 0, 1$, we define Γ'_i recursively as follows; see Fig. 1. Let $x_{\Gamma'_i}(r(T)) = 0$ and let $y_{\Gamma'_i}(r(T)) = y_{\Gamma_i}(r(T))$. If the left subtree L of $r(T)$ is non-empty, then recursively construct a drawing of it. Let x_M be the maximum x-coordinate of a node in the constructed drawing of L; horizontally translate such a drawing by subtracting $x_M + 1$ from the x-coordinate of every node in L, so that the maximum x-coordinate of any node in L is now -1. Symmetrically, if the right subtree R of $r(T)$ is non-empty, then recursively construct a drawing of it. Let x_m be the minimum x-coordinate of a node in the constructed drawing of R; horizontally translate such a drawing by subtracting $x_m - 1$ from the x-coordinate of every node in R, so that the minimum x-coordinate of any node in R is now 1.

Theorem 2. *Let T be an n-node rooted ordered binary tree, and let Γ_0 and Γ_1 be two order-preserving strictly-upward straight-line planar grid drawings of T. There exists a 3-step upward planar morph $\mathcal{M}$ from Γ_0 to Γ_1 with $h(\mathcal{M}) = \max\{h(\Gamma_0), h(\Gamma_1)\}$ and $w(\mathcal{M}) = \max\{w(\Gamma_0), w(\Gamma_1), n\}$.*

The algorithm presented before Theorem 2 can be easily generalized to rooted ordered trees with unbounded degree. Thus, there exists a 3-step upward planar morph between any two order-preserving strictly-upward straight-line planar grid drawings of an n-node rooted ordered tree. However, the generalized version of the algorithm does not guarantee polynomial bounds on the width of the morph.

4   Planar Morphs of Tree Drawings

In this section we show how to construct small-area morphs between straight-line planar grid drawings of trees. In particular, we prove the following result.

Theorem 3. *Let T be an n-node ordered tree and let Γ_0 and Γ_1 be two order-preserving straight-line planar grid drawings of T. There exists an $O(n)$-step planar morph $\mathcal{M}$ from Γ_0 to Γ_1 with $h(\mathcal{M}) \in O(D^3 n \cdot H)$ and $w(\mathcal{M}) \in O(D^3 n \cdot W)$, where $H = \max\{h(\Gamma_0), h(\Gamma_1)\}$, $W = \max\{w(\Gamma_0), w(\Gamma_1)\}$, and $D = \max\{H, W\}$.*

The rest of this section is devoted to the proof of Theorem 3. We are going to use the following definition (see Fig. 2).

Definition 1. *An upward canonical drawing of a rooted ordered tree T is an order-preserving strictly-upward straight-line planar grid drawing Γ of T satisfying the following properties:*

- *if $|V(T)| = 1$, then Γ is a grid point in the plane, representing $r(T)$;*
- *otherwise, let $\Gamma_1, \ldots, \Gamma_k$ be upward canonical drawings of the subtrees $T_1, \ldots, T_k$ of $r(T)$ (in their left-to-right order), respectively; then Γ is such that:*
  - *$r(T)$ is one unit to the left and one unit above the top-left corner of the bounding box of Γ_1;*
  - *the top sides of the bounding boxes of $\Gamma_1, \ldots, \Gamma_k$ have the same y-coordinate; and*
  - *the right side of the bounding box of Γ_i is one unit to the left of the left side of the bounding box of Γ_{i+1}, for $i = 1, \ldots, k-1$.*

By counter-clockwise rotating an upward canonical drawing of T by $\frac{\pi}{2}$, π, and $\frac{3\pi}{2}$ radians, we obtain a *leftward*, a *downward*, and a *rightward canonical drawing* of T, respectively. A *canonical drawing* of T is an upward, leftward, downward, or rightward canonical drawing of T. In an upward, leftward, downward, or rightward canonical drawing Γ of T, $r(T)$ is placed at the top-left, bottom-left, bottom-right, and top-right corner of the bounding box of Γ, respectively.

Remark 4. If T has n nodes, then a canonical drawing of T lies in the $2n$-box centered at $r(T)$.

The following lemma allows us to morph one canonical drawing into another in a constant number of morphing steps.

Lemma 1 (Pinwheel). *Let Γ and Γ' be two canonical drawings of a rooted ordered tree T, where $r(T)$ is at the same point in Γ and Γ'. If Γ and Γ' are upward and leftward, or leftward and downward, or downward and rightward, or rightward and upward, then the morph $\langle \Gamma, \Gamma' \rangle$ is planar and lies in the interior of the right, top, left, or bottom half of the $2n$-box centered at $r(T)$, respectively.*

We now describe the proof of Theorem 3. Let T be an n-node ordered tree and let Γ_0 and Γ_1 be two order-preserving straight-line planar grid drawings of T. In order to compute a morph $\mathcal{M}$ from Γ_0 to Γ_1, we root T at any leaf $r(T)$. Since T is ordered, this determines a left-to-right order of the children of each node.

We construct three morphs: a morph $\mathcal{M}^0$ from Γ_0 to a canonical drawing Γ_0^* of T, a morph $\mathcal{M}^1$ from Γ_1 to a canonical drawing Γ_1^* of T, and a morph $\mathcal{M}^{0,1}$ from Γ_0^* to Γ_1^*. Then $\mathcal{M}$ is composed of $\mathcal{M}^0$, of $\mathcal{M}^{0,1}$, and of the reverse of $\mathcal{M}^1$. The morph $\mathcal{M}^{0,1}$ consists of $O(1)$ steps and can be constructed by applying Lemma 1. We describe how to construct $\mathcal{M}^0$; the construction of $\mathcal{M}^1$ is analogous.

Let $T[0]$ be the tree T together with a labeling of each of the k internal nodes of T as unvisited and of each leaf as visited. We perform a bottom-up visit of

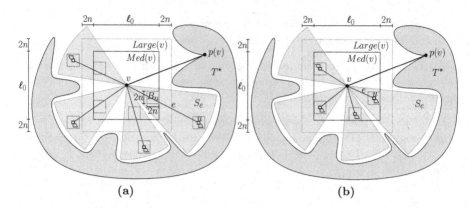

Fig. 3. (a) A partially-canonical drawing Δ_{i-1} of tree $T[i-1]$; the subtree T^* lies in the gray region, `visited` and `unvisited` nodes are represented as squares and circles, respectively. (b) Drawing Δ' of the morph $\langle \Delta_{i-1}, \Delta' \rangle$ of Claim 3.1.

T, labeling one-by-one the internal nodes of T as `visited`. We label a node v as `visited` only after all of its children have been labeled as `visited`. We denote by $T[i]$ the tree T once i of its internal nodes have been labeled as `visited`.

Let $D_0 = \max\{w(\Gamma_0), h(\Gamma_0)\}$. Let Γ be a drawing of T and let v be a node of T. We denote by $Large(v)$, $Med(v)$, and $Small(v)$ the $(\ell_0 + 4n)$-box, the ℓ_0-box, and the $2n$-box centered at v in Γ, respectively, where $\ell_0 = k_0 D_0^2 n$ for some constant $k_0 > 1$ to be determined later. We have the following definition.

Definition 2. *An order-preserving straight-line planar grid drawing Γ of T is a partially-canonical drawing of $T[i]$ if it satisfies the following properties (Fig. 3a):*

- *(a) for each `visited` node u of T, the drawing Γ_u of T_u in Γ is upward canonical or downward canonical; further, if $u \neq r(T)$, then Γ_u is upward canonical, if $y_\Gamma(u) \leq y_\Gamma(p(u))$, or downward canonical, if $y_\Gamma(u) > y_\Gamma(p(u))$;*
- *(b) for each edge $e = (v, u)$ of T, where v is the parent of u and v is `unvisited`, there exists a sector S_e of a circumference centered at v such that:*
- *(b.i) S_e encloses $Small(u)$;*
- *(b.ii) S_e contains no node with the exception of v and of, possibly, the nodes of T_u, and no edge with the exception of (u, v) and of, possibly, the edges of T_u;*
- *(b.iii) the intersection between S_e and $Med(v)$ contains a $2n$-box B_u whose corners have integer coordinates and whose center c_u is such that $y_\Gamma(c_u) \leq y_\Gamma(v)$ if and only if $y_\Gamma(u) \leq y_\Gamma(v)$; and*
- *(b.iv) for any edge $e' \neq e$ incident to v, the sectors S_e and $S_{e'}$ are internally disjoint;*
- *(c) for any two `unvisited` nodes v and w, it holds $Large(v) \cap Large(w) = \emptyset$; and*

(d) *for each* **unvisited** *node v of T, $Large(v)$ contains no node different from v, and any edge e or any sector S_e intersecting $Large(v)$ is such that e is incident to v.*

Note that, by Property (a), a partially-canonical drawing of $T[k]$ is a canonical drawing.

The algorithm to construct $\mathcal{M}^0$ is as follows. First, we scale Γ_0 up by a factor in $O(D_0^3 n)$ so that the resulting drawing Δ_0 is a partially-canonical drawing of $T[0]$ (see Lemma 2). Clearly, the morph $\mathcal{M}_0 = \langle \Gamma_0, \Delta_0 \rangle$ is planar, $w(\mathcal{M}_0) = w(\Delta_0)$, and $h(\mathcal{M}_0) = h(\Delta_0)$.

For $i = 1, \ldots, k$, let v_i be the node that is labeled as **visited** at the i-th step of the bottom-up visit of T. Starting from a partially-canonical drawing Δ_{i-1} of $T[i-1]$, we construct a partially-canonical drawing Δ_i of $T[i]$ and a morph $\mathcal{M}_{i-1,i}$ from Δ_{i-1} to Δ_i with $O(\deg(v_i))$ steps, with $w(\mathcal{M}_{i-1,i}) = w(\Delta_{i-1})$ and $h(\mathcal{M}_{i-1,i}) = h(\Delta_{i-1})$ (see Lemma 3).

Composing $\mathcal{M}_0, \mathcal{M}_{0,1}, \mathcal{M}_{1,2}, \ldots, \mathcal{M}_{k-1,k}$ yields the desired morph $\mathcal{M}^0$ from Γ_0 to a canonical drawing $\Delta_k = \Gamma_0^*$ of T. The morph has $\sum_i \deg(v_i) \in O(n)$ steps (by Lemma 3). Further, $w(\mathcal{M}^0) = w(\Delta_0)$ and $h(\mathcal{M}^0) = h(\Delta_0)$ (by Lemma 3), hence $w(\mathcal{M}^0) \in O(D_0^3 n \cdot w(\Gamma_0))$ and $h(\mathcal{M}^0) \in O(D_0^3 n \cdot h(\Gamma_0))$ (by Lemma 2).

Lemma 2. *There is an integer $B_0 \in O(D_0^3 n)$ such that the drawing Δ_0 obtained by scaling the drawing Γ_0 of T up by B_0 is a partially-canonical drawing of $T[0]$.*

Lemma 3. *For any $i \in \{1, \ldots, k\}$, let Δ_{i-1} be a partially-canonical drawing of $T[i-1]$. There exists a partially-canonical drawing Δ_i of $T[i]$ and an $O(\deg(v_i))$-step planar morph $\mathcal{M}_{i-1,i}$ from Δ_{i-1} to Δ_i such that $w(\mathcal{M}_{i-1,i}) \leq w(\Delta_0) + \ell_0 + 4n$ and $h(\mathcal{M}_{i-1,i}) \leq h(\Delta_0) + \ell_0 + 4n$.*

The rest of the section is devoted to the proof of Lemma 3. We denote by T^* the tree obtained by removing from T the nodes of T_{v_i} and their incident edges. Let Δ_i be the straight-line drawing of T obtained from Δ_{i-1} by redrawing T_{v_i} so that it is upward canonical, if $y_{\Delta_{i-1}}(v_i) \leq y_{\Delta_{i-1}}(p(v_i))$, or downward canonical, otherwise, while keeping the placement of v_i and of every node of T^* unchanged.

Lemma 4. *The drawing Δ_i is a partially-canonical drawing of $T[i]$.*

We show how to construct a morph $\mathcal{M}_{i-1,i}$ from Δ_{i-1} to Δ_i satisfying the properties of the statement of the lemma. This is done in several stages as follows.

First, consider the drawing Δ' of T obtained as described next; refer to Fig. 3b. Initialize $\Delta' = \Delta_{i-1}$. Then, for each child u of v_i, translate the drawing of T_u so that u is at the center of a $2n$-box B_u that lies in the intersection between S_e and $Med(v_i)$, whose corners have integer coordinates, and whose center c_u is such that $y_{\Delta_{i-1}}(c_u) \leq y_{\Delta_{i-1}}(v_i)$ if and only if $y_{\Delta_{i-1}}(u) \leq y_{\Delta_{i-1}}(v_i)$; such a box exists by Property (b.iii) of Δ_{i-1}. Also, redraw the edge (v_i, u) as a straight-line segment in Δ'.

Claim 3.1 *The morph* $\langle \Delta_{i-1}, \Delta' \rangle$ *is planar.*

Second, we show how to move the subtrees rooted at the children of v_i in the interior of $Large(v_i)$, so that they land in the position they have in Δ_i. The way we deal with such subtrees depends on their placement with respect to v_i and to the drawing of edge $(v_i, p(v_i))$. We consider the case in which $y(p(v_i)) \geq y(v_i)$ and $x(p(v_i)) \geq x(v_i)$; the other cases can be treated similarly. In particular, we distinguish four regions $\mathcal{R}_1$, $\mathcal{R}_2$, $\mathcal{R}_3$, and $\mathcal{R}_4$ defined as follows; refer to Fig. 4. Let $h_\rightarrow(v)$ and $h_\leftarrow(v)$ be the horizontal rays originating at a node v and directed rightward and leftward, respectively. Further, let $h_\uparrow(v)$ be the horizontal ray originating at a node v and directed upward.

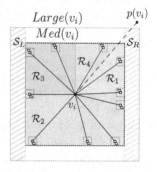

Fig. 4. Regions for v_i.

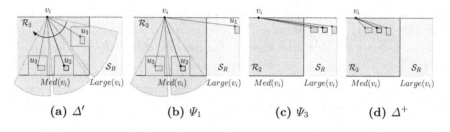

(a) Δ' (b) Ψ_1 (c) Ψ_3 (d) Δ^+

Fig. 5. Illustrations for Lemma 3, focused on the children of v_i that lie in $\mathcal{R}_2$.

Region $\mathcal{R}_1$ is defined as the intersection of $Med(v_i)$ with the wedge centered at v_i obtained by counter-clockwise rotating $h_\rightarrow(v_i)$ until it passes through $p(v_i)$; note that, if $(v_i, p(v_i))$ is a horizontal segment, then $\mathcal{R}_1 = \emptyset$.

Region $\mathcal{R}_2$ is the rectangular region that is the lower half of $Med(v_i)$;

Region $\mathcal{R}_3$ is defined as the intersection of $Med(v_i)$ with the wedge centered at v_i obtained by clockwise rotating $h_\leftarrow(v_i)$ until it coincides with $h_\uparrow(v_i)$; and

Region $\mathcal{R}_4$ is defined as the intersection of $Med(v_i)$ with the wedge centered at v_i obtained by clockwise rotating $h_\uparrow(v_i)$ until it passes through $p(v_i)$; note that, if $(v_i, p(v_i))$ is a vertical segment, then $\mathcal{R}_4 = \emptyset$.

Note that $Med(v_i) = \mathcal{R}_1 \cup \mathcal{R}_2 \cup \mathcal{R}_3 \cup \mathcal{R}_4$.

We define two more regions (see Fig. 4), which will be exploited as "buffers" that allow us to rotate subtrees via Lemma 1 without introducing crossings. Let $\mathcal{S}_L$ and $\mathcal{S}_R$ be the rectangular regions in Δ' containing all the points in $Large(v_i) - Med(v_i)$ to the left of the left side of $Med(v_i)$ and to the right of the right side of $Med(v_i)$, respectively. Observe that, since Δ_{i-1} satisfies Property (d) of a partially-canonical drawing and by the construction of Δ', the region $\mathcal{S}_L$ is empty, while the region $\mathcal{S}_R$ may only be traversed by the edge $(v_i, p(v_i))$.

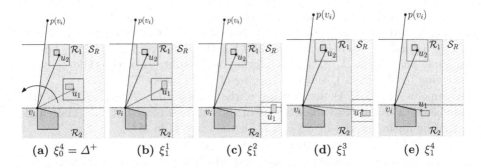

Fig. 6. Illustrations for Lemma 3, focused on the children of v_i that lie in $\mathcal{R}_1$.

We start by dealing with the children u_j of v_i that lie in the interior of $\mathcal{R}_2$; refer to Fig. 5. Consider the edges (v_i, u_j) in the order $(v_i, u_1), (v_i, u_2), \ldots, (v_i, u_m)$ in which such edges are encountered while clockwise rotating $h_\rightarrow(v_i)$; see Fig. 5a. Let Ψ_1 be the drawing obtained from Δ' by translating the drawing of the tree T_{u_1} so that u_1 lies in the interior of $\mathcal{S}_R$ and one unit below v_i and so that the right side of the bounding box of the drawing of T_{u_1} lies upon the right side of $Large(v_i)$, and by redrawing the edge (v_i, u_1) as a straight-line segment.

Claim 3.2 *The morph* $\langle \Delta', \Psi_1 \rangle$ *is planar.*

For $j = 2, \ldots, m$, let Ψ_j be the drawing obtained from Ψ_{j-1} by translating the drawing of the tree T_{u_j} so that u_j lies in the interior of $\mathcal{S}_R$ and one unit below v_i and so that the right side of the bounding box of the drawing of T_{u_j} lies one unit to the left of u_{j-1}, and by redrawing the edge (v_i, u_j) as a straight-line segment.

Claim 3.3 *For* $j = 2, \ldots, m$, *the morph* $\langle \Psi_{j-1}, \Psi_j \rangle$ *is planar.*

Let Δ^+ be the drawing obtained from Ψ_m by horizontally translating T_{u_j} so that u_j lands at its final position in Δ_i, and by redrawing the edge (v_i, u_j) as a straight-line segment, for $j = 1, 2, \ldots, m$; see Fig. 5c and d.

Claim 3.4 *The morph* $\langle \Psi_m, \Delta^+ \rangle$ *is planar.*

Next, we deal with the children u_j of v_i that lie in the interior of $\mathcal{R}_1$. Consider the edges (v_i, u_j) in the order $(v_i, u_1), (v_i, u_2), \ldots, (v_i, u_\ell)$ in which such edges are encountered while counter-clockwise rotating $h_\rightarrow(v_i)$ around v_i; refer to Fig. 6. We are going to move the subtrees rooted at the children of v_i in $\mathcal{R}_1$, one by one in the order $T_{u_1}, T_{u_2}, \ldots, T_{u_\ell}$, so that they land in the position that they have in Δ_i. Such a movement consists of four linear morphs. First, we rotate the drawing of T_{u_j} so that it becomes leftward canonical (see Fig. 6b). Second, we translate the drawing of T_{u_j} so that u_j lies in the interior of $\mathcal{S}_R$ and one unit below v_i (see Fig. 6c). Third, we rotate the drawing of T_{u_j} so that it becomes upward canonical (see Fig. 6d). Finally, we horizontally translate the drawing of T_{u_j} to its final position in Δ_i (see Fig. 6e).

We now provide the details of the above four linear morphs. For $j = 1, \ldots, \ell$, let ξ^4_{j-1} be a drawing of T with the following properties, where $\xi^4_0 = \Delta^+$ (refer to Fig. 6a and e): (P1) the drawing of T^* is the same as in Δ_i; (P2) v_i lies at the same point as in Δ_i; (P3) the drawing of the subtrees of the children of v_i belonging to $\mathcal{R}_2$ (in Δ'), and the drawing of the subtrees $T_{u_1}, T_{u_2}, \ldots, T_{u_{j-1}}$ is the same as in Δ_i; (P4) the drawing of the subtrees $T_{u_j}, T_{u_{j+1}}, \ldots, T_{u_\ell}$ is the same as in Δ^+; and (P5) the drawing of the subtrees rooted at the children of v_i that lie in the interior of $\mathcal{R}_3$ and $\mathcal{R}_4$ is the same as in Δ^+.

For $j = 1, \ldots, \ell$, we construct a drawing ξ^1_j from ξ^4_{j-1} by rotating tree T_{u_j} so that it is leftward canonical in ξ^1_j, and by leaving the position of the nodes not in T_{u_j} unaltered. This rotation can be accomplished via a linear morph $\langle \xi^4_{j-1}, \xi^1_j \rangle$ by Lemma 1; see Fig. 6b.

Claim 3.5 *For $j = 1, \ldots, \ell$, the morph $\langle \xi^4_{j-1}, \xi^1_j \rangle$ is planar.*

For $j = 1, \ldots, \ell$, let ξ^2_j be the drawing obtained from ξ^1_j by translating the drawing of T_{u_j} so that $Small(u_j)$ lies in the interior of $\mathcal{S}_R$ and so that u_j lies one unit below v_i, and by redrawing the edge (v_i, u_j) as a straight-line segment; see Fig. 6c.

Claim 3.6 *For $j = 1, \ldots, \ell$, the morph $\langle \xi^1_j, \xi^2_j \rangle$ is planar.*

For $j = 1, \ldots, \ell$, let ξ^3_j be the drawing obtained from ξ^2_j by rotating tree T_{u_j} so that it is upward canonical in ξ^3_j, and by leaving the position of the nodes not in T_{u_j} unaltered. This rotation can be accomplished via a linear morph $\langle \xi^2_j, \xi^3_j \rangle$, by Lemma 1; see Fig. 6d.

Claim 3.7 *For $j = 1, \ldots, \ell$, the morph $\langle \xi^2_j, \xi^3_j \rangle$ is planar.*

Finally, for $j = 1, \ldots, \ell$, let ξ^4_j be the drawing obtained from ξ^3_j by horizontally translating T_{u_j} so that u_j lies at its final position in Δ_i, and by leaving the position of the nodes not in T_{u_j} unaltered; see Fig. 6e.

Claim 3.8 *For $j = 1, \ldots, \ell$, the morph $\langle \xi^3_j, \xi^4_j \rangle$ is planar.*

Note that the drawing ξ^4_ℓ coincides with Δ_i, except for the drawing of the subtrees lying in the interior of $\mathcal{R}_3$ and $\mathcal{R}_4$.

Subtrees in $\mathcal{R}_3$ are treated symmetrically to the ones in $\mathcal{R}_1$. In particular, the subtrees of the children of v_i that lie in $\mathcal{R}_3$ are processed according to the clockwise order of the edges from v_i to their roots, while the role played by $\mathcal{S}_R$ is now assumed by $\mathcal{S}_L$.

The treatment of the subtrees in $\mathcal{R}_4$ is similar to the one of the subtrees in $\mathcal{R}_3$. However, when a subtree is considered, it is first horizontally translated in the interior of $\mathcal{R}_3$ and then processed according to the rules for such a region.

Altogether, we have described a morph $\mathcal{M}_{i-1,i}$ from the partially-canonical drawing Δ_{i-1} of $T[i-1]$ to Δ_i, which is a partially-canonical drawing of $T[i]$ by Lemma 4. Next, we argue about the properties of $\mathcal{M}_{i-1,i}$.

We deal with the area requirements of $\mathcal{M}_{i-1,i}$. Consider the drawing Δ_0 and place the boxes $Large(v)$ around the nodes v of T; the bounding box of the arrangement of such boxes has width $w(\Delta_0)+\ell_0+4n$ and height $h(\Delta_0)+\ell_0+4n$. We claim that the drawings of $\mathcal{M}_{i-1,i}$ lie inside such a bounding box. Assume this is true for Δ_{i-1} (this is indeed the case when $i = 1$); all subsequent drawings of $\mathcal{M}_{i-1,i}$ coincide with Δ_{i-1}, except for the placement of the subtrees rooted at the children of v_i, which however lie inside $Large(v_i)$ in each of such drawings. Since v_i has the same position in Δ_i as in Δ_0 and since $Large(v_i)$ has width and height equal to $\ell_0 + 4n$, the claim follows.

Finally, we deal with the number of linear morphs composing $\mathcal{M}_{i-1,i}$. The morph $\mathcal{M}_{i-1,i}$ consists of the morph $\langle \Delta_{i-1}, \Delta' \rangle$, followed by the morphs needed to drive the subtrees rooted at the children of v_i to their final positions in Δ_i. Since the number of morphing steps needed to deal with each of such subtrees is constant, we conclude that $M_{i-1,i}$ consists of $O(\deg(v_i))$ linear morphing steps. This concludes the proof of Lemma 3.

5 Conclusions and Open Problems

We presented an algorithm that, given any two order-preserving straight-line planar grid drawings Γ_0 and Γ_1 of an n-node ordered tree T, constructs a morph $\langle \Gamma_0 = \Delta_0, \Delta_1, \ldots, \Delta_k = \Gamma_1 \rangle$ such that k is in $O(n)$ and such that the area of each intermediate drawing Δ_i is polynomial in n and in the area of Γ_0 and Γ_1. Better bounds can be achieved if T is rooted and Γ_0 and Γ_1 are also strictly-upward drawings, especially in the case in which T is a binary tree.

We make a remark about the generality of the model that we adopted. At a first glance, our assumption that Γ_0 and Γ_1 are grid drawings seems restrictive, and it seems more general to consider drawings that have bounded resolution. However, by using an observation from [9], one can argue that two morphing steps suffice to transform a drawing with resolution r in a grid drawing whose area is polynomial in r. Namely, it suffices to scale each input drawing so that the smallest distance between any pair of geometric objects (points representing vertices or segments representing edges) is 2; this is a single morphing step which does not change the resolution of the drawing, hence the largest distance between any pair of geometric objects is in $O(r)$. Then each node can be moved to the nearest grid point; this is another morphing step, which is ensured to be planar by the fact that each node moves by at most $\sqrt{2}/2$, hence this motion only brings any two geometric objects closer by $\sqrt{2}$, while their distance is at least 2. Thus, this results in a grid drawing on an $O(r) \times O(r)$ grid.

Several problems are left open. Is it possible to generalize our results to graph classes richer than trees? Is it possible to improve our area bounds for morphs of straight-line planar grid drawings of trees or even just of paths? Is there a trade-off between the number of steps and the area required by a morph? Are there other relevant tree drawing standards for which it makes sense to consider the morphing problem?

References

1. Alamdari, S., et al.: How to morph planar graph drawings. SIAM J. Comput. **46**(2), 824–852 (2017). https://doi.org/10.1137/16M1069171
2. Alamdari, S., et al.: Morphing planar graph drawings with a polynomial number of steps. In: Khanna, S. (ed.) Proceedings of the Twenty-Fourth Annual ACM-SIAM Symposium on Discrete Algorithms, SODA 2013, New Orleans, Louisiana, USA, 6–8 January 2013, pp. 1656–1667. SIAM (2013). https://doi.org/10.1137/1.9781611973105.119
3. Angelini, P., Da Lozzo, G., Di Battista, G., Frati, F., Patrignani, M., Roselli, V.: Morphing planar graph drawings optimally. In: Esparza, J., Fraigniaud, P., Husfeldt, T., Koutsoupias, E. (eds.) ICALP 2014. LNCS, vol. 8572, pp. 126–137. Springer, Heidelberg (2014). https://doi.org/10.1007/978-3-662-43948-7_11
4. Angelini, P., Da Lozzo, G., Frati, F., Lubiw, A., Patrignani, M., Roselli, V.: Optimal morphs of convex drawings. In: Symposium on Computational Geometry. LIPIcs, vol. 34, pp. 126–140. Schloss Dagstuhl - Leibniz-Zentrum fuer Informatik (2015)
5. Angelini, P., Frati, F., Patrignani, M., Roselli, V.: Morphing planar graph drawings efficiently. In: Wismath, S., Wolff, A. (eds.) GD 2013. LNCS, vol. 8242, pp. 49–60. Springer, Cham (2013). https://doi.org/10.1007/978-3-319-03841-4_5
6. Barrera-Cruz, F., Haxell, P., Lubiw, A.: Morphing schnyder drawings of planar triangulations. In: Duncan, C., Symvonis, A. (eds.) GD 2014. LNCS, vol. 8871, pp. 294–305. Springer, Heidelberg (2014). https://doi.org/10.1007/978-3-662-45803-7_25
7. Biedl, T.C., Lubiw, A., Petrick, M., Spriggs, M.J.: Morphing orthogonal planar graph drawings. ACM Trans. Algorithms **9**(4), 29:1–29:24 (2013)
8. Cairns, S.S.: Deformations of plane rectilinear complexes. Am. Math. Monthly **51**(5), 247–252 (1944)
9. Chambers, E.W., Eppstein, D., Goodrich, M.T., Löffler, M.: Drawing graphs in the plane with a prescribed outer face and polynomial area. J. Graph Algorithms Appl. **16**(2), 243–259 (2012). https://doi.org/10.7155/jgaa.00257
10. Da Lozzo, G., Di Battista, G., Frati, F., Patrignani, M., Roselli, V.: Upward planar morphs. In: Biedl, T., Kerren, A. (eds.) GD 2018. LNCS, vol. 11282, pp. 92–105. Springer, Cham (2018). https://doi.org/10.1007/978-3-030-04414-5_7
11. Di Battista, G., Eades, P., Tamassia, R., Tollis, I.G.: Graph Drawing: Algorithms for the Visualization of Graphs. Prentice-Hall, Upper Saddle River (1999)
12. van Goethem, A., Verbeek, K.: Optimal morphs of planar orthogonal drawings. In: Speckmann, B., Tóth, C.D. (eds.) 34th International Symposium on Computational Geometry, SoCG 2018, 11–14 June 2018, Budapest, Hungary. LIPIcs, vol. 99, pp. 42:1–42:14. Schloss Dagstuhl - Leibniz-Zentrum fuer Informatik (2018). https://doi.org/10.4230/LIPIcs.SoCG.2018.42
13. Thomassen, C.: Deformations of plane graphs. J. Combin. Theory Ser. B **34**(3), 244–257 (1983)

Approximating Robust Bin Packing
with Budgeted Uncertainty

Aniket Basu Roy[1], Marin Bougeret[1], Noam Goldberg[2], and Michael Poss[1(✉)]

[1] LIRMM, University of Montpellier, CNRS, Montpellier, France
{aniket.basu-roy,marin.bougeret,michael.poss}@lirmm.fr
[2] Department of Management, Bar-Ilan University, 5290002 Ramat Gan, Israel
noam.goldberg@biu.ac.il

Abstract. We consider robust variants of the bin-packing problem where the sizes of the items can take any value in a given uncertainty set $U \subseteq \times_{i=1}^{n}[\bar{a}_i, \bar{a}_i + \hat{a}_i]$, where $\bar{a} \in [0,1]^n$ represents the nominal sizes of the items and $\hat{a} \in [0,1]^n$ their possible deviations. We consider more specifically two uncertainty sets previously studied in the literature. The first set, denoted U^{Γ}, contains scenarios in which at most $\Gamma \in \mathbb{N}$ items deviate, each of them reaching its peak value $\bar{a}_i + \hat{a}_i$, while each other item has its nominal value $\bar{a}_i$. The second set, denoted U^{Ω}, bounds by $\Omega \in [0,1]$ the total amount of deviation in each scenario. We show that a variant of the next-fit algorithm provides a 2-approximation for model U^{Ω}, and a $2(\Gamma+1)$ approximation for model U^{Γ} (which can be improved to 2 approximation for $\Gamma = 1$). This motivates the question of the existence of a constant ratio approximation algorithm for the U^{Γ} model. Our main result is to answer positively to this question by providing a 4.5 approximation for U^{Γ} model based on dynamic programming.

Keywords: Bin-packing · Robust optimization ·
Approximation algorithm · Next-fit · Dynamic programming

1 Introduction

Bin packing is the problem of assigning a given set of n items, each item of a specified size, to the smallest number of unit capacity bins. The problem has been the subject of study in an extensive body of research initiated by several publications in the 1970s including the work of Johnson et al. [11]. The problem is $\mathcal{NP}$-hard and in fact a straightforward reduction from the partition decision problem implies that it is $\mathcal{NP}$-hard to determine whether a bin-packing instance has a solution using only two bins. This also shows that the problem cannot be approximated within a factor less than 3/2. An approximation factor guarantee of 3/2 has been proven for the first-fit decreasing algorithm by Simchi-Levi [16].

This research has benefited from the support of the ANR project ROBUST [ANR-16-CE40-0018].

© Springer Nature Switzerland AG 2019
Z. Friggstad et al. (Eds.): WADS 2019, LNCS 11646, pp. 71–84, 2019.
https://doi.org/10.1007/978-3-030-24766-9_6

Much of the research has concentrated on the asymptotic setting where n tends to infinity, and in the online setting where the instance is not given in advance but each item is revealed and packed one at a time. A fully polynomial-time approximation scheme for the offline asymptotic problem is due to Karmarkar and Karp [12]. The best asymptotic and absolute online competitive ratios of 1.578 and 5/3, respectively, are due to Balogh et al. in [4] and [3], respectively.

In many applications, the sizes of the items to be packed are not fully known at the time that the packing is carried out. In cargo shipping, for example, the actual weight of a container may deviate from its declared weight or its measurements may be inaccurate. Bin packing has also been used to model the assignment of elective surgeries to operating room in hospitals [8]. Here a bin is a shift of a properly equipped and staffed operating room for performing a certain type of elective surgeries. The room scheduler has to fit in the bins as many cases (patients) as possible. In this setting clearly the length of time of performing each surgery is subject to uncertainty for example in the event of complications. One way to model the uncertainty that falls into the framework of robust optimization is to assume that the sizes are uncertain parameters taking any value in a given set $U \subset \mathbb{R}^n$, where each $a \in U$ represents a possible scenario. This leads to the following problem (where the description of U is sometimes not explicit to avoid exponential length in n)

RBP (Robust bin-packing)
Input: $U \subset \mathbb{R}^n$
Output: A solution is a partition of $[n]$ into k bins $b_1, \ldots, b_k$ such that $\max_{a \in U} \sum_{i \in b_j} a_i \leq 1$ for each $j \in [k]$
Minimize: k

Classically, robust combinatorial optimization has dealt with uncertain objective, meaning that the cost vector c can take any value in set U, unlike RBP where the uncertainty affects the feasibility of the solutions. In that context, it is well-known that arbitrary uncertainty sets U lead to robust counterparts that are hardly approximable. For instance, the robust knapsack is not approximable at all [1], while the shortest path, the spanning tree, the minimum cut, and the assignment problem do not admit constant-ratios approximation algorithms, e.g. [13,14]. Furthermore, describing U by an explicit list of scenarios runs the risk of over-fitting so the optimal solutions may become infeasible for small variations outside U. These two drawbacks are usually tackled by using more specific uncertainty sets, defined by simple budget constraints. One of these widely used uncertainty sets, U^Γ, supposes that the size of each item is either its given nominal size $\bar{a}_i$, or its peak value $\bar{a}_i + \hat{a}_i$. Furthermore, in any scenario, at most $\Gamma \in \mathbb{N}$ of the items may assume their peak value simultaneously. Formally, U^Γ can be defined as $U^\Gamma = \{a | \forall i \in [n], a_i \in \{\bar{a}_i, \bar{a}_i + \hat{a}_i\} \text{ and } \sum_{i \in [n]} (a_i - \bar{a}_i)/\hat{a}_i \leq \Gamma\}$.[1] Set U^Γ has been widely used in robust combinatorial optimization with a constant

---

[1] U^Γ is often defined alternatively in the literature, as the polytope $\{a \in \times_{i \in [n]} [\bar{a}_i, \bar{a}_i + \hat{a}_i] \mid \sum_{i \in [n]} (a_i - \bar{a}_i)/\hat{a}_i \leq \Gamma\}$. For the bin-packing problem, one readily verifies using classical arguments that the two definitions lead to the same optimization problem.

number of constraints because the set essentially preserves the complexity and approximability properties of the nominal problem. The result was initially proposed for min-max problems in [6], and was independently extended to uncertain constraints in [2,9], contrasting with the aformentionned uncertain objective. We also consider a second uncertainty set (used in [10,18], among others), characterized again by $\bar{a}$ and $\hat{a}$, as well as the number $\Omega \in [0,1]$ stating how much deviation can be spread among all sizes, formally $U^\Omega = \{a \in \times_{i \in [n]} [\bar{a}_i, \bar{a}_i + \hat{a}_i] \mid \sum_{i \in [n]} (a_i - \bar{a}_i) \leq \Omega\}$. From the approximability viewpoint, set U^Ω benefits from similar positive results as U^Γ, see [15].

The above positive complexity results (e.g. [9,15]) imply, for instance, that there exists a fully-polynomial time approximation scheme (FPTAS) for the robust knapsack problem with uncertain profits and uncertain weights belonging to U^Ω and/or U^Γ. Interestingly, these positive results do not extend to most scheduling problems (because they involve non-linearities) and to the bin-packing problem (because it involves a non-constant numbers of robust constraints). While in a previous paper [7] (with authors in common) we provided approximability results on robust scheduling, no such results have yet been proposed for the bin-packing problem, the only previous work focusing on numerical algorithms [17]. The purpose of this paper is to fill these gaps, as we present constant-ratio approximation algorithms the bin-packing problem, both for U^Ω and U^Γ.

Notations, Problems Definitions, and Next-Fit Algorithm. In this paper we consider two special cases of RBP. In the first one, ΓRBP, the input is $\mathcal{I} = (n, \bar{a}, \hat{a}, \Gamma)$ where $n \in \mathbb{N}$, and we assume that $U = U^\Gamma$. In the second one, ΩRBP, the input is $\mathcal{I} = (n, \bar{a}, \hat{a}, \Omega)$ where $n \in \mathbb{N}$, $\bar{a} \in [0,1]^n$, $\hat{a} \in [0,1]^n$, and $\Omega \in [0,1]$, and we assume that $U = U^\Omega$.

Let us now provide some important notations that will allow us to restate ΓRBP and ΩRBP in a more convenient way. Given $n \in \mathbb{N}$, sets $\{0, 1, \ldots, n\}$ and $\{1, \ldots, n\}$ are respectively denoted $[n]_0$ and $[n]$. Set $\{i, \ldots, j\}$ is denoted by $[\![i, j]\!]$. Given a vector $v \in [0,1]^n$ and a subset $X \subseteq [n]$, we define $v(X) = \sum_{i \in X} v_i$. Given two vectors $\bar{a} \in [0,1]^n$, $\hat{a} \in [0,1]^n$ and a subset of items $X \subseteq [n]$, we define $\hat{a}_\Omega(X) = \min\{\hat{a}(X), \Omega\}$, $\Gamma(X)$ as the set of Γ items in X with largest $\hat{a}$ values (ties broken by taking smallest indices), or $\Gamma(X) = X$ if $|X| < \Gamma$, and $\hat{a}_\Gamma(X) = \hat{a}(\Gamma(X))$. Accordingly, we define the fill of a bin $b \subseteq [n]$ as $f_\Gamma(b) = \bar{a}(b) + \hat{a}_\Gamma(b)$ for set U^Γ, and $f_\Omega(b) = \bar{a}(b) + \hat{a}_\Omega(b)$ for set U^Ω. The fill of a bin for a general uncertainty set U is denoted as $f_U(b) = \max_{a \in U} a(b)$.

Consider the following example. We are given an ordered set of pairs $(\bar{a}_i, \hat{a}_i)$, $X = \{(0.3, 0.2), (0.4, 0.2), (0.3, 0.1), (0.2, 0.5)\}$ with $\Gamma = 2$ and $\Omega = 0.3$. Thus, $\Gamma(X) = \{(0.3, 0.2), (0.2, 0.5)\}$, $\bar{a}(X) = 1.2$, $\hat{a}_\Gamma(X) = 0.7$, and $f_\Gamma(X) = 1.9$. Similarly, $\hat{a}_\Omega(X) = 0.3$ and $f_\Omega(X) = 1.0$.

Now, observe that $\max_{a \in U} \sum_{i \in b_j} a_i \leq 1$ (the constraint required in RBP) is equivalent to $f_U(b) \leq 1$, and thus to $f_\Gamma(b_j) \leq 1$ for ΓRBP and $f_\Omega(b_j) \leq 1$ for ΩRBP. For example in ΓRBP, $f_\Gamma(b_j) \leq 1$ simply means that the total nominal ($\bar{a}$) size of the items plus the deviating size ($\hat{a}$) of the Γ largest (in $\hat{a}$ values) items must not exceed one. Thus, the two optimization problems studied in this paper can be equivalently formulated in the following way.

ΓRBP (Γ-robust bin-packing)
Input: $\mathcal{I} = (n, \overline{a}, \hat{a}, \Gamma)$ where $n \in \mathbb{N}$, $\overline{a} \in [0,1]^n$, $\hat{a} \in [0,1]^n$, and $\Gamma \in \mathbb{N}$.
Output: A solution is a partition of $[n]$ into k bins $b_1, \ldots, b_k$ such that $f_\Gamma(b_j) \leq 1$ for each $j \in [k]$
Minimize: k

ΩRBP (Ω-robust bin-packing)
Input: $\mathcal{I} = (n, \overline{a}, \hat{a}, \Omega)$ where $n \in \mathbb{N}$, $\overline{a} \in [0,1]^n$, $\hat{a} \in [0,1]^n$, and $\Omega \in [0,1]$.
Output: A solution is a partition of $[n]$ into k bins $b_1, \ldots, b_k$ such that $f_\Omega(b_j) \leq 1$ for each $j \in [k]$
Minimize: k

The optimal solution value or cost of either problem is denoted by $\mathsf{OPT}(\mathcal{I}) = k^*$ ($\mathcal{I}$ may be omitted when the instance is clear from the context) and a corresponding optimal solution is denoted by $s^* = \{b_1^*, b_2^*, \ldots, b_{k^*}^*\}$. We introduce in Algorithm 1 a variant of the standard next fit algorithm.

---

initialization: $j = 1$
1 Pack items (with smaller index first) in b_j until $f_U(b_j) > 1$ or $n \in b_j$. If $n \notin b_j$ then $j \leftarrow j + 1$ and repeat Step 1. Otherwise, $k' \leftarrow j$ proceed to Step 2.
2 Pack the last item of each bin in a new bin: for any j, let $i = \max(b_j)$, $b_j^1 = b_j \setminus \{i\}$, and $b_j^2 = \{i\}$
   return   : $\bigcup_{j=1}^{k'} \{b_j^1, b_j^2\}$

Algorithm 1. NEXT-FIT($\mathcal{I}$)

Structure of the Paper. In Sects. 2 and 3, we analyze the ratio provided by NEXT-FIT for ΩRBP and ΓRBP, respectively. For ΩRBP, using ordering (1) (non-increasing ordering on $\frac{\hat{a}_i}{\overline{a}_i}$) the ratio is equal to 2. For ΓRBP, using ordering (2) (non-increasing ordering on $\hat{a}_i$), the ratio is bounded by $2(\Gamma + 1)$ (and can be improved to 2 for $\Gamma = 1$). As Theorem 4 shows that neither ordering (1) or (2) leads to a constant ratio using NEXT-FIT, this raises the question the existence of a constant approximation for ΓRBP. In Sect. 4 we first review some basic ideas and explain why they are not sufficient. Then, we introduce the key elements necessary to develop our dynamic programming algorithm (DP) in Sect. 5. The latter gives a ratio of 4.5 for ΓRBP and any $\Gamma \in \mathbb{N}$, which is our main result. The complete proofs of Theorems and Lemmas with a ($\star$) symbol can be found in the full version of this paper [5].

2 Next-Fit for ΩRBP

Unlike the classical bin-packing problem, executing Next-Fit on arbitrarily ordered items can lead to arbitrarily bad solutions. For example, given ϵ with $0 < \epsilon \le \frac{1}{2n}$, consider an instance with $\Omega = 1 - \epsilon$, and items $((2\epsilon, 0), (0, 1 - \epsilon), \ldots, (2\epsilon, 0), (0, 1 - \epsilon))$, where item $i \in [n]$ is denoted by the pair $(\bar{a}_i, \hat{a}_i)$. Using this ordering, Next-Fit will create $n/2$ bins b_j with $f_\Omega(b_j) > 1$ for any $j \in [n]$ (which will be turned into n bins $\{b_j^1, b_j^2\}$), whereas the optimal solution uses 2 bins. This example also illustrates that, unlike in the standard bin-packing, the total size argument no longer apply to the robust counterpart as having $f_\Omega(b_j) > 1$ for any $j \in [n]$ does not imply a large (depending on n) lower bound on the optimal.

Next, we consider an ordering of the items such that

$$\hat{a}_1/\bar{a}_1 \ge \cdots \ge \hat{a}_n/\bar{a}_n. \tag{1}$$

Lemma 1. *Suppose that the items are ordered according to (1). Then $k' \le k^*$.*

Proof. Consider an optimal solution $b_1^*, \ldots, b_{k^*}^*$ and the subset of optimal bins given by $G^* = \{j \in [k^*] \mid \hat{a}(b_j^*) > \Omega\}$. Let

$$A = \sum_{i \in [n]} (\bar{a}_i + \hat{a}_i) = \sum_{j=1}^{k'} (\bar{a}(b_j) + \hat{a}(b_j)) = \sum_{j=1}^{k^*} (\bar{a}(b_j^*) + \hat{a}(b_j^*)).$$

Let G denote the first $|G^*|$ bins opened in Step 1 of Next-Fit. If $k' \in G$ then clearly $k' \le k^*$. Otherwise, it can be observed that for each $l \in G$, $\bar{a}(b_l) > 1 - \Omega$ (as $\bar{a}(b_l) + \hat{a}_\Omega(b_l) > 1$ and $\hat{a}_\Omega(b_l) \le \Omega$) and $1 - \Omega \ge \max_{j \in G^*} \bar{a}(b_j^*)$ (as $f_\Omega(b_j) \le 1$). Thus, $\sum_{j \in G} \bar{a}(b_j) > \sum_{j \in G^*} \bar{a}(b_j^*)$ and so by the assumed ordering (1) of the items, following a standard knapsack argument, $\sum_{j \in G} \hat{a}(b_j) > \sum_{j \in G^*} \hat{a}(b_j^*)$. Letting $\bar{G} = [k'] \setminus G$ and $\bar{G}^* = [k^*] \setminus G^*$, it follows that

$$\sum_{j \in \bar{G}} (\bar{a}(b_j) + \hat{a}(b_j)) = A - \sum_{j \in G} (\bar{a}(b_j) + \hat{a}(b_j)) \le$$

$$A - \sum_{j \in G^*} (a(b_j^*) + \hat{a}(b_j^*)) = \sum_{j \in \bar{G}^*} (\bar{a}(b_j^*) + \hat{a}(b_j^*))$$

(equality may hold throughout if $G^* = \emptyset$). Further, for each $j \in \bar{G} \setminus \{k'\}$, $\bar{a}(b_j) + \hat{a}(b_j) \ge f(b_j) > 1$ and for each $j \in \bar{G}^*$, $\bar{a}(b_j^*) + \hat{a}(b_j^*) \le 1$. Therefore, $|\bar{G}| \le \left\lceil \sum_{j \in \bar{G}} (\bar{a}(b_j) + \hat{a}(b_j)) \right\rceil \le \left\lceil \sum_{j \in \bar{G}^*} (\bar{a}(b_j^*) + \hat{a}(b_j^*)) \right\rceil \le |\bar{G}^*|$ and $k' \le k^*$ as claimed. $\square$

The lemma combined with Step 2 of Next-Fit immediately imply the following theorem.

Theorem 1. *If the items are ordered according to (1) then Next-Fit is a 2-approximation algorithm for ΩRBP.*

3   Next-Fit for ΓRBP

From now on, we focus on problem ΓRBP. Remark first that using an arbitrary ordering leads to arbitrarily bad solutions, considering $\Gamma = 1$ and the same items $((2\epsilon, 0), (0, 1 - \epsilon), \ldots, (2\epsilon, 0), (0, 1 - \epsilon))$ as in the previous section. Thus, we consider here an ordering of the items such that

$$\hat{a}_1 \geq \cdots \geq \hat{a}_n. \tag{2}$$

The main result of this Section is the following.

Theorem 2 ($\star$). *If the items are ordered according to (2) then* NEXT-FIT *is a $2(\Gamma + 1)$-approximation algorithm for ΓRBP.*

The proof of Theorem 2 can be found in the full version of this paper [5].

We show here a simplified analysis showing that for $\Gamma = 1$, NEXT-FIT with ordering (2) is a 2-approximation.

The deviating item of bin j in a fixed optimal solution s^* and in the solution of NEXT-FIT are denoted by singleton sets $\{i_j^*\} = \Gamma(b_j^*)$ and $\{i_j\} = \Gamma(b_j)$, respectively. We order the bins of s^* such that $i_j^* \geq i_{j+1}^*$. Notice that by definition of NEXT-FIT and ordering (2) we also have $i_j \geq i_{j+1}$.

Lemma 2. *Suppose that the items are ordered according to (2) and that $\Gamma = 1$. Then $k' \leq k^*$.*

Proof. Suppose by contradiction that $k' > k^*$. Let $b_1, \ldots, b_{k'}$ be the bins opened at Step 1 of NEXT-FIT and notice that $f_\Gamma(b_j) = \overline{a}(b_j) + \hat{a}_\Gamma(b_j) > 1$ for each $j \in [k'-1]$, while $\overline{a}(b_j^*) + \hat{a}_\Gamma(b_j^*) \leq 1$ for each $j \in [k^*]$. We prove next by induction on $\ell \in [k^*]$ that

$$\sum_{j=1}^{\ell} \overline{a}(b_j) > \sum_{j=1}^{\ell} \overline{a}(b_j^*). \tag{3}$$

For $\ell = 1$, we have $i_1 = i_1^* = 1$ and (3) follows immediately from $\hat{a}_\Gamma(b_1) = \hat{a}_\Gamma(b_1^*)$. Suppose now that induction hypothesis is true for $\ell - 1$. By definition of i_ℓ^* and i_ℓ, we know that $[i_\ell^* - 1] \subseteq \bigcup_{j=1}^{\ell-1} b_j^*$ and $[i_\ell - 1] = \bigcup_{j=1}^{\ell-1} b_j$. Using induction hypothesis, we get that $i_\ell \geq i_\ell^*$, and accordingly $\hat{a}_\Gamma(b_\ell) \leq \hat{a}_\Gamma(b_\ell^*)$. As $l \leq k^* < l$, we have $f_\Gamma(b_l) > 1$, leading to $\overline{a}(b_\ell) > \overline{a}(b_\ell^*)$.

Thus, for $l = k^*$ we get $\sum_{j=1}^{k^*} \overline{a}(b_j) > \sum_{j=1}^{k^*} \overline{a}(b_j^*) = \sum_{i \in [n]} \overline{a}_i$, which is impossible. $\qquad\square$

As in the previous section, we obtain the following theorem.

Theorem 3. *If the items are ordered according to (2) and $\Gamma = 1$ then* NEXT-FIT *is a 2-approximation algorithm for ΓRBP.*

To complete the analysis, we establish the following lower bound on the ratio of NEXT-FIT.

Theorem 4 (⋆). *If the items are ordered according to (2) or (1), then the approximation ratio of* NEXT-FIT *for ΓRBP is at least $\frac{2\Gamma}{3}$.*

Proof. Let us define an instance where the ordering (2) can lead to Step 1 of NEXT-FIT using $k' = \Gamma$ bins while OPT $= 3$. Every row of the $\Gamma \times \Gamma$ matrix below corresponds to the set of items in a bin (after the Step 1) of NEXT-FIT algorithm

$$
\begin{matrix}
(\epsilon, 1/\Gamma - \delta_1) & (0, 1/\Gamma - \delta_1) & \ldots & (0, 1/\Gamma - \delta_1) \\
\vdots & \vdots & \ddots & \vdots \\
(\epsilon, 1/\Gamma - \delta_\Gamma) & (0, 1/\Gamma - \delta_\Gamma) & \ldots & (0, 1/\Gamma - \delta_\Gamma)
\end{matrix}
\tag{4}
$$

where $\epsilon \le 1/\Gamma$ and $\delta_1 \le \cdots \le \delta_\Gamma < \epsilon/\Gamma$. On the one hand, $\epsilon + \Gamma \cdot (1/\Gamma - \delta_l) > 1$ for each $l \in [\Gamma]$, so step 1 of NEXT-FIT outputs Γ bins. On the other hand, an optimal solution can pack all the items above except the ones in the first column into a single bin because $\Gamma \cdot 1/\Gamma - \delta_1 \le 1$. Further, the total weight of the first $\Gamma/2$ items of the first column sums up to $\Gamma/2 \cdot (1/\Gamma + \epsilon) - \sum_{l=1}^{\Gamma/2} \delta_l \le 1 - \sum_{l=1}^{\Gamma/2} \delta_l \le 1$, and similarly for the last $\Gamma/2$ items, so an optimal solution may pack the first column using two bins. Finally, instance (4) shows that NEXT-FIT produces a solution $2\Gamma/3$ times worse than the optimal one.

This instance can be adapted to establish a lower bound for the approximation ratio of NEXT-FIT when items are ordered according to (1); see [5]. □

4 First Ideas to Get a Constant Ratio for ΓRBP

We maintain the assumption that the items are ordered according to (2).

4.1 Attempts to Get a Constant Ratio

We discuss below some natural arguments to get constant ratios.

Attempt 1: Using a Classical Size Argument. NEXT-FIT without a particular ordering applied to instance of Sect. 3 leads to a solution with $k' = n/2$ bins (at the end of Step 1) where $f_\Gamma(b_j) > 1$ for each bin, while OPT $= 2$. This example shows that even if all bins are "full" (relatively to f_Γ), it does not provide a lower bound on the optimal number of bins. Moreover, as shown in Theorem 4, none of the two orders considered in the previous section leads to a constant ratio using NEXT-FIT.

Attempt 2: Using the Duality with Makespan Minimization. Given input $\mathcal{I}$, we could guess $k^* = $ OPT$(\mathcal{I})$, and then consider the input $(\mathcal{I}, k^*)$ as an input of robust makespan minimzation (which was studied in [7]). Using any ρ-approximation for the later problem (for example $\rho = 3$ in [7]), we could get in polynomial time a solution with k^* bins an such that $f_\Gamma(b_j) \le \rho$. The last step would be to convert this solution into a solution of ΓRBP by unpacking each bin (with $f_\Gamma(b_j) \leq 3$) into several bins b_j^l with $f_\Gamma(b_j^l) \leq 1$. However, even if ρ were arbitrarily close to 1, it is not possible to bound (for a fixed j) the number of bins b_j^l by a constant as showed in the instance containing n items $(\frac{\epsilon}{n}, 1 - \frac{\epsilon}{n})$ and $\Gamma = 1$. While all items fit into a single bin with capacity lower than $1 + \epsilon$, they require n bins of capacity 1 to be packed.

Attempt 3: Guessing the Profile of an Optimal Solution. Let $\mathcal{I}$ be a input of ΓRBP. Given a solution $s = \{b_j, j \in [k]\}$ for this input, we define $P(s) = \{\Gamma(b_j), j \in [k]\}$ as the profile of s and $\tilde{P}_j(s) = \{i \mid i \in \Gamma(b_\ell)$ for some $\ell \in [j]\}$ as all deviating items in the first j bins. Let $s^* = \{b_j^*, j \in [k^*]\}$ be an optimal solution. To get some insight on the problem, let us assume that we know $P(s^*)$ (even if this cannot be guessed in polynomial time). We show how we can use $P(s^*)$ to get a 2-approximation algorithm. Without loss of generality, we can always assume that $|\Gamma(b_j^*)| = \Gamma$ for any j, as otherwise we can add $\Gamma - |\Gamma(b_j^*)|$ dummy items of size $(0,0)$ to b_j^*. Remember that the items are sorted in non-increasing order of their deviating values ($\hat{a}_i \geq \hat{a}_{i+1}$). For any $j \in [k^*]$, let $i_j^* = \max(\Gamma(b_j^*))$ be the smallest (in term of $\hat{a}$ value) deviating item of bin j (when $\Gamma = 1$, $\{i_j^*\} = \Gamma(b_j^*)$ as in the previous section). Without loss of generality, let us assume that bins are sorted such that $i_j^* \geq i_{j+1}^*$. Now, given $P(s^*)$, in the first phase we construct a solution s by packing items of $P(s^*)$ as they were packed in s^*, meaning that we define $b_j = \Gamma(b_j^*)$ for $j \in [k^*]$. Let $X = [n] \setminus \bigcup_{j \in [k^*]} \Gamma(b_j^*)$ be the set of remaining items. We now pack X in the following second phase, starting with $j = 1$. Notice that in the description of the algorithm below, we consider that for $j \in [k^*]$, b_j already contains $\Gamma(b_j^*)$, whereas for any $j > k^*$, b_j is initially empty.

Step 1 pack items of X (by decreasing $\hat{a}$ values) in b_j until $f_\Gamma(b_j) > 1$ or $X = \emptyset$
Step 2 if $X \neq \emptyset$, $j = j + 1$, and go to step 1.

Let j be the bin such that X is empty after filling b_j. Let k' be the number of bins used by this algorithm. Notice that if $j \leq k^*$ then $k' = k^*$ (because of the pre-packing of item of $P(s^*)$), and otherwise $k' = j$.

Lemma 3. $k' \leq k^*$, *implying a 2-approximation as we can convert the solution of* NEXT-FIT *into a feasible solution of $2k'$ bins by repacking the last added item in each bin in a separate bin.*

Proof. Assume by contradiction that $k' > k^*$. Informally, as an item $i \in [n] \setminus \tilde{P}_{k^*}(s^*)$ does not deviate in s^*, we need to ensure that this is also the case in s. Let us prove by induction on j that the items packed greedily in Step 1 satisfy

$$\hat{a}_i \leq \hat{a}_{i_j^*}, \forall i \in b_j \setminus \tilde{P}_{k^*}(s^*), j \in [k^*]. \tag{5}$$

Let $j = 1$, and suppose there is $i \in b_1 \setminus \tilde{P}_{k^*}(s^*)$ such that $\hat{a}_i > \hat{a}_{i_1^*}$. Then, because $\hat{a}_{i_1^*} \geq \hat{a}_{i_j^*}$ for $j > 1$, $\hat{a}_i > \hat{a}_{i_j^*}$ for each j so $i \in \tilde{P}_{k^*}(s^*)$, a contradiction.

Now, consider bin b_{j+1}. By induction, we have that $\sum_{\ell \in [j]} \overline{a}(b_\ell) + \hat{a}(\tilde{P}_j(s)) > j \geq \sum_{\ell \in [j]} \overline{a}(b_\ell^*) + \hat{a}(\tilde{P}_j(s^*))$, so $\tilde{P}_j(s) = \tilde{P}_j(s^*)$ implies

$$\sum_{\ell \in [j]} \overline{a}(b_\ell) > \sum_{\ell \in [j]} \overline{a}(b_\ell^*). \tag{6}$$

Let X_j be the set of items of X left after packing bin b_j by the above procedure and X_j^* be the the the set of items of X left after the optimal solution packs bin b_j^*. Inequality (6) and the ordering used in Step 1 imply that $\lambda \geq \lambda^*$, where $\lambda = \min(X_j)$ and $\lambda^* = \min(X_j^*)$. Therefore, if there exists $i \in b_{j+1} \setminus \tilde{P}_{k^*}(s^*)$ such that $\hat{a}_i > \hat{a}_{i^*_{j+1}}$, then $\hat{a}_{\lambda^*} \geq \hat{a}_\lambda \geq \hat{a}_i > \hat{a}_{i^*_{j+1}}$, and thus $\hat{a}_{\lambda^*} > \hat{a}_{i^*_\ell}$ for any $l \in [\![j+1, k^*]\!]$, which is a contradiction as item λ^* is in X and thus does not deviate in the considered optimal.

Now that Property (5) is proved, let us get our contradiction from $k' > k^*$. Indeed, if $k' > k^*$ then $\sum_{i \in [n]} \hat{a}_i > k^* - \hat{a}(\tilde{P}_{k^*}(s^*)) \geq \sum_{j \in [k^*]} \overline{a}(b_j^*)$ where the first inequality follows from $f_\Gamma(b_j) > 1$ for $j \in [k^*]$ and Property (5), and the second one follows from $\sum_{j \in [k^*]} \overline{a}(b_j^*) + \hat{a}(\tilde{P}_{k^*}(s^*)) \leq k^*$. This implies a contradiction. $\qquad\square$

Even if the above procedure relies on a guessing step which is not polynomial, its core idea has similarities with both the analysis of Next-Fit in the proof of Theorem 2 (see [5]) and with the DP algorithm detailed later in this paper, where we only guess the deviating item with the smallest deviation of each bin (one at a time), and we pack $\Gamma - 1$ items "better" than the one packed in $P(s^*)$, at the expense of a few extra bins.

4.2 Restricting Our Attention to Small Items

We define ΓRBP with small values as the ΓRBP problem restricted to inputs where for any $i \in [n]$, $\hat{a}_i \leq \frac{1}{\Gamma}$ and $\hat{a}_i \leq \frac{1}{\Gamma}$. Below we give a justification for restricting our attention to ΓRBP with small values.

Lemma 4. *Any polynomial ρ-approximation for ΓRBP with small values implies a polynomial $(\rho + \rho_{bp})$-approximation for ΓRBP, where ρ_{bp} is the best known ratio of a polynomial time approximation for classical bin-packing.*[2]

Proof. Given an instance $\mathcal{I}$ of ΓRBP, we define the small items $\mathcal{S} = \{i \in [n] : \overline{a}_i \leq 1/\Gamma$ and $\hat{a}_i \leq 1/\Gamma\}$ and the large item as $\mathcal{B} = [n] \setminus \mathcal{S}$. We use our ρ-approximation algorithm to pack $\mathcal{S}$ into $k_\mathcal{S}$ bins, implying $k_\mathcal{S} \leq \rho\mathsf{OPT}(\mathcal{S}) \leq \rho\mathsf{OPT}(\mathcal{I})$. Then, we observe that in any packing of $\mathcal{B}$, each bin contains no more than Γ items, so that all items deviate in these bins. Hence, ΓRBP for instance $(\mathcal{B}, \Gamma)$ is equivalent to the classical bin-packing problem for items $\mathcal{B}'$

---

[2] In general, if we have a polynomial time additive approximation algorithm using $OPT + f(OPT)$ bins and polynomial time ρ-approximation algorithm for ΓRBP with small values then our algorithm uses $OPT(\rho + 1) + f(OPT)$ bins for ΓRBP in polynomial time.

where the weight of each item $i \in \mathcal{B}'$ is given by $\bar{a}_i + \hat{a}_i$. This implies that $\mathsf{OPT}_{bp}(\mathcal{B}') = \mathsf{OPT}(\mathcal{B})$ (where OPT_{bp} denotes the optimal value in classical bin-packing), and that any solution for $\mathcal{B}'$ is a solution of $\mathcal{B}$. Thus, we use a ρ_{bp}-approximation algorithm for classical bin-packing to pack $\mathcal{B}'$ in $k_{\mathcal{B}}$ bins, and use the same packing for items in $\mathcal{B}$. Note that $k_{\mathcal{B}} \leq \rho_{bp}\mathsf{OPT}_{bp}(\mathcal{B}) = \rho_{bp}\mathsf{OPT}(\mathcal{B}) \leq \rho_{bp}\mathsf{OPT}(\mathcal{I})$. We obtain a packing of $\mathcal{I}$ with cost $k_{\mathcal{S}} + k_{\mathcal{B}} \leq (\rho + \rho_{bp})\mathsf{OPT}(\mathcal{I})$. □

Observation 1. *Given an instance $\mathcal{I}$ to the ΓRBP with small values, any subset $X \subseteq [n]$ can be packed in $\lceil \frac{|X|}{\Gamma/2} \rceil$ bins.*

Notice that instances with small items are not easier to approximate by NEXT-FIT because instance (4) from Sect. 3 uses small items.

4.3   Guessing of the Full Profile and Considering Only Small Items

Let us now explain why mixing the two previous ideas is promising. As in attempt 3 where we know the full profile, we want to construct for any j bins $\{b_1, \ldots, b_j\}$ such that their total $\bar{a}$ is larger than the total value of $\bar{a}$ packed by the first j bins of s^* (the considered optimal solution), as in inequality (6). Instead of guessing the full profile $P(s^*)$, we want to design a DP algorithm (that guesses i_{j^*} one at the time) with the following intuitive outline. Start with $j = 1$.

- guess item i_j^*, the smallest (in $\hat{a}$ value) deviating item of b_j^*, and pack it in b_j
- then, as the $\Gamma - 1$ other deviating items in b_j^* are unknown and we want to pack more of the nominal size $\bar{a}$, packs separately $\Gamma - 1$ items with larger $\bar{a}$ values (among items with $\hat{a}$ values greater than $\hat{a}_{i_j^*}$). Consider that these $\Gamma - 1$ items are put in the "trash" (at the very end we will pack all items of the trash in a few additional bins)
- keep filling bin b_j greedily (by non-increasing $\hat{a}$ values) until exceeding 1
- make a recursive call with $j + 1$

If s^* uses k^* bins, we wish to output a solution s with k^* bins exceeding one, and $(\Gamma - 1)k^*$ items in the trash. This almost feasible solution can be converted into a regular one with $3k^*$ bins by removing one item from each bin and adding them to the trash, and packing the Γk^* items of the trash into $2k^*$ bins, which is possible according to Observation 1. This sketches the core ideas of the DP. However, the actual DP presented below needs to be more involved for the following reasons. Consider $j = 1$ for convenience and let $B = [\![1, i_1^* - 1]\!]$.

First, notice that items of B could be packed (as deviating items) in a bin other than b_1^* in s^*, and we may have $|B| > \Gamma - 1$. Thus, instead of trashing only $\Gamma - 1$ items of B, we have to trash all of them, and count the number of trashed items to ensure that at the end at most $(\Gamma - 1)k^*$ items are trashed. To summarize, the trash will represent the union of the $(\Gamma - 1)$ larger (in $\hat{a}$ values) deviating items of each bin. Moreover, we want to maintain that the accumulated nominal ($\bar{a}$) size of trashed items in s is larger than the accumulated nominal size of deviating items in s^*.

Second, notice that in s^*, items of $[\![i_1^* + 1, i_2^* - 1]\!]$ are either in b_1^* as non-deviating items or in a b_j^*, $j \geq 2$ as deviating items (meaning that they are trashed items in s). Thus, if we incorrectly pack some of these items in b_1 instead of trashing them, these items will not be available when considering b_2, and we may not be able to ensure then that trashed items in s have a larger $\overline{a}$ value than the deviating items in s^*.

In the next section we describe the full version of the DP. To that end, we first need to introduce formally the notion of trash.

5   Approximating ΓRBP with Small Values

Bin-Packing with Trash. For any $X \subseteq [n]$, we define $\tilde{a}_\Gamma(X) = \Gamma \hat{a}_1(X)$ ($\tilde{a}_\Gamma(X)$ is Γ times the largest deviating value of an item in X) and $\tilde{f}(X) = \overline{a}(X) + \tilde{a}_\Gamma(X)$. We introduce next a decision problem ΓRBP-T related to ΓRBP.

ΓRBP-T (Robust bin-packing with trash)
Input: $(\mathcal{I}, k, t)$ where $\mathcal{I}$ is an input of ΓRBP (where each item $(\overline{a}_i, \hat{a}_i)$ satisfies $\hat{a}_i \leq 1/\Gamma$ and $\overline{a}_i \leq 1/\Gamma$), and k, t are two integers.
Output: Decide if a solution exists, where a solution is a partition of the set of items into $k + 1$ sets $b_1, \ldots, b_k$ and T (called the trash) such that:
- $\tilde{f}(b_j) \leq 1$ for each $j = 1, \ldots, k$
- $|T| \leq t$

Notice that although each item is small in ΓRBP-T, it is possible to have an item i such that $\tilde{f}(\{i\}) > 1$, implying that i must be put in the trash. We show below how deciding ΓRBP-T is enough to approximate ΓRBP.

Lemma 5. *For any input $\mathcal{I}$ of ΓRBP and $k^* = OPT(\mathcal{I})$, $(\mathcal{I}, k^*, (\Gamma - 1)k^*)$ is a yes input of ΓRBP-T.*

Proof. Given an optimal solution of size k^* of ΓRBP problem we create a solution to ΓRBP-T problem as follows. Let b_j^* be a bin of the considered optimum. Let N_j be the non-deviating items of b_j^*, i.e., $b_j^* = N_j \cup \Gamma(b_j^*)$. Let $X = \max(\Gamma(b_j^*))$ (the smallest deviating item of b_j^*) if $|\Gamma(b_j^*)| = \Gamma$ and $X = \emptyset$ otherwise. We define $b_j' = N_j \cup X$, and add items of $Y = b_j^* \setminus b_j'$ into the trash. Notice Y is either the set of $\Gamma - 1$ largest deviating object of b_j^*, or is equal to $\Gamma(b_j^*)$ when $|\Gamma(b_j^*)| < \Gamma$. This is a feasible solution for ΓRBP-T problem as $\tilde{f}(b_j') = \overline{a}(b_j') + \tilde{a}_\Gamma(b_j')$, where $\overline{a}(b_j') \leq \overline{a}(b_j^*)$ and $\tilde{a}_\Gamma(b_j') = \tilde{a}_\Gamma(X) \leq \hat{a}_\Gamma(b_j^*)$, and as there are at most $(\Gamma - 1)k^*$ items in the trash.   □

Lemma 6. *For any input $\mathcal{I}$ of ΓRBP and integer k, given a solution of $(\mathcal{I}, k, \Gamma k)$ of ΓRBP-T, we can compute in polynomial time a solution of $3k$ bins for $\mathcal{I}$.*

Proof. Given a solution $b_1, \ldots, b_k$, T for $(\mathcal{I}, k, \Gamma k)$ of ΓRBP-T the bins remain feasible in ΓRBP as $f_\Gamma(b_j) = \overline{a}(b_j) + \hat{a}_\Gamma(b_j) \leq \overline{a}(b_j) + \tilde{a}(b_j) = \tilde{f}(b_j)$. Then, Observation 1 implies that the trash T can be packed into $\lceil k\Gamma/(\Gamma/2) \rceil \leq 2k$ additional bins.   □

A DP Algorithm for ΓRBP-T. The objective of this section is to define a DP algorithm that will be used to decide the ΓRBP-T problem. To this aim, we define G-ΓRBP-T (generalized robust bin-packing with trash), an optimization problem that the DP algorithm will solve in a relaxed way. To define G-ΓRBP-T, we consider a fixed instance $\mathcal{I}$ of ΓRBP with items ordered according to (2) and an integer k.

G-ΓRBP-T (generalized robust bin-packing with trash)

Input: $\mathcal{I} = (q, t, \ell)$, where $q \in [n]_0$, $t \in [(\Gamma - 1)k]_0$, and $\ell \in [k+1]$.

Output: A feasible solution s is a partition of $[\![q, n]\!]$ into $k - \ell + 3$ sets (b_j for $j \in [\![\ell, k]\!]$, b_0 and T), such that

- for any $j \in [\![\ell, k]\!]$, $\tilde{f}(b_j) \leq 1$ (the $k - \ell + 1$ regular bins must respect the constraint of ΓRBP-T)
- $|T| \leq t$ (we only allow t items in the trash)
- $\min(b_\ell) = q$ (meaning that the deviating item of b_ℓ is q as items are sorted in non-increasing order of $\hat{a}$ values)

Minimize: $c(s) = \bar{a}(b_0)$ (in bin b_0 we only count $\bar{a}$ values)

The objective of G-ΓRBP-T is to pack a part (defined by $[\![q, k]\!]$) of an ΓRBP-T instance given a fixed budget of resources (the number of bins and the size of the trash) while minimizing the total nominal size of items in the dummy bin b_0. The last constraint (the deviating item of b_ℓ is q) may appear artificial at first sight, but comes from the fact that the DP will guess at each new bin the largest items that should be packed in it, and therefore this constraint ensures that every optimal solution must pack q in b_ℓ as well.

Definition 1 (almost feasible solution). *We say that a bin b **exceeds by at most one item** iff $\tilde{f}(b) > 1$ and $\tilde{f}(b \setminus \{i\}) \leq 1$ where $i = \max(b)$. Given an input (q, t, ℓ) of G-ΓRBP-T, we say that a solution is **almost feasible** iff all the above constraints of G-ΓRBP-T are respected, except that for any $j \in [\![\ell, k]\!]$, we allow that b_j exceeds by at most one item instead of $\tilde{f}(b_j) \leq 1$.*

The relation between G-ΓRBP-T and ΓRBP-T is characterized in the two following lemmas whose proofs can be found in [5].

Lemma 7 ($\star$). *For any $\mathcal{I}$ input of ΓRBP and k such that $(\mathcal{I}, k, (\Gamma - 1)k)$ is a yes input of ΓRBP-T, there exists q and t such that $\mathsf{OPT}(q, t, 1) = 0$.*

Lemma 8 ($\star$). *Let us fix $\mathcal{I}$ an input of ΓRBP and k an integer. For any $q \in [(\Gamma - 1)k], t = (\Gamma - 1)k - (q - 1)$, given an almost feasible solution of $\mathcal{I}' = (q, t, 1)$ of cost 0 for G-ΓRBP-T, we can compute in polynomial time a solution of $(\mathcal{I}, k, \Gamma k)$ of ΓRBP-T.*

Thus, Lemmas 6 and 8 show that providing an almost feasible solution for $(q, t, 1)$ of cost 0 for G-ΓRBP-T implies a solution of size $3k$ for ΓRBP.

Let us now define a DP algorithm $DP(\mathcal{I})$ ($\mathcal{I}$ is an input of G-ΓRBP-T) that provides an **almost feasible** solution s with $c(s) \leq \mathsf{OPT}(\mathcal{I})$ (where $\mathsf{OPT}(\mathcal{I})$ is by definition the optimal cost of a **feasible** solution). We provide below a gentle description of the DP. Given an instance $\mathcal{I} = (q, t, \ell)$, the DP starts by guessing (q^*, t^*), where

- $q^* = \min(b_{l'}^*)$ for a bin $b_{l'}^*$ with $l' \in [\![l+1, k^*]\!]$ of an optimal solution s^*
- t^* is the number of items trashed from X^* in s^*, where $X^* = [\![q, q^* - 1]\!]$
- Notice that in s^* items of X^* must by placed in b_l^*, b_0^* or T^*. We mimic the optimal in the current call of the DP by packing X^* in b_l, b_0 and T.
- To that end, the DP:
  - packs q in b_l (as required by the corresponding constraint of G-$\varGamma$RBP-T),
  - packs the t^* largest (in terms of $\bar{a}$) remaining items of X^* to the trash
  - packs the remaining items of X^* into b_ℓ until $\tilde{f}(b_\ell) > 1$ or $X^* = \emptyset$
  - packs the remaining items of X^* into b_0 until $X^* = \emptyset$

We discuss next where the other items (of $[\![q^*, n]\!]$) are packed. Notice that in s^*, bin b_l^* may contain items of $[\![q^*, n]\!]$, and thus the DP may also have to pack items of $[\![q^*, n]\!]$ into b_l. The key is that the decision of which items of $[\![q^*, n]\!]$ to pack into b_l is not taken at this step of the algorithm but only later (to avoid packing in b_l items of large $\bar{a}$ value that are in the trash in s^*). To allow this decision to be taken later, let Δ_b be the size of the empty space in b_l after packing X^* as described above, and let $b_0^{X^*} = b_0 \cap X^*$. After the previous steps, the DP makes a recursive call to get a solution $\tilde{s}$ that packs $[\![q^*, n]\!]$ into regular bins, the trash, and a dummy bin $\tilde{b}_0$. So far solution $\tilde{s}$ has not used any of the empty space Δ_b. However, we can unpack items from $\tilde{b}_0$ to b_ℓ while ensuring that these items do not deviate in b_ℓ (as all these items have index greater than q).

The formal description of $DP(q, t, \ell)$ and its correctness, stated formally in the following two results, are provided in [5].

Lemma 9 ($\star$). *For any $\mathcal{I}$ input of G-$\varGamma$RBP-T, $DP(\mathcal{I})$ provides an almost feasible solution of cost at most $OPT(\mathcal{I})$.*

Lemma 10 ($\star$). *There is a 3-approximation for $\varGamma$RBP with small values running in $\mathcal{O}(n^6 log(n))$.*

By Lemma 4, the following theorem is now immediate using a $\frac{3}{2}$-approximation for classical bin-packing (see for example in [16]) as a black box.

Theorem 5. *There is a 4.5-approximation for $\varGamma$RBP running in $\mathcal{O}(n^6 log(n))$.*

References

1. Aissi, H., Bazgan, C., Vanderpooten, D.: Min-max and min-max regret versions of combinatorial optimization problems: a survey. Eur. J. Oper. Res. **197**(2), 427–438 (2009)
2. Álvarez-Miranda, E., Ljubic, I., Toth, P.: A note on the Bertsimas & Sim algorithm for robust combinatorial optimization problems. 4OR **11**(4), 349–360 (2013). https://doi.org/10.1007/s10288-013-0231-6
3. Balogh, J., Békési, J., Dósa, G., Sgall, J., van Stee, R.: The optimal absolute ratio for online bin packing. In: Proceedings of the Twenty-Sixth Annual ACM-SIAM Symposium on Discrete Algorithms (2015)

4. Balogh, J., Békési, J., Dósa, G., Epstein, L., Levin, A.: A new and improved algorithm for online bin packing. In: Azar, Y., Bast, H., Herman, G. (eds.) ESA. LIPIcs, vol. 112, pp. 5:1–5:14. Dagstuhl, Germany (2018)
5. Basu Roy, A., Bougeret, M., Goldberg, N., Poss, M.: Approximating the robust bin-packing with budget uncertainty (2019). https://hal.archives-ouvertes.fr/hal-02119351
6. Bertsimas, D., Sim, M.: Robust discrete optimization and network flows. Math. Program. **98**(1–3), 49–71 (2003)
7. Bougeret, M., Pessoa, A.A., Poss, M.: Robust scheduling with budgeted uncertainty. Discrete Appl. Math. **261**(31), 93–107 (2019)
8. Dexter, F., Macario, A., Traub, R.D.: Which algorithm for scheduling add-on elective cases maximizes operating room utilization? Use of bin packing algorithms and fuzzy constraints in operating room management. Anesthesiology **91**, 1491–1500 (1999)
9. Goetzmann, K.-S., Stiller, S., Telha, C.: Optimization over integers with robustness in cost and few constraints. In: Solis-Oba, R., Persiano, G. (eds.) WAOA 2011. LNCS, vol. 7164, pp. 89–101. Springer, Heidelberg (2012). https://doi.org/10.1007/978-3-642-29116-6_8
10. Gounaris, C.E., Wiesemann, W., Floudas, C.A.: The robust capacitated vehicle routing problem under demand uncertainty. Oper. Res. **61**(3), 677–693 (2013)
11. Johnson, D.S., Demers, A., Ullman, J.D., Garey, M.R., Graham, R.L.: Worst-case performance bounds for simple one-dimensional packing algorithms. SIAM J. Comput. **3**(4), 299–325 (1974)
12. Karmarkar, N., Karp, R.M.: An efficient approximation scheme for the one-dimensional bin-packing problem. In: Proceedings of the 23rd Annual Symposium on Foundations of Computer Science (SFCS 1982), pp. 312–320, November 1982
13. Kasperski, A., Zieliński, P.: On the approximability of minmax (regret) network optimization problems. Inform. Process. Lett. **109**(5), 262–266 (2009)
14. Kasperski, A., Zielinski, P.: On the approximability of robust spanning tree problems. Theor. Comput. Sci. **412**(4–5), 365–374 (2011). https://doi.org/10.1016/j.tcs.2010.10.006
15. Poss, M.: Robust combinatorial optimization with knapsack uncertainty. Discrete Optim. **27**, 88–102 (2018). https://doi.org/10.1016/j.disopt.2017.09.004
16. Simchi-Levi, D.: New worst-case results for the bin-packing problem. Naval Res. Logist. **41**, 579–585 (1994)
17. Song, G., Kowalczyk, D., Leus, R.: The robust machine availability problem–bin packing under uncertainty. IISE Trans. **50**(11), 997–1012 (2018). https://doi.org/10.1080/24725854.2018.1468122
18. Tadayon, B., Smith, J.C.: Algorithms and complexity analysis for robust single-machine scheduling problems. J. Sched. **18**(6), 575–592 (2015)

Rank-Select Indices Without Tears

Tim Baumann[1] and Torben Hagerup[2(✉)] [iD]

[1] TNG Technology Consulting GmbH, 85774 Unterföhring, Germany
tim@timbaumann.info
[2] Institut für Informatik, Universität Augsburg, 86135 Augsburg, Germany
hagerup@informatik.uni-augsburg.de

Abstract. A *rank-select index* for a sequence $B = (b_1, \ldots, b_n)$ of n bits, where $n \in \mathbb{N} = \{1, 2, \ldots\}$, is a data structure that, if provided with a constant-time operation to access (the integer whose binary representation is) the subsequence of B in $\Theta(\log n)$ specified consecutive positions (thus B is stored outside of the data structure), can compute $rank_B(j) = \sum_{i=1}^{j} b_i$ for given $j \in \{0, \ldots, n\}$ and $select_B(k) = \min\{j \in \mathbb{N} \mid rank_B(j) \geq k\}$ for given $k \in \{1, \ldots, \sum_{i=1}^{n} b_i\}$. We describe a new rank-select index that, like previous rank-select indices, occupies $O(n \log \log n / \log n)$ bits and executes *rank* and *select* queries in constant time. Its derivation is intended to be largely free of tedious low-level detail, its operations are given by straight-line code, and it can be constructed in $O(n / \log n)$ time if B can be accessed as above.

1 Introduction

When S is a finite multiset of integers and $j \in \mathbb{Z} = \{\ldots, -1, 0, 1, \ldots\}$, we write $rank_S(j)$ for the rank of j in S, i.e., $rank_S(j) = |\{i \in S : i \leq j\}|$. Moreover, for each $k \in \{1, \ldots, |S|\}$, $select_S(k) = \min\{j \in \mathbb{Z} \mid rank_S(j) \geq k\}$. If the elements of S are arranged in nondecreasing order in positions $1, \ldots, |S|$, then $rank_S(j)$ is the largest position of an element $\leq j$ (0 if there is no such element), for $j \in \mathbb{Z}$, and $select_S(k)$ is the element in position k, for $k \in \{1, \ldots, |S|\}$.

The operations *rank* and *select* are defined also for bit sequences. If $B = (b_1, \ldots, b_n)$ is a sequence of n bits, for some $n \in \mathbb{N} = \{1, 2, \ldots\}$, then $rank_B(j) = \sum_{i=1}^{j} b_i$ for $j \in \{0, \ldots, n\}$ and, again, $select_B(k) = \min\{j \in \mathbb{N} \mid rank_B(j) \geq k\}$ for $k \in \{1, \ldots, \sum_{i=1}^{n} b_i\}$. The connection between the two definitions is close: If a simple set S is a subset of $\{a, \ldots, a + n - 1\}$ for some known $a \in \mathbb{Z}$ and $n \in \mathbb{N}$, S can be represented via the bit sequence $B = (b_1, \ldots, b_n)$ with $b_i = 1 \Leftrightarrow a - 1 + i \in S$, for $i = 1, \ldots, n$. In this case we say that S is given by its *bit-vector representation* over the universe $\{a, \ldots, a + n - 1\}$ or with *offset* a and *span* n. Clearly $rank_S(j) = rank_B(j - (a - 1))$ for $j \in \{a - 1, \ldots, a - 1 + n\}$ and $select_S(k) = a - 1 + select_B(k)$ for $k \in \{1, \ldots, |S|\}$. Answering *rank* and *select* queries about a simple set of integers therefore reduces to answering *rank* and *select* queries about its bit-vector representation with some known offset and span.

A (static) *rank-select structure* for a sequence $B = (b_1, \ldots, b_n)$ of n bits, for some $n \in \mathbb{N}$, is a data structure capable of returning $rank_B(j)$ for arbitrary

© Springer Nature Switzerland AG 2019
Z. Friggstad et al. (Eds.): WADS 2019, LNCS 11646, pp. 85–98, 2019.
https://doi.org/10.1007/978-3-030-24766-9_7

given $j \in \{0, \ldots, n\}$ and $select_B(k)$ for arbitrary given $k \in \{1, \ldots, \sum_{i=1}^{n} b_i\}$. A data structure that can answer the same queries, but only if provided with a constant-time operation to access (the integer whose binary representation is) the subsequence of B in $\Theta(\log n)$ specified consecutive positions (thus B is stored outside of the data structure), is known as a *rank-select index* for B. We call B the *client sequence* of a rank-select structure or index for B. Rank-select structures and indices are of fundamental importance in space-efficient computing, have been studied extensively since the 1970s, and have many and diverse applications. Elias [3] considered the representation of multisets of integers in the context of data retrieval. For a multiset S, his *direct* or *table-lookup* question corresponds to $select_S$, and his *inverse* question is closely related to $rank_S$. Jacobson, who introduced the terms *rank* and *select* [9,10], used rank-select structures to represent trees and graphs in little space while still permitting their efficient traversal. Along the way, he solved the problem of finding matching parentheses in a balanced sequence of parentheses, again with rank-select structures as crucial components of the overall data structure. Rank-select structures and indices have also found applications in areas such as string processing [6], computational geometry [11] and graph algorithms [8].

Jacobson designed a rank-select index for bit sequences of length $n \in \mathbb{N}$ that occupies $O(n \log \log n/\log n)$ bits and answers *rank* queries in constant time. While he was unable to obtain a constant-time *select* operation, this was remedied by Clark [2] at the price of a somewhat higher space bound. From now on we will be interested only in rank-select structures and indices that answer all queries in constant time; for ease of discussion, consider this property to be part of their definition. A rank-select index that uses $O(n \log \log n/\log n)$ bits was described by Raman, Raman and Satti [13, Lemma 4.1]. A matching lower bound of $\Omega(n \log \log n/\log n)$ on the number of bits needed by a rank-select index that accesses only $O(\log n)$ bits of its client sequence during the processing of a query was proved by Golynski [5]. Thus a rank-select structure for a sequence B of n bits that consists of B plus a suitable index must occupy $n + \Omega(n \log \log n/\log n)$ bits. While this is a natural way of organizing a rank-select structure, Pătraşcu [12] showed the interesting fact that there are smaller rank-select structures that do not store B in its "raw" form.

In this paper we describe another $O(n \log \log n/\log n)$-bit rank-select index. While previous descriptions of rank-select structures and indices abound with ad-hoc and rather tedious low-level detail, we aim for a more systematic, modular and high-level approach based largely on pictures that leads to the optimal result with comparatively little effort on the part of the reader. Our rank-select index offers the first *select* operation that can be formulated as a piece of straight-line code, i.e., its implementation is free of tests and branching (in one place, fulfilling this promise involves a small amount of "cheating", as will be explained later). We also consider the problem of efficient construction of rank-select indices, an aspect that was ignored in much previous research but is essential to many applications. Our main result is the following:

Theorem 1. *For every $n \in \mathbb{N}$ and for every sequence B of n bits, given in the form of a stream of $O(n/\log n)$ chunks of $O(n \log \log n/\log n)$ consecutive bits each, a rank-select index for B that executes $rank_B$ and $select_B$ in constant time and occupies $O(n \log \log n/\log n)$ bits can be constructed in $O(n/\log n)$ time using $O(n \log \log n/\log n)$ bits of working memory.*

The theorem insists that the client sequence B be provided in several chunks because the available working space does not allow us to store B in its entirety. Typically B would be provided in $\Theta(n/\log n)$ chunks of $O(\log n)$ bits each. If B is stored in random-access read-only memory, of course, it is trivial to produce the necessary chunks, but the theorem implies that the construction of the rank-select index does not require random access to B and can make do with a single pass over B. In some applications, e.g., to the subgraph stack of [8], it is essential that the construction time of the rank-select index is $O(n/\log n)$ and not just $O(n)$.

Our model of computation is the standard word RAM [1,7] with a word length of $w = \Omega(\log n)$ bits, where w is assumed large enough to allow all memory words in use to be addressed. The word RAM has constant-time operations for addition, subtraction and multiplication modulo 2^w, division with truncation $((x,y) \mapsto \lfloor x/y \rfloor$ for $y > 0)$, left shift modulo 2^w $((x,y) \mapsto (x \ll y) \bmod 2^w$, where $x \ll y = x \cdot 2^y)$, right shift $((x,y) \mapsto x \gg y = \lfloor x/2^y \rfloor)$, and bitwise Boolean operations (AND, OR and XOR (exclusive or)).

2   Ingredients of the New Rank-Select Index

Our overall approach, shared with earlier solutions, is to break down a given instance of the rank-select problem, i.e., the problem of answering *rank* and *select* queries for a given bit sequence or multiset, into still smaller instances, eventually arriving at instances so tiny that they can be solved by brute force, i.e., table lookup. We provide a bottom-up description, proceeding from table lookup via basic reductions of instances of the rank-select problem to simpler instances and ending with the complete rank-select index that reduces *rank* and *select* queries about the client sequence all the way to table lookup. We prefer to phrase much of the discussion in terms of (multi)sets rather than bit sequences. The tables needed by the rank-select index and their computation are discussed in the next subsection.

2.1   Table Lookup

This subsection describes three different variants, denoted T1–T3, of the table-lookup method, as applied to the rank-select problem. In our applications, the parameters N and M for variants T2 and T3 will be so small as to render negligible the space occupied by the tables and the time needed to compute them.

T1. In order to answer $rank_S$ and $select_S$ queries about one particular subset S of $\{1, \dots, N\}$, where $N \in \mathbb{N}$ is known, we can simply store a table of $rank_S(j)$ for $j = 0, \dots, N$ and $select_S(k)$ for $k = 1, \dots, |S|$ and answer a query by returning the appropriate table entry. The number of bits needed is $O(N \log N)$, and the table can be computed in $O(N)$ time from a bit-vector representation of S.

T2. If the goal is to answer $rank_S$ and $select_S$ queries, where S now is also specified in the query and can be an arbitrary subset of $\{1, \dots, N\}$, we can create a subtable as for variant T1 for each of the 2^N possible subsets S and store the 2^N subtables, each indexed by the bit-vector representation over $\{1, \dots, N\}$ of the corresponding set S, in a table of $O(2^N N \log N)$ bits whose computation takes $O(2^N N)$ time.

T3. If S is a variable subset of $\{1, \dots, N\}$ but known to be of size at most M for some given $M \in \mathbb{N}$, S can be represented as an M-tuple of integers in $\{1, \dots, N\}$ by first listing the elements of S and then, if $|S| < M$, repeating the last element. Each of the M integers can in turn be represented in $\lceil \log_2 N \rceil$ bits. Since $2^{\lceil \log_2 N \rceil} \leq 2N$, this gives us an alternative to variant T2 with a table of $O((2N)^{M+1} \log N)$ bits that can be computed in $O((2N)^{M+1})$ time.

2.2   Three Basic Reductions

Let g be a nondecreasing function from $\mathbb{Z}$ to $\mathbb{Z}$ (informally, the *grouping function*) with the property that $g^{-1}(q) = \{i \in \mathbb{Z} \mid g(i) = q\}$ is finite for all $q \in \mathbb{Z}$ and let S be a finite multiset of integers. For $q \in \mathbb{Z}$, we write $S \cap g^{-1}(q)$ for the multiset $\{i \in S \mid g(i) = q\}$ of those elements of S that are mapped to q by g. E.g., if $g(j) = \lfloor j/10 \rfloor$ for all $j \in \mathbb{Z}$,

$$\{47, 47, 63, 68, 72, 76, 76, 79, 85\} \cap g^{-1}(7) = \{72, 76, 76, 79\}.$$

While denoting by $g(S)$ the simple set $\{q \in \mathbb{Z} \mid \exists i \in S : g(i) = q\} = \{q \in \mathbb{Z} \mid S \cap g^{-1}(q) \neq \emptyset\}$, we write $g((S))$ for the multiset $\{g(i) \mid i \in S\}$, in which each $q \in \mathbb{Z}$ occurs with multiplicity $|S \cap g^{-1}(q)|$. E.g., if $S = \{47, 47, 63, 68, 72, 76, 76, 79, 85\}$ and $g(j) = \lfloor j/10 \rfloor$ for all $j \in \mathbb{Z}$ as above, then $g(S) = \{4, 6, 7, 8\}$ and $g((S)) = \{4, 4, 6, 6, 7, 7, 7, 7, 8\}$. We will argue that if $j \in \mathbb{Z}$ and $q = g(j)$, then

$$rank_S(j) = rank_{g((S))}(q - 1) + rank_{S \cap g^{-1}(q)}(j). \tag{1}$$

E.g, with S and g as above, $j = 76$ and $q = g(j) = 7$,

$$rank_S(76) = rank_{\{4,4,6,6,7,7,7,7,8\}}(7 - 1) + rank_{\{72,76,76,79\}}(76) \quad (= 4 + 3 = 7).$$

The validity of (1) is easy to see: The terms $rank_{g((S))}(q-1)$ and $rank_{S \cap g^{-1}(q)}(j)$ count the elements i of $\{i \in S \mid i \leq j\}$ with $g(i) < q$ and with $g(i) = q$, respectively. Let $k \in \{1, \dots, |S|\}$. If we think of S as a sorted list, then $g(select_S(k))$ is the value under g of the kth element of S, whereas $select_{g((S))}(k)$ is the kth element of the list obtained from S by applying g to each of its elements. Thus $g(select_S(k)) = select_{g((S))}(k)$. Now, with $q = select_{g((S))}(k)$,

$$select_S(k) = select_{S \cap g^{-1}(q)}(k - rank_{g((S))}(q - 1)), \tag{2}$$

since $rank_{g((S))}(q - 1)$ again is the number of elements $i \in S$ with $g(i) < q$, so that, if S is presented in sorted order in $|S|$ positions, $select_S(k)$ is the element in position $k - rank_{g((S))}(q - 1)$ among the elements with the same value under g as itself, i.e., within $S \cap g^{-1}(q)$. As an example, if S and g are as above, $k = 6$ and $q = select_{g((S))}(k) = select_{\{4,4,6,6,7,7,7,7,8\}}(6) = 7$,

$$select_{\{47,47,63,68,72,76,76,79,85\}}(6) = select_{\{72,76,76,79\}}(6 - rank_{\{4,4,6,6,7,7,7,7,8\}}(6))$$
$$(= select_{\{72,76,76,79\}}(6 - 4) = 76 \ ).$$

Let us use $(S|g)$ as a convenient notation for the function $(g_{|S})^{-1}$ that maps each $q \in \mathbb{Z}$ to $S \cap g^{-1}(q)$. E.g., with our usual S and g, $(S|g)(4) = \{47, 47\}$ and $(S|g)(2) = (S|g)(5) = \emptyset$. Then, by Eqs. (1) and (2) above, answering $rank$ and $select$ queries about S in constant time reduces to answering $rank$ and $select$ queries about $g((S))$ and about values of $(S|g)$ in constant time. For brevity, we express this by saying that S reduces to $g((S))$ and $(S|g)$. We will use this only with $g = g_\lambda$ for some $\lambda \in \mathbb{N}$, where $g_\lambda(j) = \lfloor j/\lambda \rfloor$ for all $j \in \mathbb{Z}$, and call λ the *parameter* of the reduction. Variations of this reduction have been used since the early days of rank-select indices [9,10]. We call it BR1 ("basic reduction 1") and denote it symbolically with a triangular shape, as shown in the left subfigure of Fig. 1: S, at the apex of the triangle, reduces to $g_\lambda((S))$ and $(S|g_\lambda)$ at its base. The double line serves as a reminder that $g_\lambda((S))$ in general is a multiset even if S is not. We refrain from connecting S to the triangle with a double line because we will use the reduction only for simple sets S. A bar through the line to $(S|g_\lambda)$ indicates that $(S|g_\lambda)$ is a set-valued function rather than just a set.

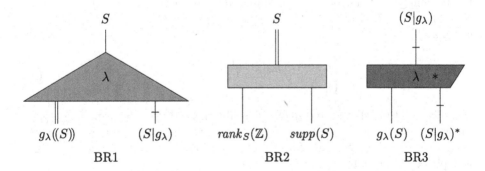

Fig. 1. The three basic reductions BR1–BR3.

Denote by $supp(S)$ the support of the multiset S, i.e., the simple set that contains exactly the same values as S, but each value only once, and let $rank_S(\mathbb{Z})$

be the image of the $rank_S$ function. For example, with $S = \{1, 1, 1, 5, 7, 7\}$ we have $supp(S) = \{1, 5, 7\}$ and $rank_S(\mathbb{Z}) = \{0, 3, 4, 6\}$. A second reduction is given by the formulas

$$rank_S(j) = select_{rank_S(\mathbb{Z})}(rank_{supp(S)}(j) + 1) \quad \text{and} \tag{3}$$

$$select_S(k) = select_{supp(S)}(rank_{rank_S(\mathbb{Z})}(k - 1)), \tag{4}$$

for $j \in \mathbb{Z}$ and $k \in \{1, \ldots, |S|\}$. E.g.,

$$rank_{\{1,1,1,5,7,7\}}(6) = select_{\{0,3,4,6\}}(rank_{\{1,5,7\}}(6) + 1)$$
$$(= select_{\{0,3,4,6\}}(2 + 1) = 4) \quad \text{and}$$
$$select_{\{1,1,1,5,7,7\}}(4) = select_{\{1,5,7\}}(rank_{\{0,3,4,6\}}(4 - 1))$$
$$(= select_{\{1,5,7\}}(2) = 5).$$

To see the validity of Eq. (3), whose origins can be traced back to Fano [4, Step 2], note that the $(q + 1)$st smallest element of $rank_S(\mathbb{Z})$, for $q = 0, \ldots, |supp(S)|$, is the total number of occurrences in S of the q smallest distinct values in S. With $q = rank_{supp(S)}(j)$, this is precisely $rank_S(j)$. Equation (4) is implied by the following observation: If S, presented in sorted order in $|S|$ positions, is thought of as partitioned into maximal *ranges* of occurrences of the same value, then $rank_{rank_S(\mathbb{Z})}(k - 1)$ is one more than the number of ranges that end strictly before the kth position, i.e., is the number of the range that contains the kth position. We may conclude that S also reduces to $rank_S(\mathbb{Z})$ and $supp(S)$. We call this reduction BR2 and depict it as shown in the middle subfigure of Fig. 1.

When S is a subset of $\mathbb{Z}$ and $\lambda \in \mathbb{N}$, we denote by $\lambda + S$ the set $\{\lambda + i \mid i \in S\}$ and by $S \bmod \lambda$ the set $\{i \bmod \lambda \mid i \in S\}$ of remainders modulo λ of elements of S. Let $(S|g_\lambda)^*$ be the function defined on $\{1, \ldots, |g_\lambda(S)|\}$ that maps q to $(S|g_\lambda)(select_{g_\lambda(S)}(q)) \bmod \lambda$, for $q = 1, \ldots, |g_\lambda(S)|$. Informally, if $(S|g_\lambda)$ is thought of as a list of subsets of S, then $(S|g_\lambda)^*$ is the sublist that contains only the nonempty subsets, but each normalized to lie in $\{0, \ldots, \lambda - 1\}$. E.g., with $S = \{47, 63, 68, 72, 76, 79, 85\}$, $(S|g_{10})^*(1) = \{7\}$ and $(S|g_{10})^*(2) = \{3, 8\}$. For $q \in \mathbb{Z}$ and $\lambda \in \mathbb{N}$,

$$(S|g_\lambda)(q) = \begin{cases} q\lambda + (S|g_\lambda)^*(rank_{g_\lambda(S)}(q)), & \text{if } q \in g_\lambda(S), \\ \emptyset, & \text{otherwise.} \end{cases} \tag{5}$$

Since we can test whether $q \in g_\lambda(S)$ by evaluating $rank_{g_\lambda(S)}(q) - rank_{g_\lambda(S)}(q - 1)$, which is 1 if $q \in g_\lambda(S)$ and 0 otherwise, $(S|g_\lambda)$ (i.e., $(S|g_\lambda)(q)$ for each $q \in \mathbb{Z}$) reduces to $g_\lambda(S)$ and $(S|g_\lambda)^*$. This third and last basic reduction, BR3, is depicted in the right subfigure of Fig. 1.

2.3 Two Combined Reductions

We can combine BR1 and BR2 as illustrated in Fig. 2. This yields a reduction, CR1, of S to $rank_{g_\lambda((S))}(\mathbb{Z})$, $supp(g_\lambda((S))) = g_\lambda(S)$, and $(S|g_\lambda)$. Incorporating

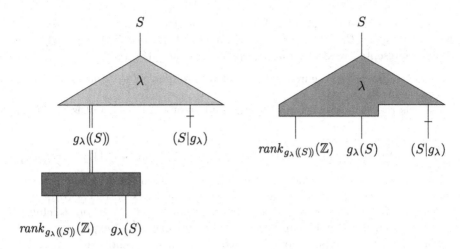

Fig. 2. The first combined reduction CR1. Left: internal structure. Right: pictorial representation.

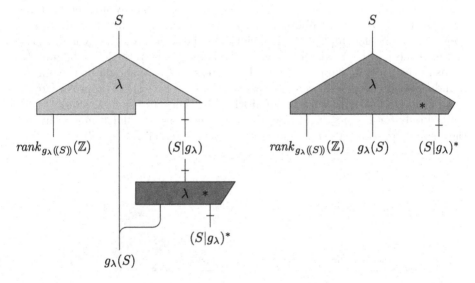

Fig. 3. The reduction CR2. Left: internal structure. Right: pictorial representation.

also BR3, we obtain a second combined reduction, CR2, shown in Fig. 3. CR1 and CR2 are very similar. Both reduce S to $rank_{g_\lambda((S))}(\mathbb{Z})$, $g_\lambda(S)$ and a third quantity, which is $(S|g_\lambda)$ in the case of CR1 and $(S|g_\lambda)^*$ in the case of CR2.

In the concrete rank-select index, sets of integers are represented as bit vectors with convenient offsets and spans. The offsets and spans can be calculated in parallel with the application of reductions according to the following rules:

The client sequence B, prefixed by a 0, is viewed as representing a set with offset 0 and span $|B| + 1$, where $|B|$ denotes the length of B. Recursively, if a set S is represented with offset a and span n and $m = |S|$, then

- $rank_{g_\lambda((S))}(\mathbb{Z})$ is represented with offset 0 and span $m + 1$ and is of size at most $\lceil n/\lambda \rceil + 1$.
- $g_\lambda(S)$ is represented with offset $g_\lambda(a)$ and span $\lceil n/\lambda \rceil + 1$ and is of size at most m.
- $(S|g_\lambda)(q)$ is represented with offset $q\lambda$ and span λ for all $q \in \mathbb{Z}$.
- $(S|g_\lambda)^*(q)$ is represented with offset 0 and span λ for all $q \in \{1, \ldots, |g_\lambda(S)|\}$.

Using the rules to keep track of offsets enables us, at the bottom of a recursive application of reductions, to translate queries about sets correctly to queries about their bit-vector representations with the given offsets. So as not to clutter the description, this simple translation will not be formulated explicitly.

The effect of each type of combined reduction on spans and sizes is depicted in Fig. 4. A pair of the form $\langle n, m \rangle$ indicates that a set has span n and size at most m, except that the rounding to integer values and the occasional $+1$ were ignored. When we refrain from bounding the size of a set by anything better than its span, $\langle n \rangle$ is used as an abbreviation for $\langle n, n \rangle$. If S has offset a and span n, $(S|g_\lambda)(q)$ can be nonempty only if q belongs to the set $\{g_\lambda(a), \ldots, g_\lambda(a+n-1)\}$ of size at most $\lceil \frac{n}{\lambda} \rceil + 1$, which motivates the label $\frac{n}{\lambda} \cdot \langle \lambda \rangle$ in the left subfigure. In terms of the concrete data structure, the expression $\frac{n}{\lambda} \cdot \langle \lambda \rangle$ should be thought of as indicating an array with index set $\{g_\lambda(a), \ldots, g_\lambda(a+n-1)\}$ of (approximately) $\frac{n}{\lambda}$ subordinate data structures, each of which is for a set of span λ. $(S|g_\lambda)^*(q)$ is defined only for $1 \le q \le |g_\lambda(S)| \le |S|$, which motivates the label $m \cdot \langle \lambda \rangle$ in the right subfigure; the corresponding array has index set $\{1, \ldots, |g_\lambda(S)|\}$.

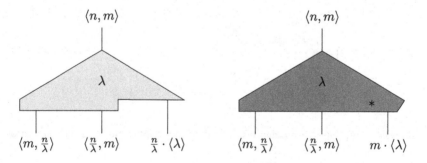

Fig. 4. The approximate effect of the two combined reductions CR1 and CR2 on $\langle \text{span}, \text{size} \rangle$.

The new optimal rank-select index can be pieced together with little effort from a constant number of instances of the combined reductions CR1 and CR2. As a warm-up and to familiarize the reader with the approach and the notation, we first develop a simpler rank-select structure that occupies $\Theta(n)$ bits for client sequences of n bits.

3   A Simplified $O(n)$-Bit Rank-Select Structure

The simplified rank-select structure is best thought of as the tree T_S shown in Fig. 5 annotated with ⟨span, size⟩ pairs suitable for a client sequence of n bits. Each inner node in T_S corresponds to an instance of the composite reduction CR1 (for brevity: is a CR1-node) and is drawn with the characteristic shape of that reduction. Each leaf in T_S corresponds to an instance or an array of instances of one of the variants of the table-lookup method and is drawn as a rectangle labeled with the name of the relevant variant. During the execution of a query, each inner node in T_S applies its associated reduction and each leaf answers queries using its table-lookup variant.

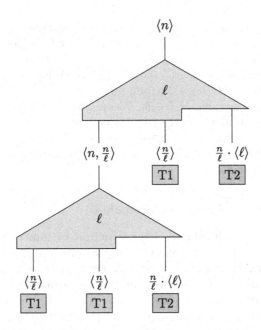

Fig. 5. A simplified rank-select structure that occupies $O(n)$ bits.

The reductions in T_S both use a parameter $\ell \in \mathbb{N}$. Here and in the following we choose $\ell = \Theta(\log n)$ such that the cost, in terms of time and space, of the table needed by table-lookup variant T2 with $N = \ell$ is negligible. In concrete terms, we take this to mean that $2^\ell \ell \log \ell = O(n/\log n)$, which is certainly satisfied if $\ell \leq (1/2) \log_2 n$.

By construction, the simplified rank-select structure is correct and executes queries in constant time. Let us analyze its storage requirements, for the time being ignoring rounding issues and pretending that the expressions for spans and sizes in Fig. 4 are exact. The first step is to verify that the ⟨span, size⟩ pairs in Fig. 5 have indeed be calculated in accordance with Fig. 4. Each leaf that uses table-lookup variant T1 (is a T1-leaf, say) stores a table of *rank* and *select* for a sequence of $\frac{n}{\ell}$ bits, which needs $O(\frac{n}{\ell} \log n) = O(n)$ bits. Similarly, each T2-leaf stores an array of $\frac{n}{\ell}$ sequences, each of ℓ bits, again for a total of $O(n)$ bits. Adding the $O(n/\log n)$ bits occupied by a global table for variant T2 and $O(\log n)$ bits for storing ℓ, links between the nodes in T_S and various other bits and pieces, we arrive at a grand total of $O(n)$ bits. It is easy to see that the error incurred by the approximation involved in Fig. 4 amounts to less than a constant factor—this is because additive error terms bounded by constants affect only quantities that are $\Omega(\log n)$—so that the true number of bits used by the simplified rank-select structure is also $O(n)$.

4   The New Optimal Rank-Select Index

The new optimal rank-select index has much in common with the simplified rank-select structure of the previous section. In order to achieve a better space bound, however, the optimal index must comprise a few additional reductions. Its structure is given by the tree T shown in Fig. 6. While most reductions use the parameter ℓ with $\ell = \Theta(\log n)$ introduced in Sect. 3, the reduction at the root of T employs a larger parameter $L \in \mathbb{N}$. We choose $L = \Theta((\log n)^2/\log \log n)$ as a multiple of ℓ such that the cost of the table needed by table-lookup variant T3 with $N = L + 1$ and $M = L/\ell + 1$ is negligible. In concrete terms, we take this to mean that $(2(L + 1))^{L/\ell+2} \log L = O(n/\log n)$, which is ensured if $L/\ell \le (1/4)\log_2 n/\log_2 \log_2 n$.

4.1   Analysis of the Space Requirements

As in the case of the simplified rank-select structure, by construction the optimal rank-select index is correct and executes queries in constant time. Because we now aim for a space bound of $o(n)$ bits, we must pay special attention to the rightmost leaf in the tree T with its 2-dimensional array A of (approximately) $\frac{n}{L} \cdot \frac{L}{\ell} = \frac{n}{\ell}$ sequences, each of (at most) ℓ bits. As is not difficult to see, each of the relevant bit sequences is a subsequence of the client sequence B. Moreover, the position of the first bit of the relevant subsequence within B is a simple function of the two integers used to index into A. Therefore there is no need to store any data in the rightmost leaf—it suffices to provide it with constant-time access to arbitrary subsequences of at most $\ell = \Theta(\log n)$ consecutive bits in B, which is precisely what the rank-select index is allowed to rely on.

The remaining part of the analysis of the space requirements parallels what was done in Sect. 3 for the simplified rank-select structure, and again we can pretend that the expressions given in Fig. 4 are exact—as a minor exception, one should observe that table-lookup variant T3 is indeed used with $N = L + 1$ and $M = L/\ell + 1$. Again, the first step is to verify that the $\langle$span, size$\rangle$ pairs in Fig. 6 have been calculated in accordance with Fig. 4.

Each T1-leaf holds tables of *rank* and *select* for a bit sequence of length $\frac{n}{L}$ or $\frac{n}{\ell^2}$. The number of bits needed for the tables is therefore $O((\frac{n}{L} + \frac{n}{\ell^2}) \log n) = O(n \log \log n/\log n)$. The total length of the bit sequences stored in T2-leaves is

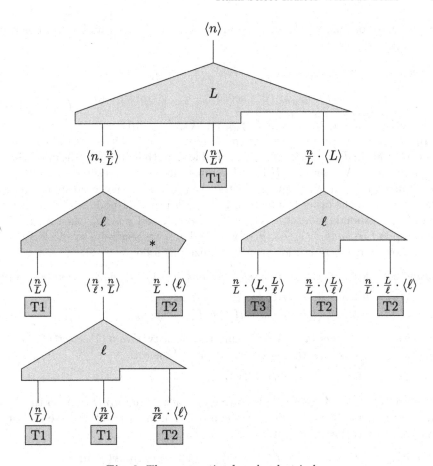

Fig. 6. The new optimal rank-select index.

$O((\frac{n}{L} + \frac{n}{\ell^2})(\ell + \frac{L}{\ell})) = O(\frac{n}{L} \cdot \ell) = O(n \log\log n/\log n)$ (recall that the rightmost leaf in T has already been considered). Finally, the single T3-leaf needs space for $\frac{n}{L}$ bit sequences, each of which represents a tuple of $\frac{L}{\ell}$ integers of $O(\log L)$ bits each, for a total of $O(\frac{n}{L} \cdot \frac{L}{\ell} \cdot \log L) = O(n \log\log n/\log n)$ bits. In summary, the number of bits occupied by the rank-select index is $O(n \log\log n/\log n)$.

4.2 The Execution of Queries

A combination of Eqs. (1)–(4) in Subsect. 2.2 yields the two formulas below, which closely mirror the derivation of the combined reduction CR1 from BR1 and BR2. S is a finite subset of $\mathbb{Z}$, and in the first formula $j \in \mathbb{Z}$ and $q = g(j)$.

$$rank_S(j) \overset{(1)}{=} rank_{g(\!(S)\!)}(q-1) + rank_{S\cap g^{-1}(q)}(j)$$

$$\overset{(3)}{=} select_{rank_{g(\!(S)\!)}(\mathbb{Z})}(rank_{supp(g(\!(S)\!))}(q-1)+1) + rank_{S\cap g^{-1}(q)}(j)$$

$$= select_{rank_{g(\!(S)\!)}(\mathbb{Z})}(rank_{g(S)}(q-1)+1) + rank_{(S|g)(q)}(j).$$

In the second formula $k \in \{1, \ldots, |S|\}$ and $q = select_{g(\!(S)\!)}(k)$, and we can rewrite $rank_{g(\!(S)\!)}(q-1)$ as above.

$$select_S(k) \overset{(2)}{=} select_{S \cap g^{-1}(q)}(k - rank_{g(\!(S)\!)}(q-1))$$

$$= select_{(S|g)(q)}(k - select_{rank_{g(\!(S)\!)}(\mathbb{Z})}(rank_{g(S)}(q-1)+1)).$$

Moreover, $q = select_{g(\!(S)\!)}(k) \overset{(4)}{=} select_{g(S)}(rank_{rank_{g(\!(S)\!)}(\mathbb{Z})}(k-1))$.

In a concrete implementation of the rank-select index, it is natural to represent each node in T by an instance of a suitable class that supports the operations *rank* and *select*. Assume that U is such an instance that corresponds to an inner node u in T and that X, Y and Z are the class instances that correspond to the children of u, in the order from left to right. If u is a CR1-node with parameter λ and U corresponds to a set S (i.e., realizes $rank_S$ and $select_S$), then X, Y and Z correspond to $rank_{g_\lambda(\!(S)\!)}(\mathbb{Z})$, $g_\lambda(S)$ and $(S|g_\lambda)$, respectively, so the formulas above show that U's operations can be realized as follows:

$U.rank(j) = $ **let** $q = g_\lambda(j)$ **in** $X.select(Y.rank(q-1)+1) + Z[q].rank(j)$ and

$U.select(k) = $ **let** $q = Y.select(X.rank(k-1))$ **in**
$\qquad\qquad Z[q].select(k - X.select(Y.rank(q-1)+1)).$

Computing $r = X.rank(k-1)$ and using the identity $Y.rank(Y.select(r)-1) = r-1$, we can streamline the implementation of $U.select$:

$U.select(k) = $ **let** $r = X.rank(k-1)$ **in** $Z[Y.select(r)].select(k - X.select(r)).$

If u is a CR2-node, Z corresponds to $(S|g_\lambda)^*$ rather than to $(S|g_\lambda)$, so the formulas must be modified. By Eq. (5), the discussion following it and the identity $Y.rank(Y.select(r)) = r$, we obtain:

$U.rank(j) = $ **let** $q = g_\lambda(j)$, $r = Y.rank(q-1)$, $s = Y.rank(q)$ **in**
$\qquad\qquad X.select(r+1) + Z[s].rank(j \bmod \lambda) \cdot (s-r)$ and

$U.select(k) = $ **let** $r = X.rank(k-1)$ **in**
$\qquad\qquad Y.select(r) \cdot \lambda + Z[r].select(k - X.select(r)).$

The formulas can clearly be expressed as straight-line code. In the case of *rank* for a CR2-node it could be argued that it would be more natural to replace the multiplication by a zero test followed by a branch. This is why claiming that our *rank* and *select* operations are straight-line involves a small amount of "cheating". Class instances that correspond to leaves in T, of course, realize their *rank* and *select* operations by a single access to a 1- or 2-dimensional table, handled in accordance with the relevant offset.

4.3   The Construction of the Index

In order to construct the optimal rank-select index for a client sequence B of n bits, we interpret each node in the tree T of Fig. 6 as a process whose task is to communicate with adjacent nodes and, in the case of leaves, to compute and store a table or an array for use in subsequent queries. Each inner node in

T receives a stream of bits from its parent and sends streams of bits or integers derived from its input stream to its children. An exception concerns the root of T, which receives the client sequence B in the stream that feeds the overall construction and prepends a single 0 to B before processing B, which is now considered to represent the same set as before, but with offset 0 and span $n+1$.

In a preprocessing phase that proceeds top-down in the tree and takes constant time, each node uses the rules formulated in Subsect. 2.3 to compute the offsets and spans of the sets that it will handle. This enables each leaf to acquire the space needed to hold its table or array.

Consider an inner node u in T with parameter λ and assume that u's input stream is a bit-vector representation of a set S whose offset a is a multiple of λ—by the 0 prepended to the client sequence as described above, this assumption, which we call the *alignment assumption*, is satisfied for the root of T. If u receives a sequence of bit-vector representations, the assumption as well as the following arguments should be applied independently to each element in the sequence. The task of u is to send either a stream of the elements of $rank_{g_\lambda((S))}(\mathbb{Z})$ (if u's left child is a T3-leaf) or a bit-vector representation of this set with offset 0 (otherwise) to its left child, a bit-vector representation of $g_\lambda(S)$ with offset $g_\lambda(a)$ to its middle child, and bit-vector representations of either $(S|g_\lambda)(q)$ with offset $q\lambda$ for $q = g_\lambda(a), g_\lambda(a)+1, \ldots$ or $(S|g_\lambda)^*(q)$ with offset 0 for $q = 1, \ldots, |g_\lambda(S)|$ to its right child.

The node u processes its input stream in *batches* of λ consecutive bits each, except that the last batch may be smaller, in which case it is filled up to size λ with 0s. Note that by the alignment assumption, a batch corresponds exactly to $g_\lambda^{-1}(q)$ for some $q \in \mathbb{Z}$. Before the processing of the first batch, u initializes a variable s to 0 and sends the integer 0 (if u's left child is a T3-leaf) or a single 1 (otherwise) to its left child. The processing of a batch begins by determining the number k of 1s in the batch. If $\lambda = \ell$, this is done in constant time by lookup in a table whose construction time $(O(2^\ell))$ and space requirements $(O(2^\ell \log \ell)$ bits) are negligible. If $\lambda = L$, k is instead found by consulting the table L/ℓ times and summing the values found there. If $k > 0$, u adds k to s and proceeds to send the current value of s (if u's left child is a T3-leaf) or $k - 1$ 0s followed by a 1 (otherwise) to its left child, a 1 to its middle child and the whole batch to its right child as the next sequence element. If $k = 0$, u instead sends nothing to its left child, a 0 to its middle child and, only if u is a CR1-node, the whole batch to its right child, again as the next sequence element.

Each T1-leaf in T constructs and stores a table of *rank* and *select* for the bit stream that it receives, and each T2-leaf, except for the rightmost leaf, receives a sequence of bit streams and simply stores these in successive cells of an array. Finally, the single T3-leaf, for each of the (approximately) $\frac{n}{L}$ sets that it receives, stores the concatenation of the $\lceil \log_2(L + 1) \rceil$-bit binary representations of the at most $L/\ell + 1$ elements of the set in the next cell of an array. It is not difficult to see that the index is constructed correctly.

If we introduce a buffer of $L + \ell$ bits between each pair of adjacent nodes in T (except between the single T3-leaf and its parent, where a buffer of $L/\ell + 1$

integers of $\lceil \log_2(L+1) \rceil$ bits each is suitable), we can repeatedly execute a top-down sweep over T in which the root processes the next batch of L bits of the client sequence in $O(L/\ell)$ time, thereby adding bits to its outbuffers, and every other node in T processes as many integers or complete batches of ℓ bits, each in constant time, as available in its inbuffer, again adding integers or bits to its outbuffers, if any. It is easy to see that each sweep can be executed in $O(L/\ell)$ time. Then the whole process finishes in $O((n/L)\cdot(L/\ell)) = O(n/\ell) = O(n/\log n)$ time, and it uses a total of $O(n \log \log n/\log n)$ bits. This concludes the proof of Theorem 1.

References

1. Angluin, D., Valiant, L.G.: Fast probabilistic algorithms for Hamiltonian circuits and matchings. J. Comput. Syst. Sci. **18**(2), 155–193 (1979). https://doi.org/10.1016/0022-0000(79)90045-X
2. Clark, D.: Compact pat trees. Ph.D. thesis, University of Waterloo (1996)
3. Elias, P.: Efficient storage and retrieval by content and address of static files. J. ACM **21**(2), 246–260 (1974). https://doi.org/10.1145/321812.321820
4. Fano, R.M.: On the number of bits required to implement an associative memory. Computation Structures Group Memo 61, MIT Project MAC, August 1971
5. Golynski, A.: Optimal lower bounds for rank and select indexes. Theor. Comput. Sci. **387**(3), 348–359 (2007). https://doi.org/10.1016/j.tcs.2007.07.041
6. Grossi, R., Gupta, A., Vitter, J.S.: High-order entropy-compressed text indexes. In: Proceedings of the 14th Annual ACM-SIAM Symposium on Discrete Algorithms, pp. 841–850. SIAM (2003)
7. Hagerup, T.: Sorting and searching on the word RAM. In: Proceedings of the 15th Annual Symposium on Theoretical Aspects of Computer Science (STACS 1998). LNCS, vol. 1373, pp. 366–398. Springer, Heidelberg (1998). https://doi.org/10.1007/BFb0028575
8. Hagerup, T., Kammer, F., Laudahn, M.: Space-efficient Euler partition and bipartite edge coloring. Theor. Comput. Sci. **754**, 16–34 (2019). https://doi.org/10.1016/j.tcs.2018.01.008
9. Jacobson, G.: Succinct static data structures. Ph.D. thesis, Carnegie Mellon University (1988)
10. Jacobson, G.: Space-efficient static trees and graphs. In: Proceedings of the 30th Annual IEEE Symposium on Foundations of Computer Science (FOCS 1989), pp. 549–554. IEEE Computer Society (1989). https://doi.org/10.1109/SFCS.1989.63533
11. Mäkinen, V., Navarro, G.: Rank and select revisited and extended. Theor. Comput. Sci. **387**(3), 332–347 (2007). https://doi.org/10.1016/j.tcs.2007.07.013
12. Pătraşcu, M.: Succincter. In: Proceedings of the 49th Annual IEEE Symposium on Foundations of Computer Science (FOCS 2008), pp. 305–313. IEEE Computer Society (2008). https://doi.org/10.1109/FOCS.2008.83
13. Raman, R., Raman, V., Satti, S.R.: Succinct indexable dictionaries with applications to encoding k-ary trees, prefix sums and multisets. ACM Trans. Algorithms **3**(4), 43 (2007). https://doi.org/10.1145/1290672.1290680

A PTAS for Bounded-Capacity Vehicle Routing in Planar Graphs

Amariah Becker[1]([✉]), Philip N. Klein[1], and Aaron Schild[2]

[1] Brown University, Providence, USA
amariah_becker@brown.edu
[2] University of California, Berkeley, USA

Abstract. The CAPACITATED VEHICLE ROUTING problem is to find a minimum-cost set of tours that collectively cover clients in a graph, such that each tour starts and ends at a specified depot and is subject to a capacity bound on the number of clients it can serve. In this paper, we present a polynomial-time approximation scheme (PTAS) for instances in which the input graph is planar and the capacity is bounded. Previously, only a quasipolynomial-time approximation scheme was known for these instances. To obtain this result, we show how to embed planar graphs into bounded-treewidth graphs while preserving, in expectation, the client-to-client distances up to a small additive error proportional to client distances to the depot.

Keywords: Capacitated Vehicle Routing ·
Approximation algorithms · Metric embeddings

1 Introduction

The CAPACITATED VEHICLE ROUTING problem with capacity $Q > 0$ for a graph G with client set S and depot r is to find a minimum-cost set of tours that collectively visit every client, such that each tour visits the depot and at most Q clients. This problem arises very naturally in both public and commercial settings including planning school bus routes and package delivery. In general metrics, CAPACITATED VEHICLE ROUTING is APX-hard, even when Q is a fixed capacity as small as three [1]. In this paper, we show that this hardness result does not extend to planar graphs. Specifically, we give the first polynomial-time approximation scheme (PTAS) for CAPACITATED VEHICLE ROUTING with fixed capacities in planar graphs.

An *embedding* of a guest graph G in a host graph H is a mapping $\phi : V(G) \longrightarrow V(H)$. One seeks embeddings in which, for each pair u, v of vertices of G, the u-to-v distance in G is in some sense approximated by the $\phi(u)$-to-$\phi(v)$ distance in H. One algorithmic strategy for addressing a metric problem is as

Research supported by NSF grant CCF-1409520.
Research supported by NSF Grant CCF-1816861.

© Springer Nature Switzerland AG 2019
Z. Friggstad et al. (Eds.): WADS 2019, LNCS 11646, pp. 99–111, 2019.
https://doi.org/10.1007/978-3-030-24766-9_8

follows: find an embedding ϕ from the input graph G to a graph H with simple structure; find a good solution in H; lift the solution to a solution in G. The success of this strategy depends on how easy it is to find a good solution in H and how well distances in H approximate corresponding distances in G.

In this paper, we give a randomized method for embedding a planar graph G into a bounded-treewidth host graph H so as to achieve a certain expected distance approximation guarantee. There is a polynomial-time algorithm to find an optimal solution to BOUNDED-CAPACITY VEHICLE ROUTING in bounded-treewidth graphs. This algorithm is used to find an optimal solution to the problem induced in H. This solution in the host graph is then lifted to obtain a near-optimal solution in G.

1.1 Related Work

Capacitated Vehicle Routing. There is a substantial body of work on approximation algorithms for CAPACITATED VEHICLE ROUTING. As the problem generalizes the TRAVELING SALESMAN PROBLEM (TSP), for general metrics and values of Q, CAPACITATED VEHICLE ROUTING is also APX-hard [16]. Haimovich and Rinnoy Kan [14] observe the following lower bound.

$$\frac{2}{Q} \sum_{v \in S} d(v, r) \leq cost(OPT) \tag{1}$$

which they use to give a $1 + (1 - \frac{1}{Q})\alpha$-approximation, where α denotes the approximation ratio of TSP. Using Christofides 1.5-approximation for TSP [9], this gives a $2.5 - \frac{1}{Q}$ approximation ratio. For general metrics and values of Q this result has not been substantially improved upon. Even for tree metrics, the best known approximation ratio for arbitrary values of Q is 4/3, due to Becker [3]. While no polynomial-time approximation schemes are known for arbitrary Q for *any* nontrivial metric, recently Becker and Paul [7] gave a bicriteria $(1, 1 + \epsilon)$ approximation scheme for tree metrics. It returns a solution of at most the optimal cost, but in which each tour is responsible for at most $(1 + \epsilon)Q$ clients.

One reasonable relaxation is to consider restricted values of Q. Even for Q as small as 3, CAPACITATED VEHICLE ROUTING is APX-hard in general metrics [1]. On the other hand, for fixed values of Q, the problem can be solved in polynomial time on trees and bounded-treewidth graphs.

Much attention has been given to approximation schemes for Euclidean metrics. In the Euclidean plane $\mathbb{R}^2$, PTASs are known for instances in which the value of Q is constant [14], $O(\log n / \log \log n)$ [1], and $\Omega(n)$ [1]. For $\mathbb{R}^3$, a PTAS is known for $Q = O(\log n)$ and for higher dimensions $\mathbb{R}^d$, a PTAS is known for $Q = O(\log^{1/d} n)$ [15]. For arbitrary values of Q, Mathieu and Das designed a quasi-polynomial time approximation scheme (QPTAS) for instances in $\mathbb{R}^2$ [10]. No PTAS is known for arbitrary values of Q.

Because algorithms for CAPACITATED VEHICLE ROUTING could be applied to logistics problems in road maps, it is particularly interesting to consider the complexity of approximating the problem in metrics that model road networks.

Becker, Klein, and Saulpic [5] gave a QPTAS for bounded-capacity instances in planar and bounded-genus graphs. The same authors gave a PTAS for graphs of bounded highway dimension [6].

Metric Embeddings. There has been much work on metric embeddings. In particular, Bartal [2] gave a randomized algorithm for selecting an embedding ϕ of the input graph into a tree so that, for any vertices u and v of G, the expected $\phi(u)$-to-$\phi(v)$ distance in the tree approximates the u-to-v distance in G to within a polylogarithmic factor. Fakcharoenphol, Rao, and Talwar [11] improved the factor to $O(\log n)$.

Talwar [17] gave a randomized algorithm for selecting an embedding of a metric space of bounded doubling dimension and aspect ratio Δ into a graph whose treewidth is bounded by a function that is polylogarithmic in Δ; the distances are approximated to within a factor of $1+\epsilon$. Feldman, Fung, Könemann, and Post [12] built on this result to obtain a similar embedding theorem for graphs of bounded highway dimension.

What about planar graphs? Chakrabarti et al. [8] showed a result that implies that unit-weight planar graphs cannot be embedded into distributions over $o(\sqrt{n})$-treewidth graphs so as to achieve approximation to within an $o(\log n)$ factor.

Let us consider distance approximation guarantees with absolute (rather than relative) error. Becker, Klein, and Saulpic [6] gave a deterministic algorithm that, given a constant $\epsilon > 0$, finds an embedding from a graph G of bounded highway dimension to a bounded-treewith graph H such that, for each pair u, v of vertices of G, the $\phi(u)$-to-$\phi(v)$ distance in H is at least the u-to-v distance in G and exceeds that distance by at most ϵ times the u-to-r distance plus the v-to-r distance, where r is a given vertex of G. This embedding was used to obtain the previously mentioned PTAS for CAPACITATED VEHICLE ROUTING with bounded capacity on graphs of bounded highway dimension.

Recently, Fox-Epstein, Klein, and Schild [13] showed how to embed planar graphs into graphs of bounded treewidth, such that distances are preserved up to a small additive error of ϵD, where D is the diameter of the graph. They show how such an embedding can be used to achieve efficient bicriteria approximation schemes for k-CENTER and d-INDEPENDENT SET.

1.2 Main Contributions

In this paper we present the first known PTAS for CAPACITATED VEHICLE ROUTING on planar graphs. We formally state the result as follows.

Theorem 1. *For any $\epsilon > 0$ and capacity Q, there is a polynomial-time algorithm for* CAPACITATED VEHICLE ROUTING *on planar graphs that returns a solution whose cost is at most $1 + \epsilon$ times optimal.*

Prior to this work, only a QPTAS was known [5] for planar graphs. As described in Sect. 1.1, PTASs for CAPACITATED VEHICLE ROUTING are known only for very few metrics. Our result expands this small list to include planar graphs—a graph class that is quite relevant to vehicle-routing problems as road networks tend to be nearly planar.

The basis for our new PTAS is a new metric-embedding theorem. For a graph G and vertices u and v, let $d_G(u, v)$ denote the u-to-v distance in G.

Theorem 2. *There is a constant c and a randomized polynomial-time algorithm that, given a planar graph G with specified root vertex r and given $0 < \epsilon < 1$, computes a graph H with treewidth at most $(\frac{1}{\epsilon})^{c\epsilon^{-1}}$ and an embedding ϕ of G into H, such that, for every pair of vertices u, v of G, $d_G(u, v) \leq d_H(\phi(u), \phi(v))$ with probability 1, and*

$$E[d_H(\phi(u), \phi(v))] \leq d_G(u, v) + \epsilon[d_G(u, r) + d_G(v, r)] \tag{2}$$

The expectation $E[\cdot]$ is over the random choices of the algorithm.

Why does this metric-embedding result give rise to an approximation scheme for CAPACITATED VEHICLE ROUTING? We draw on the following observation, which was also used in previous approximation schemes [5,6]: tours with clients far from the depot can accommodate a larger error. In particular, each client can be charged error that is proportional to its distance to the depot. In designing an appropriate embedding, we can afford a larger *error allowance* for the clients farther from the depot.

Our new embedding result builds on that of Fox-Epstein et al. [13]. The challenge in directly applying their embedding result is that it gives an *additive* error bound, proportional to the diameter of the graph. This error is too large for those clients close to the depot. Instead, we divide the graph into annuli (*bands*) defined by distance ranges from the depot and apply the embedding result to each induced subgraph independently, with an increasingly large error tolerance for the annuli farthest from the depot. In this way, each client *can* afford an error proportional to the diameter of the *subgraph* it belongs to.

How can these subgraph embeddings be combined into a global embedding with the desired properties? In particular, clients that are close to each other in the input graph may be separated into different annuli. How can we ensure that the embedding approximately preserves these distances while still achieving bounded treewidth?

We draw on a technique that has often been used, e.g. in metric embeddings. We show that by randomizing the choice of where to define the annuli boundaries, and connecting all vertices of all subgraph embeddings to a new, global depot, client distances are approximately preserved (to within their error allowance) *in expectation* by the overall embedding, without substantially increasing the treewidth. To do so we must ensure that the annuli are *wide* enough that the probability of nearby clients being separated (and thus generating large error) is small. Simultaneously, the annuli must be *narrow* enough that, within a given annulus, the clients closest to the depot can afford an error proportional to the distance of the farthest clients from the depot.

A dynamic-programming algorithm can then be used to find an optimal solution to CAPACITATED VEHICLE ROUTING in the bounded-treewidth host graph, and the solution can be lifted to obtain a solution in the input graph that in expectation is near-optimal.

Finally we describe how this result can be derandomized by trying all possible (relevant) choices for defining annuli and noting that for *some* such choice, the resulting solution cost must be near-optimal.

1.3   Outline

In Sect. 2 we describe preliminary notation and definitions. Section 3 describes the details of the embedding and provides an analysis of the desired properties. In Sect. 4 we outline our algorithm and prove Theorem 1. We conclude with some remarks in Sect. 5.

2   Preliminaries

2.1   Basics

Let $G = (V, E)$ denote a graph with vertex set V and edge set E, and let $n = |V|$. As mentioned earlier, for any two vertices $u, v \in V$, we use $d_G(u, v)$ to denote the length of the shortest u-to-v path in G. We might omit the subscript when the choice of graph is unambiguous. The *diameter* of a graph G is the maximum distance $d_G(u, v)$ over all choices of u and v.

We say that a graph is *planar* if it can be drawn in the plane without any edge crossings.

We use OPT to denote an optimal solution. For a minimization problem, an α-*approximation algorithm* is one that returns a solution whose cost is at most α times the cost of OPT. An *approximation scheme* is a family of $(1 + \epsilon)$-approximation algorithms, indexed by $\epsilon > 0$. A *polynomial-time approximation scheme* (PTAS) is an approximation scheme such that, for each $\epsilon > 0$, the corresponding algorithm runs in $O(n^c)$ time, where c is a constant independent of n but may depend on ϵ. A *quasi-polynomial-time approximation scheme* (QPTAS) is an approximation scheme such that, for each $\epsilon > 0$, the corresponding algorithm runs in $O(n^{\log^c n})$ time, where c is a constant independent of n but may depend on ϵ.

An *embedding* of a guest graph G into a host graph H is a mapping $\phi : V_G \to V_H$ of the vertices of G to the vertices of H.

A *tree decomposition* of a graph G is a tree T whose nodes (called *bags*) correspond to subsets of V with the following properties:

1. For each $v \in V$, v appears in some bag in T
2. For each $(u, v) \in E$, u and v appear *together* in some bag in T
3. For each $v \in V$, the subtree induced by the bags of T containing v is connected

The *width* of a tree decomposition is the size of the largest bag minus one, and the *treewidth* of a graph G is the minimum width over all tree decompositions of G.

2.2 Problem Statement

A *tour* in a graph G is a closed path $v_0, v_1, v_2, ..., v_L$ such that $v_0 = v_L$ and for all $i \in \{1, 2, ..., L\}$, (v_{i-1}, v_i) is an edge in G.

Given a capacity $Q > 0$ and a graph $G = (V, E)$ with specified client set $S \subseteq V$ and depot vertex $r \in V$, the CAPACITATED VEHICLE ROUTING problem is to find a set of tours $\Pi = \{\pi_1, \pi_2, ...\pi_{|\Pi|}\}$ that collectively cover all clients and such that each tour includes r and covers at most Q clients. The cost of a solution is the sum of the tour lengths, and the objective is to minimize this sum.

If a client s is covered by a tour π, we say that π *visits* s. Note that π may *pass* many other vertices (including other clients) that it does not cover.

As stated, the problem assumes that each client has unit demand. In fact, the more general case, where clients have integral demand (assumed to be polynomially bounded) that is allowed to be covered across multiple tours (demand is *divisible*) reduces to the unit-demand case as follows: For each client $s \in S$ with demand $dem(s) = k$, add k new vertices $\{v_1, v_2, ..., v_k\}$ each with unit demand and edges (s, v_i) of length zero, and set $dem(s)$ to zero. Note that this modification does not affect planarity. Additionally, since demand is assumed to be polynomially-bounded, the increase in graph size is negligible for the purpose of a PTAS.

For CAPACITATED VEHICLE ROUTING with *indivisible* demands, each client's demand must be covered by a single tour, and a tour can cover at most Q units of client demand.

We assume all non-zero distances in G are at least one. If not, the graph can be rescaled. We also assume values of ϵ are less than one. If not, any $\epsilon \geq 1$ can be replaced with a number ϵ' slightly less than one. This only helps the approximation guarantee and does not significantly increase runtime. Of course for very large values of ϵ, an efficient constant-factor approximation can be used instead (see Sect. 1.1).

3 Embedding

In this section, we prove Theorem 2, which we restate for convenience:

Theorem 2. *There is a constant c and a randomized polynomial-time algorithm that, given a planar graph G with specified root vertex r and given $0 < \epsilon < 1$, computes a graph H with treewidth at most $(\frac{1}{\epsilon})^{c\epsilon^{-1}}$ and an embedding ϕ of G into H, such that, for every pair of vertices u, v of G, $d_G(u, v) \leq d_H(\phi(u), \phi(v))$ with probability 1, and*

$$E[d_H(\phi(u), \phi(v))] \leq d_G(u, v) + \epsilon[d_G(u, r) + d_G(v, r)] \tag{3}$$

The proof uses as a black box the following result from [13]:

Lemma 1 ([13]). *There is a number c and a polynomial-time algorithm that, given a planar graph G with specified root vertex r and diameter D, computes a graph H of treewidth at most $(\frac{1}{\epsilon})^c$ and an embedding ϕ of G into H such that, for all vertices u and v,*

$$d_G(u,v) \le d_H(\phi(u),\phi(v)) \le d_G(u,v) + \epsilon D$$

For notational convenience, instead of Inequality 3 of Theorem 2, we prove

$$E[d_H(\phi(u),\phi(v))] \le d_G(u,v) + 3\epsilon[d_G(u,r) + d_G(v,r)] \qquad (4)$$

from which Theorem 2 can be proved by taking $\epsilon' = \epsilon/3$.

Our embedding partitions vertices of G into *bands* of vertices defined by distances from r. Choose $x \in [0,1]$ uniformly at random. Let B_0 be the set of vertices v such that $d_G(r,v) < \frac{1}{\epsilon}^{(x)\frac{1}{\epsilon}}$, and for $i \in \{1,2,3,...\}$ let B_i be the set of vertices v such that $\frac{1}{\epsilon}^{(i+x-1)\frac{1}{\epsilon}} \le d_G(r,v) < \frac{1}{\epsilon}^{(i+x)\frac{1}{\epsilon}}$ (see Fig. 1). Let G_i be the subgraph induced by B_i, together with all u-to-v and v-to-r shortest paths for all $u,v \in B_i$. Note that although the B_i partition V, the G_i do not partition G. Note also that the diameter of G_i is at most $4\frac{1}{\epsilon}^{(i+x)\frac{1}{\epsilon}}$. The factor of 4 addresses the fact that for $u,v \in B_i$, the u-to-v shortest path is included in G_i and may contain a vertex $w \notin B_i$. But for any such w, it must be that $d_G(r,w) \le 2\frac{1}{\epsilon}^{(i+x)\frac{1}{\epsilon}}$.

For each G_i, let H_i be the host graph resulting from applying Lemma 1 using $\epsilon' = \epsilon^{\frac{1}{\epsilon}+1}$ and let ϕ_i be the corresponding embedding. Let H be the graph resulting from adding a new vertex r' and for all i and all $v \in B_i$ adding an edge $(\phi_i(v), r')$ of length $d_G(v,r)$. That is, H is formed by connecting (all vertices of) all the H_i to r' (see Fig. 2). Finally, set $\phi(v) = \phi_i(v)$ for all $v \in B_i - \{r\}$ and set $\phi(r) = r'$.

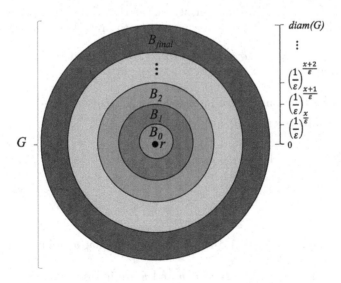

Fig. 1. G is divided into bands $B_0, B_1, ..., B_{final}$ based on distance from r.

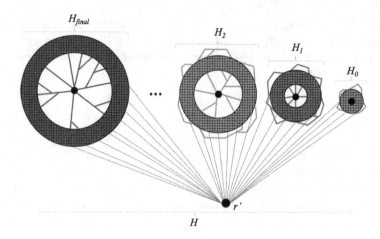

Fig. 2. Each subgraph G_i of G is embedded into a host graph H_i. These graphs are joined via edges to a new depot r' to form a host graph for G.

We can assume that there are at most n bands, since empty bands would not contribute to the embedding. The runtime for constructing H is dominated by the construction of the H_i, which by Lemma [13] is polynomial.

Let H^- be the graph obtained from H by deleting r'. The connected components of H^- are $\{H_i\}_i$. By Lemma 1, the treewidth of each host graph H_i is at most $(\frac{1}{\epsilon})^{c_0} = (\frac{1}{\epsilon})^{c_0(\epsilon^{-1}+1)}$ for some constant c_0. This also bounds the treewidth of H^-. Adding a single vertex to a graph increases the treewidth by at most one, so after adding r' back, the treewidth of H is $(\frac{1}{\epsilon})^{c_0(\epsilon^{-1}+1)} + 1 = (\frac{1}{\epsilon})^{c_1\epsilon^{-1}}$ for some constant c_1.

As for the metric approximation, it is clear that $d_G(u,v) \leq d_H(\phi(u),\phi(v))$ with probability 1. We use the following lemma to prove Inequality 4.

Lemma 2. *If $\epsilon d_G(v,r) \leq d_G(u,r) \leq d_G(v,r)$, then the probability that u and v are in different bands is at most ϵ.*

Proof. Let i be the nonnegative integer such that $d_G(u,r) = \frac{1}{\epsilon}^{(i+a)\frac{1}{\epsilon}}$ for some $a \in [0,1]$. Let b be the number such that $d_G(v,r) = \frac{1}{\epsilon}^{(i+b)\frac{1}{\epsilon}}$.

$$\frac{1}{\epsilon} \geq \frac{d_G(v,r)}{d_G(u,r)} = \frac{\frac{1}{\epsilon}^{(i+b)\frac{1}{\epsilon}}}{\frac{1}{\epsilon}^{(i+a)\frac{1}{\epsilon}}} = \frac{1}{\epsilon}^{(b-a)\frac{1}{\epsilon}}.$$

Therefore

$$b - a \leq \epsilon$$

Consider two cases. If $b \leq 1$, then the probability that u and v are in different bands is $Pr[a \leq x < b] \leq \epsilon$.

If $b > 1$ then the probability that u and v are in different bands is $Pr[x \geq a \text{ or } x \leq b-1] \leq 1 - a + b - 1 = b - a \leq \epsilon$.

We now prove Inequality 4. Let u and v be vertices in G. Without loss of generality, assume $d_G(u,r) \leq d_G(v,r)$. First we address the case where $d_G(u,r) \leq \epsilon d_G(v,r)$. Since $\phi(u)$ and $\phi(v)$ are both adjacent to r' in H, $d_H(\phi(u),\phi(v)) \leq d_H(\phi(u),r') + d_H(\phi(v),r') = d_G(u,r) + d_G(v,r) \leq 2d_G(u,r) + d_G(u,v) \leq d_G(u,v) + 2\epsilon d_G(v,r)$. Therefore $E[d_H(\phi(u),\phi(v))] \leq d_G(u,v) + 3\epsilon[d_G(u,r) + d_G(v,r)]$

Now, suppose $d_G(u,r) > \epsilon d_G(v,r)$. If u and v are in the same band B_i, then by Lemma 1,

$$d_H(\phi(u),\phi(v)) \leq d_{H_i}(\phi(u),\phi(v)) \leq d_G(u,v) + \epsilon' diam(G_i)$$
$$\leq d_G(u,v) + \epsilon'4\frac{1}{\epsilon}^{(i+x)\frac{1}{\epsilon}} = d_G(u,v) + \epsilon^{\frac{1}{\epsilon}+1}4\frac{1}{\epsilon}^{(i+x)\frac{1}{\epsilon}}$$
$$= d_G(u,v) + \epsilon4\frac{1}{\epsilon}^{(i+x-1)\frac{1}{\epsilon}} \leq d_G(u,v) + 2\epsilon(d_G(u,r) + d_G(v,r))$$

In the final inequality, when $i = 0$, we use the fact that all nonzero distances are at least one to give a lower bound on $d_G(u,r)$ and $d_G(v,r)$.

If u and v are in different bands, then since $\phi(u)$ and $\phi(v)$ are both adjacent to r' in H, $d_H(\phi(u),\phi(v)) \leq d_H(\phi(u),r') + d_H(\phi(v),r') = d_G(u,r) + d_G(v,r)$. By Lemma 2, this case occurs with probability at most ϵ.

Therefore $E[d_H(\phi(u),\phi(v))] \leq (d_G(u,v) + 2\epsilon(d_G(u,r) + d_G(v,r))) + \epsilon[d_G(u,r) + d_G(v,r)] \leq d_G(u,v) + 3\epsilon[d_G(u,r) + d_G(v,r)]$, which proves Inequality 4 and completes the proof of Theorem 2.

The construction depends on planarity only via Lemma 1. For the sake of future uses of the construction with other graph classes, we state a lemma.

Lemma 3. *Let $\mathcal{F}$ be a family of graphs closed under vertex-induced subgraphs. Suppose that there is a function f and a polynomial-time algorithm that, for any graph G in $\mathcal{F}$, computes a graph H of treewidth at most $f(\epsilon)$ and an embedding ϕ of G into H such that, for all vertices u and v,*

$$d_G(u,v) \leq d_H(\phi(u),\phi(v)) \leq d_G(u,v) + \epsilon D$$

Then there is a function g and a randomized polynomial-time algorithm that, for any graph G in $\mathcal{F}$, computes a graph H with treewidth at most $g(\epsilon)$ and an embedding ϕ of G into H, such that, for every pair of vertices u,v of G, with probability 1 $d_G(u,v) \leq d_H(\phi(u),\phi(v))$, and

$$E[d_H(\phi(u),\phi(v))] \leq d_G(u,v) + \epsilon[(d_G(u,r) + d_G(v,r)]$$

4   PTAS for Capacitated Vehicle Routing

In this section, we show how to use the embedding of Sect. 3 to give a PTAS for Capacitated Vehicle Routing, proving Theorem 1.

4.1   Randomized Algorithm

We first prove a slight relaxation of Theorem 1 in which the algorithm is randomized, and the solution value is near-optimal *in expectation*. We then show in Sect. 4.2 how to derandomize the result.

Theorem 3. *For any $\epsilon > 0$ and capacity Q, there is a randomized algorithm for* Capacitated Vehicle Routing *on planar graphs that in polynomial time returns a solution whose expected value is at most $1 + \epsilon$ times optimal.*

Our result depends on the following lemma, which is proved in the full version [4] of [6].

Lemma 4 (Lemma 20 in [6], Lemma 15 in [4]). *Given an instance of* Capacitated Vehicle Routing *with capacity Q on a graph G with treewidth w, there is a dynamic-programming algorithm that finds an optimal solution in $n^{O(wQ)}$ time.*

Given the dynamic program of Lemma 4 and the embedding of Theorem 2 as black boxes, the algorithm is as follows. First, the graph G is embedded as in Theorem 2 using $\hat{\epsilon} = \epsilon/Q$ into a host graph H with treewidth $(\frac{1}{\hat{\epsilon}})^{c\hat{\epsilon}^{-1}}$ for some constant c, and $d_G(u, v) \le E[d_H(\phi(u), \phi(v))] \le d_G(u, v) + \hat{\epsilon}(d_G(u, r) + d_G(v, r))$ for all vertices u and v. The dynamic program of Lemma 4 is then applied to H. The resulting solution SOL_H in H is then mapped back to a solution SOL_G in G which is returned by the algorithm.

Note that the tours in any vehicle-routing solution can be defined by specifying the order in which clients are visited. In particular, we use $(u, v) \in SOL$ to denote that u and v are consecutive elements of $\{clients\} \cup \{depot\}$ visited by the solution. In this way, a solution in H is easily mapped back to a corresponding solution in G, as $(u, v) \in SOL_G$ if and only if $(\phi(u), \phi(v)) \in SOL_H$. We use $cost_G(SOL)$ (resp. $cost_H(SOL)$) to denote the cost of a solution SOL in G (resp. H).

We now prove Theorem 3 by analyzing this algorithm.

Lemma 5. *For any $\epsilon > 0$ the algorithm described above finds a solution whose expected value is at most $1 + \epsilon$ times optimal.*

Proof. Let OPT be the optimal solution in G and let OPT_H be the corresponding induced solution in H. Since the dynamic program finds an optimal solution in H, we have $cost_H(SOL_H) \le cost_H(OPT_H)$. Additionally, since distances in H are no shorter than distances in G, $cost_G(SOL_G) \le cost_H(SOL_H)$. Putting these pieces together, we have,

$$E[cost_G(SOL_G)] \le E[cost_H(SOL_H)] \le E[cost_H(OPT_H)]$$

$$= E\Big[\sum_{(u,v)\in OPT} d_H(\phi(u), \phi(v)) \Big] = \sum_{(u,v)\in OPT} E[d_H(\phi(u), \phi(v))]$$

$$\le \sum_{(u,v)\in OPT} d_G(u, v) + \hat{\epsilon}(d_G(u, r) + d_G(v, r)) = \sum_{(u,v)\in OPT} d_G(u, v) + 2\hat{\epsilon}\sum_{v\in S} d_G(v, r)$$

$$\le cost_G(OPT) + 2\hat{\epsilon}\frac{Q}{2}cost_G(OPT) = (1 + \epsilon)cost_G(OPT)$$

where the final inequality comes from Lower Bound 1 (see Sect. 1.1).

The following lemma completes the proof of Theorem 3.

Lemma 6. *For any* $Q, \epsilon > 0$*, the algorithm described above runs in polynomial time.*

Proof. By Lemma 1, computing H and the embedding of G into H takes polynomial time. By Lemma 4, the dynamic program runs in $|V_H|^{O(wQ)}$ time, where w is the treewidth of H. By Theorem 2, $w = (\frac{1}{\epsilon})^{c\hat{\epsilon}^{-1}} = (\frac{Q}{\epsilon})^{c'Q\epsilon^{-1}}$, where c and c' are constants independent of $|V_H|$.

The algorithm therefore runs in $|V_H|^{(Q\epsilon^{-1})^{O(Q\epsilon^{-1})}}$ time. Finally, since $|V_H|$ is polynomial in the size of G, for fixed Q and ϵ, the running time is polynomial.

4.2 Derandomization

The algorithm can be derandomized using a standard technique. The embedding of Theorem 2 partitions the vertices of the input graph into rings depending on a value x chosen uniformly at random from $[0, 1]$. However, the partition depends on the distances of vertices from the root r. It follows that the number of partitions that can arise from different choices of x is at most the number of vertices. The deterministic algorithm tries each of these partitions, finding the corresponding solution, and returns the least costly of these solutions.

In particular, consider the optimum solution OPT. As shown in Sect. 4.1,

$$E[\sum_{(u,v)\in OPT} d_H(\phi(u), \phi(v))]$$

$$= \sum_{(u,v)\in OPT} E[d_H(\phi(u), \phi(v))]$$

$$\leq (1 + \epsilon)cost_G(OPT).$$

So for some choice of x, the induced cost of OPT in H is nearly optimal, and the dynamic program will find a solution that costs at most as much. This completes the proof of Theorem 1.

5 Conclusion

In this paper, we present the first PTAS for CAPACITATED VEHICLE ROUTING in planar graphs. Although the approximation scheme takes polynomial time, it is not an *efficient* PTAS (one whose running time is bounded by a polynomial whose degree is independent of the value of ϵ). It is an open question as to whether an efficient PTAS exists. It is also open whether a PTAS exists when the capacity Q is unbounded.

References

1. Asano, T., Katoh, N., Tamaki, H., Tokuyama, T.: Covering points in the plane by k-tours: towards a polynomial time approximation scheme for general k. In: Proceedings of the Twenty-Ninth Annual ACM Symposium on Theory of Computing, pp. 275–283. ACM (1997)
2. Bartal, Y.: Probabilistic approximations of metric spaces and its algorithmic applications. In: 37th Annual Symposium on Foundations of Computer Science, FOCS 1996, Burlington, Vermont, USA, 14–16 October 1996, pp. 184–193 (1996)
3. Becker, A.: A tight 4/3 approximation for capacitated vehicle routing in trees. In: Approximation, Randomization, and Combinatorial Optimization. Algorithms and Techniques (APPROX/RANDOM 2018). Schloss Dagstuhl-Leibniz-Zentrum fuer Informatik (2018)
4. Becker, A., Klein, P.N., Saulpic, D.: Polynomial-time approximation schemes for k-center and bounded-capacity vehicle routing in metrics with bounded highway dimension. CoRR abs/1707.08270 (2017). http://arxiv.org/abs/1707.08270
5. Becker, A., Klein, P.N., Saulpic, D.: A quasi-polynomial-time approximation scheme for vehicle routing on planar and bounded-genus graphs. In: LIPIcs-Leibniz International Proceedings in Informatics, vol. 87. Schloss Dagstuhl-Leibniz-Zentrum fuer Informatik (2017)
6. Becker, A., Klein, P.N., Saulpic, D.: Polynomial-time approximation schemes for k-center, k-median, and capacitated vehicle routing in bounded highway dimension. In: 26th Annual European Symposium on Algorithms (ESA 2018). Schloss Dagstuhl-Leibniz-Zentrum fuer Informatik (2018)
7. Becker, A., Paul, A.: A PTAS for minimum makespan vehicle routing in trees. arXiv preprint arXiv:1807.04308 (2018)
8. Chakrabarti, A., Jaffe, A., Lee, J.R., Vincent, J.: Embeddings of topological graphs: lossy invariants, linearization, and 2-sums. In: 49th Annual IEEE Symposium on Foundations of Computer Science, pp. 761–770 (2008)
9. Christofides, N.: Worst-case analysis of a new heuristic for the travelling salesman problem. Technical report , Carnegie-Mellon Univ Pittsburgh Pa Management Sciences Research Group (1976)
10. Das, A., Mathieu, C.: A quasi-polynomial time approximation scheme for Euclidean capacitated vehicle routing. In: Proceedings of the Twenty-first Annual ACM-SIAM Symposium on Discrete Algorithms, pp. 390–403. SIAM (2010)
11. Fakcharoenphol, J., Rao, S., Talwar, K.: A tight bound on approximating arbitrary metrics by tree metrics. J. Comput. Syst. Sci. **69**(3), 485–497 (2004)
12. Feldmann, A.E., Fung, W.S., Könemann, J., Post, I.: A $(1+\varepsilon)$-embedding of low highway dimension graphs into bounded treewidth graphs. In: 42nd International Colloquium on Automata, Languages, and Programming, pp. 469–480 (2015)
13. Fox-Epstein, E., Klein, P.N., Schild, A.: Embedding planar graphs into low-treewidth graphs with applications to efficient approximation schemes for metric problems. In: Proceedings of the Thirteenth Annual ACM-SIAM Symposium on Discrete Algorithms, pp. 1069–1088. SIAM (2019)
14. Haimovich, M., Rinnooy Kan, A.: Bounds and heuristics for capacitated routing problems. Math. Oper. Res. **10**(4), 527–542 (1985)

15. Khachay, M., Dubinin, R.: PTAS for the Euclidean capacitated vehicle routing problem in R^d. In: Kochetov, Y., Khachay, M., Beresnev, V., Nurminski, E., Pardalos, P. (eds.) DOOR 2016. LNCS, vol. 9869, pp. 193–205. Springer, Cham (2016). https://doi.org/10.1007/978-3-319-44914-2_16
16. Papadimitriou, C.H., Yannakakis, M.: The traveling salesman problem with distances one and two. Math. Oper. Res. **18**(1), 1–11 (1993)
17. Talwar, K.: Bypassing the embedding: algorithms for low dimensional metrics. In: 36th Annual ACM Symposium on Theory of Computing, pp. 281–290 (2004)

A Framework for Vehicle Routing Approximation Schemes in Trees

Amariah Becker[1](✉) and Alice Paul[2]

[1] Computer Science Department, Brown University, Providence, USA
amariah_becker@brown.edu
[2] Data Science Initiative, Brown University, Providence, USA
alice_paul@brown.edu

Abstract. We develop a general framework for designing polynomial-time approximation schemes (PTASs) for various vehicle routing problems in trees. In these problems, the goal is to optimally route a fleet of vehicles, originating at a depot, to serve a set of clients, subject to various constraints. For example, in MINIMUM MAKESPAN VEHICLE ROUTING, the number of vehicles is fixed, and the objective is to minimize the longest distance traveled by a single vehicle. Our main insight is that we can often greatly restrict the set of potential solutions without adding too much to the optimal solution cost. This simplification relies on partitioning the tree into clusters such that there exists a near-optimal solution in which every vehicle that visits a given cluster takes on one of a few forms. In particular, only a small number of vehicles serve clients in any given cluster. By using these coarser building blocks, a dynamic programming algorithm can find a near-optimal solution in polynomial time. We show that the framework is flexible enough to give PTASs for many problems, including MINIMUM MAKESPAN VEHICLE ROUTING, DISTANCE-CONSTRAINED VEHICLE ROUTING, CAPACITATED VEHICLE ROUTING, and SCHOOL BUS ROUTING, and can be extended to the multiple depot setting.

Keywords: Approximation algorithms · Vehicle routing ·
Rooted tree cover

1 Introduction

Vehicle routing problems address the fundamental problem of routing a fleet of vehicles from a common depot to visit a set of clients. These problems arise naturally in many real world settings, and are well-studied across computer science and operations research. We generalize a class of vehicle routing problems by introducing the notions of *vehicle load*, the problem-specific vehicle constraint (e.g. vehicle capacity, distance traveled by the vehicle, client regret, etc.), and

Research funded by NSF grant CCF-14-09520.

© Springer Nature Switzerland AG 2019
Z. Friggstad et al. (Eds.): WADS 2019, LNCS 11646, pp. 112–125, 2019.
https://doi.org/10.1007/978-3-030-24766-9_9

fleet budget, the problem-specific fleet constraint (e.g. number of vehicles, sum of distances traveled, etc.).

Most vehicle routing problems can then be framed as either MIN-MAX VEHICLE LOAD: minimize the maximum vehicle load, given a bound k on fleet budget (e.g. MINIMUM MAKESPAN VEHICLE ROUTING) or MINIMUM FLEET BUDGET: minimize the required fleet budget, given a bound D on vehicle load (e.g. DISTANCE-CONSTRAINED VEHICLE ROUTING). In fact, these are two optimization perspectives of the same decision problem: does there exist a solution with maximum vehicle load D and fleet budget k?

1.1 Main Contributions

We present a framework for designing polynomial time approximation schemes (PTASs) for MIN-MAX VEHICLE LOAD and MINIMUM FLEET BUDGET in trees. Tree (and treelike) transportation networks occur in building and warehouse layouts, mining and logging industries, and along rivers and coastlines [11,12]. Our framework applies directly to MIN-MAX VEHICLE LOAD problems and generates results of the following form.

Theorem 1. *For every $\epsilon > 0$, there is a polynomial-time algorithm that, given an instance of* MIN-MAX VEHICLE LOAD *on a tree, finds a feasible solution whose maximum vehicle load is at most $1 + \epsilon$ times optimum.*

An immediate corollary of Theorem 1 is the following result for the associated MINIMUM FLEET BUDGET problem.

Theorem 2. *Given an instance of* MINIMUM FLEET BUDGET *on a tree, if there exists a solution with fleet budget k and vehicle load D, then for any $\epsilon > 0$, there is a polynomial-time algorithm that finds a solution with fleet budget k and vehicle load at most $(1 + \epsilon)D$.*

The input to the framework is a rooted tree $T = (V, E)$ with root $r \in V$ and edge lengths $\ell(u, v) \geq 0$ for all $(u, v) \in E$. Without loss of generality, the root r represents the *depot* at which all vehicles start and the set of clients corresponds to the set of leaves in the tree (we can add zero cost edges to ensure that every client is a leaf and any subtree without a client can be safely removed from the instance). Since every edge must then be traversed by at least one vehicle, the problems are equivalent to corresponding tree-cover problems.

As stated, the framework can be customized to a wide range of problems. In Sect. 4, we illustrate in detail how to customize the framework to give a PTAS for the MINIMUM MAKESPAN VEHICLE ROUTING problem of finding k *tours* each starting and ending at a depot r that serve all clients in T such that the *makespan*, the maximum length of any tour, is minimized. Here, vehicle load is the tour length, and fleet budget is the number of vehicles. A bicriteria PTAS for the associated MINIMUM FLEET BUDGET problem, DISTANCE-CONSTRAINED VEHICLE ROUTING, follows as a corollary.

Our framework can be applied to give similar results for other vehicle-routing variants, including CAPACITATED VEHICLE ROUTING and SCHOOL BUS ROUTING, and can also be generalized to the multiple-depot setting. We state these results in Sect. 5 and refer the reader to the full version of our paper for details. The breadth of the problems listed highlights the real flexibility and convenience of the presented framework.

At a high level, the framework partitions the tree into *clusters* such that there exists a near-optimal solution that within each cluster has a very simple form, effectively coarsening the solution space. Then, given this simplified structure, a dynamic program can be designed to find such a near-optimal solution.

The clusters are designed to be *small* enough so that simplifying vehicle routes at the cluster level does not increase the optimal load by too much, but also *large* enough that the (coarsened) solutions can be enumerated efficiently. To bound the error introduced by this simplification we design a load-reassignment tool that makes cluster coverage adjustments *globally* in the tree.

Finally, standard dynamic programming techniques can result in a large accumulation of rounding error. To limit the number of times that the load of any single route is rounded, we introduce a *route projection* technique that essentially pays in advance for load that the vehicle *anticipates* accumulating, allowing the dynamic program to round only once instead of many times for this projected load.

1.2   Related Work

For trees, MINIMUM MAKESPAN VEHICLE ROUTING is equivalent to MINIMUM MAKESPAN ROOTED TREE COVER: the minimum makespan for rooted tree cover is exactly half the minimum makespan for vehicle routing, since tours traverse edges twice. MINIMUM MAKESPAN ROOTED TREE COVER is NP-hard even on star instances but admits an FPTAS if the number, k, of subtrees is constant [15] and a PTAS for general k [10]. For covering a *general graph* with rooted subtrees, [6] provides a 4-approximation; this bound was later improved to a 3-approximation by [13]. For tree metrics, an FPTAS is known for constant k [16], and a $(2 + \epsilon)$-approximation is known for general k [13]. In this paper, we improve this to a PTAS. Although a recent paper [4] also claimed to present a PTAS, in the full version of our paper we show that their result is incorrect and cannot be salvaged using the authors' proposed techniques. Additionally, we compare their approach to our own and describe how we successfully overcome the challenges where their approach fell short.

The associated DISTANCE-CONSTRAINED VEHICLE ROUTING problem is to minimize the number of tours of length at most D required to cover all client demand. Even restricted to star instances, this problem is NP-hard, and for tree instances it is hard to approximate to better than a factor of $3/2$ [14]. A 2-approximation is known for tree instances, and $O(\log D)$ and $O(\log |S|)$-approximations are known for general metrics, where S is the set of clients [14]. Allowing a multiplicative stretch in the distance constraint, a $(O(\log 1/\epsilon), 1 + \epsilon)$ bicriteria approximation is also known, which finds a solution of at most

$O(\log 1/\epsilon)OPT_D$ tours each of length at most $(1 + \epsilon)D$ [14], where OPT_D is the minimum number of tours of length at most D required to cover all clients. We give a $(1, 1 + \epsilon)$ bicriteria PTAS for trees, and note that the hardness results for trees described above [14] imply that without allowing this $(1 + \epsilon)$ stretch in D, a PTAS is unlikely to exist.

In the classic CAPACITATED VEHICLE ROUTING each vehicle can cover at most Q clients, and the objective is to minimize the *sum* of tour lengths. This problem is also NP-hard, even in star instances [12]. For tree metrics, a $4/3$-approximation is known [2], which improves upon the previous best-known approximation ratio of $(\sqrt{41} - 1)/4$ by [1] and is tight with respect to the best known lower bound. In this paper, we give a $(1, 1 + \epsilon)$ bicriteria PTAS for trees. For general metrics, a $(2.5 - \frac{1.5}{Q})$-approximation is known [9](using [5]).

The *regret* of a path is the difference between the path length and the distance between the path endpoints. The MIN-MAX REGRET ROUTING problem is to cover all clients with k *paths* starting from the depot, such that the maximum regret is minimized. For trees, there is a known 13.5-approximation algorithm [3], which we improve to a PTAS in this paper. For general graphs there is a $O(k^2)$-approximation algorithm [7].

In the related SCHOOL BUS ROUTING problem, there is a bound R on the regret of each path and the goal is to find the minimum number of paths required to cover all client demand. For general graphs, [8] provides an LP-based 15-approximation algorithm, improving upon the authors' previous 28.86-approximation algorithm [7]. In trees, there exists a 3-approximation algorithm for the uncapacitated version of this problem and a 4-approximation algorithm for the capacitated version [3]. Additionally, there is a $(3/2)$ inapproximability bound [3]. A PTAS is therefore unlikely to exist for trees. Instead, we give a $(1, 1 + \epsilon)$ bicriteria PTAS that allows a $(1 + \epsilon)$ stretch in the regret constraint.

2   Preliminaries

Let OPT denote the value of an optimum solution. For a minimization problem, a polynomial-time α-approximation algorithm is an algorithm that finds a solution of value at most $\alpha \cdot OPT$ and runs in time that is polynomial in the size of the input. A polynomial-time approximation scheme (PTAS) is a family of $(1 + \epsilon)$-approximation algorithms indexed by $\epsilon > 0$ such that for each ϵ, the algorithm runs in time polynomial in the input size, but may depend arbitrarily on ϵ.

In a rooted tree, the *parent* of a vertex v, denoted $p(v)$, is the vertex adjacent to v in the shortest path from v to r (the parent of r is undefined). If $u = p(v)$ then v is a *child* of u. The parent edge of a vertex v is the edge $(p(v), v)$ (undefined for $v = r$). The *ancestors* of vertex v are all vertices (including v and r) in the shortest v-to-r path and the *descendants* of v are all vertices u such that v is an ancestor of u. We assume every vertex has at most two children. If vertex v has $l > 2$ children $v_1, ..., v_l$, add vertex v' and edge (v, v') of length zero and replace edges $(v, v_1), (v, v_2)$ with edges $(v', v_1), (v', v_2)$ of the same lengths.

Further, the *subtree rooted at* v is the subgraph induced by the descendants of v and is denoted T_v. If $u = p(v)$, the v-*branch* at u consists of the subtree rooted at v together with the edge (u, v). We define the *length* of a subgraph $A \subseteq E$ to be $\ell(A) = \sum_{(u,v) \in A} \ell(u, v)$. For vertices u, v, we use $d_T(u, v)$ to denote the shortest-path distance in T between u and v.

Our framework applies to vehicle routing problems that can be framed as MIN-MAX VEHICLE LOAD problems, in which the objective is to minimize the maximum vehicle load, subject to a fleet budget. Given a MIN-MAX VEHICLE LOAD problem, a trivial n-approximation can be used to obtain an upper bound D_{high} for OPT. An overarching algorithm takes as input a load value $D \geq 0$ and provides the following guarantee: if there exists a solution with max load D, the algorithm will find a solution with max load at most $(1 + \epsilon)D$. A PTAS follows from using binary search between $\frac{D_{high}}{n}$ and D_{high} for the smallest value D_{low} such that the algorithm returns a solution of max load at most $(1+\epsilon)D_{low}$. This implies $D_{low} \leq OPT$. For the rest of the paper, we assume D is fixed.

3   Framework Overview

Optimization problems on trees are often well suited for dynamic programming algorithms. In fact, the following dynamic programming strategy can solve MIN-MAX VEHICLE LOAD problems on trees exactly: at each vertex v, for each value $0 \leq i \leq D$, *guess* the number of solution route segments of load exactly i in the subtree rooted at v. Such an algorithm would be exponential in D. Instead of considering every possible load value, route segment loads can be *rounded* up to the nearest θD, for some value $\theta \in (0, 1]$ that depends only on ϵ, so that only $O(\theta^{-1})$ segment load values need to be considered. In order to achieve a PTAS, we must show that this rounding does not incur too much error. Rounding the load of a route at *every* vertex accumulates too much error, but if the number of times that any given route is rounded is at most ϵ/θ, then at most ϵD error accumulates, as desired.

One main insight underlying our algorithm is that a route only needs to incur rounding error when it branches. The challenge in bounding the rounding error then becomes bounding the number of times a route branches. While a route in the optimal solution may have an arbitrary amount of branching, we show that we can greatly limit the scope of candidate solutions to those with a specific structure while only incurring an ϵD error in the maximum load. Rather than having to make decisions for covering every leaf in the tree (of which there may be arbitrarily many—each with arbitrarily small load), we partition the tree into *clusters* and then address covering the clusters.

By *reassigning* small portions of routes within a cluster, we show that there exists a near-optimal solution in which all clients (leaves) within a given cluster are covered by only one or two vehicles. These clusters are chosen to be small enough that the error incurred by the reassignment is small, but large enough that any given route covers clients in a bounded number of clusters. This *coarsens* the solutions considered by the algorithm, as vehicles must commit to covering larger fractions of load at a time. A dynamic program then finds the optimal such coarse solution using these simple building blocks within each cluster.

3.1 Simplifying the Solution Structure

Let $\hat{\epsilon}$ and δ be problem-specific values that depend only on ϵ. Let $\mathcal{H}_T$ denote the set of all subgraphs of T, and let $g : \mathcal{H} \to \mathbb{Z}^{\geq 0}$ be a problem-specific *load function*. We require g to be monotonic and subadditive. Intuitively, for all $H \in \mathcal{H}_T$, $g(H)$ is the load accumulated by a vehicle for covering H.

Condensing the Input Tree. The first step in the framework is to CONDENSE all small branches into leaf edges. Specifically, let $\mathcal{B}$ be the set of all maximal branches of load at most δD. That is, for every v-branch $b \in \mathcal{B}$, $g(b) \leq \delta D$ and for b's parent $p(v)$-branch, b_p, $g(b_p) > \delta D$. For convenience, if $b_1 \in \mathcal{B}$ is a v_1 branch at u and $b_2 \in \mathcal{B}$ is a *sibling* v_2 branch at u such that $g(b_1) + g(b_2) \leq \delta D$, we add a vertex u' and an edge (u, u') of length zero and replace (u, v_i) with edge (u', v_i) of length $\ell(u, v_i)$ for $i \in \{1, 2\}$. The u' branch at u then replaces the two branches b_1 and b_2 in $\mathcal{B}$. This ensures that any two branches in $\mathcal{B}$ with the same parent cannot be combined into a subtree of load $\leq \delta D$.

Then, for every $b \in \mathcal{B}$, we *condense* b by replacing it with a leaf edge of length $\ell(b)$ and load $g(b)$. All clients in b are now assumed to be co-located at the leaf. Though it is easier to think of these condensed branches as leaf edges, the algorithm need not actually modify the input tree; condensing a branch is equivalent to requiring a single vehicle to cover the entire branch.

Clustering the Condensed Tree. After condensing all small branches, we partition the condensed tree into clusters and define every leaf edge whose load is at least $\frac{\delta}{2}D$ to be a *leaf cluster*. The leaf-cluster-to-root paths define what we call the *backbone* of T. By construction, every edge that is not on this backbone is either a leaf cluster (of load $\geq \frac{\delta}{2}D$) or a leaf edge (of load $< \frac{\delta}{2}D$). That is, every vertex is at most one edge away from the backbone (see Fig. 1a).

We can think of the condensed tree as a binary tree whose root is the depot, whose leaves are the leaf clusters, and whose internal vertices are the branching points of the backbone. Each edge of this binary tree corresponds to a maximal path of the backbone between these vertices, together with the small leaf edges off of this path (see Fig. 1a). To avoid confusion with tree edges, we call these path and leaf subgraphs *woolly edges*. A *woolly subedge* of a woolly edge consists of a subpath of the backbone and all incident leaf edges.

A woolly edge e whose load $g(e)$ is less than $\frac{\hat{\epsilon}\delta}{2}D$ is called a *small cluster*. The remaining woolly edges have load at least $\frac{\hat{\epsilon}\delta}{2}D$. We partition each such woolly edge into one or more woolly subedges, which we call *edge clusters*, each with load in $[\frac{\hat{\epsilon}\delta}{2}D, \frac{\delta}{2}D]$. Backbone edges do not contain clients and can be subdivided as needed to ensure enough granularity in the tree edge lengths so that such a partition is always possible (see Fig. 1b).

For convenience, we label the components of edge clusters. Let $\mathcal{C}$ be the set of edge clusters. For any edge cluster $C \in \mathcal{C}$, let P_C denote the backbone path in C and let L_C denote the leaf edges in C. We order the backbone edges along P_C as $p_{C,1}, p_{C,2}, ..., p_{C,m}$ in increasing distance from the depot and similarly label

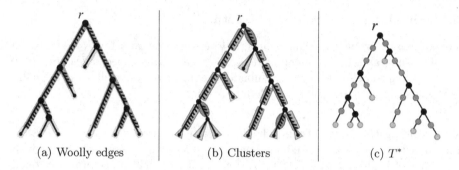

(a) Woolly edges   (b) Clusters   (c) T^*

Fig. 1. (a) Leaf clusters in yellow and woolly edges in red; (b) The tree partitioned into leaf clusters (yellow triangles), small clusters (blue ovals), and edge clusters (green rectangles); (c) The corresponding T^* for clustering from (b). (Color figure online)

the leaf edges $e_{C,1}, e_{C,2}, ..., e_{C,m-1}$ such that $e_{C,i}$ is the leaf incident to $p_{C,i}$ and $p_{C,i+1}$ for all $1 \le i < m$ (see Fig. 2). If no such incident leaf exists for some i, we can add a leaf of length zero. Likewise P_C can be *padded* with edges of length zero to ensure that each edge cluster 'starts' and 'ends' with a backbone edge.

Solution Structure. Consider the intersection of a solution with an edge cluster C. There are three different *types* of routes that visit C (see Fig. 2). A C-*passing* route traverses C without covering any clients, and thus includes all of P_C but no leaf edges in L_C. A C-*collecting* route traverses *and* covers clients in C, and thus includes all of P_C and some edges in L_C. Last, a C-*ending* route covers clients in, but does not traverse C, and thus includes backbone edges $p_{C,1}, p_{C,2}, ..., p_{C,i}$ for some $i < m$ and some leaves in L_C, but does not include all of P_C. Note that any C-*ending* route can be assumed to cover some leaves in L_C because otherwise, removing any such redundancy would only improve a solution.

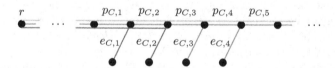

Fig. 2. Three types of route within an edge cluster C; the red tour is a C-passing route, the green tour is a C-collecting route, and the blue tour is a C-ending route. (Color figure online)

We say that a cluster C has *single coverage* if a single vehicle covers *all* clients in C. We say that an edge cluster C has *split coverage* if there is one C-ending route that covers leaf edges $e_{C,1}, e_{C,2}, ..., e_{C,i}$ for some $i < m-1$ and one C-collecting route that covers leaf edges $e_{C,i+1}, e_{C,i+2}, ..., e_{C,m-1}$ (see Fig. 2).

Finally, we say that a feasible solution has a *simple structure* if:

- Leaf clusters and small clusters have single coverage,
- Edge clusters have single or split coverage, and
- Each vehicle covers clients in $O(\frac{1}{\bar{\epsilon}^2 \delta})$ clusters

Customization of the framework requires proving a *structure theorem* stating that there exists a near-optimal solution (i.e. a feasible solution with maximum load at most $(1 + \epsilon)D$) with simple structure. Such a theorem proves that it is safe to reduce the set of potential solutions to those with simple structure.

3.2   Dynamic Program

After proving a structure theorem, the framework uses a dynamic programming algorithm (DP) to actually find a near-optimal solution with simple structure. We define the *cluster tree* T^* to be the tree that results from contracting each cluster of T to a single vertex. That is, the cluster tree has a vertex for each cluster and each branching point of the backbone (See Fig. 1c). The DP traverses T^* starting at the leaves and moving rootward, and enumerates the possible route structures within each cluster. Namely, the DP considers all ways edge cluster coverage can be split and how routes are merged at branching points.

At each vertex in this tree the algorithm stores a set of *configurations*. A configuration is interpreted as a set of routes in T that cover all clusters in the subtree of T^* rooted at v. Let $\theta \in (0, 1]$ be a problem-specific value that depends only on ϵ. A configuration at a vertex v specifies, for each multiple i of θD between 0 and $(1 + \epsilon)D$ the number of routes whose rounded load is i at the time they reach v. Because θ depends only on ϵ, the number of configurations and runtime of the DP is polynomially bounded. After traversing the entire cluster tree, the solution is found at the root. If there exists a configuration at the root such that all of the rounded route loads are at most $(1 + \epsilon)D$, the algorithm returns this solution.

To ensure that the DP actually finds a near-optimal solution, we must bound the number of times that a given route is rounded to ϵ/θ, which gives a rounding error of at most ϵD. In particular, we design the DP so that the number of times that any one route is rounded is proportional to the number of clusters that it covers clients in. Then, using the structure theorem, there exists a near-optimal solution that covers clients in $O(\frac{1}{\bar{\epsilon}^2 \delta})$ clusters and gets rounded by the DP $O(\frac{1}{\bar{\epsilon}^2 \delta})$ many times. Finally, θ is set to $c_\theta \epsilon \bar{\epsilon}^2 \delta$ for some constant c_θ.

For loads involving distance, C-passing routes pose a particular challenge for bounding rounding error. These routes may accumulate load while passing through clusters without covering any clients, yet the DP cannot afford to update the load at every such cluster. Instead, the DP *projects* routes to predetermined destinations up the tree, so that they accumulate rounding error only once while passing many clusters. The configuration then stores the (rounded) loads of the *projected* routes, and the DP need not update these load values for clusters passed through along the projection.

3.3 Reassignment Lemma

We now present a lemma that will serve as a general-purpose tool for our framework. This tool is used to *reassign* small route segments. That is, if some subgraph H is covered by several small route segments from distinct vehicles $h_1, h_2, ..., h_m$, then for some $1 \leq i \leq m$, the entire subgraph H is assigned to be covered by h_i. This *increases* load on h_i so as to cover all of H, and *decreases* load on h_j for all $j \neq i$ which are no longer required to cover H (see Fig. 3). We show that this assignment process can be performed simultaneously for many such subgraphs such that the net load increase of any one route is small.

Let $G = (A, B, E)$ be an edge-weighted bipartite graph where A is a set of *facilities*, B is a set of *clients*, and $w(a, b) \geq 0$ is the weight of edge $(a, b) \in E$. For any vertex v, we use $N(v)$ to denote the *neighborhood* of v, namely the set of vertices u such that there is an edge $(u, v) \in E$. Each facility $a \in A$ has capacity $q(a) = \sum_{b \in N(a)} w(a, b)$ and each client $b \in B$ has weight $w(b) \leq \sum_{a \in N(b)} w(a, b)$. A feasible *assignment* is a function $f : B \rightarrow A$, such that each client b is assigned to an adjacent facility $f(b) \in N(b)$. We can think of the weights $w(a, b)$ representing fractional assignment costs while weight $w(b)$ corresponds to a "discounted" cost of wholly serving client b. Ideally, the total weight of clients assigned to any facility a would not exceed the capacity $q(a)$; however, this is not always possible. We define the *overload* $h_f(a)$ of a facility a to be $w(f^{-1}(a)) - q(a) = \sum_{b | f(b) = a} w(b) - \sum_{b \in N(a)} w(a, b)$ and the *overload* h_f of an assignment to be $\max_{a \in A} h_f(a)$. The BIPARTITE WEIGHT-CAPACITATED ASSIGNMENT problem is to find an assignment with minimum overload.

Lemma 1. *Given an instance of the* BIPARTITE WEIGHT-CAPACITATED ASSIGNMENT *problem, an assignment with overload at most* $\max_{b \in B} w(b)$ *can be found efficiently.*

In our application of Lemma 1, facilities represent tours and clients represent subgraphs of T. Assignment of a client b to a facility a represents assigning subgraph b to be covered by tour a (see the proof of Lemma 2 for an example).

4   Customizing the Framework: MINIMUM MAKESPAN VEHICLE ROUTING

In this section, we demonstrate how to apply the general framework to a specific problem, MINIMUM MAKESPAN VEHICLE ROUTING. In particular, we use the framework to achieve the following:

Theorem 3. *For every* $\epsilon > 0$, *there is a polynomial-time algorithm that, given an instance of* MINIMUM MAKESPAN VEHICLE ROUTING *on a tree, finds a solution whose makespan is at most* $1 + \epsilon$ *times optimum.*

Recall that the problem is to find k *tours* that serve all clients in T such that the maximum *length* of any tour is minimized. The vehicle routes are tours, and the vehicle load is tour length, so the load $g(H)$ of subgraph H is twice the length of edges in the subgraph. The CONDENSE operation is then applied to the input tree, with $\delta = \hat{\epsilon} = \epsilon/c$ for some constant c we will define later. Leaf clusters therefore correspond to branches of length at least $\frac{\hat{\epsilon}}{4}D$ (load at least $\frac{\hat{\epsilon}}{2}D$), small clusters have total length less than $\frac{\hat{\epsilon}^2}{4}D$, and edge clusters have total length in $[\frac{\hat{\epsilon}^2}{4}D, \frac{\hat{\epsilon}}{4}D]$. As described in Sect. 3, the two steps in applying the framework are proving a structure theorem and designing a dynamic program.

4.1 MINIMUM MAKESPAN VEHICLE ROUTING Structure Theorem

We prove the following for MINIMUM MAKESPAN VEHICLE ROUTING.

Theorem 4. *If there exists a solution with makespan D, then there exists a solution with makespan at most $1 + O(\hat{\epsilon})D$ that has simple structure.*

We prove the above by starting with some optimal solution of makespan at most D and show that after a series of steps that transforms the solution into one with simple structure, the makespan is still near-optimal.

To ensure that each step maintains solution feasibility, we introduce the following notion of independence. Let T' be a connected subgraph of T containing the depot r, and let X be a set of subgraphs of T. We say that X is a *tour-independent* set with respect to T' if $T' \cup X'$ is connected for all $X' \subseteq X$. In particular, if T' is the subgraph covered by a single tour then adding any subgraphs in X' creates a new feasible tour.

Lemma 2. *The CONDENSE operation adds at most $\hat{\epsilon}D$ to the optimal makespan.*

Proof. The CONDENSE operation is equivalent to requiring every branch in $\mathcal{B}$ to be covered by a single tour. We show that there is such a solution of makespan at most $OPT + \hat{\epsilon}D$. Fix an optimal solution, and let A be the set of tours in the optimal solution that (at least partially) cover branches in $\mathcal{B}$. We define an edge-weighted bipartite graph $G = (A, \mathcal{B}, E)$ where there is an edge (a, b) if and only if tour a contains edges of branch b, and $w(a, b)$ is the length of the tour segment of a in branch b, namely twice the length of the edges covered by tour a. Note that $\forall a \in A, b \in \mathcal{B}$, $w(a, b) \leq \hat{\epsilon}D$. For each $b \in \mathcal{B}$, we define the weight $w(b)$ to be $2\ell(b)$, and for each $a \in A$, we define the capacity $q(a)$ to be the sum $\sum_{b:a \cap b \neq \emptyset} w(a, b)$ of all tour segments of a in branches of $\mathcal{B}$. Clearly, $w(b) \leq \sum_{a:a \cap b \neq \emptyset} w(a, b)$, since these tour segments collectively cover b.

Essentially, $q(a)$ represents tour a's *budget* for buying whole branches and is defined by the length of its tour segments in the branches that it partially covers. Further, we will only assign a branch to a tour that already covers some edges in the branch so there is no additional cost to connect the tour to the branch.

Applying Lemma 1 to G, we can achieve an assignment of branches to tours such that each branch is assigned to one tour and the capacity of each tour is exceeded by at most $\max_{b \in B} w(b) \leq \hat{\epsilon}D$. Further, for any tour $a \in A$, let T_a' be the corresponding subgraph visited by a excluding any branches in $\mathcal{B}$.

T'_a contains r and is connected, so $N_G(a) \subseteq \mathcal{B}$ is a tour-independent set with respect to T'_a. Thus, the reassignment of branches creates a feasible solution in which the extra distance traveled by each tour is at most $\hat{\epsilon}D$.

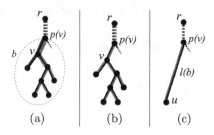

(a)   (b)   (c)

Fig. 3. (a) depicts a branch $b \in \mathcal{B}$ covered by several small tour segments; (b) shows the entire branch b being assigned to the blue tour; (c) shows the result of the condense operation. (Color figure online)

Lemma 3. *Requiring all leaf clusters and small clusters to have single coverage increases the makespan by at most $4\hat{\epsilon}D$.*

Proof. After condensing the tree, all leaf clusters have single coverage, and the effect on makespan was covered in Lemma 2. Because of the binary tree structure, we can *assign* each small cluster to a descendant leaf cluster in such a way that each leaf cluster is assigned at most two small clusters. Since each leaf cluster is covered by a single tour, we can require this tour to also cover the clients of the small cluster(s) assigned to that leaf cluster. This is feasible since small clusters are only assigned to *descendant* leaf clusters. Furthermore, since leaf clusters have length at least $\frac{\hat{\epsilon}}{4}D$, we can *charge* this error to the length of the leaf clusters. In particular, since any given tour covers at most $D/(2 \cdot \frac{\hat{\epsilon}}{4}D) = \frac{2}{\hat{\epsilon}}$ leaf clusters, this assignment adds at most $2 \cdot \frac{2}{\hat{\epsilon}} \cdot (2 \cdot \frac{\hat{\epsilon}^2}{2}D) = 4\hat{\epsilon}D$ to the makespan.

Lemma 4. *Requiring every edge cluster to have single or split coverage adds at most $3\hat{\epsilon}D$ to the optimal makespan.*

After proving Lemma 4, all that remains in proving Theorem 4 is to bound the number of clusters that a single vehicle covers clients in. See the full version of our paper for proofs.

4.2   MINIMUM MAKESPAN VEHICLE ROUTING Dynamic Program

Having proven a structure theorem, we now present a dynamic programming algorithm (DP) that actually finds a near-optimal solution with simple structure.

Recall, the DP traverses cluster tree T^* starting at the leaves and moving rootward. A configuration is a vector in $\{0, 1, 2, ..., k\}^{2\hat{\epsilon}^{-4}}$. A configuration $\boldsymbol{x}$ at a vertex v is interpreted as a set of tours *projected up to r* in T that cover all clusters in the subtree of T^* rooted at v. For $i \in \{1, 2, ..., 2\hat{\epsilon}^{-4}\}$, $\boldsymbol{x}[i]$ is the number of tours in the set that have *rounded* length $i\hat{\epsilon}^4 D$. That is, the actual tours that correspond to the $\boldsymbol{x}[i]$ tours represented in the vector each have length that may be less than $i\hat{\epsilon}^4 D$.

The algorithm categorizes the vertices into three different cases and handles them separately. The *base cases* are the leaves of T^*. Let $v \in T^*$ be such a leaf, let L_v be the corresponding leaf cluster in T, and let u be the vertex at which L_v meets the backbone. When the algorithm determines the configuration for v it addresses covering both L_v as well as covering any small clusters $C_1, ..., C_h$ that are assigned to L_v. Let ℓ_{small} be the length of all of the *leaves* of these small clusters, namely $\ell_{small} = \ell(\bigcup_{1 \le i \le h} C_i \setminus \text{backbone})$. Let ℓ_0 be $2(\ell(L_v) + \ell_{small} + d_T(u, r))$ rounded up to the nearest $\hat{\epsilon}^4 D$. The only configuration stored at v is $\boldsymbol{x}$ such that $\boldsymbol{x}[\ell_0] = 1$ and $\boldsymbol{x}[j] = 0, \forall j \neq \ell_0$. All cluster lengths and distances to the depot can be precomputed in linear time, after which each base case can be computed in constant time.

The *grow cases* are the vertices in T^* that correspond to edge clusters in T. Let $v \in T^*$ be such a vertex, and let C_v be the corresponding edge cluster in T. Let u be the root-most vertex in C_v, and let $v' \in T^*$ be the lone child vertex of v. Note that v' may correspond to a branching backbone vertex, a leaf cluster or another edge cluster, but by construction, v has exactly one child. Since C_v has single or split coverage, at most two tours in any configuration at v are involved in covering the leaves of C_v: all other tours in the configuration are C_v-passing tours, and their representation in the configuration remains unchanged. The algorithm considers all possible rounded tour lengths ℓ_1 for a C_v-ending tour t_1 for the configuration (including not having such a tour) and for each such t_1, the algorithm considers all possible (rounded) lengths ℓ_2 for an incoming C_v-collecting tour t_2, *before* the remaining length from covering leaves in C_v is added to the tour. Given ℓ_1 and ℓ_2, the algorithm can easily compute the resulting rounded length ℓ_3 of t_2 *after* covering its share of C_v leaves. For each configuration $\boldsymbol{x}'$ for child vertex v', the algorithm determines configuration $\boldsymbol{x}$ for v such that $\boldsymbol{x}[\ell_1] = \boldsymbol{x}'[\ell_1] + 1$, $\boldsymbol{x}[\ell_2] = \boldsymbol{x}'[\ell_2] - 1$, $\boldsymbol{x}[\ell_3] = \boldsymbol{x}'[\ell_3] + 1$, and $\boldsymbol{x}[i] = \boldsymbol{x}'[i]$ otherwise. If the resulting $\boldsymbol{x}$ is feasible, it is stored at v. Since there are at most $2\hat{\epsilon}^{-4}$ options for ℓ_1 and ℓ_2 and at most $k^{2\hat{\epsilon}^{-4}}$ configurations at v', the runtime for each grow case is $k^{O(\hat{\epsilon}^{-4})}$.

Finally, the *merge cases* are the vertices in T^* that correspond to branching backbone vertices in T as well as the depot. Let $v \in T^*$ be such a vertex, and let u be the corresponding vertex in T. Let $v_1, v_2 \in T^*$ be the two children of v in T^*. Every tour t in a configuration at v will either be a directly inherited tour t_i of rounded length ℓ_i from a configuration at v_i for $i \in \{1, 2\}$, or will be a *merging* of some tour t_1 from v_1 and some t_2 from v_2 with resulting length $\ell_1 + \ell_2 - 2\ell(u, r)$ rounded up to the nearest $\hat{\epsilon}^4 D$ (recall that t_1 and t_2 are tours from the depot so the subtracted amount addresses over-counting the path to the depot). For every possible (ℓ_1, ℓ_2) (including lengths of zero to account for tours inherited by children), the algorithm considers how many tours at v could have resulted from merging a tour of length ℓ_1 from v_1 with a tour of length ℓ_2 from v_2. Each of these possibilities corresponds to a configuration $\boldsymbol{x}_i$ at v_i for $i \in \{1, 2\}$ and to a merged configuration $\boldsymbol{x}$ at v. If $\boldsymbol{x}_1$ and $\boldsymbol{x}_2$ are valid configurations stored at v_1 and v_2, respectively, then the algorithm stores $\boldsymbol{x}$ at v. There are $k^{4\hat{\epsilon}^{-8}}$ such possibilities, so the runtime of each merge case is $k^{O(\hat{\epsilon}^{-8})}$. Note that the dynamic program only considers storing feasible configurations $\boldsymbol{x}$ at vertex v so the algorithm maintains that there are at most k tours total.

Since for any $\epsilon > 0$ the DP has a polynomial runtime, the following lemma, which we prove in the full version of our paper, completes the proof of Theorem 3.

Lemma 5. *The dynamic program described above finds a tour with maximum makespan at most* $(1 + \epsilon)D$.

4.3 Distance-Constrained Vehicle Routing

Recall that the DISTANCE-CONSTRAINED VEHICLE ROUTING problem is to minimize the number of tours of length at most D required to cover all clients. Since it is the MINIMUM FLEET BUDGET problem associated with MINIMUM MAKESPAN VEHICLE ROUTING, the following bicriteria PTAS follows as a corollary to Theorem 3.

Theorem 5. *Given an instance of* DISTANCE-CONSTRAINED VEHICLE ROUTING *on a tree, if there exists a solution with k tours of length at most D, then for any $\epsilon > 0$, there is a polynomial-time algorithm that finds a solution with k tours of length at most* $(1 + \epsilon)D$.

5 Framework Applications

In this section we give theorem statements for several other problems and extensions that can be solved using our framework. See the full version of our paper for details and proofs.

Theorem 6. *Given an instance of* CAPACITATED VEHICLE ROUTING *on a tree, if there exists a solution of total length k and capacity Q, then for any $\epsilon > 0$, there is a polynomial-time algorithm that finds a solution of total length k and capacity at most* $(1 + \epsilon)Q$.

Theorem 7. *Given an instance of the* SCHOOL BUS ROUTING *problem on a tree, if there exists a solution consisting of k paths of regret at most R, then for any $\epsilon > 0$, there is polynomial-time algorithm that finds a solution consisting of k paths of regret at most* $(1 + \epsilon)R$.

Theorem 8. *There is a polynomial-time 2-approximation for the* SCHOOL BUS ROUTING *problem in trees.*

Theorem 9. *For every $\epsilon > 0$ and $\rho > 0$, there is a polynomial-time algorithm that, given an instance of ρ-*DEPOT MINIMUM MAKESPAN VEHICLE ROUTING *on a tree, finds a solution whose makespan is at most $1 + \epsilon$ times optimum.*

References

1. Asano, T., Katoh, N., Kawashima, K.: A new approximation algorithm for the capacitated vehicle routing problem on a tree. J. Comb. Optim. **5**(2), 213–231 (2001). https://doi.org/10.1023/A:1011461300596
2. Becker, A.: A tight 4/3 approximation for capacitated vehicle routing in trees. CoRR abs/1804.08791 (2018). http://arxiv.org/abs/1804.08791
3. Bock, A., Grant, E., Könemann, J., Sanità, L.: The school bus problem on trees. Algorithmica **67**(1), 49–64 (2013). https://doi.org/10.1007/s00453-012-9711-x
4. Chen, L., Marx, D.: Covering a tree with rooted subtrees-parameterized and approximation algorithms. In: Proceedings of the Twenty-Ninth Annual ACM-SIAM Symposium on Discrete Algorithms, pp. 2801–2820. SIAM (2018)
5. Christofides, N.: Worst-case analysis of a new heuristic for the travelling salesman problem. Technical report, Carnegie-Mellon University Pittsburgh, PA, Management Sciences Research Group (1976)
6. Even, G., Garg, N., Könemann, J., Ravi, R., Sinha, A.: Min-max tree covers of graphs. Oper. Res. Lett. **32**(4), 309–315 (2004). https://doi.org/10.1016/j.orl.2003.11.010
7. Friggstad, Z., Swamy, C.: Approximation algorithms for regret-bounded vehicle routing and applications to distance-constrained vehicle routing. In: Proceedings of the Forty-Sixth Annual ACM Symposium on Theory of Computing, STOC 2014, pp. 744–753. ACM, New York (2014). https://doi.org/10.1145/2591796.2591840, http://doi.acm.org/10.1145/2591796.2591840
8. Friggstad, Z., Swamy, C.: Compact, provably-good LPs for orienteering and regret-bounded vehicle routing. In: Eisenbrand, F., Koenemann, J. (eds.) IPCO 2017. LNCS, vol. 10328, pp. 199–211. Springer, Cham (2017). https://doi.org/10.1007/978-3-319-59250-3_17
9. Haimovich, M., Rinnooy Kan, A.: Bounds and heuristics for capacitated routing problems. Math. Oper. Res. **10**(4), 527–542 (1985)
10. Hochbaum, D.S., Shmoys, D.B.: Using dual approximation algorithms for scheduling problems theoretical and practical results. J. ACM (JACM) **34**(1), 144–162 (1987)
11. Karuno, Y., Nagamochi, H., Ibaraki, T.: Vehicle scheduling on a tree with release and handling times. Ann. Oper. Res. **69**, 193–207 (1997)
12. Labbé, M., Laporte, G., Mercure, H.: Capacitated vehicle routing on trees. Oper. Res. **39**(4), 616–622 (1991)
13. Nagamochi, H., Okada, K.: Approximating the minmax rooted-tree cover in a tree. Inf. Process. Lett. **104**(5), 173–178 (2007)
14. Nagarajan, V., Ravi, R.: Approximation algorithms for distance constrained vehicle routing problems. Networks **59**(2), 209–214 (2012)
15. Sahni, S.K.: Algorithms for scheduling independent tasks. J. ACM (JACM) **23**(1), 116–127 (1976)
16. Xu, L., Xu, Z., Xu, D.: Exact and approximation algorithms for the min-max k-traveling salesmen problem on a tree. Eur. J. Oper. Res. **227**(2), 284–292 (2013)

Avoidable Vertices and Edges in Graphs

Jesse Beisegel[1]([⊠]), Maria Chudnovsky[2], Vladimir Gurvich[3], Martin Milanič[4,5], and Mary Servatius[6]

[1] BTU Cottbus-Senftenberg, Cottbus, Germany
beisegel@b-tu.de
[2] Princeton University, Princeton, NJ, USA
mchudnov@math.princeton.edu
[3] Higher School of Economics, National Research University, Moscow, Russia
vgurvich@hse.ru
[4] University of Primorska, IAM, Muzejski trg 2, 6000 Koper, Slovenia
martin.milanic@upr.si
[5] University of Primorska, FAMNIT, Glagoljaška 8, 6000 Koper, Slovenia
[6] Koper, Slovenia

Abstract. A vertex v in a graph G is said to be *avoidable* if every induced two-edge path with midpoint v is contained in an induced cycle. Generalizing Dirac's theorem on the existence of simplicial vertices in chordal graphs, Ohtsuki et al. proved in 1976 that every graph has an avoidable vertex. In a different generalization, Chvátal et al. gave in 2002 a characterization of graphs without long induced cycles based on the concept of simplicial paths. We introduce the concept of avoidable induced paths as a common generalization of avoidable vertices and simplicial paths. We propose a conjecture that would unify the results of Ohtsuki et al. and of Chvátal et al. The conjecture states that every graph that has an induced k-vertex path also has an avoidable k-vertex path. We prove that every graph with an edge has an avoidable edge, thus establishing the case $k = 2$ of the conjecture. Furthermore, we point out a close relationship between avoidable vertices in a graph and its minimal triangulations and identify new algorithmic uses of avoidable vertices. More specifically, applying Lexicographic Breadth First Search and bisimplicial elimination orderings, we derive a polynomial-time algorithm for the maximum weight clique problem in a class of graphs generalizing the class of 1-perfectly orientable graphs and its subclasses chordal graphs and circular-arc graphs.

Part of the work for this paper was done in the framework of two bilateral projects between Germany and Slovenia, financed by DAAD and the Slovenian Research Agency (BI-DE/17-19-18 and BI-DE/19-20-04). The third named author was partially funded by Russian Academic Excellence Project '5-100'. The work of the fourth named author is supported in part by the Slovenian Research Agency (I0-0035, research program P1-0285 and research projects N1-0032, N1-0102, J1-7051, J1-9110). Part of the work was done while the fourth named author was visiting Osaka Prefecture University in Japan, under the operation Mobility of Slovene higher education teachers 2018–2021, co-financed by the Republic of Slovenia and the European Union under the European Social Fund.

© Springer Nature Switzerland AG 2019
Z. Friggstad et al. (Eds.): WADS 2019, LNCS 11646, pp. 126–139, 2019.
https://doi.org/10.1007/978-3-030-24766-9_10

1   Introduction

A graph G is *chordal* if every cycle in G of length at least four has a chord. Chordal graphs are well-known to possess many good structural and algorithmic properties. The main goal of this paper is to study certain concepts related to chordal graphs in the framework of more general graph classes. The starting point for our research is a result due to Dirac [12] stating that every minimal cutset in a chordal graph is a clique, which implies that every chordal graph has a *simplicial vertex*, that is, a vertex whose neighborhood is a clique [14]. Denoting by P_k the k-vertex path, this result can be formulated as follows.

Theorem 1.1. *Every chordal graph has a vertex that is not the midpoint of any induced P_3.*

This theorem was generalized in the literature in various ways. Two particular ways of generalizing Theorem 1.1 include:

(i) proving a property of general graphs that, when specialized to chordal graphs, results in the existence of a simplicial vertex, and

(ii) generalizing the 'simpliciality' property from vertices, which are paths of length 0, to longer induced paths, and proving the existence of such paths for graphs excluding suitably longer cycles.

Let us explain the corresponding results in more detail.

1.1   First Generalization – From Chordal Graphs to All Graphs

A generalization of the first kind is given by the following theorem, which follows from [21, Theorem 3] as well as from [7, Main Theorem 4.1] and [1, Lemma 2.3].

Theorem 1.2. *Every graph G has a vertex v such that every induced P_3 having v as its midpoint is contained in an induced cycle in G.*

We formalize this property as follows.

Definition 1.1. *A vertex v in a graph G is said to be* avoidable *if between any pair x and y of neighbors of v there exists an x,y-path all the internal vertices of which avoid v and all neighbors of v. Equivalently, a vertex v is* avoidable *if every induced P_3 with midpoint v closes to an induced cycle.*

This terminology is motivated by considering a setting where G represents a symmetric acquaintance relation on a group of people. In this setting, the property of person (equivalently, vertex) a being avoidable can be interpreted as follows: whenever two acquaintances of a need to share some information that they would not like to share with a, they can do so by passing the information along a path completely avoiding both a and all her other acquaintances. Thus, a is in a sense avoidable, as information can be passed around in her immediate proximity without her knowledge.

Note that every simplicial vertex in a graph is avoidable. If we analyze avoidable vertices in graph classes, rather than in general graphs, we see that this definition is a generalization of many well known concepts. For example, in a tree a vertex is avoidable if and only if it is a leaf, while in a chordal graph a vertex is avoidable if and only if it is simplicial. With this terminology, Theorem 1.2 can be equivalently stated as follows.

Theorem 1.3. *Every graph has an avoidable vertex.*

The notion of avoidable vertices has appeared in the literature (with different terminology) in a variety of settings. To our knowledge, the earliest appearance was in the paper from 1976 by Ohtsuki et al. [21], where avoidable vertices were characterized as exactly the vertices from which a minimal elimination ordering can start. Here, a *minimal elimination ordering* of a graph $G = (V, E)$ is a procedure of eliminating vertices one at a time so that before each vertex is removed, its neighborhood is turned into a clique, and the resulting set F of edges added throughout the procedure is an inclusion-minimal set of non-edges of G such that $(V, E \cup F)$ is a chordal graph (in other words, $(V, E \cup F)$ is a *minimal triangulation* of G). Given a graph G, an avoidable vertex in G can be found in linear time using graph search algorithms such as Lexicographic Breadth First Search (LBFS) [25] (see also [16]) or Maximum Cardinality Search (MCS) [6]. The presentation closest to our setting is the one used by Ohtsuki et al. [21]. In fact, Berry et al. [5,6] named avoidable vertices *OCF-vertices*, after the initials of the three authors of [21].

1.2   Second Generalization – From Vertices to Longer Paths

In order to generalize the notion of simplicial vertices to longer paths, the next definition, partially following Chvátal et al. [11], will be useful.

Definition 1.2. *Given an induced path P in a graph G, an* extension *of P is any induced path in G obtained by adding to P one edge at each end. An induced path is said to be* simplicial *if it has no extension.*

In this terminology, Theorem 1.1 can be stated as follows: every graph containing at least one vertex without induced cycles of length more than 3 has a simplicial induced P_1. In 2002, Chvátal et al. [11] generalized this result as follows.

Theorem 1.4. *For each $k \geq 1$, every $\{C_{k+3}, C_{k+4}, \ldots\}$-free graph that contains an induced P_k also contains a simplicial induced P_k.*

1.3   A Common Generalization?

Theorems 1.3 and 1.4 suggest that a further common generalization might be possible, based on the following generalization of Definition 1.1 (definition of avoidable vertices) to longer paths.

Definition 1.3. *An induced path P in a graph G is said to be* avoidable *if every extension of P is contained in an induced cycle.*

Thus, in particular, a vertex v in a graph G is avoidable if and only if the corresponding one-vertex path is avoidable. Moreover, every simplicial induced path is (vacuously) avoidable. We conjecture that the following common generalization of Theorems 1.3 and 1.4 holds.

Conjecture 1.1. For every $k \geq 1$, every graph that contains an induced P_k also contains an avoidable induced P_k.

Theorem 1.3 implies the conjecture for $k = 1$, while Theorem 1.4 implies it for every positive integer k provided we restrict ourselves to the class of graphs without induced cycles of length more than $k + 2$. Indeed, if G is a $\{C_{k+3}, C_{k+4}, \ldots\}$-free graph that contains an induced P_k, then by Theorem 1.4 graph G contains a simplicial induced P_k, and every simplicial induced path is avoidable.

1.4 Our Results

The results of this paper can be summarized as follows.

1. Characterization, existence, and computation of avoidable vertices. Following the work of Ohtsuki et al. [21], we revisit the connection between avoidable vertices and minimal triangulations of graphs by characterizing avoidable vertices in a graph G as exactly the simplicial vertices in some minimal triangulation of G (Theorem 3.1). Using properties of LBFS that follow from works of Berry and Bordat [7] and Aboulker et al. [1], we show that every graph with at least two vertices contains a diametral pair of avoidable vertices (Theorem 4.1). The same approach shows that a pair of distinct avoidable vertices (though not necessarily a diametral pair) in a given graph G with at least two vertices can be computed in linear time (Theorem 4.2).

2. New polynomially solvable cases of the maximum weight clique problem. A graph is 1-perfectly orientable if its edges can be oriented so that the out-neighborhood of every vertex induces a tournament, and hole-cyclically orientable if its edges can be oriented so that each induced cycle of length at least four is oriented cyclically. We connect the structural and algorithmic properties of avoidable vertices with the concept of bisimplicial vertices to develop an efficient algorithm for the maximum weight clique problem in the class of 1-perfectly orientable graphs and more generally in the class of hole-cyclically orientable graphs (Theorem 5.3). This result generalizes the well-known fact of polynomial-time solvability of the maximum weight clique problem for the classes of chordal graphs and circular-arc graphs.

3. Existence of avoidable edges. We show that for every graph G and every non-universal vertex $v \in V(G)$ there exists an avoidable vertex in the non-neighborhood of v (Theorem 3.2). We then adapt this approach to prove the existence of two avoidable edges in any graph with at least two edges (Theorem 6.1). This settles in the affirmative the case $k = 2$ of Conjecture 1.1 and generalizes the case $k = 2$ of Theorem 1.4.

Structure of the Paper. In Sect. 2 we summarize the main notations and definitions of some of the most frequently used notions in the paper. In Sect. 3 we discuss structural aspects of avoidable vertices in graphs, including a characterization of avoidable vertices as simplicial vertices in some minimal triangulation of the graph and a new proof of the existence result. Section 4 is devoted to algorithmic issues regarding the problem of efficient computation of avoidable vertices in a given graph. In Sect. 5 we present an algorithmic application of this concept to the maximum weight clique problem, by identifying a rather general class of graphs in which every avoidable vertex is bisimplicial, which leads to a polynomial-time algorithm for the maximum weight clique problem in this class of graphs. In Sect. 6, we settle in the affirmative the case $k = 2$ of Theorem 1.3. We conclude the paper in Sect. 7 with some open problems. Due to space restrictions, most proofs are omitted.

2   Preliminaries

All graphs in this paper will be finite but may be either undirected or directed, and will contain at least one vertex. We will refer to an undirected graph simply as a *graph* and denote it as $G = (V, E)$ where V is the vertex set and E the edge set. For a graph $G = (V, E)$, we also write $V(G)$ for V and $E(G)$ for E. A directed graph will be called a *digraph* and denoted as $D = (V, A)$ where V is again the set of vertices and A the set of arcs. Graphs and digraphs in this paper will always be simple, that is, without loops or multiple edges (but pairs of oppositely oriented arcs in digraphs are allowed). Unless stated otherwise we use standard graph and digraph terminology and notation. The set of all vertices adjacent to a vertex v in G, i.e., its *neighborhood*, is denoted by $N_G(v)$ (or simply $N(v)$ if the graph is clear from the context). The cardinality of $N_G(v)$ is the *degree* of v, denoted by $d_G(v)$. Similarly, the *closed neighborhood* $N_G(v) \cup \{v\}$ is written as $N_G[v]$ (or simply $N[v]$ if the graph is clear from the context). Given a digraph $D = (V, A)$, the *in-neighborhood* of a vertex v in D, denoted by $N_D^-(v)$, is the set of all vertices w such that $(w, v) \in A$. Similarly, the *out-neighborhood* of v in D, denoted by $N_D^+(v)$, is the set of all vertices w such that $(v, w) \in A$. An *orientation* of a graph $G = (V, E)$ is a digraph obtained by assigning to each edge of G a direction. A *tournament* is an orientation of the complete graph. The *distance* between two vertices s and t in a graph G is the length of a shortest path between these two vertices and will be denoted by $\text{dist}_G(s, t)$. A vertex x with largest distance from s is called *eccentric* to s and its distance to s is the *eccentricity* $\text{ecc}_G(s)$ of s. The *diameter* of G, denoted $\text{diam}(G)$, is the largest such value among all vertices.

A permutation $\sigma = (v_1, \ldots, v_n)$ of the vertices of G will be called a *vertex ordering*. For a vertex ordering $\sigma = (v_1, \ldots, v_n)$, we write $v_i \prec_\sigma v_j$ if $i < j$. Following Heggernes [19], a graph $H = (V, E \cup F)$ is called a *triangulation* of $G = (V, E)$ if H is chordal and we say that it is a *minimal triangulation* if for every proper subset F' of F, the graph $(V, E \cup F')$ is not chordal. An *elimination ordering* σ of the vertices of G is a vertex ordering given as input of the Elimination Procedure, as defined in Algorithm 1, to compute greedily a triangulation of G called G_σ^+. If G_σ^+ is a minimal triangulation we call σ a *minimal elimination ordering*. If G_σ^+ is equal to G, then σ is a *perfect elimination ordering* and G is chordal by definition. The *deficiency* of a vertex v is defined as the set $D_G(v) = \{uw \notin E \mid u, w \in N_G(v)\}$.

Input: A graph $G = (V, E)$ and an ordering $\sigma = (v_1, \ldots, v_n)$.
Output: The filled graph G_σ^+.

$G^0 = G$;
for $i = 1$ **to** n **do**
  | Let $F^i = D_{G^{i-1}}(v)$;
  | Obtain G^i by adding the edges in F^i to G^{i-1} and removing v_i;
$G_\sigma^+ = (V, E \cup \bigcup_{i=1}^n F^i)$

Algorithm 1. Elimination Procedure

By P_n, C_n, and K_n we denote the path, cycle, and the complete graph with n vertices, respectively. By $2K_2$ we denote the graph consisting of two disjoint copies of K_2. A graph is *bipartite* if its vertex set can be partitioned into two independent sets, which are then said to form a *bipartition* of the graph. By $K_{m,n}$ we denote the complete bipartite graph with m vertices in one part of the bipartition and n in the other one. Given a family of graphs $\mathcal{F}$, we say that a graph is $\mathcal{F}$-*free* if no induced subgraph of G is isomorphic to a graph in $\mathcal{F}$. A graph is *cobipartite* if its complement is bipartite and *circular arc* if it is the intersection graph of arcs on a circle. For undefined notation and terms related to graphs and graph classes, we refer to [9,15,27,30].

3   Characterization and Existence of Avoidable Vertices

The proof of Theorem 3 in the paper [21] by Ohtsuki, Cheung, and Fujisawa (which itself relied on earlier works of Rose [22–24]) leads to the characterization of avoidable vertices given by the following theorem. Since we are not aware of any explicit statement of this result in the literature, we state it here and give a short self-contained proof that does not rely on the concept of minimal elimination orderings.

Theorem 3.1. *Let $G = (V, E)$ be a graph and let $v \in V$. Then v is avoidable in G if and only if v is a simplicial vertex in some minimal triangulation of G.*

Proof sketch. Suppose that $v \in V$ is a simplicial vertex in some minimal triangulation $G' = (V, E \cup F)$ of G such that v is not avoidable in G. Let $S = N_G[v] \setminus \{x, y\}$, let F^* be the set of all pairs $\{u, w\} \in F$ such that u and w are in different connected components of the graph $G - S$, and let G^* be the graph $(V, E \cup (F \setminus F^*))$. Since v is not avoidable in G, the set F^* is non-empty. It can be shown that G^* is a triangulation of G, contradicting the minimality of G'. For the converse direction, assume that $|V| \geq 2$, let v be an avoidable vertex in G, let $S = N_G(v)$, and let $G' = (V', E')$ be the graph obtained from $G - v$ by turning S into a clique. Moreover, let $G'_1 = (V', E' \cup F')$ be a minimal triangulation of G' and let $G_1 = (V, E \cup E' \cup F')$. Then v is a simplicial vertex in G_1 and it can be shown that G_1 is a minimal triangulation of G.   □

Theorem 3.1 reveals a close connection with *potential maximal cliques*, sets of vertices of a graph G that are maximal cliques in some minimal triangulation of G [8]. The theorem states that given a vertex $v \in V(G)$, its closed neighborhood $N_G[v]$ is a potential maximal clique in G if and only if v is avoidable.

Since every graph has a minimal triangulation, Theorems 1.1 and 3.1 imply Theorem 1.3, which coincides with the statement of Conjecture 1.1 for the case $k = 1$. The following slightly stronger form of the same result can be proved using a direct approach, which can be adapted to prove the case $k = 2$ of the conjecture (cf. Sect. 6). A vertex in a graph is said to be *universal* if it is adjacent to every other vertex, and *non-universal* otherwise.

Theorem 3.2. *For every graph G and every non-universal vertex $v \in V(G)$ there exists an avoidable vertex $a \in V(G) \setminus N[v]$.*

4   Computing Avoidable Vertices

Knowing that every graph has an avoidable vertex, the next question is how to compute one efficiently. The obvious polynomial-time method would be to decide for each vertex v of the graph G whether it is avoidable. For this we have to check for each pair of nonadjacent neighbors x and y of v, if they are in the same connected component of $(G - N[v]) \cup \{x, y\}$. If we use breadth- or depth-first search to compute the connected components, this gives a running time of $\mathcal{O}(|V(G)||E(\overline{G})|(|V(G)| + |E(G)|))$. The same method can be used to compute the set of all avoidable vertices. However, if we are only interested in computing one or two avoidable vertices, we show next that this can be done in linear time.

We have already seen that in chordal graphs the avoidable vertices are exactly the same as the simplicial vertices. Therefore, any graph search algorithm that can compute simplicial vertices in a chordal graph is a good candidate for computing avoidable vertices. In 1976, Rose et al. [25] defined a linear-time algorithm (Lex-P), which computes a perfect elimination ordering if there is one, and is thus a recognition algorithm for chordal graphs. This is a recognition algorithm for chordal graphs. This algorithm, since named *Lexicographic Breadth First Search* (LBFS), exhibits many interesting structural properties and has been used as an ingredient in many other recognition and optimization algorithms on graphs. Any vertex ordering of G that can be produced by LBFS is called an *LBFS ordering* (of G).

Input: A connected n-vertex graph $G = (V, E)$ and a distinguished vertex $s \in V$.

Output: A vertex ordering σ.

$label(s) \leftarrow n$;
foreach *each vertex* $v \in V \setminus \{s\}$ **do** $label(v) \leftarrow \emptyset$;
for $i \leftarrow 1, \ldots, n$ **do**
    pick an unnumbered vertex v with lexicographically largest label;
    $\sigma(i) \leftarrow v$;
    foreach *unnumbered vertex* $u \in N(v)$ **do** append $(n - i)$ to $label(w)$;

Algorithm 2. Lexicographic Breadth First Search

The pseudocode of Lexicographic Breadth First Search given above is presented for connected graphs. However, the method can be generalized to work for arbitrary graphs, by executing the search component after component (in an arbitrary order) and concatenating the resulting vertex orderings.

In this context we will be mainly interested in the properties of the vertices of a given graph G visited *last* by some execution of LBFS (and for a suitable choice of s), also called *end vertices*. The essential claim of the following lemma due to Aboulker et al. [1] can also be found in [7].

Lemma 4.1. *Let $G = (V, E)$ be a graph and let $\sigma = (v_1, \ldots, v_n)$ be an LBFS ordering of G. Then for all triples of vertices $a, b, c \in V$ such that $a \prec_\sigma b \prec_\sigma c$ and $ac \in E$, there exists a path from a to b whose internal vertices are disjoint from $N[v_n]$.*

Corollary 4.1. *Let $G = (V, E)$ be a graph and let $\sigma = (v_1, \ldots, v_n)$ be an LBFS ordering of G. Then v_n is avoidable in G. In fact, for any $i \in \{1, \ldots, n\}$, the vertex v_i is avoidable in $G[v_1, \ldots, v_i]$.*

Note that Lexicographic Breadth First Search is a breadth-first search, that is, when LBFS runs from a vertex s, it orders the vertices of G according to their distance from the starting vertex s. In particular, this implies the following strengthening of Theorem 3.2.

Corollary 4.2. *For every graph $G = (V, E)$ and every vertex $v \in V$ there is an avoidable vertex $a \in V$ is eccentric to v.*

This corollary generalizes the fact that for every vertex v in a chordal graph G, there is a simplicial vertex in G that is eccentric to v [13,29]. Moreover, with Corollary 4.2 at hand we can strengthen Theorem 1.3 to the following generalization of Dirac's theorem on chordal graphs (Theorem 1.1).

Theorem 4.1. *Every graph G with at least two vertices contains two avoidable vertices whose distance to each other is the diameter of G.*

Proof. Let $s \in V$ be a vertex of maximum eccentricity in G and let $\sigma = (s = v_1, \ldots, v_n = a)$ be the ordering given by an LBFS starting in s. By Corollary 4.1, vertex a is avoidable. On the other hand, if $\tau = (a = w_1, \ldots, w_n = b)$ is an LBFS of G starting in a, then b is avoidable due to Corollary 4.1. Moreover, $a \neq b$ and $\mathrm{dist}_G(a, b) = \mathrm{ecc}_G(a) = \mathrm{ecc}_G(s) = \mathrm{diam}(G)$. □

Since LBFS can be implemented to run in linear time (see, e.g., [15]), employing the same approach as in the proof of Theorem 4.1, except that vertex s is chosen arbitrarily, we obtain the announced consequence for the computation of avoidable vertices.

Theorem 4.2. *Given a graph G with at least two vertices, two distinct avoidable vertices in G can be computed in linear time.*

5   Implications for the Maximum Weight Clique Problem

In this section, we present an application of the concept of avoidable vertices to the maximum weight clique problem: given a graph $G = (V, E)$ with a vertex weight function $w : V \to \mathbb{R}_+$, find a clique in G of maximum total weight, where the weight of a set $S \subseteq V$ is defined as $w(S) := \sum_{x \in S} w(x)$. We will show that this problem, which is generally NP-hard, is solvable in polynomial time in the class of 1-perfectly orientable graphs, and even more generally in the class of hole-cyclically orientable graphs. The importance of these two graph classes, the definitions of which will be given shortly, is due to the fact that they form a common generalization of two well studied graph classes, the chordal graphs and the circular-arc graphs.

The link between avoidable vertices and the classes of 1-perfectly orientable or hole-cyclically orientable graphs will be given by considering particular orientations of the input graph. Barot et al. [4] introduced the class of *cyclically orientable graphs* as the class of graphs that admit an orientation such that every chordless cycle is oriented cyclically. If we allow triangles to be oriented arbitrarily, while all other chordless cycles must be oriented cyclically, we obtain the class of hole-cyclically orientable graphs. More formally, we say that a *hole* in a graph is a chordless cycle of length at least four, that an orientation D of a graph G is *hole-cyclic* if all holes of G are oriented cyclically in D, and that a graph is *hole-cyclically orientable* if it admits a hole-cyclic orientation.

While the class of hole-cyclically orientable graphs does not seem to have been studied in the literature, it generalizes the previously studied class of 1-perfectly orientable graphs, defined as follows. We say that an orientation of a graph is an *out-tournament*, or *1-perfect* [20], if the out-neighborhood of every vertex induces a tournament. A graph is said to be *1-perfectly orientable* if it admits a 1-perfect orientation. The class of 1-perfectly orientable graphs forms a common generalization of the classes of chordal graphs and of circular-arc graphs [26,28]. While 1-perfectly orientable graphs can be recognized in polynomial time via a reduction to a 2-SAT [3], their structure is not understood (except for some special cases, see [3,10,17,18]) and the complexity of many classical optimization problems such as maximum clique, maximum independent set, or k-coloring for fixed $k \geq 3$ is still open for this class of graphs.

In this section, we show that the maximum weight clique problem is solvable in polynomial time in the class of 1-perfectly orientable graphs. Moreover, we do this even in the more general setting of hole-cyclically orientable graphs. The fact that every 1-perfectly orientable graph is hole-cyclically orientable is a consequence of the following simple lemma (see, e.g., [17]).

Lemma 5.1. *Every 1-perfect orientation of a graph G is hole-cyclic.*

Our algorithm for the maximum weight clique problem in the class of hole-cyclically orientable graphs will be based on the fact that the classes of 1-perfectly orientable and hole-cyclically orientable graphs coincide within the class of cobipartite graphs, where they also coincide with circular-arc graphs. The equivalence between properties 1, 3 and 4 in the lemma below was already observed in [17]. Due to Lemma 5.1, the list can be trivially extended with the hole-cyclically orientable property.

Lemma 5.2. *For every cobipartite graph G, the following properties are equivalent:*

1. *G is 1-perfectly orientable.*
2. *G is hole-cyclically orientable.*
3. *G has an orientation in which every induced 4-cycle is oriented cyclically.*
4. *G is a circular-arc graph.*

Another important notion for the algorithm is that of bisimplicial elimination orderings. A vertex v in a graph G is *bisimplicial* if its neighborhood is the union of two cliques in G (or, equivalently, if the graph $G[N(v)]$ is cobipartite). Let $G = (V, E)$ be a graph and let $\sigma = (v_1, \ldots, v_n)$ be a vertex ordering of G. We say that σ is a *bisimplicial elimination ordering* of G if v_i is bisimplicial in the graph $G[\{v_1, \ldots, v_i\}]$ for every $i \in \{1, \ldots, n\}$. Ye and Borodin [31] proved the following.

Theorem 5.1. *The maximum weight clique problem is solvable in polynomial time in the class of graphs having a bisimplicial elimination ordering.*

The algorithm can be summarized as follows.

Input: A graph $G = (V, E)$, a weight function $w : V \to \mathbb{R}_+$, and a
        bisimplicial elimination ordering $\sigma = (v_1, \ldots, v_n)$
Output: A maximum weight clique C^* of G

```
C* := ∅;
for i = 0 to n − 1 do
    v := σ(n − i);
    Compute a maximum weight clique C_v of G[N(v)];
    if w(C_v) > w(C*) then C* := C_v;
    G := G − v;
```

Algorithm 3. Solving the maximum weight clique problem in graphs with a bisimplicial elimination ordering

As shown by Addario-Berry et al. [2], every graph containing at least one vertex without even holes has a bisimplicial vertex. It turns out that this property also holds for hole-cyclically orientable graphs. This fact, instrumental to the polynomial-time solvability of the maximum weight clique problem in this class of graphs, is based on a simple argument involving avoidable vertices.

Lemma 5.3. *Every avoidable vertex in a hole-cyclically orientable graph is bisimplicial.*

Lemma 5.3 and Corollary 4.1 lead to the following.

Theorem 5.2. *Every hole-cyclically orientable graph has a bisimplicial vertex. Moreover, a bisimplicial elimination ordering of a hole-cyclically orientable graph can be computed in linear time.*

Theorems 5.1 and 5.2 imply that the maximum weight clique problem is solvable in polynomial time in the class of hole-cyclically orientable graphs. The degree of the polynomial involved in the running time of the algorithm given in [31] was not estimated; it was based on polynomial-time solvability of the maximum weight clique problem in the class of perfect graphs. It is not difficult to show that the algorithm can be implemented to run in time $\mathcal{O}(|V(G)|^4)$. For the class of hole-cyclically orientable graphs, we can improve the running time further using the structure of cobipartite graphs in this class given by Lemma 5.2.

Theorem 5.3. *The maximum weight clique problem is solvable in time* $\mathcal{O}(n(n \log n + m \log \log n))$ *in the class of hole-cyclically orientable graphs (and, in particular, in the class of 1-perfectly orientable graphs) with n vertices and m edges.*

6   Avoidable Edges in Graphs

We will call an edge e in a graph G *avoidable* (resp., *simplicial*) if the path P_2 induced by its endpoints is avoidable (resp., simplicial). The case $k = 2$ of Conjecture 1.1 states that every graph with an edge has an avoidable edge. Theorem 1.4 settles this case of the conjecture for $\{C_5, C_6, \ldots\}$-free graphs; in fact, it asserts that every $\{C_5, C_6, \ldots\}$-free graph with an edge has a simplicial edge. We prove the case $k = 2$ of Conjecture 1.1 for all graphs. Given a graph G, two edges will be called *independent* in G if their endpoints form an induced $2K_2$ in G. We first consider the case when the graph contains no two independent edges.

Lemma 6.1. *Let G be a graph with at least two edges but with no two independent edges. Then G contains at least two avoidable edges.*

Corollary 6.1. *Every graph with at least one edge but with no two independent edges contains an avoidable edge.*

Two distinct edges will be called *weakly adjacent* if they are not independent. An edge $e \in E(G)$ will be called *universal* in G if every edge of G other than e is weakly adjacent to e. The case not considered by Lemma 6.1 is settled in the next lemma.

Lemma 6.2. *For every graph G and every non-universal edge $e \in E(G)$ there is an edge $f \in E(G)$ independent of e which is avoidable.*

Lemmas 6.1 and 6.2 imply the following.

Theorem 6.1. *Every graph with an edge has an avoidable edge. Every graph with at least two edges has two avoidable edges.*

7 Conclusion

We introduced the notion of avoidability in graphs, a concept that has been implicitly used in a variety of contexts in algorithmic graph theory. We discussed both structural and algorithmic aspects of avoidable vertices, including a characterization of avoidable vertices as simplicial vertices in some minimal triangulation of the graph and the fact that one or two avoidable vertices in a graph can be found in linear time using a simple application of LBFS. This approach was then used to construct a polynomial-time algorithm for the maximum weight clique problem in the class of graphs admitting an orientation in which every hole is oriented cyclically. We suggested a generalization of the concept of avoidability from vertices to nontrivial induced paths and proposed a conjecture about their existence (Conjecture 1.1). We proved the conjectures for edges, that is, two-vertex paths.

Many interesting questions remain. The main open question related to this work is to resolve the status of Conjecture 1.1. Theorems 1.3 and 6.1 imply that the conjecture is true for $k \in \{1, 2\}$. In turn, this fact and Theorem 1.4 imply that the conjecture is true for the class of $\{C_6, C_7, \ldots\}$-free graphs, which includes several well studied graph classes such as weakly chordal graphs, cocomparability graphs, and AT-free graphs. While we gave a linear-time algorithm to compute two distinct avoidable vertices in any nontrivial graph (Theorem 4.2), it would also be of interest to devise an algorithm to compute *all* avoidable vertices that is more efficient than the naïve approach. Finally, having introduced the class of hole-cyclically orientable graphs as a generalization of 1-perfectly orientable graphs, we can ask for structural properties of these graphs. In particular, it is not known whether they can be recognized in polynomial time. The complexity of the maximum independent set and k-coloring problems (for fixed $k \geq 3$) is also open both for 1-perfectly orientable and for hole-cyclically orientable graphs.

Acknowledgement. The authors are grateful to Ekkehard Köhler, Matjaž Krnc, Irena Penev, and Robert Scheffler for interest in their work and helpful remarks.

References

1. Aboulker, P., Charbit, P., Trotignon, N., Vušković, K.: Vertex elimination orderings for hereditary graph classes. Discrete Math. **338**(5), 825–834 (2015)
2. Addario-Berry, L., Chudnovsky, M., Havet, F., Reed, B., Seymour, P.: Bisimplicial vertices in even-hole-free graphs. J. Comb. Theory Ser. B **98**(6), 1119–1164 (2008)
3. Bang-Jensen, J., Huang, J., Prisner, E.: In-tournament digraphs. J. Comb. Theory Ser. B **59**(2), 267–287 (1993)
4. Barot, M., Geiss, C., Zelevinsky, A.: Cluster algebras of finite type and positive symmetrizable matrices. J. Lond. Math. Soc. **73**(3), 545–564 (2006)
5. Berry, A., Blair, J.R.S., Bordat, J.-P., Simonet, G.: Graph extremities defined by search algorithms. Algorithms **3**(2), 100–124 (2010)
6. Berry, A., Blair, J.R.S., Heggernes, P., Peyton, B.W.: Maximum cardinality search for computing minimal triangulations of graphs. Algorithmica **39**(4), 287–298 (2004)
7. Berry, A., Bordat, J.-P.: Separability generalizes Dirac's theorem. Discrete Appl. Math. **84**(1–3), 43–53 (1998)
8. Bouchitté, V., Todinca, I.: Treewidth and minimum fill-in: grouping the minimal separators. SIAM J. Comput. **31**(1), 212–232 (2001)
9. Brandstädt, A., Le, V.B., Spinrad, J.P.: Graph Classes: A Survey. SIAM Monographs on Discrete Mathematics and Applications. Society for Industrial and Applied Mathematics (SIAM), Philadelphia (1999)
10. Brešar, B., Hartinger, T.R., Kos, T., Milanič, M.: 1-perfectly orientable K_4-minor-free and outerplanar graphs. Discrete Appl. Math. **248**, 33–45 (2018)
11. Chvátal, V., Rusu, I., Sritharan, R.: Dirac-type characterizations of graphs without long chordless cycles. Discrete Math. **256**(1–2), 445–448 (2002)
12. Dirac, G.A.: On rigid circuit graphs. Abh. Math. Sem. Univ. Hamburg **25**, 71–76 (1961)
13. Farber, M., Jamison, R.E.: Convexity in graphs and hypergraphs. SIAM J. Algebraic Discrete Methods **7**(3), 433–444 (1986)
14. Fulkerson, D.R., Gross, O.A.: Incidence matrices and interval graphs. Pac. J. Math. **15**, 835–855 (1965)
15. Golumbic, M.C.: Algorithmic Graph Theory and Perfect Graphs. Annals of Discrete Mathematics, vol. 57, 2nd edn. Elsevier Science B.V., Amsterdam (2004)
16. Habib, M., McConnell, R., Paul, C., Viennot, L.: Lex-BFS and partition refinement, with applications to transitive orientation, interval graph recognition and consecutive ones testing. Theoret. Comput. Sci. **234**(1–2), 59–84 (2000)
17. Hartinger, T.R., Milanič, M.: Partial characterizations of 1-perfectly orientable graphs. J. Graph Theory **85**(2), 378–394 (2017)
18. Hartinger, T.R., Milanič, M.: 1-perfectly orientable graphs and graph products. Discrete Math. **340**(7), 1727–1737 (2017)
19. Heggernes, P.: Minimal triangulations of graphs: a survey. Discrete Math. **306**(3), 297–317 (2006)
20. Kammer, F., Tholey, T.: Approximation algorithms for intersection graphs. Algorithmica **68**(2), 312–336 (2014)
21. Ohtsuki, T., Cheung, L.K., Fujisawa, T.: Minimal triangulation of a graph and optimal pivoting order in a sparse matrix. J. Math. Anal. Appl. **54**(3), 622–633 (1976)
22. Rose, D.J.: Symmetric elimination on sparse positive definite systems and the potential flow network problem. ProQuest LLC, Ann Arbor, MI. Thesis (Ph.D.), Harvard University (1970)

23. Rose, D.J.: Triangulated graphs and the elimination process. J. Math. Anal. Appl. **32**, 597–609 (1970)
24. Rose, D.J.: A Graph-Theoretic Study of the Numerical Solution of Sparse Positive Definite Systems of Linear Equations, pp. 183–217. Academic Press, New York (1972)
25. Rose, D.J., Tarjan, R.E., Lueker, G.S.: Algorithmic aspects of vertex elimination on graphs. SIAM J. Comput. **5**(2), 266–283 (1976)
26. Skrien, D.J.: A relationship between triangulated graphs, comparability graphs, proper interval graphs, proper circular-arc graphs, and nested interval graphs. J. Graph Theory **6**(3), 309–316 (1982)
27. Spinrad, J.P.: Efficient Graph Representations. Fields Institute Monographs, vol. 19. American Mathematical Society, Providence (2003)
28. Urrutia, J., Gavril, F.: An algorithm for fraternal orientation of graphs. Inf. Process. Lett. **41**(5), 271–274 (1992)
29. Voloshin, V.I.: Properties of triangulated graphs. In: Operations Research and Programming, pp. 24–32. Shtiintsa, Kishinev (1982)
30. West, D.B.: Introduction to Graph Theory. Prentice Hall Inc., Upper Saddle River (1996)
31. Ye, Y., Borodin, A.: Elimination graphs. ACM Trans. Algorithms **8**(2), 23 (2012). Art. 14

Plane Hop Spanners for Unit Disk Graphs

Ahmad Biniaz[✉]

University of Waterloo, Waterloo, Canada
ahmad.biniaz@gmail.com

Abstract. The unit disk graph (UDG) is a widely employed model for
the study of wireless networks. In this model, wireless nodes are represented by points in the plane and there is an edge between two points if
and only if their Euclidean distance is at most one. A *hop spanner* for
the UDG is a spanning subgraph H such that for every edge (p, q) in the
UDG the topological shortest path between p and q in H has a constant
number of edges. The *hop stretch factor* of H is the maximum number
of edges of these paths. A hop spanner is *plane* (i.e. embedded planar) if
its edges do not cross each other.

The problem of constructing hop spanners for the UDG has received
considerable attention in both computational geometry and wireless ad
hoc networks. Despite this attention, there has not been significant
progress on getting hop spanners that (i) are plane, and (ii) have low
hop stretch factor. Previous constructions either do not ensure the planarity or have high hop stretch factor. The only construction that satisfies
both conditions is due to Catusse, Chepoi, and Vaxès [5]; their plane hop
spanner has hop stretch factor at most 449.

Our main result is a simple algorithm that constructs a plane hop
spanner for the UDG. In addition to the simplicity, the hop stretch factor of the constructed spanner is at most 341. Even though the algorithm
itself is simple, its analysis is rather involved. Several results on the plane
geometry are established in the course of the proof. These results are of
independent interest.

Keywords: Unit disk graph · Plane graph · Hop spanner ·
Hop stretch factor · Delaunay triangulation · Square grid

1 Introduction

Computational geometry techniques are widely used to solve problems, such as
topology construction, routing, and broadcasting, in wireless ad hoc networks. A
wireless ad hoc network is usually modeled as a unit disk graph (UDG). In this
model wireless devices are represented by points in the plane and assumed to
have identical unit transmission radii. There exists an edge between two points
if their Euclidean distance is at most one unit; this edge indicates that the corresponding devices are in each other's transmission range and can communicate.

Supported by NSERC Postdoctoral Fellowship.

© Springer Nature Switzerland AG 2019
Z. Friggstad et al. (Eds.): WADS 2019, LNCS 11646, pp. 140–154, 2019.
https://doi.org/10.1007/978-3-030-24766-9_11

A *geometric graph* is a graph whose vertices are points in the plane and whose edges are straight-line segments between the points. A geometric graph is *plane* if its edges do not cross each other. Let G be a geometric graph. A *topological shortest path* between any two vertices u and v in G is a path that connects u and v and has the minimum number of edges. The *hop distance* $h_G(u, v)$ between u and v is the number of edges of a topological shortest path between them.

For a point set P in the plane, the *unit disk graph* $UDG(P)$ is a geometric graph with vertex set P that has an edge between two points p and q if and only if their Euclidean distance $|pq|$ is at most 1. A *hop spanner* for $UDG(P)$ is a spanning subgraph H such that for any edge $(p, q) \in UDG(P)$ it holds that $h_H(p, q) \leqslant t$, where t is some positive constant. The constant t is called the *hop stretch factor* of H. This definition of hop spanner H implies that for any two points p and q in P (not necessarily connected by an edge of $UDG(P)$) it holds that $h_H(p, q) \leqslant t \cdot h_{UDG(P)}(p, q)$. In this paper we study the problem of constructing UDG hop spanners that are plane and have low hop stretch factor.

For general graphs we cannot hope to always get a planar hop spanner. For example take a complete graph with five vertices, which is not planar, and replace each of its edges by an arbitrary long path. The resulting graph—which is not planar—does not have any planar hop spanner (with constant stretch factor).

The *Euclidean spanner* and *Euclidean stretch factor* are defined in a similar way, but for the distance measure they use the total Euclidean length of path edges. Both hop spanners and Euclidean spanners have received considerable attention in computational geometry and wireless ad hoc networks; see e.g. the surveys by Eppstein [8], Bose and Smid [4], Li [13], and the book by Narasimhan and Smid [16]. Unit disk graph spanners have been used to reduce the size of a network and the amount of routing information. They are also used in topology control for maintaining network connectivity, improving throughput, and optimizing network lifetime; see the surveys by Li [13] and Rajaraman [17]. Constructions of UDG spanners, both centralized and distributed, also with additional properties like planarity and power saving have been widely studied [10,12,14,15]. Researchers also studied the construction of spanners for general disk graphs [9] and for quasi unit disk graphs [6].

1.1   Related Work

Gao et al. [10] proposed a randomized algorithm for constructing a UDG spanner. First they create several clusters of points each containing one point as the clusterhead. Then they connect the clusters by a restricted Delaunay graph, and then connect the remaining points to clusterheads. The restricted Delaunay graph can be maintained in a distributed manner when points move around. Although the underlying restricted Delaunay graph is plane, the entire spanner is not. This spanner has constant Euclidean stretch factor in expectation, and constant hop stretch factor for some unspecified constant.

Alzoubi et al. [1] proposed a distributed algorithm for the construction of a hop spanner for the UDG. Their algorithm integrates the connected dominating set and the local Delaunay graph of [14] to form a backbone for the spanner.

Although the backbone is plane, the entire spanner is not. The hop stretch factor of this spanner is at most 15716 (around 15000 as estimated in [5]).

Yan, Xiang, and Dragan [18] showed how to obtain for any n-vertex unit disk graph G, a system of $O(\log n)$ spanning trees such that for any two vertices p and q there exists a tree T with $h_T(p,q) \leqslant 3 \cdot h_G(p,q) + 12$. This immediately gives a hop spanner with $O(n \log n)$ edges for G. Although this spanner is not planar, it has some interesting properties, e.g., has small hop stretch factor (which is 15) and gives a compact routing scheme with guaranteed delivery for G.

To the best of our knowledge, the only construction that guarantees the planarity of the entire hop spanner is due to Catusse, Chepoi, and Vaxès [5]. First they use a regular square-grid to partition input points into clusters. Then they add edges between points in different clusters, and also between points in the same cluster to obtain a hop spanner, which is not necessarily plane. Then they go through several steps and in each step they remove some edges to ensure planarity, and add some new edges to maintain constant hop stretch factor. At the end they obtain a plane hop spanner with hop stretch factor at most 449. This spanner can be obtained by a localized distributed algorithm.

1.2   Our Contribution

Our main contribution in this paper is a polynomial-time simple algorithm that constructs a plane hop spanner, with hop stretch factor at most 341, for unit disk graphs. Our algorithm works as follows: Given a set P of points in the plane, we first select a subset S of P (in a clever way), then compute a plane graph $DT_1(S)$ (which is the Delaunay triangulation of S minus edges of length more than 1), and then connect every remaining point of P to its closest visible vertex of $DT_1(S)$. In addition to improving the hop stretch factor, this algorithm is straightforward and the planarity proof is simple, in contrast to that of Catusse et al. [5]. Our analysis of hop stretch factor is still rather involved. Towards the correctness proof of our algorithm, we prove several results on the plane geometry, which are of independent interest. Our construction uses only local information and can be implemented as a localized distributed algorithm.

Catusse et al. [5] also showed a construction of a hop spanner, with stretch factor 5 and at most $10n$ edges, for any n-vertex UDG. By a simple modification to their construction we obtain such a spanner with at most $9n$ edges.

2   Preliminaries and Some Geometric Results

We say that a set of points in the plane is in *general position* if no three points lie on a straight line and no four points lie on a circle. Throughout this paper, every given point set is assumed to be in general position. For a set P of points in the plane, we denote by $DT(P)$ the Delaunay triangulation of P. Let p and q be any two points in the plane. We denote by pq the straight-line segment between p and q, and by $\overrightarrow{pq}$ the ray that emanates from p and passes through q. The *diametral disk* $D(p,q)$ between p and q is the disk with diameter $|pq|$ that has p and q on its boundary. Every disk considered in this paper is closed, i.e., the disk contains its boundary circle.

Consider the Delaunay triangulation of a point set P. In their seminal work, Dobkin, Friedman, and Supowit [7] proved that for any two points $p, q \in P$ there exists a path, between p and q in $DT(P)$, that lies in the diametral disk between p and q. Their proof makes use of Voronoi cells (of the Voronoi diagram of P) that intersect the line segment pq. In the following theorem we give a simple inductive proof for a more general claim that shows the existence of such a path in any disk (not only the diametral disk) between p and q.

Theorem 1. *Let P be a set of points in the plane in general position and let $DT(P)$ be the Delaunay triangulation of P. Let p and q be any two points of P and let D be any disk in the plane that has p and q on its boundary. There exists a path, between p and q in $DT(P)$, that lies in D.*

Proof. We prove this theorem by induction on the number of points in D. If D does not contain any point of $P \setminus \{p, q\}$ in its interior then (p, q) is an edge of $DT(P)$, and thus (p, q) is a desired path. Assume that D contains a point $r \in P \setminus \{p, q\}$ in its interior. Let c be the center of D. Consider the ray $\vec{pc}$. Fix D at p and shrink it along $\vec{pc}$ until r becomes on its boundary circle; see the figure to the right. Denote the resulting disk by D_{pr}; this disk lies fully in D. Com-

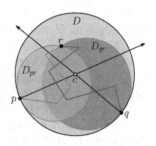

pute the disk D_{qr} in a similar fashion by shrinking D along $\vec{qc}$. Since r is in the interior of D, the disk D_{pr} does not contain q and the disk D_{qr} does not contain p. Thus, the number of points in each of D_{pr} and D_{qr} is smaller than that of D. Therefore, by induction hypothesis there exists a path, between p and r in $DT(P)$, that lies in D_{pr}, and similarly there exists a path, between q and r in $DT(P)$, that lies in D_{qr}. The union of these two paths contains a path, between p and q in $DT(P)$, that lies in D. $\square$

Let G be a plane geometric graph and let $p \notin G$ be any point in the plane. We say that a vertex $q \in G$ is *visible* from p if the straight-line segment pq does not cross any edge of G. One can simply verify that for every p such a vertex q exists. Among all vertices of G that are visible from p, we refer to the one that is closest to p by the *closest visible vertex* of G from p.

The following theorem (though simple) turns out to be crucial in the planarity proof of our hop spanner; this theorem is of independent interest. Although it answers a basic question, we were unable to find such a result in the literature; there exist however related results, see e.g. [3,11].

Theorem 2. *Let G be a plane geometric graph, and let Q be a set of points in the plane that is disjoint from G. The graph, that is obtained by connecting every point of Q to its closest visible vertex of G, is plane.*

Proof. Let E be the set of edges that connect every point of Q to its closest visible vertex of G. To prove the theorem, it suffices to show that the edges of

$G \cup E$ do not cross each other. The edges of G do not cross each other because G is plane. It is implied from the definition of visibility that the edges of E do not cross the edges of G.

We prove by contradiction that the edges of E do not cross each other. To that end consider two crossing edges (p, s) and (q, r) of E where p, q are two points of Q and s, r are two vertices of G. Let c be their intersection point of (p, s) and (q, r). By the triangle inequality we have $|pr| < |ps|$ or $|qs| < |qr|$. After a suitable relabeling assume that $|pr| < |ps|$, and thus

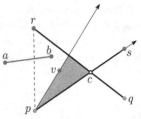

p is closer to r than to s. The reason that p was not connected to r, is that r is not visible from p. Therefore there are edges of G that block the visibility of r from p. Take any such edge (a, b). The edge (a, b) does not intersect any of (p, s) and (q, r) because otherwise (a, b) blocks the visibility of s from p or the visibility of r from q, and as such we wouldn't have these edges in E; see the figure to the right. Therefore, exactly one endpoint of (a, b), say b, lies in the triangle $\triangle pcr$. Rotate the ray $\vec{ps}$ towards b and stop as soon as hitting a vertex of G in $\triangle pcr$. This vertex is visible from p. Denote this vertex by v (it might be that $v = b$). Since v lies in $\triangle pcr$, it turns out that $|pv| \leqslant \max\{|pr|, |pc|\} < |ps|$. Thus, v is a closer visible vertex of G from p. This contradicts the fact that s is a closest visible vertex from p.   □

We refer to a hop spanner with hop stretch factor t as a t-*hop spanner*. Catusse et al. [5] showed a simple construction of a sparse 5-hop spanner with at most $10n$ edges, for any n-vertex unit disk graph. With a simple modification to their construction we obtain a 5-hop spanner with at most $9n$ edges. See the full version of this paper [2] for proofs of Theorem 3 and Lemma 1.

Theorem 3. *Every n-vertex UDG has a 5-hop spanner with at most $9n$ edges.*

Lemma 1. *Let C be a convex shape of diameter d in the plane, and let pq be a straight-line segment that intersects C. Then for any points $r \in C$, we have $\min\{|rp|, |rq|\} \leqslant \sqrt{d^2 + |pq|^2/4}$.*

3   Plane Hop Spanner Algorithm

This section presents our main contribution which is a polynomial-time algorithm for construction of plane hop spanners for unit disk graphs.

Let P be a set of points in the plane in general position, and let $UDG(P)$ be the unit disk graph of P. Our algorithm first partitions P into some clusters by using a regular square-grid; this is a standard initial step in many UDG algorithms, see e.g. [5,6]. We use this partition to select a subset S of P that satisfies some properties, which we will describe later. Then we compute the Delaunay triangulation of S and remove every edge that has length more than 1. We denote the resulting graph by $DT_1(S)$. Then we connect every point of $P \setminus S$ to its closest visible vertex of $DT_1(S)$. Let H denote the final resulting graph. We claim that H is a plane hop spanner, with hop stretch factor at most 341, for

$UDG(P)$. In Sect. 3.1 we show how to compute S. The points of S are distributed with constant density, i.e., there are $O(1)$ points of S in any unit disk in the plane. Based on this and the fact that $DT_1(S)$ has only edges of length at most 1, $DT_1(S)$ can be computed by a localized distributed algorithm. In Sect. 3.2 we prove the correctness of the algorithm that H is plane and H is a subgraph of $UDG(P)$. In Sect. 3.3 we analyze the stretch factor of H. The following theorem summarizes our result in this section.

Theorem 4. *There exists a plane 341-hop spanner for the unit disk graph of any set of points in the plane in general position. Such a spanner can be computed in polynomial time.*

3.1 Computation of S

In this section, we compute the subset S; we will see properties of S at the end of this section. Let Γ be a regular square-grid on the plane with squares of diameter 1. The side-length of these squares is $1/\sqrt{2}$. Without loss of generality we assume that no point of P lies on a grid line (this can be achieved by moving the grid by a small amount horizontally and vertically). Let E be the edge set containing the shortest edge of $UDG(P)$ that runs between any two nonempty cells of Γ if such an edge exists. Since every edge of $UDG(P)$ has length at most 1, for every cell π there are at most 20 edges in E going from π to other cells $\pi_1, \ldots, \pi_{20}$ as depicted in Fig. 1. Let $V(E)$ be the set of endpoints of E, i.e., endpoints of the edges of E. The set $V(E)$ has the following two properties:

- Every cell of Γ contains at most 20 points of $V(E)$.
- For every cell $\pi \in \Gamma$ and every $i \in \{1, \ldots, 20\}$ if there is an edge in $UDG(P)$ between π and π_i, then there are two points $s_i, t_i \in V(E)$ such that $s_i \in \pi$, $t_i \in \pi_i$, and (s_i, t_i) is the shortest edge of $UDG(P)$ that runs between π and π_i.

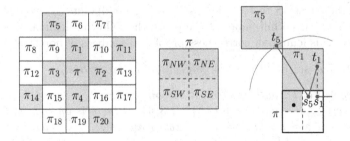

Fig. 1. Illustration of the computation of S.

We want to modify the edge set E and also compute a point set T such that $V(E) \cup T$ satisfies some more properties that we will see later. To that end we partition every cell π of Γ into four sub-cells of diameter $1/2$, namely $\pi_{NW}, \pi_{NE}, \pi_{SW}, \pi_{SE}$ as in Fig. 1. For each cell π, consider four triplets (π_{NW},

π_1, π_5), ($\pi_{NE}, \pi_2, \pi_{11}$), ($\pi_{SW}, \pi_3, \pi_{14}$), and ($\pi_{SE}, \pi_4, \pi_{20}$); these triplets are colored in Fig. 1. Let T be the empty set. We perform the following three-step process on each of the four triplets of every cell π. We describe the process only for (π_{NW}, π_1, π_5); the processes of other triplets are analogous. In our description "a point of π_{NW}" refers to a point of P that lies in π_{NW}.

1. If π_{NW} is empty (has no point of P) then do nothing and stop the process. Assume that π_{NW} contains some points of P. If it contains an endpoint of E, i.e. an endpoint of some edge of E, then do nothing and stop the process.
2. Assume now that π_{NW} contains some points of P but does not contain any endpoint of E. If there is no edge in E that runs between π and π_1 or between π and π_5 then we take a point of π_{NW} arbitrary and add it to T, and then stop the process.
3. Assume that E contains an edge between π and π_1, and an edge between π and π_5. We are now in the case where π_{NW} contains some points of P but not any endpoint of E, and both s_1 and s_5 exist. If $s_1 = s_5$, then we add a point of π_{NW} to T, and then stop the process. Assume that $s_1 \neq s_5$. Since π_{NW} does not contain any endpoint of E, the points s_1 and s_5 do not lie in π_{NW}. In particular, s_5 must be in sub-cell π_{NE} because the distance between π_5 and each of π_{SE} and π_{SW} is more than 1; however s_1 can lie in other sub-cells. In this setting the disk with center s_5 and radius $|s_5 t_5|$ contains the entire π_1 (see Fig. 1), and thus the distance between s_5 and any point in π_1 is at most 1. We replace the edge (s_1, t_1) of E by the edge (s_5, t_1), which has length at most one; see Fig. 1. Then we add a point of π_{NW} to T, and then stop the process. (We note that in the future steps when we process triplets of the cell π_1, the edge (s_5, t_1) might be replaced by another edge that runs between π_1 and π. This is okay for the purpose of Lemma 2 because we replace an edge between two cells by another edge between the same two cells).

This is the end of process for triplet (π_{NW}, π_1, π_5). After performing this process on all triplets of all cells, we obtain an edge set E and a point set T. We define the subset S to be union of T and the endpoints of edges of E, i.e., $S = V(E) \cup T$. In the full paper [2] we prove the following lemma. We will use properties of this lemma in correctness proof and analysis of hop stretch factor.

Lemma 2. *The set S satisfies the following four properties:*

(P1) *Every cell of Γ contains at most 20 points of S.*
(P2) *For every cell π and every $i \in \{1, 2, 3, 4\}$, if there is an edge in $UDG(P)$ between π and π_i, then there are two points $s_i, t_i \in S$ such that $s_i \in \pi$, $t_i \in \pi_i$, and $|s_i t_i| \leq 1$.*
(P3) *For every cell π and every $i \in \{5, \ldots, 20\}$, if there is an edge in $UDG(P)$ between π and π_i, then there are two points $s_i, t_i \in S$ such that $s_i \in \pi$, $t_i \in \pi_i$, $|s_i t_i| \leq 1$, and (s_i, t_i) is the shortest edge of $UDG(P)$ that runs between π and π_i.*
(P4) *The set S contains at least one point from every nonempty sub-cell π_{NW}, π_{NE}, π_{SW}, π_{SE} of each cell π.*

3.2 Correctness Proof

In this section we prove the correctness of our algorithm. Recall the grid Γ, and the subset S of P that is computed in Sect. 3.1. Recall that our algorithm computes the Delaunay triangulation $DT(S)$ and removes every edge of length more than 1 to obtain $DT_1(S)$, and then connects every point of $P \setminus S$ to its closest visible vertex of $DT_1(S)$. Let H denotes the resulting graph. One can simply verify that this algorithm takes polynomial time. Since $DT(S)$ is plane, its subgraph $DT_1(S)$ is also plane. It is implied from Theorem 2 (where $DT_1(S)$ and $P \setminus S$ play the roles of G and Q) that H is plane. As we stated at the outset, except for the computation of S which is a little more involved, the algorithm and the planarity proof are straightforward.

To finish the correctness proof it remains to show that every edge of H has length at most 1. Consider any edge e of H. By our construction, either the two endpoints of e belong to S, or one endpoint of e belongs to S and its other endpoint belongs to $P \setminus S$. If both endpoints of e are in S, then e belongs to $DT_1(S)$ and hence has length at most 1. If one endpoint of e is in S and its other endpoint is in $P \setminus S$, then by following lemma the length of e is at most $1/\sqrt{2}$.

Lemma 3. *The length of every edge of H, that has an endpoint in S and an endpoint in $P \setminus S$, is at most $1/\sqrt{2}$.*

Proof. Consider any edge $(p, s) \in H$ with $p \in P \setminus S$ and $s \in S$. By our construction, s is the closest visible vertex of $DT_1(S)$ from p. Thus, to prove the lemma, it suffices to show the existence of a vertex $v \in DT_1(S)$ that is visible from p and for which $|pv| \leqslant 1/\sqrt{2}$; this would imply that the distance between p and s, which is the closest visible vertex from p, is at most $1/\sqrt{2}$. In the rest of the proof we show the existence of such vertex v.

Let π be the cell that contains p (the dashed cell in Fig. 2). After a suitable rotation we assume that p lies in sub-cell π_{NW}. Since π_{NW} is nonempty, by property (P4) in Lemma 2 the set S contains at least one point from π_{NW}. Let S' be the set of points of π_{NW} that are in S. Notice that $S' \subseteq S$ and $S' \neq \emptyset$. If any point of S' is visible from p, then this point is a desired vertex v with $|pv| \leqslant 1/2$ because the diameter of π_{NW} is $1/2$.

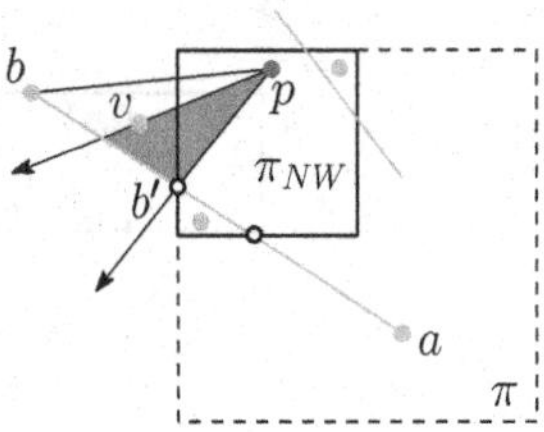

Fig. 2. Proof of Lemma 3. The red points belong to S. (Color figure online)

Assume that no point of S' is visible from p. The visibility of (points of) S' from p is blocked by some edges of $DT_1(S)$; these edges properly cross π_{NW} and separate p from points of S' (the red edges in Fig. 2). Among these edges take one whose intersection points with the boundary of π_{NW} are visible from p (observe that such an edge always exists). Denote this edge by (a, b). Since the diameter of π_{NW} is $1/2$ and $|ab| \leqslant 1$, it is implied from Lemma 1 that the distance from p to a or to b is at most $1/\sqrt{2}$; after a suitable relabeling assume that $|pb| \leqslant 1/\sqrt{2}$. Of the two intersection points of (a, b) with the boundary of π_{NW}, denote by b' the one that is closer to b. By our choice of (a, b), b' is visible from p. We rotate the ray $\overrightarrow{pb'}$

towards b and stop as soon as hitting a vertex $v \in S$ in triangle $\triangle pbb'$ (it might be that $v = b$). The vertex v is visible from p. Since v is in triangle $\triangle pbb'$ it holds that $|pv| \leqslant \max\{|pb|, |pb'|\}$. Since $|pb| \leqslant 1/\sqrt{2}$ and $|pb'| \leqslant 1/2$ it turns out that $|pv| \leqslant 1/\sqrt{2}$. □

3.3 Hop Stretch Factor

In this section we prove that the hop stretch factor of H is at most 341. We show that for any edge $(u, v) \in UDG(P)$ there exists a path of length at most 341 between u and v in H.

In this section a "cell" refers to the interior of a square of Γ, a "grid point" refers to the intersection point of a vertical and a horizontal grid line, and a "corner of π" refers to a grid point on the boundary of a cell π. We define *neighbors* of a cell π to be the set of eight cells that share sides or corners with π. We partition the neighbors of π into +-*neighbors* and ×-*neighbors*, where +-neighbors are the four cells that share sides with π, and ×-neighbors are the four cells each sharing exactly one grid point with π. In Fig. 1 the cells $\pi_1, \pi_2, \pi_3, \pi_4$ are the +-neighbors of π, and the cells $\pi_9, \pi_{10}, \pi_{15}, \pi_{16}$ are the ×-neighbors of π.

Consider any two points $p, q \in S$. If $|pq| \leqslant 1$ then every edge of $DT(S)$, that lies in $D(p, q)$, has length at most 1, and thus all these edges are present in $DT_1(S)$. Combining this with Theorem 1 we get the following corollary.

Corollary 1. *For any two points $p, q \in S$, with $|pq| \leqslant 1$, there exists a path, between p and q in $DT_1(S)$, that lies in $D(p, q)$.*

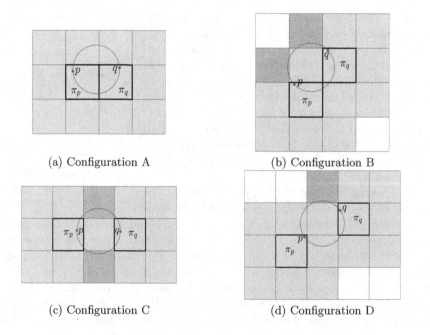

(a) Configuration A   (b) Configuration B

(c) Configuration C   (d) Configuration D

Fig. 3. Relative positions of the cells π_p and π_q where $|pq| \leqslant 1$.

Consider any two points p and q in the plane that lie in different cells, say π_p and π_q. If $|pq| \leqslant 1$ then the relative positions of π_p and π_q is among four configurations A, B, C, and D that are shown in Fig. 3. In the rest of this section we consider different configurations of a disk intersecting some cells of Γ. Although mentioned before, we emphasis that a "cell" refers to the interior of a square of grid (and hence a cell is open and does not contain its boundary) while a "disk" is closed (and hence contains its boundary).

3.3.1   Disk-Cell Intersections

To cope with the number of cases that appear in the analysis of hop stretch factor we use Lemmas 4, 5, and 6 about disk-cell intersections. These lemmas enable us to reduce the number of cases in our analysis. See the full paper [2] for the proofs of these lemmas; in the proofs of Lemmas 5 and 6 we look at relative positions of two cells, as in Fig. 3. We say that an element x is "outside" a set X if $x \notin X$.

Lemma 4. *Let p and q be any two points in the plane with $|pq| \leqslant 1$.*

1. *If p and q are in different cells, then $D(p,q)$ intersects at most 7 cells.*
2. *If p and q are in the same cell π, then $D(p,q)$ can intersect only π and its four $+$-neighbors.*

Lemma 5. *Let p and q be any two points in the plane that are in different cells π_p and π_q. Let X be the set containing the cells π_p and π_q and their $+$-neighbors.*

1. *If $|pq| \leqslant 1$, then $D(p,q)$ does not intersect any cell outside the neighborhoods of π_p and π_q.*
2. *If $|pq| \leqslant 1$, then $D(p,q)$ intersects at most two cells outside X.*
3. *If $|pq| \leqslant 1/\sqrt{2}$, then $D(p,q)$ does not intersect any cell outside X.*

Lemma 6. *Consider two cells π and π'. Let p_1 and p_2 be any two points in π', and let p_3 and p_4 be any two points in π. Let $\mathcal{D}$ be the union of three disks $D(p_1, p_2)$, $D(p_2, p_3)$, and $D(p_3, p_4)$. Then the following statements hold:*

1. *If $|p_2 p_3| \leqslant 1/\sqrt{2}$ then $\mathcal{D}$ intersects at most 8 cells.*
2. *If $|p_2 p_3| \leqslant 1$, and π and π' are $+$-neighbors, then $\mathcal{D}$ intersects at most 8 cells.*
3. *If $|p_2 p_3| \leqslant 1$, and π and π' are $\times$-neighbors, then $\mathcal{D}$ intersects at most 10 cells.*
4. *If $|p_2 p_3| \leqslant 1$, and π and π' are not neighbors, then $\mathcal{D}$ intersects at most 11 cells.*

3.3.2   Analysis of Hop Stretch Factor

With lemmas in the previous section, we have all tools for proving the hop stretch factor of H. Recall that no point of P lies on a grid line of Γ, and thus every point of P is in the interior of some square of Γ. Consider any edge $(u, v) \in UDG(P)$, and notice that $|uv| \leqslant 1$. In this section we prove the existence of a path, of length at most 341, between u and v in H. Depending on whether u or v belong to S, we have three cases: (1) $u \notin S$ and $v \notin S$, (2) $u \in S$ and $v \in S$, and (3) $u \notin S$ and $v \in S$, or vice versa. These cases are treated using similar arguments.

We give a detailed description of case (1) which gives rise to the worst stretch factor for our algorithm. We give a brief description of other cases at the end of this section. We denote by "(p, q)-path" a simple path between two points p and q.

Case (1): In this case $u, v \in P \setminus S$. Recall that, in H, u and v are connected to their closest visible vertices of $DT_1(S)$; let u_1 and v_1 denote these vertices respectively. Therefore, in H, there is a (u, v)-path that consists of the edge (u, u_1), a (u_1, v_1)-path in $DT_1(S)$, and the edge (v_1, v); see Fig. 4-top. In the following description we prove the existence of a (u_1, v_1)-path in $DT_1(S)$ of desired length. By Lemma 3 we have $|uu_1| \leqslant 1/\sqrt{2}$ and $|vv_1| \leqslant 1/\sqrt{2}$; we will use these inequalities in our description.

Let π_u, π_v, π_u' and π_v' denote the cells containing u, v, u_1 and v_1 respectively. Depending on the identicality of these cells we can have—up to symmetry—the following five sub-cases: (i) $\pi_u = \pi_v$ or (ii) $\pi_u' = \pi_u$ or (iii) $\pi_u' = \pi_v$ or (iv) $\pi_u' = \pi_v'$ or (v) all four cells are pairwise distinct. These sub-cases are treated using similar arguments. We give a detailed description of sub-case (v) which gives rise to the worst stretch factor for our algorithm. We give a brief description of other sub-cases at the end of this section.

Assume that π_u, π_v, π_u' and π_v' are pairwise distinct. Since $|uu_1| \leqslant 1/\sqrt{2}$, π_u and π_u' are neighbors; similarly π_v and π_v' are neighbors. Since $(u, u_1) \in UDG(P)$, by properties (P2) and (P3) in Lemma 2 there exist two points $u_2, u_3 \in S$ such that $u_2 \in \pi_u'$, $u_3 \in \pi_u$, and $|u_2u_3| \leqslant 1$. Similarly, there exist two points $v_2, v_3 \in S$ such that $v_2 \in \pi_v'$, $v_3 \in \pi_v$, and $|v_2v_3| \leqslant 1$. Moreover, since $(u, v) \in UDG(P)$, there exist two points $u_4, v_4 \in S$ such that $u_4 \in \pi_u$, $v_4 \in \pi_v$, and $|u_4v_4| \leqslant 1$. See Fig. 4. It might be the case that $u_1 = u_2$, $u_3 = u_4$, $v_3 = v_4$, or $v_1 = v_2$. Since u_1 and u_2 are in the same cell, $|u_1u_2| \leqslant 1$; similarly $|u_3u_4| \leqslant 1$, $|v_1v_2| \leqslant 1$, and $|v_3v_4| \leqslant 1$. Having these distance constraints, Corollary 1 implies that in $DT_1(S)$ there exists a walk between u_1 and v_1 that consists of a (u_1, u_2)-path in $D(u_1, u_2)$, a (u_2, u_3)-path in $D(u_2, u_3)$, a (u_3, u_4)-path in $D(u_3, u_4)$, a (u_4, v_4)-path in $D(u_4, v_4)$, a (v_4, v_3)-path in $D(v_4, v_3)$, a (v_3, v_2)-path in $D(v_3, v_2)$, and a (v_2, v_1)-path in $D(v_2, v_1)$. Thus, there is a (u_1, v_1)-path in $DT_1(S)$ that lies in the union of these seven disks; see Fig. 4.

Let $\mathcal{D}$ denote the union of the seven disks. We want to obtain an upper bound on the number of cells intersected by $\mathcal{D}$. To that end, set $\mathcal{D}_u = D(u_1, u_2) \cup D(u_2, u_3) \cup D(u_3, u_4)$, and $\mathcal{D}_v = D(v_4, v_3) \cup D(v_3, v_2) \cup D(v_2, v_1)$. Define X_u as the set containing the cells π_u and π_u' and their +-neighbors. Since π_u and π_u' are neighbors, their relative positions is among configurations A and B (Fig. 3(a) and (b)); in these configurations X_u contains 8 cells. Analogously, define X_v with respect to π_v and π_v', and notice that X_v also contains 8 cells.

Claim. Each of $\mathcal{D}_u$ and $\mathcal{D}_v$ intersects at most 8 cells. Moreover, the cells that are intersected by $\mathcal{D}_u$ and $\mathcal{D}_v$ belong to X_u and X_v, respectively.

Proof. Because of symmetry, we prove this claim only for $\mathcal{D}_u$. Recall that π_u and π_u' are neighbors. If π_u and π_u' are +-neighbors, then $\mathcal{D}_u$ intersects at most 8 cells by statement 2 in Lemma 6. The proof of statement 2 also implies that these (at most 8) cells belong to X_u. If π_u and π'_u are ×-neighbors, then by property (P3) in Lemma 2, (u_2, u_3) is the shortest edge of $UDG(P)$ that runs between π'_u and π_u. Since (u_1, u) is also an edge between π'_u and π_u, we have $|u_2 u_3| \leqslant |u_1 u| \leqslant 1/\sqrt{2}$. In this case $\mathcal{D}_u$ intersects at most 8 cells by statement 1 in Lemma 6. The proof of statement 1 implies that these cells belong to X_u. □

Notice that $\mathcal{D} = \mathcal{D}_u \cup \mathcal{D}_v \cup D(u_4, v_4)$. Based on this and the above claim, in order to obtain an upper bound on the number of cells that are intersected by $\mathcal{D}$ it suffices to obtain an upper bound on the number of cells, outside $X_u \cup X_v$, that are intersected by $D(u_4, v_4)$. To that end, define X as the set containing the cells π_u and π_v and their +-neighbors, and notice that $X \subseteq X_u \cup X_v$. By statement 2 of Lemma 5, the disk $D(u_4, v_4)$ intersects at most 2 cells outside X, and hence at most 2 cells outside $X_u \cup X_v$. Therefore, the number of cells intersected by $\mathcal{D}$ is at most $|X_u \cup X_v| + 2 \leqslant 8 + 8 + 2 = 18$. Since by property (P1) in Lemma 2 each cell contains at most 20 points of S, the set $\mathcal{D}$ contains at most 360 points of S. Therefore, the (u_1, v_1)-path in $DT_1(S)$ has at most 360 vertices, and hence at most 359 edges. Thus, the (u, v)-path in H has at most 361 edges (including (u, u_1) and (v_1, v)).

With a closer look at relative positions of π_u and π_v we show that $\mathcal{D}$ in fact intersects at most 17 cells. This would imply that the (u, v)-path has at most 341 edges as claimed. To that end we consider four configurations A, B, C, and D for π_u and π_v, which we refer to them as sub-cases (v)-A, (v)-B, (v)-C, and (v)-D, respectively.

- (v)-A. In this case X_u and X_v share π_u and π_v and thus $|X_u \cup X_v| \leqslant 14$. Moreover, by the proof of Lemma 5 the disk $D(u_4, v_4)$ does not intersect any cell outside $X_u \cup X_v$; see Fig. 3(a). Thus $\mathcal{D}$ intersects at most 14 cells.
- (v)-B. In this case X_u and X_v share at least two cells (two +-neighbors of π_u and π_v) and thus $|X_u \cup X_v| \leqslant 14$. Moreover, by the proof of Lemma 5 the disk $D(u_4, v_4)$ intersects at most two cells outside $X_u \cup X_v$; see Fig. 3(b). Thus $\mathcal{D}$ intersects at most 16 cells.
- (v)-C. In this case X_u and X_v share at least one cell (one +-neighbor of π_u and π_v), and by the proof of Lemma 5 the disk $D(u_4, v_4)$ intersects at most two cells outside $X_u \cup X_v$; see Fig. 3(c). Thus $\mathcal{D}$ intersects at most 17 cells (the shaded cells in Fig. 4-top).
- (v)-D. In this case X_u and X_v may not share any cell, but by the proof of Lemma 5 the disk $D(u_4, v_4)$ intersects at most one cell outside $X_u \cup X_v$; see Fig. 3(d). Thus $\mathcal{D}$ intersects at most 17 cells (shaded in Fig. 4-bottom).

Even though $\mathcal{D}$ may intersect exactly 17 cells, in the full version of this paper [2] we show a possibility of decreasing the upper bound on the length of (u, v)-path even further.

Other Cases and Sub-cases: We gave a detailed analysis for sub-case (v) (of case (1)) where u, v, u_1, v_1 lie in distinct cells $\pi_u, \pi_v, \pi'_u, \pi'_v$. Our analysis shows the existence of a (u_1, u_4)-path in $\mathcal{D}_u$ which intersects at most 8 cells (which belong to X_u), and the existence of a (v_1, v_4)-path in $\mathcal{D}_v$ which intersects at

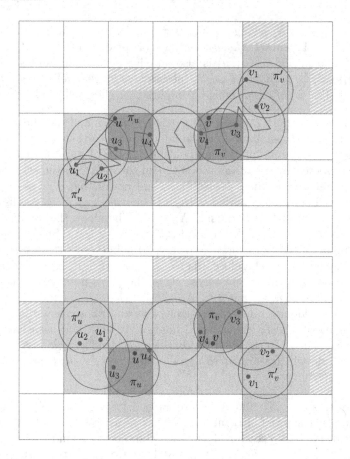

Fig. 4. Points u, v, u_1, v_1 belong to distinct cells. The cells π_u, π_v are in configurations C (top) and D (bottom). The red path corresponds to a (u_1, v_1)-path in $DT_1(S)$. (Color figure online)

most 8 cells (which belong to X_v). In the sequel we give short descriptions of case (2), case (3), and remaining sub-cases of case (1).

Recall case (1) where $u, v \in P \setminus S$. In sub-case (i) where $\pi_u = \pi_v$, the sets X_u and X_v share at least 5 cells (π_u and its four +-neighbors). Therefore, $\mathcal{D}$ intersects at most $|X_u \cup X_v| + 2 \leqslant 8 + 8 - 5 + 2 = 13$ cells. Similarly, in each of sub-cases (iii) where $\pi'_u = \pi_v$ and (iv) where $\pi'_u = \pi'_v$, the sets X_u and X_v share at least 5 cells, and thus $\mathcal{D}$ intersects at most 13 cells. In sub-case (ii) where $\pi'_u = \pi_u$, the set X_u contains 5 cells (π_u and its four +-neighbors), and thus $\mathcal{D}$ intersects at most $5 + 8 + 2 = 15$ cells. Thus, in all these remaining sub-cases, the (u, v)-path has at most 301 edges (including (u, u_1) and (v_1, v)).

Now consider case (3) where $v \in S$ or $u \in S$ but not both. By symmetry we assume that $v \in S$, and thus $u \in P \setminus S$. In this case we do not have the point v_1 nor the cell π'_v; one may assume that $v_4 = v_3 = v_2$. Thus, X_v contains at most 5 cells (π_v and its +-neighbors). By an argument similar to that of case (1), there exists a (u_1, v)-path in $DT_1(S)$ that lies in $\mathcal{D}$ which intersects at most $|X_u \cup X_v| + 2 \leqslant 8 + 5 + 2 = 15$ cells. Therefore, there is a (u, v)-path in H that has at most 300 edges (including the edge (u, u_1)).

Consider case (2) where $u, v \in S$. Since $|uv| \leqslant 1$, by Corollary 1 there is a (u, v)-path in $DT_1(S)$ that lies in $D(u, v)$. By Lemma 4, $D(u, v)$ intersects at most seven cells, and thus contains at most 140 points of S. Therefore, the (u, v)-path has at most 139 edges.

References

1. Alzoubi, K.M., Li, X., Wang, Y., Wan, P., Frieder, O.: Geometric spanners for wireless ad hoc networks. IEEE Trans. Parallel Distrib. Syst. **14**(4), 408–421 (2003)
2. Biniaz, A.: Plane hop spanners for unit disk graphs: simpler and better. arXiv:1902.10051 (2019). http://arxiv.org/abs/1902.10051
3. Bose, P., Fagerberg, R., van Renssen, A., Verdonschot, S.: On plane constrained bounded-degree spanners. In: Fernández-Baca, D. (ed.) LATIN 2012. LNCS, vol. 7256, pp. 85–96. Springer, Heidelberg (2012). https://doi.org/10.1007/978-3-642-29344-3_8
4. Bose, P., Smid, M.: On plane geometric spanners: a survey and open problems. Comput. Geom. **46**(7), 818–830 (2013)
5. Catusse, N., Chepoi, V., Vaxès, Y.: Planar hop spanners for unit disk graphs. In: Scheideler, C. (ed.) ALGOSENSORS 2010. LNCS, vol. 6451, pp. 16–30. Springer, Heidelberg (2010). https://doi.org/10.1007/978-3-642-16988-5_2
6. Chen, J., Jiang, A., Kanj, I.A., Xia, G., Zhang, F.: Separability and topology control of quasi unit disk graphs. Wirel. Netw. **17**(1), 53–67 (2011). Also in INFOCOM 2007
7. Dobkin, D.P., Friedman, S.J., Supowit, K.J.: Delaunay graphs are almost as good as complete graphs. Discrete Comput. Geom. **5**, 399–407 (1990). Also in FOCS 1987
8. Eppstein, D.: Spanning trees and spanners. Technical report, Information and Computer Science, University of California, Irvine (1996)
9. Fürer, M., Kasiviswanathan, S.P.: Spanners for geometric intersection graphs with applications. J. Comput. Geom. **3**(1), 31–64 (2012)
10. Gao, J., Guibas, L.J., Hershberger, J., Zhang, L., Zhu, A.: Geometric spanners for routing in mobile networks. IEEE J. Sel. Areas Commun. **23**(1), 174–185 (2005). Also in MobiHoc 2001
11. Hurtado, F., Kano, M., Rappaport, D., Tóth, C.D.: Encompassing colored planar straight line graphs. Comput. Geom.: Theory Appl. **39**(1), 14–23 (2008). Also in CCCG 2004
12. Kanj, I.A., Perkovic, L.: On geometric spanners of Euclidean and unit disk graphs. In: Proceedings of the 25th Annual Symposium on Theoretical Aspects of Computer Science (STACS), pp. 409–420 (2008)
13. Li, X.: Algorithmic, geometric and graphs issues in wireless networks. Wirel. Commun. Mob. Comput. **3**(2), 119–140 (2003)
14. Li, X., Călinescu, G., Wan, P.: Distributed construction of planar spanner and routing for ad hoc wireless networks. In: Proceedings of the 21st Annual Joint Conference of the IEEE Computer and Communications Societies (INFOCOM), pp. 1268–1277 (2002)

15. Li, X., Wang, Y.: Efficient construction of low weighted bounded degree planar spanner. Int. J. Comput. Geom. Appl. **14**(1–2), 69–84 (2004). Also in COCOON 2003
16. Narasimhan, G., Smid, M.: Geometric Spanner Networks. Cambridge University Press, Cambridge (2007)
17. Rajaraman, R.: Topology control and routing in ad hoc networks: a survey. SIGACT News **33**(2), 60–73 (2002)
18. Yan, C., Xiang, Y., Dragan, F.F.: Compact and low delay routing labeling scheme for Unit Disk Graphs. Comput. Geom.: Theory Appl. **45**(7), 305–325 (2012). Also in WADS 2009

On the Minimum Consistent
Subset Problem

Ahmad Biniaz[1]([⊠]), Sergio Cabello[2], Paz Carmi[3], Jean-Lou De Carufel[4],
Anil Maheshwari[5], Saeed Mehrabi[5], and Michiel Smid[5]

[1] David R. Cheriton School of Computer Science,
University of Waterloo, Waterloo, Canada
ahmad.biniaz@gmail.com
[2] Department of Mathematics, IMFM and FMF,
University of Ljubljana, Ljubljana, Slovenia
[3] Department of Computer Science,
Ben-Gurion University of the Negev, Beer-Sheva, Israel
[4] School of Electrical Engineering and Computer Science,
University of Ottawa, Ottawa, Canada
[5] School of Computer Science, Carleton University, Ottawa, Canada

Abstract. Let P be a set of n colored points in the plane. Introduced
by Hart [7], a *consistent subset* of P, is a set $S \subseteq P$ such that for every
point p in $P \setminus S$, the closest point of p in S has the same color as p.
The consistent subset problem is to find a consistent subset of P with
minimum cardinality. This problem is known to be NP-complete even
for two-colored point sets. Since the initial presentation of this problem,
aside from the hardness results, there has not been significant progress
from the algorithmic point of view. In this paper we present the following
algorithmic results:

1. The first subexponential-time algorithm for the consistent subset
   problem.
2. An $O(n \log n)$-time algorithm that finds a consistent subset of size
   two in two-colored point sets (if such a subset exists). Towards our
   proof of this running time we present a deterministic $O(n \log n)$-time
   algorithm for computing a variant of the compact Voronoi diagram;
   this improves the previously claimed expected running time.
3. An $O(n \log^2 n)$-time algorithm that finds a minimum consistent subset in two-colored point sets where one color class contains exactly
   one point; this improves the previous best known $O(n^2)$ running
   time which is due to Wilfong (SoCG 1991).
4. An $O(n)$-time algorithm for the consistent subset problem on
   collinear points that are given from left to right; this improves the
   previous best known $O(n^2)$ running time.
5. A non-trivial $O(n^6)$-time dynamic programming algorithm for the
   consistent subset problem on points arranged on two parallel lines.

To obtain these results, we combine tools from paraboloid lifting, planar
separators, additively-weighted Voronoi diagrams with respect to convex
distance functions, point location in farthest-point Voronoi diagrams,
range trees, minimum covering of a circle with arcs, and several geometric
transformations.

© Springer Nature Switzerland AG 2019
Z. Friggstad et al. (Eds.): WADS 2019, LNCS 11646, pp. 155–167, 2019.
https://doi.org/10.1007/978-3-030-24766-9_12

Keywords: Consistent subset · Colored points · Planar separator · Voronoi diagram · Paraboloid lifting · Range tree · Circle covering

1   Introduction

One of the important problems in pattern recognition is to classify new objects according to the current objects using the nearest neighbor rule. Motivated by this problem, in 1968, Hart [7] introduced the notion of *consistent subset* as follows. For a set P of colored points[1] in the plane, a set $S \subseteq P$ is a consistent subset if for every point $p \in P \setminus S$, the closest point of p in S has the same color as p. The *consistent subset problem* asks for a consistent subset with minimum cardinality. Formally, we are given a set P of n points in the plane that is partitioned into $P_1, \ldots, P_k$, with $k \geqslant 2$, and the goal is to find a smallest set $S \subseteq P$ such that for every $i \in \{1, \ldots, k\}$ it holds that if $p \in P_i$ then the nearest neighbor of p in S belongs to P_i. It is implied by the definition that S should contain at least one point from every P_i. To keep the terminology consistent with some recent works on this problem we will be dealing with colored points instead of partitions, that is, we assume that the points of P_i are colored i. Following this terminology, the consistent subset problem asks for a smallest subset S of P such that the color of every point $p \in P \setminus S$ is the same as the color of its closest point in S. The notion of consistent subset has a close relation with Voronoi diagrams, a well-known structure in computational geometry. Consider the Voronoi diagram of a subset S of P. Then, S is a consistent subset of P if and only if for every point $s \in S$ it holds that the points of P, that lie in the Voronoi cell of s, have the same color as s.

Since the initial presentation of this problem in 1968, there has not been significant progress from the algorithmic point of view. Although there were several attempts for developing algorithms, they either did not guarantee the optimality [5,7,13] or had exponential running time [12]. In SoCG 1991, Wilfong [13] proved that the consistent subset problem is NP-complete if the input points are colored by at least three colors—the proof is based on the NP completeness of the disc cover problem [11]. He further presented a technically-involved $O(n^2)$-time algorithm for a special case of two-colored input points where one point is red and all other points are blue; his elegant algorithm transforms the consistent subset problem to the problem of covering points with disks which in turn is transformed to the problem of covering a circle with arcs. It has been recently proved, by Khodamoradi et al. [9], that the consistent subset problem with two colors is also NP-complete—the proof is by a reduction from the planar rectilinear monotone 3-SAT [2]. Observe that the one color version of the problem is trivial because every single point is a consistent subset. More recently, Banerjee et al. [1] showed that the consistent subset problem on collinear points, i.e., points that lie on a straight line, can be solved optimally in $O(n^2)$ time.

Recently, Gottlieb et al. [6] studied a two-colored version of the consistent subset problem—referred to as the nearest neighbor condensing problem—where

---

[1] In some previous works the points have labels, as opposed to colors.

the points come from a metric space. They prove a lower bound for the hardness of approximating a minimum consistent subset; this lower bound includes two parameters: the doubling dimension of the space and the ratio of the minimum distance between points of opposite colors to the diameter of the point set. Moreover, for this two-colored version of the problem, they give an approximation algorithm whose ratio almost matches the lower bound.

In a related problem, called the selective subset problem, the goal is to find the smallest subset S of P such that for every $p \in P_i$ the nearest neighbor of p in $S \cup (P \setminus P_i)$ belongs to P_i. Wilfong [13] showed that this problem is NP-complete even with two colors. See [1] for some recent progress on this problem.

In this paper we study the consistent subset problem. We improve some previous results and present some new results. To obtain these results, we combine tools from planar separators, additively-weighted Voronoi diagrams with respect to a convex distance function, point location in farthest-point Voronoi diagrams, range trees, paraboloid lifting, minimum covering of a circle with arcs, and several geometric transformations. In the full version of this paper [4] we present the first subexponential-time algorithm for the consistent subset problem. We use a recursive separator-based technique that was introduced in 1993 by Hwang et al. [8], and then extended by Marx and Pilipczuk [10]. The application of this technique in our setting is not straightforward and requires technical details. The following theorem summarizes this result.

Theorem 1. *A minimum consistent subset of n colored points in the plane can be computed in $n^{O(\sqrt{k})}$ time, where k is the size of the minimum consistent subset.*

In Sect. 2 we present an $O(n \log n)$-time algorithm that finds a consistent subset of size two in two-colored point sets (if such a subset exists); this is obtained by transforming the problem into a point-cone incidence problem in dimension three. Towards our proof of this running time, we present a deterministic $O(n \log n)$-time algorithm for computing a variant of the compact Voronoi diagram (see the full paper [4]); this improves the $O(n \log n)$ expected running time of the randomized algorithm of Bhattacharya et al. [3]. In particular, we prove the following theorem.

Theorem 2. *Let C be a cone in $\mathbb{R}^3$ with non-empty interior that is given as the intersection of n halfspaces. Given n translations of C and a set of n points in $\mathbb{R}^3$, we can decide in $O(n \log n)$ time whether or not there is a point-cone incidence.*

In Sect. 3 we revisit the case where one point is red and all other points are blue. We give an $O(n \log^2 n)$-time algorithm for this case, thereby improving the previous $O(n^2)$ running time of [13]. In Sect. 4 we present an $O(n)$-time algorithm for collinear points; this improves the previous running time by a factor of $\Theta(n)$. We also present a non-trivial $O(n^6)$-time dynamic programming algorithm for points arranged on two parallel lines.

2   Consistent Subset of Size Two

In this section we investigate the existence of a consistent subset of size two in a set of bichromatic points where every point is colored by one of the two colors, say red and blue. Before stating the problem formally we introduce some terminology. For a set P of points in the plane, we denote the convex hull of P by $CH(P)$. For two points p and q in the plane, we denote the straight-line segment between p and q by pq, and the perpendicular bisector of pq by $\beta(p, q)$.

Let R and B be two disjoint sets of total n points in the plane such that the points of R are colored red and the points of B are colored blue. We want to decide whether or not $R \cup B$ has a consistent subset of size two. Moreover, if the answer is positive, then we want to find such points, i.e., a red point $r \in R$ and a blue point $b \in B$ such that all red points are closer to r than to b, and all blue points are closer to b than to r. Alternatively, we want to find a pair of points $(r, b) \in R \times B$ such that $\beta(r, b)$ separates $CH(R)$ and $CH(B)$. This problem can be solved in $O(n^2 \log n)$ time by trying all the $O(n^2)$ pairs $(r, b) \in R \times B$; for each pair (r, b) we can verify, in $O(\log n)$ time, whether or not $\beta(r, b)$ separates $CH(R)$ and $CH(B)$. In this section we show how to solve this problem in time $O(n \log n)$. To that end, we assume that $CH(R)$ and $CH(B)$ are disjoint, because otherwise there is no such pair (r, b).

It might be tempting to believe that a solution of this problem contains points only from the boundaries of $CH(R)$ and $CH(B)$. However, this is not necessarily the case; in Fig. 1, the only minimum consistent subset contains r and b which are in the interiors of $CH(R)$ and $CH(B)$. Also, due to the close relation between Voronoi diagrams and Delaunay triangulations, one may believe that a solution is defined by the two endpoints of an edge in the Delaunay triangulation of $R \cup B$. This is not necessarily the case either; the endpoints of green edges in Fig. 1, which are the Delaunay edges between R and B, do not introduce any solution.

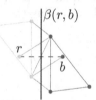

Fig. 1. The set $\{r, b\}$ is the minimum consistent subset of the given eight points. (Color figure online)

Let R' and B' be the subsets of R and B on the boundaries of $CH(R)$ and $CH(B)$, respectively; see Fig. 2. For two points p and q in the plane, let $D(p, q)$ be the closed disk that is centered at p and has q on its boundary.

Lemma 1. *For every two points $r \in R$ and $b \in B$, the bisector $\beta(r, b)$ separates R and B if and only if*

(i) $\forall r' \in R' :$ $b \notin D(r', r)$, and
(ii) $\forall b' \in B' :$ $b \in D(b', r)$.

Proof. For the direct implication since $\beta(r, b)$ separates R and B, every red point r' (and in particular every point in R') is closer to r than to b; this implies that $D(r', r)$ does not contain b and thus (i) holds. Also, every blue point b' (and in particular every point in B') is closer to b than to r; this implies that $D(b', r)$ contains b and thus (ii) holds. See Fig. 2.

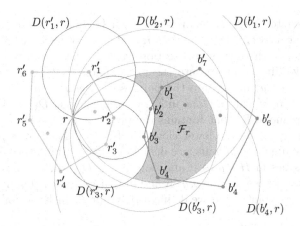

Fig. 2. $R' = \{r'_1, \ldots, r'_6\}$ and $B' = \{b'_1, \ldots, b'_7\}$ are the points of R and B on boundaries of $CH(R)$ and $CH(B)$. The feasible region F_r for point r is shaded. (Color figure online)

Now we prove the converse implication by contradiction. Assume that both (i) and (ii) hold for some $r \in R$ and some $b \in B$, but the bisector $\beta(r, b)$ does not separate R and B. After a suitable rotation we may assume that $\beta(r, b)$ is vertical, r is to the left side of $\beta(r, b)$ and b is to the right side of $\beta(r, b)$. Since $\beta(r, b)$ does not separate R and B, there exists either a point of R to the right side of $\beta(r, b)$, or a point of B to the left side of $\beta(r, b)$. If there is a point of R to the right side of $\beta(r, b)$ then there is also a point $r' \in R'$ to the right side of $\beta(r, b)$. In this case r' is closer to b than to r, and thus the disk $D(r', r)$ contains b which contradicts (i). If there is a point of B to the left side of $\beta(r, b)$ then there is also a point $b' \in B'$ to the left side of $\beta(r, b)$. In this case b' is closer to r than to b and thus $D(b', r)$ does not contain b which contradicts (ii). $\quad\square$

Lemma 1 implies that a pair $(r, b) \in R \times B$ is a consistent subset of $R \cup B$ if and only if every point of R' is closer to r than to b, and every point of B' is closer to b than to r. This lemma does not imply that r and b are necessarily in R' and B'. By symmetry, Lemma 1 holds even if we swap the roles of r, r', R' with b, b', B' in (i) and (ii), however we do not use this in the rest of our proof.

For every red point $r \in R$ we define a *feasible region* $\mathcal{F}_r$ as follows:

$$\mathcal{F}_r = \left(\bigcap_{b' \in B'} D(b', r) \right) \setminus \left(\bigcup_{r' \in R'} D(r', r) \right).$$

See Fig. 2 for an illustration of a feasible region. Lemma 1, together with this definition, imply the following corollary.

Corollary 1. *For every two points $r \in R$ and $b \in B$, the bisector $\beta(r, b)$ separates R and B if and only if $b \in \mathcal{F}_r$.*

This corollary reduces our original decision problem to the following question.

Question 1. Is there a blue point $b \in B$ such that b lies in the feasible region $\mathcal{F}_r$ of some red point $r \in R$?

If the answer to Question 1 is positive then $\{r, b\}$ is a consistent subset for $R \cup B$, and if the answer is negative then $R \cup B$ does not have a consistent subset with two points. In the rest of this section we show how to answer Question 1. To that end, we lift the plane onto the paraboloid $z = x^2 + y^2$ by projecting every point $s = (x, y)$ in $\mathbb{R}^2$ onto the point $\hat{s} = (x, y, x^2 + y^2)$ in $\mathbb{R}^3$. This lift projects a circle in $\mathbb{R}^2$ onto a plane in $\mathbb{R}^3$. Consider a disk $D(p, q)$ in $\mathbb{R}^2$ and let $\pi(p, q)$ be the plane in $\mathbb{R}^3$ that contains the projection of the boundary circle of $D(p, q)$. Let $H^-(p, q)$ be the lower closed halfspace defined by $\pi(p, q)$, and let $H^+(p, q)$ be the upper open halfspace defined by $\pi(p, q)$. For every point $s \in \mathbb{R}^2$, its projection $\hat{s}$ lies in $H^-(p, q)$ if and only if $s \in D(p, q)$, and lies in $H^+(p, q)$ otherwise. Moreover, $\hat{s}$ lies in $\pi(p, q)$ if and only if s is on the boundary circle of $D(p, q)$. For every point $r \in R$ we define a polytope $\mathcal{C}_r$ in $\mathbb{R}^3$ as follows:

$$\mathcal{C}_r = \left(\bigcap_{b' \in B'} H^-(b', r) \right) \cap \left(\bigcap_{r' \in R'} H^+(r', r) \right).$$

By the above discussion, Corollary 1 can be translated to the following corollary.

Corollary 2. *For every two points $r \in R$ and $b \in B$, the bisector $\beta(r, b)$ separates R and B if and only if $\hat{b} \in \mathcal{C}_r$.*

This corollary, in turn, translates Question 1 to the following question.

Question 2. Is there a blue point $b \in B$ such that its projection $\hat{b}$ lies in the polytope $\mathcal{C}_r$ for some red point $r \in R$?

Now, we are going to answer Question 2. The polytope $\mathcal{C}_r$ is the intersection of some halfspaces, each of which has $\hat{r}$ on its boundary plane. Therefore, $\mathcal{C}_r$ is a cone in $\mathbb{R}^3$ with apex $\hat{r}$; see Fig. 4. Recall that $|R \cup B| = n$, however, for the purposes of worst-case running-time analysis and to simplify indexing, we will index the red points, and also the blue points, from 1 to n. Let $r_1, r_2, , \ldots, r_n$ be the points of R. For every point $r_i \in R$, let τ_i be the translation that brings $\hat{r}_1$ to $\hat{r}_i$. Notice that τ_1 is the identity transformation. In the rest of this section we will write $\mathcal{C}_i$ for $\mathcal{C}_{r_i}$.

Lemma 2. *For every point $r_i \in R$, the cone $\mathcal{C}_i$ is the translation of $\mathcal{C}_1$ with respect to τ_i.*

Proof. For a circle C in $\mathbb{R}^2$, let π_C denote the plane in $\mathbb{R}^3$ that C translates onto. For every two concentric circles C_1 and C_i in $\mathbb{R}^2$ it holds that π_{C_1} and π_{C_i} are parallel; see Fig. 3. It follows that, if C_1 passes through the point r_1, and C_i passes through the point r_i, then π_{C_i} is obtained from π_{C_1} by the translation τ_i that brings $\hat{r}_1$ to $\hat{r}_i$, that is $\tau_i(\pi_{C_1}) = \pi_{C_i}$. A similar argument holds also for the halfspaces defined by π_{C_1} and π_{C_i}. Since for every $a \in R' \cup B'$ the disks $D(a, r_1)$ and $D(a, r_i)$ are concentric and the boundary of

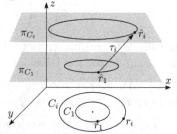

Fig. 3. Parallel planes π_{C_1} and π_{C_i} in $\mathbb{R}^3$ for two concentric circles C_1 and C_i in $\mathbb{R}^2$.

$D(a, r_1)$ passes through r_1 and the boundary of $D(a, r_i)$ passes through r_i, it follows that $\tau_i(H^+(a, r_1)) = H^+(a, r_i)$ and $\tau_i(H^-(a, r_1)) = H^-(a, r_i)$. Since a translation of a polytope is obtained by translating each of the halfspaces defining it, we have $\tau_i(\mathcal{C}_1) = \mathcal{C}_i$ as depicted in Fig. 4. □

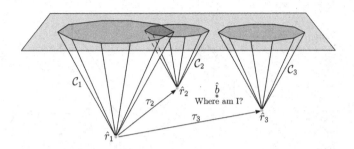

Fig. 4. The cones $\mathcal{C}_2$ and $\mathcal{C}_3$ are the translations of $\mathcal{C}_1$ with respect to τ_2 and τ_3. (Color figure online)

Based on Lemma 2, to answer Question 2 it suffices to solve the following problem: Given a cone $\mathcal{C}_1$ that is defined by n halfspaces, n translations of $\mathcal{C}_1$, and a set of n points, we want to decide whether or not there is a point in some cone (see Fig. 4). This can be verified in $O(n \log n)$ time by Theorem 2. This is the end of our constructive proof. The following theorem summarizes our result in this section.

Theorem 3. *Given a set of n bichromatic points in the plane, in $O(n \log n)$ time, we can compute a consistent subset of size two (if such a set exists).*

3 One Red Point

In this section we revisit the consistent subset problem for the case where one input point is red and all other points are blue. Let P be a set of n points in the plane consisting of a red point and $n - 1$ blue points. Observe that any consistent subset of P contains the only red point and some blue points. In his seminal work in SoCG 1991, Wilfong [13] showed that P has a consistent subset of size at most seven (including the red point); this implies an $O(n^6)$-time brute force algorithm for this problem. Wilfong showed how to solve this problem in $O(n^2)$-time; his elegant algorithm transforms the consistent subset problem to the problem of covering points with disks which in turn is transformed to the problem of covering a circle with arcs. The running time of his algorithm is dominated by the transformation to the circle covering problem which involves computation of $n - 1$ arcs in $O(n^2)$ time; all other transformations together with the solution of the circle covering problem take $O(n \log n)$ time ([13, Lemma 19 and Theorem 9]).

We first introduce the circle covering problem, then we give a summary of Wilfong's transformation to this problem, and then we show how to perform this transformation in $O(n \log^2 n)$ time which implies the same running time for the entire algorithm. We emphasis that the most involved part of the algorithm, which is the correctness proof of this transformation, is due to Wilfong.

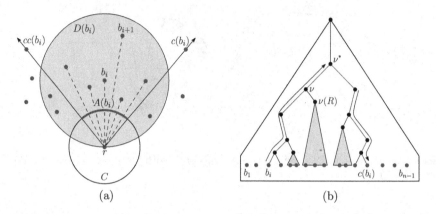

Fig. 5. (a) Transformation to the circle covering problem. (b) The range tree T. (Color figure online)

Let C be a circle and let $\mathcal{A}$ be a set of arcs covering the entire C. The *circle covering* problem asks for a subset of $\mathcal{A}$, with minimum cardinality, that covers the entire C.

Wilfong's algorithm starts by mapping input points to the projective plane, and then transforming (in two stages) the consistent subset problem to the circle covering problem. Let P denote the set of points after the mapping, and let r denote the only red point of P. The transformation, which is depicted in Fig. 5(a), proceeds as follows. Let C be a circle centered at r that does not contain any blue point. Let $b_1, b_2, \ldots, b_{n-1}$ be the blue points in clockwise circular order around r (b_1 is the first clockwise point after b_{n-1}, and b_{n-1} is the first counterclockwise point after b_1). For each point b_i, let $D(b_i)$ be the disk of radius $|rb_i|$ centered at b_i. Define $cc(b_i)$ to be the first counterclockwise point (measured from b_i) that is not in $D(b_i)$, and similarly define $c(b_i)$ to be the first clockwise point that is not in $D(b_i)$. Denote by $A(b_i)$ the open arc of C that is contained in the wedge with counterclockwise boundary ray from r to $cc(b_i)$ and the clockwise boundary ray from r to $c(b_i)$.[2] Let $\mathcal{A}$ be the set of all arcs $A(b_i)$; since blue points are assumed to be in circular order, $\mathcal{A}$ covers the entire C. Wilfong proved that our instance of the consistent subset problem is equivalent to the problem of covering C with

---

[2] Wilfong shrinks the endpoint of $A(b_i)$ that corresponds to $cc(b_i)$ by half the clockwise angle from $cc(b_i)$ to the next point, and shrinks the other endpoint of $A(b_i)$ by half the counterclockwise angle from $c(b_i)$ to the previous point.

$\mathcal{A}$. The running time of his algorithm is dominated by the computation of $\mathcal{A}$ in $O(n^2)$ time. We show how to compute $\mathcal{A}$ in $O(n \log^2 n)$ time.

In order to find each arc $A(b_i)$ it suffices to find the points $cc(b_i)$ and $c(b_i)$. Having the clockwise ordering of points around r, one can find these points in $O(n)$ time for each b_i, and consequently in $O(n^2)$ time for all b_i's. In the rest of this section we show how to find $c(b_i)$ for all b_i's in $O(n \log^2 n)$ time; the points $cc(b_i)$ can be found in a similar fashion.

By the definition of $c(b_i)$ all points of the sequence $b_{i+1}, \ldots, c(b_i)$, except $c(b_i)$, lie inside $D(b_i)$. Therefore among all points $b_{i+1}, \ldots, c(b_i)$, the point $c(b_i)$ is the farthest from b_i. This implies that in the farthest-point Voronoi diagram of $b_{i+1}, \ldots, c(b_i)$, the point b_i lies in the cell of $c(b_i)$. To exploit this property of $c(b_i)$, we construct a 1-dimensional range tree T on all blue points based on their clockwise order around r; blue points are stored at the leaves of T as in Fig. 5(b). At every internal node ν of T we store the farthest-point Voronoi diagram of the blue points that are stored at the leaves of the subtree rooted at ν; we refer to this diagram by $\text{FVD}(\nu)$. This data structure can be computed in $O(n \log^2 n)$ time because T has $O(\log n)$ levels and in each level we compute farthest-point Voronoi diagrams of total $n - 1$ points. To simplify our following description, at the moment we assume that $b_1, \ldots, b_{n-1}$ is a linear order. At the end of this section, in Remark 1, we show how to deal with the circular order.

We use the above data structure to find each point $c(b_i)$ in $O(\log^2 n)$ time. To that end, we walk up the tree from the leaf containing b_i (first phase), and then walk down the tree (second phase) as described below; also see Fig. 5(b). For every internal node ν, let $\nu(L)$ and $\nu(R)$ denote its left and right children, respectively. In the first phase, for every internal node ν in the walk, we locate the point b_i in $\text{FVD}(\nu(R))$ and find the point b_f that is farthest from b_i. If b_f lies in $D(b_i)$ then also does every point stored at the subtree of $\nu(R)$. In this case we continue walking up the tree and repeat the above point location process until we find, for the first time, the node ν^* for which b_f does not lie in $D(b_i)$. To this end we know that $c(b_i)$ is among the points stored at $\nu^*(R)$. Now we start the second phase and walk down the tree from $\nu^*(R)$. For every internal node ν in this walk, we locate b_i in $\text{FVD}(\nu(L))$ and find the point b_f that is farthest from b_i. If b_f lies in $D(b_i)$, then also does every point stored at $\nu(L)$, and hence we go to $\nu(R)$, otherwise we go to $\nu(L)$. At the end of this phase we land up in a leaf of T, which stores $c(b_i)$. The entire walk has $O(\log n)$ nodes and at every node we spend $O(\log n)$ time for locating b_i. Thus the time to find $c(b_i)$ is $O(\log^2 n)$. Therefore, we can find all $c(b_i)$'s in $O(n \log^2 n)$ total time.

Theorem 4. *A minimum consistent subset of n points in the plane, where one point is red and all other points are blue, can be computed in $O(n \log^2 n)$ time.*

Remark 1. To deal with the circular order $b_1, \ldots, b_{n-1}$, we build the range tree T with $2(n-1)$ leaves $b_1, \ldots, b_{n-1}, b_1, \ldots, b_{n-1}$. For a given b_i, the point $c(b_i)$ can be any of the points $b_{i+1}, \ldots, b_{n-1}, b_1, \ldots, b_{i-1}$. To find $c(b_i)$, we first follow the path from the root of T to the leftmost leaf that stores b_i, and then from that leaf we start looking for $c(b_i)$ as described above.

4   Restricted Point Sets

In this section we present polynomial-time algorithms for the consistent subset problem on collinear points and on points that are placed on two parallel lines.

4.1   Collinear Points

Let P be a set of n colored points on the x-axis, and let $p_1, \ldots, p_n$ be the sequence of these points from left to right. We present a dynamic programming algorithm that solves the consistent subset problem on P in $O(n)$-time; this improves the previous quadratic-time algorithm of Banerjee et al. [1]. To simplify the description of our algorithm we add a point p_{n+1} very far (at distance at least $|p_1p_n|$) to the right of p_n. We set the color of p_{n+1} to be different from that of p_n. Observe that every solution for $P \cup \{p_{n+1}\}$ contains p_{n+1}. Moreover, by removing p_{n+1} from any optimal solution of $P \cup \{p_{n+1}\}$ we obtain an optimal solution for P. Therefore, to compute an optimal solution for P, we first compute an optimal solution for $P \cup \{p_{n+1}\}$ and then remove p_{n+1}.

Our algorithm maintains a table T with $n+1$ entries $T(1), \ldots, T(n+1)$. Each table entry $T(k)$ represents the number of points in a minimum consistent subset of $P_k = \{p_1, \ldots, p_k\}$ provided that p_k is in this subset. The number of points in an optimal solution for P will be $T(n+1) - 1$; the optimal solution itself can be recovered from T. In the rest of this section we show how to solve a subproblem with input P_k provided that p_k should be in the solution (thereby in the rest of this section the phrase "solution of P_k" refers to a solution that contains p_k). In fact, we show how to compute $T(k)$, by a bottom-up dynamic programming algorithm that scans the points from left to right. If P_k is monochromatic, then the optimal solution contains only p_k, and thus, we set $T(k) = 1$. Hereafter assume that P_k is not monochromatic. Consider the partition of P_k into maximal blocks of consecutive points such that the points in each block have the same color. Let $B_1, B_2, \ldots, B_{m-1}, B_m$ denote these blocks from left to right, and notice that p_k is in B_m. Assume that the points in B_m are red and the points in B_{m-1} are blue. Let p_y be the leftmost point in B_{m-1}; see Fig. 6(a). Any optimal solution for P_k contains at least one point from $\{p_y, \ldots, p_{k-1}\}$; let p_i be the rightmost such point (p_i can be either red or blue). Then, $T(k) = T(i) + 1$. Since we do not know the index i, we try all possible values in $\{y, \ldots, k-1\}$ and select one that produces a *valid* solution, and that minimizes $T(k)$:

$$T(k) = \min\{T(i) + 1 \mid i \in \{y, \ldots, k-1\} \text{ and } i \text{ produces a valid solution}\}.$$

The index i produces a valid solution (or p_i is valid) if one of the following conditions hold:

(i) p_i is red, or

(ii) p_i is blue, and for every $j \in \{i+1, \ldots, k-1\}$ it holds that if p_j is blue then p_j is closer to p_i than to p_k, and if p_j is red then p_j is closer to p_k than to p_i.

If (i) holds then p_i and p_k have the same color. In this case the validity of our solution for P_k is ensured by the validity of the solution of P_i. If (ii) holds then p_i and p_k have distinct colors. In this case the validity of our solution for P_k depends on the colors of points $p_{i+1}, \ldots, p_{k-1}$. To verify the validity in this case, it suffices to check the colors of only two points that are to the left and to the right of the mid-point of the segment $p_i p_k$. This can be done in $O(|B_{m-1}|)$ time for all blue points in B_{m-1} while scanning them from left to right. Thus, $T(k)$ can be computed in $O(k)$ time because $|B_{m-1}| = O(k)$. Therefore, the total running time of the above algorithm is $O(n^2)$.

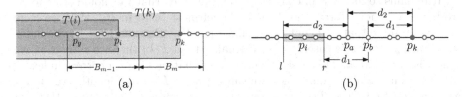

Fig. 6. (a) Illustration of the computation of $T(k)$ from $T(i)$. (b) Any blue point in the range $[l, r]$ is valid. (Color figure online)

We are now going to show how to compute $T(k)$ in constant time, which in turn improves the total running time to $O(n)$. To that end we first prove the following lemma.

Lemma 3. *Let $s \in \{1, \ldots, m\}$ be an integer, $p_i, p_{i+1}, \ldots, p_j$ be a sequence of points in B_s, and $x \in \{i, \ldots, j\}$ be an index for which $T(x)$ is minimum. Then, $T(j) \leqslant T(x) + 1$.*

Proof. To verify this inequality, observe that by adding p_j to the optimal solution of P_x we obtain a valid solution (of size $T(x) + 1$) for P_j. Therefore, any optimal solution of P_j has at most $T(x) + 1$ points, and thus $T(j) \leqslant T(x) + 1$. □

At every point p_j, in every block B_s, we store the index i of the first point p_i to the left of p_j where $p_i \in B_s$ and $T(i)$ is strictly smaller than $T(j)$; if there is no such point p_i then we store j at p_j. These indices can be maintained in linear time while scanning the points from left to right. We use these indices to compute $T(k)$ in constant time as described below.

Notice that if the minimum, in the above calculation of $T(k)$, is obtained by a red point in B_m then it always produces a valid solution, but if the minimum is obtained by a blue point then we need to verify its validity. In the former case, it follows from Lemma 3 that the smallest $T(\cdot)$ for red points in $B_m \setminus \{p_k\}$ is obtained either by p_{k-1} or by the point whose index is stored at p_{k-1}. Therefore we can find the smallest $T(\cdot)$ in constant time. Now consider the latter case where the minimum is obtained by a blue point in B_{m-1}. Let p_a be the rightmost point of B_{m-1}, and let p_b be the leftmost endpoint of B_m. Set $d_1 = |p_b p_k|$ and $d_2 = |p_a p_k|$ as depicted in Fig. 6(b). Set $l = x(p_a) - d_2$ and $r = x(p_b) - d_1$, where

$x(p_a)$ and $x(p_b)$ are the x-coordinates of p_a and p_b. Any point $p_i \in B_{m-1}$ that is to the right of r is invalid because otherwise p_b would be closer to p_i than to p_k. Any point $p_i \in B_{m-1}$ that is to the left of l is also invalid because otherwise p_a would be closer to p_k than to p_i. However, every point $p_i \in B_{m-1}$, that is in the range $[l, r]$, is valid because it satisfies condition (ii) above. Thus, to compute $T(k)$ it suffices to find a point of B_{m-1} in range $[l, r]$ with the smallest $T(\cdot)$. By slightly abusing notation, let p_r be the rightmost point of B_{m-1} in range $[l, r]$. It follows from Lemma 3 that the smallest $T(\cdot)$ is obtained either by p_r or by the point whose index is stored at p_r. Thus, in this case also, we can find the smallest $T(\cdot)$ in constant time.

It remains to identify, in constant time, the index that we should store at p_k (to be used in next iterations). If p_k is the leftmost point in B_m, then we store k at p_k. Assume that p_k is not the leftmost point in B_m, and let x be the index stored at p_{k-1}. In this case, if $T(x)$ is smaller than $T(k)$ then we store x at p_k, otherwise we store k. This assignment ensures that p_k stores a correct index.

Based on the above discussion we can compute $T(k)$ and identify the index at p_k in constant time. Therefore, our algorithm computes all values of $T(\cdot)$ in $O(n)$ total time. The following theorem summarizes our result in this section.

Theorem 5. *A minimum consistent subset of n collinear colored points can be computed in $O(n)$ time, provided that the points are given from left to right.*

4.2   Points on Two Parallel Lines

Given a set of n colored points on two parallel lines, in the full version of this paper [4] we present an algorithm that finds a minimum consistent subset in $O(n^6)$ time. Our algorithm uses a top-down dynamic programming technique. Although the technique is standard, its application to our problem is non-trivial and rather involved. In several places, our algorithm produces collinear instances of the problem that have been discussed in Sect. 4.1. The following theorem summarizes this result.

Theorem 6. *A minimum consistent subset of n colored points on two parallel lines can be computed in $O(n^6)$ time.*

In this setting (points on two parallel lines), we also study the bichromatic version of the problem where each of the n points is colored either red or blue, and where all red points lie on one line and all blue points lie on the other line. We give another involved dynamic programming algorithm that finds a minimum consistent subset, for this version of the problem, in $O(n^4)$ time (see the full paper [4]).

Theorem 7. *Let P be a set of n bichromatic points on two parallel lines, such that all points on the same line have the same color. Then, a minimum consistent subset of P can be computed in $O(n^4)$ time.*

Acknowledgement. This work initiated at the *Sixth Annual Workshop on Geometry and Graphs*, March 11–16, 2018, at the Bellairs Research Institute of McGill University, Barbados. The authors are grateful to the organizers and to the participants of this workshop.

Ahmad Biniaz was supported by NSERC Postdoctoral Fellowship. Sergio Cabello was supported by the Slovenian Research Agency, program P1-0297 and projects J1-8130, J1-8155. Paz Carmi was supported by grant 2016116 from the United States – Israel Binational Science Foundation. Jean-Lou De Carufel, Anil Maheshwari, and Michiel Smid were supported by NSERC. Saeed Mehrabi was supported by NSERC and by Carleton-Fields Postdoctoral Fellowship.

References

1. Banerjee, S., Bhore, S., Chitnis, R.: Algorithms and hardness results for nearest neighbor problems in bicolored point sets. In: Bender, M.A., Farach-Colton, M., Mosteiro, M.A. (eds.) LATIN 2018. LNCS, vol. 10807, pp. 80–93. Springer, Cham (2018). https://doi.org/10.1007/978-3-319-77404-6_7

2. de Berg, M., Khosravi, A.: Optimal binary space partitions for segments in the plane. Int. J. Comput. Geom. Appl. **22**(3), 187–206 (2012)

3. Bhattacharya, B., Bishnu, A., Cheong, O., Das, S., Karmakar, A., Snoeyink, J.: Computation of non-dominated points using compact Voronoi diagrams. In: Rahman, M.S., Fujita, S. (eds.) WALCOM 2010. LNCS, vol. 5942, pp. 82–93. Springer, Heidelberg (2010). https://doi.org/10.1007/978-3-642-11440-3_8

4. Biniaz, A., Cabello, S., Carmi, P., De Carufel, J.-L., Maheshwari, A., Mehrabi, S., Smid, M.: On the minimum consistent subset problem (2018). http://arxiv.org/abs/1810.09232

5. Gates, G.: The reduced nearest neighbor rule. IEEE Trans. Inform. Theory **18**(3), 431–433 (1972)

6. Gottlieb, L., Kontorovich, A., Nisnevitch, P.: Near-optimal sample compression for nearest neighbors. IEEE Trans. Inform. Theory **64**(6), 4120–4128 (2018). Also in NIPS 2014

7. Hart, P.E.: The condensed nearest neighbor rule. IEEE Trans. Inform. Theory **14**(3), 515–516 (1968)

8. Hwang, R.Z., Lee, R.C.T., Chang, R.C.: The slab dividing approach to solve the Euclidean p-center problem. Algorithmica **9**(1), 1–22 (1993)

9. Khodamoradi, K., Krishnamurti, R., Roy, B.: Consistent subset problem with two labels. In: Panda, B.S., Goswami, P.P. (eds.) CALDAM 2018. LNCS, vol. 10743, pp. 131–142. Springer, Cham (2018). https://doi.org/10.1007/978-3-319-74180-2_11

10. Marx, D., Pilipczuk, M.: Optimal parameterized algorithms for planar facility location problems using Voronoi diagrams. In: Bansal, N., Finocchi, I. (eds.) ESA 2015. LNCS, vol. 9294, pp. 865–877. Springer, Heidelberg (2015). https://doi.org/10.1007/978-3-662-48350-3_72. Full version in arXiv:1504.05476

11. Masuyama, S., Ibaraki, T., Hasegawa, T.: Computational complexity of the m-center problems in the plane. Trans. Inst. Electron. Commun. Eng. Jpn. Section E **E64**(2), 57–64 (1981)

12. Ritter, G., Woodruff, H., Lowry, S., Isenhour, T.: An algorithm for a selective nearest neighbor decision rule. IEEE Trans. Inform. Theory **21**(6), 665–669 (1975)

13. Wilfong, G.T.: Nearest neighbor problems. Int. J. Comput. Geom. Appl. **2**(4), 383–416 (1992). Also in SoCG 1991

Parameterized Complexity
of Conflict-Free Graph Coloring

Hans L. Bodlaender[1], Sudeshna Kolay[2], and Astrid Pieterse[3(✉)]

[1] Utrecht University, Utrecht, Netherlands
`h.l.bodlaender@uu.nl`
[2] Ben Gurion University of Negev, Beersheba, Israel
`sudeshna.kolay@gmail.com`
[3] Eindhoven University of Technology, Eindhoven, Netherlands
`astridpieterse@outlook.com`

Abstract. Given a graph G, a q-open neighborhood conflict-free coloring or q-ONCF-coloring is a vertex coloring $c: V(G) \rightarrow \{1, 2, \ldots, q\}$ such that for each vertex $v \in V(G)$ there is a vertex in $N(v)$ that is uniquely colored from the rest of the vertices in $N(v)$. When we replace $N(v)$ by the closed neighborhood $N[v]$, then we call such a coloring a q-closed neighborhood conflict-free coloring or simply q-CNCF-coloring. In this paper, we study the NP-hard decision questions of whether for a constant q an input graph has a q-ONCF-coloring or a q-CNCF-coloring. We will study these two problems in the parameterized setting. First of all, we study running time bounds on FPT-algorithms for these problems, when parameterized by treewidth. We improve the existing upper bounds, and also provide lower bounds on the running time under ETH and SETH. Secondly, we study the kernelization complexity of both problems, using vertex cover as the parameter. We show that both $(q \geq 2)$-ONCF-coloring and $(q \geq 3)$-CNCF-coloring cannot have polynomial kernels when parameterized by the size of a vertex cover unless NP $\subseteq$ coNP/poly. On the other hand, we obtain a polynomial kernel for 2-CNCF-coloring parameterized by vertex cover. We conclude the study with some combinatorial results. Denote $\chi_{ON}(G)$ and $\chi_{CN}(G)$ to be the minimum number of colors required to ONCF-color and CNCF-color G, respectively. Upper bounds on $\chi_{CN}(G)$ with respect to structural parameters like minimum vertex cover size, minimum feedback vertex set size and treewidth are known. To the best of our knowledge only an upper bound on $\chi_{ON}(G)$ with respect to minimum vertex cover size was known. We provide tight bounds for $\chi_{ON}(G)$ with respect to minimum vertex cover size. Also, we provide the first upper bounds on $\chi_{ON}(G)$ with respect to minimum feedback vertex set size and treewidth.

This research was done with support by the NWO Gravitation grant NETWORKS. The research was partially done when the first and second author were associated with Eindhoven University of Technology. Part of this work was done at the Lorentz center workshop on Fixed-Parameter Computational Geometry, May 14–18, 2018, in Leiden, the Netherlands.

© Springer Nature Switzerland AG 2019
Z. Friggstad et al. (Eds.): WADS 2019, LNCS 11646, pp. 168–180, 2019.
https://doi.org/10.1007/978-3-030-24766-9_13

Keywords: Conflict-free coloring · Kernelization ·
Fixed-parameter tractability · Combinatorial bounds

1   Introduction

Often, in frequency allocation problems for cellular networks, it is important
to allot a unique frequency for each client, so that at least one frequency is
unaffected by cancellation. Such problems can be theoretically formulated as a
coloring problem on a set system, better known as conflict-free coloring [5].
Formally, given a set system $\mathcal{H} = (U, \mathcal{F})$, a q-conflict-free coloring $c: U \rightarrow$
$\{1, 2, \ldots, q\}$ is a function where for each set $f \in \mathcal{F}$, there is an element $v \in f$
such that for all $w \neq v \in f$, $c(v) \neq c(w)$. In other words, each set f has at
least one element that is uniquely colored in the set. This variant of coloring
has also been extensively studied for set systems induced by various geometric
regions [1,8,14].

A natural step to study most coloring problems is to study them in graphs.
Given a graph G, $V(G)$ denotes the set of n vertices of G while $E(G)$
denotes the set of m edges in G. A q-coloring of G, for $q \in \mathbb{N}$ is a function
$c: V(G) \rightarrow \{1, 2, \ldots, q\}$. The most well-studied coloring problem on graphs
is proper-coloring. A q-coloring c is called a proper-coloring if for each edge
$\{u, v\} \in E(G)$, $c(u) \neq c(v)$. In this paper, we study two specialized variants
of q-conflict-free coloring on graphs, known as q-ONCF-coloring and q-CNCF-
coloring, which are defined as follows.

Definition 1. *Given a graph G, a q-coloring $c: V(G) \rightarrow \{1, 2, \ldots, q\}$ is called
a q-ONCF-coloring, if for every vertex $v \in V(G)$, there is a vertex u in the open
neighborhood $N(v)$ such that $c(u) \neq c(w)$ for all $w \neq u \in N(v)$. In other words,
every open neighborhood in G has a uniquely colored vertex.*

Definition 2. *Given a graph G, a q-coloring $c: V(G) \rightarrow \{1, 2, \ldots, q\}$ is called
a q-CNCF-coloring, if for for every vertex $v \in V(G)$, there is a vertex u in the
closed neighborhood $N[v]$ such that $c(u) \neq c(w)$ for all $w \neq u \in N[v]$. In other
words, every closed neighborhood in G has a uniquely colored vertex.*

Observe that by the above definitions, the q-ONCF-coloring (or q-CNCF-
coloring) problem is a special case of the conflict-free coloring of set systems.
Given a graph G, we can associate it with the set system $\mathcal{H} = (V(G), \mathcal{F})$, where
$\mathcal{F}$ consists of the sets given by open neighborhoods $N(v)$ (respectively, closed
neighborhoods $N[v]$) for $v \in V(G)$. A q-ONCF-coloring (or q-CNCF-coloring)
of G then corresponds to a q-conflict-free coloring of the associated set system.

Notationally, let $\chi_{CF}(\mathcal{H})$ denote the minimum number of colors required for
a conflict-free coloring of a set system $\mathcal{H}$. Similarly, we denote by $\chi_{ON}(G)$ and
$\chi_{CN}(G)$ the minimum number of colors required for an ONCF-coloring and a
CNCF-coloring of a graph G, respectively. The study of conflict-free coloring
was initially restricted to combinatorial studies. This was first explored in [5]
and [13]. Pach and Tardos [12] gave an upper bound of $\mathcal{O}(\sqrt{m})$ on $\chi_{CF}(\mathcal{H})$ for

a set system $\mathcal{H} = (U, \mathcal{F})$ when the size of $\mathcal{F}$ is m. In [12], it was also shown that for a graph G with n vertices $\chi_{CN}(G) = \mathcal{O}(\log^2 n)$. This bound was shown to be tight in [7]. Similarly, [3] showed that $\chi_{ON}(G) = \Theta(\sqrt{n})$.

However, computing $\chi_{ON}(G)$ or $\chi_{CN}(G)$ is NP-hard. This is because deciding whether a 2-ONCF-coloring or a 2-CNCF-coloring of G exists is NP-hard [6]. This motivates the study of the following decision problems under the lens of parameterized complexity.

---

q-ONCF-COLORING
Input: A graph G.
Question: Is there a q-ONCF-coloring of G?

The q-CNCF-COLORING problem is defined analogously.

Note that because of the NP-hardness for q-ONCF-COLORING or q-CNCF-COLORING even when $q = 2$, the two problems are para-NP-hard under the natural parameter q. Thus, the problems were studied under structural parameters. Gargano and Rescigno [6] showed that both q-ONCF-COLORING and q-CNCF-COLORING have FPT algorithms when parameterized by (i) the size of a vertex cover of the input graph G, (ii) and the neighborhood diversity of the input graph. Gargano and Rescigno also mention that due to Courcelle's theorem, for a non-negative constant q, the two decision problems are FPT with the treewidth of the input graph as the parameter.

Our Results and Contributions. In this paper, we extend the parameterized study of the above two problems with respect to structural parameters. Our first objective is to provide both upper and lower bounds for FPT algorithms when using treewidth as the parameter (Sect. 3). We show that both q-ONCF-COLORING and q-CNCF-COLORING parameterized by treewidth t can be solved in time $(2q^2)^t n^{\mathcal{O}(1)}$. On the other hand, for $q \geq 3$, both problems cannot be solved in time $(q - \epsilon)^t n^{\mathcal{O}(1)}$ under Strong Exponential Time Hypothesis (SETH). For $q = 2$, both problems cannot be solved in time $2^{o(t)} n^{\mathcal{O}(1)}$ under Exponential Time Hypothesis (ETH).

We also study the polynomial kernelization question (Sect. 4). Observe that both q-ONCF-COLORING and q-CNCF-COLORING cannot have polynomial kernels under treewidth as the parameter, as there are straightforward AND-cross-compositions from each problem to itself.[1] Therefore, we will study the kernelization question by a larger parameter, namely the size of a vertex cover in the input graph. The kernelization complexity of the q-COLORING problem (asking for a proper-coloring of the input graph) is very well-studied for this parameter, the problem admits a kernel of size $\widetilde{\mathcal{O}}(k^{q-1})$ [10] which is known to be tight unless NP $\subseteq$ coNP/poly [9]. From this perspective however, q-CNCF-COLORING and q-ONCF-COLORING turn out to be much harder: q-CNCF-COLORING for $q \geq 3$ and q-ONCF-COLORING for $q \geq 2$ do not have polynomial kernels under

---

[1] This is true for a number of graph problems when parameterized by treewidth. For more information, see [4, Theorem 15.12] and the example given for TREEWIDTH (parameterized by solution size) in [4, p. 534].

the standard complexity assumptions, when parameterized by the size of a vertex cover. Interestingly, 2-CNCF-COLORING parameterized by vertex cover size *does* have a polynomial kernel and we obtain an explicit polynomial compression for the problem. Although this does not lead to a polynomial kernel of reasonable size, we study a restricted version called 2-CNCF-COLORING-VC-EXTENSION (Sect. 4.1) and show that this problem has a $\mathcal{O}(k^2 \log k)$ kernel where k is the vertex cover size. Therefore, 2-CNCF-COLORING behaves significantly differently from the other problems.

Finally, we obtain a number of combinatorial results regarding ONCF-colorings of graphs. Denote by $\chi(G)$ the minimum q for which a q-proper-coloring for G exists. While $\chi_{CN}(G) \leq \chi(G)$, the same upper bound does not hold for $\chi_{ON}(G)$ [6]. For a graph G, let $\mathsf{vc}(G)$, $\mathsf{fvs}(G)$ and $\mathsf{tw}(G)$ denote the size of a minimum vertex cover, the size of a minimum feedback vertex set and the treewidth of G, respectively. From the known result that $\chi(G) \leq \mathsf{tw}(G) + 1 \leq \mathsf{fvs}(G) + 1 \leq \mathsf{vc}(G) + 1$, we could immediately obtain the fact that the same behavior holds for $\chi_{CN}(G)$. However, to show that $\chi_{ON}(G)$ behaves similarly more work needs to be done. To the best of our knowledge no upper bounds on $\chi_{ON}(G)$ with respect to $\mathsf{fvs}(G)$ and $\mathsf{tw}(G)$ were known, while a loose upper bound was provided with respect to $\mathsf{vc}(G)$ in [6]. We give a tight upper bound on $\chi_{ON}(G)$ with respect to $\mathsf{vc}(G)$ and also provide the first upper bounds on $\chi_{ON}(G)$ with respect to $\mathsf{fvs}(G)$ and $\mathsf{tw}(G)$ (Sect. 5).

Our main contributions in this work are structural results for the conflict-free coloring problem, which we believe gives more insight into the decision problems on graphs. Firstly, the gadgets we build for the ETH-based lower bounds could be useful for future lower bounds, but are also useful to understand difficult examples for conflict-free coloring which have not been known in abundance so far. We are able to reuse these gadgets in the constructions needed to prove the kernelization lower bounds. Secondly, our combinatorial results also give constructible conflict-free colorings of graphs and therefore provide more insight into conflict-free colored graphs. Finally, the kernelization dichotomy we obtain for q-ONCF-COLORING and q-CNCF-COLORING under vertex cover size as a parameter is a very surprising one.

2   Preliminaries

For a positive integer n, we denote the set $\{1, 2, \ldots, n\}$ in short with $[n]$. For a graph G, given a q-coloring $c \colon V(G) \to [q]$ and a subset $S \subseteq V(G)$, we denote by $c|_S$ the restriction of c to the subset S. For a graph G that is q-ONCF-colored by a coloring c, for a vertex $v \in V(G)$, suppose $w \in N(v)$ is such that $c(w) \neq c(w')$ for each $w' \neq w \in N(v)$; then $c(w)$ is referred to as the ONCF-color of v. Similarly, for a graph G that is q-CNCF-colored by a coloring c, for a vertex $v \in V(G)$, a unique color in $N[v]$ is referred to as the CNCF-color of v.

An *edge-star graph* is a generalization of a star graph where there is a central edge $\{u, v\}$ and all other vertices w have $N(w) = \{u, v\}$. A triangle is an example of an edge-star graph.

For statements marked with a star (⋆), the (complete) proof can be found in the full version of the paper [2].

2.1 Parameterized Complexity

Let Σ be a finite alphabet. A parameterized problem $\mathcal{Q}$ is a subset of $\Sigma^* \times \mathbb{N}$.

Definition 3 (Kernelization). *Let $\mathcal{Q}, \mathcal{Q}'$ be two parameterized problems and let $h \colon \mathbb{N} \to \mathbb{N}$ be some computable function. A generalized kernel from $\mathcal{Q}$ to $\mathcal{Q}'$ of size $h(k)$ is an algorithm that given an instance $(x, k) \in \Sigma^* \times \mathbb{N}$, outputs $(x', k') \in \Sigma^* \times \mathbb{N}$ in time $poly(|x| + k)$ such that (i) $(x, k) \in \mathcal{Q}$ if and only if $(x', k') \in \mathcal{Q}'$, and (ii) $|x'| \le h(k)$ and $k' \le h(k)$.*
The algorithm is a kernel if $\mathcal{Q} = \mathcal{Q}'$. It is a polynomial (generalized) kernel if $h(k)$ is a polynomial in k.

3 Algorithmic Results Parameterized by Treewidth

In this section, we state the algorithmic results obtained for the ONCF-COLORING and CNCF-COLORING problems parameterized by treewidth. On the algorithmic side, we have the following theorem.

Theorem 1 (⋆). q-ONCF-COLORING *and* q-CNCF-COLORING *parameterized by treewidth t admits a $(2q^2)^t n^{\mathcal{O}(1)}$ time algorithm.*

We also obtain algorithmic lower bounds for the problems under standard assumptions.

Theorem 2 (⋆). *The following algorithmic lower bounds can be obtained:*

1. *For $q \ge 3$, q-ONCF-COLORING or q-CNCF-COLORING parameterized by treewidth t cannot be solved in $(q - \varepsilon)^t n^{\mathcal{O}(1)}$ time, under SETH.*
2. *2-ONCF-COLORING or 2-CNCF-COLORING parameterized by treewidth t cannot be solved in $2^{o(t)} n^{\mathcal{O}(1)}$ time, under ETH.*

Due to paucity of space, the full proofs of the Theorems above have been omitted from this extended abstract. As a brief overview of our lower bound techniques, in the remainder of section we will show the running time lower bound on 2-ONCF-COLORING under ETH claimed in Theorem 2. The bound will be obtained by giving a reduction from 3-SAT, to give the reduction we will need the following type of gadget.

Definition 4. *An ONCF-gadget is a gadget on ten vertices, as depicted in Fig. 1.*

The objective of this gadget is the following. The vertices $\{g_1, g_2, g_3, g_{10}\}$ in Fig. 1 will be the interaction points of the ONCF-gadget with the outside world. As will be proved in the following two lemmas, the gadget is designed so as to (i) disallow certain 2-ONCF-colorings and (ii) allow certain 2-ONCF-colorings on its interaction points.

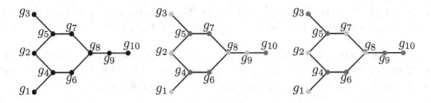

Fig. 1. The ONCF-gadget (left). Observe that if g_1, g_2, and g_3 are all red, then g_9 must also be red (middle), and if one of g_1, g_2, or g_3 is blue, then g_9 may be blue (right). (Color figure online)

Lemma 1 ($\star$). *Let G be a ONCF-gadget with a coloring $c\colon V(G) \to \{\mathrm{red}, \mathrm{blue}\}$ such that for all $4 \leq i \leq 9$ the neighborhood of g_i is ONCF-colored by c. If $c(g_1) = c(g_2) = c(g_3) = \mathrm{red}$, then $c(g_9) = \mathrm{red}$.*

Lemma 2 ($\star$). *Let G be a ONCF-gadget. Let $c'\colon \{g_1, g_2, g_3\} \to \{\mathrm{red}, \mathrm{blue}\}$ be a partial 2-ONCF-coloring of G. If there exists $i \in [3]$ such that $c'(g_i) = \mathrm{blue}$, then c' can be extended to a coloring c satisfying*

1. *For every $4 \leq i \leq 9$, the neighborhood of vertex g_i is ONCF-colored by c (contains at most one red, or at most one blue vertex), and*
2. *$c(g_9) = \mathrm{blue}$, $c(g_8) = \mathrm{red}$, $c(g_4) = c(g_5) = \mathrm{blue}$, and $c(g_{10}) = \mathrm{blue}$.*

Now that we have introduced the necessary gadgets, we can prove the running time lower bound for 2-ONCF-COLORING.

Lemma 3 ($\star$). *2-ONCF-COLORING parameterized by treewidth t cannot be solved in $2^{o(t)}n^{\mathcal{O}(1)}$ time, under ETH.*

Proof (Proof sketch). We show this by giving a reduction from 3-SAT. Given an instance of 3-SAT with variables $x_1, \ldots, x_n$ and clauses $C_1, \ldots, C_m$, create a graph G as follows. Start by creating palette vertices R, R', and B, and edges $\{R, R'\}$ and $\{R', B\}$. For each variable $i \in [n]$, create vertices u_i, v_i, w_i and add edges $\{u_i, v_i\}$ and $\{v_i, w_i\}$. For each $j \in [m]$, add an ONCF-gadget G_j and connect g_{10} of this gadget to R. Add vertices s_j^1, s_j^2, and s_j^3 and connect s_j^b to g_b in G_j for $b \in [3]$. Let clause $C_j := (\ell_1, \ell_2, \ell_3)$. Now if $\ell_b = x_i$ for some $i \in [n], b \in [3]$, connect s_j^b to u_i. Similarly, if $\ell_b = \neg x_i$, connect s_j^b to w_i. This concludes the construction of G. The main idea towards showing that φ is satisfiable if and only if G is 2-ONCF-colorable is to let the situation where u_i is *red* and w_i is *blue* mean that the corresponding variable is set to *true*. A more detailed explanation can be found in the full version of this paper [2].

Note that the graph induced by $V(G) \setminus (\{R, R', B\} \cup \{u_i, v_i, w_i \mid i \in [n]\})$ is a disjoint union of ONCF-gadgets for which every g_b for $b \in [3]$ has an additional degree-1 vertex attached to it. It is easy to see that every connected component of this graph has treewidth two. Thus, G has treewidth at most $3n + 5$. This implies that a 3-SAT formula ϕ on n variables and m clauses is reduced to a graph G with treewidth at most $3n + 5$. Since 3-SAT cannot be solved in $2^{o(n)}n^{\mathcal{O}(1)}$

time under ETH, this also implies that 2-CNCF-COLORING parameterized by treewidth t cannot be solved in $2^{o(t)}n^{\mathcal{O}(1)}$ time, under ETH.   □

Note that a reduction from 3-SAT to 2-ONCF-COLORING was given in Theorem 2 of [6]. However, that reduction led to a quadratic blow-up in the input size. Hence, the need for the alternative reduction given above.

4   Kernelization

In this section, we will study the kernelizability of the ONCF- and CNCF-coloring problems, when parameterized by the size of a vertex cover. We prove the following two theorems to obtain a dichotomy on the kernelization question.

Theorem 3 ($\star$). q-ONCF-COLORING *for $q \geq 2$ and q-CNCF-COLORING for $q \geq 3$, parameterized by vertex cover size do not have polynomial kernels, unless* NP $\subseteq$ coNP/poly.

Theorem 4. 2-CNCF-COLORING *parameterized by vertex cover size k has a generalized kernel of size $\mathcal{O}(k^{10})$.*

Note that by using an NP-completeness reduction, this results in a polynomial kernel for 2-CNCF-COLORING parameterized by vertex cover size. We also obtain an $\mathcal{O}(k^2 \log k)$ kernel for an extension problem of 2-CNCF-COLORING and this is described in Sect. 4.1.

In the remainder of this section we will prove Theorem 4, by obtaining a polynomial generalized kernel for 2-CNCF-COLORING parameterized by vertex cover size. This result is in contrast to the kernelization results we obtain for q-CNCF-COLORING for $q \geq 3$ as well as q-ONCF-COLORING for $q \geq 2$. We will start by transforming an instance of 2-CNCF-COLORING to an equivalent instance of another problem, namely d-POLYNOMIAL ROOT CSP. We will then carefully rephrase the d-POLYNOMIAL ROOT CSP instance such that it uses only a limited number of variables, such that we can use a known kernelization result for d-POLYNOMIAL ROOT CSP to obtain our desired compression. We start by introducing the relevant definitions.

Define d-POLYNOMIAL ROOT CSP over a field F as follows [11].

---

d-POLYNOMIAL ROOT CSP
Input: A list L of polynomial equalities over variables $V = \{x_1, \ldots, x_n\}$. An equality is of the form $f(x_1, \ldots, x_n) = 0$, where f is a multivariate polynomial over F of degree at most d.
Question: Does there exist an assignment of the variables $\tau\colon V \to \{0, 1\}$ satisfying all equalities (over F) in L?

A field F is said to be *efficient* if both the field operations and Gaussian elimination can be done in polynomial time in the size of a reasonable input encoding. In particular, $\mathbb{Q}$ is an efficient field by this definition. The following theorem was shown by Jansen and Pieterse.

Theorem 5 ([11, Theorem 5]). *There is a polynomial-time algorithm that, given an instance (L, V) of d-*POLYNOMIAL ROOT CSP *over an efficient field F, outputs an equivalent instance (L', V) with at most $n^d + 1$ constraints such that $L' \subseteq L$.*

Using the theorem introduced above, we can now prove Theorem 4.

Proof (Proof of Theorem 4). Given an input instance G with vertex cover S of size k, we start by preprocessing G. For each set $X \subseteq S$ with $|X| \leq 2$, mark 3 vertices in $v \in G \setminus S$ with $N(v) = X$ (if there do not exist 3 such vertices, simply mark all). Let $S' \subseteq V(G) \setminus S$ be the set of all marked vertices. Remove all $w \in V(G) \setminus (S \cup S')$ with $deg(w) \leq 2$ from G. Let the resulting graph be G'.

Claim 1. *G' is 2-CNCF-colorable if and only if G is 2-CNCF-colorable.*

Proof. In one direction, suppose G' has a 2-CNCF coloring c using colors $\{r, b\}$. Consider a vertex $w \in V(G) \setminus V(G')$. Let $X_w \subseteq S$ be the neighborhood of w. Note that $|X_w|$ is at most 2. Consider $N(X_w) \cap S'$. Since w was deleted, there are 3 vertices in $N(X_w) \cap S'$. Consider the color from $\{r, b\}$ that appears in majority on the vertices of $N(X_w) \cap S'$. If we color w with the same color, it is easy to verify that this extension of c to G is a 2-CNCF coloring of G.

In the reverse direction, suppose G has a 2-CNCF coloring c using colors $\{r, b\}$. We describe a new coloring c' for G as follows. Consider a subset $X \subseteq S$ of size at most 2 and let N be the set of vertices in $G \setminus S$ that have X as their neighborhood. If $|N| > 3$ and $N \setminus S'$ has a vertex w that is uniquely colored in the set N, then we arbitrarily choose a vertex $w' \in N \cap S'$. We define $c'(w') = c(w)$ and $c'(w) = c(w')$. All other vertices have the same color in c and c'. It is easy to verify that c' is also a 2-CNCF coloring of G and the restriction of c' to G' is a 2-CNCF coloring of G'.   ⌐

We continue by creating an instance of 2-POLYNOMIAL ROOT CSP that is satisfiable if and only if G' is 2-CNCF-colorable. Let $V := \{r_v, b_v \mid v \in V(G)\}$ be the variable set. We create L over $\mathbb{Q}$ as follows.

1. For each $v \in V(G')$, add the constraint $r_v + b_v - 1 = 0$ to L.
2. For all $v \in V(G')$, add the constraint $(-1 + \sum_{u \in N[v]} r_u) \cdot (-1 + \sum_{u \in N[v]} b_v) = 0$.
3. For each $v \in V(G') \setminus (S \cup S')$ of degree $d_v = |N(v)|$ add the constraint

$$(\sum_{u \in N(v)} r_u)(-1 + \sum_{u \in N(v)} r_u)(-(d_v - 1) + \sum_{u \in N(v)} r_u)(-d_v + \sum_{u \in N(v)} r_u) = 0.$$

Note that such a constraint is a quadratic polynomial.

Intuitively, the first constraint ensures that every vertex is either *red* or *blue*. The second constraint ensures that in the closed neighborhood of every vertex, exactly one vertex is *red* or exactly one is *blue*. The third constraint is seemingly redundant, saying that the open neighborhood of every vertex outside the vertex cover does not have two *red* or two *blue* vertices, which is clearly forbidden. The requirement for these last constraints is made clear in the proof of Claim 4.

We show that this results in an instance that is equivalent to the original input instance, in the following sense.

Claim 2 ($\star$). (L, V) *is a yes-instance of* 2-POLYNOMIAL ROOT CSP *if and only if* G' *is* 2-*CNCF-colorable*.

Clearly, $|V| = 2n$ if n is the number of vertices of G'. We will now show how to modify L, such that it uses only variables for the vertices in $S \cup S'$. To this end, we introduce the following function. For $v \notin (S \cup S')$, let $f_v(V) := g\left(\sum_{u \in N(v)} r_u, |N(v)|\right)$, where

$$g(x, N) = -\frac{(N - x)(x - 1)(N - 2(x + 1))}{N(N - 2)}.$$

Note that for any fixed $N > 2$, $g(x, N)$ describes a degree-3 polynomial in x over $\mathbb{Q}$. The following is easy to verify.

Observation 1. $g(0, N) = g(N - 1, N) = 1$, *and* $g(N, N) = g(1, N) = 0$ *for all* $N \in \mathbb{Z} \setminus \{0, 2\}$.

Observe that f_v only uses variables defined for vertices that are in S. As such, let $V' := \{r_v, b_v \mid v \in S\} \cup \{r_v, b_v \mid v \in S'\}$, and let L' be equal to L with every occurrence of r_v for $v \notin (S \cup S')$ substituted by f_v and every occurrence of b_v for $v \notin (S \cup S')$ substituted by $(1 - f_v(V))$.

Claim 3 ($\star$). *If* $\tau : V \rightarrow \{0, 1\}$ *is a satisfying assignment for* (L, V), *then* $\tau|_{V'}$ *is a satisfying assignment for* (L', V').

The next claim shows the equivalence between (L', V') and (L, V).

Claim 4 ($\star$). *If* $\tau : V' \rightarrow \{0, 1\}$ *is a satisfying assignment for* (L', V'), *then there exists a satisfying assignment* $\tau' : V \rightarrow \{0, 1\}$ *for* (L, V) *such that* $\tau'|_{V'} = \tau$.

Using the method described above, we obtain an instance (L, V) of 2-POLYNOMIAL ROOT CSP such that (L, V) has a satisfying assignment if and only if G is 2-CNCF-colorable by Claims 1 and 2. Then we obtain an instance (L', V') such that (L', V') is satisfiable if and only if (L, V) is satisfiable by Claims 3 and 4. As such, (L', V') is a yes-instance if and only if G is 2-CNCF-colorable and it suffices to give a kernel for (L', V'). Observe that $|V'| = \mathcal{O}(k^2)$.

We start by partitioning L' into three sets L_S', L_1' and L_2'. Let L_S' contain all equalities created for a vertex $v \in S$. Let L_1' contain all equations that contain at least one of the variables in $\{r_v, b_v \mid v \in S'\}$ and let L_2 contain the remaining equalities. Observe that $|L_S'| = k$ by definition. Furthermore, the polynomials in L_1' have degree at most 2, as they were created for vertices in $V(G') \setminus S$, and these are not connected. As such, we use Theorem 5 to obtain $L_1'' \subseteq L_1'$ such that $|L_1''| = \mathcal{O}((k^2)^2) = \mathcal{O}(k^4)$ and any boolean assignment satisfying all equalities in L_1'' satisfies all equalities in L_1'.

Similarly, we observe that L_2' by definition contains none of the variables in $\{r_v, b_v \mid v \in S'\}$, implying that the equations in L_2' are equations over only k variables. Since the polynomials in L_2' have degree at most 6, we can apply Theorem 5 to obtain $L_2'' \subseteq L_2'$ such that $|L_2''| \leq \mathcal{O}(k^6)$ and any assignment satisfying all equations in L_2'' satisfies all equalities in L_2'.

We now define $L'' := L_1'' \cup L_2'' \cup L_S'$, and the output of our polynomial generalized kernel will be (L'', V'). The correctness of the procedure is proven above, it remains to bound the number of bits needed to store instance (L'', V').

By this definition, $|L''| \leq \mathcal{O}(k^6)$. To represent a single constraint, it is sufficient to store the coefficients for each variable in V'. The storage space needed for a single coefficient is $\mathcal{O}(\log(n))$, as the coefficients are bounded by a polynomial in n. Thereby, (L'', V') can be stored in $\mathcal{O}(k^6 \cdot k^2 \log n)$ bits. To bound this in terms of k, we observe that it is easy to solve 2-CNCF-COLORING in time $\mathcal{O}(2^{k^2} \cdot \text{poly}(n))$. This is done by guessing the coloring of S, extending this coloring to the entire graph (observe $G \setminus S$ has no vertices of degree less than three) and verifying whether this results in a CNCF-coloring. Therefore, we can assume that $\log(n) \leq k^2$, as otherwise we can solve the 2-CNCF-COLORING problem in $\mathcal{O}(2^{k^2} \text{poly}(n))$ time, which is then polynomial in n. Thereby we conclude that (L'', V') can be stored in $\mathcal{O}(k^{10})$ bits.   □

4.1   Kernelization Bounds for Conflict-Free Coloring Extension

We furthermore provide kernelization bounds for the following extension problems.

---

q-CNCF-COLORING-VC-EXTENSION
Input: A graph G with vertex cover S and partial q-coloring $c \colon S \to [q]$.
Question: Does there exist a q-CNCF-coloring of G that extends c?

We define q-ONCF-COLORING-VC-EXTENSION analogously.

We obtain the following kernelization results when parameterized by vertex cover size, thereby classifying the situations where the extension problem has a polynomial kernel. The extension problem turns out to have a polynomial kernel in the same case as the normal problem. However, we manage to give a significantly smaller kernel. Observe that the kernelization result is non-trivial, since 2-CNCF-COLORING-VC-EXTENSION is NP-hard ($\star$).

Theorem 6 ($\star$). *The following results hold.*

1. 2-CNCF-COLORING-VC-EXTENSION *has a kernel with* $\mathcal{O}(k^2)$ *vertices and edges that can be stored in* $\mathcal{O}(k^2 \log k)$ *bits. Here k is the size of the input vertex cover S.*
2. q-CNCF-COLORING-VC-EXTENSION *for any $q \geq 3$, and* 2-ONCF-COLORING-VC-EXTENSION *parameterized by the size of a vertex cover do not have a polynomial kernel, unless* NP $\subseteq$ coNP/poly.

5   Combinatorial Bounds

Given a graph G, it is easy to prove that $\chi_{CN}(G) \leq \chi(G)$. However, there are examples that negate the existence of such bounds with respect to χ_{ON} [6]. In this section, we prove combinatorial bounds for χ_{ON} with respect to common graph parameters like treewidth, feedback vertex set and vertex cover.

First, note that if G is a graph with isolated vertices then the graph can have no ONCF-coloring. Therefore, in all the arguments below we assume that G does not have any isolated vertices. We obtain the following result. Recall that for a graph G, $\mathsf{vc}(G)$, $\mathsf{fvs}(G)$ and $\mathsf{tw}(G)$ denote the size of a minimum vertex cover, the size of a minimum feedback vertex set and the treewidth of G, respectively.

Theorem 7 ($\star$). *Given a connected graph G,*

1. $\chi_{ON}(G) \leq 2\mathsf{tw}(G) + 1$,
2. $\chi_{ON}(G) \leq \mathsf{fvs}(G) + 3$,
3. $\chi_{ON}(G) \leq \mathsf{vc}(G) + 1$. *Furthermore, if G is not a star graph or an edge-star graph, then $\chi_{ON}(G) \leq \mathsf{vc}(G)$.*

Here, we only give a proof sketch of the result with respect to $\mathsf{vc}(G)$ and relegate the other two combinatorial results to the full version of the paper. The next lemma bounds the value of $\chi_{ON}(G)$ for graphs with a vertex cover of size k. In particular, we improve the bound given by Gargano and Rescigno [6, Lemma 4], who showed that $\chi_{ON}(G) \leq 2k + 1$.

Lemma 4 ($\star$). *Let G be a connected graph with $\mathsf{vc}(G) = k$. Then $\chi_{ON}(G) \leq k + 1$. Furthermore, if G is not a star graph or an edge-star graph, then $\chi_{ON}(G) \leq k$.*

Proof (Proof sketch). See Fig. 2 for a sketch of the colorings described in the proof. We start by proving the bounds for the case where G is not a star and not an edge-star. Let S be a minimum vertex cover of G and let k be the size of S. We do a case distinction on the size and connectedness of S.

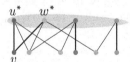 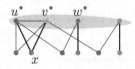 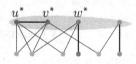

Fig. 2. (left) A coloring of the graph when all vertices in $G[S]$ are isolated. (middle) The case where $G[S]$ contains an edge and the endpoints have a common neighbor. (right) The case where $G[S]$ contains an edge and the endpoints have no common neighbors.

($k = 2$ **and S connected**) First, we prove the bounds for $k = 2$ and $G[S]$ is an edge $\{u^*, v^*\}$. Note that G is not an edge-star graph. Therefore at least one of u^* or v^* have neighbors with degree exactly 1 in $G \setminus S$. As shown in the full proof, it is possible to ONCF-color such a graph with 2 colors, namely r and b.

($G[S]$ **disconnected or** $k \geq 3$) We now prove the bounds for $k = 2$ and $G[S]$ is disconnected, and $k \geq 3$. We consider a number of cases.

(Suppose $G[S]$ contains a connected component C of size at least three.) Let $v^* \in C$ be a vertex such that $G[C \setminus \{v\}]$ remains connected. We color the vertices in G as follows. For every vertex $u \in S$, let $c(u) := c_u$. For every vertex $u \in S$ that is isolated in $G[S]$, pick an arbitrary neighbor $v \notin S$ and (re)color v such that $c(v) := c_u$. Notice that a vertex v in $G \setminus S$ may be picked multiple times as the candidate for an arbitrary neighbor for an isolated vertex in S, and in this case the color of this vertex v is set to the last color it is assigned. For every vertex v that is not yet colored, let $c(v) := c_{v^*}$. It can be shown that c is the required coloring.

(Suppose $G[S]$ only contains connected components of size one.) Note that $|S| > 1$. Start by letting $c(v) := c_v$ for every vertex $v \in S$. Since G is connected, there exists $v \notin S$ such that $|N(v)| \geq 2$. Pick two vertices $u^*, w^* \in N(v)$ with $u^* \neq w^*$. Let $c(v) := c_{u^*}$. For every vertex $u \in S \setminus \{u^*, w^*\}$ pick an arbitrary neighbor $v \notin S$ and recolor v to c_u. Color the vertices that remained uncolored by this procedure with c_{w^*}. It can be shown that c is the required coloring.

(Otherwise.) In this case $G[S]$ has size at least 3, contains multiple connected components, and at least one such component has size two. This leads to two further cases, that have been analyzed in the complete proof in the full version of this paper.

If G is not a star and not an edge-star, we are in one of the cases above. Otherwise, it is easy to observe that stars have a vertex cover of size one and can always be colored with two colors, and edge-stars can be colored with three colors while having a minimum vertex cover size of two.   □

Observe that the bounds of Lemma 4 are tight. First, a star graph requires 2 colors and has vertex cover size 1 while an edge-star graph requires 3 colors and has vertex cover size 2. On the other hand, given an $q \geq 3$, taking the complete graph K_q and subdividing each edge once results in a graph that requires q colors [6] for an ONCF-coloring and has a vertex cover of size q.

6   Open Problems

The study in this paper leads to some interesting open questions. In this paper we only exhibit a generalized kernel of size $\mathcal{O}(k^{10})$ for 2-CNCF-COLORING and it remains to resolve the size of tight polynomial kernels for the problem. On the combinatorial side, with respect to minimum vertex cover, we obtain tight upper bounds on $\chi_{ON}(G)$. It would be interesting to obtain corresponding tight bounds for $\chi_{ON}(G)$ with respect to feedback vertex set and treewidth.

References

1. Ajwani, D., Elbassioni, K., Govindarajan, S., Ray, S.: Conflict-free coloringfor rectangle ranges using $O(n^{.382})$ colors. Discrete Comput. Geom. **48**(1), 39–52 (2012). https://doi.org/10.1007/s00454-012-9425-5
2. Bodlaender, H.L., Kolay, S., Pieterse, A.: Parameterized complexity of conflict-free graph coloring. CoRR abs/1905.00305 (2019). https://arxiv.org/abs/1905.00305
3. Cheilaris, P.: Conflict-free coloring. Ph.D. thesis, City University of New York (2009)
4. Cygan, M., et al.: Parameterized Algorithms. Springer, Heidelberg (2015). https://doi.org/10.1007/978-3-319-21275-3
5. Even, G., Lotker, Z., Ron, D., Smorodinsky, S.: Conflict-free colorings of simple geometric regions with applications to frequency assignment in cellular networks. In: Proceedings of the 43rd FOCS, pp. 691–700 (2002). https://doi.org/10.1109/SFCS.2002.1181994
6. Gargano, L., Rescigno, A.A.: Complexity of conflict-free colorings of graphs. Theor. Comput. Sci. **566**, 39–49 (2015). https://doi.org/10.1016/j.tcs.2014.11.029
7. Glebov, R., Szabó, T., Tardos, G.: Conflict-free colouring of graphs. Combin. Probab. Comput. **23**(3), 434–448 (2014)
8. Har-Peled, S., Smorodinsky, S.: Conflict-free coloring of points and simpleregions in the plane. Discrete Comput. Geom. **34**(1), 47–70 (2005). https://doi.org/10.1007/s00454-005-1162-6
9. Jansen, B.M.P., Kratsch, S.: Data reduction for graph coloring problems. Inf. Comput. **231**, 70–88 (2013). https://doi.org/10.1016/j.ic.2013.08.005
10. Jansen, B.M.P., Pieterse, A.: Optimal data reduction for graph coloring using low-degree polynomials. In: Proceedings of the 12th IPEC, pp. 22:1–22:12 (2017). https://doi.org/10.4230/LIPIcs.IPEC.2017.22
11. Jansen, B.M.P., Pieterse, A.: Optimal sparsification for some binary CSPs using low-degree polynomials. CoRR abs/1606.03233v2 (2018)
12. Pach, J., Tardos, G.: Conflict-free colourings of graphs and hypergraphs. Combin. Probab. Comput. **18**(05), 819–834 (2009). https://doi.org/10.1017/S0963548309990290
13. Smorodinsky, S.: Combinatorial problems in computational geometry. Ph.D. thesis, School of Computer Science, Tel-Aviv University (2003)
14. Smorodinsky, S.: Conflict-free coloring and its applications. In: Bárány, I., Böröczky, K.J., Tóth, G.F., Pach, J. (eds.) Geometry—Intuitive, Discrete, and Convex. BSMS, vol. 24, pp. 331–389. Springer, Heidelberg (2013). https://doi.org/10.1007/978-3-642-41498-5_12

Graph Isomorphism for (H_1, H_2)-Free Graphs: An Almost Complete Dichotomy

Marthe Bonamy[1], Konrad K. Dabrowski[2]([✉]), Matthew Johnson[2], and Daniël Paulusma[2]

[1] CNRS, LaBRI, Université de Bordeaux, Bordeaux, France
marthe.bonamy@u-bordeaux.fr
[2] Department of Computer Science, Durham University, Durham, UK
{konrad.dabrowski,matthew.johnson2,daniel.paulusma}@durham.ac.uk

Abstract. We almost completely resolve the computational complexity of GRAPH ISOMORPHISM for classes of graphs characterized by two forbidden induced subgraphs H_1 and H_2. Schweitzer settled the complexity of this problem restricted to (H_1, H_2)-free graphs for all but a finite number of pairs (H_1, H_2), but without explicitly giving the number of open cases. Grohe and Schweitzer proved that GRAPH ISOMORPHISM is polynomial-time solvable on graph classes of bounded clique-width. By combining known results with a number of new results, we reduce the number of open cases to seven. By exploiting the strong relationship between GRAPH ISOMORPHISM and clique-width, we simultaneously reduce the number of open cases for boundedness of clique-width for (H_1, H_2)-free graphs to five.

Keywords: Hereditary graph class · Induced subgraph · Clique-width · Graph Isomorphism

1 Introduction

The GRAPH ISOMORPHISM problem, which is that of deciding whether two given graphs are isomorphic, is a central problem in Computer Science. It is not known if this problem is polynomial-time solvable, but it is not NP-complete unless the polynomial hierarchy collapses [24]. Analogous to the use of the notion of NP-completeness, we can say that a problem is GRAPH ISOMORPHISM-complete (abbreviated to GI-complete). Babai [1] proved that GRAPH ISOMORPHISM can be solved in quasi-polynomial time.

In order to increase understanding of the computational complexity of GRAPH ISOMORPHISM, it is natural to place restrictions on the input. This approach has yielded many graph classes on which GRAPH ISOMORPHISM is

Research supported by the London Mathematical Society (SC7-1718-04), ANR project HOSIGRA (ANR-17-CE40-0022), EPSRC (EP/K025090/1) and the Leverhulme Trust (RPG-2016-258).

© Springer Nature Switzerland AG 2019
Z. Friggstad et al. (Eds.): WADS 2019, LNCS 11646, pp. 181–195, 2019.
https://doi.org/10.1007/978-3-030-24766-9_14

polynomial-time solvable, and many other graph classes on which the problem remains GI-complete. We refer to [23] for a survey, but some recent examples include a polynomial-time algorithm for unit square graphs [20] and a complexity dichotomy for H-induced-minor-free graphs [2] for every graph H.

In this paper we consider the GRAPH ISOMORPHISM problem for hereditary graph classes, which are the classes of graphs that are closed under vertex deletion. It is readily seen that a graph class $\mathcal{G}$ is hereditary if and only if there exists a family of graphs $\mathcal{F}_\mathcal{G}$, such that the following holds: a graph G belongs to $\mathcal{G}$ if and only if G does not contain any graph from $\mathcal{F}_\mathcal{G}$ as an induced subgraph. We implicitly assume that $\mathcal{F}_\mathcal{G}$ is a family of minimal forbidden induced subgraphs, in which case $\mathcal{F}_\mathcal{G}$ is unique. We note that $\mathcal{F}_\mathcal{G}$ may have infinite size. For instance, if $\mathcal{G}$ is the class of bipartite graphs, then $\mathcal{F}_\mathcal{G}$ consists of all odd cycles.

A natural direction for a *systematic* study of the computational complexity of GRAPH ISOMORPHISM is to consider graph classes $\mathcal{G}$, for which $\mathcal{F}_\mathcal{G}$ is small, starting with the case where $\mathcal{F}_\mathcal{G}$ has size 1. A graph is H-*free* if it does not contain H as induced subgraph; conversely, we write $H \subseteq_i G$ to denote that H is an induced subgraph of G. The classification for H-free graphs [4] is due to an unpublished manuscript of Colbourn and Colbourn (see [16] for a proof).

Theorem 1 (see [4,16]). *Let H be a graph. Then* GRAPH ISOMORPHISM *on H-free graphs is polynomial-time solvable if $H \subseteq_i P_4$ and* GI-*complete otherwise.*

Later, it was shown that GRAPH ISOMORPHISM is polynomial-time solvable even for the class of permutation graphs [7], which form a superclass of the class of P_4-free graphs. Classifying the case where $\mathcal{F}_\mathcal{G}$ has size 2 is much more difficult than the size-1 case. Kratsch and Schweitzer [16] initiated this classification. Schweitzer [25] extended the results of [16] and proved that only a finite number of cases remain open. A graph is (H_1, H_2)-*free* if it has no induced subgraph isomorphic to H_1 or H_2. This leads to our research question:

> *Is it possible to determine the computational complexity of* GRAPH ISO-MORPHISM *for (H_1, H_2)-free graphs for all pairs H_1, H_2?*

We recall that the analogous research question for H-induced-minor-free graphs was fully answered by Belmonte, Otachi and Schweitzer [2], who also determined all graphs H for which the class of H-induced-minor-free graphs has bounded clique-width. Similar classifications for GRAPH ISOMORPHISM [22] and boundedness of clique-width [12] are also known for H-free minor graphs.

Lokshtanov et al. [17] recently gave an FPT algorithm for GRAPH ISOMORPHISM with parameter k on graph classes of treewidth at most k, and this has since been improved by Grohe et al. [13]. Whether an FPT algorithm exists when parameterized by clique-width is still open. Grohe and Schweitzer [14] proved membership of XP.

Theorem 2 ([14]). *For every c,* GRAPH ISOMORPHISM *is polynomial-time solvable on graphs of clique-width at most c.*

Our Results. Combining known results [16, 25] with Theorem 2, we narrow the list of open cases for GRAPH ISOMORPHISM on (H_1, H_2)-free graphs to 14. Of these 14 cases, we prove that two are polynomial-time solvable (Sect. 3) and five others are GI-complete (Sect. 4). Thus we reduce the number of open cases to seven. In Sect. 5 we provide an explicit list of all known and open cases.

Besides Theorem 2, there is another reason why results for clique-width are of importance for GRAPH ISOMORPHISM. Namely, Schweitzer [25] pointed out great similarities between proving unboundedness of clique-width of some graph class $\mathcal{G}$ and proving that GRAPH ISOMORPHISM stays GI-complete for $\mathcal{G}$. We will illustrate these similarities by noting that our construction demonstrating that GRAPH ISOMORPHISM is GI-complete for (gem, $P_1 + 2P_2$)-free graphs can also be used to show that this class has unbounded clique-width. This reduces the number of pairs (H_1, H_2) for which we do not know if the class of (H_1, H_2)-free graphs has bounded clique-width from six [11] to five. As such, our paper also continues a project [3, 6, 8, 9, 11, 12] aiming to classify the boundedness of clique-width of (H_1, H_2)-free graphs for all pairs (H_1, H_2) (see [10] for a summary).

2 Preliminaries

We consider only finite, undirected graphs without multiple edges or self-loops. The *disjoint union* $(V(G) \cup V(H), E(G) \cup E(H))$ of two vertex-disjoint graphs G and H is denoted by $G + H$ and the disjoint union of r copies of a graph G is denoted by rG. For a subset $S \subseteq V(G)$, we let $G[S]$ denote the subgraph of G *induced* by S, which has vertex set S and edge set $\{uv \mid u, v \in S, uv \in E(G)\}$. If $S = \{s_1, \ldots, s_r\}$, then we may write $G[s_1, \ldots, s_r]$ instead of $G[\{s_1, \ldots, s_r\}]$. Recall that for two graphs G and G' we write $G' \subseteq_i G$ to denote that G' is an induced subgraph of G. For a set of graphs $\{H_1, \ldots, H_p\}$, a graph G is $(H_1, \ldots, H_p)$-*free* if it has no induced subgraph isomorphic to a graph in $\{H_1, \ldots, H_p\}$; recall that if $p = 1$, we may write H_1-free instead of (H_1)-free. For a graph G, the set $N(u) = \{v \in V \mid uv \in E\}$ denotes the *(open) neighbourhood* of $u \in V(G)$ and $N[u] = N(u) \cup \{u\}$ denotes the *closed neighbourhood* of u. The *degree* $d_G(v)$ of a vertex v in a graph G is the number of vertices in G that are adjacent to v.

A *(connected) component* of a graph G is a maximal subset of vertices that induces a connected subgraph of G; it is *non-trivial* if it has at least two vertices, otherwise it is *trivial*. The *complement* $\overline{G}$ of a graph G has vertex set $V(\overline{G}) = V(G)$ such that two vertices are adjacent in $\overline{G}$ if and only if they are not adjacent in G.

The graphs C_r, K_r, $K_{1, r-1}$ and P_r denote the cycle, complete graph, star and path on r vertices, respectively. Let $K_{1,n}^+$ and $K_{1,n}^{++}$ be the graphs obtained from $K_{1,n}$ by subdividing one edge once or twice, respectively. The graphs $K_{1,3}$, $\overline{2P_1 + P_2}$, $\overline{P_1 + P_3}$, $\overline{P_1 + P_4}$ and $\overline{2P_1 + P_3}$ are also called the claw, diamond, paw, gem and crossed house, respectively. We need the following result.

Lemma 1 ([25]). *For every fixed t, GRAPH ISOMORPHISM is polynomial-time solvable on $(2K_{1,t}, K_t)$-free graphs.*

The graph $S_{h,i,j}$, for $1 \leq h \leq i \leq j$, denotes the *subdivided claw*, that is, the tree that has only one vertex x of degree 3 and exactly three leaves, which are at distance h, i and j from x, respectively. Observe that $S_{1,1,1} = K_{1,3}$. A *subdivided star* is a graph obtained from a star by subdividing its edges an arbitrary number of times. A graph is a *path star forest* if all of its connected components are subdivided stars.

Let G be a graph and let $X, Y \subseteq V(G)$ be disjoint sets. The edges between X and Y form a *perfect matching* if every vertex in X is adjacent to exactly one vertex in Y and vice versa. A vertex $x \in V(G) \setminus Y$ is *complete* (resp. *anti-complete*) to Y if it is adjacent (resp. non-adjacent) to every vertex in Y. Similarly, X is complete (resp. anti-complete) to Y if every vertex in X is complete (resp. anti-complete) to Y. A graph is *split* if its vertex set can be partitioned into a clique and an independent set. A graph is *complete multipartite* if its vertex set can be partitioned into independent sets $V_1, \ldots, V_k$ such that V_i is complete to V_j whenever $i \neq j$.

Lemma 2 ([21]). *Every connected $(\overline{P_1 + P_3})$-free graph is either complete multipartite or K_3-free.*

Given two graphs G and H, an *isomorphism* from G to H is a bijection $f : V(G) \to V(H)$ such that $vw \in E(G)$ if and only if $f(v)f(w) \in E(H)$. For a function $f : X \to Y$, if $X' \subseteq X$, we define $f(X') := \{f(x) \in Y \mid x \in X'\}$. The GRAPH ISOMORPHISM problem is defined as follows.

GRAPH ISOMORPHISM
   Instance: Graphs G and H.
   Question: Is there an isomorphism from G to H?

The *clique-width* of a graph G, denoted by $\mathrm{cw}(G)$, is the minimum number of labels needed to construct G using the following four operations:

(i)   create a new graph consisting of a single vertex v with label i;
(ii) take the disjoint union of two labelled graphs G_1 and G_2;
(iii) join each vertex with label i to each vertex with label j ($i \neq j$);
(iv) rename label i to j.

A class of graphs $\mathcal{G}$ has bounded clique-width if there is a constant c such that the clique-width of every graph in $\mathcal{G}$ is at most c; otherwise the clique-width of $\mathcal{G}$ is *unbounded*.

Let G be a graph. For an induced subgraph $G' \subseteq_i G$, the *subgraph complementation* operation (acting on G with respect to G') replaces every edge present in G' by a non-edge, and vice versa, that is, the resulting graph has vertex set $V(G)$ and edge set $(E(G) \setminus E(G')) \cup \{xy \mid x, y \in V(G'), x \neq y, xy \notin E(G')\}$. Similarly, for two disjoint vertex subsets S and T in G, the *bipartite complementation* operation with respect to S and T acts on G by replacing every edge with one end-vertex in S and the other in T by a non-edge and vice versa.

Let $k \geq 0$ be a constant and let γ be some graph operation. We say that a graph class $\mathcal{G}'$ is (k, γ)-*obtained* from a graph class $\mathcal{G}$ if the following two conditions hold:

(i) every graph in $\mathcal{G}'$ is obtained from a graph in $\mathcal{G}$ by performing γ at most k times, and

(ii) for every $G \in \mathcal{G}$ there exists at least one graph in $\mathcal{G}'$ obtained from G by performing γ at most k times.

We say that γ *preserves* boundedness of clique-width if for any finite constant k and any graph class $\mathcal{G}$, any graph class $\mathcal{G}'$ that is (k, γ)-obtained from $\mathcal{G}$ has bounded clique-width if and only if $\mathcal{G}$ has bounded clique-width.

Fact 1. Vertex deletion preserves boundedness of clique-width [19].

Fact 2. Subgraph complementation preserves boundedness of clique-width [15].

Fact 3. Bipartite complementation preserves boundedness of clique-width [15].

We need the following lemmas on clique-width.

Lemma 3 ([5]). *The class of* $\overline{2P_1 + P_3}$-*free split graphs has bounded clique-width.*

Lemma 4 ([18]). *The class of* $(P_2 + P_3)$-*free bipartite graphs has bounded clique-width.*

We also need the special case of [12, Theorem 3] when $V_{0,i} = V_{i,0} = \emptyset$ for $i \in \{1, \ldots, n\}$.

Lemma 5 ([12]). *For* $m \geq 1$ *and* $n > m + 1$ *the clique-width of a graph G is at least* $\lfloor \frac{n-1}{m+1} \rfloor + 1$ *if* $V(G)$ *has a partition into sets* $V_{i,j}$ *(*$i, j \in \{1, \ldots, n\}$*) with the following properties:*

1. $|V_{i,j}| \geq 1$ *for all* $i, j \geq 1$.
2. $G[\cup_{j=1}^{n} V_{i,j}]$ *is connected for all* $i \geq 1$.
3. $G[\cup_{i=1}^{n} V_{i,j}]$ *is connected for all* $j \geq 1$.
4. *For* $i, j, k, \ell \geq 1$, *if a vertex of* $V_{i,j}$ *is adjacent to a vertex of* $V_{k,\ell}$, *then* $|k - i| \leq m$ *and* $|\ell - j| \leq m$.

3   New Polynomial-Time Results

In this section we prove Theorem 3, which states that GRAPH ISOMORPHISM is polynomial-time solvable on $(\overline{2P_1 + P_3}, P_2 + P_3)$-free graphs (see also Fig. 1). The complexity of GRAPH ISOMORPHISM on $(\overline{2P_1 + P_3}, 2P_2)$-free graphs was previously unknown, but since this class is contained in the class of $(\overline{2P_1 + P_3}, P_2 + P_3)$-free graphs, Theorem 3 implies that GRAPH ISOMORPHISM is also polynomial-time solvable on this class. Before proving Theorem 3, we first prove a useful lemma.

Lemma 6. *Let G be a* $\overline{2P_1 + P_3}$-*free graph containing an induced K_5 with vertex set* K^G. *Then* $V(G)$ *can be partitioned into sets* $A_1^G, \ldots, A_p^G, N_1^G, \ldots, N_p^G, B^G$ *for some* $p \geq 5$ *such that:*

(i) $K^G \subseteq \bigcup A_i^G$;

(ii) $G[\bigcup A_i^G]$ *is a complete multipartite graph, with partition* $A_1^G, \ldots, A_p^G$;

(iii) *For every* $i \in \{1, \ldots, p\}$, *every vertex of* N_i^G *has a neighbour in* A_i^G, *but is anti-complete to* A_j^G *for every* $j \in \{1, \ldots, p\} \setminus \{i\}$; *and*

(iv) B^G *is anti-complete to* $\bigcup A_i^G$.

Furthermore, given K^G, *this partition is unique (up to permuting the indices on the* A_i^Gs *and corresponding* N_i^Gs*) and can be found in polynomial time.*

Proof. Let G be a $\overline{2P_1 + P_3}$-free graph containing an induced K_5 with vertex set K^G. If a vertex $v \in V(G) \setminus K^G$ has two neighbours $x, x' \in K^G$ and two non-neighbours $y, y' \in K^G$, then $G[x, x', y, v, y']$ is a $\overline{2P_1 + P_3}$, a contradiction. Therefore every vertex in $V(G) \setminus K^G$ has either at most one non-neighbour in K^G or at most one neighbour in K^G. Let L^G denote the set of vertices that are either in K^G or have at most one non-neighbour in K^G and note that L^G is uniquely defined by the choice of K^G.

We claim that $G[L^G]$ is a complete multipartite graph. Suppose, for contradiction, that $G[L^G]$ is not complete multipartite. Then $G[L^G]$ contains an induced $P_1 + P_2 = \overline{P_3}$, say on vertices v, v', v'' (note that some of these vertices may be in K^G). Now each of v, v', v'' has at most one non-neighbour in K^G and if a vertex $w \in \{v, v', v''\}$ is in K^G, then it is adjacent to every vertex in $K^G \setminus \{w\}$. Therefore, since $|K^G| = 5$, there must be vertices $u, u' \in K^G \setminus \{v, v', v''\}$ that are complete to $\{v, v', v''\}$. Now $G[u, u', v', v, v'']$ is a $\overline{2P_1 + P_3}$. This contradiction completes the proof that $G[L^G]$ is complete multipartite.

We let $A_1^G, \ldots, A_p^G$ be the partition classes of the complete multipartite graph $G[L^G]$. Note that $p \geq 5$, since each A_i^G contains at most one vertex of K^G. We claim that each vertex not in L^G has neighbours in at most one set A_i^G. Suppose, for contradiction, that there is a vertex $v \in V(G) \setminus L^G$ with neighbours in two distinct sets A_i^G, say v is adjacent to $u \in A_1^G$ and $u' \in A_2^G$. Since $v \notin L^G$, the vertex v has at most one neighbour in K^G. Since $|K^G| = 5$, there must be two vertices $y, y' \in K^G \setminus (A_1^G \cup A_2^G)$ that are non-adjacent to v. Now $G[u, u', y, v, y']$ is a $\overline{2P_1 + P_3}$, a contradiction. Therefore every vertex not in L^G has neighbours in at most one set A_i^G. Let N_i^G be the set of vertices in $V(G) \setminus L^G$ that have neighbours in A_i^G and let B^G be the set of vertices in $V(G) \setminus L^G$ that are anti-complete to L^G. Finally, note that the partition of $V(G)$ into sets $A_1^G, \ldots, A_p^G, N_1^G, \ldots, N_p^G, B^G$ can be found in polynomial time and is unique (up to permuting the indices on the A_i^Gs and corresponding N_i^Gs).   □

Theorem 3. GRAPH ISOMORPHISM *is polynomial-time solvable on* $(\overline{2P_1 + P_3}, P_2 + P_3)$-*free graphs.*

Proof. As GRAPH ISOMORPHISM can be solved component-wise, we need only consider connected graphs. Therefore, as GRAPH ISOMORPHISM is polynomial-time solvable on $(K_5, P_2 + P_3)$-free graphs by Lemma 1, and we can test whether a graph is K_5-free in polynomial time, it only remains to consider the class of connected $(\overline{2P_1 + P_3}, P_2 + P_3)$-free graphs G that contain an induced K_5. Let K^G

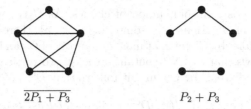

$$\overline{2P_1 + P_3} \qquad\qquad P_2 + P_3$$

Fig. 1. Forbidden induced subgraphs from Theorem 3.

be the vertices of such a K_5 in G (note that such a set K^G can be found in polynomial time, but it is not necessarily unique). Let $A_1^G, \ldots, A_p^G, N_1^G, \ldots, N_p^G, B^G$ be defined as in Lemma 6 and let $L^G = \bigcup A_i^G$ and $D^G = V(G) \setminus L^G$.

Suppose G and H are connected $(\overline{2P_1 + P_3}, P_2 + P_3)$-free graphs that each contain an induced K_5. If G and H have bounded clique-width (which happens in Case 1 below), then by Theorem 2 we are done. Otherwise, note that if K^G and K^H are vertex sets that induce a K_5 in G and H, respectively, then Lemma 6 implies that L^G, D^G, L^H and D^H are uniquely defined. Therefore, we fix one choice of K^G and, for each choice of K^H, test whether there is an isomorphism $f : G \to H$ such that $f(L^G) = L^H$ (we use this approach in Cases 2 and 3 below). Clearly, we may assume that the vertex partitions given by Lemma 6 for G and H have the same value of p and that $|A_i^G| = |A_i^H|$ and $|N_i^G| = |N_i^H|$ for all $i \in \{1, \ldots, p\}$ and $|B^G| = |B^H|$. Furthermore, for any claims we prove about G and its vertex sets, we may assume that the same claims hold for H (otherwise such an isomorphism f does not exist). We start by proving the following four claims.

Claim 1. $G[D^G]$ is P_3-free

Indeed, suppose, for contradiction, that $G[D^G]$ contains an induced P_3, say on vertices u, u', u''. Since $|K^G| = 5$ and each vertex in D^G has at most one neighbour in K^G, there must be vertices $v, v' \in K^G$ that are anti-complete to $\{u, u', u''\}$. Then $G[v, v', u, u', u'']$ is a $P_2 + P_3$, a contradiction. ◇

Claim 2. If $v \in N_j^G$ for some $j \in \{1, \ldots, p\}$ and there are two adjacent vertices $u, u' \in D^G \setminus N_j^G$, then v is complete to $\{u, u'\}$.

Since $G[D^G]$ is P_3-free by Claim 1, the vertex v must be either complete or anti-complete to $\{u, u'\}$. Suppose, for contradiction, that v is anti-complete to $\{u, u'\}$. Since $v \in N_j^G$, v has a neighbour $v' \in A_j^G$. Since $|K^G \setminus A_j^G| \geq 4$ and each vertex in D^G has at most one neighbour in K^G, there is a vertex $v'' \in K^G \setminus A_j^G$ that is non-adjacent to both u and u'. Since $v'' \notin A_j^G$, v'' is also non-adjacent to v, but is adjacent to v'. Now $G[u, u', v, v', v'']$ is a $P_2 + P_3$, a contradiction. ◇

Claim 3. If $G[D^G]$ has at least two components and one of these components C has at least three vertices, then there is an $i \in \{1, \ldots, p\}$ such that $D^G \setminus C \subset N_i^G \cup B^G$ and all but at most one vertex of C belongs to N_i^G.

By Claim 1, $G[D^G]$ is a disjoint union of cliques. As G is connected, $D^G \setminus C$ cannot be a subset of B^G. Hence, for some $i \in \{1, \ldots, p\}$, there must be a vertex $x \in N_i^G \setminus C$. Therefore, by Claim 2, at most one vertex of C can lie outside of N_i^G. As $|C| \geq 3$, it follows that $C \cap N_i^G$ contains at least two vertices. As the vertices in C are pairwise adjacent, by Claim 2 it follows that $D^G \setminus C \subset N_i^G \cup B^G$. ◇

Claim 4. *Let $i \in \{1, \ldots, p\}$. If $G[D^G]$ contains at least two non-trivial components and there is a vertex v in A_i^G with two non-neighbours in the same component of $G[D^G]$, then v is anti-complete to D^G. Furthermore, there is at most one vertex in A_i^G with this property.*

Suppose $v \in A_i^G$ has two non-neighbours x, x' in some component C of $G[D^G]$. By Claim 1, $G[D^G]$ is a disjoint union of cliques, so x must be adjacent to x'. We claim that v is anti-complete to $D^G \setminus C$. Suppose, for contradiction, that v has a neighbour $y \in D^G \setminus C$. Since every vertex of D^G has at most one neighbour in K^G, there must be a vertex $z \in K^G \setminus A_i^G$ that is non-adjacent to x, x' and y and so $G[x, x', y, v, z]$ is a $P_2 + P_3$. This contradiction implies that v is indeed anti-complete to $D^G \setminus C$. Now $G[D^G \setminus C]$ contains another non-trivial component C' and we have shown that v is anti-complete to C'. Repeating the same argument with C' taking the place of C, we find that v is anti-complete to $D^G \setminus C'$, and therefore v is anti-complete to D^G. Finally, suppose, for contradiction, that there are two vertices $v, v' \in A_i^G$ that are both anti-complete to D^G. Let x, x' be adjacent vertices in D^G and let $z \in K^G \setminus A_i^G$ be a vertex non-adjacent to x and x'. Then $G[x, x', v, z, v']$ is a $P_2 + P_3$, a contradiction. ◇

We now start a case distinction and first consider the following case.

Case 1. *$G[D^G]$ contains at most one non-trivial component.*

In this case we will show that G has bounded clique-width, and so we will be done by Theorem 2. By Claim 1, every component of $G[D^G]$ is a clique. Since $G[D^G]$ contains at most one non-trivial component, we may partition D^G into a clique C and an independent set I (note that C or I may be empty). If $|C| \geq 3$ and $|I| \geq 1$, then by Claim 3 there is an $i \in \{1, \ldots, p\}$ such that at most one vertex of $C \cup I$ is outside N_i^G; if such a vertex exists, then by Fact 1 we may delete it. Now if $|C| \leq 3$, then by Fact 1 we may delete the vertices of C. Thus we may assume that either $C = \emptyset$ or $|C| \geq 4$ and furthermore, if $|C| \geq 4$ and $|I| \geq 1$, then $C \cup I \subseteq N_i^G$ for some $i \in \{1, \ldots, p\}$. Note that $I \cap B^G = \emptyset$ since G is connected, so $B^G \subset C$. Hence $G[B^G]$ is a complete graph, so it has clique-width at most 2. Applying a bipartite complementation between B^G and $C \setminus B^G$ removes all edges between B^G and $V(G) \setminus B^G$. By Fact 3, we may thus assume that $B^G = \emptyset$.

Let M be the set of vertices in L^G that have neighbours in I. We claim that M is complete to all but at most one vertex of C. We may assume that $|C| \geq 4$ and $|I| \geq 1$, otherwise the claim follows trivially. Therefore, as noted above, $C \cup I \subseteq N_i^G$ for some $i \in \{1, \ldots, p\}$. Suppose $u \in M$ has a neighbour $u' \in I$ and note that this implies $u \in A_i^G$, $u' \in N_i^G$. Suppose, for contradiction, that u has two non-neighbours $v, v' \in C$ and let $w \in K^G \setminus A_i^G$. Then $G[v, v', u', u, w]$

is a $P_2 + P_3$, a contradiction. Therefore if $u \in M$, then u has at most one non-neighbour in C. Now suppose that there are two vertices $u, u' \in M$. It follows that $u, u' \in A_i^G$, so these vertices must be non-adjacent. Furthermore, each of these vertices has at most one non-neighbour in C. If u and u' have different neighbourhoods in C, then without loss of generality we may assume that there are vertices $x, y, y' \in C$ such that u is adjacent to x, y and y' and u' is adjacent to y and y', but not to x. Now $G[y, y', u, u', x]$ is a $\overline{2P_1 + P_3}$, a contradiction. Hence every vertex in M has the same neighbourhood in C, which consists of all but at most one vertex of C and the claim holds. If the vertices of M are not complete to C, then we delete one vertex of C (we may do so by Fact 1), after which M will be complete to C. Hence we may assume that M is complete to C.

Now note that for all $i \in \{1, \ldots, p\}$, the graph $G_i = G[(A_i^G \setminus M) \cup (N_i^G \cap C)]$ is a $\overline{2P_1 + P_3}$-free split graph, so it has bounded clique-width by Lemma 3. Furthermore $G_i' = G[(A_i^G \cap M) \cup (N_i^G \cap I)]$ is a $(P_2 + P_3)$-free bipartite graph, so it has bounded clique-width by Lemma 4. Let G_i'' be the graph obtained from the disjoint union $G_i + G_i'$ by complementing A_i^G and $(N_i^G \cap C)$. By Fact 2, G_i'' also has bounded clique-width. Therefore the disjoint union G^* of all the G_i''s has bounded clique-width. Now G can be constructed from G^* by complementing L^G, complementing C and applying a bipartite complementation between C and M. Hence, by Facts 2 and 3, G has bounded clique-width. This completes Case 1.

We may now assume that Case 1 does not apply, that is, $G[D^G]$ has at least two non-trivial components. This leads us to our second and third cases.

Case 2. $G[D^G]$ *contains at least two non-trivial components, but is K_4-free.*

Recall that $G[D^G]$ is P_3-free by Claim 1, so every component of $G[D^G]$ is a clique. Let C be a non-trivial component of $G[D^G]$ and let $x, y \in C$. Then x is adjacent to y and $x, y \in N_i^G \cup N_j^G \cup B^G$ for some (not necessarily distinct) $i, j \in \{1, \ldots, p\}$. By Claim 2, every vertex z in a component of $G[D^G]$ other than C must also be in $N_i^G \cup N_j^G \cup B^G$. As $G[D^G]$ contains at least two non-trivial components, repeating this argument with another non-trivial component implies that every vertex of D^G lies in $N_i^G \cup N_j^G \cup B^G$. Without loss of generality, we may therefore assume that $N_k^G = \emptyset$ for $k \geq 3$.

Since $G[D^G]$ is K_4-free, for each $i \in \{1, \ldots, p\}$ the graph $G[D^G \cup A_i^G]$ is K_5-free. This means that every K_5 in G is entirely contained in L^G. By Claim 4, for $i \geq 3$, $|A_i^G| = 1$ and so $L^G \setminus (A_1^G \cup A_2^G)$ must be a clique. The vertices of $L^G \setminus (A_1^G \cup A_2^G)$ have no neighbours outside L^G and are adjacent to every other vertex of L^G, so these vertices are in some sense interchangeable. Indeed, $N[v] = L^G$ for every $v \in L^G \setminus (A_1^G \cup A_2^G)$, and so every bijection that permutes the vertices of $L^G \setminus (A_1^G \cup A_2^G)$ and leaves the other vertices of G unchanged is an isomorphism from G to itself. Let G' be the graph obtained from G by deleting all vertices in A_i^G for $i \geq 6$ (if any such vertices are present). Now G' is K_6-free and thus $(K_6, P_2 + P_3)$-free. We can test isomorphism of such graphs G' in polynomial time by Lemma 1. If there is an isomorphism between two such graphs G' and H', then, as the vertices of $L^G \setminus (A_1^G \cup A_2^G)$ are interchangeable, we can extend it to a full isomorphism of G and H by mapping the remaining vertices of $L^G \setminus (A_1^G \cup A_2^G)$ to $L^H \setminus (A_1^H \cup A_2^H)$ arbitrarily. This completes Case 2.

Case 3. $G[D^G]$ *contains at least two non-trivial components and contains an induced* K_4.

Recall that $G[D^G]$ is P_3-free by Claim 1, so every component of $G[D^G]$ is a clique. We claim that $D^G \subseteq N_i^G \cup B^G$ for some $i \in \{1, \ldots, p\}$. Let C be a component of $G[D^G]$ that contains at least four vertices, and let C' be a component of $G[D^G]$ other than C, and note that such components exist by assumption. By Claim 3, there is an $i \in \{1, \ldots, p\}$ such that $D^G \setminus C \subset N_i^G \cup B^G$ and all but at most one vertex of C belongs to N_i^G. In particular, this implies that $C' \subset N_i^G \cup B^G$. By Claim 2, it follows that C cannot have a vertex in N_j^G for some $j \in \{1, \ldots, p\} \setminus \{i\}$, and so $C \subset N_i^G \cup B^G$. Without loss of generality, we may therefore assume that $N_j^G = \emptyset$ for $j \in \{2, \ldots, p\}$ and so $D^G = N_1^G \cup B^G$. Now if $j \in \{2, \ldots, p\}$, then the vertices of A_j^G are anti-complete to D^G, so Claim 4 implies that $|A_j^G| = 1$. This implies that $L^G \setminus A_1^G$ is a clique.

By Claim 4 there is at most one vertex $x^G \in A_1^G$ that has two non-neighbours in the same non-trivial component C of $G[D^G]$ and if such a vertex exists, then it must be anti-complete to D^G. Let $A_1^{*G} = A_1^G \setminus \{x^G\}$ if such a vertex x^G exists and $A_1^{*G} = A_1^G$ otherwise. Then every vertex in A_1^{*G} has at most one non-neighbour in each component of $G[D^G]$. Note that A^{*G} is non-empty, since D^G is non-empty and G is connected.

Suppose C is a component of $G[D^G]$ on at least four vertices. Now suppose, for contradiction, that there are two vertices $y, y' \in A_1^{*G}$ with different neighbourhoods in C. Then without loss of generality there is a vertex $x \in C$ that is adjacent to y, but not to y'. Since $|C| \geq 4$ and every vertex in A_1^{*G} has at most one non-neighbour in C, there must be two vertices $z, z' \in C$ that are adjacent to both y and y'. Now $G[z, z', x, y', y]$ is a $\overline{2P_1 + P_3}$, a contradiction. We conclude that every vertex in A_1^{*G} has the same neighbourhood in C. This implies that every vertex of C is either complete or anti-complete to A_1^{*G}. If a vertex of C is anti-complete to A_1^{*G}, then it is anti-complete to A_1^G, and so it lies in B^G.

Let D^{*G} be the set of vertices in D^G that are in components of $G[D^G]$ that have at most three vertices. Then every vertex of $D^G \setminus D^{*G}$ is complete or anti-complete to A_1^{*G} and anti-complete to $A_1^G \setminus A_1^{*G}$.

Now let $G' = G[D^{*G} \cup L^G \setminus (A_1^G \setminus A_1^{*G})]$ and note that this graph is uniquely defined by G and K^G. Then $G'[D^{*G}]$ is K_4-free, so $G'[D^{*G} \cup A_1^{*G}]$ is K_5-free, so every induced K_5 in G' is entirely contained in $L^G \setminus (A_1^G \setminus A_1^{*G})$. Furthermore, since $p \geq 5$, every vertex in $L^G \setminus (A_1^G \setminus A_1^{*G})$ is contained in an induced K_5 in G'. Therefore every isomorphism q from G' to H' satisfies $q(L^G \setminus (A_1^G \setminus A_1^{*G})) = L^H \setminus (A_1^H \setminus A_1^{*H})$. Therefore a bijection $f : V(G) \to V(H)$ is an isomorphism from G to H such that $f(L^G) = L^H$ if and only if all of the following hold:

1. The restriction of f to $V(G')$ is an isomorphism from G' to H' such that $f(A_1^{*G}) = A_1^{*H}$.
2. $f(A_1^G \setminus A_1^{*G}) = A_1^H \setminus A_1^{*H}$.

3. For every component C of $G[D^G]$ with at least four vertices, $f(C)$ is a component of $H[D^H]$ on the same number of vertices and $|C \cap B^G| = |f(C) \cap B^H|$.

It is therefore sufficient to test whether there is a bijection from G to H with the above properties. Note that these properties are defined on pairwise disjoint vertex sets, and the edges in G and H between these sets are completely determined by the definition of the sets. Thus it is sufficient to independently test whether there are bijections satisfying each of these properties. If D^{*G} is empty, then G' is a complete multipartite graph, so we can easily test if Property 1 holds in this case. Otherwise, since A_j^G has no neighbours outside L^G for $j \in \{2, \ldots, p\}$, every isomorphism from G' to H' satisfies $f(A_1^{*G}) = A_1^{*H}$, so it is sufficient to test if G' and H' are isomorphic, and we can do this by applying Case 1 or Case 2. The sets $A_1^G \setminus A_1^{*G}$ and $A_1^H \setminus A_1^{*H}$ consist of at most one vertex, so we can test if Property 2 can be satisfied in polynomial time. To satisfy Property 3, we only need to check whether there is a bijection q from the components of $G[D^{*G} \setminus D^G]$ to the components of $H[D^{*H} \setminus D^H]$ such that $|q(C)| = |C|$ and $|q(C) \cap B^H| = |C \cap B^G|$ for every component of $G[D^{*G} \setminus D^G]$ and this can clearly be done in polynomial time. This completes the proof of Case 3. □

4   New GI-Complete Results

We state Theorems 4, 5 and 6, which establish that GRAPH ISOMORPHISM is GI-complete on (diamond, $2P_3$)-free, (diamond, P_6)-free and (gem, $P_1 + 2P_2$)-free graphs, respectively. The complexity of GRAPH ISOMORPHISM on $(\overline{2P_1 + P_3}, 2P_3)$-free graphs and (gem, P_6)-free graphs was previously unknown, but as these classes contain the classes of (diamond, $2P_3$)-free graphs and (diamond, P_6)-free graphs, respectively, Theorems 4 and 5, respectively, imply that GRAPH ISOMORPHISM is also GI-complete on these classes. In Theorems 4 and 5, GI-completeness follows from the fact that the constructions used in our proofs (which we omit) fall into the framework of so-called simple path encodings (see [25]). The construction used in the proof of Theorem 6 does not fall into this framework and we give a direct proof of GI-completeness in this case.

Theorem 4. GRAPH ISOMORPHISM *is* GI-*complete on* (*diamond*, $2P_3$)-*free graphs.*

Theorem 5. GRAPH ISOMORPHISM *is* GI-*complete on* (*diamond*, P_6)-*free graphs.*

Theorem 6. GRAPH ISOMORPHISM *is* GI-*complete on* (*gem*, $P_1 + 2P_2$)-*free graphs. Furthermore,* (*gem*, $P_1 + 2P_2$)-*free graphs have unbounded clique-width.*

Proof Sketch. Let G be a graph. Let $v_1^G, \ldots, v_n^G$ be the vertices of G and let $e_1^G, \ldots, e_m^G$ be the edges of G. We construct a graph $q(G)$ from G as follows:

1. Create a complete multipartite graph with partition $(A_1^G, \ldots, A_n^G)$, where $|A_i^G| = d_G(v_i^G)$ for $i \in \{1, \ldots, n\}$ and let $A^G = \bigcup A_i^G$.

2. Create a complete multipartite graph with partition $(B_1^G, \ldots, B_m^G)$, where $|B_i^G| = 2$ for $i \in \{1, \ldots, m\}$ and let $B^G = \bigcup B_i^G$.
3. Take the disjoint union of the two graphs above, then for each edge $e_i^G = v_{i_1}^G v_{i_2}^G$ in G in turn, add an edge from one vertex of B_i^G to a vertex of $A_{i_1}^G$ and an edge from the other vertex of B_i^G to a vertex of $A_{i_2}^G$. Do this in such a way that the edges added between A^G and B^G form a perfect matching.

It can be checked that $q(G)$ is (gem, $P_1 + 2P_2$)-free for every graph G. Let G and H be graphs. Let G^* and H^* be the graphs obtained from G and H, respectively, by adding four pairwise adjacent vertices that are adjacent to every vertex of G and H, respectively. Note that every vertex of G^* and H^* has degree at least 3. We claim that G is isomorphic to H if and only if $q(G^*)$ is isomorphic to $q(H^*)$. As the latter two graphs are (gem, $P_1 + 2P_2$)-free, this proves the first result.

Let H_n be the $n \times n$ grid. We use Lemma 5 with $m = 1$ combined with Fact 2 to prove that the set of graphs $\{q(H_n) \mid n \in \mathbb{N}\}$, which are (gem, $P_1 + 2P_2$)-free as stated above, has unbounded clique-width. $\square$

5   Classifying the Complexity of GRAPH ISOMORPHISM for (H_1, H_2)-Free Graphs

Given four graphs H_1, H_2, H_3, H_4, the classes of (H_1, H_2)-free graphs and (H_3, H_4)-free graphs are *equivalent* if the unordered pair H_3, H_4 can be obtained from the unordered pair H_1, H_2 by some combination of the operations:

  (i) complementing both graphs in the pair, and
 (ii) if one of the graphs in the pair is K_3, replacing it with the paw or vice versa.

Note that two graphs G and H are isomorphic if and only if their complements $\overline{G}$ and $\overline{H}$ are isomorphic. Therefore, for every pair of graphs H_1, H_2, the GRAPH ISOMORPHISM problem is polynomial-time solvable or GI-complete for (H_1, H_2)-free graphs if and only if the same is true for $(\overline{H_1}, \overline{H_2})$-free graphs. Since GRAPH ISOMORPHISM can be solved component-wise, and it can easily be solved on complete multipartite graphs in polynomial time, Lemma 2 implies that for every graph H_1, the GRAPH ISOMORPHISM problem is polynomial-time solvable or GI-complete for (H_1, K_3)-free graphs if and only if the same is true for (H_1, paw)-free graphs. Thus if two classes are equivalent, then the complexity of GRAPH ISOMORPHISM is the same on both of them. Here is the summary of known results for the complexity of GRAPH ISOMORPHISM on (H_1, H_2)-free graphs (see Sect. 2 for notation; we omit the proof).

Theorem 7. *For a class $\mathcal{G}$ of graphs defined by two forbidden induced subgraphs, the following holds:*

1. GRAPH ISOMORPHISM *is solvable in polynomial time on $\mathcal{G}$ if $\mathcal{G}$ is equivalent to a class of (H_1, H_2)-free graphs such that one of the following holds:*

   (i) H_1 or $H_2 \subseteq_i P_4$
   (ii) $\overline{H_1}$ and $H_2 \subseteq_i K_{1,t} + P_1$ for some $t \geq 1$
  (iii) $\overline{H_1}$ and $H_2 \subseteq_i tP_1 + P_3$ for some $t \geq 1$
  (iv) $H_1 \subseteq_i K_t$ and $H_2 \subseteq_i 2K_{1,t}, K_{1,t}^+$ or P_5 for some $t \geq 1$
   (v) $H_1 \subseteq_i$ paw and $H_2 \subseteq_i P_2 + P_4, P_6, S_{1,2,2}$ or $K_{1,t}^{++} + P_1$ for some $t \geq 1$
  (vi) $H_1 \subseteq_i$ diamond and $H_2 \subseteq_i P_1 + 2P_2$
 (vii) $H_1 \subseteq_i$ gem and $H_2 \subseteq_i P_1 + P_4$ or P_5
(viii) $H_1 \subseteq_i \overline{2P_1 + P_3}$ and $H_2 \subseteq_i P_2 + P_3$.

2. GRAPH ISOMORPHISM *is* GI-*complete on* $\mathcal{G}$ *if* $\mathcal{G}$ *is equivalent to a class of* (H_1, H_2)-*free graphs such that one of the following holds:*
   (i) *neither* H_1 *nor* H_2 *is a path star forest*
  (ii) *neither* $\overline{H_1}$ *nor* $\overline{H_2}$ *is a path star forest*
  (iii) $H_1 \supseteq_i K_3$ *and* $H_2 \supseteq_i 2P_1 + 2P_2, P_1 + 2P_3, 2P_1 + P_4$ *or* $3P_2$
  (iv) $H_1 \supseteq_i K_4$ *and* $H_2 \supseteq_i K_{1,4}^{++}, P_1 + 2P_2$ *or* $P_1 + P_4$
   (v) $H_1 \supseteq_i K_5$ *and* $H_2 \supseteq_i K_{1,3}^{++}$
  (vi) $H_1 \supseteq_i C_4$ *and* $H_2 \supseteq_i K_{1,3}, 3P_1 + P_2$ *or* $2P_2$
 (vii) $H_1 \supseteq_i$ diamond *and* $H_2 \supseteq_i K_{1,3}, P_2 + P_4, 2P_3$ *or* P_6
(viii) $H_1 \supseteq_i$ gem *and* $H_2 \supseteq_i P_1 + 2P_2$.

Open Problem 1. *What is the complexity of* GRAPH ISOMORPHISM *on* (H_1, H_2)-*free graphs in the following cases?*

   (i) $H_1 = K_3$ *and* $H_2 \in \{P_7, S_{1,2,3}\}$
  (ii) $H_1 = K_4$ *and* $H_2 = S_{1,1,3}$
  (iii) $H_1 =$ diamond *and* $H_2 \in \{P_1 + P_2 + P_3, P_1 + P_5\}$
  (iv) $H_1 = \overline{\text{gem}}$ *and* $H_2 = P_2 + P_3$
   (v) $H_1 = \overline{2P_1 + P_3}$ *and* $H_2 = P_5$

    Note that all of the classes of (H_1, H_2)-free graphs in Open Problem 1 are incomparable. We omit the proof of the next theorem.

Theorem 8. *Let* $\mathcal{G}$ *be a class of graphs defined by two forbidden induced subgraphs. Then* $\mathcal{G}$ *is not equivalent to any of the classes listed in Theorem 7 if and only if it is equivalent to one of the seven cases listed in Open Problem 1.*

6   Conclusions

By combining known and new results we determined the complexity of GRAPH ISOMORPHISM in terms of polynomial-time solvability and GI-completeness for (H_1, H_2)-free graphs for all but seven pairs (H_1, H_2). This also led to a new class of (H_1, H_2)-free graphs whose clique-width is unbounded. In particular, we developed a technique for showing polynomial-time solvability for $(\overline{2P_1 + P_3}, H)$-free graphs, which we illustrated for the case $H = P_2 + P_3$. For future work we have some preliminary results for the case where $H = P_5$.

References

1. Babai, L.: Graph isomorphism in quasipolynomial time [extended abstract]. In: Proceedings of STOC 2016, pp. 684–697 (2016)
2. Belmonte, R., Otachi, Y., Schweitzer, P.: Induced minor free graphs: isomorphism and clique-width. Algorithmica **80**(1), 29–47 (2018)
3. Blanché, A., Dabrowski, K.K., Johnson, M., Lozin, V.V., Paulusma, D., Zamaraev, V.: Clique-width for graph classes closed under complementation. In: Proceedings of MFCS 2017. LIPIcs, vol. 83, pp. 73:1–73:14 (2017)
4. Booth, K.S., Colbourn, C.J.: Problems polynomially equivalent to graph isomorphism. Technical report CS-77-04, Department of Computer Science, University of Waterloo (1979)
5. Brandstädt, A., Dabrowski, K.K., Huang, S., Paulusma, D.: Bounding the clique-width of H-free split graphs. Discrete Appl. Math. **211**, 30–39 (2016)
6. Brandstädt, A., Dabrowski, K.K., Huang, S., Paulusma, D.: Bounding the clique-width of H-free chordal graphs. J. Graph Theory **86**(1), 42–77 (2017)
7. Colbourn, C.J.: On testing isomorphism of permutation graphs. Networks **11**(1), 13–21 (1981)
8. Dabrowski, K.K., Dross, F., Paulusma, D.: Colouring diamond-free graphs. J. Comput. Syst. Sci. **89**, 410–431 (2017)
9. Dabrowski, K.K., Huang, S., Paulusma, D.: Bounding clique-width via perfect graphs. J. Comput. Syst. Sci. **104**, 202–215
10. Dabrowski, K.K., Johnson, M., Paulusma, D.: Clique-width for hereditary graph classes. In: Proceedings of BCC 2019. London Mathematical Society Lecture Note Series, vol. 456, 1–56 (2019)
11. Dabrowski, K.K., Lozin, V.V., Paulusma, D.: Clique-width and well-quasi-ordering of triangle-free graph classes. In: Bodlaender, H.L., Woeginger, G.J. (eds.) WG 2017. LNCS, vol. 10520, pp. 220–233. Springer, Cham (2017). https://doi.org/10. 1007/978-3-319-68705-6_17
12. Dabrowski, K.K., Paulusma, D.: Clique-width of graph classes defined by two forbidden induced subgraphs. Comput. J. **59**(5), 650–666 (2016)
13. Grohe, M., Neuen, D., Schweitzer, P., Wiebking, D.: An improved isomorphism test for bounded-tree-width graphs. In: Proceedings of ICALP 2018. LIPIcs, vol. 107, pp. 67:1–67:14 (2018)
14. Grohe, M., Schweitzer, P.: Isomorphism testing for graphs of bounded rank width. In: Proceedings of FOCS 2015, pp. 1010–1029 (2015)
15. Kamiński, M., Lozin, V.V., Milanič, M.: Recent developments on graphs of bounded clique-width. Discrete Appl. Math. **157**(12), 2747–2761 (2009)
16. Kratsch, S., Schweitzer, P.: Graph isomorphism for graph classes characterized by two forbidden induced subgraphs. Discrete Appl. Math. **216**(Part 1), 240–253 (2017)
17. Lokshtanov, D., Pilipczuk, M., Pilipczuk, M., Saurabh, S.: Fixed-parameter tractable canonization and isomorphism test for graphs of bounded treewidth. SIAM J. Comput. **46**(1), 161–189 (2017)
18. Lozin, V.V.: Bipartite graphs without a skew star. Discrete Math. **257**(1), 83–100 (2002)
19. Lozin, V.V., Rautenbach, D.: On the band-, tree-, and clique-width of graphs with bounded vertex degree. SIAM J. Discrete Math. **18**(1), 195–206 (2004)
20. Neuen, D.: Graph isomorphism for unit square graphs. In: Proceedings of ESA 2016. LIPIcs, vol. 57, pp. 70:1–70:17 (2016)

21. Olariu, S.: Paw-free graphs. Inf. Process. Lett. **28**(1), 53–54 (1988)
22. Ponomarenko, I.N.: Isomorphism problem for classes of graphs closed under contractions. Zapiski Nauchnykh Seminarov (LOMI) **174**, 147–177 (1988). (in Russian, English translation in Journal of Soviet Mathematics, 55(2), 1621–1643, (1991))
23. de Ridder, H.N., et al.: Information system on graph classes and their inclusions (2001–2019). http://www.graphclasses.org
24. Schöning, U.: Graph isomorphism is in the low hierarchy. J. Comput. Syst. Sci. **37**(3), 312–323 (1988)
25. Schweitzer, P.: Towards an isomorphism dichotomy for hereditary graph classes. Theory Comput. Syst. **61**(4), 1084–1127 (2017)

Hamiltonicity for Convex Shape Delaunay and Gabriel Graphs

Prosenjit Bose[1], Pilar Cano[1,2(✉)], Maria Saumell[3,4], and Rodrigo I. Silveira[2]

[1] School of Computer Science, Carleton University, Ottawa, Canada
jit@scs.carleton.ca
[2] Departament de Matemàtiques, Universitat Politècnica de Catalunya,
Barcelona, Spain
{m.pilar.cano,rodrigo.silveira}@upc.edu
[3] Institute of Computer Science, The Czech Academy of Sciences, Prague, Czechia
[4] Department of Theoretical Computer Science, Faculty of Information Technology,
Czech Technical University in Prague, Prague, Czechia
maria.saumell@fit.cvut.cz

Abstract. We study Hamiltonicity for some of the most general variants of Delaunay and Gabriel graphs. Instead of defining these proximity graphs using circles, we use an arbitrary convex shape $\mathcal{C}$. Let S be a point set in the plane. The k-order Delaunay graph of S, denoted $k\text{-}DG_{\mathcal{C}}(S)$, has vertex set S and edge pq provided that there exists *some* homothet of $\mathcal{C}$ with p and q on its boundary and containing at most k points of S different from p and q. The k-order Gabriel graph $k\text{-}GG_{\mathcal{C}}(S)$ is defined analogously, except for the fact that the homothets considered are restricted to be *smallest* homothets of $\mathcal{C}$ with p and q on its boundary. We provide upper bounds on the minimum value of k for which $k\text{-}GG_{\mathcal{C}}(S)$ is Hamiltonian. Since $k\text{-}GG_{\mathcal{C}}(S) \subseteq k\text{-}DG_{\mathcal{C}}(S)$, all results carry over to $k\text{-}DG_{\mathcal{C}}(S)$. In particular, we give upper bounds of 24 for every $\mathcal{C}$ and 15 for every point-symmetric $\mathcal{C}$. We also improve the bound to 7 for squares, 11 for regular hexagons, 12 for regular octagons, and 11 for even-sided regular t-gons (for $t \geq 10$). These constitute the first general results on Hamiltonicity for convex shape Delaunay and Gabriel graphs.

1 Introduction

The study of the combinatorial properties of geometric graphs has played an important role in the area of Discrete and Computational Geometry. One of the

P.B. was partially supported by NSERC. P.C. was supported by CONACyT. M.S. was supported by the Czech Science Foundation, grant number GJ19-06792Y, and by institutional support RVO:67985807. R.S. was supported by MINECO through the Ramón y Cajal program. P.C. and R.S. were also supported by projects MINECO MTM2015-63791-R and Gen. Cat. 2017SGR1640. This project has received funding from the European Union's Horizon 2020 research and innovation programme under the Marie Skłodowska-Curie grant agreement No 734922.

© Springer Nature Switzerland AG 2019
Z. Friggstad et al. (Eds.): WADS 2019, LNCS 11646, pp. 196–210, 2019.
https://doi.org/10.1007/978-3-030-24766-9_15

fundamental structures that has been studied intensely is the Delaunay triangulation of a planar point set and some of its spanning subgraphs, such as the Gabriel Graph, the Relative Neighborhood Graph and the Minimum Spanning Tree. Delaunay triangulations possess many interesting properties. For example, among all triangulations of a given planar point set, the Delaunay triangulation maximizes the minimum angle. It is also a 1.99-spanner [19] (i.e., for any pair of vertices x, y, the shortest path between x and y in the Delaunay triangulation has length that is at most 1.99 times $|xy|$). See [16] for an encyclopedic treatment of this structure and its many properties.

Shamos [18] conjectured that the Delaunay triangulation contains a Hamiltonian cycle. This conjecture sparked a flurry of research activity. Although Dillencourt [11] disproved this conjecture, he showed that Delaunay triangulations are *almost* Hamiltonian [12], in the sense that they are 1-tough.[1] Focus then shifted on determining how much the definition of the Delaunay triangulation has to be loosened to achieve Hamiltonicity. One such direction is to relax the *empty disk* requirement. Given a planar point set S and two points $p, q \in S$, the k-Delaunay graph (k-DG) with vertex set S has an edge pq provided that there exists a disk with p and q on the boundary containing at most k points of S different from p and q.[2] If the disk with p and q on its boundary is restricted to disks with pq as diameter, then the graph is called the k-Gabriel graph (k-GG). For the k-Relative Neighborhood graph (k-RNG), pq is an edge provided that there are at most k points of S whose distance to both p and q is less than $|pq|$. Note that k-$RNG \subseteq k$-$GG \subseteq k$-DG. Chang et al. [9] showed that 19-RNG is Hamiltonian.[3] Abellanas et al. [1] lowered this bound to 15-GG. Currently, the lowest known bound is by Kaiser et al. [15] who showed that 10-GG is Hamiltonian. All of these results are obtained by studying properties of *bottleneck Hamiltonian cycles*. Given a planar point set, a bottleneck Hamiltonian cycle is a cycle whose maximum edge length is minimum among all Hamiltonian cycles of the point set. Biniaz et al. [6] showed that there exist point sets such that its 7-GG does not contain a bottleneck Hamiltonian cycle, implying that this approach cannot yield an upper bound lower than 8. Despite this, it is conjectured that 1-DG is Hamiltonian [1].

Another avenue that has been explored is through the relaxation of the shape defining the Delaunay triangulation. Delaunay graphs where the disks have been replaced by various convex shapes have been studied in the literature. For instance, Chew [10] showed that the $\triangle$-Delaunay graph (i.e., where the shape is an equilateral triangle), denoted $DG_{\triangle}$, is a 2-spanner and that the $\square$-Delaunay graph (i.e., where the shape is a square), denoted $DG_{\square}$, is a $\sqrt{10}$-spanner. Bose et al. [8] proved that the convex-Delaunay graph (i.e., where the shape is an arbitrary convex shape) is a c-spanner where the constant c depends only on the perimeter and width of the convex shape.

As for Hamiltonicity in convex shape Delaunay graphs, not much is known. Bonichon et al. [7] proved that every plane triangulation is Delaunay-realizable

---

[1] A graph is *1-tough* if removing k vertices from it results in $\leq k$ connected components.

[2] Note that this implies that the standard Delaunay triangulation is the 0-DG.

[3] According to the definition of k-RNG in [9], they showed Hamiltonicity for 20-RNG.

Table 1. Bounds on the minimum k for which $k\text{-}DG_\mathcal{C}(S)$ is Hamiltonian and for which $k\text{-}GG_\mathcal{C}(S)$ contains a $d_\mathcal{C}-$bottleneck Hamiltonian cycle.

Type of shape $\mathcal{C}$	$k \leq$	$k \geq$	Bottleneck-$k \geq$
Circles	10 [15]	1 [11]	8 [6]
Equilateral triangles	7 [5]	1 [5]	6 [5]
Squares	7 [Theorem 10]	1 [17]	3 [Lemma 5]
Regular hexagons	11 [Theorem 11]	1 [Lemma 4]	6 [Lemma 6]
Regular octagons	12 [Theorem 13]	1 [Lemma 4]	-
Regular t-gons (t even, $t \geq 10$)	11 [Theorem 12]	-	-
Point-symmetric convex	15 [Theorem 8]	-	-
Convex	24 [Theorem 5]	-	-

where homothets of a triangle act as the empty convex shape. This implies that there exist such triangulations that do not contain Hamiltonian paths or cycles. For the special case of $\triangle$-Delaunay graph, Biniaz et al. [5] showed that $7\text{-}DG_\triangle$ contains a bottleneck Hamiltonian cycle and that there exist points sets where $5\text{-}DG_\triangle$ does not contain a bottleneck Hamiltonian cycle. Ábrego et al. [2] showed that the $DG_\square$ admits a Hamiltonian path, while Saumell [17] showed that the $DG_\square$ is not necessarily 1-tough, and therefore does not necessarily contain a Hamiltonian cycle.

Results. In this article, we generalize the above results by replacing the disk with an arbitrary convex shape. We show that the k-Gabriel graph, and hence also the k-Delaunay graph, is Hamiltonian for any convex shape $\mathcal{C}$ when $k \geq 24$. Furthermore, we give improved bounds for point-symmetric shapes, as well as for even-sided regular polygons. Table 1 summarizes the bounds obtained. Together with the results of Bose et al. [8], our results are the first results on graph-theoretic properties of generalized Delaunay graphs that apply to any convex shape.

The beauty of our results relies on the use of normed metrics and packing lemmas. In fact, in contrast to previous work on Hamiltonicity for generalized Delaunay graphs, our results are the first to use properties of normed metrics to obtain simple proofs for general and specific convex shape Delaunay graphs.

Due to space limitations, some proofs are omitted.

2 Convex Distances and the $\mathcal{C}$-Gabriel Graph

Let p and q be two points in the plane. Let $\mathcal{C}$ be a compact convex set that contains the origin, denoted $\bar{o}$, in its interior. We denote *the boundary of* $\mathcal{C}$ by $\partial\mathcal{C}$. The *convex distance* $d_\mathcal{C}(p,q)$ is defined as follows: If $p = q$, then $d_\mathcal{C}(p,q) = 0$. Otherwise, let $\mathcal{C}_p$ be the convex $\mathcal{C}$ translated by the vector $\overrightarrow{p}$ and let q' be the intersection of the ray from p through q and $\partial\mathcal{C}_p$. Then, $d_\mathcal{C}(p,q) = \frac{d(p,q)}{d(p,q')}$ (see Fig. 1) where d denotes the Euclidean distance.

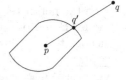

Fig. 1. Convex distance from p to q.

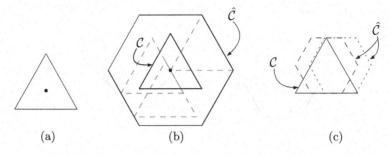

Fig. 2. (a) A triangle is a non-symmetric shape $\mathcal{C}$. (b) $\hat{\mathcal{C}}$ for this triangle is a hexagon. (c) The shape $\hat{\mathcal{C}}$ with radius $\frac{1}{2}$ does not contain $\mathcal{C}$.

The convex set $\mathcal{C}$ is the *unit $\mathcal{C}$-disk* of $d_{\mathcal{C}}$ with center $\bar{o}$, i.e., every point p in $\mathcal{C}$ satisfies that $d_{\mathcal{C}}(\bar{o}, p) \leq 1$. The *$\mathcal{C}$-disk with center c and radius r* is defined as the homothet of $\mathcal{C}$ centered at c and with scaling factor r. The *triangle inequality* holds: $d_{\mathcal{C}}(p, q) \leq d_{\mathcal{C}}(p, z) + d_{\mathcal{C}}(z, q), \forall p, q, z \in \mathbb{R}^2$. However, this distance may not define a metric when $\mathcal{C}$ is not *point-symmetric* about the origin,[4] since there may be points p, q for which $d_{\mathcal{C}}(p, q) \neq d_{\mathcal{C}}(q, p)$. When $\mathcal{C}$ is point-symmetric with respect to the origin, $d_{\mathcal{C}}$ is called a *symmetric convex distance function*. We will refer to such distance functions as *symmetric convex*. Such a distance defines a metric; moreover, $d_{\mathcal{C}}(\bar{o}, p)$ defines a *norm*[5] of a *metric space*. In addition, if a point p is on the line segment ab, then $d_{\mathcal{C}}(a, b) = d_{\mathcal{C}}(a, p) + d_{\mathcal{C}}(p, b)$ (see [3, Chapter 7]).

Let S be a set of points in the plane satisfying the following general position assumption: For each pair $p, q \in S$, any minimum homothet of $\mathcal{C}$ having p and q on its boundary does not contain any other point of S on its boundary. The *k-order $\mathcal{C}$-Delaunay graph* of S, denoted k-$DG_{\mathcal{C}}(S)$, is the graph with vertex set S such that, for each pair of points $p, q \in S$, the edge pq is in k-$DG_{\mathcal{C}}(S)$ if there exists a $\mathcal{C}$-disk that has p and q on its boundary and contains at most k points of S different from p and q. When $k = 0$ and $\mathcal{C}$ is a circle, k-$DG_{\mathcal{C}}(S)$ is the standard *Delaunay triangulation*.

Unlike the definition of Delaunay graphs, the one for Gabriel graphs requires a distance function. To circumvent the problem that $d_{\mathcal{C}}$ might not be symmetric, Aurenhammer and Paulini [4] showed how to define, from any convex shape $\mathcal{C}$, another distance function that is always symmetric: The distance from p to q is given by the scaling factor of a smallest homothet containing p and q on its boundary, which is equivalent to $\min_{v \in \mathcal{C}} d_{\mathcal{C}_{-v}}(p, q) = d_{\hat{\mathcal{C}}}(p, q)$ where $\hat{\mathcal{C}} = \bigcup_{v \in \mathcal{C}} \mathcal{C}_{-v}$. The set $\hat{\mathcal{C}}$ is a symmetric convex set that is the Minkowski sum[6]

---

[4] A shape $\mathcal{C}$ is point-symmetric with respect to a point $x \in \mathcal{C}$ provided that for every point $p \in \mathcal{C}$ there is a corresponding point $q \in \mathcal{C}$ such that $pq \in \mathcal{C}$ and x is the midpoint of pq.

[5] A function $\rho(x)$ is a norm if: (a) $\rho(x) = 0$ if and only if $x = \bar{o}$, (b) $\rho(\lambda x) = |\lambda| \rho(x)$ where $\lambda \in \mathbb{R}$, and (c) $\rho(x + y) \leq \rho(x) + \rho(y)$.

[6] The Minkowski sum of two sets A and B is defined as $A \oplus B = \{a + b : a \in A, b \in B\}$.

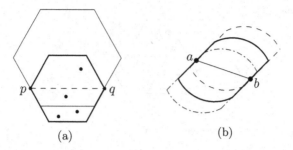

Fig. 3. (a) $\mathcal{C}$ is a regular hexagon. Edge pq is in 2-$DG_{\mathcal{C}}(S)$ but it is not in 2-$GG_{\mathcal{C}}(S)$. (b) Many $\mathcal{C}$-disks $\mathcal{C}(a,b)$ may exist for a and b.

of $\mathcal{C}$ and its shape reflected about its center. For an example, see Fig. 2. The diameter and width of $\hat{\mathcal{C}}$ is twice the diameter and width of $\mathcal{C}$, respectively. Moreover, if $\mathcal{C}$ is point-symmetric, $d_{\hat{\mathcal{C}}}(p,q) = \frac{d_{\mathcal{C}}(p,q)}{2}$.

We define the k-order $\mathcal{C}$-*Gabriel graph* of S, denoted k-$GG_{\mathcal{C}}(S)$, as the graph with vertex set S such that, for every pair of points $p, q \in S$, the edge pq is in k-$GG_{\mathcal{C}}(S)$ if and only if there exists a $\mathcal{C}$-disk with radius $d_{\hat{\mathcal{C}}}(p,q)$ that has p and q on its boundary and contains at most k points of S different from p and q. From the definition of k-$GG_{\mathcal{C}}(S)$ and k-$DG_{\mathcal{C}}(S)$ we note that k-$GG_{\mathcal{C}}(S) \subseteq k$-$DG_{\mathcal{C}}(S)$, and it can be a proper subgraph. See Fig. 3a for an example. Further, $\hat{\mathcal{C}}$ always contains $\mathcal{C}$ in its interior. However, for some non point-symmetric convex $\mathcal{C}$ it is not true that the $\hat{\mathcal{C}}$-disk with radius $\frac{1}{2}$ contains $\mathcal{C}$ (refer to Fig. 2c). Thus, for non point-symmetric shapes $\mathcal{C}$, in general $GG_{\hat{\mathcal{C}}} \nsubseteq GG_{\mathcal{C}}$.

3   Hamiltonicity for General Convex Shapes

Define $\mathcal{H}$ to be the set of all Hamiltoninan cycles of the point set S. Define the $d_{\hat{\mathcal{C}}}$-length sequence of $h \in \mathcal{H}$, denoted $dsc(h)$, as an edge sequence sorted in decreasing order with respect to the length of the edges in $d_{\hat{\mathcal{C}}}$-metric. Sort the elements of $\mathcal{H}$ in lexicographic order with respect to their $d_{\hat{\mathcal{C}}}$-length sequence, breaking ties arbitrarily. This order is strict. For $h_1, h_2 \in \mathcal{H}$, if h_1 is smaller than h_2 in this order, we write $h_1 \prec h_2$.

For simplicity, denote by $\mathcal{C}_r(a,b)$ a $\mathcal{C}$-disk with radius r containing the points a and b on its boundary. For the special case of a *diametral disk*, i.e., when the radius of $\mathcal{C}_r(a,b)$ is $d_{\hat{\mathcal{C}}}(a,b)$, we denote it as $\mathcal{C}(a,b)$. Note that $\mathcal{C}(a,b)$ may not be unique, see Fig. 3b. In addition, we denote by $D_{\mathcal{C}}(c,r)$ the $\mathcal{C}$-disk centered at point c with radius r.

Claim 1. *Let $\mathcal{C}$ be a point-symmetric convex shape. Let u be a point in the plane different from the origin $\bar{o}$. Let $r < d_{\mathcal{C}}(u, \bar{o})$. Let p be the intersection point of $\partial D_{\mathcal{C}}(u,r)$ and line segment $\bar{o}u$. Let $u' = \lambda u$, with $\lambda > 1 \in \mathbb{R}$, be a point defined by vector u scaled by a factor of λ. Then $D_{\mathcal{C}}(u,r) \subset D_{\mathcal{C}}(u', d_{\mathcal{C}}(u',p))$. (See Fig. 4a).*

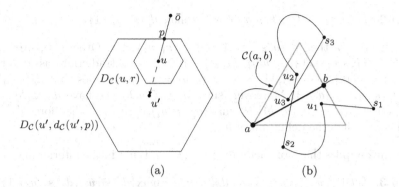

Fig. 4. (a) $D_{\mathcal{C}}(u,r)$ is contained in $D_{\mathcal{C}}(u', d_{\mathcal{C}}(u',p))$ where $u' = \lambda u$ with $\lambda > 1$. (b) Example of U in $\mathcal{C}(a,b)$.

Proof. Let $q \in D_{\mathcal{C}}(u,r)$; then $d_{\mathcal{C}}(u,q) \leq d_{\mathcal{C}}(u,p)$. Since u is on the line segment $u'p$, we have that $d_{\mathcal{C}}(u',p) = d_{\mathcal{C}}(u',u) + d_{\mathcal{C}}(u,p)$. Hence $d_{\mathcal{C}}(u',q) \leq d_{\mathcal{C}}(u',u) + d_{\mathcal{C}}(u,q) \leq d_{\mathcal{C}}(u',u) + d_{\mathcal{C}}(u,p) = d_{\mathcal{C}}(u',p)$. Therefore, $D_{\mathcal{C}}(u,r)$ is contained in $D_{\mathcal{C}}(u', d_{\mathcal{C}}(u',p))$. □

Let h be the minimum element in $\mathcal{H}$, often called *bottleneck Hamiltonian cycle*. The approach we follow to prove our bounds, which is similar to the approach in [1,9,15], is to show that h is contained in $k\text{-}GG_{\mathcal{C}}(S)$ for a small value of k. The strategy for proving that h is contained in $24\text{-}GG_{\mathcal{C}}(S)$ is by showing that for every edge $ab \in h$ there are at most 24 points in any $\mathcal{C}(a,b)$. In order to do this, we associate each point in the interior of an arbitrary fixed $\mathcal{C}(a,b)$ with another point. Later, we show that the $d_{\hat{\mathcal{C}}}$-distances between such associated points and a is at least $d_{\hat{\mathcal{C}}}(a,b)$. Finally, we use a packing lemma for showing that there are at most 24 associated points, which leads to 24 points contained in $\mathcal{C}(a,b)$.

Let $ab \in h$; we can assume without loss of generality that $d_{\hat{\mathcal{C}}}(a,b) = 1$. Let $U = \{u_1, u_2, \ldots, u_k\}$ be the set of points in S different from a and b that are in the interior of an arbitrary fixed $\mathcal{C}(a,b)$.[7] We assume that, when traversing h from b to a, we visit the points of U in order $u_1, \ldots, u_k$. For each point u_i, we define s_i to be the point preceding u_i in h. See Fig. 4b.

Note that if a point p is in the interior of $\mathcal{C}(a,b)$, then for any q on the boundary of $\mathcal{C}(a,b)$ there exists a $\mathcal{C}$-disk (not necessarily diametral) through p and q contained in $\mathcal{C}(a,b)$, and any diametral disk through p and q is smaller than or equal (in size) to this disk. Therefore, $d_{\hat{\mathcal{C}}}(a,u_i) < 1$ and $d_{\hat{\mathcal{C}}}(b,u_i) < 1$ for any $i \in \{1, \ldots, k\}$. Furthermore, we have the following:

---

[7] Since S is in general position, only a and b can lie on the boundary of $\mathcal{C}(a,b)$.

Claim 2. *Let $1 \leq i \leq k$. Then $d_{\hat{\mathcal{C}}}(a, s_i) \geq \max\{d_{\hat{\mathcal{C}}}(s_i, u_i), 1\}$*

Proof. If $s_1 = b$, then $d_{\hat{\mathcal{C}}}(a, s_1) = 1$ and $d_{\hat{\mathcal{C}}}(s_1, u_1) < 1$. Otherwise, we define $h' = (h \setminus \{ab, s_i u_i\}) \cup \{as_i, u_i b\}$. For the sake of a contradiction, assume that $d_{\hat{\mathcal{C}}}(a, s_i) < \max\{d_{\hat{\mathcal{C}}}(s_i, u_i), 1\}$. Since $d_{\hat{\mathcal{C}}}(a, b) = 1$, this implies that $d_{\hat{\mathcal{C}}}(a, s_i) < \max\{d_{\hat{\mathcal{C}}}(s_i, u_i), d_{\hat{\mathcal{C}}}(a, b)\}$. Moreover, since $u_i \in \mathcal{C}(a, b)$, we have $d_{\hat{\mathcal{C}}}(u_i, b) < 1$. Thus, $\max\{d_{\hat{\mathcal{C}}}(a, s_i), d_{\hat{\mathcal{C}}}(u_i, b)\} < \max\{d_{\hat{\mathcal{C}}}(s_i, u_i), d_{\hat{\mathcal{C}}}(a, b)\}$. Therefore $h' \prec h$, which contradicts the definition of h.   □

Claim 2 implies that, for each $i \in \{1, \ldots, k\}$, s_i is not in the interior of $\mathcal{C}(a, b)$.

Claim 3. *Let $1 \leq i < j \leq k$. Then $d_{\hat{\mathcal{C}}}(s_i, s_j) \geq \max\{d_{\hat{\mathcal{C}}}(s_i, u_i), d_{\hat{\mathcal{C}}}(s_j, u_j), 1\}$.*

Proof. For the sake of contradiction, we assume that $d_{\hat{\mathcal{C}}}(s_i, s_j) < \max\{d_{\hat{\mathcal{C}}}(s_i, u_i), d_{\hat{\mathcal{C}}}(s_j, u_j), 1\}$. Consider the Hamiltonian cycle $h' = h \setminus \{(a, b), (s_i, u_i), (s_j, u_j)\} \cup \{(s_i, s_j), (u_i, a), (u_j, b)\}$. Similarly as in Claim 2 we have that $d_{\hat{\mathcal{C}}}(u_i, a) < 1$ and $d_{\hat{\mathcal{C}}}(u_j, b) < 1$. So, $\max\{d_{\hat{\mathcal{C}}}(s_i, s_j), d_{\hat{\mathcal{C}}}(u_i, a), d_{\hat{\mathcal{C}}}(u_j, b)\} < \max\{d_{\hat{\mathcal{C}}}(s_i, u_i), d_{\hat{\mathcal{C}}}(s_j, u_j), d_{\hat{\mathcal{C}}}(a, b)\}$. Therefore, $h' \prec h$ which contradicts the minimality of h.   □

Without loss of generality we assume that a is the origin $\bar{o}$. Then, by the definition of $\hat{\mathcal{C}}$, we have that $D_{\hat{\mathcal{C}}}(\bar{o}, 1)$ contains $\mathcal{C}(a, b)$. Also, from Claim 2, we have that s_i is not in the interior of $D_{\hat{\mathcal{C}}}(\bar{o}, 1)$ for all $i \in \{1, \ldots, k\}$. Let $D_{\hat{\mathcal{C}}}(\bar{o}, 2)$ be the $\hat{\mathcal{C}}$-disk centered at a with radius 2. For each $s_i \notin D_{\hat{\mathcal{C}}}(\bar{o}, 2)$, define s_i' as the intersection of $\partial D_{\hat{\mathcal{C}}}(\bar{o}, 2)$ with the ray $\overrightarrow{as_i}$. We let $s_i' = s_i$ when s_i is inside $D_{\hat{\mathcal{C}}}(\bar{o}, 2)$. See Fig. 5. The *distance from a point v to a $\mathcal{C}$-disk C* is given by the minimum $\mathcal{C}$-distance from v to any point u in C. Notice that if $v \notin C$, then the distance from v to C is defined by a point on ∂C. This can be seen by taking an $\epsilon \in \mathbb{R}^+$ small enough such that $D_{\mathcal{C}}(v, \epsilon)$ does not intersect C and by making ϵ grow until $D_{\mathcal{C}}(v, \epsilon)$ hits C.

Observation 4. *Let p be the intersection point of $\partial D_{\hat{\mathcal{C}}}(\bar{o}, 1)$ and $\bar{o}s_j$. If $s_j \notin D_{\hat{\mathcal{C}}}(\bar{o}, 2)$ (with $1 \leq j \leq k$), the $d_{\hat{\mathcal{C}}}$-distance from s_j' to $D_{\hat{\mathcal{C}}}(\bar{o}, 1)$ is 1. Moreover, $d_{\hat{\mathcal{C}}}(s_j', p) = 1$.*

Lemma 1. *For any pair s_i and s_j with $i \neq j$, we have that $d_{\hat{\mathcal{C}}}(s_i', s_j') \geq 1$.*

Proof. If both s_i and s_j are in $D_{\hat{\mathcal{C}}}(\bar{o}, 2)$, then from Claim 3 we have that $d_{\hat{\mathcal{C}}}(s_i', s_j') = d_{\hat{\mathcal{C}}}(s_i, s_j) \geq 1$. In the following, we assume, without loss of generality, that $d_{\hat{\mathcal{C}}}(\bar{o}, s_j) \geq d_{\hat{\mathcal{C}}}(\bar{o}, s_i)$. Since s_j' is on the line segment $\bar{o}s_j$, we have $s_j = \lambda s_j'$ for some $\lambda > 1 \in \mathbb{R}$. Let p be the intersection point of $\partial D_{\hat{\mathcal{C}}}(\bar{o}, 1)$ and $\bar{o}s_j$. Since $d_{\hat{\mathcal{C}}}$ defines a norm, we have $d_{\hat{\mathcal{C}}}(\lambda s_j', \bar{o}) = \lambda d_{\hat{\mathcal{C}}}(s_j', \bar{o})$. By Observation 4 we have that $d_{\hat{\mathcal{C}}}(s_j, p) = d_{\hat{\mathcal{C}}}(s_j, \bar{o}) - d_{\hat{\mathcal{C}}}(p, \bar{o}) = \lambda d_{\hat{\mathcal{C}}}(s_j', \bar{o}) - 1 = 2\lambda - 1$, which is the distance from s_j to $D_{\hat{\mathcal{C}}}(\bar{o}, 1)$. Otherwise, there exists a point $v \in \partial D_{\hat{\mathcal{C}}}(\bar{o}, 1)$ such that $d_{\hat{\mathcal{C}}}(s_j, v) < d_{\hat{\mathcal{C}}}(s_j, p)$. Thus, $2\lambda = d_{\hat{\mathcal{C}}}(\bar{o}, s_j) \leq d_{\hat{\mathcal{C}}}(\bar{o}, v) + d_{\hat{\mathcal{C}}}(v, s_j) < d_{\hat{\mathcal{C}}}(\bar{o}, v) + d_{\hat{\mathcal{C}}}(p, s_j) = 1 + 2\lambda - 1 = 2\lambda$, which is a contradiction. Further, $d_{\hat{\mathcal{C}}}(s_j, s_j') = d_{\hat{\mathcal{C}}}(s_j, \bar{o}) - d_{\hat{\mathcal{C}}}(s_j', \bar{o}) = 2\lambda - 2$. Let us prove that $d_{\hat{\mathcal{C}}}(s_i', s_j') \geq 1$. For the sake of a contradiction, assume that $d_{\hat{\mathcal{C}}}(s_i', s_j') \leq 1$. Let $D_{s_j'} = D_{\hat{\mathcal{C}}}(s_j', 1)$.

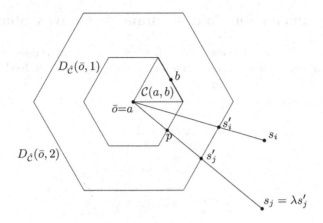

Fig. 5. The points s_i' and s_j' are projections of s_i and s_j on $\partial D_{\hat{\mathcal{C}}}(\bar{o}, 2)$, respectively.

By Observation 4, $d_{\hat{\mathcal{C}}}(s_j', p) = 1$. Therefore, p is on $\partial D_{s_j'}$. Now, we consider two cases:

Case (1) $s_i \in D_{\hat{\mathcal{C}}}(\bar{o}, 2)$. Then $d_{\hat{\mathcal{C}}}(\bar{o}, s_i) \le 2$. Since $d_{\hat{\mathcal{C}}}(s_i', s_j') \le 1$, we have $s_i \in D_{s_j'}$. From Claim 1 it follows that $D_{s_j'}$ is contained in $D_{\hat{\mathcal{C}}}(s_j, d_{\hat{\mathcal{C}}}(s_j, p))$. Thus, $s_i' \in D_{\hat{\mathcal{C}}}(s_j, d_{\hat{\mathcal{C}}}(s_j, p))$. Hence, $d_{\hat{\mathcal{C}}}(s_j, s_i') = d_{\hat{\mathcal{C}}}(s_j, s_i) \le d_{\hat{\mathcal{C}}}(s_j, p)$. Since S is in general position, u_j is in the interior of $D_{\hat{\mathcal{C}}}(\bar{o}, 1)$. Hence, $d_{\hat{\mathcal{C}}}(s_j, s_i) \le d_{\hat{\mathcal{C}}}(s_j, p) < d_{\hat{\mathcal{C}}}(s_j, u_j)$, which contradicts Claim 3.

Case (2) $s_i \notin D_{\hat{\mathcal{C}}}(\bar{o}, 2)$. Then $d_{\hat{\mathcal{C}}}(\bar{o}, s_i) > 2$. Thus, $s_i = \delta s_i'$ for some $\delta > 1 \in \mathbb{R}$. Moreover, since $d_{\hat{\mathcal{C}}}(\bar{o}, s_j) \ge d_{\hat{\mathcal{C}}}(\bar{o}, s_i)$ and s_i', s_j' are on $\partial D_{\hat{\mathcal{C}}}(\bar{o}, 2)$, $\delta \le \lambda$. Hence, s_i is on the line segment $s_i'(\lambda s_i')$. Let $D_{s_j} = D_{\hat{\mathcal{C}}}(s_j, 2\lambda - 1)$. Note that $\lambda < 2\lambda - 1$ because $\lambda > 1$. Since $d_{\hat{\mathcal{C}}}$ defines a norm, $d_{\hat{\mathcal{C}}}(s_j, \lambda s_i') = d_{\hat{\mathcal{C}}}(\lambda s_j', \lambda s_i') = \lambda d_{\hat{\mathcal{C}}}(s_j', s_i') \le \lambda < 2\lambda - 1$. Hence, $\lambda s_i' \in D_{s_j}$. In addition, from Claim 1 it follows that $D_{s_j'} \subseteq D_{s_j}$. Therefore, $s_i' \in D_{s_j}$. Thus, the line segment $s_i'(\lambda s_i')$ is contained in D_{s_j}. Hence, $s_i \in D_{s_j}$. Then, $d_{\hat{\mathcal{C}}}(s_j, s_i) \le 2\lambda - 1 = d_{\hat{\mathcal{C}}}(s_j, p) < d_{\hat{\mathcal{C}}}(s_j, u_j)$ which contradicts Claim 3. $\qquad \square$

Theorem 5. *For any set S of points in general position and convex shape $\mathcal{C}$, the graph $24\text{-}GG_{\mathcal{C}}(S)$ is Hamiltonian.*

Proof. For each s_i we define the $\hat{\mathcal{C}}$-disk $D_i = D_{\hat{\mathcal{C}}}(s_i', \frac{1}{2})$. We also set $D_0 := D_{\hat{\mathcal{C}}}(a, \frac{1}{2})$. By Lemma 1, each pair of $\hat{\mathcal{C}}$-disks D_i and D_j $(0 < i < j \le k)$ are internally disjoint. Note that if s_i' is on $\partial D_{\hat{\mathcal{C}}}(\bar{o}, 2)$ then D_0 and D_i are internally disjoint. On the other hand, If s_i' is in the interior of $D_{\hat{\mathcal{C}}}(\bar{o}, 2)$, then by definition $s_i' = s_i$. Thus, by Claim 2 D_0 is internally disjoint from D_i. See Fig. 6. Since $s_i' \in D_{\hat{\mathcal{C}}}(\bar{o}, 2)$ for all i, each disk D_i is inside $D_{\hat{\mathcal{C}}}(a, \frac{5}{2})$. There can be at most $\frac{Area(D_{\hat{\mathcal{C}}}(\bar{o}, \frac{5}{2}))}{Area(D_0)} = \frac{(\frac{5}{2})^2 Area(\hat{\mathcal{C}})}{(\frac{1}{2})^2 Area(\hat{\mathcal{C}})} = 25$ disjoint disks in $D_{\hat{\mathcal{C}}}(\bar{o}, 2)$. Thus, there are at most 24 points s_i' in $D_{\hat{\mathcal{C}}}(\bar{o}, 1)$, since D_0 is centered at a. Hence, there are at most 24 points in the interior of $\mathcal{C}(a, b)$. $\qquad \square$

4   Hamiltonicity for Point-Symmetric Convex Shapes

Using the fact that $d_\mathcal{C}$ defines a metric when $\mathcal{C}$ is point-symmetric, we can improve the upper bound for point-symmetric convex shapes. Indeed, given that $d_\mathcal{C} = 2d_{\hat{\mathcal{C}}}$ we can prove the following.

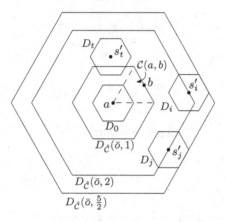

Fig. 6. The $\hat{\mathcal{C}}$-disks D_i, D_j and D_t are contained in $D_{\hat{\mathcal{C}}}(a, \frac{5}{2})$.

Claim 6. $d_\mathcal{C}(s_i, a) \geq \max\{d_\mathcal{C}(s_i, u_i), 2\}$.

Claim 7. Let $1 \leq i < j \leq k$, then $d_\mathcal{C}(s_i, s_j) \geq \max\{d_\mathcal{C}(s_i, u_i), d_\mathcal{C}(s_j, u_j), 2\}$.

By using $\mathcal{C}(a, b)$ instead of $D_{\hat{\mathcal{C}}}(\bar{o}, 1)$, $D_\mathcal{C}(\bar{o}, 3)$ instead of $D_{\hat{\mathcal{C}}}(\bar{o}, 2)$, and $D_\mathcal{C}(\bar{o}, 4)$ instead of $D_{\hat{\mathcal{C}}}(\bar{o}, \frac{5}{2})$, in combination with Claims 6 and 7, and with arguments similar to those in the previous section, we obtain the following results.

Lemma 2. For any pair s_i and s_j with $i \neq j$, we have that $d_\mathcal{C}(s_i', s_j') \geq 2$. Moreover, if at least one of s_i and s_j is not in $D_\mathcal{C}(\bar{o}, 3)$, then $d_\mathcal{C}(s_i', s_j') > 2$.

Theorem 8. For any set S of points in general position and point-symmetric convex shape $\mathcal{C}$, the graph $15\text{-}GG_\mathcal{C}(S)$ is Hamiltonian.

4.1   Hamiltonicity for Regular Polygons

An important family of point-symmetric convex shapes is that of regular even-sided polygons. When $\mathcal{C}$ is a regular polygon with t sides $\mathcal{P}_t$, for t even, we can improve the previous bound by analyzing the geometry for different values of t.

First, we consider the case when the polygon is a square. In this case, we divide $D_\square(\bar{o}, 3)$ into 9 disjoint squares of radius 1 and show that there can be at most one point of $\{a, s_1', \ldots, s_k'\}$ in each such square. We use lines $x = -1, x = 1, y = -1$, and $y = 1$ to split $D_\square(3, \bar{o})$ into 9 squares of radius 1. Refer to Fig. 7. Let $D_0, D_1, \ldots, D_7$ be the squares of radius 1 in $D_\square(\bar{o}, 3)$ different from $C(a, b)$,

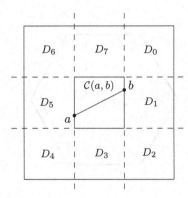

Fig. 7. Lines $x = -1, x = 1, y = -1$, and $y = 1$ split $D_\square(3, \bar{o})$ into nine unit squares: $\mathcal{C}(a, b), D_0, \ldots, D_7$.

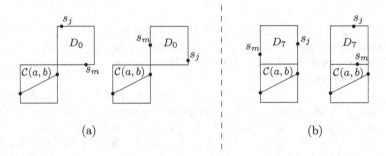

(a)   (b)

Fig. 8. (a) Points s_j and s_m can be either on the horizontal sides of D_0 or on the vertical sides of D_0. In both cases, the distance from s_j to $\mathcal{C}(a, b)$ is 2. (b) Either s_j and s_m belong to different squares or the distance from s_j to $\mathcal{C}(a, b)$ is 2.

ordered clockwise where D_0 is the top-right corner square. In the following lemma we prove that there is at most one point in each D_i. Note that each D_i shares a side with D_{i-1}, i modulo 8 and for each odd i, D_i shares a side with $\mathcal{C}(a, b)$. Moreover, there exists a D_i that contains a on its boundary. We will associate a point p in $D_\square(\bar{o}, 3)$ to a unique square D_i in the following fashion. Let indices be taken modulo 8. Let p be a point in D_i. If p is on $D_i \cap D_{i-1}$, we say that p is associated to D_{i-1}. If i is odd and p is the intersection point $D_i \cap D_{i-1} \cap D_{i-2}$, then p is associated to D_{i-2}. Otherwise, p is associated to D_i.

Observation 9. *If there are two points at $d_\square$-distance 2 in a unit square, then such points are in opposite sides of the square.*

Lemma 3. *There is at most one s'_j associated to each D_i. Moreover, the D_i containing a on its boundary has no s'_j associated to it.*

Proof. Assume there are two points s'_j and s'_m in D_i. From Lemma 2 we have that $d_\square(s'_j, s'_m) \geq 2$. Also, since D_i is a unit square, $d_\square(s'_j, s'_m) \leq 2$. Therefore,

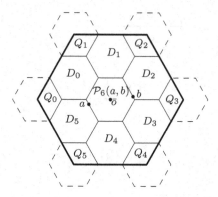

Fig. 9. The bold hexagon is the boundary of $D_{\mathcal{P}_6}(\bar{o}, 3)$. Such hexagon is divided into 13 interior-disjoint regions: 6 quadrangles—a third of a unit $\mathcal{P}_6$-disk—and 7 unit $\mathcal{P}_6$-disks.

$d_\square(s'_j, s'_m) = 2$. Then Lemma 2 implies that s_j and s_m must be inside $D_{\mathcal{C}}(\bar{o}, 3)$. In addition, by Observation 9, the points s_i and s_j are on ∂D_i. For simplicity we will assume that $d_\square(\bar{o}, s_j) \geq d_\square(\bar{o}, s_m)$. By Observation 9, the points s_j and s_m are on opposite sides of D_i.

If i is even, then the $d_\square$-distance of s_j to $\mathcal{C}(a, b)$ is exactly 2. We refer to Fig. 8a. Recall that by our definition of points in general position, u_j is in the interior of $\mathcal{C}(a, b)$. Thus, the distance from s_j to $\mathcal{C}(a, b)$ is less than $d_\square(u_j, s_j)$; i.e., $d_\square(u_j, s_j) > 2$. Hence, $d_\square(s_j, s_m) = 2 < d_\square(s_j, u_j)$ which contradicts Claim 7. Therefore, if i is even, there is at most one point in D_i which is associated to it.

If i is odd, then s_j is either on $D_i \cap D_{i-1}$ or $D_i \cap D_{i+1}$, or on $D_i \cap D_{\mathcal{C}}(\bar{o}, 3)$ (indices are taken modulo 8). We refer to Fig. 8b. If s_j is on $D_i \cap D_{i-1}$ or $D_i \cap D_{i+1}$, then only one of s_j and s_m is associated to D_i. If s_j is on $D_i \cap D_{\mathcal{C}}(\bar{o}, 3)$, then by Observation 9, s_m is on $\mathcal{C}(a, b)$, which contradicts our general position assumption. Therefore, there is only one point associated to D_i.

Finally, if D_i contains a, then there is no point s'_j in D_i. Indeed, assume for the sake of contradiction that $s'_j \in D_i$. Then, s_j is not in D_i, otherwise, $d_\square(a, s_j) < d_\square(s_j, u_j)$, contradicting Claim 6. Thus, s'_j is on $D_i \cap D_\square(\bar{o}, 3)$ and $s_j = \lambda s'_j$ for some $\lambda > 1$. Hence, $d_\square(s'_j, a) = 2$, which means that $a \in D_\square(s'_j, 2)$. Let p be the point $\bar{o}s'_j \cap \partial \mathcal{C}(a, b)$. By Claim 1, $D_\square(s'_j, 2) \subset D_\square(s_j, d_\square(s_j, p))$. So, $a \in D_\square(s_j, d_\square(s_j, p))$. Thus, $d_\square(s_j, a) < d_\square(s_j, u_j)$, contradicting Claim 6. $\square$

Theorem 10. *For any set S of points in general position, the graph $7\text{-}GG_\square(S)$ is Hamiltonian.*

Proof. From Lemma 3 we have that for each $0 \leq i \leq 7$ there is at most one point associated to D_i, and any square containing a has no s'_i associated to it. Since there is at least one D_i containing a, there are at most 7 points in $D_\square(\bar{o}, 3)$. Therefore, there are at most 7 points of S in the interior of $\mathcal{C}(a, b)$. $\square$

The analysis for the case of hexagons is similar to the previous one. First we divide the hexagon $D_{\mathcal{P}_6}(\bar{o}, 3)$ into 13 different regions $\mathcal{C}(a, b), D_0, \ldots, D_5$,

$Q_0, \ldots, Q_5$, shown in Fig. 9. Then, we show that there is at most one point s'_j associated to each region D_i and Q_i with i modulo 6. Moreover, there is no point s'_j in the hexagon D_i that contains a for some i modulo 6. The hexagon $D_{\mathcal{P}_6}(\bar{o}, 3)$ contains at most 11 points $s'_1, \ldots, s'_k$. Consequently, the following theorem holds.

Theorem 11. *The graph 11-$GG_{\mathcal{P}_6}$ is Hamiltonian.*

Finally, for the remaining regular polygons with even sides we use the ex-circle of $D_{\mathcal{P}_t}(\bar{o}, 3)$ in order to give an upper bound on the number of points in $D_{\mathcal{P}_t}(\bar{o}, 3)$ at pairwise Euclidean distance at least 2. Without loss of generality we assume that the incircle of the unit $\mathcal{P}_t$-disk has Euclidean radius 1.

Theorem 12. *For any set S of points in general position and regular polygon $\mathcal{P}_t$ with even $t \geq 10$, the graph 11-$GG_{\mathcal{P}_t}(S)$ is Hamiltonian.*

Proof. Let $\mathcal{P}_t$ be a polygon with $t \geq 10$ sides and t even. Then $D_{\mathcal{P}_t}(\bar{o}, 3)$ is inscribed in a circle of radius $r = \frac{3}{\cos(\frac{\pi}{t})}$. Since the function $\cos(\frac{\pi}{t})$ is an increasing function for $t \geq 2$ we have that $r \leq \frac{3}{\cos(\frac{\pi}{10})}$. Therefore, $D_{\mathcal{P}_t}(\bar{o}, 3)$ is inside the excircle of a decagon with incircle of radius 3. In addition, from Lemma 2 we know that for any pair of points s'_i, s'_j in $D_{\mathcal{P}_t}(\bar{o}, 3)$, $d_{\mathcal{P}_t}(s'_i, s'_j) \geq 2$. Since the incircle of the 2-unit $\mathcal{P}_t$-disk has Euclidean radius 2, we have that $d(s'_i, s'_j) \geq 2$. So, it suffices to show that there are at most 12 points in $D_{\mathcal{P}_t}(\bar{o}, 3)$ at pairwise Euclidean distance at least 2. Fodor [13] proved that the minimum radius R of a circle having 13 points at pairwise Euclidean distance at least 2 is $R \approx 3.236$, which is greater than $\frac{3}{\cos(\frac{\pi}{10})} \approx 3.154$. Thus, $D_{\mathcal{P}_t}(\bar{o}, 3)$ contains at most 12 points at pairwise distance at least 2. Since a is also at distance at least 2 from all s'_i's, there are at most 11 points inside $\mathcal{P}_t(a, b)$. □

In the case of octagons, the ex-circle of $D_{\mathcal{P}_8}(\bar{o}, 3)$ is greater than 3.236. Thus we cannot use the result of Fodor that we apply for Theorem 12. However, we can use another result from Fodor [14] to show that $D_{\mathcal{P}_8}(\bar{o}, 3)$ contains at most 13 points at pairwise Euclidean distance at least 2, leading to the following result.

Theorem 13. *The graph 12-$GG_{\mathcal{P}_8}(S)$ is Hamiltonian.*

5 Non-Hamiltonicity for Regular Polygonal Shapes

Until now we have discussed upper bounds for k, so that k-$GG_\mathcal{C}$ is Hamiltonian. As mentioned in Sect. 2, k-$GG_\mathcal{C} \subseteq k$-$DG_\mathcal{C}$, thus all upper bounds given in the previous sections hold for k-order $\mathcal{C}$-Delaunay graphs as well. In this section we present point sets for which $DG_{\mathcal{P}_t}$ is not Hamiltonian, for $t = 5, 6, \ldots 11$ (we note that these point sets can be generalized to larger values of t inductively, considering separately the cases of even and odd k). See Fig. 10. For the case of squares, Saumell [17] showed that for any $n \geq 9$ there exists a point set S such that $DG_\square(S)$ is non-Hamiltonian.

In order to prove the following lemma we need to recall that a graph is *1-tough* if removing any k vertices from it results in at most k connected components. As mentioned in the introduction, every Hamiltonian graph is 1-tough.

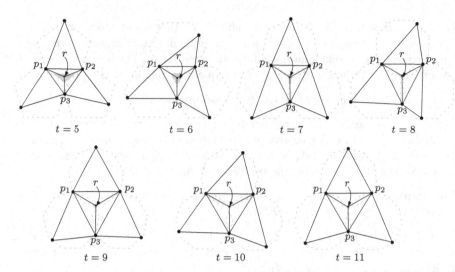

Fig. 10. For each $t \in \{5, 6, 7, 8, 9, 10, 11\}$ the graph $DG_{\mathcal{P}_t}(S)$ is non-Hamiltonian.

Lemma 4. *For any $n \geq 7$, there exists an n-point set S such that $DG_{\mathcal{P}_t}(S)$ is a non-Hamiltonian graph for any $t \in \{5, 6, \ldots, 11\}$.*

Proof. Let $t \in \{5, 6, \ldots, 11\}$. Consider the graph $DG_{\mathcal{P}_t}(S)$ in Fig. 10 for such t. Note that such graph is indeed a $\mathcal{P}_t$-Delaunay graph, since for each edge there exists a $\mathcal{P}_t$-disk that contains its vertices and at most 3 points of S are on its boundary. Also, note that some edges from the convex hull of S do not appear in such graphs. This is because any $\mathcal{P}_t$-disk that contains the vertices of such edge contains one point p_i in its interior with $i \in \{1, 2, 3\}$. Now, let $G' = DG_{\mathcal{P}_t}(S) \setminus \{p_1, p_2, p_3\}$. The graph G' consists of 4 connected components. Thus, $DG_{\mathcal{P}_t}(S)$ is not 1-tough. Hence, it is non-Hamiltonian. Finally, notice that there exists an area r which can have any number of points in its interior. $\qquad \square$

The proofs of the following results are very similar to those in [5, 6, 15], but adapted to squares and hexagons.

Lemma 5. *There exists a point set S with $n \geq 17$ points such that 2-$GG_{\square}$ does not contain any $d_{\square}$-bottleneck Hamiltonian cycle of S.*

Lemma 6. *There exist a point set S with $n \geq 22$ points such that 5-$GG_{\mathcal{P}_6}$ does not contain any $d_{\mathcal{P}_6}$-bottleneck Hamiltonian cycle of S.*

6   Conclusions

In this paper we have presented the first general results on Hamiltonicity for higher-order convex-shape Delaunay and Gabriel graphs. By combining properties of metrics and packings, we have achieved general bounds for any convex shape, and improved bounds for point-symmetric shapes, as well as for even-sided regular polygons. For future research, we point out that our results are based on bottleneck Hamiltonian cycles, in the same way as all previously obtained bounds [1,9,15]. However, in several cases this technique is reaching its limit. Therefore a major challenge to effectively close the existing gaps will be to devise a different approach to prove Hamiltonicity of Delaunay graphs.

References

1. Abellanas, M., Bose, P., García-López, J., Hurtado, F., Nicolás, C.M., Ramos, P.: On structural and graph theoretic properties of higher order Delaunay graphs. Int. J. Comput. Geom. Appl. **19**(6), 595–615 (2009). https://doi.org/10.1142/S0218195909003143

2. Ábrego, B.M., et al.: Matching points with squares. Discrete Comput. Geom. **41**(1), 77–95 (2009). https://doi.org/10.1007/s00454-008-9099-1

3. Aurenhammer, F., Klein, R., Lee, D.T.: Voronoi Diagrams and Delaunay Triangulations. World Scientific Publishing Company, Hackensack (2013)

4. Aurenhammer, F., Paulini, G.: On shape Delaunay tessellations. Inf. Process. Lett. **114**(10), 535–541 (2014). https://doi.org/10.1016/j.ipl.2014.04.007

5. Biniaz, A., Maheshwari, A., Smid, M.: Higher-order triangular-distance Delaunay graphs: graph-theoretical properties. Comput. Geom. **48**(9), 646–660 (2015). https://doi.org/10.1016/j.comgeo.2015.07.003

6. Biniaz, A., Maheshwari, A., Smid, M.: Bottleneck matchings and Hamiltonian cycles in higher-order Gabriel graphs. In: Proceedings of the 32nd European Workshop on Computational Geometry (EuroCG16), pp. 179–182 (2016)

7. Bonichon, N., Gavoille, C., Hanusse, N., Ilcinkas, D.: Connections between theta-graphs, Delaunay triangulations, and orthogonal surfaces. In: Thilikos, D.M. (ed.) WG 2010. LNCS, vol. 6410, pp. 266–278. Springer, Heidelberg (2010). https://doi.org/10.1007/978-3-642-16926-7_25

8. Bose, P., Carmi, P., Collette, S., Smid, M.: On the stretch factor of convex Delaunay graphs. J. Comput. Geom. **1**(1), 41–56 (2010). https://doi.org/10.1007/978-3-540-92182-0_58

9. Chang, M., Tang, C.Y., Lee, R.C.T.: 20-relative neighborhood graphs are Hamiltonian. J. Graph Theory **15**(5), 543–557 (1991). https://doi.org/10.1002/jgt.3190150507

10. Chew, L.P.: There are planar graphs almost as good as the complete graph. J. Comput. System Sci. **39**(2), 205–219 (1989). https://doi.org/10.1016/0022-0000(89)90044-5

11. Dillencourt, M.B.: A non-Hamiltonian, nondegenerate Delaunay triangulation. Inf. Process. Lett. **25**(3), 149–151 (1987). https://doi.org/10.1016/0020-0190(87)90124-4

12. Dillencourt, M.B.: Toughness and Delaunay triangulations. Discrete Comput. Geom. **5**, 575–601 (1990). https://doi.org/10.1007/BF02187810

13. Fodor, F.: The densest packing of 13 congruent circles in a circle. Beitr. Algebra Geom. **44**(2), 431–440 (2003)

14. Fodor, F.: Packing of 14 congruent circles in a circle. Stud. Univ. Zilina Math. Ser **16**, 25–34 (2003)

15. Kaiser, T., Saumell, M., Cleemput, N.V.: 10-Gabriel graphs are Hamiltonian. Inf. Process. Lett. **115**(11), 877–881 (2015). https://doi.org/10.1016/j.ipl.2015.05.013

16. Okabe, A., Boots, B., Sugihara, K., Chiu, S.N.: Spatial Tessellations: Concepts and Applications of Voronoi Diagrams. Wiley, Chichester (2000)
17. Saumell Mendiola, M.: Some problems on proximity graphs. Ph.D. thesis, Universitat Politècnica de Catalunya (2011)
18. Shamos, M.: Computational Geometry. Ph.D. thesis, Yale University (1978)
19. Xia, G.: The stretch factor of the Delaunay triangulation is less than 1.998. SIAM J. Comput. **42**(4), 1620–1659 (2013). https://doi.org/10.1137/110832458

Computing Maximum Independent Set on Outerstring Graphs and Their Relatives

Prosenjit Bose[3], Paz Carmi[1], Mark J. Keil[2], Anil Maheshwari[3],
Saeed Mehrabi[3(✉)], Debajyoti Mondal[2], and Michiel Smid[3]

[1] Ben-Gurion University of the Negev, Beer-Sheva, Israel
carmip@cs.bgu.ac.il
[2] University of Saskatchewan, Saskatoon, Canada
keil@cs.usask.ca, d.mondal@usask.ca
[3] Carleton University, Ottawa, Canada
{jit,anil,michiel}@scs.carleton.ca, saeed.mehrabi@carleton.ca

Abstract. A graph G with n vertices is called an *outerstring graph* if it has an intersection representation of a set of n curves inside a disk such that one endpoint of every curve is attached to the boundary of the disk. Given an outerstring graph representation, the *Maximum Independent Set* (MIS) problem of the underlying graph can be solved in $O(s^3)$ time, where s is the number of segments in the representation (Keil et al., Comput. Geom., 60:19–25, 2017). If the strings are of constant size (e.g., line segments, L-shapes, etc.), then the algorithm takes $O(n^3)$ time.

In this paper, we examine the fine-grained complexity of the MIS problem on some well-known outerstring representations. We show that solving the MIS problem on grounded segment and grounded square-L representations is at least as hard as solving MIS on circle graph representations. Note that no $O(n^{2-\delta})$-time algorithm, $\delta > 0$, is known for the MIS problem on circle graphs. For the grounded string representations where the strings are y-monotone simple polygonal paths of constant length with segments at integral coordinates, we solve MIS in $O(n^2)$ time and show this to be the best possible under the strong exponential time hypothesis (SETH). For the intersection graph of n L-shapes in the plane, we give a $(4 \cdot \log \mathsf{OPT})$-approximation algorithm for MIS (where OPT denotes the size of an optimal solution), improving the previously best-known $(4 \cdot \log n)$-approximation algorithm of Biedl and Derka (WADS 2017).

1 Introduction

Let $G = (V, E)$ be an undirected graph with $|V(G)| = n$; graph G is *weighted* if each edge in $E(G)$ is associated with a non-negative value, called its *weight*. A set $S \subseteq V(G)$ is an *independent set* if no two vertices in S are adjacent.

The research of Prosenjit Bose, Anil Maheshwari, Debajyoti Mondal and Michiel Smid is supported in part by NSERC. Part of this work was done when Saeed Mehrabi was visiting the University of Saskatchewan.

© Springer Nature Switzerland AG 2019
Z. Friggstad et al. (Eds.): WADS 2019, LNCS 11646, pp. 211–224, 2019.
https://doi.org/10.1007/978-3-030-24766-9_16

The objective of the *Maximum Independent Set* (MIS) problem is to compute a maximum-cardinality independent set of G. The MIS problem is NP-complete and it is known that no approximation algorithm with approximation factor within $|V(G)|^{1-\epsilon}$ is possible for any $\epsilon > 0$, unless $\mathsf{P} = \mathsf{NP}$ [20]. The inapproximability of the MIS problem has motivated a rich body of research to study the MIS problem on the intersection graph of geometric objects. Let O be a set of n geometric objects in the plane. Then, the *intersection graph* of O has the objects in O as its vertices and two vertices $o_i, o_j \in O$ are adjacent in the graph if and only if $o_i \cap o_j \neq \emptyset$. If O is a set of curves in the plane (resp., a set of chords of a circle), then the intersection graph of O is called a *string graph* (resp., *circle graph*); see Fig. 1(b–c) for an example.

Ehrlich et al. [13] showed in 1976 that every planar graph has a string representation. Moreover, the longstanding Scheinerman's conjecture [30], stating that all planar graphs can be represented as intersection graphs of line segments was proved affirmatively only in 2009 by Chalopin and Gonçalves [10]. For the MIS problem, Fox and Pach [16] gave an algorithm with an approximation factor of n^ϵ when the input consists of a set of curves, any two intersecting at most a constant number of times. The MIS problem has been studied on the intersection graph of other geometric objects such as line segments [1], disks and squares [14], rectangles [9] and pseudo-disks [11].

We study the MIS problem on outerstring graphs and their relatives with respect to the time-complexity of solving MIS in circle graph representations.

Definition 1 (Outerstring Graph [24]). *Graph G is called an* outerstring *graph if it is the intersection graph of a set of curves that lie inside a disk such that each curve intersects the boundary of the disk in one of its endpoints.*

Figure 1(d) shows an example of an outerstring graph. A string representation of a graph is called *grounded*, if one endpoint of each string is attached to a *grounding line* ℓ and all strings lie on one side of ℓ. For example, a graph G is called a *grounded segment graph*, if it is the intersection graph of a set of segments such that each segment is attached to a grounding line ℓ at one of its endpoints and all segments lie on one side of ℓ; see Fig. 1(e).

Gavril [18] presented an $O(n^3)$ algorithm for solving the MIS problem on circle graphs. Subsequent improvement reduced the complexity to $O(n^2)$ [3,31]. Several algorithms exist with running time sensitive to various graph parameters, e.g., $O(nd)$ time [2,32], or $O(n \min\{d, \alpha\})$ time [29]. Here d is a parameter known as the density of the circle graph, and α is the independence number of the circle graph. However, no truly subquadratic-time algorithm (i.e., an $O(n^{2-\delta})$-time algorithm where $\delta > 0$) is known for the MIS problem on circle graphs.

Although recognizing an intersection graph may require $\Theta(n^2)$ time (since there could be $\Theta(n^2)$ edges), the MIS problem can be solved faster if an intersection representation is given. For example, MIS in an interval graph representation can be solved in $O(n)$ time [17]. Moreover, recognizing grounded segment graphs is ∃ℝ-complete [8], but given an outerstring representation, one can solve the weighted MIS problem in $O(s^3)$ time, where s is the number of segments in the

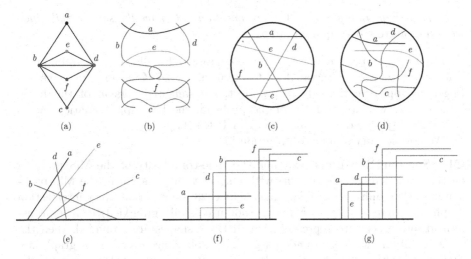

Fig. 1. (a) A graph G with six vertices. (b) A string graph, (c) a circle graph, (d) an outerstring graph, (e) a segment graph, (f) a grounded L, and (g) a grounded square-L representation of G.

representation [23]. For grounded segment graphs, this yields a time complexity of $O(n^3)$, where n is the number of vertices in the grounded segment graph. Although the strings in a grounded segment graph are straight line segments, no faster algorithm is known for this case. Thus a natural question is to ask whether one can prove non-trivial lower bounds on the time complexity of the MIS problem for outerstring graphs or simpler variants of such graphs.

An L-shape is the union of a vertical segment and a horizontal segment that share an endpoint; hence, there are four possible types of L-shapes: $\{\ulcorner, \urcorner, \llcorner, \lrcorner\}$. A graph is called a B_1-*VPG graph* if it is the intersection graph of a set of L-shapes in the plane. This class of string graphs belongs to a larger class called the Vertex intersection of Paths on a Grid (VPG) and denoted by B_k-VPG, where k indicates the maximum number of bends each path can have in the grid representation [4]. These graphs and their relatives have been studied extensively in terms of recognition problems (e.g., see [4,12,15,19]). Recently, there has been an increasing attention on studying optimization problems on these graphs; see [5,6,26,27] and the references therein. For the MIS problem, it is known that the problem is NP-complete on B_k-VPG graphs even when $k = 1$ [25], and the previously best-known approximation algorithms have factor $4 \cdot \log n$ [6,27]. Combining B_1-VPG and grounded string graphs, we consider the MIS problem on *grounded* L and *grounded square-*L graphs.

Definition 2 (Grounded L and Grounded Square-L Graphs). *Graph G is called a* grounded L *graph if G is the intersection of a set of* L*-shapes such that each* L*-shape is of type* $\ulcorner$ *and the lower endpoint of the vertical segment of each* L*-shape is attached to a grounding line ℓ. If the vertical and horizontal segments* *of every L-shape in a grounded L representation of G have the same length, then we call G a* grounded square-L *graph.*

See Fig. 1(f–g) for examples of these graphs. Finally, for the MIS problem on a set of n rectangles, Chalermsook and Chuzhoy [9] gave an $(\log \log n)$-approximation algorithm for the unweighted version of the problem. For the weighted version of the problem, the best approximation factor is $O(\log n / \log \log n)$ due to Chan and Har-Peled [11].

We now summarize our contribution in C_1–C_3.

C1. (Section 2): We first examine the time-complexity of the MIS problem on the grounded segment graphs with respect to its relation to the MIS problem in circle graphs. Middendorf and Pfeiffer [28] showed that every intersection graph of L-shapes of types ⌐ and ∟ (not necessarily grounded) can be transformed into a segment representation. If the L-shapes are grounded, then the transformation yields a grounded segment graph. Since every circle graph is a grounded L graph [22], they are also grounded segment graphs. However, the transformation [28] into the grounded segment representation is by an inductive proof, and it is unclear whether the constructed representation can be encoded in a subquadratic number of bits. We show that the MIS problem in a circle graph representation is $O(n \log n)$-time reducible to the MIS problem in an implicit representation of a grounded segment graph, where the representation takes $O(n \log n)$ bits. This indicates that solving MIS in such grounded segment representations is as hard as solving MIS in circle graph representations.

C2. (Sections 3, 4): Since grounded L graphs include circle graphs, we examined a simpler variant: grounded square-L graphs. We show that there exist grounded square-L graphs (resp., grounded L graphs) that are not circle graphs (resp., grounded square-L graphs). Although grounded square-L is a simpler variant, we prove that it includes the circle graphs. In fact, we give an $O(n \log n)$-time reduction, showing that MIS in grounded square-L representations is at least as hard as MIS in circle graph representations. In contrast, for the grounded string representations where the strings are y-monotone simple polygonal paths of constant length with segments at integral coordinates, we can solve MIS in $O(n^2)$ time. Assuming the strong exponential time hypothesis (SETH) [21], we show that an $O(s^{2-\delta})$-time algorithm, where $\delta > 0$, for computing MIS in outerstring representations of size $O(s)$ is unlikely, even when each string has one bend.

C3. (Section 5): We give a $(4 \cdot \max\{1, \log \mathsf{OPT}\})$-approximation algorithm for the weighted MIS problem on the intersection graph of a set of n L-shapes in the plane. This improves the previously best-known algorithm, which has an approximation factor of $4 \cdot \log n$ [6,27]. Moreover, our algorithm can be used to obtain a simple $(4 \cdot \max\{1, \log \mathsf{OPT}\})$-approximation algorithm for the weighted MIS problem on a set of n axis-parallel rectangles in the plane.

Throughout the paper, the complete proofs of lemmas and theorems marked with (∗) appear in the full version of the paper [7] due to space constraints.

2   MIS on Grounded Segment Representations

In this section, we show that the MIS problem in a circle graph representation is $O(n \log n)$-time reducible to the MIS problem in a representation of a grounded segment graph, where the representation takes $O(n \log n)$ bits. This indicates that solving MIS on grounded segment representations could be as hard as solving MIS on circle graph representations.

An *overlap graph* is an intersection graph of intervals, where two vertices are adjacent if and only if their corresponding intervals properly intersects (i.e., the intersection is non-empty but neither contains the other). Gavril [18] showed that a graph is a circle graph if and only if it is an overlap graph. Given the circle graph representation, one can find an overlap representation in linear time by computing the shadow of each chord on a horizontal line below the circle, assuming the point light source is at the apex of the circle as illustrated in Fig. 2(a–c). It now suffices to show that the overlap representation can be transformed into a grounded segment representation in linear time.

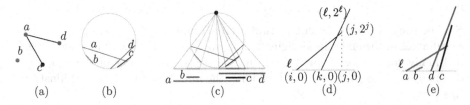

Fig. 2. (a) A circle graph G. (b) A circle graph representation of G. (c) Transformation into an overlap graph. (d)-(e) Transformation into a grounded segment graph. We only show a schematic representation for space constraints.

We assume that the circle graph representation is non-degenerate, i.e., no two chords share a common endpoint. Consequently, the overlap representation is also non-degenerate. We now sort the endpoints of the intervals and relabel them with integral coordinates. For each interval $[i, j]$ in the overlap graph, we define a line segment with coordinates $(i, 0), (j, 2^j)$. Note that all the segments are grounded at the line $y = 0$; i.e., line ℓ in Fig. 2(d). Moreover, it is straightforward to encode the representation implicitly in $O(n \log n)$ bits (note that an explicit representation would require $O(n^2)$ bits). Let the resulting representation be $\mathcal{R}$. In the proof of the following theorem we show that $\mathcal{R}$ is the required grounded segment representation.

Theorem 1 (∗).   *Given a circle graph representation with n chords, in $O(n \log n)$ time one can transform it into an implicit grounded segment representation, which uses $O(n \log n)$ bits. Thus, the MIS problem on grounded segment representations is at least as hard as the MIS problem on circle graph representations.*

3   MIS on Grounded Square-L Representations

In this section, we show that solving MIS in a circle graph representation is $O(n \log n)$-time reducible to solving MIS in a grounded square-L representation.

Given a circle graph representation, we first compute the corresponding overlap graph in the same way as we did in Sect. 2, and relabel the endpoints with integral coordinates from 0 to $2n$. We now transform this into a grounded square-L representation. The idea is to process the intervals in the order of their endpoints, and sometimes shifting the endpoints by a certain offset γ to avoid unnecessary crossings. We now give formal description of the steps of the construction by S_1–S_3.

S_1. Initialize an empty list Q, and then process the intervals in the increasing order of the x-coordinates of their left endpoints. While processing an interval $I = [I_\ell, I_r]$, we first find the closest non-intersecting interval $J = [J_\ell, J_r]$ to the left of I. If no such interval exists, then we continue processing the next interval. Otherwise, let (X, γ) be the tuple at the end of the list Q (assume a dummy tuple $(\Phi, 0)$ if the list is empty). If $J \neq X$, then append a new tuple $(J, J_\ell + \gamma)$ to Q.

S_2. For each pair of consecutive tuples (A, α) and (B, β) in Q, update the x-coordinates of the endpoints originally lying in $[A_r + 1, B_\ell]$ by adding the integer α. Finally, for the last tuple (X, γ), update the x-coordinates of the endpoints originally lying in $[X_r + 1, +\infty]$, by adding the integer γ. Figure 3(a–b) illustrate this step.

S_3. For each interval $[I_\ell, I_r]$ in the increasing order of their left endpoints, construct a square-L shape with endpoints $(\frac{I_\ell}{2}, -\frac{I_\ell}{2})$ and $(I_r + \frac{I_\ell}{2}, -\frac{I_\ell}{2})$, and create the bend point at $(I_\ell + \frac{I_r - I_\ell}{2}, \frac{I_r - I_\ell}{2})$. See Fig. 3(c).

By S_3, it is straightforward to see that all the shapes are grounded on the line $x + y = 0$. Let Γ be the resulting grounded square-L representation. The following lemma claims the correctness of the representation.

Lemma 1. *The graph represented by Γ is the same as the graph represented by the overlap representation.*

Proof. Let G be the graph corresponding to the input overlap representation. While processing the kth interval B in S_3, it suffices to verify the invariant that the subgraph H_k of G induced by B and the intervals with left endpoints smaller than B_ℓ has been correctly represented with a grounded square-L representation.

The invariant is trivial for the first interval, and assume that it holds for $H_1, \ldots, H_{k-1}$, where $k > 1$. Consider now the kth interval B. Let b be the vertex corresponding to interval B, and let a be another vertex in H_k, and denote by A, the interval of a. Let A', B' be the modified intervals (computed in S_2). For any interval I, let $L(I)$ be the square-L shape constructed as in S_3. We now consider the following cases.

Case 1 (a and b are adjacent in H_k): In this case A and B properly intersect; i.e., neither contains the other. Let J' be the closest non-overlapping interval

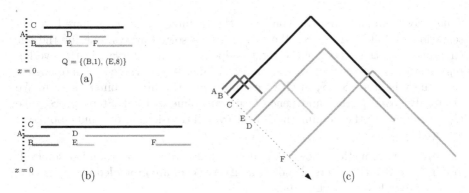

Fig. 3. (a) An overlap representation. (b) Modification after step S_2. (c) The grounded square-L representation constructed at S_3; A is grounded at $(0,0)$.

to the left of B'. Any interval having an endpoint between $[J'_r + 1, B'_\ell]$ will be shifted together with B. Thus all these intersections remain valid, along with all the other intervals who were intersecting B but did not have an endpoint in $[J'_r + 1, B'_\ell]$. We now show that the shift in S_2 keeps the ordering of the intervals in H_{k-1} intact.

Consider a pair of vertices in p, q in H_{k-1}, and let P and Q be their corresponding intervals. Let P' and Q' be the modified intervals in S_2. If p and q are adjacent, then P and Q must properly intersect. Since the endpoints of P and Q can only get extended to the right (i.e., the intervals don't shrink), $L(P')$ and $L(Q')$ must intersect. Now consider the case when p and q are not adjacent. If one of P and Q contains the other, then the same argument holds. If neither contains the other, then the offset may only increase their distance. Therefore, if $L(P)$ and $L(Q)$ do not intersect, then $L(P')$ and $L(Q')$ cannot intersect.

Case 2 (a and b are non-adjacent in H_k): In this case either A and B do not intersect, or A contains B (note that B cannot contain A). If A contains B, then by the same argument as in Case 1, we can see that $L(A')$ and $L(B')$ will not intersect.

Assume now that A and B do not intersect. Recall that B has been processed after A. While we processed B in S_1, we first computed the closest interval J to the left of B. Hence $A_r \leq J_r$. In S_2, we ensured that the endpoints of B are shifted to the right by at least an amount of $J_\ell + \gamma$. Here, γ corresponds to the overall shift for J to accommodate the segments that were processed before J, and the term J_ℓ is to avoid the crossing between $L(J')$ and $L(B')$. Since $A_r \leq J_r$, the shapes $L(A')$ and $L(B')$ cannot intersect. Using the argument of Case 1, observe that such shifting still maintains a valid representation for H_{k-1}. □

Theorem 2. *Given a circle graph representation with n chords, in $O(n \log n)$ time one can transform it into a grounded square-L representation. Thus, the* MIS *problem on grounded square-L representations is at least as hard as the* MIS *problem on circle graph representations.*

Proof. By Lemma 1, one can construct the required grounded square-L representation by following S_1–S_3. We compute two sorted arrays, one for the left endpoints and the other for the right endpoints of the intervals in the overlap representation. The sorting takes $O(n \log n)$ time. We use these arrays to answer each query in steps S_1–S_3 in $O(\log n)$ time by performing a binary search. We need only $O(n)$ queries, and hence $O(n \log n)$ time in total. Steps S_2–S_3 take $O(n)$ time. Hence the running time of the overall transformation can be bounded by $O(n \log n)$.   □

Remark. Our reduction shows that every circle graph is a grounded square-L graph. However, the reverse is not true. Even, there are grounded L graphs that are not grounded square-L graphs.

Theorem 3 (∗). *There are grounded square-L graphs that are not circle graphs. Moreover, there are grounded L graphs that are not grounded square-L graphs.*

The strong exponential time hypothesis (SETH), introduced by Impagliazzo, Paturi, and Zane [21], has been used to analyze fine-grained time-complexity of problems that lie in P. Under SETH, CNF-SAT on n variables cannot be solved in $O(2^{n(1-\epsilon)} poly(n))$ time for any $\epsilon > 0$. The following theorem sates that under SETH, finding MIS in outerstring graphs requires $\Omega(n^{2-\epsilon})$ time.

Theorem 4 (∗). *Assuming the strong exponential time hypothesis (SETH), computing an MIS in an outerstring representation with n strings requires $\Omega(n^{2-\epsilon})$ time, even when each string contains $O(1)$ bends.*

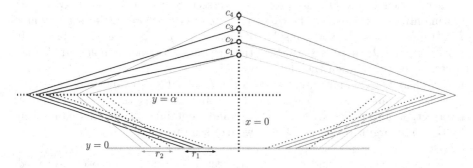

Fig. 4. Illustration for the proof of Theorem 4.

Proof. Given an instance of CNF-SAT, the idea is to partition its n variables into two sets A, B. For each of the $2^{n/2}$ truth assignments for the variables in A, we construct a set of α outerstrings that correspond to the α clauses that it satisfies. For example, an interval r_i, $1 \le i \le 2^{n/2}$, in Fig. 4, corresponds to a truth assignment of the variables of the variables in A, and the strings (solid lines) grounded in r_i correspond to the clauses that the assignment satisfies. We construct the strings for the set B symmetrically. We show that an MIS of size m, where m is the number of clauses, would correspond to an affirmative solution to the CNF-SAT instance, and vice versa. See the full version of the paper [7] for more details.   □

4 Representations with Bounded-Length Integral Shapes

In this section, we consider string representations where the strings are y-monotone (not necessarily strict) polygonal paths, the length of each string is bounded by a constant κ, and all the bends and endpoints are on integral coordinates. We show that the MIS problem on such representations can be solved in $O(n^2)$ time. For simplicity, we first examine the case when each string is an L-shape of type $\ulcorner$. Denote by M_p, an axis-aligned simple y-monotone (not necessarily strictly monotone) polygonal path that satisfies the following three constraints: (a) M_p starts at point p, and ends at a point on the line $y = \kappa$. (b) M_p contains at most 2κ bends, and (c) the length of each line segment in M_p is bounded by κ. Then the number of such distinct strings can be at most $f(\kappa) \in O(1)$ (since κ is a constant). Denote the set of such strings by $\mathcal{M}_p$.

We employ a dynamic programming technique, where we express a subproblem with two points a, b on the grounding line and two monotone paths M_a and M_b. Figure 5(a) illustrates a subproblem $\mathsf{MIS}(a, b, M_a, M_b)$. The subproblem contains all the L-shapes of the given representation that are in the region between M_a and M_b. The left side of the region is open and the right side is closed, hence the L-shape that starts at a must be excluded. While constructing subproblems, we will ensure that a and b belong to the set of grounding points on the grounding line. The initial problem can be expressed as $\mathsf{MIS}(i, j, M_i, M_j)$, where i is a grounding point of a dummy L-shape I lying to the left of all the L-shapes, and j is the grounding point of the rightmost L-shape. M_i and M_j are two strings that bound all the L-shapes in between.

Given a problem of the form $\mathsf{MIS}(a, b, M_a, M_b)$, we first find a grounding point q at the median position among the distinct grounding points between a and b, as illustrated in Fig. 5(b). Note that L-shapes can share grounding points, and we only consider the distinct points while considering the median point. If q coincides with b, then we have the base case where all the L-shapes starts at b. We thus return 1 or 0 depending on whether there exists a L-shape in the region between M_a and M_b (this takes $O(n)$ time). Otherwise, we compute the solution using the following recurrence relation.

$$\mathsf{MIS}(a, b, M_a, M_b) = \max_{M \in \mathcal{M}_q} \mathsf{MIS}(a, q, M_a, M) + \mathsf{MIS}(q, b, M, M_b).$$

To verify the correctness of the recurrence relation, observe that any independent set of $\mathsf{MIS}(a, b, M_a, M_b)$ can be partitioned by a string in M_q. The size of the dynamic programming table is bounded by $O(n^2) \times O(1)$, where the first term comes from the choices for a and b, and the $O(1)$ term corresponds to the possible choices for M_a and M_b. Computing a base case requires $O(n)$ time. In the base case, a and b are consecutive on the ground line, and hence there can be at most $O(n) \times f(\kappa) \times f(\kappa)$ distinct base cases, requiring $O(n^2)$ time in total. Computing an entry in the general case requires $f(\kappa) \in O(1)$ time (using constant time table look-up). Hence the running time for the general case is also bounded by $O(n^2)$ in total.

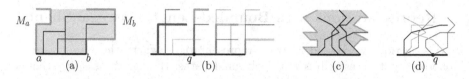

Fig. 5. Illustration for the dynamic programming. (a) A subproblem. (b) Splitting into subproblems. (c)–(d) General y-monotone strings.

Although we described the algorithm for L-shapes, it is straightforward to generalize the algorithm for y-monotone strings, as illustrated in Fig. 5(c)–(d). The only difference is that we need to define M_p as a simple y-monotone path. The following theorem summarizes the results of this section.

Theorem 5. *Let R be a string representation such that the strings are y-monotone (not necessarily strict), the length of each string is bounded by a constant, and all the bends and endpoints are on integral coordinates. Then, the MIS problem in R can be solved in $O(n^2)$ time.*

5   A $(4 \cdot \log \mathsf{OPT})$-Approximation Algorithm

In this section, we give a $(4 \cdot \max\{1, \log \mathsf{OPT}\})$-approximation algorithm for the MIS problem on the intersection graph of a set of n L-shapes. To this end, we first give a $(\max\{1, \log \mathsf{OPT}\})$-approximation algorithm for the problem when the input consist of only L-shapes of type $\ulcorner$. We discuss the generalization of our algorithm to the weighted version of the MIS problem and for approximating the MIS problem on rectangles at the end of this section.

Consider the input L-shapes from left to right in the increasing order of the x-coordinate of their vertical segment; we denote the ith L-shape in this ordering by L_i. For any $1 \le i < j \le n$, we define $I[i, j]$ as the set of L-shapes L_x such that (i) $i \le x \le j$, and (ii) L_x does not intersect the line through the vertical segment of L_{j+1}. We add a dummy L-shape L_{n+1} far to the right such that no input L-shape intersects the line through the vertical segment of L_{n+1}; thus, $I[1, n]$ is the set of all input L-shapes. Moreover, let $\mathsf{OPT}[i, j]$ denote the size of an optimal solution for the MIS problem on the set of L-shapes in $I[i, j]$; we denote $\mathsf{OPT}[1, n]$ simply by OPT. For any such i, j and some $i < k < j$, let I_k denote the set of L-shapes L_y such that (i) $i \le y \le j$ and (ii) L_y intersects the line through the vertical segment of L_k. Moreover, let $\mathsf{OPT}(I_k)$ be the size of an optimal solution for the MIS problem on the intersection graph induced by the L-shapes in I_k.

We define $S[i, j]$ as the solution returned by our algorithm on the L-shapes in $I[i, j]$, for all $1 \le i < j \le n$. Initially, for every pair $1 \le i < j \le n$, if $I[i, j] = \emptyset$, then we set $S[i, j] = 0$. Then, for every pair $1 \le i < j \le n$, we check to see if $\mathsf{OPT}[i, j] \le 4$; if so, then we directly store $\mathsf{OPT}[i, j]$ in $S[i, j]$. Otherwise, we compute $S[i, j]$ as follows.

$$S[i, j] = \max\{ \max_{i<k<j} S[i, k-1] + S[k+1, j], \mathsf{OPT}(I_k)\}.$$

The algorithm returns $S[1, n]$ as the solution. Computing the actual solution can be done in the standard manner; to this end, we also store the corresponding value of k in $S[i, j]$.

Approximation Factor. To show the approximation factor, let $\mathsf{S}^{\mathsf{OPT}}[i, j]$ be the set of L-shapes in $\mathsf{OPT}[i, j]$. If $\mathsf{OPT}[i, j] \leq 4$, then we have $S[i, j] = \mathsf{OPT}[i, j]$. We now prove by induction that for all $1 \leq i < j \leq n$, if $\mathsf{OPT}[i, j] > 4$, then $S[i, j] \geq \mathsf{OPT}[i, j]/\log \mathsf{OPT}[i, j]$. Suppose that $S[i, j] \geq \mathsf{OPT}[i, j]/\log \mathsf{OPT}[i, j]$ for all $1 \leq i < j \leq n$ for which $4 < \mathsf{OPT}[i, j] < m$. Take any pair $1 \leq i < j \leq n$ for which $\mathsf{OPT}[i, j] = m$, and let k_j^i be the index such that $\mathsf{L}_{k_j^i}$ is the median of the L-shapes in $\mathsf{S}^{\mathsf{OPT}}[i, j]$ (i.e., each $\mathsf{S}^{\mathsf{OPT}}[i, k_j^i - 1]$ and $\mathsf{S}^{\mathsf{OPT}}[k_j^i + 1, j]$ contains at most $\mathsf{OPT}[i, j]/2$ L-shapes). Notice that

$$\mathsf{OPT}(I_{k_j^i}) \geq |\mathsf{S}^{\mathsf{OPT}}[i, j] \cap I_{k_j^i}|. \tag{1}$$

Now, if $\mathsf{OPT}[i, k_j^i - 1] \leq 4$, then we know that $S[i, k_j^i - 1] = \mathsf{OPT}[i, k_j^i - 1]$. Otherwise, by the induction hypothesis, we have

$$S[i, k_j^i - 1] \geq \frac{\mathsf{OPT}[i, k_j^i - 1]}{\log \mathsf{OPT}[i, j]/2} \geq \frac{|\mathsf{S}^{\mathsf{OPT}}[i, j] \cap I[i, k_j^i - 1]|}{\log \mathsf{OPT}[i, j] - 1}. \tag{2}$$

Similarly, if $\mathsf{OPT}[k_j^i + 1, j] \leq 4$, then we know that $S[k_j^i + 1, j] = \mathsf{OPT}[k_j^i + 1, j]$. Otherwise, by the induction hypothesis, we have

$$S[k_j^i + 1, j] \geq \frac{\mathsf{OPT}[k_j^i + 1, j]}{\log \mathsf{OPT}[i, j]/2} \geq \frac{|\mathsf{S}^{\mathsf{OPT}}[i, j] \cap I[k_j^i + 1, j]|}{\log \mathsf{OPT}[i, j] - 1}. \tag{3}$$

Therefore,

$$S[i, j] = \max\{\max_{i < k < j} S[i, k - 1] + S[k + 1, j], \mathsf{OPT}(I_k)\}$$

$$\geq \max\{S[i, k_j^i - 1] + S[k_j^i + 1, j], \mathsf{OPT}(I_{k_j^i})\}$$

$$\geq \max\{\frac{|\mathsf{S}^{\mathsf{OPT}}[i, j] \cap I[i, k_j^i - 1]| + |\mathsf{S}^{\mathsf{OPT}}[i, j] \cap I[k_j^i + 1, j]|}{\log \mathsf{OPT}[i, j] - 1}, |\mathsf{S}^{\mathsf{OPT}}[i, j] \cap I_{k_j^i}|\}$$

$$\geq \max\{\frac{\mathsf{OPT}[i, j] - |\mathsf{S}^{\mathsf{OPT}}[i, j] \cap I_{k_j^i}|}{\log \mathsf{OPT}[i, j] - 1}, |\mathsf{OPT}[i, j] \cap I_{k_j^i}|\}.$$

The first inequality is because our algorithm tries all values of $i < k < j$, which includes k_j^i. Moreover, the second inequality is because of (3), (2) and (1). Now, if $|\mathsf{S}^{\mathsf{OPT}}[i, j] \cap I_{k_j^i}| \geq \mathsf{OPT}[i, j]/\log \mathsf{OPT}[i, j]$, then we are done. Otherwise,

$$\frac{\mathsf{OPT}[i, j] - |\mathsf{S}^{\mathsf{OPT}}[i, j] \cap I_{k_j^i}|}{\log \mathsf{OPT}[i, j] - 1} \geq \frac{\mathsf{OPT}[i, j] - \mathsf{OPT}[i, j]/\log \mathsf{OPT}[i, j]}{\log \mathsf{OPT}[i, j] - 1}$$

$$= \frac{\mathsf{OPT}[i, j]}{\log \mathsf{OPT}[i, j]}.$$

This completes the proof of the induction step. By setting $i = 1$ and $j = n$, we have $S[1, n] \geq \mathsf{OPT}/\log \mathsf{OPT}$.

Running Time. For a fixed triple i, j and k, we can compute $\mathsf{OPT}(I_k)$ in $O(n^3)$ time because the corresponding graph is an outerstring graph for which MIS can be solved in $O(n^3)$ time [23]. Since there are $O(n)$ choices for k for a fixed pair of i and j, and $O(n^2)$ entries in the table for i and j, the overall running time of the algorithm is $O(n^6)$. We show in the full version of the paper [7] how to improve the running time to $O(n^5)$ time, and so we have the following lemma.

Lemma 2 (∗). *There exists an $O(n^5)$-time $(\max\{1, \log \mathsf{OPT}\})$-approximation algorithm for the MIS problem on a set of n L-shapes of type $\ulcorner$, where OPT denotes the size of an optimal solution.*

When the input consists of all four types of L-shapes, we run the algorithm of Lemma 2 four times (once for each type of the input L-shapes), and then return the largest solution as the final answer. Clearly, this gives us a $(4 \cdot \log \mathsf{OPT})$-approximation algorithm for the original problem and so we have the main result of this section.

Theorem 6. *There exists an $O(n^5)$-time $(4 \cdot \max\{1, \log \mathsf{OPT}\})$-approximation algorithm for the MIS problem on any set of n L-shapes, where OPT denotes the size of an optimal solution.*

Generalizations. Our algorithm can be generalized in two ways: for the weighted version of the MIS problem on L-shapes, and for the weighted MIS problem on axis-parallel rectangles.

Theorem 7 (∗). *There exists an $O(n^5)$-time $(4 \cdot \max\{1, \log \mathsf{OPT}\})$-approximation algorithm (resp., an $O(n^3)$-time $(\max\{1, \log \mathsf{OPT}\})$-approximation algorithm) for the weighted MIS problem on any set of n L-shapes (resp., a set of n axis-parallel rectangles in the plane), where OPT is the size of an optimal solution.*

We note that for the case of rectangles, our $(\max\{1, \log \mathsf{OPT}\})$-approximation algorithm provides a somewhat simpler algorithm than the one that can be obtained (with the same approximation factor) from the $O(\log n / \log \log n)$-approximation algorithm of Chan and Har-Peled [11].

6   Conclusion

In this paper, we studied the time-complexity and approximability of the MIS problem on outerstring graphs and their relatives. Our work gives rise to some natural open questions:

1. Does there exist a quadratic-time algorithm that can solve the MIS problem on grounded segment or grounded square-L graphs?
2. Can we improve the approximation factor of the algorithm of Theorem 6?
3. Can we find an $\Omega(n^{2-\epsilon})$-time lower bound under SETH for finding MIS in grounded segment representations?

References

1. Agarwal, P.K., Mustafa, N.H.: Independent set of intersection graphs of convex objects in 2D. Comput. Geom. **34**(2), 83–95 (2006)
2. Apostolico, A., Atallah, M.J., Hambrusch, S.E.: New clique and independent set algorithms for circle graphs. Discrete Appl. Math. **36**(1), 1–24 (1992)
3. Asano, T., Imai, H., Mukaiyama, A.: Finding a maximum weight independent set of a circle graph. IEICE Trans. **E74**(4), 681–683 (1991)
4. Asinowski, A., Cohen, E., Golumbic, M.C., Limouzy, V., Lipshteyn, M., Stern, M.: Vertex intersection graphs of paths on a grid. J. Graph Algorithms Appl. **16**(2), 129–150 (2012)
5. Bandyapadhyay, S., Maheshwari, A., Mehrabi, S., Suri, S.: Approximating dominating set on intersection graphs of rectangles and L-frames. In: Proceedings of the 43rd International Symposium on Mathematical Foundations of Computer Science (MFCS 2018), Liverpool, UK, pp. 37:1–37:15 (2018)
6. Biedl, T., Derka, M.: Splitting B_2-VPG graphs into outer-string and cocomparability graphs. Algorithms and Data Structures. LNCS, vol. 10389, pp. 157–168. Springer, Cham (2017). https://doi.org/10.1007/978-3-319-62127-2_14
7. Bose, P., et al.: Computing maximum independent set on outerstring graphs and their relatives. CoRR abs/1903.07024 (2019). http://arxiv.org/abs/1903.07024
8. Cardinal, J., Felsner, S., Miltzow, T., Tompkins, C., Vogtenhuber, B.: Intersection graphs of rays and grounded segments. J. Graph Algorithms Appl. **22**(2), 273–295 (2018)
9. Chalermsook, P., Chuzhoy, J.: Maximum independent set of rectangles. In: Proceedings of the 20th Annual ACM-SIAM Symposium on Discrete Algorithms (SODA 2009), New York, NY, USA. pp. 892–901 (2009)
10. Chalopin, J., Gonçalves, D.: Every planar graph is the intersection graph of segments in the plane: extended abstract. In: Proceedings of the 41st Annual ACM Symposium on Theory of Computing (STOC 2009), Bethesda, MD, USA. pp. 631–638 (2009)
11. Chan, T.M., Har-Peled, S.: Approximation algorithms for maximum independent set of pseudo-disks. Discrete Comput. Geom. **48**(2), 373–392 (2012)
12. Chaplick, S., Jelínek, V., Kratochvíl, J., Vyskočil, T.: Bend-bounded path intersection graphs: sausages, noodles, and waffles on a grill. In: Golumbic, M.C., Stern, M., Levy, A., Morgenstern, G. (eds.) WG 2012. LNCS, vol. 7551, pp. 274–285. Springer, Heidelberg (2012). https://doi.org/10.1007/978-3-642-34611-8_28
13. Ehrlich, G., Even, S., Tarjan, R.E.: Intersection graphs of curves in the plane. J. Comb. Theory, Ser. B **21**(1), 8–20 (1976)
14. Erlebach, T., Jansen, K., Seidel, E.: Polynomial-time approximation schemes for geometric intersection graphs. SIAM J. Comput. **34**(6), 1302–1323 (2005)
15. Felsner, S., Knauer, K.B., Mertzios, G.B., Ueckerdt, T.: Intersection graphs of L-shapes and segments in the plane. Discrete Appl. Math. **206**, 48–55 (2016)
16. Fox, J., Pach, J.: Computing the independence number of intersection graphs. In: Proceedings of the 22nd Annual ACM-SIAM Symposium on Discrete Algorithms (SODA 2011), San Francisco, CA, USA, pp. 1161–1165 (2011)
17. Frank, A.: Some polynomial algorithms for certain graphs and hypergraphs. In: Proceedings of the 5th British Combinatorial Conference (1975)
18. Gavril, F.: Algorithms for a maximum clique and a maximum independent set of a circle graph. Networks **3**, 261–273 (1973)

19. Gonçalves, D., Isenmann, L., Pennarun, C.: Planar graphs as L-intersection or L-contact graphs. In: Proceedings of the 29th Annual ACM-SIAM Symposium on Discrete Algorithms (SODA 2018), New Orleans, LA, USA, pp. 172–184 (2018)
20. Håstad, J.: Clique is hard to approximate within $n^{1-\epsilon}$. In: Proceedings of the 37th Annual Symposium on Foundations of Computer Science (FOCS 1996), Burlington, Vermont, USA, pp. 627–636 (1996)
21. Impagliazzo, R., Paturi, R., Zane, F.: Which problems have strongly exponential complexity? J. Comput. Syst. Sci. **63**(4), 512–530 (2001)
22. Jelínek, V., Töpfer, M.: On grounded L-graphs and their relatives. CoRR abs/1808.04148 (2018)
23. Keil, J.M., Mitchell, J.S.B., Pradhan, D., Vatshelle, M.: An algorithm for the maximum weight independent set problem on outerstring graphs. Comput. Geom. **60**, 19–25 (2017)
24. Kratochvíl, J.: String graphs. I. the number of critical nonstring graphs is infinite. J. Comb. Theory, Ser. B **52**(1), 53–66 (1991)
25. Lahiri, A., Mukherjee, J., Subramanian, C.R.: Maximum independent set on B_1-VPG graphs. In: Proceedings of the 9th International Conference Combinatorial Optimization and Applications (COCOA 2015), Houston, TX, USA, pp. 633–646 (2015)
26. Mehrabi, S.: Approximating domination on intersection graphs of paths on a grid. In: Solis-Oba, R., Fleischer, R. (eds.) WAOA 2017. LNCS, vol. 10787, pp. 76–89. Springer, Cham (2018). https://doi.org/10.1007/978-3-319-89441-6_7
27. Mehrabi, S.: Approximation algorithms for independence and domination on B_1-VPG and B_1-EPG graphs. CoRR abs/1702.05633 (2017)
28. Middendorf, M., Pfeiffer, F.: Weakly transitive orientations, Hasse diagrams and string graphs. Discrete Math. **111**(1–3), 393–400 (1993)
29. Nash, N., Gregg, D.: An output sensitive algorithm for computing a maximum independent set of a circle graph. Inf. Process. Lett. **110**(16), 630–634 (2010)
30. Scheinerman, E.R.: Intersection Classes and Multiple Intersection Parameters of Graphs. Ph.D. thesis, Princeton University (1984)
31. Supowit, K.J.: Finding a maximum planar subset of a set of nets in a channel. IEEE Trans. CAD Integr. Circ. Syst. **6**(1), 93–94 (1987)
32. Valiente, G.: A new simple algorithm for the maximum-weight independent set problem on circle graphs. In: Ibaraki, T., Katoh, N., Ono, H. (eds.) ISAAC 2003. LNCS, vol. 2906, pp. 129–137. Springer, Heidelberg (2003). https://doi.org/10.1007/978-3-540-24587-2_15

Online Bin Covering with Advice

Joan Boyar[1], Lene M. Favrholdt[1], Shahin Kamali[2(✉)], and Kim S. Larsen[1]

[1] University of Southern Denmark, Odense, Denmark
{joan,lenem,kslarsen}@imada.sdu.dk
[2] University of Manitoba, Winnipeg, Manitoba, Canada
shahin.kamali@umanitoba.ca

Abstract. The bin covering problem asks for covering a maximum number of bins with an online sequence of n items of different sizes in the range $(0, 1]$; a bin is said to be covered if it receives items of total size at least 1. We study this problem in the advice setting and provide tight bounds for the size of advice required to achieve optimal solutions. Moreover, we show that any algorithm with advice of size $o(\log \log n)$ has a competitive ratio of at most 0.5. In other words, advice of size $o(\log \log n)$ is useless for improving the competitive ratio of 0.5, attainable by an online algorithm without advice. This result highlights a difference between the bin covering and the bin packing problems in the advice model: for the bin packing problem, there are several algorithms with advice of constant size that outperform online algorithms without advice. Furthermore, we show that advice of size $O(\log \log n)$ is sufficient to achieve a competitive ratio that is arbitrarily close to $0.53\bar{3}$ and hence strictly better than the best ratio 0.5 attainable by purely online algorithms. The technicalities involved in introducing and analyzing this algorithm are quite different from the existing results for the bin packing problem and confirm the different nature of these two problems. Finally, we show that a linear number of bits of advice is necessary to achieve any competitive ratio better than 15/16 for the online bin covering problem.

1 Introduction

In the bin covering problem [3], the input is a multi-set of items of different sizes in the range $(0, 1]$ which need to be placed into a set of bins. A bin is said to be covered if the total size of items in it is at least 1. The goal of the bin covering problem is to place items into bins so that a maximum number of bins is covered. In the online setting, items form a sequence which is revealed in a piece-by-piece manner; that is, at each given time, one item of the sequence is revealed and an online algorithm has to place the item into a bin without any information about the forthcoming items. The decisions of the algorithm are irrevocable.

The first, second, and fourth authors were supported in part by the Danish Council for Independent Research, Natural Sciences, grant DFF-1323-00247.

© Springer Nature Switzerland AG 2019
Z. Friggstad et al. (Eds.): WADS 2019, LNCS 11646, pp. 225–238, 2019.
https://doi.org/10.1007/978-3-030-24766-9_17

Bin covering is closely related to the classic bin packing problem and is sometimes called the dual bin packing problem[1]. The input to both problems is the same. In the bin packing problem, however, the goal is to place items into a minimum number of bins so that the total size of items in each bin is at most 1. Online algorithms for bin packing can naturally be extended to bin covering. For example, Next-Fit is a bin packing algorithm which keeps one "open" bin at any time: To place an incoming item x, if the size of x is smaller than the remaining capacity of the open bin, x is placed in the open bin; otherwise, the bin is closed (never used again) and a new bin is opened. Dual-Next-Fit [3] is a bin covering algorithm that behaves similarly, except that it closes the bin when the total size of items in it becomes at least 1.

In the offline setting, the bin packing and bin covering problems are NP-hard. There is an asymptotic fully polynomial-time approximation scheme (AFPTAS) for bin covering [17]. There are also bin packing algorithms which open $\text{OPT}(\sigma) + o(\text{OPT}(\sigma))$ bins [16,18,23], where $\text{OPT}(\sigma)$ is the number of bins in the optimal packing. The additive term was improved in [23] and further, to $O(\log \text{OPT}(\sigma))$, in [16].

Online algorithms are often compared under the framework of competitive analysis. An algorithm, $\mathbb{A}$, for bin covering (respectively, bin packing) is *c-competitive*, if there exists a constant b such that, for any input sequence, σ, $\mathbb{A}(\sigma) \geq c \cdot \text{OPT} - b$ (respectively, $\mathbb{A}(\sigma) \leq c \cdot \text{OPT} + b$). The *competitive ratio* of a bin covering (respectively, bin packing) algorithm, $\mathbb{A}$, is $\sup\{c \mid \mathbb{A} \text{ is } c\text{-competitive}\}$ (respectively, $\inf\{c \mid \mathbb{A} \text{ is } c\text{-competitive}\}$).

Despite similarities between bin covering and bin packing, the status of these problems are different in the online setting. In the case of bin covering, it is known that no online algorithm can achieve a competitive ratio better than $1/2$ [13], while bin covering algorithms such as Dual-Next-Fit [3] have the best possible competitive ratio of $1/2$. Hence, we have a clear picture of the complexity of deterministic bin covering under competitive analysis. The situation is more complicated for the bin packing problem. It is known that no deterministic algorithm can achieve a competitive ratio of 1.54278 [5] while the best existing deterministic algorithm has a competitive ratio of 1.5783 [4]. Note there is a gap between the best known upper and lower bounds.

Advice complexity is a formalized way of measuring how much knowledge of the future is required for an online algorithm to obtain a certain level of performance, as measured by the competitive ratio. When such advice is available, algorithms with advice could lead to semi-online algorithms. Unlike related approaches such as "lookahead" [15] (in which some forthcoming items are revealed to the algorithm) and "closed bin packing" [2] (where the length of the input is revealed), *any* information can be encoded and sent to the algorithm under the advice setting. This generality means that lower bound results under the advice model also imply strong lower bound results on semi-online

---

[1] There is another problem, also sometimes referred to as "dual bin packing", which asks for maximizing the number of items packed into a fixed number of bins; for the advice complexity of that dual bin packing problem, see [9,21].

algorithms, where one can infer impossibility results simply from the length of an encoding of the information a semi-online algorithm is provided with. Advice complexity is also closely related to randomization; complexity bounds from advice complexity can be transferred to the randomization case and vice versa [6,8,14,20].

The advice is generated by a benevolent oracle with unlimited computational power. The advice is written on a tape and the algorithm knows its meaning. This general approach has been studied for many problems (we refer the reader to a recent survey on advice complexity of online problems [11]). In particular, bin packing has been studied under the advice complexity [1,12,22].

Contributions

In this article, we provide the first results with respect to the advice complexity of the bin covering problem. To obtain an optimal result, advice essentially corresponding to an encoding of an entire optimal solution is necessary and sufficient. Not surprisingly, this follows from a similar proof for bin packing, since for both problems, bins filled to size one in an optimal solution are at the core of the proof. However, unlike the bin packing problem, advice of constant size cannot help improve the competitive ratio of algorithms. We establish this result by showing that any algorithm with advice of size $o(\log \log n)$ has a competitive ratio of at most 0.5, which is the competitive ratio of online algorithms without advice. We prove a tight result that advice of size $O(\log \log n)$ suffices to achieve a competitive ratio arbitrarily close to $0.53\bar{3}$. Finally, using a reduction from the binary string guessing problem [7], we show that advice of linear size is necessary to achieve any competitive ratio larger than 15/16. This is similar to, but more intricate, than the corresponding result for bin packing.

2   Optimal Covering and Advice

It is not hard to see that advice of size $O(n \log(\text{OPT}(\sigma)))$ is sufficient to achieve an optimal covering for an input σ of length n; note that $\text{OPT}(\sigma)$ denotes the number of bins in an optimal covering of σ. Provided with $O(\log(\text{OPT}(\sigma)))$ bits of advice for each item, the offline oracle can indicate in which bin the item is placed in the optimal packing. Provided with this advice, the online algorithm just needs to pack each item in the bin indicated by the advice. Clearly, the size of the advice is $O(n \log(\text{OPT}(\sigma)))$ and the outcome is an optimal packing. Note that it is always assumed that the oracle that generates the advice has unbounded computational power. However, if the time complexity of the oracle is a concern, we can use the AFPTAS of [17] to generate an almost-optimal packing and encode it in the advice. Similarly, if the input is assumed to have only m distinct known sizes, one can encode the entire request sequence, specifying for each distinct size how many of that size occur in the sequence. This only requires $O(m \log(n))$ bits of advice. The following theorem shows that the above naive solutions are asymptotically tight.

Theorem 1. *For online bin covering on sequences σ of length n, advice of size $\Theta(n \log \text{OPT}(\sigma))$ is required and sufficient to achieve an optimal solution, assuming $2\,\text{OPT}(\sigma) \leq (1 - \varepsilon)n$ for some positive value of ε. When the input is formed by n items with $m \in o(n)$ distinct, known item sizes, advice of size $\Theta(m \log n)$ is required and sufficient to achieve an optimal solution.*

Proof. The lower bounds follow immediately from the corresponding results for bin packing [12, Theorems 1, 3]. Since the optimal result in those proofs have all bins filled to size 1, any non-optimal bin packing would also lead to a non-optimal bin covering. ☐

3   Advice of Size $o(\log \log n)$ Is Not Helpful

In this section, we show that advice of size $o(\log \log n)$ does not help for improving the competitive ratio of bin covering algorithms. This result is in contrast to bin packing where advice of constant size can improve the competitive ratio. Our lower bound sequence is similar to the one in [13], where the authors proved a lower bound on the competitive ratio of purely online algorithms.

Theorem 2. *There is no algorithm with advice of size $o(\log \log n)$ and competitive ratio better than $1/2$.*

Proof. Consider a family of sequences formed as follows:

$$\sigma_j = \langle \underbrace{\varepsilon, \varepsilon, \dots \varepsilon}_{n \text{ items}}, \underbrace{1 - j\varepsilon, 1 - j\varepsilon, \dots, 1 - j\varepsilon}_{n/j \text{ items}} \rangle$$

Here, j takes a value between 1 and n and hence there are n sequences in the family. All sequences start with the same prefix of n items of size ε. We assume that $\varepsilon < \frac{1}{2n}$ to ensure that, even if all these items are placed in the same bin, the level of that bin is still less than $1/2$. Note that the suffix, formed by items of size $1 - j\varepsilon$ has length $O(n)$, and hence the length of all sequences is $\Theta(n)$.

Clearly, for packing σ_j, an optimal algorithm places j items of size ε in each bin and covers n/j bins. So we have $\text{OPT}(\sigma_j) = n/j$.

The proof is by contradiction, so assume there is an algorithm, $\mathbb{A}$, using $o(\log \log n)$ advice bits and having competitive ratio $1/2 + \mu$ for some constant $\mu > 0$. Thus, there exists a fixed constant d such that for any sequence σ_j we have

$$\mathbb{A}(\sigma_j) \geq (1/2 + \mu)\text{OPT}(\sigma_j) - d = \frac{n}{2j} + \frac{\mu n}{j} - d \qquad (1)$$

We say two sequences belong to the same *sub-family* if they receive the same advice string. Since the advice has size $o(\log \log n)$, there are $o(\log n)$ sub-families. Let $\sigma_{a_1}, \dots, \sigma_{a_w}$ be the sequences in one sub-family. Since the advice and the first n items (of size ε) are the same for any two members of this sub-family, $\mathbb{A}$ will place these n items identically. Let m_i denote the number of bins receiving at least i items in such a placement. So, we have $\sum_{i=1}^{n} m_i = n$ (a bin with exactly x items is counted x times). Moreover, for any σ_j, we have

$$\mathbb{A}(\sigma_j) \leq m_j + (n/j - m_j)/2 = \frac{n}{2j} + \frac{m_j}{2} \tag{2}$$

This follows since any bin with at least j items of size ε can be covered using only one item of size $1 - j\varepsilon$, while the other bins require two such items. From Eqs. 1 and 2, we get $\mu\frac{n}{j} \leq \frac{m_j}{2} + d$. Summing over $j \in \{a_1, \ldots, a_w\}$, we get that

$$\mu n \left(\frac{1}{a_1} + \frac{1}{a_2} + \ldots + \frac{1}{a_w}\right) \leq \frac{1}{2}(m_{a_1} + m_{a_2} + \ldots + m_{a_w}) + wd$$

Since $\frac{1}{2}(m_{a_1} + m_{a_2} + \ldots + m_{a_w}) + dw \leq \frac{1}{2} \cdot \sum_{i=1}^{n} m_i + dn = (d + \frac{1}{2})n$, we have

$$\frac{1}{a_1} + \frac{1}{a_2} + \ldots + \frac{1}{a_w} \in O(1)$$

Summing the left-hand side over all families, we include every sequence once and obtain $\Sigma_{i=1}^{n} \frac{1}{i}$. Since there are $o(\log n)$ sub-families, it follows that $\Sigma_{i=1}^{n} \frac{1}{i} \in o(\log n)$. This is a contradiction since the Harmonic number $\Sigma_{i=1}^{n} \frac{1}{i} \in \Theta(\log n)$.

Thus, our initial assumption is wrong and with advice of size $o(\log \log n)$, no algorithm with competitive ratio strictly better than $1/2$ can exist.   □

4   An Algorithm with Advice of Size $O(\log \log n)$

In this section, we show that advice of size $O(\log \log n)$ is sufficient to achieve a competitive ratio arbitrarily close to $0.53\bar{3}$. Throughout this section, we call an item *small* if it has size less than $1/2$ and *large* otherwise.

Consider a packing of the input sequence σ. We partition the bins in this packing into three groups. A *large-small (LS)* bin includes one large item and some small items, a *large-large (LL)* bin includes only two large items, and a *small (S)* bin includes only small items. We assume there is no small item in the LL bins of OPT (such small items can be moved to another bin without decreasing the number of covered bins). We also assume that, in OPT's packing, large items in LS bins are larger than those in LL bins (otherwise, we can move them around and LL bins will be still covered while the level of LS bins will increase). We use m and m', respectively, to denote the number of LS and LL bins in the optimal packing. For $m \geq 1$, we let $\beta \geq 1$ satisfy $m + m' = \beta m$. See Table 1 for a summary of notation used in this section.

In the following lemma and later, we use the algorithm Dual-Worst-Fit, which, given a fixed number of bins, places an item in a least full bin.

Lemma 1. *Given an integer q, assume we apply Dual-Worst-Fit to cover q bins. Let S denote the total size of packed items and d denote the maximum size of any item in the sequence. The level of any bin is at least $S/q - d$.*

Table 1. Notation used in Sect. 4

Notation	Meaning
n	The length of the input
m	The number of LS bins in the optimal packing
m'	The number of LL bins in the optimal packing
S_l	An integer representing the total size of small items in the LS bins of the optimal packing (rounded down)
S_s	An integer representing the total size of small items in the S bins of the optimal packing (rounded down)
β	The value of $\frac{m+m'}{m}$. The algorithm behaves differently when $\beta \geq 15/14$ compared to when $\beta < 15/14$
α	A parameter of the algorithm when $\beta < 15/14$. Approximately $\lfloor \alpha m \rfloor$ of covered bins include exactly one large item. We assume $\alpha < \frac{7-6\beta}{15} < \frac{4}{105}$
k	An integer representing the precision of approximate encodings in $O(\log \log n)$ bits. We assume k is a large constant and we have $k \geq 6$

Proof. The level of any two bins cannot differ by more than d; otherwise the last item placed in the bin with the larger level had to be placed in the bin with the smaller level. Let $B_{\min}$ and $B_{\max}$ be the two bins with minimum and maximum levels, respectively. From the above observation, we have level$(B_{\min}) \geq$ level$(B_{\max}) - d$. The maximum level of any bin is no less than the average level of all bins. That is, level$(B_{\max}) \geq S/q$ which gives level$(B_{\min}) \geq S/q - d$.   □

The following lemma shows that sequences with relatively few LS bins in an optimal packing are "easy" instances. The lemma is used in the case where $\beta \geq 15/14$, i.e., when there are at most 14 times as many LS bins as LL bins in the optimal packing. Note that, in this case, $(2\beta - 1)/(2\beta) \geq 8/15$.

Lemma 2. *There is an online bin covering algorithm with competitive ratio at least* $\min\{2/3, \frac{2\beta-1}{2\beta}\}$.

Proof. Consider a simple algorithm, $\mathbb{A}$, that places large and small items separately. Each pair of large items cover one bin and small items are placed using the Dual-Next-Fit strategy, that is, they are placed in the same bin until the bin is covered (and then a new bin is started). Let S denote the total size of small items. Note that the number of large items is $m + 2m'$. The number of bins covered by $\mathbb{A}$ is at least $\lfloor (m + 2m')/2 \rfloor + \lfloor 2S/3 \rfloor$. The number of bins covered by OPT is at most $m + m' + \lfloor S \rfloor$. Thus, for any input sequence, σ,

$$\mathbb{A}(\sigma) \geq \frac{\lfloor (m + 2m')/2 \rfloor + \lfloor 2S/3 \rfloor}{m + m' + \lfloor S \rfloor} \cdot \text{OPT}(\sigma)$$

$$= \frac{(m + 2m')/2 + 2S/3}{m + m' + S} \cdot \text{OPT}(\sigma) - O(1).$$

This proves a competitive ratio of at least $\min\{2/3, \frac{2\beta-1}{2\beta}\}$, since

$$\frac{(m+2m')/2}{m+m'} = \frac{2m+2m'-m}{2m+2m'} = \frac{2\frac{m+m'}{m}-1}{2\frac{m+m'}{m}} = \frac{2\beta-1}{2\beta}.$$

$\square$

Recall that among the large items, we assume that the largest m items form LS bins in the optimal packing. Let S_l and S_s be two integers that denote the floor of the total size of small items placed in respectively the LS and SS bins. So, the number of bins covered by OPT is at most $\beta m + S_s$. In what follows, we define (α, k)-desirable packings, which act as reference packings for our algorithm. Here α and k are two parameters of the algorithm that we will introduce later.

For the following definition, it may be helpful to confer with Fig. 1.

Definition 1. *A covering is (α, k)-desirable, where α is a real number in the range $(0, 1]$ and k is a positive integer, if and only if all the following hold:*

I *The covering has at least $\lfloor \alpha m \rfloor$ LS bins. All LS bins, except possibly a constant number of them, are covered.*
II *The large items not in LS bins appear in pairs, with each pair covering one bin (except one item when there are an odd number of such large items).*
III *The small items not in LS bins cover at least $\lfloor (1-\frac{1}{2^k})\frac{2S_s}{3} \rfloor - 1$ bins.*

Lemma 3. *For any input sequence, σ, the number of bins covered in an (α, k)-desirable packing is at least $\min\{\frac{\alpha+2\beta-1}{2\beta}, (1-\frac{1}{2^k})\frac{2}{3}\} \cdot \text{OPT}(\sigma) - O(1)$.*

Proof. The number of bins covered in the optimal packing is at most $m' + m + S_s = \beta m + S_s$. The number of bins covered by an (α, k)-desirable packing is at least $\lfloor \alpha m \rfloor - c$ (for covered LS bins; c is a constant) plus $\lfloor (2\beta-1-\alpha)m/2 \rfloor$ (for bins covered by pairs of large items) plus at least $\lfloor (1-\frac{1}{2^k})\frac{2S_s}{3} \rfloor - 1$ bins (covered by small items). So, the number, d, of bins covered in the (α, k)-desirable packing of σ will be

$$d \geq \frac{\alpha m/2 + (2\beta-1)m/2 + (1-\frac{1}{2^k})2S_s/3}{\beta m + S_s} \cdot \text{OPT}(\sigma) - O(1)$$

$$\geq \min\left\{\frac{\alpha+2\beta-1}{2\beta}, \left(1-\frac{1}{2^k}\right)\frac{2}{3}\right\} \cdot \text{OPT}(\sigma) - O(1).$$

$\square$

In the remainder of this section, we describe an algorithm that achieves an (α, k)-desirable covering for certain values of α and k. Here, k is used as a parameter to encode approximate values of a few numbers passed to the algorithm. Before describing these numbers, we explain how the approximate encodings work. Given a positive integer x, we can write the length of the binary encoding of x in $O(\log \log x)$ bits, using self-delimited encoding as in [19]. The approximate value of x will be represented by the binary encoding of the length of x,

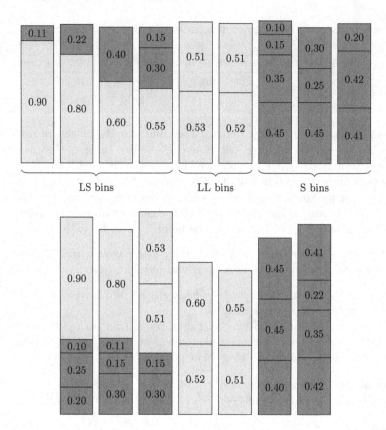

Fig. 1. (left) an optimal packing with $m = 4, m' = 2, S_l = \lfloor 1.18 \rfloor$ and $S_s = \lfloor 3.08 \rfloor$ (right) an (α, k)-desirable packing with $\alpha = 1/2$ and $k = 6$.

plus the k most significant bits of x after the high-order 1. Setting the unknown lower order bits to zero gives an approximation to x which we denote by $\bar{x}$. We can bound $\bar{x}$ as follows: If $\bar{x} = y \cdot 2^\ell$ for some y represented by $k + 1$ bits, where the high-order bit is a one, then $2^k \le y < 2^{k+1}$. Given $\bar{x}$, the largest x could be is $y \cdot 2^\ell + (2^\ell - 1)$. Thus, $(1 - \frac{1}{2^k})x < \bar{x} \le x$.

In the remaining more technical part of the section, it may be beneficial to consider that if we had $O(\log n)$ bits of advice instead of $O(\log \log n)$, many arguments would be simplified, and it could be helpful on a first reading to ignore multiplicative terms such as $1 - \frac{1}{2^k}$ that are there because we know only approximate as opposed to exact values of the parameters we receive information about in the advice.

First, we describe how the algorithm treats the small items and then discuss the large items. The algorithm receives $\bar{S}_s$ and $\bar{m}$, i.e., the approximate values of S_s and m, in $O(\log \log n + k)$ bits of advice. It places small items using the Dual-Next-Fit strategy until a point at which the sum of small items observed so far becomes larger than $\bar{S}_s$. Let p be the small item that causes the sum to exceed $\bar{S}_s$. The algorithm places p and any other small item that follows it using the Dual-Worst-Fit strategy in $\lceil \bar{m}/3 \rceil$ bins, ignoring any large items when calculating the levels of the bins. In what follows, we refer to these $\lceil \bar{m}/3 \rceil$ bins as *reserved bins*. The items before p have a total size of more than $\bar{S}_s - 1$ and hence cover at least $\lfloor (2/3)(\bar{S}_s - 1) \rfloor \geq \lfloor 2\bar{S}_s/3 - 1 \rfloor \geq \lfloor (2/3)(1 - 1/2^k)S_s \rfloor - 1$. So, Property III of an (α, k)-desirable covering holds.

Next, we describe how the algorithm places large items so that properties I and II also hold. For that, the algorithm will need the approximate value of m (which was also required for small items) and m'. As before, these values can be encoded in $O(\log\log n + k)$ bits of advice. We call the largest $\lceil m/3 \rceil$ items in the input sequence *good items*. The algorithm aims at placing $\lfloor \alpha m \rfloor$ of the good items in the reserved bins. Before describing how the algorithm detects good items, we prove the following lemma, showing that the reserved bins with one good item will be covered.

Lemma 4. *A reserved bin that includes any good item will be covered in the final solution (covering) of the algorithm.*

Proof. Define the *desired level* to be $d = 1 - \text{size}(x)$ where x is the smallest good item. Consider an LS-bin B in the optimal packing that does not include a good item (that is, it has one large item smaller than any good item). The total size of small items in B will be at least d. As there are at least $\lceil 2m/3 \rceil$ such bins, we have that $S_l \geq (2m/3) \cdot d$, so $d \leq \frac{3S_l}{2m}$. On the other hand, the total size of small items placed in the reserved bins is at least $S_s + S_l - \bar{S}_s \geq S_l$. Since we use the Dual-Worst-Fit strategy to place these items into $m/3$ reserved bins, by Lemma 1, the total size of small items in any reserved bin is at least $\frac{S_l}{m/3} - y$ where y is the largest small item in those bins. Now, if a reserved bin includes a small item of size at least d, its level is already at least d; otherwise, $\frac{S_l}{m/3} - x$ will be at least $3S_l/m - d$ and since $d \leq \frac{3S_l}{2m}$, the level of the bins is at least d. $\square$

So, in order to achieve an (α, k)-desirable packing, our algorithm needs to select $\lfloor \alpha m \rfloor$ good items and place them in the reserved bins; the above lemma indicates that these bins will be covered (Property I holds). Meanwhile, the algorithm ensures that other large items are paired and hence each pair of them covers a bin (Property II holds). In order to provide the above guarantees, the algorithm considers three cases depending on the location of good items (advice will be used to select the correct case).

Lemma 5. *When $\beta < 15/14$, there exists an (α, k)-desirable packing for $\alpha \leq \frac{7-6\beta}{15} - \frac{176-18\beta}{75 \cdot 2^k + 120}$ and k sufficiently large.*

Proof. Throughout the proof, $1 \leq \beta < 15/14$ and $\alpha < \frac{7-6\beta}{15} - \frac{176-18\beta}{75 \cdot 2^k + 120} < \frac{7-6\beta}{15} < \frac{1}{15}$. Note that under the description of the algorithm, we established Property III, and just prior to the statement of the lemma, we established Property I. The proof to establish Property II is a case analysis on where good items appear in the request sequence.

Case 1: Assume there are $\lfloor \alpha \bar{m}/(1 - 1/2^k) \rfloor$ good items among the first $A = \lfloor \bar{m}/3 \rfloor$ large items in the sequence. In this case, the algorithm places the first A large items into the reserved bins. After seeing all these A items, the algorithm chooses the largest $\lfloor \alpha \bar{m}/(1 - 1/2^k) \rfloor \geq \lfloor \alpha m \rfloor$ of them and declares them to be good items, which by Lemma 4 are guaranteed to be covered. The remaining $A - \lfloor \alpha m/(1-1/2^k) \rfloor$ large items in the reserved bins will be paired with forthcoming large items. Since there are at least $m - A \geq 2m/3$ forthcoming large items and fewer than $m/3$ large items in the reserved bins waiting to be paired, all these large items (except possibly one) can be paired (Property II holds). In summary, in the final covering, there are $\lfloor \alpha \bar{m}/(1 - 1/2^k) \rfloor \geq \lfloor \alpha m \rfloor$ bins covered by a large item (and some small items) while the remaining large items are paired (except possibly one). Hence, the result will be an (α, k)-desirable packing.

Case 2: Assume there are fewer than $\lfloor \alpha \bar{m}/(1 - 1/2^k) \rfloor$ good items among the first A large items in the sequence (Case 1 does not apply). Furthermore, assume there are $\lfloor \alpha \bar{m}/(1-1/2^k) \rfloor$ good items among the $A = \lfloor \bar{m}/3 \rfloor$ large items that follow the first A large items. In this case, the algorithm places the first A large items pairwise in $\lceil A/2 \rceil$ bins. The A large items that follow are placed in the reserved bins. After placing the last of these items in the reserved bins, the algorithm considers these A items and declares the $\lfloor \alpha \bar{m}/(1 - 1/2^k) \rfloor \geq \lfloor \alpha m \rfloor$ largest to be good items, which by Lemma 4 are guaranteed to be covered. The remaining $A - \lfloor \alpha \bar{m} \rfloor$ reserved bins (with large items) will need to be covered by forthcoming large items. We know there are at least $m - 2A \geq \lfloor m/3 \rfloor$ forthcoming large items and fewer than $\lfloor m/3 \rfloor$ large items in reserved bins waiting to be paired, so all these large items (except possibly one) can be paired (Property II holds). Thus, the result is an (α, k)-desirable packing.

Case 3: Assume there are fewer than $\lfloor \alpha \bar{m}/(1 - 1/2^k) \rfloor$ good items among the first $A = \lfloor \bar{m}/3 \rfloor$ large items and also fewer than $\lfloor \alpha \bar{m}/(1-1/2^k) \rfloor$ good items among the following A items (Cases 1 and 2 do not apply). In what follows, we assume $\beta < 15/14$ and let α be some positive value such that $\alpha \leq \frac{7 - 6\beta}{15} - \frac{176 - 18\beta}{75 \cdot 2^k + 120}$. We will later choose k sufficiently large; here $k \geq 6$ will ensure that α is positive. Note also that we have $\alpha < \frac{7 - 6\beta}{15}$. In this case, the algorithm places the first $2A$ large items in pairs. There are $C = 2m' + m - 2A = 2m' + m - 2\lfloor \bar{m}/3 \rfloor$ remaining large items. The algorithm places the first $F = \bar{m}' + \lfloor \bar{m}/6 \rfloor + \lfloor \frac{\alpha \bar{m}}{2} \rfloor - 1$ of the last C large items in the reserved bins (note that this is roughly half of the last C large items when α is small). For this to be possible, we show that the number of reserved bins is at least F, and that at least $\alpha m - 6$ of the F items placed in the reserved bins are good items (see the full paper [10] for the calculations). After placing these F items, the algorithm declares the largest $\lfloor \alpha m \rfloor - 6$ among them to be good items. By Lemma 4, these items (along with small items in the reserved bins) will cover their respective bins. There are $F - \lfloor \alpha m \rfloor + 6$ (positive for $k \geq 1$ and $\alpha \leq \frac{1}{15}$) large items in reserved bins which have not been declared good, and the $C - F$ large items which have not arrived at this point will be paired with them. Calculations show that the number of large items in the reserved bins which are not paired will be at most 6 (see the full paper [10]). The bins in which these items are placed, along with the $\lfloor \alpha m \rfloor - 6$ bins that include good items, will be the LS bins in the final (α, k)-desirable packing. Note that the number of these bins is $\lfloor \alpha m \rfloor$ minus an additive constant which is allowed in desirable packings. All large items placed in bins other than LS bins are paired and hence, Property II also holds.   □

Theorem 3. *There is an algorithm that, provided with $O(\log \log n)$ bits of advice, achieves a competitive ratio of at least $\frac{12\beta-4}{15} - \frac{88-9\beta}{75\beta 2^k + 120\beta}$, where k is a large but constant parameter of the algorithm. Since $\beta \geq 1$, for any $\varepsilon > 0$, there exists an algorithm using a sufficiently large k with competitive ratio at least $\frac{8}{15} - \varepsilon$.*

Proof. The advice indicates the values of $\bar{m}$, $\bar{m}'$, and $\bar{S}_s$. These values can all be encoded in $O(\log \log n)$ bits of advice. Note that one cannot calculate β exactly, since m and m' are not known exactly. Thus, the advice also includes 1 bit to indicate if Lemma 2 should be used because β is larger than $15/14$. If not, the advice also indicates one of the three cases described above; this requires two more bits. Thus, the size of advice is $O(\log \log n)$.

If Lemma 2 is used, the competitive ratio is at least $\min\{2/3, \frac{2\beta-1}{2\beta}\}$, which for $\beta \geq 15/14$ is at least $8/15$. Otherwise, provided with this advice and a sufficiently large integer parameter k, the algorithm can create an (α, k)-packing of the input sequence for any $\alpha \leq \frac{7-6\beta}{15} - \frac{176-18\beta}{75 \cdot 2^k + 120}$. By Lemma 3, the resulting packing has a competitive ratio of at least $\frac{\alpha+2\beta-1}{2\beta}$. Choosing $\alpha = \frac{7-6\beta}{15} - \frac{176-18\beta}{75 \cdot 2^k + 120}$ gives a scheme with competitive ratio at least $\frac{12\beta-4}{15\beta} - \frac{88-9\beta}{75\beta 2^k + 120\beta}$. Since this is an increasing function of β and $\beta \geq 1$, the competitive ratio approaches $\frac{8}{15}$ for large values of k.   □

5   Impossibility Result for Advice of Sub-linear Size

This section uses what is normally referred to as lower bound techniques, but since our ratios are smaller than 1, an upper bound is a negative result, and we refer to such results as negative or impossibility results. In what follows, we show that, in order to achieve any competitive ratio larger than $15/16$, advice of linear size is necessary. We use a reduction from the binary separation problem:

Definition 2. *The* Binary Separation Problem *is the following online problem. The input $I = (n_1, \sigma = \langle y_1, y_2, \ldots, y_n \rangle)$ consists of $n = n_1 + n_2$ positive values which are revealed one by one. There is a fixed partitioning of the set of items into a subset of n_1 large items and a subset of n_2 small items, so that all large items are larger than all small items. Upon receiving an item y_i, an online algorithm must guess if y belongs to the set of small or large items. After the algorithm has made a guess, it is revealed to the algorithm which class y_i belongs to.*

A reduction from a closely related problem named "binary string guessing with known history" shows that, in order to guess more than half of the items correctly, advice of linear size is required:

Lemma 6. *[12] For any fixed $\beta > 0$, any deterministic algorithm for the Binary Separation Problem that is guaranteed to guess correctly on more than $(1/2+\beta)n$ input items on an input of length n needs at least $\Omega(n)$ bits of advice.*

The following lemma provides the actual reduction from the Binary Separation Problem to bin covering. The complete proof can be found in the full paper [10].

Lemma 7. *Consider the bin covering problem on sequences of length $2n$ for which OPT covers n bins. Assume that there is an online algorithm $\mathbb{A}$ that solves the problem on these instances using $b(n)$ bits of advice and covers at least $n - r(n)/8$ bins. Then there is also an algorithm BSA that solves the Binary Separation Problem on sequences of length n using $b(n)$ bits of advice and guessing incorrectly at most $r(n)$ times.*

Proof. (sketch) Given an instance of the Binary Separation Problem formed by $n = n_1 + n_2$ values, we create an instance of the bin covering problem that starts with n_1 "huge" items of size $1 - \varepsilon$ for some $\varepsilon < \frac{1}{2n}$. Any "reasonable" bin covering algorithm has to place these items in separate bins. The next n items are created in an online manner, and each is associated with a value x in the Binary Separation Problem. The size of the item created for x will be an increasing function of x in the range $(\varepsilon, 2\varepsilon)$. We call the item associated with x "small" if x is small and "large" otherwise. If the bin covering algorithm places the item associated with x in a bin with a huge item, we guess that x is "small"; otherwise, we guess that x is "large". The last n_2 items of the bin covering instance are defined as complements of the large items. An optimal algorithm places small items in the bins opened for huge items and covers one bin with each large item and its complement. So, the number of covered bins in an optimal solution is n. We say an algorithm "makes a mistake" when it places a large item in a bin with a huge item or places a small item in a bin without a huge item. A detailed analysis shows that, for each 8 mistakes, the algorithm covers at least 1 bin fewer. Hence, if the number of covered bins is at least $n - r(n)/8$, then the number of binary separation errors must be at most $r(n)$.   □

It turns out that reducing the Binary Separation Problem to bin covering (the above lemma) is more involved than a similar reduction to the bin packing problem [12]. The difference roots in the fact that there are more ways to place items into bins in the bin covering problem compared to bin packing; this is because many arrangements of items are not allowed in bin packing due to the capacity constraint.

Theorem 4. *Consider the bin covering problem on sequences of length n. To achieve a competitive ratio of $15/16 + \delta$, in which δ is a small, but fixed positive constant, an online algorithm needs to receive $\Omega(n)$ bits of advice.*

Proof. Suppose for the sake of contradiction that there is a bin covering algorithm $\mathbb{A}$ with competitive ratio $15/16 + \delta$ using $o(n)$ bits of advice. Consider sequences of length $2n$ for which OPT covers n bins. $\mathbb{A}$ covers $(15/16 + \delta)n = n - r(n)/8$ bins for $r(n) = (1/2 - 8\delta)n$. Applying Lemma 7, we conclude that there is an algorithm that solves the Binary Separation Problem on sequences of length n using $o(n)$ bits of advice, while making at most $(1/2 - 8\delta)n$ errors. By Lemma 6, we know that such an algorithm requires $\Omega(n)$ bits of advice. So, our initial assumption that $\mathbb{A}$ required only $o(n)$ bits of advice is wrong. $\square$

6   Concluding Remarks

We have established that $\Theta(\log \log n)$ bits of advice are necessary and sufficient to improve the competitive ratio obtainable by purely online algorithms.

Obvious questions are: How much better than our bound of $8/15 = 0.53\bar{3}$ can one do with $O(\log \log n)$ bits of advice? Can one do better with $O(\log n)$ bits of advice?

References

1. Angelopoulos, S., Dürr, C., Kamali, S., Renault, M.P., Rosén, A.: Online bin packing with advice of small size. Theory Comput. Syst. **62**(8), 2006–2034 (2018)
2. Ásgeirsson, E.I., et al.: Closed on-line bin packing. Acta Cybernetica **15**(3), 361–367 (2002)
3. Assmann, S.F., Johnson, D.S., Kleitman, D.J., Leung, J.Y.-T.: On a dual version of the one-dimensional bin packing problem. J. Algorithms **5**(4), 502–525 (1984)
4. Balogh, J., Békési, J., Dósa, G., Epstein, L., Levin, A.: A new and improved algorithm for online bin packing. In: 26th Annual European Symposium on Algorithms (ESA), volume 112 of Leibniz International Proceedings in Informatics (LIPIcs), pp. 5:1–5:14. Schloss Dagstuhl-Leibniz-Zentrum fuer Informatik (2018)
5. Balogh, J., Békési, J., Dósa, G., Epstein, L., Levin, A.: A new lower bound for classic online bin packing. ArXiv, arXiv:1807.05554 [cs:DS] (2018)
6. Böckenhauer, H.-J., Hromkovič, J., Komm, D.: A technique to obtain hardness results for randomized online algorithms – a survey. In: Calude, C.S., Freivalds, R., Kazuo, I. (eds.) Computing with New Resources. LNCS, vol. 8808, pp. 264–276. Springer, Cham (2014). https://doi.org/10.1007/978-3-319-13350-8_20
7. Böckenhauer, H.-J., Hromkovič, J., Komm, D., Krug, S., Smula, J., Sprock, A.: The string guessing problem as a method to prove lower bounds on the advice complexity. Theor. Comput. Sci. **554**, 95–108 (2014)
8. Böckenhauer, H.-J., Komm, D., Královic, R., Královic, R.: On the advice complexity of the k-server problem. J. Comput. Syst. Sci. **86**, 159–170 (2017)
9. Borodin, A., Pankratov, D., Salehi-Abari, A.: A simple PTAS for the dual bin packing problem and advice complexity of its online version. In: 1st Symposium on Simplicity in Algorithms (SOSA), LIPIcs, pp. 8:1–8:12. Schloss Dagstuhl - Leibniz-Zentrum fuer Informatik (2018)
10. Boyar, J., Favrholdt, L.M., Kamali, S., Larsen, K.S.: Online bin covering with advice. ArXiv, arXiv:1905.00066 [cs:DS] (2019)

11. Boyar, J., Favrholdt, L.M., Kudahl, C., Larsen, K.S., Mikkelsen, J.W.: Online algorithms with advice: a survey. ACM Comput. Surv. **50**(2), 19:1–19:34 (2017)
12. Boyar, J., Kamali, S., Larsen, K.S., López-Ortiz, A.: Online bin packing with advice. Algorithmica **74**(1), 507–527 (2016)
13. Csirik, J., Totik, V.: Online algorithms for a dual version of bin packing. Discrete Appl. Math. **21**(2), 163–167 (1988)
14. Dürr, C., Konrad, C., Renault, M.P.: On the power of advice and randomization for online bipartite matching. In: 24th Annual European Symposium on Algorithms (ESA), volume 57 of LIPIcs, pp. 37:1–37:16. Schloss Dagstuhl - Leibniz-Zentrum fuer Informatik (2016)
15. Grove, E.F.: Online bin packing with lookahead. In: 6th Annual ACM-SIAM Symposium on Discrete Algorithms (SODA). SIAM, pp. 430–436 (1995)
16. Hoberg, R., Rothvoss, T.: A logarithmic additive integrality gap for bin packing. In: 28th Annual ACM-SIAM Symposium on Discrete Algorithms (SODA). SIAM, pp. 2616–2625 (2017)
17. Jansen, K., Solis-Oba, R.: An asymptotic fully polynomial time approximation scheme for bin covering. Theoret. Comput. Sci. **306**(1–3), 543–551 (2003)
18. Karmarkar, N., Karp, R.M.: An efficient approximation scheme for the one-dimensional bin-packing problem. In: 23rd Annual Symposium on Foundations of Computer Science (FOCS), pp. 312–320. IEEE Computer Society (1982)
19. Komm, D.: An Introduction to Online Computation - Determinism, Randomization, Advice. Texts in Theoretical Computer Science. An EATCS Series. Springer, Switzerland (2016). https://doi.org/10.1007/978-3-319-42749-2
20. Mikkelsen, J.W.: Randomization can be as helpful as a glimpse of the future in online computation. In: 43rd International Colloquium on Automata, Languages, and Programming (ICALP), pp. 39:1–39:14. Springer, Heidelberg (2016)
21. Renault, M.P.: Online algorithms with advice for the dual bin packing problem. CEJOR **25**(4), 953–966 (2017)
22. Renault, M.P., Rosén, A., van Stee, R.: Online algorithms with advice for bin packing and scheduling problems. Theoret. Comput. Sci. **600**, 155–170 (2015)
23. Rothvoss, T.: Approximating bin packing within o(log OPT * log log OPT) bins. In: 54th Annual IEEE Symposium on Foundations of Computer Science (FOCS), pp. 20–29. IEEE Computer Society (2013)

Stackelberg Packing Games

Toni Böhnlein[1]($\boxtimes$), Oliver Schaudt[2]($\boxtimes$), and Joachim Schauer[3]($\boxtimes$)

[1] Institut für Informatik, Universität zu Köln, Weyertal 80, 50931 Köln, Germany
boehnlein@zpr.uni-koeln.de
[2] Institut für Mathematik, RWTH Aachen, Pontdriesch 10, 52062 Aachen, Germany
schaudt@mathc.rwth-aachen.de
[3] Department of Statistics and Operations Research, University of Graz,
Universitätsstr. 15, 8010 Graz, Austria
joachim.schauer@uni-graz.at

Abstract. In a Stackelberg pricing game a distinguished player, the *leader*, chooses prices for a set of items, and the other player, the *follower*, seeks to buy a minimal cost feasible subset of the items. The goal of the leader is to maximize her revenue, which is determined by the sold items and their prices. Typically, the follower's feasible subsets are given by a combinatorial covering problem. In the Stackelberg shortest path game, for example, the items are edges in a network graph and the follower's feasible subsets are *s-t*-paths. This game has been used to model road-toll setting problems by Labbé et al. [14].

We initiate the study of pricing problems where the follower's feasible subsets are given by a packing problem, e.g., a matching or an independent set problem. We introduce a model that naturally extends packing problems to Stackelberg pricing games. The resulting pricing games have applications related to scheduling.

Our interest is the complexity of computing leader-optimal prices depending on different types of followers. As the main result, we show that the Stackelberg pricing game where the follower is given by the well-known interval scheduling problem is solvable in polynomial time. The interval scheduling problem is equivalent to the independent set problem on interval graphs.

As a complementary result, we prove APX-hardness when the follower is given by the bipartite matching problem. This result also shows APX-hardness for the case where the follower is given by the independent set problem on perfect graphs. On a more general note, we prove Σ_2^p-completeness if the follower is given by a particular packing problem that is NP-complete. In this case, the leader's pricing problem is hard even if she has an NP-oracle at hand.

Keywords: Stackelberg games · Algorithmic pricing · Revenue maximization

ⓒ Springer Nature Switzerland AG 2019
Z. Friggstad et al. (Eds.): WADS 2019, LNCS 11646, pp. 239–253, 2019.
https://doi.org/10.1007/978-3-030-24766-9_18

1   Introduction

Assume an agent seeks to complete a set of jobs. If job i is completed, the agent receives a fixed benefit $b(i)$, i.e., the selling value of a good that is produced. However, the agent is not able to complete the jobs by himself but relies on a manufacturer to complete them for him. The agent offers a payment (or price) $p(i)$ to the manufacturer to execute job i and if the manufacturer finishes a job, the agent pays the price. The agent's objective is to maximize his revenue which is the sum of the margins $b(i) - p(i)$ over those jobs i that are finished by the manufacturer. Higher prices are more appealing to the manufacturer but yield less revenue for the agent.

Whether it is profitable to complete a job or not, depends on the price for which the manufacturer is willing to execute it. In our model, this price depends on payments that competitors offer to the manufacturer to carry out their jobs. Additionally, the manufacturer's schedule has to obey a number of constraints. For instance, two jobs might not be scheduled together because they have to be executed in the same time window, and there is only one machine available. The manufacturer selects a feasible schedule that maximizes his income. The agent's task is to decide which jobs should be offered and for what price. This can be seen as a *make or buy* decision.

As a special case, we introduce a so-called Stackelberg pricing game that is based on the well-known *interval scheduling problem*. In this problem, there is one machine and a set of weighted jobs I. Each job i has a fixed starting time $s_i \in \mathbb{R}$ and terminating time $t_i \in \mathbb{R}$. Hence, a job can be represented by an interval $[s_i, t_i]$ on the line. We say that two intervals *overlap* if their intersection is non-empty. The objective is to find a subset of non-overlapping intervals of maximum total weight. On the left-hand side of Fig. 1, we have an instance of an interval scheduling problem if we only consider the solid intervals a, b, c, d. An optimal solution is the set $\{a, b\}$ with a total weight of 7. Such an optimal set of jobs can be computed in polynomial time.

To fall in line with the terminology of Stackelberg pricing games, we call the agent *leader* and the manufacturer *follower*. In our game, the solid intervals a, b, c, d represent the jobs of the competitors. The dashed lines x, z, y are the jobs of the leader with their respective benefits 4, 3 and 5. In the first step of our game, the leader sets the prices p_x, p_y and p_z. On the right hand side of Fig. 1, prices $p_x = 3$, $p_y = 0$, $p_z = 4$ are set. For this price vector, the follower selects the jobs c, x, z with total (maximum) weight 9. Note that the prices p_x and p_z are optimal in the following sense: if either of p_x or p_z is decreased by some $\varepsilon > 0$, the intervals x or z are not selected by the follower. The leader obtains revenue 2 under these prices.

The optimal prices $p_x = 1$, $p_y = 2$, $p_z = 0$ yield revenue 4. Under these prices the solutions $\{b, c, x, y\}$ and $\{a, b\}$ are optimal for the follower; both have a weight of 7. A common assumption for Stackelberg pricing games is that the follower is cooperative: he always chooses the optimal solution which is most profitable for the leader. This assumption is made to avoid technicalities with ε-values. When setting the prices, the leader is aware of the competitors' jobs, the constraints they imply and the offered payments. After prices are set, the follower selects a feasible subset of jobs–offered by the leader and her competitors–that maximizes his income, i.e, the follower solves an interval scheduling instance. Computing leader-optimal prices is a special case of revenue maximization in combinatorial auctions where it is common to assume knowledge of the customers' preference. In our case, the customers' preferences correspond to the follower's offer situation.

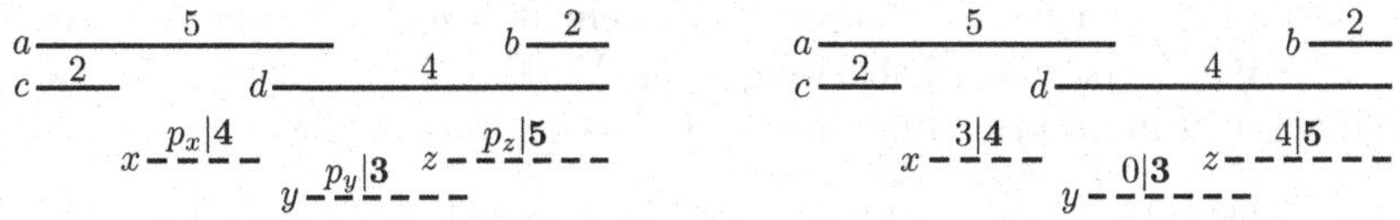

Fig. 1. An instance of the Stackelberg interval scheduling game.

In the literature, such problems are captured by a game-theoretic model called *Stackelberg pricing games* (cf. [17]). Originally, in a Stackelberg pricing game the leader chooses prices for a number of items. After that, one or several followers are interested in buying these items. The goal of the leader is to maximize her revenue while followers want to minimize their costs. Depending on the followers' preferences, computing optimal prices can be a highly non-trivial problem.

A major line of research studies Stackelberg pricing games where the follower's preferences are given by a combinatorial optimization problem. Labbé et al. [14] model road-toll setting problems by a Stackelberg pricing game based on the shortest path problem. In this game, the leader sets prices for a subset of priceable edges of a network graph while the remaining edges have fixed costs. Each follower has a pair of vertices (s, t) and buys a minimum cost path from s to t. The costs of a path depend on both the fixed costs and the prices set by the leader. Roche et al. [16] show that the problem is NP-hard, even if there is only one follower. More recently, other combinatorial optimization problems were studied in their Stackelberg pricing game version. Cardinal et al. [10,11] investigate the Stackelberg minimum spanning tree game. They show APX-hardness and give a logarithmic approximation algorithm.

1.1 Our Results

The Stackelberg pricing games which were previously studied are based on covering problems. The scenario above gives us a framework to formulate Stackelberg pricing games that are based on packing problems. We are interested in the complexity of computing leader-optimal prices depending on a single follower that is given by different packing problems.

Our main result is a polynomial time algorithm that solves the Stackelberg interval scheduling game. Since it will be more handy later on, we state this game in terms of an independent set problem on an interval graph. Associated with an instance of the interval scheduling problem I, there is a corresponding *interval graph* $G = (V, E)$ with vertex weights. For each interval $i \in I$ there is a corresponding vertex $v_i \in V$ with weight $w(v_i) = w(i)$. If two intervals i, j overlap, there is an edge $(v_i, v_j) \in E$. Recall that an *independent set* in a graph is a subset of mutually non-adjacent vertices. The interval scheduling problem on I is equivalent to finding a maximum weight independent set of G.

STACKELBERG INTERVAL SCHEDULING (SIS)
Input: An interval graph $G = (V, E)$ with priceable vertices $P \subseteq V$ and $|P| = k$. For every $v \in P$ there is a benefit $b(v) \in \mathbb{R}$, and for every $u \in F = V \setminus P$ there is a fixed weight $w(u) \in \mathbb{R}$.
Objective: Find a price function $p : P \to \mathbb{R}_{\geq 0}$ maximizing

$$\max_S \{b(S \cap P) - p(S \cap P) \mid S \text{ is a maximum weight independent set}\},$$

where the weight of an independent set S is defined as $w(S \cap F) + p(S \cap P)$.

The objective reflects the follower's cooperative behavior: From his optimal solutions he chooses a solution which maximizes the leader's revenue. This is a common assumption for such pricing games and is made to avoid technicalities with small ε-values (cf. [15]).

Theorem 1. *Given a SIS instance (G, P, w, b) with $|P| = k$, optimal leader prices can be computed in time $\mathcal{O}(k^3 (|V(G)| + |E(G)|))$.*

We present a sketch of the proof in Sect. 2. It builds on the analysis of a special case where priceable vertices are not allowed to be adjacent. For this special case, we formulate a combinatorial algorithm. Its analysis relies on a linear program, and the computed price vectors have a special structure which is important for the general case.

For the general case, this combinatorial algorithm allows us to compute optimal leader-prices for any subset of non-adjacent, priceable vertices. However, enumerating all subsets is not efficient. An ordering-property that is inherent to interval graphs allows us to show that we only need to look at $\mathcal{O}(k^2)$ many subsets. We express this property as a recursion and formulate a dynamic programming algorithm. We remark that our theorem is one of the few cases in which one can solve a Stackelberg pricing game based on a non-trivial optimization problem to optimality.

As a complementary result, we show that the Stackelberg matching game is hard to approximate. Analogously, we formulate the pricing problem as follows.

STACKELBERG MATCHING
Input: A graph $G = (V, E)$ with priceable edges $P \subseteq E$ where $k = |P|$. For every $e \in P$ there is a benefit $b(e) \in \mathbb{R}$, and for every $e \in F = E \setminus P$ there are fixed weights $w(e) \in \mathbb{R}$.
Objective: Find a price function $p : P \to \mathbb{R}_{\geq 0}$ maximizing

$$\max_M \{b(M \cap P) - p(M \cap P) \mid M \text{ is a maximum weight matching}\},$$

where the weight of a matching M is defined as $w(M \cap F) + p(M \cap P)$.

Theorem 2. STACKELBERG MATCHING *is APX-hard even when the graph G is bipartite,* $w(f) \in \{1, 2\}$ *for all* $f \in F$, *and* $b(e) = 4$ *for all* $e \in P$.

In Sect. 3, we sketch the theorem's proof. The theorem rules out that STACKELBERG MATCHING is fixed-parameter tractable when parameterized by the number of fixed weight values. Moreover, the theorem shows that STACKELBERG INTERVAL SCHEDULING on perfect graphs–instead of interval graphs–is APX-hard. The matching problem on a graph is equivalent to the independent set problem on its line graph. And it is known that the line graph of a bipartite graph is a perfect graph. This complements Theorem 1 since perfect graphs are a super-class of interval graphs.

Corollary 1. STACKELBERG INDEPENDENT SET ON PERFECT GRAPHS *is APX-hard even if* $w(f) \in \{1, 2\}$ *for all* $f \in F$ *and* $b(e) = 4$ *for all* $e \in P$.

In the full version of the paper, we show that the Stackelberg maximum spanning tree game is APX-hard. This follows readily with the proof by Cardinal et al. [10] for the minimization version. Yet, it shows that a Stackelberg pricing game based on a packing problem is hard if the follower optimizes over a matroid.

Additionally, we take a step back from our example problems and study the complexity of the Stackelberg pricing game in its own right. It turns out that there are combinatorial optimization problems in NP such that the corresponding Stackelberg pricing game is Σ_2^p-complete. In other words, such a pricing problem is computationally difficult even if an NP oracle is provided, unless the polynomial hierarchy collapses to the second level.

Theorem 3. *There is a linear combinatorial maximization problem* Π *in NP such that the* STACKELBERG Π *is* Σ_2^p-*complete.*

We present the proof in the full version of the paper.

1.2 Related Work

As mentioned above, Labbé et al. [14] use the Stackelberg shortest path game to model road-toll problems. They establish NP-hardness and use LP bi-level formulations to solve smaller instances. Roche et al. [16] formulate a combinatorial approximation algorithm with logarithmic approximation guarantee. The best lower bound is due to Briest et al. [6]: they show that the problem is NP-hard to approximate within a factor of less than 2. This is an improvement above previous results by Joret [13] showing APX-hardness. For a survey on the Stackelberg shortest path game and bi-level programming see Hoesel [12] or Labbé and Violin [15].

Bilo et al. [4] show that the Stackelberg shortest path tree game is NP-hard, and give an efficient algorithm in case the number of priceable edges is constant; their algorithm was improved by Cabello [9]. Briest et al. [8] develop an efficient algorithm for a special case of the Stackelberg bipartite vertex cover game. Their algorithm is based on max-flow computations. An improved algorithm, building on the preflow-push algorithm, was later given by Baïou and Barahona [1].

Cardinal et al. [10,11] investigate the Stackelberg minimum spanning tree game. They show APX-hardness and prove several positive approximation results. The game remains NP-hard if the instances are planar graphs, and it becomes polynomial-time solvable on graphs of bounded treewidth. Bilo et al. [3] consider further variations.

Briest et al. [8] analyze general Stackelberg network pricing games, and give a logarithmic approximation algorithm. This algorithm uses a single price strategy. Independently, a slightly more general result was obtained by Balcan et al. [2]. Böhnlein et al. [5] study the single price strategy in a non-discrete setting as well as the parameterized complexity of such pricing problems.

Briest et al. [7] study Stackelberg pricing games based on the knapsack and vertex cover problem. Here, the follower runs a known approximation algorithm.

2   Stackelberg Interval Scheduling Games

Recall that an instance of the interval scheduling problem is given by a set I of intervals. Each interval is specified by a tuple $(s_i, t_i) \in \mathbb{R}^2$ with $s_i < t_i$ and a weight $w(i) \in \mathbb{R}$. As mentioned in the introduction, associated to I there is an interval graph $G = (V, E)$. It is well-known that the interval representation of an interval graph is not unique. We say that $(v_1, \ldots, v_k)$ is an *interval order* of $\{v_1, \ldots, v_k\} \subseteq V$ if there exists an interval representation such that $s_{v_1} \leq \ldots \leq s_{v_k}$.

Before we can turn to the general problem, we need to analyze a special case of STACKELBERG INTERVAL SCHEDULING. In this case, we set aside the benefits on the priceable vertices P. Instead, we require that $P \subseteq S$ where S is the follower's solution. The goal of the leader is to minimize the total price under which all her vertices are bought. If all priceable vertices have to be part of S, P itself has to be an independent set, i.e., in terms of interval scheduling priceable intervals are non-overlapping.

2.1   Stackelberg Interval Scheduling with Non-overlapping Intervals

STACKELBERG INTERVAL SCHEDULING WITH NON-OVERLAPPING INTERVALS (NOSIS)
Input: An interval graph $G = (V, E)$ with priceable vertices $P \subseteq V$ and fixed weights $w(u)$ for $u \in F = V \setminus P$. The set P is an independent set of G and $|P| = k$.
Objective: Find a price function $p : P \to \mathbb{R}_{\geq 0}$ such that there is a maximum weight independent set S with $P \subseteq S$ and $p(P)$ is minimum. The weight of an independent set S is defined as $w(S \cap F) + p(S \cap P)$.

Briest et al. [8] formulate a linear program for so-called Stackelberg Network Pricing Games which generalize NOSIS. Their linear program has exponentially many constraints but it can be solved in polynomial time using a separation oracle. We formulate a linear program for NOSIS where the number of constraints is of order $\mathcal{O}(k^2)$. More importantly, this linear program is crucial for results we give later.

A Linear Program for NOSIS. Given a NOSIS instance (G, P, w) and a price function p, we say G *under* p when we take the weights of vertices in P determined by p. A price function p is *feasible* if there exists a maximum weight independent set S in G *under* p such that $P \subseteq S$.

To set up the linear program, several independent sets of subsets of F have to be computed. For a subset $U' \subseteq F$, let $opt(U')$ be the weight of a maximum weight independent set of the induced graph $G[U']$.

If a price function p is feasible, the weight of maximum weight independent set that contains all priceable vertices must be at least $opt(F)$–the maximum weight of an independent without priceable vertices. Let $U := \{u \in F \mid (u, v) \notin E, \forall v \in P\}$ be the set of vertices with fixed weights that are not adjacent to any priceable vertex. Hence, $opt(F) - opt(U)$ is a lower bound on the total price of a feasible price function. This lower bound yields the first constraint of our linear program: every feasible price function p necessarily satisfies

$$p(P) \geq opt(F) - opt(U).$$

However, this constraint is not sufficient to ensure that p is feasible. For example, the price vector $p' = (opt(F) - opt(U), 0, \ldots, 0)$ satisfies the constraint but it is unlikely feasible. Analogous constraints for subset of priceable vertices are necessary. With Proposition 1, we show that it is sufficient to consider constraints for subsets of vertices that are consecutive in an interval order.

Let $I = (v_1, v_2, \ldots, v_k)$ be an interval order of the priceable vertices P.[1] For a consecutive sub-series $v_i, \ldots, v_j$ of I, we derive a lower bound on $p(v_i) + \ldots + p(v_j)$. The constraint is set up under the condition that the vertices $v_1, \ldots, v_{i-1}$ and $v_{j+1}, \ldots, v_k$ are included in every optimal solution of the follower. Hence, we want to consider only the vertices of F that interfere with $v_i, \ldots, v_j$ but not with the remaining priceable vertices. The relevant subset of F is denoted by $G[i, j]$. A formal definitions is given in the full version. To define $\bar{G}[i, j]$, we also remove the vertices of F that are adjacent to $v_i, \ldots, v_j$. The weights of maximum weight independent sets are denoted as $opt(G[i, j])$ and $opt(\bar{G}[i, j])$, respectively. In terms of interval scheduling, think of the instance that is between the intervals corresponding to v_{i-1} and v_{j+1}. It holds that $opt(G[1, k]) = opt(F)$ and that $opt(\bar{G}[1, k]) = opt(U)$. With this notation, the NOSIS-LP has constraints

$$p(v_i) + \ldots + p(v_j) \geq opt(G[i, j]) - opt(\bar{G}[i, j])$$

for all $1 \leq i \leq j \leq k$ and the objective is to minimize $p(P)$. Note that these are $\mathcal{O}(k^2)$ constraints in total.

Proposition 1. *Given a NOSIS instance (G, P, w) with the interval order $I = (v_1, \ldots, v_k)$ of P, a price vector $p \in \mathbb{R}^k_{\geq 0}$ is feasible if and only if it satisfies the constraints of the corresponding NOSIS-LP.*

---

[1] Any interval representation is suitable for our purpose, but we need to fix one.

Proposition 1 implies that an optimal price vector can be computed using the NOSIS-LP. Its proof is contained in the full version of the paper.

A Combinatorial Algorithm for NOSIS. In this section, we introduce a combinatorial algorithm for NOSIS. It computes a price vector based on an interval order $(v_1, \ldots, v_k)$ of P in k iterations. In iteration $i \in [k]$ the price for v_i is computed based on one modified constraint of the NOSIS-LP. We use the subset $G[1, i]$ of F in iteration i where we add the priceable vertices $v_1, \ldots, v_{i-1}$ with the previously computed prices. We denote this by $G[1, i](p)$ and the weight of a maximum weight independent set as $opt(G[1, i](p))$ where p is a price vector for $v_1, \ldots, v_{i-1}$.

Algorithm 1. STACKELBERG INTERVAL SCHEDULING WITH NON-OVERLAPPING INTERVALS

1: **Input:** a NOSIS instance (G, P, w) and an interval order $(v_1, \ldots, v_k)$ of P
2: **for** $i := 1, \ldots, k$ **do**
3:     $p(v_i) = opt(G[1, i](p(v_1), \ldots, p(v_{i-1}))) - opt(\bar{G}[1, i]) - \sum_{l=1}^{i-1} p(v_l)$
4: **return** $(p(v_1), \ldots, p(v_k))$

Proposition 2. *Given a NOSIS instance (G, P, w) and an interval order $I = (v_1, \ldots, v_k)$ of P, if $p \in \mathbb{R}_{\geq 0}^k$ is a price vector computed by Algorithm 1, then p is feasible.*

In the proof, we show that the computed price vector satisfies all constraints of the NOSIS-LP.

Proposition 3. *Given a NOSIS instance (G, P, w) and an interval order $I = (v_1, \ldots, v_k)$ of P, Algorithm 1 computes a feasible price vector p with minimum total weight $p(P)$ in time of order $\mathcal{O}(k(|V| + |E|))$.*

To prove the proposition, we consider the set O of feasible price vectors with minimum total weight. We compare p to a price vector $q = \arg\max_{o \in O}\{l \mid p(v_l') = o(v_l')$ for $l' < l\}$. That is, q is the optimal price vector which agrees with p on as many positions as possible, starting at position 1. It is possible to turn q into p by 'shifting weight' to higher indices while keeping the total weight constant. Hence, we show that p has minimum total weight.

Moreover, the proof hints that the first entry of p is special. The following lemma shows that p without the first entry is feasible for the remaining priceable vertices. This property is essential for the next section.

Lemma 1. *Given a NOSIS instance (G, P, w) and an interval order $I = (v_1, \ldots, v_k)$ of P, if $p \in \mathbb{R}_{\geq 0}^k$ is a price vector computed by Algorithm 1, then there exists an maximum weight independent set S under $(0, p(v_2), \ldots, p(v_k))$ such that $\{v_2, \ldots, v_k\} \subseteq S$.*

2.2   Stackelberg Interval Scheduling

In this section, we formulate a polynomial time algorithm that solves STACK-ELBERG INTERVAL SCHEDULING (SIS). Here, priceable vertices can be adjacent. The input of the algorithm is a SIS instance (G, P, w, b) and an interval order $I = (v_1, \ldots, v_k)$ of P. Two techniques characterize the algorithm.

The first ingredient uses the NOSIS machinery to compute optimal prices for an independent set P' of G where $P' \subseteq P$. For this, we compute the induced subgraph $G' = G[F \cup P']$. Now (G', P', w) is an instance of NOSIS. Algorithm 1 allows us to compute an optimal price vector p'. In turn, the price vector p' can be extended to a price vector p for the original SIS instance by setting the price of vertices in the set $P \setminus P'$ to zero. For the fixed set P', the benefits are constant, and p has a minus sign in the leader's objective function of the SIS instance. Consequently, the price vector p yields maximum revenue under the assumption that P' is contained in the follower's solution. We denote this maximum revenue by

$$rev_G(P').$$

On this account, the SIS algorithm selects several independent sets of P and determines their maximum revenue. However, it cannot explicitly enumerate all independent sets of P since it is supposed to run in polynomial time.

The second ingredient is dynamic programming: the algorithm solves a SIS instance from a series of sub-instances. The first sub-instance contains only the priceable vertex v_1. In each iteration, one priceable vertex is added according to the interval order I. Define $P_i := \{v_1, v_2, \ldots, v_i\}$ and $G_i := G[F \cup P_i]$ for $i \in [k]$. Note that in each induced subgraph all vertices with fixed weights are present. The sub-instance of iteration i is $(G_i, P_i, w, b|_{P_i})$. In iteration i our algorithm computes an independent set $S_i \subseteq \{v_1, v_2, \ldots, v_i\}$ with $v_i \in S_i$. The set S_i is the optimal selection to form an independent set of P_{i-1} and v_i. Formally, for any independent set $Q \subseteq P_{i-1}$ we have $rev_{G_i}(Q \cup v_i) \leq rev_{G_i}(S_i)$. In Proposition 4, we show that the optimal solution in iteration i can be computed with the following recursion:

$$S_i = \underset{j < i,\, (v_i, v_j) \notin E}{\arg\max} \; rev_{G_i}(S_j \cup \{v_i\})$$

Note that $S_j \cup \{v_i\}$ is an independent set since we require that $(v_i, v_j) \notin E$. To allow our algorithm to test sets where $S_j = \emptyset$, a priceable vertex v_0 with $S_0 = \emptyset$ is added. The vertex v_0 is isolated in G and $b(v_0) = 0$. The interval order I is extended to $(v_0, v_1, \ldots, v_k)$ and we put $P_i = P_i \cup \{v_0\}$ for all $i \in [k]$.

The vertex v_0 together with $S_0 = \emptyset$ is the base case for the recursion. After computing optimal independent sets for $v_0, v_1, v_2, \ldots, v_k$ we can select the globally optimal set and thus solve the instance to optimality. The SIS algorithm is formulated around the recursion above. The full version contains a formulation in pseudo-code. The algorithm has a running time of order $\mathcal{O}(k^3 (|V| + |E|))$.

Proposition 4. *Given a SIS instance (G, P, w, b) and an interval order $I = (v_0, v_1, \ldots, v_k)$ of $P \cup \{v_0\}$, the SIS algorithm computes an optimal price vector.*

Proof. To prove the proposition, we show that it is not necessary to explicitly enumerate all independent sets to determine an optimal selection of priceable vertices. If a new priceable vertex is considered, it suffices to test the pre-computed independent sets. We show that a different selection of vertices cannot yield more revenue. Let $i, j \in [k]$ such that $j < i$ and $(v_i, v_j) \notin E$.

Suppose that we are in the i-th iteration. Let us compare the set S_j to a different selection of vertices. For this, let $Q \subseteq \{v_1, v_2, \ldots v_j\}$ be another independent set with $v_j \in Q$. Furthermore, let p and q be optimal price vectors for the NOSIS instances $(G[F \cup S_j], S_j, w)$ and $(G[F \cup Q], Q, w)$, respectively. By induction we may assume that

$$b(S_j) - p(S_j) \geq b(Q) - q(Q). \tag{1}$$

Now, we show that S_j is an optimal selection of an independent set if v_j and v_i are fixed. For ease of notation, let $G_S = G[F \cup S_j \cup \{v_i\}]$ and $G_Q = G[F \cup Q \cup \{v_i\}]$. Let $\hat{p}$ and $\hat{q}$ be optimal price vectors for the NOSIS instances $(G_S, S_j \cup \{v_i\}, w)$ and $(G_Q, Q \cup \{v_i\}, w)$, respectively. Suppose that

$$b(S_j) - \hat{p}(S_j) + b(v_i) - \hat{p}(v_i) < b(Q) - \hat{q}(Q) + b(v_i) - \hat{q}(v_i). \tag{2}$$

From (1) and (2) we derive a contradiction. More precisely, we show that $\hat{q}$ (without $\hat{q}(v_i)$) yields more revenue than q for $(G[F \cup Q], Q, w)$.

Let $\bar{I} = (v_k, v_{k-1}, \ldots, v_1)$ be the reversed interval order of I. This interval order induces interval orders $\bar{I}_S = \bar{I}[S_j \cup \{v_i\}]$ and $\bar{I}_Q = \bar{I}[Q \cup \{v_i\}]$ for the priceable vertices of $S_j \cup \{v_i\}$ and $Q \cup \{v_i\}$, respectively. We choose the two price vectors $\hat{p}$ and $\hat{q}$ as computed by Algorithm 1 for the interval orders $\bar{I}_S$ and $\bar{I}_Q$, respectively. Due to the special structure of price vectors computed by the combinatorial algorithm, the following claim holds.

Claim. $\hat{p}(v_i) = \hat{q}(v_i)$

This allows us to mix up the price vectors. Consider the NOSIS instance $(G_S, S_j \cup \{v_i\}, w)$. With p and $\hat{q}(v_i)$ we form the price vector $p' = (p(v_1), \ldots, p(v_j), \hat{q}(v_i))$.

Claim. The price vector p' is feasible for $(G_S, S_j \cup \{v_i\}, w)$.

This claim holds because $\hat{q}(v_i)$ was computed together with v_j (recall that $v_j \in Q$). Vertex v_j is also contained in S_j. Any vertex $v \in F$ that is adjacent to v_i an some priceable vertex in either Q or S_j (that is not v_j) must also be adjacent to v_j; this is due to the ordering-property of interval graphs. Intuitively, vertex v_j cancels influences that a different selection Q can have on the price of v_i.

Since $\hat{p}$ is an optimal price vector for $(G_S, S_j \cup v_i, w)$,

$$b(S_j) - \hat{p}(S_j) + b(v_i) - \hat{p}(v_i) \geq b(S_j) - p(S_j) + b(v_i) - \hat{q}(v_i).$$

With (2) it follows that

$$b(Q) - \hat{q}(Q) + b(v_i) - \hat{q}(v_i) > b(S_j) - p(S_j) + b(v_i) - \hat{q}(v_i)$$

which simplifies to
$$b(Q) - \hat{q}(Q) > b(S_j) - p(S_j)$$

With (1) it follows that

$$b(Q) - \hat{q}(Q) > b(Q) - q(Q). \tag{3}$$

Claim. Price vector $\hat{q}$ (without $\hat{p}(v_i)$) is feasible for $(G[F \cup Q], Q, w)$.

Proof. As $\hat{q}$ was computed with $\bar{I}_Q$, the claim follows from Lemma 1.

Equation 3 states that vector $\hat{q}$ (without $\hat{p}(v_i)$) yields more revenue than q for $(G[F \cup Q], Q, w)$. This contradicts that q is an optimal price vector. As a consequence, there can be no independent set like Q that yields strictly more revenue than S_j when v_j and v_i are fixed. Therefore, it is valid to test only the pre-computed independent sets. After the computation, our algorithm chooses the optimal selection of priceable vertices and the corresponding price vector is returned.

3   Stackelberg Matching Games

One of the most common scheduling problems involves n jobs and m machines; each machine can execute at most one job, and the task is to decide which machine executes which job. This situation can be modeled by a bipartite graph $G = (U \cup V, E)$. The two blocks U and V correspond to the jobs and machines, respectively. To set up a Stackelberg pricing game, we say that the follower receives a payment to execute job i on machine j which is represented by an weighted edge connecting the corresponding vertices. Hence, he solves a maximum weight matching problem to maximize his income. Let a subset of jobs or vertices belong to the leader. An edge e which is incident to such a vertex is a priceable edge and has a benefit $b(e)$. The leader's objective is to set prices $p(e)$ that maximize her revenue $b(e) - p(e)$ over the priceable edges that are part of the follower's matching.

In Sect. 1, we formalized this scenario as the Stackelberg matching game. In this section, we sketch the proof of the following theorem.

Theorem 4. *The* STACKELBERG MATCHING *is NP-hard even when the graph G is bipartite, $w(f) \in \{1, 2\}$ for all $f \in F$, and $b(e) = 4$ for all $e \in P$.*

As we will see in the proof, the graph used for the reduction covers the model of the scheduling problem just described. Moreover, the reduction implies the stronger statement of Theorem 2. This extension is similar to the APX-hardness proof for the Stackelberg minimum spanning tree game by Cardinal et al. [10]. We do not present it in this extended abstract.

Proof (Sketch of Theorem 4's proof). We provide a reduction from the NP-complete SET COVER problem. Shorthand, we write $\bigcup C$ for $\bigcup_{c \in C} c$ where C is a family of subsets.

SET COVER
Input: A ground set $\mathcal{B} = \{\hat{u}_1, \ldots, \hat{u}_n\}$ and a family $\mathcal{S} = \{S_1, \ldots, S_m\}$ of subsets of $\mathcal{B}$ where $\bigcup \mathcal{S} = \mathcal{B}$.
Objective: Find a subfamily $C \subseteq \mathcal{S}$ with minimum cardinality such that $\bigcup C = \mathcal{B}$.

Given a SET COVER instance $(\mathcal{B}, \mathcal{S})$ with $n = |\mathcal{B}|$ and $m = |\mathcal{S}|$, we construct a STACKELBERG MATCHING instance (G, P, b, w). The graph $G = (V, E = P \cup F)$ is set up as follows.

– There is the set $B \subseteq V$ that contains a vertex $u_j \in B$ for each $\hat{u}_j \in \mathcal{B}$.

A *subset-gadget* is constructed for each subset of $\mathcal{S}$. The set of priceable edges consists of two disjoint subsets $P = P_{sub} \cup P_{cov}$.
Let $S_i \in \mathcal{S}$ and $\hat{u}_j \in S_i$, then

– there is a vertex $u_{i,j} \in V \setminus B$ and a priceable edge $\sigma_{i,j} = (u_{i,j}, u_j) \in P_{cov}$ with benefit $b(\sigma_{i,j}) = 4$. An edge $\sigma_{i,j}$ is called a *covering-edge* of u_j. The subset $P_{cov} \subset P$ contains all covering-edges.
– there are two vertices $v_i, w_i \in V \setminus B$ as well as the priceable edge $s_i = (v_i, w_i) \in P_{sub}$ with benefit $b(s_i) = 4$. The edge s_i is called the *subset-edge* of S_i. The subset $P_{sub} \subset P$ contains all subset-edges.
– there are paths of length two connecting v_i to each $u_{i,j}$. Such a path consists of a vertex $\bar{u}_{i,j} \in V \setminus B$ as well as the edge $(u_{i,j}, \bar{u}_{i,j}) \in F$ with weight $w((u_{i,j}, \bar{u}_{i,j})) = 1$ and the edge $(\bar{u}_{i,j}, v_i) \in F$ with weight $w((\bar{u}_{i,j}, v_i)) = 2$.

Given a price function p there can be several maximum weight matchings. We call the maximum weight matchings that yield maximum revenue for the leader *maximum revenue matchings* under p. The idea of the construction is to enforce certain matchings as maximum revenue matchings by having large enough benefits on all priceable edges.

Claim. A maximum revenue matching contains m subset-edges and n covering-edges.

Our proof relies on establishing bounds on the prices of edges. We show that under an optimal price function all covering-edges in a maximum revenue matching have price 1. If a subset-gadget has a covering-edge with price 1, then its subset-edge has to have price 2. Without such a covering-edge it suffices to set price 1 for the subset-edge. Consequently, a subset-gadget is more "expensive" if it is covering some ground elements and finding a minimum set cover coincides with computing an optimal price function. More formally, we show the following claim.

Claim. If p is an optimal price function, $p(s) \in \{1, 2\}$ for all $s \in P_{sub}$.

To prove the theorem, we have to show that a minimum set cover corresponds to an optimal price function. With the following two claims, we can convert a price function to a cover and vice-versa.

Claim. Given a set cover $C \subseteq S$ where $t = |C|$. Consider a price function p_C set up as follows.

(i) For $S_i \in C$ set $p_C(s_i) = 2$ and for $\sigma \in \sigma_i$ set $p_C(\sigma) = 1$, and
(ii) for $S_i \notin C$ set $p_C(s_i) = 1$ and for $\sigma \in \sigma_i$ set $p_C(\sigma) = 0$.

Then a maximum revenue matching of G under p_C yields $n(4 - 1) + t(4 - 2) + (m - t)(4 - 1)$ revenue for the leader.

Claim. Let p be an optimal price function. Then the family of subsets $D = \{S_i \in S \mid p(s_i) = 2\}$ is a set cover of $\mathcal{B}$ i.e. $\bigcup D = \mathcal{B}$.

Given an optimal price function or minimum cover, we convert it to a corresponding cover or price function, respectively. If we suppose that these corresponding objects are not optimal, a contradiction can be reached by using the transformations again.

Note that, there is a subset $U \subseteq V(G)$ of the vertices that all belong to the same block of G, and all priceable edges are incident to some vertex in U. Hence, the pricing problem based on the scheduling problem from the beginning of this section is APX-hard.

4   Conclusion and Future Work

We aim to advance the study of Stackelberg pricing games where the follower solves a combinatorial optimization problem. We contributed a model to handle packing problems whereas up to now only covering problems were studied. Given this model, we studied the complexity of such pricing games depending on the complexity of the underlying follower problem. We showed that STACKELBERG INTERVAL SCHEDULING can be solved in polynomial time. This stands in contrast to the fact that most previously studied Stackelberg pricing games are NP-hard even if the follower's problem is solvable in polynomial time. This turned out to be the case for STACKELBERG MATCHING and STACKELBERG MAXIMUM SPANNING TREE. In addition, we showed that Stackelberg pricing games can be Σ_2^p-complete even if the underlying problem is in NP.

Pricing Independent Sets. Recall that for STACKELBERG INTERVAL SCHEDULING the follower computes a maximum independent set in an interval graph. It is natural to ask for the complexity of this pricing problem on more general graph classes. Corollary 1 says that pricing independent sets on perfect graphs is APX-hard. It might be intriguing to study the complexity of the Stackelberg pricing game where the follower computes a maximum weight independent set in a chordal graph since these graphs generalize interval graphs and are perfect.

Σ_2^p-**Hardness.** Theorem 3 states that there is a problem in NP whose Stackelberg pricing problem is Σ_2^p-complete. However, the problem we use is rather artificial and not covered by any well-known problem to the best of our knowledge. In the context of this paper, the question arises whether STACKELBERG Π is Σ_2^p-complete if Π is some classical NP-complete maximization problem. To stimulate further research on this matter, we formulate the following conjecture.

Conjecture 1. STACKELBERG Π is Σ_2^p-complete if Π is the general independent set problem or if Π is the knapsack problem.

References

1. Baïou, M., Barahona, F.: Stackelberg bipartite vertex cover and the preflow algorithm. Algorithmica **74**(3), 1174–1183 (2016)
2. Balcan, M., Blum, A., Mansour, Y.: Item pricing for revenue maximization. In: Proceedings 9th ACM Conference on Electronic Commerce (EC-2008), Chicago, IL, USA, pp. 50–59, 8–12 June 2008. https://doi.org/10.1145/1386790.1386802, http://doi.acm.org/10.1145/1386790.1386802
3. Bilò, D., Gualà, L., Leucci, S., Proietti, G.: Specializations and generalizations of the stackelberg minimum spanning tree game. Theoret. Comput. Sci. **562**, 643–657 (2015)
4. Bilò, D., Gualà, L., Proietti, G., Widmayer, P.: Computational aspects of a 2-player stackelberg shortest paths tree game. In: Papadimitriou, C., Zhang, S. (eds.) WINE 2008. LNCS, vol. 5385, pp. 251–262. Springer, Heidelberg (2008). https://doi.org/10.1007/978-3-540-92185-1_32
5. Böhnlein, T., Kratsch, S., Schaudt, O.: Revenue maximization in stackelberg pricing games: beyond the combinatorial setting. In: 44th International Colloquium on Automata, Languages, and Programming, ICALP 2017, Warsaw, Poland. pp. 46:1–46:13, 10–14 July 2017. https://doi.org/10.4230/LIPIcs.ICALP.2017.46
6. Briest, P., Chalermsook, P., Khanna, S., Laekhanukit, B., Nanongkai, D.: Improved hardness of approximation for stackelberg shortest-path pricing. In: Saberi, A. (ed.) WINE 2010. LNCS, vol. 6484, pp. 444–454. Springer, Heidelberg (2010). https://doi.org/10.1007/978-3-642-17572-5_37
7. Briest, P., Hoefer, M., Gualà, L., Ventre, C.: On stackelberg pricing with computationally bounded consumers. In: Leonardi, S. (ed.) WINE 2009. LNCS, vol. 5929, pp. 42–54. Springer, Heidelberg (2009). https://doi.org/10.1007/978-3-642-10841-9_6
8. Briest, P., Hoefer, M., Krysta, P.: Stackelberg network pricing games. Algorithmica **62**(3–4), 733–753 (2012). https://doi.org/10.1007/s00453-010-9480-3, http://dx.doi.org/10.1007/s00453-010-9480-3
9. Cabello, S.: Stackelberg Shortest Path Tree Game. arXiv preprint arXiv:1207.2317 (2012)
10. Cardinal, J., et al.: The stackelberg minimum spanning tree game. Algorithmica **59**, 129–144 (2011)
11. Cardinal, J., Demaine, E.D., Fiorini, S., Joret, G., Newman, I., Weimann, O.: The stackelberg minimum spanning tree game on planar and bounded-treewidth graphs. J. Comb. Optim. **25**(1), 19–46 (2013)
12. van Hoesel, S.: An overview of stackelberg pricing in networks. Eur. J. Oper. Res. **189**, 1393–1402 (2008)

13. Joret, G.: Stackelberg network pricing is hard to approximate. Networks **57**(2), 117–120 (2011)
14. Labbé, M., Marcotte, P., Savard, G.: A bilevel model of taxation and its application to optimal highway pricing. Manage. Sci. **44**, 1608–1622 (1998)
15. Labbé, M., Violin, A.: Bilevel programming and price setting problems. Annals OR **240**(1), 141–169 (2016). https://doi.org/10.1007/s10479-015-2016-0
16. Roche, S., Savard, G., Marcotte, P.: An approximation algorithm for stackelberg network pricing. Networks **46**, 57–67 (2005)
17. von Stackelberg, H.: Marktform und Gleichgewicht. Springer, Berlin (1934)

FRESH: Fréchet Similarity with Hashing

Matteo Ceccarello[1] , Anne Driemel[2], and Francesco Silvestri[3(✉)]

[1] IT University, Copenhagen, Denmark
mcec@itu.dk
[2] University of Bonn, Bonn, Germany
driemel@cs.uni-bonn.de
[3] University of Padova, Padova, Italy
silvestri@dei.unipd.it

Abstract. This paper studies the r-range search problem for curves under the continuous Fréchet distance: given a dataset S of n polygonal curves and a threshold $r > 0$, construct a data structure that, for any query curve q, efficiently returns all entries in S with distance at most r from q. We propose FRESH, an approximate and randomized approach for r-range search, that leverages on a locality sensitive hashing scheme for detecting candidate near neighbors of the query curve, and on a subsequent pruning step based on a cascade of curve simplifications. We experimentally compare FRESH to exact and deterministic solutions, and we show that high performance can be reached by suitably relaxing precision and recall.

Keywords: Similarity search · Range reporting ·
Locality Sensitive Hashing · Fréchet distance · Algorithm engineering

1 Introduction

The target of this paper is similarity search for time series and trajectories or, more generally, for curves: indeed, time series and trajectories can be envisioned as polygonal curves with vertices from $\mathbb{R}^d$, for a suitable dimension $d \geq 1$.[1] Similarity search of curves frequently arises in several applications, like ridesharing recommendation [27], frequent routes [25], players performance [21], and seismology [26]. In the paper, we address the *r-range search problem*: given a dataset S of n curves from a domain $\mathcal{X}$ and a threshold $r > 0$, construct a data structure that, for any query curve $q \in \mathcal{X}$, efficiently returns *all entries* in S with distance at most r from q. Range reporting is a primitive widely used for solving the similarity join and k-nearest neighbor problems.

There is no common agreement on the best distance measure for curves, for it depends on the application domain, quality of input data, and performance requirements. There are several functions to measure the distance between two curves, such as continuous Fréchet distance, Dynamic Time Warping (DTW),

---

[1] Usually, we have $d = 1$ for time series and $d > 1$ for trajectories.

© Springer Nature Switzerland AG 2019
Z. Friggstad et al. (Eds.): WADS 2019, LNCS 11646, pp. 254–268, 2019.
https://doi.org/10.1007/978-3-030-24766-9_19

Euclidean distance, and Hausdorff distance. We focus on the *continuous Fréchet distance*, that was introduced in computer science by Alt and Godau in the '90s [3]. The continuous Fréchet distance and its discrete variant, named discrete Fréchet distance [19], have been widely studied in theory (e.g. [1,9,22]) and used in different applications, like handwriting recognition [28], protein structure alignment [31] and, in particular, trajectories of moving objects (e.g., [24]). Recently, the Fréchet distance has been addressed by the ACM SIGSPATIAL Cup 2017, drawing attention to this measure from a practical domain.

The Fréchet distance[2] between two curves is traditionally explained with this metaphor: a man is walking on a curve and his dog on another curve; the man and dog follow their curves from start to end and can vary their speeds, but they cannot go backward; the minimum length of the leash necessary to connect man and dog during the walk is the continuous Fréchet distance. The Fréchet distance does not require a one-to-one mapping between points of two curves, and it is hence invariant under differences in speed: this allows, for instance, to detect the trajectories of two cars following the same street but with different speeds due to traffic conditions.

Range search is known to be computational demanding in high dimensions under different distances, including the Fréchet distance: from a worst-case point of view, there is indeed evidence that it is not possible to obtain a truly sublinear algorithm unless with a breakthrough for the Satisfiability problem [9,16]. Locality Sensitive Hashing (LSH), introduced in [23], is the most common technique for developing approximate and randomized algorithms for similarity search problems. LSH is a hashing scheme where near points have a higher collision probability than far points. Recently, [16] has introduced a family of LSH schemes for curves under the discrete Fréchet and Dynamic Time Warping distances.

1.1   Our Results

The goal of this paper is to describe and experimentally evaluate FRESH, an approximate and randomized approach for r-range search under the continuous Fréchet distance. FRESH builds on the theoretical ideas in [16] and extends it by providing a solid and efficient framework for trading precision and performance.

Algorithm Design. The core component of FRESH is a filter based on the LSH scheme for the discrete Fréchet distance in [16], which is boosted with multiply-shift hashing [15] and tensoring [4,14] for better performance. For a given input set S with n curves and a query curve q, the filter selects as candidate near neighbors all curves colliding with q under at least one of L hash functions randomly selected from the LSH scheme. This filters out a significant number of curves, without even reading them. All candidates are associated with a *score*, representing the fraction of collisions under the L hash functions. If FRESH is seen as a classifier for detecting near and far curves for a given query q, the score of a curve p represents the probability that p and q are near.

---

[2] If not differently stated, "Fréchet distance" refers to the continuous definition.

The second component of FRESH is a candidate pruning step for reducing false positives (i.e., far curves marked as near). The pruning consists in verifying that the fraction $0 \leq \tau \leq 1$ of candidates with smaller scores have continuous Fréchet distance from the query not larger than r. As verifying the Fréchet distance is a costly operation, we propose a procedure exploiting a cascade of curve simplifications from [17] and verification heuristics from [6,11]: each step can successfully show that the distance is larger or not than r, or it can fail and do not provide an answer; the procedure applies the aforementioned simplifications and heuristics until one of them succeeds.

Performance/Quality Trade-Off. FRESH trades the quality of the results with the overall performance by suitably settings the aforementioned L and τ parameters.[3]: We measure the quality of the results in terms of: (1) *recall*, that is the fraction of true positives reported by the algorithm over all the positives in the ground truth; (2) *precision*, that is the fraction of true positives over the predicted positives (i.e. the sum of true positives and false positives). By increasing the number L of hash functions used in FRESH, it is possible to increase the recall of our algorithm by increasing the query time (linear in L) and of the space requirements (equal to $L \cdot n + I$, where I is the input size). Once the recall has been fixed, it is possible to improve the precision by increasing the τ parameter at the cost of a higher query time. The recall is not affected by this step and a perfect precision is reached by setting $\tau = 1$.

Practical and Theoretical Guarantees. We have carried out an extensive experimental evaluation of the FRESH algorithm over several datasets. To evaluate FRESH, we use it as a primitive for solving a self-similarity join on each dataset D: specifically, for every curve in D, we perform an r-range search query over D. The experiments show that the scores computed under a query q provide a good indicator of the distance from q, and thus filtering points according with scores is a sound approach. From a performance point of view, we compare FRESH with the exact solutions that won the ACM SIGSPATIAL 2017 challenge [6,11,18]. When the recall is approximately 70–80% and the precision is approximately 50%, FRESH exhibits better running times with speedups above 5x for some inputs. Although the precision is low, the returned points are never too far from the query (up to a constant factor from r) by the property of the LSH scheme. With higher precision, the heuristics adopted in the exact solutions, in particular the bounding box approach in [18], are very effective with the 1-dimensional datasets (i.e., time series) considered in the experiments and highlight the limitations of FRESH in this setting. FRESH is also supported by the theoretical foundations of the LSH scheme in [16].

The FRESH algorithm is described in Sect. 3 and the experimental results in Sect. 4. The code of FRESH is available at https://github.com/Cecca/FRESH.

---

[3] In addition to parameters τ and L, the FRESH algorithm has other second order parameters that are introduced in Sect. 3, which marginally affect performance and quality. However, from an application point of view, the trade-off is mainly captured by L and τ, and the remaining parameters can be left to the default value in the implementation.

We refer to the full version [12] for a more detailed coverage of our results, including the theoretical analysis bounding the collision probability and further experiments.

1.2 Related Works

Similarity Search for Curves. Data structures for searching among curves under the Fréchet distance have been studied under different angles. One of the earlier theoretical works is [22] that proposes a nearest neighbor data structure for Fréchet distance. In 2011, [7] revived the topic motivated by the availability of high-resolution trajectories of soccer players in the emerging area of sports analytics. A comprehensive study of the complexity of range searching under the Fréchet distance appeared in [1], that also gives lower bounds on the space-query-time trade-off of range searching under the Fréchet distance. Recently, the annual data competition within the ACM SIGSPATIAL conference on geographic information science has drawn attention to the timeliness of this problem [30]. The focus of the challenge was on exact solutions and hence none of the awarded submissions [6, 11, 18] propose approximate solutions. An LSH for the discrete Fréchet distance is described in [16]. A follow-up paper [20] provides better theoretical approximation bounds using a slightly different approach, but their results do not apply to the setting that we focus on in this paper. Sketches for the Hausdorff and discrete Fréchet distances are proposed in [5], which gives an LSH scheme with similar properties of [16].

Verifying the Fréchet Distance. In order to improve the precision of the proposed LSH scheme, we suggest to filter the query results by verifying the distances for selected curves. However, verifying the distance is a non-trivial and expensive operation. It is known that the (discrete or continuous) Fréchet distance between two fixed curves cannot be decided in strictly subquadratic time in the number of vertices of the curves, unless the Strong Exponential Time Hypothesis is false [8]. The fastest algorithms for computing the continuous and discrete Fréchet distance are described in [10] and [2]. Both algorithms take roughly quadratic time. However, [17] shows that one can approximate the distance in near-linear time under certain realistic assumptions on the shape of the input curves. We use this algorithm to filter the query results, in order to improve the precision of our method.

2 Preliminaries

Continuous and Discrete Fréchet Distances. A *time series* (or *trajectory*) is a series $(p_1, t_1), \ldots, (p_m, t_m)$ of measurements $p_i \in \mathbb{R}^d$ of a signal taken at times t_i, where $0 = t_1 < t_2 < \ldots < t_m = 1$ and m is finite. A time series denotes a *polygonal curve* p of length m and defined by the sequence of *vertices* $p_1, \ldots, p_m$. A polygonal curve p may be viewed as a continuous function $p : [0, n] \to \mathbb{R}^d$ by linearly interpolating $p_1, \ldots, p_m$ in order of t_i, $i = 1, \ldots, m$. Each segment between p_i and p_{i+1} is called *edge* $\overline{p_i p_{i+1}} = \{x p_i + (1-x)p_{i+1} | x \in [0,1]\}$. We let $|p|$ denote the *length* of curve p, that is the number of vertices in p. The space of all polygonal curves in $\mathbb{R}^d$ is denoted by Δ^d. As all our curves are polygonal, we omit the term "polygonal" for the sake of simplicity.

For two vertices in $p, q \in \mathbb{R}^d$, we let $d_E(p,q) = \|p-q\|_2$ denote their Euclidean distance. Let Φ_n be the set of all continuous and non-decreasing functions ϕ from $[0,1]$ into $[1,n]$. The *continuous Fréchet distance* of two curves p and q, denoted by $d_F(p,q)$, is defined as

$$d_F(p,q) = \inf_{\substack{\phi_1 \in \Phi_{|p|} \\ \phi_2 \in \Phi_{|q|}}} \max_{t \in [0,1]} \left\| p_{\phi_1(t)} - q_{\phi_2(t)} \right\|_2 . \tag{1}$$

Each pair $(\phi_1, \phi_2) \in \Phi_{|p|} \times \Phi_{|q|}$ is called *continuous traversal*, and it can been seen as a schedule for simultaneously traversing the two curves, starting on the first vertices of both curves at time 0 and ending on the last vertices at time 1.

The problem of verifying that the Fréchet distance between two curves is less than or equal to a threshold r is usually done with the so-called *free space diagram* [3], which has quadratic cost in the worst case. However, it was shown in [17] that if the algorithm operates on simplified copies of the curves, then the complexity reduces to near-linear under certain assumptions on the shape of the curves. The simplification introduces an approximation error to the verification algorithm, but as shown in [17], the error can be bounded if the simplification parameters are wisely chosen. By exploiting the bounded error, it is possible to use the simplification for confirming or denying that two curves have distance at most r.

Range Search and LSH. Given a set $S \subseteq \mathcal{X}$ of n points in a domain $\mathcal{X}$, a distance function $d : \mathcal{X} \times \mathcal{X} \rightarrow [0, +\infty)$, and a radius $r > 0$, the *r-range search* (also known as range reporting) problem requires to construct a data structure that, for any given query point $q \in \mathcal{X}$, returns all points $p \in S$ such that $d(q,p) \leq r$. We say that a point p is a *r-near* or *r-far* point of q if $d(p,q) \leq r$ or $d(p,q) > r$, respectively; if r is clear in the context, we will just say that p is a near or far point of q.

Locality Sensitive Hashing (LSH) [23] is a common tool for r-range search in high dimensions. For a given radius $r > 0$ and approximation factor $c > 1$, an LSH is an hash scheme $\mathcal{H}$ where for a random selected map $h \in \mathcal{H}$ and two points x and y, we have that $\Pr_{h \in \mathcal{H}}[h(x) = h(y)] \geq p_1$ if $d(x,y) \leq r$, and $\Pr_h[h(x) = h(y)] \leq p_2$ if $d(x,y) > c \cdot r$. Probabilities p_1 and p_2 depend on the LSH scheme and the quality of an LSH scheme is given by $\rho = \rho(H) = \frac{\log 1/p_1}{\log 1/p_2}$ (values of ρ closer to 0 are better). Concatenation is a technique for building an LSH scheme with a small collision probability p_2 of far points: by concatenating $k \geq 1$ hash functions randomly and uniformly selected from $\mathcal{H}$, we get an LSH scheme with collision probability p_1^k for near points and p_2^k for far points.

The standard data structure based on LSH for solving the r-range search problem is the following [23]. Assume that, after concatenation, we have $p_2 \leq 1/n$. Let $\ell_1, \ldots, \ell_L$ be L functions randomly and uniformly chosen from $\mathcal{H}$. The data structure consists of L hash tables $H_1, \ldots H_L$: each hash table H_i stores the input set S, partitioned by the hash function ℓ_i. For each query q, we compute the set $S_q = \cup_{i=1}^L H_i(\ell_i(q))$, where $H_i(\ell_i(q))$ denotes the set of points in S colliding with q under the hash function ℓ_i. Then, we scan S_q and remove all points with distance larger than r from q; the remaining points are returned as r-near points of q. If $L = \Theta\left(p_1^{-1}\right) = \Theta\left(n^\rho\right)$, then the above data structure returns in expectation a constant fraction of all near points of q.

3   FRESH Algorithm

We let S denote our input set with n curves of maximum length m, and let q be a query curve. For each query q, FRESH returns a set O_q of pairs (t, s_t) where $t \in S$ is a curve and $0 \le s_t \le 1$ is its *score*. Each score s_t denotes the likelihood of t to be close to the query q: a large value of s_t implies a high probability that t is a r-near curve of q; further, if two curves t and t' have scores $s_t \le s_{t'}$, then it is more likely that t' is closer to q than t. Curves with scores equal to 0 are not reported since they are considered far from q.

The above approach can generate both false negatives and false positives. As we will later see, false negatives (i.e., near curves that are not reported in O_q) can be reduced by increasing the number of LSH functions (i.e., the parameter L) used in the score computations. On the other hand, false positives (i.e., far curves that are reported in O_q) can be reduced by verifying the distance from q of a subset of curves in O_q with small scores. Verifying that two curves have continuous Fréchet distance at most r is however an expensive operation, we thus propose a heuristic based on a cascade of curve simplifications that efficiently rules out or confirms the distance between the curves.

The section is organized as follows: Sect. 3.1 explains how scores are computed; Sect. 3.2 describes how to reduce false positives; Sect. 3.3 shows how to verify if two curves have continuous Fréchet distance at most r.

3.1   Score Computations with LSH

At a high level, the score s_p of a curve $p \in S$ with query q is given by the normalized number of collisions with q under $L \ge 1$ hash functions from the LSH scheme $\mathcal{G}_\delta^k$ described below, where δ and k are suitable parameters.

LSH Scheme $\mathcal{G}_\delta^k$. Our starting point is the LSH scheme $\hat{\mathcal{G}}_\delta$ in [16], which maps each curve into a smaller curve with vertices from a random shifted grid $G_{\delta,t} = \left\{(x_1, \ldots, x_d) \in \mathbb{R}^d \mid \forall\, i \in [d]\; \exists\, j \in \mathbb{N} : x_i = j \cdot \delta + t\right\}$ where $\delta > 0$ is the side of the grid and $t = (t_1, \ldots t_d)$ is a random variable uniformly distributed in $[0, \delta)^d$. For a curve p with vertices $p_1, \ldots, p_m$, the function $g_{\delta,t}(p)$ returns the curve obtained by: (1) replacing each vertex p_i with its closest grid vertex in $G_{\delta,t}$; (2) removing consecutive duplicates in the new curve. The LSH family $\hat{\mathcal{G}}_\delta$ is defined as $\hat{\mathcal{G}}_\delta = \{g_{\delta,t}, \forall t \in [0, \delta)^d\}$. We also define $\hat{\mathcal{G}}_\delta^k$ as the LSH family obtained by concatenating $k \ge 1$ copies of hash functions uniformly and independently selected in $\hat{\mathcal{G}}_\delta$. We have that $\Pr_{g^k \in \hat{\mathcal{G}}_\delta^k}[g^k(q) = g^k(p)] = \Pr_{g \in \hat{\mathcal{G}}_\delta}[g(q) = g(p)]^k$: the lower collision probability of far curves allows to decrease false positives.

FRESH requires the computation of a large number of hash values in $\hat{\mathcal{G}}_\delta^k$: indeed, $k \cdot L \cdot n$ hash values are computed at construction time and $k \cdot L$ hash values for each query. We speed up the hash computation with the tensoring approach. Tensoring was initially proposed in [4] and then further studied in [14]; to the best of our knowledge, it has only been used in practice in [29]. The tensoring approach generates L hash functions building on two collections of $\sqrt{L}$ hash functions, reducing the actual number of hash computations by a $\sqrt{L}$ factor. Specifically, let $\Lambda_1 = \{g_1, \ldots, g_{L'}\}$ and $\Lambda_2 = \{g'_1, \ldots, g'_{L'}\}$ be two groups of $L' = \sqrt{L}$ random hash functions from $\hat{\mathcal{G}}_\delta^{k/2}$. Then, it is possible to construct $L' \cdot L' = L$ LSH hash functions from G_δ^k by concatenating the pair (g_i, g'_j) for all possible values of i and j in $\{1, \ldots L'\}$. This technique reduces the number of hash value computations for the initial data structure construction from $k \cdot L \cdot n$ to $k \cdot \sqrt{L} \cdot n$, and for the query procedure from $k \cdot L$ to $k \cdot \sqrt{L}$.

Finally, as storing and searching signatures is quite inefficient, we map all signatures on integers with the multiply-shift hashing scheme $\mathcal{H}$ in [15]. We denote with $\mathcal{G}_\delta^k$ the LSH hash family obtained by first using the tensoring approach to construct (a subset of) $\hat{\mathcal{G}}_\delta^k$, and then by applying the multiply-shift hashing $\mathcal{H}$ on the signature. We observe that the signature of a curve does not need to be generated and stored: while we scan a curve p to compute its signature, the hash value $h(g(p))$ is built on the fly.

Data Structure. The data structure of FRESH for efficiently computing the scores leverages on the traditional approach for solving range search with LSH. $L \geq 1$ hash functions $g_1, \ldots, g_L$ are randomly chosen from the above LSH family $\mathcal{G}_\delta^k$, for suitable values of δ and k; then for each g_i, a hash table H_i is created for storing the n input curves partitioned by g_i. For each query q, we compute the multiset $T_q = \cup_{i=1}^L H_i(g_i(q))$, where $H_i(g_i(q))$ denotes the set of curves colliding with q under g_i. If $t \in T_q$ and its multiplicity in T_q is $\hat{s}_t$, then its score s_t is $\hat{s}_t/L$. Note that the hash tables do not need to store the complete curves but just their identifiers: thus, the space required by the data structure is $I + \Theta(Ln)$ memory words, where I is the number of words to store S.

3.2 Filtering False Positives

All curves with non-zero score are not too far from the query: indeed, if the hash function uses a grid of side length δ, then all colliding curves have maximum distance δ. However, as in general $\delta > r$ (in our experiments $\delta = 4dr$, where d is the point dimension), we may report some curves with distance in $(r, \delta]$. To improve the precision, a simple approach is to set a threshold Δ and verify all curves with scores less than Δ. However, the limitations of this approach are: (1) it is not clear how to select the best Δ as it might be query dependent; (2) Δ does not directly allow to trade precision and running time. The approach used in FRESH is to verify a fraction τ, with $0 \leq \tau \leq 1$, of the curves in O_q with smaller scores. The parameter τ can be used for trading performance (with $\tau = 0$ no curve in O_q is verified) with precision (with $\tau = 1$, all curves in O_q are verified which implies a 100% precision).

3.3 Verifying the Fréchet Distance

Verifying that two curves p and q are within Fréchet distance r is an expensive operation [8]: to speed up this operation, we introduce the procedure VERIFY for checking if two curves p and q have continuous Fréchet distance less than or equal to r. VERIFY consists of two procedures, named VERIFYSIMPL and VERIFYHEUR, that exploit strategies from [6,11,17]: each procedure can successfully show that $d_F(p,q) \leq r$ or $d_F(p,q) > r$, or it can fail and do not provide an answer. Procedure VERIFYHEUR exploits the heuristics *Equal-time alignment* [11], *Greedy algorithm* [6] and *Negative filter* [6], and it stops as soon as one of them succeeds. On the other hand, procedure VERIFYSIMPL is a decision procedure based on the concept of simplification in [17]: p and q are mapped on suitable smaller trajectories p' and q' through a transformation based on a parameter $\varepsilon \geq 0$ ($\varepsilon = 0$ gives the original curves). Evaluating distance predicates on p' and q' allows to answer distance predicates on p and q, by suitable setting the parameter ε. We refer to the full version of this paper [12] for a more detailed description of VERIFYSIMPL and VERIFYHEUR. Procedure VERIFY is then the following:

1. In the first stage, we only consider the first (p_1 and q_1) and last vertices ($p_{|p|}$ and $q_{|q|}$) of p and q. If $||p_1 - q_1||_2 > r$ or $||p_{|P|} - q_{|Q|}||_2 > r$, then the two curves cannot be r-near by the definition of continuous traversal. We call this heuristic ENDPOINTS.
2. In the second stage, we look at the bounding boxes of the two curves. If the ℓ_1 distance of corresponding corners of the bounding boxes is larger than r, then the two curves cannot be r-near [18]. We call this heuristic BOUNDINGBOX.
3. In the third stage, we use VERIFYSIMPL with decreasing values of ε (which will be fixed in the experimental analysis), corresponding to simplifications becoming less aggressive. For a given ε, if VERIFYSIMPL can give an answer, then we return it, otherwise we move to the next ε.
4. The fourth stage runs if none of the calls to VERIFYSIMPL could return an answer: in this case we return the result of the invocation of VERIFYHEUR on the original curve.

4   Experimental Evaluation

In this section, we present our experimental evaluation of FRESH. Section 4.1 describes the setup of our experiments, including the benchmarks and the exact baseline algorithm used as reference. Section 4.2 analyzes the performance and quality of the LSH scheme in FRESH, without the partial verification to reduce false positives: in particular, we investigate how the number of LSH repetitions (L) and of LSH concatenations (k) affect performance and quality (recall/precision). Section 4.3 examines how the partial verification affects the performance and precision under different values of the fraction τ of verified candidate curves, and it analyses the effectiveness of the various heuristics used in FRESH to prune false positives.

4.1 Experimental Setup

We implement our algorithm in C++ with OpenMP, using the gcc compiler version 4.9.2. We run the experiments on a Debian GNU/Linux machine (kernel version 3.16.0) equipped with 24 GB of RAM, and an Intel I7 Nehalem processor (clock frequency 3.07 GHz).

As benchmarks we use datasets from the UCR collection [13], which is comprised of 85 datasets of trajectories in one dimension. For brevity, we report on 2 of these datasets: the results for the other datasets can be found in the full version of this paper [12]. We also include in our benchmark a dataset of road trips in San Francisco that was used in the SIGSPATIAL 2017 challenge [30], along with the TDrive dataset [32]. Both are datasets of trajectories in 2 dimensions. We refer to full version of the paper [12] for some statistics about these datasets.

For each dataset, we perform a self-similarity join using a set of fixed Fréchet distance thresholds, by solving the r-range search problem for each curve of the dataset. The thresholds are set to the first and fifth percentiles of the pairwise distances for any given dataset, so that the output size is 1% and 5% of the number of possible pairs, respectively. Given the large number of possible pairs, these percentiles are computed on the pairwise distances of a sample of 1000 points of each dataset. Figure 1 gives the distribution of pairwise distances in the datasets we are considering. Each result is the average over at least 5 runs.

To establish a baseline, we ran the code provided by the three winners of the SIGSPATIAL 2017 challenge [6,11,18], compiled with all optimizations enabled and ran with the default parameters. Table 1 reports these results.

4.2 Evaluating the LSH Scheme

We analyze how the LSH scheme affects the performance and quality of FRESH without the partial verification. In other words, each pair colliding in at least one of the L repetitions (i.e., with a non-zero score) is reported as a positive match, without further verification. We test this setup using hash values obtained as the concatenation of $k = 1, 2, 4$ hash functions and with $L = 128, 256, 512, 1024$ repetitions, setting the grid size to $\delta = 4dr$. Figure 2 reports, for each dataset and combination of parameters, the performance in the precision-recall space. The recall is the fraction of true positives reported by the algorithm over all the positives in the ground truth, whereas the precision is the fraction of true positives over the predicted positives (i.e., the sum of true positives and false positives). Both scores range from 0 to 1, with 1 being the best, hence in the plots of Fig. 2 we have that the closer the top right corner, the better the performance. Note that we use the precision instead of the false positive rate due to the large

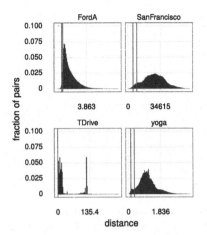

Fig. 1. Distribution of pairwise distances for all the datasets considered. The green line highlights the first percentile, the red one highlights the fifth percentile. (Color figure online)

Table 1. Baseline times (in seconds) for the two different radii, which are defined, respectively, as the first and fifth percentile of all pairwise distances. Results marked with ‡ were obtained using the code by Baldus et al. [6], the ones marked with ⋆ were obtained using the code by Dutsch et al. [18].

Dataset	Range	Best time
FordA	1.07 (first)	299 ⋆
	1.20 (fifth)	1190 ⋆
Yoga	0.14 (first)	23 ⋆
	0.33 (fifth)	87 ⋆
SanFrancisco	5213.21 (first)	413 ⋆
	9205.43 (fifth)	417 ‡
TDrive	0.17 (first)	3913 ⋆
	0.23 (fifth)	20372 ⋆

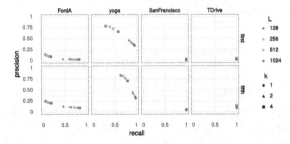

Fig. 2. Performance in terms of precision and recall of FRESH on all the datasets considered. The color of a point denotes the number of repetitions L, while the shape of a point represents the number of concatenations k.

number of negatives in the ground truth, which makes very easy to attain a small false positive rate.

In general, we have that increasing the number of repetitions L improves the recall, lowering the precision, as expected. Symmetrically, increasing k makes the LSH more selective, hence it increases the precision, at the expense of the recall. Note that on some datasets our LSH technique is more effective than on others. In general, using sufficiently many repetitions we can get good recall, while getting a good precision is harder, and may be very costly in terms of recall. We will address this problem in the next subsection.

On the SanFrancisco and TDrive datasets we get perfect recall and low precision, almost irrespective of the configuration of parameters. This is due to the distance distribution of these datasets: by setting the query range to the first and

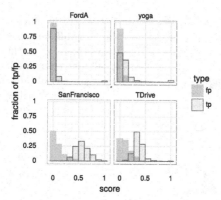

Fig. 3. The distribution of scores assigned to colliding pairs for $k = 2$ and $L = 1024$, with query radius equals to the first percentile of distances, shows that the majority of false positive pairs (fp, in orange) have lower scores than the true positive colliding pairs (tp, in blue), with some overlapping of the two distributions. The results for other configurations of parameters are similar. (Color figure online)

fifth percentiles of distances, the algorithm constructs grids with a resolution so large that almost all curves collide with the queries.

4.3   Improving the Precision by Partial Verification

In this section we verify the trade-off between precision and running time proposed in Sect. 3.2. From the previous experiments we selected a configuration of parameters striking a good balance of recall and precision on most datasets: $k = 2$ and $L = 1024$. For $\tau \in \{0, 0.1, 0.2, 0.5, 1\}$ we run the algorithm evaluating the τm pairs with lowest non-zero scores, where m is the number of pairs with non-zero scores. When $\tau = 0$, the algorithm runs in the same configuration used in the previous subsection, when $\tau = 1$ the algorithm verifies all the colliding pairs. We apply 3 simplifications in the verification pipeline, using $\varepsilon = 10, 1, 0.1$, from coarsest to finest.

First, we consider the distribution of scores before any verification happens, to assert that verifying the lowest-score pairs is actually sound (Fig. 3). We have that the false positive pairs (colored in orange) have lower scores than the true positive colliding pairs (in blue), with some overlapping of the two distributions. Therefore, verifying pairs starting from the low-score ones seems like a sensible choice, since we are likely to get rid of many false positives, which we expect to improve the recall. Note that verifying some pairs does not remove true positives (neither it can introduce them), therefore the recall remains unchanged, irrespective to the fraction of pairs τ that we verify.

We now move to assess the influence of the fraction of verified pairs τ on the precision and the runtime performance (Fig. 4). For measuring the latter, we focus on the *speedup*, defined as the ratio between the time of the baseline and LSH based algorithm. As we expect, increasing τ increases the precision, with

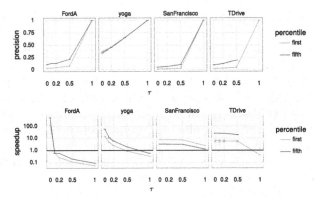

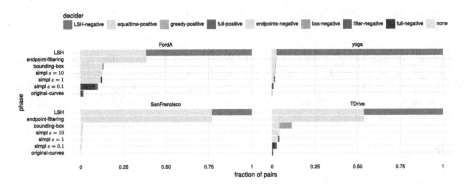

Fig. 4. Precision and speedup per pair given for varying τ, for $k = 2$, $L = 1024$. The black line on the speedup plots marks speedup 1, i.e. the performance of the best baseline algorithm.

perfect precision when $\tau = 1$, when all the pairs are verified and the algorithm reports no false positives. The speedup decreases with the increase of τ: this is because we evaluate more and more pairs, which is a costly operation. We observe that on two-dimensional trajectories, the speedup that can be obtained is larger than on one-dimensional datasets, even at higher precision values.

Fig. 5. Breakdown of the effect of the various heuristics used to decide whether a pair is a positive match or not. The hue of the colors increases with the cost of the heuristic, so *full-negative* is more expensive to compute than *endpoints-negative*. (Color figure online)

Finally, we analyze the contribution to the decision process of the LSH and the various heuristics employed (Fig. 5). We concentrate on a single run, for each dataset, with $k = 2$, $L = 1024$ and the radius set to the first percentile of distances, evaluating all pairs with nonzero score. The parts shaded in gray denote pairs for which the algorithm was not able to reach a decision and needed to move to the next stage. Then, parts in shades of green (resp. red) denote pairs for which a positive (resp. negative) decision was reached using one of the heuristics. The pairs excluded by the LSH scheme are shaded in blue rather than red, to remark that even if they are rejected as negatives they may contain some false negatives: the larger the blue bar, the more effective the filtering power of the LSH scheme. Some datasets are more amenable to be processed with the LSH strategy, and this is in line with the precision results reported in Fig. 2. Of the pairs surviving this first filtering, several can be discarded by looking at the endpoints, as shown by the *endpoint-filtering* column in the plot. The simplifications have varying degrees of effectiveness, depending on the dataset: on some datasets coarser simplifications are effective, whereas on some others we have to use finer simplifications (i.e., with a smaller ε).

5   Conclusion

As future work, it would be interesting to develop a general approach that merges the techniques in FRESH with the ones used in the exact solutions of the ACM SIGSPATIAL competition; more generally, a challenge is understanding which input features make a solution more efficient than others. The filtering approach used in FRESH can be enriched by using techniques for classifier assessment that consider the different costs that false positives and false negatives can have on the final application. Finally, we observe that the LSH scheme for the discrete Fréchet distance in [16] also holds under the DTW distance: an interesting direction is to extend and analyze FRESH to report near curves under the DTW distance and other distance measures.

Acknowledgments. The authors would like to thank M. Aumüller, K. Bringmann, F. Dütsch, R. Pagh and J. Vahrenhold for useful comments, and the developers of the UCR collection. This work has been partially supported by: ERC project "Scalable Similarity Search", NWO Veni project 10019853, SID 2018 and 2017 projects of the University of Padova.

References

1. Afshani, P., Driemel, A.: On the complexity of range searching among curves. In: Proceedings 29th Symposium on Discrete Algorithms (SODA), pp. 898–917 (2018)
2. Agarwal, P., Avraham, R., Kaplan, H., Sharir, M.: Computing the discrete Fréchet distance in subquadratic time. SIAM J. Comput. **43**(2), 429–449 (2014)
3. Alt, H., Godau, M.: Computing the Fréchet distance between two polygonal curves. Int. J. Comput. Geom. Appl. **5**, 75–91 (1995)
4. Andoni, A., Indyk, P.: Efficient algorithms for substring near neighbor problem. In: Proceedings of 17th SIAM Symposium on Discrete Algorithm (SODA), pp. 1203–1212 (2006)
5. Astefanoaei, M., Cesaretti, P., Katsikouli, P., Goswami, M., Sarkar, R.: Multi-resolution sketches and locality sensitive hashing for fast trajectory processing. In: Proceedings of International Conference on Advances in Geographic Information Systems (SIGSPATIAL) (2018)

6. Baldus, J., Bringmann, K.: A fast implementation of near neighbors queries for Fréchet distance (GIS Cup). In: Proceedings 25th International Conference on Advances in Geographic Information Systems (SIGSPATIAL), pp. 99:1–99:4 (2017)
7. de Berg, M., Cook IV, A.F., Gudmundsson, J.: Fast Fréchet queries. Comput. Geom. Theory Appl. **46**(6), 747–755 (2013)
8. Bringmann, K.: Why walking the dog takes time: Fréchet distance has no strongly subquadratic algorithms unless SETH fails. In: Proceedings 55th Symposium on Foundations of Computer Science (FOCS), pp. 661–670 (2014)
9. Bringmann, K., Mulzer, W.: Approximability of the discrete Fréchet distance. J. Comput. Geom. **7**(2), 46–76 (2016)
10. Buchin, K., Buchin, M., Meulemans, W., Mulzer, W.: Four soviets walk the dog-with an application to Alt's conjecture. In: Proceedings 25th Symposium on Discrete Algorithms (SODA), pp. 1399–1413 (2014)
11. Buchin, K., Diez, Y., van Diggelen, T., Meulemans, W.: Efficient trajectory queries under the Fréchet distance (GIS Cup). In: Proceedings 25th International Conference on Advances in Geographic Information Systems (SIGSPATIAL), pp. 101:1–101:4 (2017)
12. Ceccarello, M., Driemel, A., Silvestri, F.: FRESH: Fréchet similarity with hashing, arXiv abs/1809.02350 (2019)
13. Chen, Y., et al.: The UCR time series classification archive, July 2015. www.cs. ucr.edu/~eamonn/time_series_data/
14. Christiani, T.: Fast locality-sensitive hashing frameworks for approximate near neighbor search, arXiv:1708.07586 (2017)
15. Dietzfelbinger, M., Hagerup, T., Katajainen, J., Penttonen, M.: A reliable randomized algorithm for the closest-pair problem. J. Algorithms **25**(1), 19–51 (1997)
16. Driemel, A., Silvestri, F.: Locality-sensitive hashing of curves. In: Proceedings 33rd International Symposium on Computational Geometry (SoCG) (2017)
17. Driemel, A., Har-Peled, S., Wenk, C.: Approximating the Fréchet distance for realistic curves in near linear time. Discrete Comput. Geom. **48**(1), 94–127 (2012)
18. Dütsch, F., Vahrenhold, J.: A filter-and-refinement- algorithm for range queries based on the Fréchet distance (GIS Cup). In: Proceedings of 25th International Conference on Advances in Geographic Information Systems (SIGSPATIAL), pp. 100:1–100:4 (2017)
19. Eiter, T., Mannila, H.: Computing discrete Fréchet distance. Technical report CD-TR 91/16, TU Vienna (1994)
20. Emiris, I.Z., Psarros, I.: Products of Euclidean metrics and applications to proximity questions among curves. In: Proceedings of 34th International Symposium on Computational Geometry (SoCG). LIPIcs, vol. 99, pp. 37:1–37:13 (2018)
21. Gudmundsson, J., Horton, M.: Spatio-temporal analysis of team sports. ACM Comput. Surv. **50**(2), 22 (2017)
22. Indyk, P.: Approximate nearest neighbor algorithms for Fréchet distance via product metrics. In: Proceedings of 18th Symposium on Computational Geometry (SoCG), pp. 102–106 (2002)
23. Indyk, P., Motwani, R.: Approximate nearest neighbors: towards removing the curse of dimensionality. In: Proceedings of 30th Symposium on the Theory of Computing (STOC), pp. 604–613 (1998)
24. Konzack, M., et al.: Visual analytics of delays and interaction in movement data. Int. J. Geogr. Inf. Sci. **31**(2), 320–345 (2017)
25. Luo, W., Tan, H., Chen, L., Ni, L.M.: Finding time period-based most frequent path in big trajectory data. In: Proceedings of International Conference on Management of Data (SIGMOD) (2013)

26. Rong, K., et al.: Locality-sensitive hashing for earthquake detection: a case study of scaling data-driven science. Proc. VLDB Endow. **11**(11), 1674–1687 (2018)
27. Shang, S., Ding, R., Zheng, K., Jensen, C.S., Kalnis, P., Zhou, X.: Personalized trajectory matching in spatial networks. VLDB J. **23**(3), 449–468 (2014)
28. Sriraghavendra, E., Karthik., K., Bhattacharyya, C.: Fréchet distance based approach for searching online handwritten documents. In: Proceedings of 9th International Conference on Document Analysis and Recognition (ICDAR 2007), vol. 1 (2007)
29. Sundaram, N., et al.: Streaming similarity search over one billion tweets using parallel locality-sensitive hashing. Proc. VLDB Endow. **6**(14), 1930–1941 (2013)
30. Werner, M., Oliver, D.: ACM SIGSPATIAL GIS Cup 2017: range queries under Fréchet distance. SIGSPATIAL Spec. **10**(1), 24–27 (2018). http://sigspatial2017. sigspatial.org/giscup2017/home
31. Wylie, T., Zhu, B.: Protein chain pair simplification under the discrete Fréchet distance. IEEE/ACM Trans. Comput. Biol. Bioinf. **10**(6), 1372–1383 (2013)
32. Yuan, J., et al.: T-drive: driving directions based on taxi trajectories. In: Proceedings of 18th International Conference on Advances in Geographic Information Systems (SIGSPATIAL), pp. 99–108. ACM (2010)

Range Closest-Pair Search in Higher Dimensions

Timothy M. Chan[1], Saladi Rahul[1], and Jie Xue[2(✉)]

[1] University of Illinois at Urbana-Champaign, Urbana, IL, USA
tmc@illinois.edu, saladi.rahul@gmail.com
[2] University of Minnesota - Twin Cities, Minneapolis, MN, USA
xuexx193@umn.edu

Abstract. Range closest-pair (RCP) search is a range-search variant of the classical closest-pair problem, which aims to store a given set S of points into some space-efficient data structure such that when a query range Q is specified, the closest pair in $S \cap Q$ can be reported quickly. RCP search has received attention over years, but the primary focus was only on $\mathbb{R}^2$. In this paper, we study RCP search in higher dimensions. We give the first nontrivial RCP data structures for orthogonal, simplex, halfspace, and ball queries in $\mathbb{R}^d$ for any constant d. Furthermore, we prove a conditional lower bound for orthogonal RCP search for $d \geq 3$.

1 Introduction

The closest-pair problem is one of the most fundamental problems in computational geometry and finds numerous applications in various areas, such as collision detection, traffic control, etc. In many scenarios, instead of finding the global closest-pair, people want to know the closest pair contained in some specified ranges. This results in the notion of *range closest-pair* (RCP) search. RCP search is a range-search variant of the classical closest-pair problem, which aims to store a given set S of points into some space-efficient data structure such that when a query range Q is specified, the closest pair in $S \cap Q$ can be reported quickly. RCP search has received considerable attention over the years [1,4,10,11,17,18,20–23].

Unlike most traditional range-search problems, RCP search is *non-decomposable*. That is, if we partition the dataset S into S_1 and S_2, given a query range Q, the closest pair in $S \cap Q$ cannot be obtained efficiently from the closest pairs in $S_1 \cap Q$ and $S_2 \cap Q$. Due to the non-decomposability, many traditional range-search techniques are inapplicable to RCP search, which makes the problem quite challenging. As such, despite of much effort made on this topic, most known results are restricted to the plane case, i.e., RCP search in

---

A full version of the paper is available at [6]. Work by the first author has been partially supported by NSF Grant CCF-1814026. Work by the third author has been partially supported by a Doctoral Dissertation Fellowship from the Graduate School of the University of Minnesota.

© Springer Nature Switzerland AG 2019
Z. Friggstad et al. (Eds.): WADS 2019, LNCS 11646, pp. 269–282, 2019.
https://doi.org/10.1007/978-3-030-24766-9_20

$\mathbb{R}^2$. Beyond $\mathbb{R}^2$, only very specific query types have been studied, such as 2-sided box queries.

In this paper, we investigate RCP search in higher dimensions. We consider four widely-studied query types: orthogonal queries, simplex queries, halfspace queries, and ball queries. We are interested in designing efficient RCP data structures (in terms of space cost, query time, and preprocessing time) for these kinds of query ranges, and proving conditional lower bounds for these problems.

Related Work. The closest-pair problem and range search are both well-studied problems in computational geometry; see [2,19] for surveys of these two topics.

RCP search was for the first time introduced by Shan et al. [17] and subsequently studied in [1,4,10,11,18,20–23]. In $\mathbb{R}^2$, the query types studied include quadrants, strips, rectangles, and halfplanes. RCP search with these query ranges can be solved using near-linear space with poly-logarithmic query time. The best known data structures were given by Xue et al. [22], and we summarize the bounds in Table 1. For *fat* rectangles queries (i.e., rectangles of constant aspect ratio), Bae and Smid [4] showed an improved RCP data structure using $O(n \log n)$ space and $O(\log n)$ query time. In a recent work [20], Xue considered a colored version of RCP search in which the goal is to report the bichromatic closest pair contained in a query range, and proposed efficient data structures for orthogonal colored approximate RCP search (mainly in $\mathbb{R}^2$).

Table 1. Best known results in $\mathbb{R}^2$

Query type	Space cost	Query time	Preprocessing time
Quadrant	$O(n)$	$O(\log n)$	$O(n \log^2 n)$
Strip	$O(n \log n)$	$O(\log n)$	$O(n \log^2 n)$
Rectangle	$O(n \log^2 n)$	$O(\log^2 n)$	$O(n \log^7 n)$
Halfplane	$O(n)$	$O(\log n)$	$O(n \log^2 n)$

Beyond $\mathbb{R}^2$, the problem is quite open. To our best knowledge, the only known results are the orthogonal RCP data structure given by Gupta et al. [10] which only has guaranteed average-case performance and the approximate colored RCP data structures given by Xue [20] which can only handle restricted query types (dominance query in $\mathbb{R}^3$ and 2-sided box query in $\mathbb{R}^d$).

A key ingredient in existing solutions for RCP search in $\mathbb{R}^2$ is the *candidate-pair* method. Roughly speaking, this method tries to show that among the $\Omega(n^2)$ point pairs, only a few (called *candidate pairs*) can be the answer of some query. If this can be shown, then it suffices to store the candidate pairs and search the answer among them. Unfortunately, it is quite difficult to generalize this method to higher dimensions, as the previous approaches for proving the number of candidate pairs heavily rely on the fact that the data points are given in the plane. This might be the main reason why RCP search can be efficiently solved in $\mathbb{R}^2$, while remaining open in higher dimensions.

Our Contributions. In this paper, we give the first non-trivial RCP data structures for orthogonal, simplex, halfspace, and ball queries in $\mathbb{R}^d$, for any constant d. The performances of our new data structures are summarized in Table 2, where the notation $\tilde{O}(\cdot)$ hides $\log n$ factors. All these data structures have near-linear space cost, sub-linear query time, and sub-quadratic preprocessing time. For example, we obtain $\tilde{O}(n^{7/8})$ query time for two-dimensional triangular ranges, and $\tilde{O}(n^{2/3})$ query time for three-dimensional halfspaces and two-dimensional balls (i.e., disks).[1]

Furthermore, we complement these results by establishing a conditional lower bound, implying that our $\tilde{O}(\sqrt{n})$ query time bound for orthogonal RCP search in $\mathbb{R}^d$ for any $d \geq 3$ is likely the best possible (and in particular explaining why polylogarithmic solution seems not possible beyond two dimensions). Specifically, we show that orthogonal RCP search in $\mathbb{R}^3$ is at least as hard as the *set intersection query problem*, which is conjectured to require $\tilde{\Omega}(\sqrt{n})$ query time for linear-space structures.

Table 2. Performances of our new RCP data structures in $\mathbb{R}^d$

Query type	Source	Space cost	Query time	Preprocessing time
Orthogonal	Theorem 1	$\tilde{O}(n)$	$\tilde{O}(\sqrt{n})$	$\tilde{O}(n\sqrt{n})$
Simplex	Theorem 3	$\tilde{O}(n)$	$\tilde{O}(n^{1-1/(2d^2)})$	$\tilde{O}(n^{(3d^2+1)/(2d^2+1)})$
Halfspace	Theorem 4	$\tilde{O}(n)$	$\tilde{O}(n^{1-1/(d\lfloor d/2 \rfloor)})$	$\tilde{O}(n^{2-1/(2d^2)})$
Ball	Full version [6]	$\tilde{O}(n)$	$\tilde{O}(n^{1-1/((d+1)\lceil d/2 \rceil)})$	$\tilde{O}(n^{2-1/(2(d+1)^2)})$

Overview of Our Techniques. Our approach for designing these new data structures is quite different from those in the previous work. We avoid using the aforementioned candidate-pair method. Instead, our RCP data structures solve the problems as follows (roughly). For a given query range Q, the data structure first partitions the points in $S \cap Q$ into two subsets, say K and L. The size of L is guaranteed to be small, while K may have a large size. Then the data structure computes the closest pair ϕ in K using some pre-stored information and computes the closest pair ϕ' in L using the standard closest-pair algorithm (which can be done efficiently as L is small). If the two points of the closest pair ϕ^* in $S \cap Q$ are both in K or both in L, we are done. The only remaining case is that one point of ϕ^* is in K while the other point is in L. The data structure handles this case by finding the nearest neighbor of a in Q for every $a \in L$ via reporting all the points in Q that are "near" a. Using a packing argument, we can show that one only needs to report a constant number of points for each $a \in L$, and hence this procedure can be completed efficiently (since L is small).

To implement this strategy, we incorporate a number of existing geometric data structuring techniques. For orthogonal RCP, we use *range trees* and adapt

---

[1] Gupta et al. [10] obtained $\tilde{O}(\sqrt{n})$ query time for two-dimensional disks, but only for *uniformly distributed* point sets; the general problem was left open in their paper.

an idea from Gupta et al. [10] of classifying nodes as "heavy" and "light" (originally for solving a different problem, two-dimensional orthogonal *range diameter*, in near-linear space and $\tilde{O}(\sqrt{n})$ query time). For simplex RCP, we use *simplicial partitions* instead of range trees. For halfspace RCP, we switch to dual space and use *cuttings*, similar to an idea from Chan et al. [7] (for solving a different problem, halfspace *range mode*, in near-linear space and $\tilde{O}(n^{1-1/d^2})$ time). Overall, the combination of existing and new ideas is nontrivial (and interesting, in our opinion). Our conditional lower bound proof for three-dimensional orthogonal RCP is similar to some previous work (for example, Davoodi et al.'s conditional lower bound for two-dimensional range diameter [9]), and along the way, we introduce a new variant of colored range searching, *color uniqueness query*, which may be of independent interest.

2 Preliminaries

The first two results we need are the well-known partition lemma and cutting lemma, both of which are extensively used for solving range-search problems.

Lemma 1 (Partition lemma [13]). *Given a set S of n points in $\mathbb{R}^d$ and a parameter $1 \le r \le n^{1-\delta}$ for an arbitrarily small constant $\delta > 0$, one can compute in $O(n \log n)$ time a partition $\{S_1, \ldots, S_r\}$ of S and r simplices $\Delta_1, \ldots, \Delta_r$ in $\mathbb{R}^d$ such that (1) $S_i \subseteq \Delta_i$ for all $i \in \{1, \ldots, r\}$, (2) $|S_i| = O(n/r)$ for all $i \in \{1, \ldots, r\}$, and (3) any hyperplane in $\mathbb{R}^d$ crosses $O(r^{1-1/d})$ simplices among $\Delta_1, \ldots, \Delta_r$.*

Lemma 2 (Cutting lemma [8]). *Given a set $\mathcal{H}$ of n hyperplanes in $\mathbb{R}^d$ and a parameter $1 \le r \le n$, one can compute in $O(nr^{d-1})$ time a cutting of $\mathbb{R}^d$ into $O(r^d)$ cells each of which is a constant-complexity polytope intersecting $O(n/r)$ hyperplanes in $\mathcal{H}$. In addition, the algorithm for computing the cutting stores the cells into an $O(r^d)$-space data structure which can report in $O(\log r)$ time, for a specified point in $x \in \mathbb{R}^d$, the cell containing x.*

We shall also use the standard range-reporting data structures for orthogonal, simplex, and halfspace queries, stated in the following lemma:

Lemma 3. *Given a set S of n points in $\mathbb{R}^d$, one can build in $O(n \log^{O(1)} n)$ time an $O(n \log^{O(1)} n)$-space data structure which can*

(a) **(Orthogonal range reporting [5])** *report, for a specified orthogonal box B in $\mathbb{R}^d$, the points in $S \cap B$ in $O(\log^{O(1)} n + k)$ time where $k = |S \cap B|$;*
(b) **(Simplex range reporting [13])** *report, for a specified simplex Δ in $\mathbb{R}^d$, the points in $S \cap \Delta$ in $O(n^{1-1/d} \log^{O(1)} n + k)$ time where $k = |S \cap \Delta|$;*
(c) **(Halfspace range reporting [14])** *report, for a specified halfspace H in $\mathbb{R}^d$, the points in $S \cap H$ in $O(n^{1-1/\lfloor d/2 \rfloor} \log^{O(1)} n + k)$ query time where $k = |S \cap H|$.*

Using a multi-level data structure that combines range trees with the above structures, we can obtain range-reporting structures for query ranges that are the intersections of an orthogonal box and a simplex/halfspace (see the full version [6] for a detailed proof).

Lemma 4. *Given a set S of n points in $\mathbb{R}^d$, one can build in $O(n \log^{O(1)} n)$ time an $O(n \log^{O(1)} n)$-space data structure which can*

(a) **(Box-simplex range reporting)** *report, for a specified orthogonal box B and simplex Δ in $\mathbb{R}^d$, the points in $S \cap B \cap \Delta$ in $O(\log^{O(1)} n + m^{1-1/d} \log^{O(1)} n + k)$ time where $m = |S \cap B|$ and $k = |S \cap B \cap \Delta|$;*

(b) **(Box-halfspace range reporting)** *report, for a specified orthogonal box B and halfspace H in $\mathbb{R}^d$, the points in $S \cap B \cap H$ in $O(\log^{O(1)} n + m^{1-1/\lfloor d/2 \rfloor} \log^{O(1)} n + k)$ time where $m = |S \cap B|$ and $k = |S \cap B \cap H|$.*

3 Orthogonal RCP Queries

3.1 Data Structure

Let S be a set of n points in $\mathbb{R}^d$. In this section, we show how to build a RCP data structure on S for orthogonal queries. First, we build a (standard) d-dimensional range tree $\mathcal{T}$ on S. Each node $\mathbf{u}$ of $\mathcal{T}$ corresponds to a *canonical subset* of S, which we denote by $S(\mathbf{u})$. We say $\mathbf{u}$ is a *heavy node* if $|S(\mathbf{u})| \geq \sqrt{n}$. For every pair $(\mathbf{u}, \mathbf{v})$ of heavy nodes, we compute the closest pair $\phi_{\mathbf{u},\mathbf{v}}$ in $S(\mathbf{u}) \cup S(\mathbf{v})$; denote by Φ the set of all these pairs. Then we build an orthogonal range-reporting data structure $\mathcal{D}(S)$ on S (Lemma 3(a)). Our orthogonal RCP data structure consists of the range tree $\mathcal{T}$, the data structure $\mathcal{D}(S)$, and the pair set Φ.

Query Procedure. Consider a query box B in $\mathbb{R}^d$. Our goal is to find the closest pair in $S \cap B$ using the data structure described above. By searching in the range tree $\mathcal{T}$, we can find $t = O(\log^{O(1)} n)$ canonical nodes $\mathbf{c}_1, \dots, \mathbf{c}_t$ corresponding to B. We have $S \cap B = \bigcup_{i=1}^{t} S(\mathbf{c}_i)$. Let $I = \{i : \mathbf{c}_i \text{ is a heavy node}\}$ and $I' = \{1, \dots, t\} \backslash I$. (See Fig. 1(left).) For all $i, j \in I$, we obtain the pair $\phi_{\mathbf{c}_i, \mathbf{c}_j}$ from Φ and take the closest one $\phi \in \{\phi_{\mathbf{c}_i, \mathbf{c}_j} : i, j \in I\}$. On the other hand, we compute $L = \bigcup_{i \in I'} S(\mathbf{c}_i)$. We take the closest pair ϕ' in L. Let $\delta = \min\{|\phi|, |\phi'|\}$. For each $a \in L$, let $\square_a$ be the hypercube centered at a with side-length 2δ. We query, for each $a \in L$, the box range-reporting data structure $\mathcal{D}(S)$ with $\square_a \cap B$ to obtain the set $P_a = S \cap \square_a \cap B$. After this, for each $a \in L$, we compute a pair ψ_a consisting of a and the nearest neighbor of a in $P_a \backslash \{a\}$. We then take the closest one $\psi \in \{\psi_a : a \in L\}$. Finally, if $|\psi| < |\phi|$, then we return ψ as the answer; otherwise, we return ϕ as the answer.

We now verify the correctness of the above query procedure. Let $\phi^* = (a, b)$ be the closest pair in $S \cap B$. It suffices to show that $|\phi| \leq |\phi^*|$ or $|\psi| \leq |\phi^*|$. Suppose $a \in S(\mathbf{c}_i)$ and $b \in S(\mathbf{c}_j)$. If $i, j \in I$, then $|\phi| \leq |\phi_{\mathbf{c}_i, \mathbf{c}_j}| \leq |\phi^*|$ and we are done. Otherwise, either $i \in I'$ or $j \in I'$; assume $i \in I'$ without loss of generality. It follows that $a \in L$. Since ϕ^* is the closest pair in $S \cap B$, we have $|\phi^*| \leq |\phi|$ and $|\phi^*| \leq |\phi'|$, which implies that the distance between a and b is at most δ. Therefore, $b \in P_a$. Now we have $|\psi| \leq |\psi_a| \leq |\phi^*|$, which completes the proof of the correctness.

Analysis. We analyze the performance (space, query time, and preprocessing time) of our orthogonal RCP data structure. To this end, we first bound the

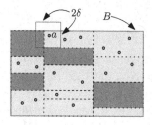 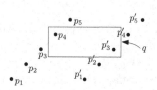

Fig. 1. (Left) The canonical nodes in the range tree $\mathcal{T}$ break the query box B into thirteen disjoint regions. The green regions correspond to set I (the heavy nodes). The orange points form the set L. For one of the points in L (denoted by a), the box $\square_a$ is shown in blue. The crucial property is that the number of points which lie in $B \cap \square_a$ is $O(1)$. (Right) Reduction from the set intersection query to the color uniqueness query. The set intersection query is to test if S_4 and S_3 are disjoint, and the query rectangle q for the color uniqueness query exactly contains points p_4 and p_3'. (Color figure online)

number of the heavy nodes. The lemma below follows immediately from the well-known fact that the sum of sizes of the canonical subsets in a range tree is $O(n \log^d n)$.

Lemma 5. *There are $O(\sqrt{n} \log^{O(1)} n)$ heavy nodes in $\mathcal{T}$.*

By the above lemma, the space of the data structure is $O(n \log^{O(1)} n)$. Indeed, the range tree $\mathcal{T}$ and the data structure $\mathcal{D}(S)$ both occupy $O(n \log^{d-1} n)$ space, and the pair-set Φ takes $O(n \log^{2d-2} n)$ space as there are $O(\sqrt{n} \log^{O(1)} n)$ heavy nodes. The preprocessing time is $O(n\sqrt{n} \log^{O(1)} n)$. Indeed, building the range tree $\mathcal{T}$ and the data structure $\mathcal{D}(S)$ takes $O(n \log^{O(1)} n)$ time. We claim that the pair-set Φ can be computed in $O(n\sqrt{n} \log^{O(1)} n)$ time. We first find the set $\mathcal{H}$ of heavy nodes, which can be done in $O(n \log^{O(1)} n)$ time by simply checking every node of $\mathcal{T}$. For two pairs $(\mathbf{u}, \mathbf{v})$ and $(\mathbf{u}', \mathbf{v}')$ of nodes in $\mathcal{H}$, we write $(\mathbf{u}, \mathbf{v}) \preceq (\mathbf{u}', \mathbf{v}')$ if $|S(\mathbf{u})| + |S(\mathbf{v})| \le |S(\mathbf{u}')| + |S(\mathbf{v}')|$. Then "$\preceq$" is a partial order on $\mathcal{H} \times \mathcal{H}$. We consider the pairs of heavy nodes in this partial order from the smallest to the greatest. For each pair $(\mathbf{u}, \mathbf{v})$, we compute $\phi_{\mathbf{u},\mathbf{v}}$ as follows. If $|S(\mathbf{u})| < 2\sqrt{n}$ and $|S(\mathbf{v})| < 2\sqrt{n}$, we explicitly compute $S(\mathbf{u}) \cup S(\mathbf{v})$ and then compute $\phi_{\mathbf{u},\mathbf{v}}$ using the standard closest-pair algorithm in $O(\sqrt{n} \log n)$ time. Otherwise, either $|S(\mathbf{u})| \ge 2\sqrt{n}$ or $|S(\mathbf{v})| \ge 2\sqrt{n}$. Without loss of generality, assume $|S(\mathbf{u})| \ge 2\sqrt{n}$. Then the two children $\mathbf{u}_1$ and $\mathbf{u}_2$ of $\mathbf{u})$ are both heavy. Note that $\phi_{\mathbf{u},\mathbf{v}}$ is the closest one among $\phi_{\mathbf{u}_1,\mathbf{v}}, \phi_{\mathbf{u}_2,\mathbf{v}}, \phi_{\mathbf{u}_1,\mathbf{u}_2}$ by construction. Also note that $(\mathbf{u}_1, \mathbf{v}) \preceq (\mathbf{u}, \mathbf{v})$, $(\mathbf{u}_2, \mathbf{v}) \preceq (\mathbf{u}, \mathbf{v})$, $(\mathbf{u}_1, \mathbf{u}_2) \preceq (\mathbf{u}, \mathbf{v})$, thus $\phi_{\mathbf{u}_1,\mathbf{v}}, \phi_{\mathbf{u}_2,\mathbf{v}}, \phi_{\mathbf{u}_1,\mathbf{u}_2}$ have already been computed when considering $(\mathbf{u}, \mathbf{v})$. With $\phi_{\mathbf{u}_1,\mathbf{v}}, \phi_{\mathbf{u}_2,\mathbf{v}}, \phi_{\mathbf{u}_1,\mathbf{u}_2}$ in hand, we can compute $\phi_{\mathbf{u},\mathbf{v}}$ in $O(1)$ time. In sum, $\phi_{\mathbf{u},\mathbf{v}}$ can be computed in $O(\sqrt{n} \log n)$ time in any case. Since $|\mathcal{H} \times \mathcal{H}| = O(n \log^{O(1)} n)$, we can compute Φ in $O(n\sqrt{n} \log^{O(1)} n)$ time. This completes the discussion of the preprocessing time. Next, we analyze the query time. Finding the canonical nodes $\mathbf{c}_1, \ldots, \mathbf{c}_t$ takes $O(\log^{O(1)} n)$ time, so does computing the index sets I and I'. Obtaining the set $\{\phi_{\mathbf{c}_i,\mathbf{c}_j} : i, j \in I\}$ and computing ϕ takes $O(\log^{O(1)} n)$ time since $|I| \leq t$ and $t = O(\log^{O(1)} n)$. Computing ϕ' requires $O(\sqrt{n} \log^{O(1)} n)$ time, because $|L| = O(t\sqrt{n}) = O(\sqrt{n} \log^{O(1)} n)$. For a point $a \in L$, reporting the points in P_a takes $O(\log^{O(1)} n + |P_a|)$ time. Therefore, computing all the P_a's can be done in $O(|L| \log^{O(1)} n + \sum_{a \in L} |P_a|)$ time. To bound this quantity, we observe the following fact.

Lemma 6. $|P_a| = O(1)$ for all $a \in L$.

Proof. We have $S \cap B = (\bigcup_{i \in I} S(\mathbf{u}_i)) \cup L$. It suffices to show that $|(\bigcup_{i \in I} S(\mathbf{u}_i)) \cap \square_a| = O(1)$ and $|L \cap \square| = O(1)$. Both facts follow from the pigeonhole principle readily. Indeed, we have $|(\bigcup_{i \in I} S(\mathbf{u}_i)) \cap \square_a| = O(1)$ because ϕ is the closest pair in $\bigcup_{i \in I} S(\mathbf{u}_i)$ and $|\phi| \geq \delta$. We have $|L \cap \square| = O(1)$ because ϕ' is the closest pair in L and $|\phi'| \geq \delta$. This completes the proof. $\square$

By the above lemma and the fact $|L| = O(\sqrt{n} \log^{O(1)} n)$, we can compute all the P_a's in $O(\sqrt{n} \log^{O(1)} n)$ time. The pair ψ can be directly obtained after knowing all the P_a's, hence the total query time is $O(\sqrt{n} \log^{O(1)} n)$. We conclude the following.

Theorem 1. *Given a set S of n points in $\mathbb{R}^d$, one can construct in $\tilde{O}(n\sqrt{n})$ time an orthogonal RCP data structure on S with $\tilde{O}(n)$ space and $\tilde{O}(\sqrt{n})$ query time.*

3.2   Conditional Hardness

In this subsection, we prove a conditional lower-bound for the orthogonal RCP query, which shows that the upper bound given in Theorem 1 is tight, ignoring $\log n$ factors. First, we define the following problem [15].

Problem 1 (**Set intersection query**). The input is a collection of sets $S_1, S_2, \ldots, S_m$ of positive reals such that $\sum_{i=1}^{m} |S_i| = n$. Given query indices i and j, report if S_i and S_j are disjoint, or not?

This problem can be viewed as a query version of Boolean matrix multiplication, and is conjectured to be *hard*: in the cell-probe model without the floor function and where the cardinality of each set S_i is upper-bounded by $\log^{O(1)} m$, any data structure to answer the set intersection problem in $\tilde{O}(\alpha)$ time requires $\tilde{\Omega}((n/\alpha)^2)$ space, for $1 \leq \alpha \leq n$ [9,15]. In particular, any linear-space structure is believed to require $\tilde{\Omega}(\sqrt{n})$ time.

Next we introduce an intermediate geometric problem, which may be of independent interest:

Problem 2 (**Color uniqueness query**). The input is a set S of n *colored* points in $\mathbb{R}^2$. Specifically, let C be a collection of distinct colors, and each point $p \in S$ is associated with some color from C. Given a query rectangle q, report if all the colors are unique in $S \cap q$? In other words, is there a color which has at least two points in $S \cap q$?

We will perform a two-step reduction: first, reduce the set intersection query to the color uniqueness query, and then reduce the two-dimensional color uniqueness query to the three-dimensional orthogonal RCP query.

Reduction from Set Intersection to Color Uniqueness in $\mathbb{R}^2$. Given an instance of the set intersection query, we will construct an instance of the color uniqueness query. Let $p_1 = (1, 1), p_2 = (2, 2), \ldots, p_m = (m, m)$, and $p'_1 = (m + 1, 1), p'_2 = (m + 2, 2), \ldots, p'_m = (2m, m)$. Next, assign a unique color to each distinct element in $S_1 \cup S_2 \cup \ldots \cup S_m$. Now replace each point p_i with $|S_i|$ *new* points such that (a) the new points are within a distance of $\varepsilon \ll 1$ from p_i, and (b) each new point picks a distinct color from the colors assigned to the elements in S_i. Perform a similar operation for points p'_i. Let P be the collection of these $2n$ new points.

To answer if S_i and S_j are disjoint ($j < i$), we ask a color uniqueness query on P with an axis-aligned rectangle $q = [i - \varepsilon, m + j + \varepsilon] \times [j - \varepsilon, i + \varepsilon]$ (see Fig. 1(right)). If there is a color which contains two points, then we report that S_i and S_j are not disjoint; otherwise, we report that S_i and S_j are disjoint. The correctness is easy to see: the key observation is that q exactly contains the points of S_i and S_j. Therefore, S_i and S_j are disjoint iff all the colors are unique in $P \cap q$. Reductions of this flavor have been performed before [3,9,12,16].

Reduction from Color Uniqueness in $\mathbb{R}^2$ to Orthogonal RCP in $\mathbb{R}^3$. Given an instance of the color uniqueness query, we will now construct an instance of the orthogonal RCP query in $\mathbb{R}^3$. Let $d_{\max}$ be the maximum Euclidean distance between any two points in S, and let $c_1, c_2, \ldots, c_{|C|}$ be the $|C|$ colors in the dataset. Then each point $p = (p_x, p_y) \in S$ with color c_i is mapped to a 3-d point $p' = (p_x, p_y, 2 \cdot i \cdot d_{\max})$. Let P be the collection of these n newly mapped points.

To answer the color uniqueness query for a rectangle q, we will ask an orthogonal RCP query on P with the query box $q \times (-\infty, \infty)$. If the closest-pair distance is less than or equal to $d_{\max}$, then we report that there is a color which contains at least two points inside q; otherwise, we report that all the colors are unique inside q. Once again, the correctness is easy to see: the key observation is that the distance between points of different colors in P is at least $2 \cdot d_{\max}$.

The above two reductions together implies our conditional lower bound, which is presented in the following theorem.

Theorem 2. *The orthogonal RCP query is at least as hard as the set intersection query.*

4   Simplex RCP Queries

Let S be a set of n points in $\mathbb{R}^d$, and r be a parameter to be specified shortly. In this section, we show how to build a RCP data structure on S for simplex queries. First, we use Lemma 1 to compute a partition $\{S_1, \ldots, S_r\}$ of S and r simplices $\Delta_1, \ldots, \Delta_r$ in $\mathbb{R}^d$ satisfying the conditions in the lemma. For every $i, j \in \{1, \ldots, r\}$, we compute the closest pair $\phi_{i,j}$ in $S_i \cup S_j$; denote by Φ the set of all these pairs. Then we build a box-simplex range-reporting data structure $\mathcal{D}'(S)$ on S (Lemma 4(a)). Our simplex RCP data structure consists of the partition $\{S_1, \ldots, S_r\}$, the simplices $\Delta_1, \ldots, \Delta_r$, the data structure $\mathcal{D}'(S)$, and the pair set Φ.

Query Procedure. Consider a query simplex Δ in $\mathbb{R}^d$. Our goal is to find the closest pair in $S \cap \Delta$ using the data structure described above. We first compute two index sets $I = \{i : \Delta_i \subseteq \Delta\}$, $I' = \{i : \Delta_i \not\subseteq \Delta \text{ and } \Delta_i \cap \Delta \neq \emptyset\}$. (See Fig. 2.) These index sets are computed by explicitly considering the r simplices $\Delta_1, \ldots, \Delta_r$. For all $i, j \in I$, we obtain the pair $\phi_{i,j}$ from Φ and take the closest one $\phi \in \{\phi_{i,j} : i, j \in I\}$. On the other hand, we compute a set $L = (\bigcup_{i \in I'} S_i) \cap \Delta$ by simply checking, for every $i \in I'$ and every $a \in S_i$, whether $a \in \Delta$. We take the closest pair ϕ' in L. Let $\delta = \min\{|\phi|, |\phi'|\}$. For each $a \in L$, let $\square_a$ be the hypercube centered at a with side length 2δ. We query, for each $a \in L$, the box-simplex range-reporting data structure $\mathcal{D}'(S)$ with $\square_a$ and Δ to obtain the set $P_a = S \cap \square_a \cap \Delta$. After this, for each $a \in L$, we compute a pair ψ_a consisting of a and the nearest neighbor of a in $P_a \backslash \{a\}$. We then take the closest one $\psi \in \{\psi_a : a \in L\}$. Finally, if $|\psi| < |\phi|$, then we return ψ as the answer; otherwise, we return ϕ as the answer.

We now verify the correctness of the above query procedure. Let $\phi^* = (a, b)$ be the closest pair in $S \cap \Delta$. It suffices to show that $|\phi| \leq |\phi^*|$ or $|\psi| \leq |\phi^*|$. Suppose $a \in S_i$ and $b \in S_j$. We first notice that $i, j \in I \cup I'$. Indeed, if $i \notin I \cup I'$ (resp., $j \notin I \cup I'$), then $\Delta_i \cap \Delta = \emptyset$ (resp., $\Delta_j \cap \Delta = \emptyset$) and hence $S_i \cap \Delta = \emptyset$ (resp., $S_j \cap \Delta = \emptyset$), which contradicts the fact that $a \in S_i \cap \Delta$ (resp., $b \in S_i \cap \Delta$). If $i, j \in I$, then $|\phi| \leq |\phi_{i,j}| \leq |\phi^*|$ and we are done. Otherwise, either $i \in I'$ or $j \in I'$; assume $i \in I'$ without loss of generality. It follows that $a \in L$. Since ϕ^* is the closest pair in $S \cap \Delta$, we have $|\phi^*| \leq |\phi|$ and $|\phi^*| \leq |\phi'|$, which implies that the distance between a and b is at most δ. Therefore, $b \in P_a$. Now we have $|\psi| \leq |\psi_a| \leq |\phi^*|$, which completes the proof of the correctness.

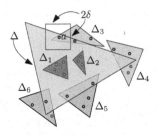

Fig. 2. $I = \{\Delta_1, \Delta_2\}$ and $I' = \{\Delta_3, \Delta_4, \Delta_5, \Delta_6\}$.

Analysis. We analyze the performance (space, query time, and preprocessing time) of our simplex RCP data structure. The space of the data structure is $O(n \log^{O(1)} n + r^2)$, because $\mathcal{D}'(S)$ occupies $O(n \log^{O(1)} n)$ space and Φ occupies $O(r^2)$ space. The preprocessing time is $O(nr \log^{O(1)} n)$. Indeed, computing the partition $\{S_1, \ldots, S_r\}$ and the simplices $\Delta_1, \ldots, \Delta_r$ takes $O(n \log n)$ time by Lemma 1. Computing $\phi_{i,j}$ for some fixed $i, j \in \{1, \ldots, r\}$ can be done in $O((n/r)\log(n/r))$ time using the standard closest-pair algorithm, because $|S_i \cup S_j| = O(n/r)$. It follows that computing Φ takes $O(nr \log n)$ time. Finally, building the data structure $\mathcal{D}'(S)$ requires $O(n \log^{O(1)} n)$ time. As such, our simplex RCP data structure can be constructed in $O(nr \log^{O(1)} n)$ time. Next, we analyze the query time. The index sets I and I' are computed in $O(r)$ time. Obtaining the set $\{\phi_{i,j} : i, j \in I\}$ and computing ϕ requires $O(r^2)$ time. The set L is computed by explicitly considering all the points in $\bigcup_{i \in I'} S_i$ in $O(\sum_{i \in I'} |S_i|)$ time. We notice that $|I'| = O(r^{1-1/d})$, since each facet of Δ only intersects $O(r^{1-1/d})$ simplices among $\Delta_1, \ldots, \Delta_r$ by Lemma 1. It follows that $\sum_{i \in I'} |S_i| = O(n/r^{1/d})$, because $|S_i| = O(n/r)$. That says, L can be computed in $O(n/r^{1/d})$ time and in particular, $|L| = O(n/r^{1/d})$. Once L is obtained, ϕ' can be computed in $O((n/r^{1/d})\log(n/r^{1/d}))$ time using the standard closest-pair algorithm. For a point $a \in L$, reporting the points in P_a takes $O(\log^{O(1)} n + m_a^{1-1/d} \log^{O(1)} m_a + |P_a|)$ time where $m_a = |S \cap \square_a|$, by Lemma 4(a). Therefore, computing all the P_a's can be done in $O(\sum_{a \in L} m_a^{1-1/d} \log^{O(1)} n + \sum_{a \in L} |P_a|)$ time. To bound this quantity, we observe the following fact.

Lemma 7. $\sum_{a \in L} m_a = O(n)$ *and* $|P_a| = O(1)$ *for all* $a \in L$.

Proof. We first prove $\sum_{a \in L} m_a = O(n)$. Consider a point $p \in S$. Let $\square_p$ be the hypercube centered at p with side-length 2δ. Note that $p \in P_a$ only if $a \in \square_p$ for all $a \in L$. Since ϕ' is the closest pair in L and $|\phi'| \geq \delta$, we have $L \cap \square_p = O(1)$ by the pigeonhole principle. Therefore, only a constant number of points in L is contained in p. In other words, any point $p \in S$ is contained in P_a for only a constant number of $a \in L$, which implies $\sum_{a \in L} m_a = O(n)$. Next, we prove that $|P_a| = O(1)$ for all $a \in L$. Clearly, $S \cap \Delta = (\bigcup_{i \in I} S_i) \cup L$. So it suffices to show that $|(\bigcup_{i \in I} S_i) \cap \square_a| = O(1)$ and $|L \cap \square_a| = O(1)$. Both facts follow from the pigeonhole principle readily. Indeed, we have $|(\bigcup_{i \in I} S_i) \cap \square_a| = O(1)$ because ϕ is the closest pair in $\bigcup_{i \in I} S_i$ and $|\phi| \geq \delta$. We have $|L \cap \square_a| = O(1)$ because ϕ' is the closest pair in L and $|\phi'| \geq \delta$. This completes the proof of $|P_a| = O(1)$. $\square$

By the above lemma and Hölder's inequality, we have

$$\sum_{a \in L} m_a^{1-1/d} \leq O(n^{1-1/d} |L|^{1/d}) = O\left(\frac{n}{r^{1/d^2}}\right),$$

which implies that computing all the P_a's takes $O((n \log^{O(1)} n)/r^{1/d^2})$ time. The pair ψ can be directly obtained after knowing all the P_a's. Hence, the total query time is $O(r^2 + (n \log^{O(1)} n)/r^{1/d^2})$. Setting $r = n^{d^2/(2d^2+1)}$ gives:

Theorem 3. *Given a set S of n points in $\mathbb{R}^d$, one can construct in $\tilde{O}(n^{(3d^2+1)/(2d^2+1)})$ time a simplex RCP data structure on S with $\tilde{O}(n)$ space and $\tilde{O}(n^{1-1/(2d^2)})$ query time.*

Note that our data structure above can also handle constant-complexity polytope RCP queries (with the same query procedure and query time). In other words, the data structure can be used to report, for specified $O(1)$ halfspaces $H_1, \ldots, H_c$ in $\mathbb{R}^d$, the closest pair in $S \cap (\bigcap_{i=1}^c H_i)$ in $\tilde{O}(n^{1-1/(2d^2)})$ time.

5 Halfspace RCP Queries

Let S be a set of n points in $\mathbb{R}^d$, and r be a parameter to be specified shortly. In this section, we show how to build an RCP data structure on S for halfspace queries. The same method can also result in an RCP data structure for ball queries, using the standard lifting argument. Since halfspace query is a special case of simplex query, the simplex RCP data structure in the last section can be directly used to answer halfspace RCP queries. But in fact, for halfspace RCP queries, we can achieve better bounds.

It suffices to consider the halfspaces which are regions below non-vertical hyperplanes, namely, halfspaces of the form $x_d \leq a_1 x_1 + \cdots + a_{d-1} x_{d-1}$. By duality, a point $a \in S$ maps to a hyperplane a^* in the dual space (which is also a copy of $\mathbb{R}^d$). Also, a non-vertical hyperplane h in the primal $\mathbb{R}^d$ maps to a point h^* in the dual space. The property of duality guarantees that a is above (resp., below) h iff h^* is above (resp., below) a^* for all $a \in S$ and all hyperplanes h (see Fig. 3). Define $\mathcal{H} = \{a^* : a \in S\}$. We use Lemma 2 to cut $\mathbb{R}^d$ (the dual space) into $R = O(r^d)$ cells $\Xi_1, \ldots, \Xi_R$ each of which is a constant-complexity polytope intersecting $O(n/r)$ hyperplanes in $\mathcal{H}$. For $i \in \{1, \ldots, R\}$, let $S_i = \{a : a^* \text{ is below } \Xi_i\}$. We associate to the cell Ξ_i the closest pair ϕ_i in S_i. Furthermore, we build a simplex range-reporting data structure $\mathcal{D}(S)$ on S (Lemma 3(b)) and a box-halfspace range-reporting data structure $\mathcal{D}'(S)$ in S (Lemma 4(b)). Our halfspace RCP data structure consists of the cells $\Xi_1, \ldots, \Xi_R$ (with the associated pairs $\phi_1, \ldots, \phi_r$) and the data structures $\mathcal{D}(S)$ and $\mathcal{D}'(S)$. The cells $\Xi_1, \ldots, \Xi_R$ are stored in the way mentioned in Lemma 2 (so that we can do point location efficiently).

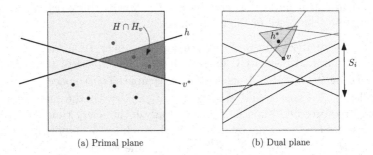

(a) Primal plane   (b) Dual plane

Fig. 3. The dataset shown in (a) consists of seven points. The dual h^* of the query hyperplane h lies inside the cell Ξ_i shown in pink in (b). The closest pair among the black points, ϕ_i, is computed in the preprocessing phase itself (since the dual of the black points is the set S_i). The red points belong to set L and are explicitly reported during the query procedure. (Color figure online)

Query Procedure. Consider a query halfspace H that is the region below a non-vertical hyperplane h. Our goal is to find the closest pair in $S \cap H$ using the data structure described above. To this end, we first find the cell Ξ_i such that $h^* \in \Xi_i$. Let V be the set of the vertices of Ξ_i. We have $V = O(1)$ by Lemma 2. For every $v \in V$, let H_v be the halfspace above the non-vertical hyperplane v^* in the primal $\mathbb{R}^d$. Using $\mathcal{D}(S)$, we find the points in $S \cap (H \cap H_v)$ for all $v \in V$ and obtain the set $L = \bigcup_{v \in V} S \cap (H \cap H_v)$. We take the closest pair ϕ' in L. Let $\delta = \min\{|\phi_i|, |\phi'|\}$ (recall that ϕ_i is the pair associated to Ξ_i). For each $a \in L$, let $\square_a$ be the hypercube centered at a with side-length 2δ. We query, for each $a \in L$, the box-halfspace range-reporting data structure $\mathcal{D}'(S)$ with $\square_a$ and H to obtain the set $P_a = S \cap \square_a \cap H$. After this, for each $a \in L$, we compute a pair ψ_a consisting of a and the nearest neighbor of a in $P_a \setminus \{a\}$. We then take the closest one $\psi \in \{\psi_a : a \in L\}$. Finally, if $|\psi| < |\phi_i|$, then we return ψ as the answer; otherwise, we return ϕ_i as the answer.

We now verify the correctness of the above query procedure. First of all, we claim that $S \cap H = S_i \cup L$. Indeed, we have $L \subseteq S \cap H$ by definition and $S_i \subseteq S \cap H$ because a^* is below Ξ_i (and hence below h^*) for all $a \in S_i$; this implies $S_i \cup L \subseteq S \cap H$. To see $S \cap H \subseteq S_i \cup L$, let $a \in S \cap H$ be a point. If a^* is below Ξ_i, then $a \in S_i$. Otherwise, there exists $v \in V$ such that a^* is above v. It follows that $a \in S \cap (H \cap H_v) \subseteq L$. Therefore, $S \cap H \subseteq S_i \cup L$ and $S \cap H = S_i \cup L$. With this observation in hand, we first show that the returned answer is a pair in $S \cap H$. It suffices to show that both ϕ_i and ψ are pairs in $S \cap H$. The two points of ϕ_i are both in S_i and hence in $S \cap H$. To see ψ is a pair in $S \cap H$, suppose $\psi = \psi_a$ for $a \in L$. By definition, ψ_a consists of a and the nearest neighbor of a in $P_a \setminus \{a\}$. We have $a \in L \subseteq S \cap H$ and $P_a \subseteq L \subseteq S \cap H$, hence ψ is a pair in $S \cap H$. Next, we show that the returned answer is the closest pair in $S \cap H$. Let $\phi^* = (a, b)$ be the closest-pair in $S \cap H$. It suffices to show that $|\phi_i| \leq |\phi^*|$ or $|\psi| \leq |\phi^*|$. If $a, b \in S_i$, then $|\phi_i| \leq |\phi^*|$ and we are done. Otherwise, assume $a \notin S_i$ and thus $a \in L$, without loss of generality. Since ϕ^* is the closest pair in $S \cap H$, we have $|\phi^*| \leq |\phi_i|$, which implies that the distance between a and b is at most δ. Therefore, $b \in P_a$. Now we have $|\psi| \leq |\psi_a| \leq |\phi^*|$, which completes the proof of the correctness.

Analysis. We analyze the performance (space, query time, and preprocessing time) of our halfspace RCP data structure. The space of the data structure is $O(n \log^{O(1)} n + R)$, because $\mathcal{D}(S)$ occupies $O(n)$ space, $\mathcal{D}'(S)$ occupies $O(n \log^{O(1)} n)$ space, and storing $\Xi_1, \ldots, \Xi_R$ (with the associated pairs $\phi_1, \ldots, \phi_R$) requires $O(R)$ space. Next, we analyze the query time. Determining the cell Ξ_i takes $O(\log r)$ time by Lemma 2. For each $v \in V$, reporting the points in $S \cap (H \cap H_v)$ takes $O(n^{1-1/d} \log^{O(1)} n + k_v)$ time where $k_v = |S \cap (H \cap H_v)|$. We claim that a^* intersects Ξ_i for any $a \in S \cap (H \cap H_v)$. Indeed, a^* is below h because $a \in H$ and is above v because $a \in H_v$. Thus, a^* intersects the segment connecting h^* and v. Since $h^*, v \in \Xi_i$, a^* intersects Ξ_i. It follows that $k_v = O(n/r)$ by Lemma 2. Furthermore, because $V = O(1)$, L can be computed in $O(n^{1-1/d} \log^{O(1)} n + \sum_{v \in V} k_v) = O(n^{1-1/d} \log^{O(1)} n + n/r)$ time and $|L| = O(\sum_{v \in V} k_v) = O(n/r)$. Once L is obtained, ϕ' can be computed in $O((n/r)\log(n/r))$ time using the standard closest-pair algorithm. For a point $a \in L$, reporting the points in P_a takes $O(\log^{O(1)} n + m_a^{1-1/\lfloor d/2\rfloor} \log^{O(1)} m_a + |P_a|)$ time where $m_a = |S \cap \square_a|$, by Lemma 4(b). By exactly the same argument in the proof of Lemma 7, we have the following observation:

Lemma 8. $\sum_{a\in L} m_a = O(n)$ and $|P_a| = O(1)$ for all $a \in L$.

By the above lemma and Hölder's inequality, we have

$$\sum_{a\in L} m_a^{1-1/\lfloor d/2\rfloor} \leq O(n^{1-1/\lfloor d/2\rfloor}|L|^{1/\lfloor d/2\rfloor}) = O\left(\frac{n}{r^{1/\lfloor d/2\rfloor}}\right),$$

which implies that computing all the P_a's takes $O(n \log^{O(1)} n/r^{1/\lfloor d/2\rfloor})$ time. The pair ψ can be directly obtained after knowing all the P_a's. Hence, the total query time is $O(\log r + n \log^{O(1)} n/r^{1/\lfloor d/2\rfloor})$. Finally, we analyze the pre-processing time. The data structures $\mathcal{D}(S)$ and $\mathcal{D}'(S)$ can both be constructed in $O(n \log^{O(1)} n)$ time by Lemmas 3(b) and 4(b). The cells $\Xi_1, \ldots, \Xi_R$ can be computed in $O(nr^{d-1})$ time by Lemma 2. So it suffices to show how to compute the pairs $\phi_1, \ldots, \phi_R$ efficiently. To this end, we build a simplex RCP data structure on S as described in Theorem 3, which takes $\tilde{O}(n^{(3d^2+1)/(2d^2+1)})$ time. Fix $i \in \{1, \ldots, R\}$ and let V be the set of the $O(1)$ vertices of Ξ_i. For $v \in V$, let H'_v be the halfspace below the hyperplane v^* in the primal space. We claim that $S_i = S \cap (\bigcap_{v\in V} H'_v)$. To see this, consider a point $a \in S$. We have $a \in S_i$ iff a^* is below Ξ_i iff v is below a^* for all $v \in V$, or equivalently, $a \in H'_v$ for all $v \in V$. Thus, $S_i = S \cap (\bigcap_{v\in V} H'_v)$. We can then compute the closest pair ϕ_i in S_i using the simplex RCP data structure with the query range $\bigcap_{v\in V} H'_v$ (as mentioned at the end of Sect. 4, our simplex RCP data structure can handle queries which are intersections of constant number of halfspaces). Computing ϕ_i takes $O(n^{1-1/(2d^2)} \log^{O(1)} n)$ time, and hence computing all pairs $\phi_1, \ldots, \phi_R$ takes $O(Rn^{1-1/(2d^2)} \log^{O(1)} n)$ time. In sum, the preprocessing time of our halfs-pace RCP data structure is $O((nr^{d-1} + n^{(3d^2+1)/(2d^2+1)} + Rn^{1-1/(2d^2)}) \log^{O(1)} n)$. Setting $r = n^{1/d}$ gives:

Theorem 4. Given a set S of n points in $\mathbb{R}^d$, one can construct in $\tilde{O}(n^{2-1/(2d^2)})$ time a halfspace RCP data structure on S with $\tilde{O}(n)$ space and $\tilde{O}(n^{1-1/(d\lfloor d/2\rfloor)})$ query time.

References

1. Abam, M.A., Carmi, P., Farshi, M., Smid, M.: On the power of the semi-separated pair decomposition. Comput. Geom. **46**(6), 631–639 (2013)
2. Agarwal, P.K., Erickson, J.: Geometric range searching and its relatives. Contemp. Math. **223**, 1–56 (1999)
3. Agarwal, P.K., Kumar, N., Sintos, S., Suri, S.: Range-max queries on uncertain data. J. Comput. Syst. Sci. **94**, 118–134 (2018)
4. Won Bae, S., Smid, M.: Closest-pair queries in fat rectangles. CoRR arXiv:1809.10531 (2018)

5. de Berg, M., Cheong, O., van Kreveld, M., Overmars, M.: Computational Geometry: Algorithms and Applications. Springer, Heidelberg (2008). https://doi.org/10.1007/978-3-662-03427-9
6. Chan, T., Rahul, S., Xue, J.: Range closest-pair search higher dimensions. CoRR arXiv:1905.01029 (2019)
7. Chan, T.M., Durocher, S., Larsen, K.G., Morrison, J., Wilkinson, B.T.: Linear-space data structures for range mode query in arrays. Theory Comput. Syst. **55**(4), 719–741 (2014)
8. Chazelle, B.: Cutting hyperplanes for divide-and-conquer. Discrete Comput. Geom. **9**(2), 145–158 (1993)
9. Davoodi, P., Smid, M., van Walderveen, F.: Two-dimensional range diameter queries. In: Fernández-Baca, D. (ed.) LATIN 2012. LNCS, vol. 7256, pp. 219–230. Springer, Heidelberg (2012). https://doi.org/10.1007/978-3-642-29344-3_19
10. Gupta, P., Janardan, R., Kumar, Y., Smid, M.: Data structures for range-aggregate extent queries. Comput. Geom. **2**(47), 329–347 (2014)
11. Gupta, P.: Range-aggregate query problems involving geometric aggregation operations. Nord. J. Comput. **13**(4), 294–308 (2006)
12. Kaplan, H., Rubin, N., Sharir, M., Verbin, E.: Counting colors in boxes. In: Proceedings of the Annual ACM-SIAM Symposium on Discrete Algorithms (SODA), pp. 785–794 (2007)
13. Matoušek, J.: Efficient partition trees. Discrete Comput. Geom. **8**(3), 315–334 (1992)
14. Matoušek, J.: Reporting points in halfspaces. Comput. Geom. **2**(3), 169–186 (1992)
15. Pătraşcu, M., Roditty, L.: Distance oracles beyond the Thorup-Zwick bound. SIAM J. Comput. **43**(1), 300–311 (2014)
16. Rahul, S., Janardan, R.: Algorithms for range-skyline queries. In: Proceedings of ACM Symposium on Advances in Geographic Information Systems (GIS), pp. 526–529 (2012)
17. Shan, J., Zhang, D., Salzberg, B.: On spatial-range closest-pair query. In: Hadzilacos, T., Manolopoulos, Y., Roddick, J., Theodoridis, Y. (eds.) SSTD 2003. LNCS, vol. 2750, pp. 252–269. Springer, Heidelberg (2003). https://doi.org/10.1007/978-3-540-45072-6_15
18. Sharathkumar, R., Gupta, P.: Range-aggregate proximity queries. Technical report TR/2007/80, IIIT Hyderabad, Telangana (2007)
19. Smid, M.: Closest point problems in computational geometry. In: Sack, J.-R., Urrutia, J. (eds.) Handbook of Computational Geometry, pp. 877–935. Elsevier Science, Amsterdam (1999)
20. Xue, J.: Colored range closest-pair problem under general distance functions. In: Proceedings of the Annual ACM-SIAM Symposium on Discrete Algorithms (SODA), pp. 373–390 (2019)
21. Xue, J., Li, Y., Janardan, R.: Approximate range closest-pair search. In: Proceedings of the Canadian Conference on Computational Geometry (CCCG), pp. 282–287 (2018)
22. Xue, J., Li, Y., Rahul, S., Janardan, R.: New bounds for range closest-pair problems. In: Proceedings of Symposium on Computational Geometry (SoCG), pp. 73:1–73:14. Schloss Dagstuhl-Leibniz-Zentrum fur Informatik GmbH, Dagstuhl Publishing (2018)
23. Xue, J., Li, Y., Rahul, S., Janardan, R.: Searching for the closest-pair in a query translate. CoRR arXiv:1807.09498 (2018)

Orthogonal Range Reporting and Rectangle Stabbing for Fat Rectangles

Timothy M. Chan[1], Yakov Nekrich[2(✉)], and Michiel Smid[3]

[1] Department of Computer Science, University of Illinois at Urbana-Champaign, Champaign, USA
`tmc@illinois.edu`
[2] Cheriton School of Computer Science, University of Waterloo, Waterloo, Canada
`yakov.nekrich@googlemail.com`
[3] School of Computer Science, Carleton University, Ottawa, Canada
`michiel@scs.carleton.ca`

Abstract. In this paper we study two geometric data structure problems in the special case when input objects or queries are fat rectangles. We show that in this case a significant improvement compared to the general case can be achieved.

We describe data structures that answer two- and three-dimensional orthogonal range reporting queries in the case when the query range is a *fat* rectangle. Our two-dimensional data structure uses $O(n)$ words and supports queries in $O(\log \log U + k)$ time, where n is the number of points in the data structure, U is the size of the universe and k is the number of points in the query range. Our three-dimensional data structure needs $O(n \log^\varepsilon U)$ words of space and answers queries in $O(\log \log U + k)$ time. We also consider the rectangle stabbing problem on a set of three-dimensional fat rectangles. Our data structure uses $O(n)$ space and answers stabbing queries in $O(\log U \log \log U + k)$ time.

1 Introduction

Orthogonal range reporting and rectangle stabbing are two fundamental problems in computational geometry. In the orthogonal range reporting problem we keep a set of points in a data structure; for any axis-parallel query rectangle Q we must report all points in Q. Rectangle stabbing is, in a sense, a dual problem. We keep a set of axis-parallel rectangles in a data structure. For a query point q we must report all rectangles that are stabbed by q, i.e., all rectangles that contain q. A rectangle is fat if its aspect ratio (the ratio of its longest and shortest edges) is bounded by a constant. In this paper we consider the range reporting problem in scenario when query rectangles are fat. We show that significant improvements can be achieved for this special case. We also describe a data structure that supports three-dimensional stabbing queries on a set of fat three-dimensional rectangles.

The range reporting problem and its variants have been studied extensively over the last four decades; see for example, [1–3,6,7,9,10,14,16,22,23,26]. We

© Springer Nature Switzerland AG 2019
Z. Friggstad et al. (Eds.): WADS 2019, LNCS 11646, pp. 283–295, 2019.
https://doi.org/10.1007/978-3-030-24766-9_21

refer to [18,24] for extensive surveys of previous results. The best known data structure for two-dimensional point reporting uses $O(n \log^{\varepsilon} n)$ words of space and supports queries in $O(\log \log U + k)$ time [7]. Henceforth n is the total number of geometric objects (points or rectangles) in the data structure, k is the number of reported objects, and ε is an arbitrarily small positive constant; we assume that all point coordinates are positive integers bounded by a parameter U. The space usage can be reduced to linear or almost-linear at the cost of paying a non-constant penalty for every reported point. Thus there is an $O(n)$-word data structure that supports queries in $O(\log \log U + (k + 1) \log^{\varepsilon} n)$ time and $O(n \log \log n)$-word data structure that answers queries in $O(\log \log U + k \log \log n)$ time. If we want to use linear space and spend constant time for every reported point, then the overall query cost is increased to polynomial: the fastest linear-space data structure requires $O(n^{\varepsilon} + k)$ time to answer a query [5]. Better results are known only in the special case when the query range is bounded on three sides [2,20]; there is a linear-space data structure that answers three-sided queries in $O(\log \log U + k)$ time (or even in $O(1 + k)$ time if $U = O(n)$) [2].

In this paper we show that two-dimensional orthogonal range reporting queries can be answered in $O(\log \log U + k)$ time using an $O(n)$-space data structure under assumption that query rectangles are fat. We also demonstrate that the fatness assumption is profitable for three-dimensional orthogonal range reporting. We show in this paper how to report all points in a three-dimensional axis-parallel fat rectangle in $O(\log \log U + k)$ time using a $O(n \log^{\varepsilon} U)$-word data structure. This is comparable to the current best results of Chan et al. [7] for general two-dimensional orthogonal range searching and three-dimensional 4-sided orthogonal range searching, which had the same $O(\log \log U + k)$ query time with $O(n \log^{1+\varepsilon} n)$ words of space. In fact, we observe (see the remark after Theorem 2) that the latter problem reduces to three-dimensional (6-sided) fat rectangles, so our result for three-dimensional fat rectangles cannot be improved unless there is a breakthrough for the latter problem. The third problem considered in this paper is the three-dimensional stabbing problem on a set of fat rectangles. For a query point q we must report all rectangles that are stabbed by q. We describe a data structure that uses $O(n)$ words of space and supports queries in $O(\log U \log \log U + k)$ time. For comparison, the best known data structures for general rectangles use $O(n \log^* n)$ words of space and support queries in $O(\log^2 n)$ time [25] (Table 1).

Our data structure for two-dimensional range reporting, described in Sects. 2 and 3, is based on quadtrees. Using a marking scheme on nodes of a quadtree, we divide the plane into $O(n/d)$ canonical rectangles, so that each rectangle contains $O(d)$ points for $d = \log n$. For any fat query rectangle Q, we can quickly find all canonical rectangles R satisfying $Q \cap R \neq \emptyset$ and report all points in $Q \cap R$ for all such R. In Sect. 4 we describe a data structure that supports three-dimensional range reporting for fat query ranges. It is based on an interesting new variant of the recursive grid approaches by Alstrup, Brodal, and Rauhe [2] (also adapted by Chan, Larsen, and Patrascu [7] and Karpinski and Nekrich [16]); these previous approaches use nonuniform grid cells, but in our scheme, grid cells are cubes.

Table 1. Space-time trade-offs for two-dimensional range reporting. Result in line 7 is a corollary from [11], but it is not stated there.

Reference	Space	Time	Query type
[7]	$O(n)$	$O(\log \log U + (k+1) \log^\varepsilon n)$	General
[7]	$O(n \log \log n)$	$O(\log \log U + k \log \log n)$	General
[7]	$O(n \log^\varepsilon n)$	$O(\log \log U + k)$	General
[5]	$O(n)$	$O(n^\varepsilon + k)$	General
[20]	$O(n$	$O(\log n + k)$	Three-sided
[2]	$O(n$	$O(\log \log U + k)$	Three-sided
[11]	$O(n)$	$O(\log n + k)$	Fat
New	$O(n)$	$O(\log \log U + k)$	Fat

Our data structure employs a "lopsided" van Emde Boas recursion, in which each node of the tree stores a known data structure for 5-sided range searching [7]. We will describe a data structure for stabbing queries on a set of three-dimensional fat rectangles in the full version [8]. This result is based on reducing a stabbing query to $O(\log U)$ three-dimensional dominance queries. The results of this paper are valid in the word RAM model of computation.

Related Work. A result about range reporting in two-dimensional fat rectangles is implicitly contained in the paper of Chazelle and Edelsbrunner [11]. In [11] the authors describe a linear-space data structure for triangular range reporting. Their data structure can report all points in an arbitrary query triangle, provided that the sides of the triangle parallel to three fixed directions; queries are supported in $O(\log n + k)$ time where k is the number of reported points. We can represent a square as a union of two such triangles and we can represent an arbitrary fat rectangle as a union of $O(1)$ squares. Hence we can answer two-dimensional range reporting queries for fat rectangles in $O(\log n + k)$ time and $O(n)$ space.

Data structures for fat convex objects are studied in e.g., [4,12,13,17]. Iacono and Langerman [15] describe a data structure that supports point location queries in a set of axis-parallel fat d-dimensional rectangles. This data structure answers queries in $O(\log \log U)$ time and uses $O(n \log \log U)$ space for any fixed dimension d.

Another related problem is k-nearest-neighbor search under the L_∞ metric. The decision version of the problem, reporting all k points of L_∞ distance smaller than a given query radius from a query point, is equivalent to range reporting for a hypercube, which is fat. We are not aware of any previous sublogarithmic results for the exact decision problem in three dimensions.

2   Quadtree-Based Rectangular Subdivision

In this section we describe a planar rectangular subdivision that is used by our two-dimensional data structure. To make the description self-contained, we start with the definition of a compressed quadtree.

A quadtree T_Q is a hierarchical data structure that divides the plane into regions. Let U denote the maximum of x- and y-coordinates of all points. We associate a square (also called a cell) $\mathtt{square}(v)$ to every quadtree node v. The root of a quadtree is associated to the square $[0, U] \times [0, U]$. W. l. o. g. we assume that U is a power of 2. If a square $\mathtt{square}(v)$ of a node v contains more than one point, then the node v has four children. We divide $\mathtt{square}(v)$ into four squares of equal size and associate them to four child nodes of v. A compressed quadtree T is a subtree of T_Q obtained by keeping only those internal nodes of T_Q that have more than one non-empty child (Fig. 1).

Marking Nodes in a Quadtree. Let T denote a compressed quadtree on a set of n points. Let $d = \log n$. We mark selected nodes in T by employing the following marking scheme: (i) every d-th leaf is marked and (ii) if an internal node u has at least two children with marked descendants, then u is marked. We can mark nodes of a given quadtree T in linear time using the following method. We will say that a node u is a special node if exactly one child of u has marked descendants. First we traverse the leaves of T in the left-to-right order and mark every d-th leaf, starting with the leftmost one. Then we visit all internal nodes of T in post-order. If a visited node u has exactly one child u_i such that u_i is either marked or special, then we declare that the node u is special. If u has two or more children that are either special or marked, then the node u is marked. Marked nodes induce a subtree T' of T. T' has n/d leaves. Since every internal node of T' has at least two children, T' has at most $n/d - 1$ internal nodes. Hence the total number of marked nodes is $O(n/d)$. Similar methods for selecting nodes were previously used in other tree-based data structures, see e.g., [19,21].

Rectangular Subdivision. When nodes are marked, we traverse T from the top to the bottom and divide it into $O(n/d)$ rectangles so that each rectangle contains $O(d)$ points. The subdivision is produced as follows. A direct marked descendant of a node u is a descendant u' of u such that u' is marked and there are no marked nodes between u and u'. Suppose that a node u is a marked node and let $u_1, \ldots, u_f$ denote its direct marked descendants. A marked node has at most 4 direct marked descendants, therefore $f \leq 4$. Let $\mathtt{square}(u)$ denote the cell of a node u. We can represent $\mathtt{square}(u) \setminus (\cup_{i=1}^{f} \mathtt{square}(u_i))$ as a union of a constant number of rectangles $R_j(u)$. We will say that $R_j(u)$ are rectangles associated to the node u. See Fig. 2. There are $O(n/d)$ marked nodes in the quadtree. Our subdivision consists of rectangles $R_i(u)$ for all marked internal nodes of T and cells $\mathtt{square}(v)$ for all marked leaves v of T. By dividing every marked node with marked descendants into rectangles as described above, we obtain a sub-division of the plane into $O(n/d)$ rectangles. Rectangles of this subdivision will be further called *canonical rectangles*.

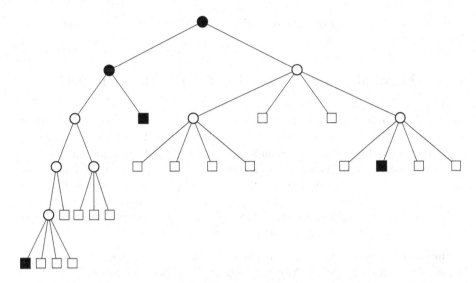

Fig. 1. Marking nodes in a compressed quadtree for $d = 8$. Leaves are shown with squares and internal nodes are shown with circles. Marked leaves and internal nodes are depicted with filled circles and filled squares respectively. Only a part of the quadtree is shown.

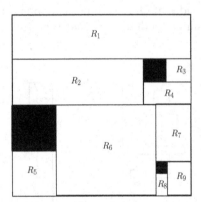

Fig. 2. Subdivision of a marked cell into rectangles. Cells corresponding to direct marked descendants are shown in black.

Lemma 1. *Every canonical rectangle contains $O(d)$ points.*

Proof: Consider a rectangle $R(u)$ associated to a node u. Let $u_1, \ldots, u_f$ denote the direct marked descendants of u. We can show that the set $P_0 = \texttt{square}(u) \setminus (\cup_{i=1}^{f} \texttt{square}(u_i))$ contains $O(d)$ points. Let L_0 denote the set of leaves in which points from P_0 are stored. There are at most d leaf nodes from L_0 between u_i and u_{i+1} for $1 \le i < f$; there are at most d leaf descendants of u to the left of u_1 and at most d leaf descendants of u to the right of u_f. Hence the total number of leaves in L_0 does not exceed $(f + 2)d$. Since $R(u) \subseteq L_0$ and $f \leq 4$, $R(u)$ contains $O(d)$ points.   □

3   Orthogonal Range Reporting for Fat Boxes in 2-D

Data Structure. We divide the plane into canonical rectangles as described in Sect. 2. For every rectangle R in this subdivision we keep the list $L_x(R)$ of points in R sorted by their x-coordinates and the list $L_y(R)$ of points in R sorted by their y-coordinates. We also keep a data structure $D(R)$ that supports two-dimensional range reporting queries on points of R. Since R contains $O(\log n)$ points, we can implement $D(R)$ in $O(\log n)$ space so that queries are supported in $O(k)$ time. The data structure $D(R)$ will be described in Sect. 5. We will denote by P the set of points stored in our data structure.

Orthogonal Range Queries. For simplicity we will consider the case when the query range is a square. Any fat rectangle can be represented as a union of $O(1)$ squares. Consider a query $Q = [a, b] \times [c, d]$. All canonical rectangles that intersect Q can be divided into three categories: (i) corner rectangles that contain a corner of Q (ii) rectangles that cut one side of Q or are completely contained in Q; such rectangles will be called internal rectangles (iii) rectangles that cross two opposite sides of Q, but do not contain corners of Q; we say that such rectangles are spanning rectangles or that type (iv) rectangles span Q. See Fig. 3.

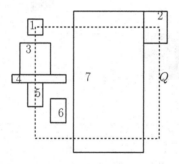

Fig. 3. Examples of different rectangles with respect to a query Q. Rectangles 1 and 2 are corner rectangles, rectangles 3, 4, 5, 6 are internal rectangles, and rectangle 7 spans Q.

Lemma 2. *If a rectangle* $Q = [a, b] \times [c, d]$ *is a square, then* Q *is spanned by* $O(1)$ *canonical rectangles.*

Proof: Suppose that a canonical rectangle $R(u)$, associated to a node u, spans Q. Then either (i) square(u) contains two corners of Q and Q is not contained in $\mathtt{square}(u')$ for any descendant u' of u, or (ii) $\mathtt{square}(u)$ contains Q but Q is not contained in $\mathtt{square}(u')$ for any descendant u' of u.

If Q is contained in $\mathtt{square}(u)$ and at least one rectangle $R(u)$ spans Q, then Q is not contained in $\mathtt{square}(u')$ for any descendant u' of u. Hence there is at most one cell that satisfies condition (i).

Suppose that $\mathtt{square}(u)$ contains two corners of Q and some rectangle $R(u)$ spans Q. Let us assume w.l.o.g. that Q crosses the left side ℓ of $\mathtt{square}(u)$. Let u' be some descendant of u. If $\mathtt{square}(u')$ for a descendant u' of u does not touch the left side of $\mathtt{square}(u)$, then the distance from ℓ to $\mathtt{square}(u')$ is greater than or equal to the size of $\mathtt{square}(u')$. Hence $\mathtt{square}(u')$ does not contain two corners of Q. If $\mathtt{square}(u')$ touches ℓ and contains two corners of Q, then there is no canonical rectangle $R(u)$ that spans Q. Thus there is at most one cell that satisfies condition (ii).

Since there is only one cell satisfying condition (i) and only one cell satisfying condition (ii), the total number of canonical rectangles that span Q is bounded by a constant. □

A query range can overlap with a large number of internal rectangles. But we can find all internal rectangles R, such that $R \cap Q \cap P \neq \emptyset$ by answering a range reporting query on a set P' (defined below) that contains $O(n/d)$ representative points for $d = \log n$. It was shown in Lemma 2 that a square is intersected by $O(1)$ spanning rectangles. There are at most four corner rectangles for any query range Q. Since there is a constant number of corner rectangles and spanning rectangles, we can process all of them in constant time. A more detailed description follows.

We can identify all internal rectangles (type (ii) rectangles) that contain at least one point from $P \cap Q$ as follows. For every canonical rectangle, we keep its topmost point, its lowermost point, its leftmost point, and its rightmost point in the set P'. P' contains $O(n/d)$ points. We keep P' in the data structure D' that supports orthogonal range reporting queries in $O(\log \log U + k)$ time [7]. D' uses space $O(n' \log^{\varepsilon} n')$, where n' is the number of points in P'. Since $n' = O(n/d)$, D' uses space $O(n)$. If $rect(u)$ is an internal rectangle and $rect(u) \cap Q \cap P \neq \emptyset$, then at least one of its extreme points is in Q. We can find all such rectangles by answering the same query Q on the set P'. For every reported point p, we examine the canonical rectangle R_p that contains p.

There are at most four corner rectangles. We can find corner rectangles by keeping all canonical rectangles in the point location data structure of Chan [6]. For each corner point q of Q, we identify the rectangle R_q that contains q in $O(\log \log U)$ time.

Rectangles that span Q are the most difficult to deal with. All points of a spanning rectangle can be outside of Q. It is not clear how we can find spanning rectangles R such that $R \cap Q \cap P \neq \emptyset$. Existence of these rectangles is the reason why our method cannot be extended to the general case of the orthogonal range reporting. However, by Lemma 2, a square query range Q is spanned by $O(1)$ canonical rectangles from our subdivision. All rectangles that span Q can be found as follows. For a rectangle R we denote by $\mathtt{left}(R)$, $\mathtt{right}(R)$, $\mathtt{bot}(R)$, and $\mathtt{top}(R)$ the lower and upper bounds of its horizontal and vertical projections; that is, $R = [\texttt{left}(R), \texttt{right}(R)] \times [\texttt{bot}(R), \texttt{top}(R)]$. If a rectangle R spans Q, then at least one side of R spans Q. That is, R satisfies one of the following conditions: (i) $\texttt{left}(R) \leq a$, $\texttt{right}(R) \geq b$, and $c \leq \texttt{top}(R) \leq d$; (ii) $\texttt{left}(R) \leq a$, $\texttt{right}(R) \geq b$, and $c \leq \texttt{bot}(R) \leq d$; (iii) $a \leq \texttt{left}(R) \leq b$, $\texttt{bot}(R) \leq c$, and $\texttt{top}(R) \geq d$; (iv) $a \leq \texttt{right}(R) \leq b$, $\texttt{bot}(R) \leq c$, and $\texttt{top}(R) \geq d$. We keep information about every rectangle in four three-dimensional data structures. The data structure $\mathcal{R}_1$ contains a tuple $(\texttt{left}(R), \texttt{right}(R), \texttt{top}(R))$ for every canonical rectangle R. $\mathcal{R}_1$ can find all R that satisfy $\texttt{left}(R) \leq a$, $\texttt{right}(R) \geq b$, and $c \leq \texttt{top}(R) \leq d$. The data structure $\mathcal{R}_2$ contains a tuple $(\texttt{left}(R), \texttt{right}(R), \texttt{bot}(R))$ for every canonical rectangle R. $\mathcal{R}_2$ can find all R that satisfy $\texttt{left}(R) \leq a$, $\texttt{right}(R) \geq b$, and $c \leq \texttt{bot}(R) \leq d$. Data structures $\mathcal{R}_3$ and $\mathcal{R}_4$ contain tuples $(\texttt{left}(R), \texttt{bot}(R), \texttt{top}(R))$ and $(\texttt{right}(R), \texttt{bot}(R), \texttt{top}(R))$ respectively for every canonical rectangle R. $\mathcal{R}_3$ supports queries $a \leq \texttt{left}(R) \leq b$, $\texttt{bot}(R) \leq c$, and $\texttt{top}(R) \geq d$; $\mathcal{R}_4$ supports queries $a \leq \texttt{right}(R) \leq b$, $\texttt{bot}(R) \leq c$, and $\texttt{top}(R) \geq d$. Queries supported by data structures $\mathcal{R}_i$ are a special case of three-dimensional orthogonal range reporting queries, called 4-sided queries (the query range is bounded on four sides). Using the result of Chan et al. [7], we can answer such queries in $O(\log \log U + k)$ time using $O(n' \log^\varepsilon n)$ space where $n' = O(n/d)$ is the number of tuples in $\mathcal{R}_i$. If a rectangle R is returned by a query to $\mathcal{R}_i$, then R spans Q or R contains two corners of Q. If Q is a square, then we can answer all queries on $\mathcal{R}_i$ described above and identify all canonical rectangles that span Q in $O(\log \log U + f) = O(\log \log U)$ time, where $f = O(1)$ is the number of canonical rectangles that span Q.

For every corner or spanning rectangle R, we find all points in $R \cap Q$ using the data structure $D(R)$. Since the total number of corner and spanning rectangles is bounded by $O(1)$, we can find all relevant points in $O(k)$ time. Using data structure D' we can find all internal rectangles in $O(\log \log U + n_I)$ time where n_I is the number of internal rectangles. For every internal rectangle R_I we traverse the list of points in $L_x(R_I)$ or $L_y(R_I)$ and report all points in $R_I \cap Q$ in time $O(k_I)$ where $k_I = |R_I \cap Q|$. The result of this section can be summed up as follows.

Theorem 1. *There is a linear-space data structure that answers orthogonal range reporting queries in $O(\log \log U + k)$ time provided the query range $Q = [a, b] \times [c, d]$ is a fat rectangle.*

4 Orthogonal Range Reporting for Fat Boxes in 3-D

In this section, we describe a data structure for 3-d orthogonal range reporting for fat query boxes, by adopting a recursive grid approach. Nonuniform grids have been used in previous range searching data structures by Alstrup, Brodal, and Rauhe [2] and Chan, Larsen, and Patrascu [7], but we use uniform grids instead. Also, the way we use recursion is a little different, and more closely resembles the recursion from van Emde Boas trees. Each node in our recursive structure is augmented with a general 5-sided range reporting structure; thus, our solution can be viewed as a reduction from fat 6-sided range searching to 5-sided range searching.

The Data Structure. Let P be a given set of n points in $[U]^3$, where $[U]$ denotes $\{0, 1, \ldots, U - 1\}$. Let r be a parameter (a function of U) to be chosen later. Divide $[U]^3$ into r^3 *grid cells*, each a cube of the form $\{(x, y, z) : (U/r)i \leq x < (U/r)(i + 1), (U/r)j \leq y < (U/r)(j + 1), (U/r)k \leq z < (U/r)(k + 1)\}$ for some $i, j, k \in [r]$. We call (i, j, k) the label of such a grid cell. A *grid slab* refers to a region of the form $\{(x, y, z) : (U/r)i \leq x < (U/r)(i + 1)\}$, $\{(x, y, z) : (U/r)j \leq y < (U/r)(j + 1)\}$, or $\{(x, y, z) : (U/r)k \leq z < (U/r)(k + 1)\}$. A *grid-aligned box* refers to a box whose x-, y-, and z-coordinates are all multiples of U/r. We construct our data structure as follows:

A. For each nonempty grid cell γ, recursively build a data structure for $P \cap \gamma$; also store $P \cap \gamma$ as a linked list.
B. Let Γ be the set of all nonempty grid cells. Recursively build a data structure for the labels of Γ.
C. For each grid slab σ, build Chan, Larsen, and Patrascu's data structure [7] for $P \cap \sigma$ for 3-d 5-sided queries, which requires $O(n \log^\varepsilon n)$ words of space and $O(\log \log U)$ query time[1].

Analysis of Space. Since we use a uniform grid, we will represent the space usage and query time as functions of the universe size U. Let $s(U)$ be the *amortized* space complexity of our data structure in bits, i.e., the total space complexity in bits divided by the number of points n. Item A of the data structure requires at most $s(U/r)$ bits per point, since after translation, each grid cell becomes $[U/r]^3$. This ignores the space for the linked lists, which require a total of $O(n \log U)$ bits. Item B requires at most $s(r)$ bits per point, since the labels lie in $[r]^3$. Item C requires a total of $O(n \log^\varepsilon n \log U) \leq O(n \log^{1+\varepsilon} U)$ bits (since $n \leq U^3$). Thus,

$$s(U) \leq s(U/r) + s(r) + O(\log^{1+\varepsilon} U).$$

Query Algorithm. We consider the case when the query range is an (axis-parallel) cube; any fat query box can be expressed as a union of $O(1)$ cubes. Given a query cube Q, we report all points of P in Q as follows:

1. If Q is completely contained in a grid cell γ, then recursively report all points of $P \cap \gamma$ in Q. Otherwise:
2. Decompose Q into (at most) one grid-aligned cube Q' and (at most) six other boxes $Q_1, \ldots, Q_6$, where each Q_i is a 5-sided box in a grid slab σ_i. (See Fig. 4 for an analogous 2-d depiction).
3. Recursively report all grid cells of Γ in Q'. For each reported grid cell $\gamma \in \Gamma$, report all points in the linked list $P \cap \gamma$.
4. For each $i \in \{1, \ldots, 6\}$, report all points of $P \cap \sigma_i$ in Q_i.

---

[1] For simplicity, we ignore the time needed to output points in this section.

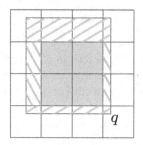   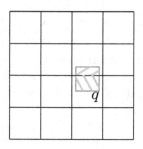

Fig. 4. In 2-d, if a query square Q is not completely contained in any grid cell, it can be decomposed into one grid-aligned square and four 3-sided rectangles in four grid slabs (shown in the left), or just two 3-sided rectangles (shown in the right). Similarly, in 3-d, if a query cube Q is not completely contained in any grid cell, it can be decomposed into one grid-aligned cube and six 5-sided rectangles in six grid slabs, or just two 5-sided rectangles.

Analysis of Query Time. Let $t(U)$ denote the running time of the query algorithm, excluding the outputting cost (which is $O(k)$ for k output points). Step 1 takes $t(U/r)$ time. The recursive call in step 3 takes $t(r)$ time. Step 2 takes $O(\log \log U)$ time. Thus,

$$t(U) \leq \max\{t(U/r) + O(1),\ t(r) + O(\log\log U)\}.$$

We can eliminate the $O(1)$ in the first term of the max by the following idea: Consider the tree formed by expanding the recursion due to item A (treating the recursive structures from item B as secondary structures at the nodes of the main tree). Then we can jump to the first node of the tree at which Q is not completely contained in a grid cell, in $O(1)$ time by an LCA operation in the tree (actually, because the tree is perfectly balanced, for our choice of r (see below), the LCA operation can be simulated by standard arithmetic and bitwise-logical operations on the coordinates of Q).

Conclusion. Setting $r = U^{1/b}$ for a fixed parameter b gives

$$s(U) \leq s(U^{1-1/b}) + s(U^{1/b}) + O(\log^{1+\varepsilon} U)$$
$$t(U) \leq \max\left\{t(U^{1-1/b}),\ t(U^{1/b}) + O(\log\log U)\right\},$$

which solves to $s(U) = O(b \log^{1+\varepsilon} U)$ and $t(U) = O(\log_b \log U \cdot \log \log U)$. The outputting cost goes up to $O(k \log_b \log U)$, since each point may be reported $O(\log_b \log U)$ times.

Setting $b = \log^\varepsilon U$ gives $O(\log^{1+2\varepsilon} U)$ amortized space in bits and $O(\log \log U + k)$ query time. Readjusting ε by a half, we conclude:

Theorem 2. *We can store n points in $[U]^3$ in a data structure with $O(n \log^\varepsilon U)$ words of space so that we can report all k points in any query fat box in $O(\log \log U + k)$ time.*

Remark. The above theorem can't be improved with current state of the art, because 3-d 4-sided range reporting reduces to our problem and the current best data structure for the former requires $O(n \log^\varepsilon n)$ space and $O(\log \log U + k)$ query time. To see the reduction, note that a 4-sided box $[x_1, x_2] \times (-\infty, y) \times (-\infty, z)$ contains the same points as the cube $[x_1, x_2] \times [y - (x_2 - x_1), y) \times [z - (x_2 - x_1), z)$, assuming that $x_2 - x_1 > U$. The assumption can be guaranteed after stretching the x-axis by a factor of U (so that the universe is now $[U^2] \times [U] \times [U]$).

5   Orthogonal Range Reporting on a Small Set of Points

In this section we show how two-dimensional orthogonal range reporting on a set of $d = O(\log n)$ points can be supported in $O(k)$ time. Our data structure uses space $O(d)$, but needs an additional universal look-up table of size $o(n)$. That is, we can keep many instances of our data structure for different point sets and all instances can use the same look-up table.

Lemma 3. *If a set P contains $d = O(\log n)$ points, then we can keep P in a linear-space data structure $D(P)$ that answers two-dimensional range reporting queries in $O(k)$ time. This data structure relies on a universal look-up table of size $o(n)$.*

Proof: First we observe that we can answer a query on a set P' that contains at most $d' = (1/4) \log n / \log \log n$ points using a look-up table of size $o(n)$. Suppose that all points in P' have positive integer coordinates bounded by d'. There are $2^{d' \log d'}$ combinatorially different sets P'. For every instance of P', we can ask $(d')^4$ different queries and the answer to each query consists of $O(d')$ points. Hence the total space needed to keep answers to all possible queries on all instances of P' is $O(2^{(\log d')d'}(d')^5) = o(n)$ points. The general case (when point coordinates are arbitrary integers) can be reduced to the case when point coordinates are bounded by d' using reduction to rank space [2,14].

A query on P can be reduced to $O(1)$ queries on sets that contain $O(d')$ points using the grid approach [2,7]. The set of points P is divided into $4 \log d$ columns C_i and $4 \log d$ rows R_j so that every row and every column contains $(1/4)d / \log d$ points. Hence we can support range reporting queries on points in a row/column using the look-up table approach described above. The top set P_t contains a meta-point (i, j) iff the intersection of the i-th column and the j-th row is not empty, $R_j \cap C_i \neq \emptyset$. Since P_t contains $O(\log^2 d) = o(d')$ points, we can also support queries on P_t in $O(k)$ time. For each meta-point (i, j) in P_t we store the list of points L_{ij} contained in the intersection of the i-th column and the j-th row, $L_{ij} = C_i \cap R_j \cap P$.

Consider a query $Q = [a, b] \times [c, d]$. If Q is contained in one column or one row, we answer the query using the data structure for that column/row. Otherwise we identify the rows R_l and R_t that contain c and d respectively (i.e., the line $y = c$ is contained in R_b and the line $y = d$ is contained in R_t). We also identify the columns C_f and C_r containing a and b. We report all points in $Q \cap C_l$, $Q \cap C_r$, $Q \cap R_b$ and $Q \cap R_t$. We find all meta-points (i, j) in P_t such that $f < i < r$ and $l < j < t$; for every found (i, j) we report all points in L_{ij}.   □

References

1. Afshani, P.: On dominance reporting in 3D. In: Halperin, D., Mehlhorn, K. (eds.) ESA 2008. LNCS, vol. 5193, pp. 41–51. Springer, Heidelberg (2008). https://doi.org/10.1007/978-3-540-87744-8_4
2. Alstrup, S., Brodal, G.S., Rauhe, T.: New data structures for orthogonal range searching. In: Proceedings of 41st Annual Symposium on Foundations of Computer Science, (FOCS 2000), pp. 198–207 (2000)
3. Arge, L., Samoladas, V., Vitter, J.S.: On two-dimensional indexability and optimal range search indexing. In: Proceedings of 18th ACM SIGACT-SIGMOD-SIGART Symposium on Principles of Database Systems (PODS), pp. 346–357 (1999)
4. Aronov, B., de Berg, M., Gray, C.: Ray shooting and intersection searching amidst fat convex polyhedra in 3-space. Comput. Geom. $41(1–2)$, 68–76 (2008)
5. Bentley, J.L., Maurer, H.A.: Efficient worst-case data structures for range searching. Acta Inf. 13, 155–168 (1980)
6. Chan, T.M.: Persistent predecessor search and orthogonal point location on the word RAM. ACM Trans. Algorithms $9(3)$, 22 (2013)
7. Chan, T.M., Larsen, K.G., Patrascu, M.: Orthogonal range searching on the RAM. In: Proceedings of 27th ACM Symposium on Computational Geometry (SoCG 2011), pp. 1–10 (2011)
8. Chan, T.M., Nekrich, Y., Smid, M.: Orthogonal range reporting and rectangle stabbing for fat rectangles. CoRR, abs/1905.02322 (2019)
9. Chazelle, B.: Filtering search: a new approach to query-answering. SIAM J. Comput. $15(3)$, 703–724 (1986)
10. Chazelle, B.: A functional approach to data structures and its use in multidimensional searching. SIAM J. Comput. $17(3)$, 427–462 (1988)
11. Chazelle, B., Edelsbrunner, H.: Linear space data structures for two types of range search. Discrete Comput. Geom. 2, 113–126 (1987)
12. de Berg, M., Gray, C.: Vertical ray shooting and computing depth orders for fat objects. SIAM J. Comput. $38(1)$, 257–275 (2008)
13. Efrat, A., Katz, M.J., Nielsen, F., Sharir, M.: Dynamic data structures for fat objects and their applications. Comput. Geom. $15(4)$, 215–227 (2000)
14. Gabow, H.N., Bentley, J.L., Tarjan, R.E.: Scaling and related techniques for geometry problems. In: Proceedings of 16th Annual ACM Symposium on Theory of Computing (STOC 1984), pp. 135–143 (1984)
15. Iacono, J., Langerman, S.: Dynamic point location in fat hyperrectangles with integer coordinates. In: Proceedings of 12th Canadian Conference on Computational Geometry (2000)
16. Karpinski, M., Nekrich, Y.: Space efficient multi-dimensional range reporting. In: Ngo, H.Q. (ed.) COCOON 2009. LNCS, vol. 5609, pp. 215–224. Springer, Heidelberg (2009). https://doi.org/10.1007/978-3-642-02882-3_22
17. Katz, M.J.: 3-D vertical ray shooting and 2-D point enclosure, range searching, and arc shooting amidst convex fat objects. Comput. Geom. $8(6)$, 299–316 (1997)
18. Kreveld, M.V., Löffler, M.: Range Searching, pp. 1–7. Springer, Heidelberg (2014). https://doi.org/10.1007/978-1-4939-2864-4
19. Lewenstein, M., Nekrich, Y., Vitter, J.S.: Space-efficient string indexing for wildcard pattern matching. In: Proceedings of 31st International Symposium on Theoretical Aspects of Computer Science (STACS 2014), pp. 506–517 (2014)
20. McCreight, E.M.: Priority search trees. SIAM J. Comput. $14(2)$, 257–276 (1985)

21. Navarro, G., Nekrich, Y.: Top-k document retrieval in optimal time and linear space. In: Proceedings of 23rd Annual ACM-SIAM Symposium on Discrete Algorithms (SODA 2012, pp. 1066–1077 (2012)
22. Nekrich, Y.: A data structure for multi-dimensional range reporting. In: Proceedings of 23rd ACM Symposium on Computational Geometry (SoCG), pp. 344–353 (2007)
23. Nekrich, Y.: Space efficient dynamic orthogonal range reporting. Algorithmica **49**(2), 94–108 (2007)
24. Nekrich, Y.: Orthogonal Range Searching on Discrete Grids, pp. 1–6. Springer, Boston (2008). https://doi.org/10.1007/978-3-642-27848-8
25. Rahul, S.: Improved bounds for orthogonal point enclosure query and point location in orthogonal subdivisions in $\mathbb{R}^3$. In: Proceedings of 26th Annual ACM-SIAM Symposium on Discrete Algorithms (SODA 2015), pp. 200–211 (2015)
26. Vengroff, D.E., Vitter, J.S.: Efficient 3-D range searching in external memory. In: Proceedings of 28th Annual ACM Symposium on the Theory of Computing (STOC 1996), pp. 192–201 (1996)

Kernelization of Graph Hamiltonicity: Proper H-Graphs

Steven Chaplick[1], Fedor V. Fomin[2], Petr A. Golovach[2], Dušan Knop[3,4(✉)], and Peter Zeman[5]

[1] Lehrstuhl für Informatik I, University of Würzburg, Würzburg, Germany
steven.chaplick@uni-wuerzburg.de
[2] Department of Informatics, University of Bergen, Bergen, Norway
{fedor.fomin,petr.golovach}@uib.no
[3] Algorithmics and Computational Complexity,
Faculty IV, TU Berlin, Berlin, Germany
dusan.knop@tu-berlin.de
[4] Department of Theoretical Computer Science, Faculty of Information Technology,
Czech Technical University in Prague, Prague, Czech Republic
[5] Department of Applied Mathematics, Faculty of Mathematics and Physics,
Charles University, Prague, Czech Republic
zeman@kam.mff.cuni.cz

Abstract. We obtain new polynomial kernels and compression algorithms for PATH COVER and CYCLE COVER, the well-known generalizations of the classical HAMILTONIAN PATH and HAMILTONIAN CYCLE problems. Our choice of parameterization is strongly influenced by the work of Biró, Hujter, and Tuza, who in 1992 introduced H-graphs, intersection graphs of connected subgraphs of a subdivision of a fixed (multi) graph H. In this work, we turn to proper H-graphs, where the containment relationship between the representations of the vertices is forbidden. As the treewidth of a graph measures how similar the graph is to a tree, the size of graph H is the parameter measuring the closeness of the graph to a proper interval graph. We prove the following results.

- PATH COVER admits a kernel of size $\mathcal{O}(\|H\|^8)$, that is, we design an algorithm that for an n-vertex graph G and an integer $k \geq 1$, in time polynomial in n and $\|H\|$, outputs a graph G' of size $\mathcal{O}(\|H\|^8)$ and $k' \leq |V(G')|$ such that the vertex set of G is coverable by k vertex-disjoint paths if and only if the vertex set of G' is coverable by k' vertex-disjoint paths.
- CYCLE COVER admits a compression of size $\mathcal{O}(\|H\|^{10})$ into another problem, called PRIZE COLLECTING CYCLE COVER, that is, we design an algorithm that, in time polynomial in n and $\|H\|$, outputs an equivalent instance of PRIZE COLLECTING CYCLE COVER of size $\mathcal{O}(\|H\|^{10})$.

The research was supported by the Research Council of Norway via the projects CLASSIS and MULTIVAL. Steven Chaplick is supported by DFG grant WO 758/11-1, Dušan Knop by grants GAČR 17-20065S and DFG "MaMu", NI 369/19, and Peter Zeman by grants GAČR 19-17314J and GAUK 1334217.

© Springer Nature Switzerland AG 2019
Z. Friggstad et al. (Eds.): WADS 2019, LNCS 11646, pp. 296–310, 2019.
https://doi.org/10.1007/978-3-030-24766-9_22

In all our algorithms we assume that a proper H-decomposition is given as a part of the input.

Keywords: Cycle Cover · Path Cover · Proper H-graphs · Kernelization

1 Introduction

HAMILTONIAN CYCLE is one of the oldest mathematical puzzles, whose study can be traced back to the 9th century, and it is still actively studied. Our results about the HAMILTONIAN CYCLE problem are in the intersection of two research areas: kernelization and algorithms on special graph classes. In both areas HAMILTONIAN CYCLE has been intensively investigated.

Parameterized Algorithms and Kernelization. The most popular generalization of the HAMILTONIAN CYCLE problem studied in parameterized complexity is known under the name LONGEST CYCLE. This problem is to decide whether a graph contains a cycle of length at least k, where k is an integer parameter. LONGEST CYCLE and its close relative LONGEST PATH are important representatives of the so-called family of "non-local" problems and this is why these problems served as testbeds in the development of various fundamental techniques in the area such as color coding [1], algebraic methods [4,28,34], or Cut & Count [15] to name a few. We refer to the book of Cygan et al. [14] for an overview of these techniques. From the perspective of kernelization, the framework developed by Bodlaender et al. [5] excludes the existence of a polynomial kernel (up to some reasonable assumption from complexity theory) for LONGEST CYCLE with the natural parameter k. This lower bound initiated the development of kernelization algorithms for HAMILTONIAN CYCLE with "structural kernelization". Fellows et al. [19] proved that HAMILTONIAN CYCLE parameterized by the *max leaf number* of the input graph G, that is the maximum number of leaves in a spanning tree of G, admits a kernel of polynomial size. A systematic approach in the study of structural kernelization of HAMILTONIAN CYCLE (and other related problems) was taken by Bodlaender et al. [6] who considered kernelization of HAMILTONIAN CYCLE parameterized by the size of the modulator to some nice graph property. More precisely, for a graph G the modulator to a graph property Π is a set of vertices or edges such that after removing this set from graph G, the resulting graph has property Π. In particular, Bodlaender et al. [6] have shown that HAMILTONIAN CYCLE admits a polynomial kernel when parameterized by the size of a minimum vertex cover (a minimum modulator to an independent set) or by the size of a minimum modulator to the cluster graph, that is, the disjoint union of complete graphs. They also provided a number of lower bounds on the structural kernelization of the problem by showing, for example, that the problem does not admit a polynomial kernel when the parameter is the minimum size of a modulator to an outerplanar graph.

Graph Classes. There is a large research area in graph algorithms, where the structural properties of graphs, like being interval or chordal, are exploited for developing of efficient algorithms problems intractable on general graphs. We refer to the books [8,24] for the introduction and survey of the known results. Without a doubt, the oldest and the most studied class of intersection graphs is the class of interval graphs and there is a long history of research on the HAMIL-TONIAN CYCLE and HAMILTONIAN PATH problems on interval, circular-arc and related graph classes. It was shown by Keil [27] in 1985 that HAMILTONIAN CYCLE can be solved in linear time for interval graphs (see also [9,10,16,30]). The problem for circular-arc graphs proved to be much more involved and the first polynomial algorithm for HAMILTONIAN CYCLE on circular-arc graphs was given by Shih et al. [33] in 1992 (see also [25]). On the other hand, for proper interval graphs, it was already shown by Bertossi [2] that every connected proper interval graph has a Hamiltonian path, and a proper interval graph has a Hamiltonian cycle if and only if it is 2-connected graph with at least three vertices (see also [13,26]). This immediately implies a linear-time algorithm for the problem. It follows from the results of Brandstädt et al. [7] that HAMILTONIAN CYCLE can be solved in linear time for circular-arc graphs. Thus, HAMILTONIAN CYCLE can be solved in linear time for (proper) interval and circular graphs. For chordal graphs, HAMILTONIAN CYCLE is well-known to be NP-complete and is even NP-complete for strongly chordal split graphs [31].

Our Results. In this paper we follow the main question of structural kernelization—if a computational problem can be solved in polynomial time on instances with some structural properties, does it admit a polynomial kernel parameterized by some "distance" to this nice structural property? In our setting the nice structural property is to be a proper interval graph. However, the "distance" we use is quite different from the commonly used the size of a modulator.

Our measure of similarity with proper interval graphs is based on the beautiful concept of H-graph introduced by Biró et al. [3] in the context of the precoloring extension problem. An *intersection representation* of a graph G assigns a set S_v to every vertex $v \in V(G)$ such that $S_u \cap S_v \neq \emptyset$ if and only if $uv \in E(G)$. When the sets S_u are intervals of the real line, this defines an interval graph. From a different perspective, every interval graph can be viewed as an intersection graph of subpaths of some (sufficiently long) path. Similarly, *circular-arc graphs*, a natural generalization of interval graphs, are the intersection graphs of subpaths of some cycle. It is also a well-known fact that a graph is *chordal* if and only if it is an intersection graph of subtrees of some tree. A natural generalization of these classes are intersection graphs of subgraphs of some subdivision of an arbitrary *underlying graph* H. For a fixed graph H, we say that a graph G is an H-*graph*, if it is an intersection graph of connected subgraphs of a subdivision of H. In this language, interval graphs are K_2-graphs, circular-arc graphs are K_3-graphs, and every chordal graph is a T-graph for some tree T.

An *intersection representation* $\{S_v\}_{v \in V(G)}$ of a graph G is a *proper* representation, if $S_u \subseteq S_v$ implies $u = v$. Then a graph G is a *proper H-graph*, if it admits a proper intersection representation by connected subgraphs of a subdivision of H. For example, proper K_2-graphs are proper interval graphs, that is the graphs admitting a proper representation by intervals of the real line. Various aspects of proper interval and proper circular-arc representations are well-studied, and our goal is again to study how these carry to general proper H-graphs. Clearly, all positive algorithmic results obtained for H-graphs in [11,12,20] are valid for proper H-graphs, but since we consider a more restricted graph class, we can hope that the tractability area could be expanded.

We consider the following fundamental generalizations of HAMILTONIAN CYCLE and HAMILTONIAN PATH problems.

CYCLE COVER (PATH COVER)

Input: A graph G and a positive integer k.

Task: Decide whether G has a cycle (path) cover $\mathcal{C}$ with at most k cycles (paths).

The main results of this paper are the following theorems about kernelization of CYCLE COVER and PATH COVER. In both theorems we assume that a proper H-representation of input graph G is given.

Theorem 1. PATH COVER *admits a kernel of size $\mathcal{O}(h^8)$, where h is the size of the graph H in a proper H-representation of the input graph G.*

For CYCLE COVER we only construct a polynomial compression of the explicitly stated size. (Roughly speaking, the difference between kernelization and compression is that kernelization algorithm outputs an equivalent instance of the same parameterized problem, while a compression algorithm maps an instance of a parameterized problem to an equivalent instance of another non-parameterized problem. We refer to Sect. 2 with preliminaries, where we define kernelization and compressing algorithms.) Let us note that since we compress into an NP-complete problem, the standard trick involving the Cook-Levin theorem, see e.g. [21], implies the existence of a polynomial in h kernel for CYCLE COVER but we are unable to give the exact size of such a kernel.

Theorem 2. CYCLE COVER *admits a compression of size $\mathcal{O}(h^{10})$, where h is the size of the graph H in a proper H-representation of the input graph G.*

However for the special case of CYCLE COVER with $k = 1$, namely HAMILTONIAN CYCLE, we also are able to obtain a kernel of size $\mathcal{O}(h^8)$.

Note that, the requirement that a proper H-representation is given in the input of the considered problems on proper H-graphs is likely unavoidable. Namely, the hardness result of Chaplick et al. [11] can be adapted to show that the recognition problem for proper H-graphs is NP-hard even for small fixed graphs H.

Organization of the Paper. Section 2 provides the notions used in the paper. Due to space constraints, we are unable to explain all the details in this extended abstract. Instead, in Sect. 3, we give an informal description of our kernelization and compression algorithms.

2   Preliminaries

Graphs. All graphs considered in this paper assumed to be simple, that is, finite undirected graphs without loops or multiple edges, unless it is said explicitly that we consider a multigraph. For each of the graph problems considered in this paper, we let $n = |V(G)|$ and $m = |E(G)|$ denote the number of vertices and edges, respectively, of the input graph G if it does not create confusion; $\|G\| = |E(G)|$ is the *size* of G. A path P (a cycle C) in G is *Hamiltonian* if $V(P) = V(G)$ ($V(C) = V(G)$ respectively). A family of paths $\mathcal{P} = \{P_1, \ldots, P_k\}$ (cycles $\mathcal{C} = \{C_1, \ldots, C_k\}$) is a *path cover* (*cycle cover*) if the paths (cycles) are pairwise disjoint and the union of their vertices is $V(G)$. The *size* of a path or cycle cover is the number of paths or cycles in it. A family $\mathcal{Q} = \{Q_1, \ldots, Q_s\}$ of cliques is said to be a *(vertex) clique cover* if the cliques are pairwise disjoint and $\bigcup_{i=1}^{s} Q_i = V(G)$. Note that we consider only vertex clique covers.

Let $\mathcal{S}$ be a collection of sets. The *intersection graph* G of $\mathcal{S}$ has $\mathcal{S}$ as its vertex set and two distinct vertices $X, Y \in \mathcal{S}$ are adjacent if and only if $X \cap Y \neq \emptyset$. For an intersection graph G, $\mathcal{S}$ is called an *(intersection) model* of G. The intersection graph of a family of intervals of the real line is called an *interval* graph; it is also said that G is an interval graph if there is a family of intervals (called *interval model* or *representation*) such that G is isomorphic to the intersection graph of this family. Throughout the paper we assume that the intervals of an interval model are closed. An interval graph is *proper* if it has an interval model such that no interval is a subinterval of another one.

Let H be a multigraph. We say that H' is obtained from H by the *subdivision of an edge* $e = xy$ if to construct H', we delete e and add a new vertex z along with two new edges zx and zy. Similarly, H' is a *subdivision* of H if H' is obtained from H by iteratively subdividing its edges.

For a multigraph H, a simple graph G is an *H-graph* if G is an intersection graph of connected subgraphs of some subdivision H' of H or, equivalently, G is an intersection graph of connected subsets of vertices of H'. Throughout the paper we only allow the H's in H-graphs to be multigraphs and all other graphs are assumed to be simple. To distinguish the vertices of H and H' from the vertices of G, we refer to the vertices of H and H' as *nodes*. We also say the nodes of H are *branching* nodes of H' and the other nodes are *subdivision* nodes. A pair $(H', \mathcal{M})$, where $\mathcal{M} = \{M_v\}_{v \in V(G)}$ is a collection of connected vertex sets of H' such that G is the intersection graph of $\mathcal{M}$, is called an *H-representation* of G. A representation $(H', \mathcal{M})$ is *proper* if for every two distinct $u, v \in V(G)$, neither $M_u \subseteq M_v$ nor $M_v \subseteq M_u$. In this sense, G is a *proper H-graph* if it has a proper H-representation. It is straightforward to see that every interval graph G is a K_2-graph and every proper interval graph is a proper K_2-graph.

Note that every graph has the following trivial model. For a graph G, let $I(G)$ be the *incidence* graph of G, that is, the result of subdividing each edge of G exactly once.

Observation 1. *Every graph G is a proper G-graph. Its trivial proper G-representation is $(I(G), \{N_{I(G)}[v]\}_{v \in V(G)})$.*

Parameterized Complexity and Kernelization. We refer to the books [14, 17, 21] for a detailed introduction to the field.

Parameterized Complexity is a two dimensional framework for studying the computational complexity of a problem. One dimension is the input size n and the other is a *parameter* k associated with the input. The main goal is to confine the combinatorial explosion in the running time of an algorithm, for an NP-hard problem, to depend only on k. In this sense, a parameterized problem is said to be *fixed parameter tractable* (or FPT) if it can be solved in time $f(k) \cdot n^{\mathcal{O}(1)}$ for some computable function f.

A *compression* of a parameterized problem Π_1 into a (non-parameterized) problem Π_2 is a polynomial algorithm that maps each instance (I, k) of Π_1 with the input I and the parameter k to an instance I' of Π_2 such that (i) (I, k) is a yes-instance of Π_1 if and only if I' is a yes-instance of Π_2, and (ii) $|I'|$ is bounded by $f(k)$ for a computable function f. The output I' is also called a *compression*. The function f is said to be the *size* of the compression. A compression is *polynomial* if f is polynomial. A *kernelization* algorithm for a parameterized problem Π is a polynomial algorithm that maps each instance (I, k) of Π to an instance (I', k') of Π such that (i) (I, k) is a yes-instance of Π if and only if (I', k') is a yes-instance of Π, and (ii) $|I'| + k'$ is bounded by $f(k)$ for a computable function f. Respectively, (I', k') is a *kernel* and f is its *size*. A kernel is *polynomial* if f is polynomial. While it can be shown that every decidable parameterized problem is FPT if and only if it admits a kernel, it is unlikely that every problem in FPT has a polynomial kernel (see, e.g., [14, 21] for the details).

For CYCLE COVER, we show that it admits a polynomial compression into a special problem called PRIZE COLLECTING CYCLE COVER, defined next.

Let G be a graph and let $\omega \colon E(G) \to \mathbb{N}_0$ be a *weight* function; note that we allow zero weights. For a cycle C, $\omega(C)$ is the sum of the weights of its edges. Let $\alpha \colon \mathbb{N} \to \mathbb{N}$ be a non-decreasing *penalty* function. For a cycle cover $\mathcal{C} = \{C_1, \ldots, C_k\}$ of G, the weight of $\mathcal{C}$ is $\omega(\mathcal{C}) = \sum_{i=1}^{k} \omega(C_i)$ and the *cost* of $\mathcal{C}$ is $c_{\alpha, \omega}(\mathcal{C}) = \omega(\mathcal{C}) - \alpha(|\mathcal{C}|)$. Observe that the cost may be negative.

PRIZE COLLECTING CYCLE COVER ———————————————————

Input: A graph G with a weight function $\omega \colon E(G) \to \mathbb{N}_0$, a non-decreasing penalty function $\alpha \colon \{1, \ldots, |V(G)|\} \to \mathbb{N}$, and an integer r.

Task: Decide whether G has a cycle cover $\mathcal{C}$ of cost $c_{\alpha, \omega}(\mathcal{C}) \geq r$.

Notice that if G is a graph with zero edge-weights and the penalty function $\alpha(x) = x$ for $x \in \mathbb{N}$, then G has a cycle cover with at most k cycles if and only if G has a cycle cover of cost at least $r = -k$, that is, PRIZE COLLECTING CYCLE COVER generalizes CYCLE COVER. We prove that CYCLE COVER admits a polynomial compression to PRIZE COLLECTING CYCLE COVER of size $\mathcal{O}(h^{10})$ when parameterized by the size h of H if a proper H-representation is given on the input.

3 Description of Algorithms

In this section, we give an informal high-level description of our kernelization and compression algorithms as to outline how they work. Our first step towards the kernelization of PATH COVER and compression of CYCLE COVER is a kernelization algorithm for CYCLE COVER, PATH COVER, and PRIZE COLLECTING CYCLE COVER being parameterized by the size of a (vertex) clique cover of the input graph. In the second step, we describe how having a proper H-model leads to kernels with small clique covers.

From small Clique Covers to Kernels. These results are of independent interest. The parameterization of HAMILTONIAN CYCLE by the clique cover size was considered by Lampis et al. [29] who proved that the problem is FPT for this parameterization. We extend their result by showing the following theorem.

Theorem 3. CYCLE COVER, PATH COVER, and HAMILTONIAN CYCLE admit kernels of size $\mathcal{O}(s^8)$, where s is the size of a clique cover. PRIZE COLLECTING CYCLE COVER admits a kernel of size $\mathcal{O}((s+\ell)^{10})$, where s is the size of a clique cover and ℓ is the number of edges of the input graph with non-zero weights. In all kernels we assume that a clique cover of size s is given in the input.

We sketch the main ideas of the kernelization for CYCLE COVER, which is the easiest among these problems, and then explain how to modify it for the other problems under consideration.

Recall that a clique cover is a collection $\mathcal{Q} = \{Q_1, \ldots, Q_s\}$ of disjoint cliques such that $V(G) = \bigcup_{i=1}^{s} Q_i$. First we show that there is always an optimal solution to CYCLE COVER with very specific properties. We call a cycle cover *regular* (for the clique cover $\mathcal{Q} = \{Q_1, \ldots, Q_s\}$) if it satisfies the following properties for each $i, j \in \{1, \ldots, s\}, i \neq j$,

(i) at most one cycle of the cover has an edge between cliques Q_i and Q_j,
(ii) every cycle of the cover has at most two edges between Q_i and Q_j.

It is possible to prove that every cycle cover can be transformed into a regular one without increasing its size. Informally, if two distinct cycles have edges between two cliques Q_i and Q_j, we can "glue" them together as it is shown in Fig. 1(left), and if a cycle has at least three edges between the cliques, then we can pick two of them that are in the "same direction" according an arbitrary orientation of the cycle and reroute the cycle, see Fig. 1(right).

Because the cycles of a regular cycle cover have a limited number of edges that are between the cliques of $\mathcal{Q}$, it is possible to modify and/or reroute them using the fact that the vertices of the same clique are pairwise adjacent. The regularity of a cycle cover allows us to apply the following reduction rules.

– If there is a clique $Q_i \in \mathcal{Q}$ and $v \in Q_i$ such that $N_G[v] = Q_i$ and $|Q_i| \geq s+3$, then set $G = G - v$ and $Q_i = Q_i \setminus \{v\}$.

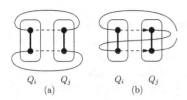

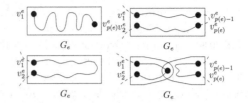

Fig. 1. Rerouting cycles; the deleted edges are shown by dashed lines and the added edges are shown by thick lines.

Fig. 2. The types of covering (up to symmetry) of G_e by a tamed path cover.

- If there are $i, j \in \{1, \ldots, s\}$, $i \neq j$, such that the bipartite graph G_{ij}, with vertex set $Q_i \cup Q_j$ and whose edges are the edges of G between Q_i and Q_j, has a matching M of size at least $4s - 3$, then select (arbitrarily) an edge $e \in M$, set $G = G - e$.
- If there is a clique $Q_i \in \mathcal{Q}$ and $v \in V(G) \setminus Q_i$ such that $|N_G(v) \cap Q_i| \geq 2s + 1$, then for an arbitrary edge $e = uv$ with $u \in Q_i$, set $G = G - e$.

The first item asserts that if a sufficiently large clique of the clique cover has a simplicial vertex, this vertex is irrelevant. Similarly, if there is a large matching between two cliques, then one edge of this matching can be deleted safely. Finally, if there is a vertex outside a clique which is adjacent to many vertices of the clique, any edge between such a vertex and the clique can be removed. We apply the rules exhaustively. We prove that any irreducible instance has $\mathcal{O}(s^4)$ vertices, that is, the size of the obtained instance of CYCLE COVER is $\mathcal{O}(s^8)$ and this implies the claim of Theorem 3 for the problem.

As HAMILTONIAN CYCLE is the special case of CYCLE COVER when $k = 1$ and the reduction rules do not modify k, the kernelization algorithm for HAMILTONIAN CYCLE is the same. For PATH COVER, we need a tiny adjustment to reroute the paths of a path cover in a slightly different way. However, PRIZE COLLECTING CYCLE COVER requires additional work.

Let (G, ω, α, r) be an instance of PRIZE COLLECTING CYCLE COVER and let S be the set of edges of G with non-zero weights, $\ell = |S|$. First, we modify the clique cover $\mathcal{Q}$ of G by making the end-vertices of the edges of S trivial cliques of size one. Thus, we obtain the clique cover $\hat{\mathcal{Q}}$ of size $t \leq s + 2\ell$. Then we observe that the modifications of the cycles of a cycle cover that were used for CYCLE COVER never affect edges of G with both end-vertices in trivial cliques. In particular, if the cycles of a cycle cover contain $e \in S$, one of the cycles of the cycle cover obtained by the reroutings still contains e. Also, we do not increase the number of cycles in cycle covers by such reroutings. This implies that we still can use our reduction rules. It is possible to show that an irreducible instance of PRIZE COLLECTING CYCLE COVER obtained by the exhaustive application of the rules has $\mathcal{O}(t^4)$ vertices.

Note that this is not a polynomial kernel yet because we still have to compress the edge weights as well as the values of the penalty function α and r. For this, we apply the approach proposed by Etscheid et al. [18] for constructing kernels for weighted problems. These techniques are based on the classical algorithm for compressing numbers given by Frank and Tardos [22]. This allows us to encode the value of the weight function for each $e \in S$ and the value of the penalty function for each of the $\mathcal{O}(t^4)$ vertices by a binary string of length $\mathcal{O}((s + \ell)^6)$. Summarizing, we obtain an instance of PRIZE COLLECTING CYCLE COVER of size $\mathcal{O}((s + \ell)^{10})$. This completes the sketch of the kernelization algorithm.

Note that Theorem 3 requires that a clique cover of the input graph is given. This may be unavoidable as it is already NP-complete to decide whether a graph has a clique cover of size 3 [23] (this is the same as 3-COLORING in the graph complement).

From Proper H-Representations to Small Clique Covers. Now we use Theorem 3 to construct kernelization and compression algorithms for PATH COVER and CYCLE COVER on proper H-graphs, i.e., we build kernels with small clique covers.

Suppose that G is a proper H-graph given together with its proper H-representation $(H', \mathcal{M})$. Notice that for every branching node $x \in V(H)$, the set $K_x = \{v \in V(G) \mid x \in M_v\}$ is a clique of G. Observe also that the graph $G - \bigcup_{x \in V(H)} K_x$ can be seen as a union of proper interval graphs G_e corresponding to the edges $e \in E(H)$. More formally, let $e = xy \in E(H)$ and consider the (x, y)-path P_e in H' obtained from e by the subdivisions. We denote by G_e the subgraph of G induced by $V_e = \{v \in V(G) \mid M_v \subseteq V(P_e) \setminus \{x, y\}\}$. Clearly, G_e is a proper interval graph and the sets M_v for $v \in V_e$ form a proper interval representation of it. This representation defines a corresponding total ordering of its vertices (see [32]). We assume that these orderings are fixed for every G_e. In particular, whenever we speak about leftmost and rightmost vertices of G_e, we mean the leftmost and the rightmost vertices with respect to this ordering. Notice that for $e = xy$, $N_G(V_e) \subseteq K_x \cup K_y$, that is, paths or cycles that cover the vertices in G_e are either completely in G_e or enter G_e via the vertices of K_x or K_y that we call the *left* and *right* cliques respectively.

The graphs G_e could be huge but, since they are proper interval graphs, they have a relatively simple structure. We exploit this structure in order to replace them by small gadget graphs while maintaining the equivalence of the instances of the considered problems. Since the vertices of $\bigcup_{x \in V(H)} K_x$ can be covered by at most $|V(H)|$ cliques and the set of vertices of each gadget replacing G_e can be covered by a constant number of cliques, we obtain a graph that has a clique cover of size $\mathcal{O}(|V(H)| + |E(H)|)$.

To formalize the proof idea sketched above, we show that we can assume that the considered H-representation $(H', \mathcal{M})$ of G has no redundancies, that is, for every node $x \in V(H')$, there is a vertex $v \in V(G)$ with $x \in M_v$ and, moreover, for every edge $xy \in E(H')$, there is $v \in V(G)$ with $x, y \in M_v$. We call such a representation *nice*. To achieve this, we first observe that if the input graph G has a component F that is a proper interval graph, we can find the minimum number of paths or cycles that cover F depending on the considered problem, and then delete F and modify the parameter k of PATH COVER or

CYCLE COVER respectively. Somehow surprisingly, to the best of our knowledge CYCLE COVER was not studied on proper interval graph. Therefore, we design a linear time algorithm for the problem. Note that it may happen that we solve the problem by applying the reduction rules. Otherwise, we obtain an induced subgraph G' of G such that every component of G' has a vertex v with M_v containing a branching node of H'. Then we modify H' by removing irrelevant nodes and edges. This procedure can create new nodes of degree one from some subdivision nodes of H' but the number of such vertices is at most $2|E(H)|$. From this, we derive that G' is an $\hat{H}$-graph for some $\hat{H}$ with at most $3|E(H)|$ nodes and at most $2|E(H)|$ edges, and we construct the corresponding nice proper $\hat{H}$-representation.

From now on we can concentrate only on nice representations. In particular, we assume that every graph G_e for $e = xy \in E(H)$ is connected and that the leftmost and the rightmost vertices of G_e have neighbors in the left and the right cliques respectively.

Recall that for PATH COVER, we prove the following theorem.

Theorem 1. PATH COVER *admits a kernel of size* $\mathcal{O}(h^8)$, *where h is the size of the graph H in a proper H-representation of the input graph G.*

Let G be a proper H-graph and $(H', \mathcal{M})$ be a nice proper representation of it. Let $\mathcal{P}$ be a path cover of G. For $e \in E(H)$, let $\mathcal{P}_e$ denote the family of inclusion-maximal subpaths of the paths $P \in \mathcal{P}$ with all their vertices in V_e. We say that $\mathcal{P}_e$ is the *projection* of $\mathcal{P}$ on G_e. Since $\mathcal{P}$ is a path cover of G, $\mathcal{P}_e$ is a path cover of G_e. It is possible to show that if G has a path cover of size at most k, then G has a path cover of size at most k such that the paths in each projection $\mathcal{P}_e$ have a very special structure in the case when the vertices of the graph G_e cannot be covered by two cliques. We call such a cover *tamed* (this is a slightly simplified definition which we use only for the high-level description of the algorithm). One possibility is that $\mathcal{P}_e$ consists of a single Hamiltonian path of G_e with its end-vertices being the leftmost and the rightmost vertices of G_e. In all other cases, $\mathcal{P}_e$ consists of at most two paths that are proper subpaths of some paths of $\mathcal{P}$ and, moreover, every path of $\mathcal{P}_e$ extends in two directions in the path of $\mathcal{P}$. We prove the following properties of $\mathcal{P}_e$.

- If G_e is 2-connected, then
  - either $\mathcal{P}_e$ consists of one Hamiltonian path of G_e such that its end-vertices are the two leftmost vertices of G_e (symmetrically, the two rightmost vertices),
  - or $\mathcal{P}_e$ consists of two paths such that each of them has one of its end-vertices among the two leftmost vertices of G_e and the second end-vertex is among the two rightmost vertices of G_e, and these paths are proper subpaths of the same path P of $\mathcal{P}$ that occur in P in the "opposite directions" for an arbitrary orientation of P.
- If G_e has a cut-vertex, then $\mathcal{P}_e$ consists of two paths such that one of them has its end-vertices in the two leftmost vertices or just in the leftmost vertex if the path is a trivial one-vertex path, and the second path behaves symmetrically.

The structure of paths in the projection of a tamed path cover is shown in Fig. 2, the vertices of G_e are denoted by $v_1^e, \ldots, v_{p(e)}^e$ in the figure according to their proper interval ordering. Note that every path of $\mathcal{P}$ that enters G_e uses the (one or two) leftmost and rightmost vertices as entry-points.

We use this structural result for our kernelization. For each G_e that cannot be covered by two cliques, we analyse the possible structure of paths in $\mathcal{P}_e$ for a tamed path cover $\mathcal{P}_e$. It appears that the types of paths in $\mathcal{P}_e$ are defined by cut-vertices of G_e and that the adjacencies of the second leftmost and the second rightmost vertices of G_e to the corresponding left and right cliques (if, say, the second leftmost vertex is not adjacent to the left clique, then the leftmost vertex "cuts" in a special sense this clique from the remaining part of G_e). Then we replace G_e by a gadget from Fig. 3 which has the same structure with respect to how they can be covered by a tamed path cover. Since each of these gadgets can be covered by at most two cliques, in the end we obtain an equivalent instance of PATH COVER such that the input graph can be covered by at most $|V(H)| + 2|E(H)|$ cliques.

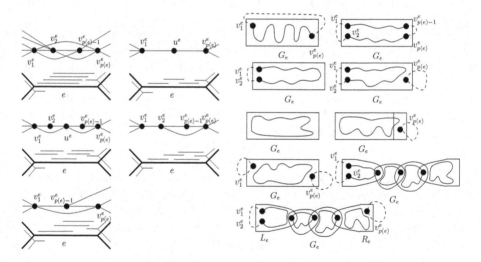

Fig. 3. Replacement gadgets (up to symmetry) and their H-representations.

Fig. 4. The types of covering (up to symmetry) of G_e by a tamed cycle cover.

Then we can apply Theorem 3 where $h \leq |V(H)| + 2|E(H)|$. Note that the kernelization from Theorem 3 can destroy the proper H-representation. Thus we have to be a bit careful here to specify the value of the parameter. We do it by using Observation 1 and output the trivial proper $\hat{G}$-representation for the obtained graph $\hat{G}$.

CYCLE COVER is more complicated. While the general idea follows the one for PATH COVER, there are several non-trivial differences which we underline below. We first recall the statement of the main result for CYCLE COVER.

Theorem 2 CYCLE COVER *admits a compression of size $\mathcal{O}(h^{10})$, where h is the size of the graph H in a proper H-representation of the input graph G.*

Let G be a proper H-graph and $(H', \mathcal{M})$ be a nice proper representation of it. Let $\mathcal{C}$ be a cycle cover of G. As for path covers, for each $e \in E(H)$, we define the *projection* $\mathcal{C}_e$ of $\mathcal{C}$ on G_e that is now a family of paths and cycles of G_e. We show that it suffices to only consider special structured cycle covers that again are called *tamed*; the structure of paths and cycles in the projection of a tamed cycle cover can be seen in Fig. 4. The crucial difference between projections of tamed path and cycle covers is that the number of elements of the projection of a tamed cycle cover is not bounded by any constant. Namely, if G has at least three blocks, then either $\mathcal{C}_e$ contains a Hamiltonian path with its end-vertices in the leftmost and the rightmost vertices of G_e or each middle block with at least three vertices should contain a cycle of $\mathcal{C}_e$. This implies that we cannot replace G_e by a gadget which both has the same number of cycles as the original projection, and can be covered by cliques whose number is any function of the size of H.

To deal with this situation we introduce weights that encode the number of cycles that we need to cover G_e if we do not use a Hamiltonian path between the leftmost and the rightmost vertices. For each G_e, we construct a gadget with at most three edges of positive weight. The remaining edges of the considered graph receive zero weights. To give a rough idea how this works, we observe that the non-zero weights are assigned to the edges of a gadget in such a way that (i) there is a Hamiltonian path between the leftmost and rightmost vertices that contains all these edges and (ii) for any cycle cover whose projection has no such path, the cycles of the cover miss some edges of non-zero weights. The simplest way to achieve this property is to use bridges in the replacement gadgets for the assignment of non-zero weights but it is not always possible and we have to use also more complicated gadgets. We replace the leftmost (the rightmost) block by a copy of K_5 if it has size at least 6 and leave it intact otherwise. The replacement gadgets are attached to the graph by the two leftmost (rightmost) vertices of G_e and the unique cut-vertices of the corresponding blocks. The same

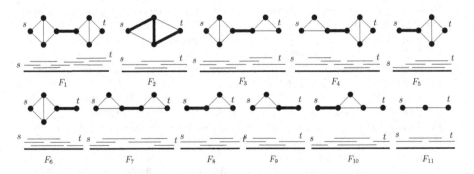

Fig. 5. The construction of F_1–F_{11} and their proper interval representations; the edges of non-zero weights are shown by thick lines.

replacement is done for the middle part if G_e has a unique middle block. If G_e has at least two middle blocks, we replace them by one of the graphs F_1–F_{11} shown in Fig. 5, the non-zero weight edges are shown by thick lines and the gadgets are attached via the vertices s and t.

This way we construct an instance of PRIZE COLLECTING CYCLE COVER, where at most $3|E(H)|$ edges have non-zero weights. Then we apply Theorem 3.

Notice that for HAMILTONIAN CYCLE, we have no such difficulties, because we are looking for a single cycle. This allows us to construct a kernel of size $\mathcal{O}(h^8)$.

4 Conclusion and Open Questions

We obtained compression and kernelization results for HAMILTONIAN PATH and HAMILTONIAN CYCLE and their generalizations for some classes of intersection graphs. We proved that HAMILTONIAN CYCLE and PATH COVER on proper H-graphs admit a polynomial kernel of size $\mathcal{O}(h^8)$ when parameterized by the size h of H if a proper H-representation is given in the input. For CYCLE COVER, it was shown that it admits a polynomial compression into PRIZE COLLECTING CYCLE COVER. As a byproduct, we also established that HAMILTONIAN CYCLE, CYCLE COVER, PATH COVER, and PRIZE COLLECTING CYCLE COVER admit polynomial kernels when parameterized by the clique cover size if a clique cover is given in the input.

It is natural to ask whether our results for proper H-graphs could be generalized to (not necessarily proper) H-graphs. While HAMILONIAN CYCLE is NP-complete on strongly chordal split graphs [31], this question is open even for special families of graphs H, like trees or stars. It could be also interesting to consider other (covering) problems on (proper) H-graphs. For example, what can be said about CLIQUE COVER?

References

1. Alon, N., Yuster, R., Zwick, U.: Color-coding. J. ACM **42**(4), 844–856 (1995)
2. Bertossi, A.A.: Finding Hamiltonian circuits in proper interval graphs. Inf. Process. Lett. **17**(2), 97–101 (1983)
3. Biro, M., Hujter, M., Tuza, Z.: Precoloring extension. I. Interval graphs. Discrete Math. **100**(1), 267–279 (1992)
4. Björklund, A., Husfeldt, T., Kaski, P., Koivisto, M.: Narrow sieves for parameterized paths and packings. CoRR abs/1007.1161 (2010)
5. Bodlaender, H.L., Downey, R.G., Fellows, M.R., Hermelin, D.: On problems without polynomial kernels. J. Comput. Syst. Sci. **75**(8), 423–434 (2009)
6. Bodlaender, H.L., Jansen, B.M.P., Kratsch, S.: Kernel bounds for path and cycle problems. Theor. Comput. Sci. **511**, 117–136 (2013)
7. Brandstädt, A., Dragan, F.F., Köhler, E.: Linear time algorithms for hamiltonian problems on (claw, net)-free graphs. SIAM J. Comput. **30**(5), 1662–1677 (2000)

8. Brandstädt, A., Le, V.B., Spinrad, J.P.: Graph Classes: A Survey. Society for Industrial and Applied Mathematics (SIAM), Philadelphia, PA, SIAM Monographs on Discrete Mathematics and Applications (1999)
9. Broersma, H., Fiala, J., Golovach, P.A., Kaiser, T., Paulusma, D., Proskurowski, A.: Linear-time algorithms for scattering number and hamilton-connectivity of interval graphs. J. Graph Theor. **79**, 282–299 (2015)
10. Chang, M., Peng, S., Liaw, J.: Deferred-query: an efficient approach for some problems on interval graphs. Networks **34**(1), 1–10 (1999)
11. Chaplick, S., Töpfer, M., Voborník, J., Zeman, P.: On H-topological intersection graphs. In: Bodlaender, H.L., Woeginger, G.J. (eds.) WG 2017. LNCS, vol. 10520, pp. 167–179. Springer, Cham (2017). https://doi.org/10.1007/978-3-319-68705-6_13
12. Chaplick, S., Zeman, P.: Combinatorial problems on H-graphs. In: EUROCOMB, vol. 61, ENDM, pp. 223–229 (2017). arXiv:1706.00575
13. Chen, C., Chang, C., Chang, G.J.: Proper interval graphs and the guard problem. Discrete Math. **170**(1–3), 223–230 (1997)
14. Cygan, M., et al.: Parameterized Algorithms. Springer, Cham (2015). https://doi.org/10.1007/978-3-319-21275-3
15. Cygan, M., Nederlof, J., Pilipczuk, M., Pilipczuk, M., van Rooij, J.M.M., Wojtaszczyk, J.O.: Solving connectivity problems parameterized by treewidth in single exponential time. In: FOCS 2011, pp. 150–159. IEEE (2011)
16. Damaschke, P.: Paths in interval graphs and circular arc graphs. Discrete Math. **112**(1–3), 49–64 (1993)
17. Downey, R.G., Fellows, M.R.: Fundamentals of Parameterized Complexity. Texts in Computer Science. Springer, London (2013). https://doi.org/10.1007/978-1-4471-5559-1
18. Etscheid, M., Kratsch, S., Mnich, M., Röglin, H.: Polynomial kernels for weighted problems. J. Comput. Syst. Sci. **84**, 1–10 (2017)
19. Fellows, M.R., Lokshtanov, D., Misra, N., Mnich, M., Rosamond, F.A., Saurabh, S.: The complexity ecology of parameters: an illustration using bounded max leaf number. Theory Comput. Syst. **45**(4), 822–848 (2009)
20. Fomin, F.V., Golovach, P.A., Raymond, J.: On the tractability of optimization problems on H-graphs. In: ESA 2018, vol. 112 of LIPIcs, Schloss Dagstuhl - Leibniz-Zentrum fuer Informatik, pp. 30:1–30:14 (2018)
21. Fomin, F.V., Lokshtanov, D., Saurabh, S., Zehavi, M.: Kernelization. In: Theory of Parameterized Preprocessing, Cambridge University Press (2019)
22. Frank, A., Tardos, É.: An application of simultaneous diophantine approximation in combinatorial optimization. Combinatorica **7**(1), 49–65 (1987)
23. Garey, M.R., Johnson, D.S.: Computers and intractability. In: Computers and Intractability: A Guide to the Theory of NP-Completeness, A Series of Books in the Mathematical Sciences. W. H. Freeman & Co. (1979)
24. Golumbic, M.C.: Algorithmic Graph Theory and Perfect Graphs, Annals of Discrete Mathematics, vol. 57, 2nd edn. Elsevier Science B.V., Amsterdam (2004). With a foreword by Claude Berge
25. Hung, R., Chang, M.: Linear-time certifying algorithms for the path cover and hamiltonian cycle problems on interval graphs. Appl. Math. Lett. **24**(5), 648–652 (2011)
26. Ibarra, L.: A simple algorithm to find Hamiltonian cycles in proper interval graphs. Inf. Process. Lett. **109**(18), 1105–1108 (2009)
27. Keil, J.M.: Finding hamiltonian circuits in interval graphs. Inf. Process. Lett. **20**(4), 201–206 (1985)

28. Koutis, I.: Faster algebraic algorithms for path and packing problems. In: Aceto, L., Damgård, I., Goldberg, L.A., Halldórsson, M.M., Ingólfsdóttir, A., Walukiewicz, I. (eds.) ICALP 2008. LNCS, vol. 5125, pp. 575–586. Springer, Heidelberg (2008). https://doi.org/10.1007/978-3-540-70575-8_47
29. Lampis, M., Makino, K., Mitsou, V., Uno, Y.: Parameterized edge hamiltonicity. Discrete Appl. Math. **248**, 68–78 (2018)
30. Manacher, G.K., Mankus, T.A., Smith, C.J.: An optimum theta (n log n) algorithm for finding a canonical hamiltonian path and a canonical hamiltonian circuit in a set of intervals. Inf. Process. Lett. **35**(4), 205–211 (1990)
31. Müller, H.: Hamiltonian circuits in chordal bipartite graphs. Discrete Math. **156**(1–3), 291–298 (1996)
32. Roberts, F.S.: Indifference graphs. In: Proof Techniques in Graph Theory (Proceedings of the Second Ann Arbor Graph Theory Conference, Ann Arbor, Michigan, 1968), pp. 139–146. Academic Press, New York (1969)
33. Shih, W., Chern, T.C., Hsu, W.: An o(n^2 log n) algorithm for the hamiltonian cycle problem on circular-arc graphs. SIAM J. Comput. **21**(6), 1026–1046 (1992)
34. Williams, R.: Finding paths of length k in $O(2^k)$ time. Inf. Process. Lett. **109**(6), 315–318 (2009)

Weighted Throughput Maximization
with Calibrations

Vincent Chau[1], Shengzhong Feng[1,2], Minming Li[3], Yinling Wang[4(✉)],
Guochuan Zhang[5], and Yong Zhang[1]

[1] Shenzhen Institutes of Advanced Technology, Chinese Academy of Sciences,
Shenzhen, China
[2] National Supercomputing Centre in Shenzhen, Shenzhen, China
[3] City University of Hong Kong, Hong Kong, China
[4] Dalian University of Technology, Dalian, China
yinling_wang@foxmail.com
[5] Zhejiang University, Hangzhou, China

Abstract. The scheduling problem with calibrations was introduced by
Bender et al. (SPAA 2013). In sensitive applications, machines need to
be periodically calibrated to ensure that they run correctly. Formally,
we are given a set of n jobs with release times, deadlines and weights.
Calibrating a machine requires a cost and remains calibrated for a period
of T time units, after which it must be recalibrated before it can resume
running jobs. Moreover, we are given a budget of K calibrations. The
objective is to schedule a set of jobs such that the total weight is maximized on m identical machines with at most K calibrations.

In this paper, we present a $(1/3)$-approximation polynomial time algorithm when jobs have unit processing time. For the arbitrary processing
time case, we give a $((1 - \varepsilon)/3)$-approximation pseudo-polynomial time
algorithm and a $((1 - \varepsilon)/18)$-approximation polynomial time algorithm.

1 Introduction

The scheduling problem of minimizing the number of calibrations has been
recently introduced by Bender et al. [5]. It is motivated by the Integrated Stock-
pile Evaluation (ISE) program at Sandia National Laboratories for testing in
contexts where safety mistakes may have serious consequences. More generally,
this problem concerns the manufacture of modern industrial products according
to their exact standards like digital cameras or processors. The machines that
make these products are precise and execute dedicated tasks. Therefore, they
need to be calibrated carefully before they can perform any job. The products

Vincent Chau, Shengzhong Feng and Yong Zhang are supported by Shen-
zhen research grant (KQJSCX20180330170311901, JCYJ20180305180840138 and
GGFW2017073114031767), NSFC (No. 61433012) and Hong Kong GRF 17210017.
Minming Li is supported by a grant from Research Grants Council of the Hong Kong
Special Administrative Region, China (Project No. CityU 11268616). Guochuan Zhang
is supported by NSFC (No. 11531014).

© Springer Nature Switzerland AG 2019
Z. Friggstad et al. (Eds.): WADS 2019, LNCS 11646, pp. 311–324, 2019.
https://doi.org/10.1007/978-3-030-24766-9_23

can be considered reliable only if the machines have been well calibrated beforehand. These calibrations can be very expensive and can be much more expensive than the running cost of the machines.

This motivation can extend to any context where machines performing jobs must be calibrated periodically. High-precision machines require periodic calibration to ensure precision. Methodology for calibrating these machines is itself an area of research; mainly, it includes [14,15,17,18]. Generally, machines are no longer considered to be accurate after a period. There are many guidelines to determine this *calibration interval*, or period of time between calibrations, both from industry and academia [4,9,11,13,16].

The scheduling problem with calibrations proposed by Bender et al. [5] is as follows. We are given a set of n jobs that need to be scheduled on a set of m identical machines. Each job j has a release time r_j, a deadline d_j and a processing time p_j. A job is scheduled if it is processed entirely inside its interval $[r_j, d_j)$. However, we can schedule jobs only if we perform calibrations beforehand. A calibration activates instantaneously the machine for a period of T time units and the machine can start to process jobs as soon as it is calibrated. After T time units, the machine cannot schedule any jobs unless we perform another calibration. The goal is to find a feasible schedule such that all jobs are scheduled with the minimum number of calibrations.

1.1 Related Works

The scheduling problem of minimizing the number of calibrations has been first studied in the seminal paper by Bender et al. [5]. They restricted their work to the unit processing time case and provided an optimal algorithm for this problem when there is only one available machine, and a 2-approximation algorithm[1] when there are multiple available machines. Moreover, they proposed several structural properties and several cases where the algorithm is optimal. Although this problem admits a polynomial time algorithm for the case of one machine, it remains open whether it is polynomial or NP-hard for the multiple machines case. Later, Fineman and Sheridan [7] generalized the calibration model. They considered the case where jobs have arbitrary processing time, preemption of jobs is not allowed and jobs must be entirely scheduled within the same calibration. The problem is NP-complete to find a feasible solution and they study the problem with resource-augmentation. More recently, Angel et al. [1] developed dynamic programming algorithms for further generalizations—where for example there are multiple kinds of calibrations, or jobs have non-unit processing times, but are preemptible and can be assigned to different calibrations. Chau et al. [6] considered the flow time problem with calibration constraints. They studied the online version whose goal is to minimize the total (weighted)

---

[1] A ρ-approximation algorithm for an optimization problem is a polynomial-time algorithm that for all instances of the problem produces a solution whose value is within a factor of ρ of the value of an optimal solution. By convention, we have $\rho > 1$ for minimization problems, while $\rho < 1$ for maximization problems.

flow time, the (weighted) time between the release time of a job and its completion as well as the calibration cost. They proposed several constant competitive online algorithms where jobs are not known in advance and also proposed a polynomial time algorithm for the offline case.

Scheduling with calibrations has similarities with some other well-known scheduling problems, such as minimizing idle periods [2], and scheduling on cloud-based machines which must be rented to perform work [12].

1.2 Our Problem

In this paper, we study the *throughput* version of the scheduling problem with calibrations. Instead of minimizing the number of calibrations such that all jobs are scheduled, we are interested to schedule a subset of jobs with a limited number of calibrations such that the total weights of the selected jobs is maximized. More formally, we are given a set of n jobs with their respective release time, deadline, processing time and each job is associated to a weight (profit). We have a limited budget K in the number of calibrations and each job has its own weight. The goal is to find a subset of jobs that can be scheduled with at most K calibrations such that the total weights of the selected jobs is maximized.

A first observation is that in the worst case, each job is assigned to a different calibration. So in any reasonable feasible schedule, there are at most n calibrations. Secondly, our problem is at least as hard as the calibration minimization problem. Indeed, let us consider the following decision problem: given a calibration budget K, is it possible to schedule a subset of jobs with total weights at least W? If we can answer this question in a positive way, we can perform a binary search on the number of calibrations we are allowed to use and aim to find the minimum number of calibrations such that all jobs are scheduled. In other words, a polynomial time algorithm that can solve our problem implies that the calibration minimization problem can be solved in polynomial time.

Furthermore, we study the variant where jobs have arbitrary processing time. Jobs can be preempted but must be resumed in the same calibration. Indeed, resuming a job in a different calibration may incur overhead cost. This work is the first to our knowledge to study the *throughput* version of the calibration scheduling problem.

1.3 Our Contributions

When jobs have unit processing time, we propose a framework that achieves a $(1 - \frac{1}{e})$-approximation (Theorem 1) when the calibration budget is less than or equal to the number of machines, a $\frac{1}{3}$-approximation otherwise (Theorem 3). On the other hand, when jobs have arbitrary processing time, we show that we are able to solve the problem in polynomial time by losing a factor of 6 (Proposition 3). Therefore, we get a polynomial time algorithm for the case when jobs have arbitrary processing time and the calibration budget is more than the number of machines, and its approximation ratio is $\frac{1}{18}(1 - \varepsilon)$ (Theorem 6).

The paper is organized as follows: we first study the case where jobs have unit processing time and the budget of calibrations is less than or equal to the number of machines in Sect. 2 as well as the complementary case (when the budget of calibrations is more than the number of machines) in Sect. 3. We then study the case where jobs have arbitrary processing time in Sect. 4: a pseudo-polynomial time approximation algorithm is given in Sect. 4.1 while a polynomial time algorithm is given in Sect. 4.2. Then we conclude in Sect. 5. Due to space limitation, some proofs are omitted.

2   Warm-Up: $K \leq m$, Unit Processing Time

We first present a simple algorithm to solve the problem of maximum throughput with calibrations when the calibration budget is less than or equal to the number of machines. In this case, we do not need to worry about whether we have enough machines or not since we can perform each calibration on a different machine. The idea is to transform our problem to a Maximum Coverage Problem.

Definition 1 (Maximum Coverage Problem). *We are given a collection of sets $S = \{S_1, S_2, ..., S_p\}$. We aim to select at most K sets such that the total weight of the covered elements is maximized, i.e. find a subset $S' \subseteq S$ such that $|S'| \leq K$ and $\sum_{S_i \in S'} w(S_i)$ is maximized.*

A greedy algorithm [8] (Algorithm 1) has been proved to have an approximation ratio of $(1 - 1/e)$. The algorithm is as follows. Select the set that contains the maximum weight of uncovered elements until we select K sets.

Algorithm 1. Greedy algorithm for the Maximum Coverage Problem

---
Require: K, $S = \{S_1, S_2, ..., S_p\}$
1: $C \leftarrow \emptyset$
2: $E \leftarrow \emptyset$
3: **while** $|C| \leq K$ **do**
4:     $S' \leftarrow \arg \max_{S_i \in S} \{w(S_i \setminus E)\}$
5:     $C \leftarrow C \cup S'$
6:     $E \leftarrow E \bigcup_{e \in S'} e$
7: **end while**
8: **return** C

We now construct an instance of the Maximum Coverage Problem from an instance of our problem. Each element corresponds to a job, and each set corresponds to a calibration containing some jobs. Depending on when the calibration starts, it may not contain the same set of jobs. From [1], we know that it is sufficient to consider a restricted number of starting times of calibrations.

Definition 2 (Definition 1 [1]). *Let $\Psi := \bigcup_i \{d_i - n, d_i - n + 1, ..., d_i\}$.*

Proposition 1 (Proposition 1 [1]). *There exists an optimal solution in which each calibration starts at a time in Ψ.*

High Level Idea of the Algorithm. At each step of the algorithm, we choose the calibration that covers the maximum number of unselected jobs. It is not necessary to consider all subsets of jobs for a given calibration among all feasible subsets. In other words, if a calibration starts at some time $t \in \Psi$, we will choose the one that contains the maximum number of jobs. Therefore, we only need to consider one schedule for each different starting time of the calibration. To get a feasible solution that contains a maximum number of jobs, we can use the algorithm for the classical scheduling problem whose aim is to maximize the number (weight) of scheduled jobs. We first construct a new instance in the following way.

- After deciding the starting time of the calibration, we know that the machine is available in the interval $[t, t + T)$.
- For each job j whose release time is before t, change it to t or the deadline d_j whichever happens first, i.e. $r_j := \min\{t, d_j\}$,
- Similarly for the deadline of a job j that is after $t + T$, change it to $t + T$ or its release time r_j, i.e. $d_j := \max\{t + T, r_j\}$.
- Remove all infeasible jobs, these jobs verify $r_j = d_j$ if they are outside the interval $[t, t + T)$.

After we create such an instance, we use the algorithm proposed by Baptiste et al. [3] in order to find the set of jobs that can be scheduled in the interval $[t, t + T)$ in polynomial time such that the total weight is maximized.

Definition 3. *An* EDF *schedule is a schedule in which the machine schedules the job with the earliest deadline among the set of available jobs at this time.*

In the sequel, we only consider EDF schedules inside each calibration. Among different starting times of calibrations, we choose the one with the maximum weight. After we choose the calibration and the associated jobs, we update the set of jobs by removing the already chosen jobs, and we repeat the procedure until we choose K calibrations. The algorithm is described in Algorithm 2. Since we have $K \leq m$, the resulting solution is a feasible schedule and we do not need to verify whether there is at most $m - 1$ calibrations for a given time t (line 6).

Theorem 1. *Algorithm 2 can solve the problem of maximizing the total weights of selected jobs with a calibration budget $K \leq m$ in $O(Kn^6)$ time and it is a $(1 - \frac{1}{e})$-approximation algorithm.*

Proof. The approximation ratio of Algorithm 2 comes from the approximation ratio of the algorithm for the maximum coverage problem [8]. If we generate all possible calibrations (for each starting time of calibration, we choose a subset of jobs in $\{0, 1\}^n$) and apply Algorithm 1, the returned solution is the same as if we use Algorithm 2. As mentioned previously, we do not need to generate all possible subsets of jobs for a given starting time of calibration. There is one calibration that dominates all others (choose the lexicographically smaller (in terms of the vector of jobs in $\{0, 1\}^n$) if tie). Hence, at each step, both algorithms will select the same calibrations as well as the same set of jobs.

Algorithm 2. Greedy algorithm for the problem of maximizing the total weights of selected jobs with a calibration budget

---

Require: K, $J = \{1, 2, \ldots, n\}$
1: $\Psi \leftarrow \bigcup_i \{d_i - n, d_i - n + 1, \ldots, d_i\}$.
2: $C \leftarrow \emptyset$ //set of calibrations
3: $E \leftarrow \emptyset$ //set of chosen jobs
4: **while** $|C| \leq K$ **do**
5:     $S \leftarrow \emptyset$ // set of choices of calibrations
6:     **for** $t \in \Psi$ such that at most $m - 1$ calibrations in interval $[t, t + T]$ **do**
7:         $S_t \leftarrow$ Solve the maximum weighted throughput scheduling problem in interval $[t, t + T]$
8:         $S \leftarrow S \cup \{S_t\}$
9:     **end for**
10:     $S' \leftarrow \arg\max_{S_t \in S}\{E \cup S_t\}$
11:     $C \leftarrow C \cup S'$
12:     $E \leftarrow E \bigcup_{e \in S'} e$
13:     $J \leftarrow J \setminus S'$
14: **end while**
15: **return** C

The set Ψ has $O(n^2)$ elements. For each different starting time of calibration in Ψ, we use the algorithm proposed by Baptiste et al. [3] in order to find the calibration that dominates all others. The running time of their algorithm is $O(n^4)$. Since we need to select K calibrations, the overall running time is $O(Kn^6)$. □

3   A Framework when $K > m$

In this section, we use the same Algorithm 2, but we need to analyze in a different way. In order to check whether there are at most m calibrations at timeslot t, we only need to check that there are at most m calibrations that start in the interval $[t - T + 1, t]$. We use a charging scheme argument to show the approximation ratio of the greedy algorithm. We compare the schedule returned by Algorithm 2 with the optimal schedule. In particular, we will compare directly the jobs scheduled in each calibration.

Definition 4. *Let* $A = \bigcup_{1 \leq i \leq K} A_i$ *be the set of jobs that Algorithm 2 chooses where* A_i *is the set of chosen jobs at the i-th step. Let* $w(A_i)$ *be the total weights of the jobs in* A_i. *From Algorithm 2, we know that* $w(A_1) \geq w(A_2) \geq \ldots \geq w(A_K)$.

Without loss of generality, we require that both schedules, the one returned by Algorithm 2 and the optimal solution, have the following properties:

- Calibrations are sorted in non-increasing order of their profit (weight).
- Following this order, calibrations are assigned to the first available machine.

In order to prove the approximation ratio, we will compare the algorithm with the optimal solution in three phases. Firstly, we check whether there are common jobs in both schedules. Secondly, we compare the calibrations that contain some overlap in both schedules, i.e., there exists a non-empty interval on some machine such that a calibration in A and a calibration in the optimal schedule occur at the same time. Finally, we analyze the remaining calibrations.

The calibrations from the optimal schedule will be mapped to the calibrations of the schedule returned by Algorithm 2. We can bound the number of times that each calibration will be mapped to, which leads to the approximation ratio.

Definition 5. *For a given machine, we say that two calibrations overlap when they have at least one common time slot.*

Observation 1. *When comparing two feasible schedules on the same machine, each calibration in one schedule overlaps with at most two calibrations in the other schedule.*

Theorem 2. *Algorithm 2 is a $\frac{1}{3}$-approximation for the problem of maximizing the throughput with calibration constraints.*

Proof. Let ALG be the set of calibrations that Algorithm 2 chooses. Each calibration contains a set of jobs that we schedule. Similarly, let OPT be the set of calibrations in the optimal solution. We denote J^A to be the set of jobs of a schedule A or a calibration A (depending on the context). Let $\mathcal{C} = J^{ALG} \cap J^{OPT}$ be the set of common jobs that are scheduled by Algorithm 2 as well as by the optimal solution. We use a charging argument to prove the approximation ratio of the algorithm. A charging from an element (in our case a job or a calibration) a to b is valid only if the profit of a is less than or equal to the profit of b. Finally, we show that each calibration of ALG receives a charge of a most 3 times its profit which leads to the approximation ratio of 3. The proof is divided into two parts. We charge each job of $\mathcal{C}$ from OPT to ALG. After we map these jobs, we do not consider them in OPT anymore in the sequel. Then, we charge each calibration in OPT (without considering jobs in $\mathcal{C}$) to a calibration in ALG (with the jobs in $\mathcal{C}$). See Fig. 1 for an illustration of the charging scheme.

First, we map each job of $\mathcal{C}$ one by one from OPT to ALG. This mapping can be done because these jobs are scheduled in both OPT and ALG, so each job of $\mathcal{C}$ in ALG receives a charge equal to its own profit.

Now, we map each calibration in OPT (without considering the jobs in $\mathcal{C}$) to a calibration in ALG (with the jobs in $\mathcal{C}$). We consider the case where there is an overlap of a calibration in ALG with a calibration in OPT, i.e. they have at least one common time slot on the same machine. The complementary case (without overlap) will be handled at the end. By Observation 1, we know that a calibration $\mathcal{O}$ in OPT overlaps with at most two calibrations in ALG. At least one of these has a profit more valuable than the profit of the current calibration. As we do not consider jobs in $\mathcal{C}$ in calibration $\mathcal{O}$, if the profit of jobs in $J^{\mathcal{O}} \setminus \mathcal{C}$ is larger than the other overlapped calibration(s) in OPT, we have a contradiction with the algorithm, because the algorithm would have chosen this calibration before the others. So we can map the jobs in $J^{\mathcal{O}}$ to a calibration that has a higher profit.

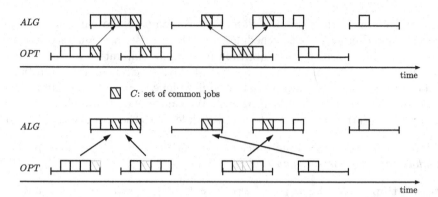

Fig. 1. Illustration of the charging scheme. In the first schedule, we map each job of $\mathcal{C}$ (the set of common jobs) from J^{OPT} to the same job in J^{ALG}. In the second schedule, the jobs in $\mathcal{C}$ are not considered any more (in gray). Each calibration in OPT is mapped to a calibration in ALG such that the profit of a calibration in OPT (without the common jobs) is smaller than the profit of a calibration in ALG (with the common jobs).

For the remaining calibrations in OPT and the calibrations in ALG that are not mapped, we can map them one by one. As previously, since we do not consider jobs in $\mathcal{C}$, it means that the remaining profit of each calibration in OPT is lower than the profit of each unmapped calibration in ALG, otherwise, the algorithm would have chosen such a calibration. Note that both schedules have K calibrations (it may have no jobs assigned to them).

Finally, we show that each calibration in ALG receives at most 3 times its profit. Each calibration may receive a mapping from the jobs of $\mathcal{C}$. Since we never schedule a job twice, then the mapping is at most the profit of the calibration. Each calibration in ALG overlaps with at most two calibrations in OPT on a machine (Observation 1). So it may be mapped by at most two calibrations that have a profit less than or equal to its profit. The calibrations that were not considered before, are mapped to the calibrations that have not been mapped before. So a calibration in ALG can be mapped at most 3 times. Thus we have $w(OPT) \leq 3w(ALG)$ and $\frac{1}{3}w(OPT) \leq w(ALG) \leq w(OPT)$. $\square$

Theorem 3. *When jobs have unit processing time, Algorithm 2 runs in time* $O(Kn^6)$.

Proof. Since jobs have unit processing time, the set Ψ has a size of $O(n^2)$. In order to solve the maximum throughput scheduling problem (line 7), we use the algorithm proposed by Baptiste et al. [3]. In particular, their algorithm can find the maximum throughput in time $O(n^4)$. Finally, we need to choose at most K calibrations. So the running time of Algorithm 2 is $O(Kn^6)$. $\square$

4   Arbitrary Processing Time

In this section, we study the case where jobs have arbitrary processing time. Preemption of jobs is allowed but must be resumed in the same calibration, i.e., a job can only be assigned to one calibration. We first show that the problem is NP-hard by reducing our problem from the KNAPSACK problem and we obtain the following theorem.

Theorem 4. *The problem of maximizing the profit with calibration constraint when jobs have arbitrary processing times and arbitrary weights is NP-hard.*

Sketch of Proof: It is sufficient to show that the problem of maximizing the profit with calibration is NP-hard on a single machine. The idea is to show that a special case of our problem is equivalent to the KNAPSACK problem. In particular, we consider the following instance: we are allowed to use only one calibration whose length T is large enough. Moreover, jobs are released at time 0 while their deadline are T. The goal is to choose the set of jobs such that they can be scheduled within one calibration and such that the total profit is maximized. This instance is indeed equivalent to the KNAPSACK problem.

4.1   A Pseudo-Polynomial Time Algorithm

The idea is similar to the unit processing time case: we find a set of potential starting times of calibrations, find the calibration that maximizes the profit of scheduled jobs, then repeat until we use K calibrations. Since jobs have arbitrary processing times, we need to consider every time slot at a distance at most $P = \sum p_j$ from every deadline which is the starting time of a set of consecutive calibrations (a set of calibrations without a time slot such that the machine is not calibrated). With this starting time, we know we can calibrate the machine every T unit times, and at most n times since in the worst case, each job is scheduled in a different calibration. We get the following Definition 6 and Proposition 2. It is clear that the running time of Algorithm 2 becomes pseudo polynomial.

Definition 6. $\Omega := \bigcup_j \{d_j - i + aT, \ i = 0, \ldots, P, \ a = -n, -n+1, \ldots, n-1, n\}$ *be the set of starting times of the calibrations.*

Proposition 2. *There exists an optimal solution in which each calibration starts at a time in Ω.*

Theorem 5. *When jobs have arbitrary processing times, Algorithm 2 runs in time $O(n^5 P W^2 K)$, where W is the sum of weights of jobs and P is the sum of processing times and has an approximation ratio of $1/3$.*

Proof. When jobs have arbitrary processing time, the set Ω has a size of $O(n^2 P)$. In order to solve the maximum throughput scheduling problem (line 7), we use the algorithm proposed by Lawler [10]. In particular, this algorithm can find the maximum throughput in time $O(n^3 W^2)$. Finally, we need to choose at most K

calibrations. Therefore the running time of Algorithm 2 is $O(n^5 PW^2 K)$. With Proposition 1 and the algorithm proposed by Lawler [10], it is also optimal. The approximation ratio of the overall algorithm (Theorem 2) is $1/3$. $\qquad\qquad$ □

Corollary 1. *When jobs have arbitrary processing times, Algorithm 2 runs in time $O(KPn^5 \lfloor n/\varepsilon \rfloor^2)$, where P is the sum of processing times, and has an approximation ratio of $\frac{1}{3}(1 - \varepsilon)$.*

Sketch of Proof: With a similar proof of the FPTAS (Fully Polynomial Time Approximation Scheme) for the Knapsack problem, we can round the weights of the jobs as follows: $\widetilde{w}_j = \lfloor \frac{w_j}{R} \rfloor$ where $R = \frac{\varepsilon W}{n}$. In this way, the solution returned by the algorithm in [10] is a $(1 - \varepsilon)$-approximated solution.

Corollary 2. *When jobs have arbitrary processing times but unit weight, Algorithm 2 runs in time $O(KPn^6)$ where P is the sum of processing times and has an approximation ratio of $\frac{1}{3}$.*

4.2 A Polynomial Time Algorithm

We consider in this section that each calibration must be of length T because we do not allow to recalibrate the machine if the current calibration is still valid. In the following, we consider the optimal schedule OPT such that each calibration starts or ends at a time in Ω. We restrict our attention to a subset of times $\Theta \subseteq \Omega$ such that Θ has a polynomial number of times by losing a constant factor on the objective. Since we are not allowed to stop a calibration, we know that we may need to calibrate the machine every T time units.

Definition 7. *Let $\Theta := \bigcup_j \{d_j + aT, a = -n, -n+1, \ldots, n-1, n\}$ be a set of restricted starting times of calibrations.*

The above definition guarantees that there is at least one time in Θ inside each calibration interval in OPT.

Proposition 3. *There exists a $\frac{1}{6}$-approximation schedule such that each calibration starts or ends at a time in Θ.*

Proof. We show how to transform an optimal schedule that satisfies Proposition 2 into a schedule that satisfies Proposition 3 by losing a factor of 6 on the profit. Without loss of generality, we prove for a single machine as we can repeat such a transformation independently on each machine. First, we number the calibrations on the machine in the increasing order of their starting times. We separate all the calibrations into two sets: the calibrations with odd index and the calibrations with even index. Let W^{odd} (resp. W^{even}) be the total profit of the odd (resp. even) calibrations. If $W^{odd} > W^{even}$, we keep the calibrations with odd index (even index otherwise). In this way, we lose at most half of the profit, but we create enough space for our modifications, i.e., there are at least T unit times between two consecutive calibrations.

Let c_i be the first calibration in the optimal schedule such that its starting time or its ending time is not in Θ. Let $[b_i, e_i)$ be the interval of the calibration c_i. Let $t \in \Theta \cap [b_i, e_i)$ be a critical time inside the calibration. The idea is to cut the calibration at a time t such that the calibration starts or ends at this time and therefore verify the proposition we are proving. Let us consider that t is the smallest release time r_j in $[b_i, e_i)$. Figure 2 shows different cases of the transformation.

- J^B: the set of jobs that are entirely processed before t;
- J^A: the set of jobs that are entirely processed after t;
- J^M: the set of remaining jobs.

Since we have $w(J^B) + w(J^A) + w(J^M) = w(c_i)$, then at least one of the following cases is true:

case (a) $w(J^B) \geq \frac{1}{3}w(c_i)$

case (b) $w(J^A) \geq \frac{1}{3}w(c_i)$

case (c) $w(J^M) \geq \frac{1}{3}w(c_i)$

We claim that there is at most one job in J^M, i.e., at most one job is scheduled before and after t. Since, we consider EDF schedule, the preemption of a job may occur only at a release time (a more urgent job become available when it is released). And since t is the smallest release time in $[b_i, e_i)$, then there is no another job that is scheduled before and after t.

Case (a) If $w(J^B) \geq \frac{1}{3}w(c_i)$, then we can stop the calibration at time t and discard all jobs from J^A and J^M. In this case, we discard at most 2/3 of its initial profit. In order to make the calibration length equal to T, we can start the calibration at time $t - T$. The schedule is feasible because it covers $[b_i, t)$, so jobs in J^B can be scheduled as before, and the new ending time of the calibration is smaller because we perform the calibration earlier. Such a modification is possible because we have $b_i - T \leq t - T \leq b_i \leq t$ and we know that the machine is not calibrated at time $b_i - T$ since we remove the previous calibration (odd or even calibrations). Therefore, the new calibration cannot start more than T time units earlier.

Case (b) If $w(J^A) \geq \frac{1}{3}w(c_i)$, then we start the calibration at time t and discard all other jobs from J^B and J^M. We have the same argument as in the first case. The calibration starts at time t and ends at time $t + T$ and schedules all the jobs from J^A.

Case (c) If $w(J^M) \geq \frac{1}{3}w(c_i)$, the set J^M has only one job. We try to schedule the job in J^M as early as possible such that its starting time is in Θ. Let e_{i-1} be the completion time of calibration c_{i-1}. By definition, we have $e_{i-1} \in \Theta$, then we can schedule the job in J^M at time $\max\{e_{i-1}, r_{JM}\}$ and the calibration ends at time $\max\{e_{i-1} + T, r_{JM} + T\}$. Note that the release time r_{JM} is before the starting time of the calibration, otherwise, we have a contradiction with the fact that we choose the smallest release time in the interval. So we always have $\max\{e_{i-1} + T, r_{JM} + T\} < e_i$. This leads to a feasible schedule because the calibration starts earlier since the job in J^M will be completed before e_i.

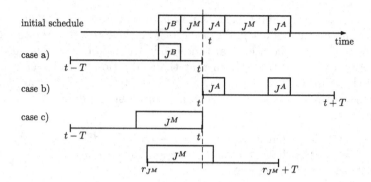

Fig. 2. Illustration of different cases of the modification. After the modification, the calibration starts or ends at time $t \in \Theta$. In the last case, the job in J^M needs to start at its release time which is also in Θ. Note that jobs are not necessarily scheduled continuously.

We repeat this modification as long as there is a calibration that does not start or end at a time in Θ. Thus, with such a modification, we can get a feasible schedule by losing a factor of 6 of the profit from the optimal schedule such that each calibration of this schedule starts and ends at a time in Θ. □

Theorem 6. *When jobs have arbitrary processing times and arbitrary weights, Algorithm 2 runs in time $O(Kn^5 \lfloor n/\varepsilon \rfloor^2)$ and has an approximation ratio of $\frac{1}{18}(1 - \varepsilon)$.*

Proof. Instead of considering the set Ψ in Algorithm 2, we use the set Θ which contains $O(n^2)$ starting times. From Proposition 3, we get a $\frac{1}{6}$-approximation by only considering the starting times of calibrations in Θ. For each potential starting time of the calibration, we calculate the maximum profit with the algorithm in [10] which runs in time $O(n^3W^2)$ where $W = \sum_j w_j$. As we show in Corollary 1, it is possible to get a $(1 - \varepsilon)$-approximation on the maximum profit and the running time is $O(n^3 \lfloor n/\varepsilon \rfloor^2)$. Finally by Theorem 2, the greedy algorithm (the choice of the calibrations) has an approximation of $\frac{1}{3}$ and has K steps. Notice that we compare the solution returned by Algorithm 2 with the optimal solution such that calibrations start and end at a time in Θ. Thus, Algorithm 2 has an approximation ratio of $\frac{1}{18}(1 - \varepsilon)$ when jobs have arbitrary processing times and arbitrary weights and runs in time $O(Kn^5 \lfloor n/\varepsilon \rfloor^2)$. □

Corollary 3. *When jobs have arbitrary processing times and unit weight, Algorithm 2 runs in time $O(Kn^6)$ and has an approximation ratio of $\frac{1}{18}$.*

5   Conclusion

This work is the first to our knowledge to study the throughput maximization scheduling problem with calibrations and the first to give polynomial time algorithms with constant approximation ratio. When jobs have arbitrary processing times, we get a $\frac{1}{18}(1 - \varepsilon)$-approximation polynomial time algorithm. On the other hand, we have a $\frac{1}{3}$-approximation polynomial time algorithm for the unit processing time case. A natural question is whether it is possible to improve the approximation ratio.

Since this problem is at least as hard as the calibration minimization scheduling problem, if we are able to propose an optimal polynomial time algorithm for the case where jobs have unit processing time and unit weight, it will answer the question of the hardness of the problem initially proposed by Bender et al. [5].

References

1. Angel, E., Bampis, E., Chau, V., Zissimopoulos, V.: On the complexity of minimizing the total calibration cost. In: Xiao, M., Rosamond, F. (eds.) FAW 2017. LNCS, vol. 10336, pp. 1–12. Springer, Cham (2017). https://doi.org/10.1007/978-3-319-59605-1_1
2. Baptiste, P.: Scheduling unit tasks to minimize the number of idle periods: a polynomial time algorithm for offline dynamic power management. In: SODA, pp. 364–367. ACM Press (2006)
3. Baptiste, P., Chrobak, M., Dürr, C., Jawor, W., Vakhania, N.: Preemptive scheduling of equal-length jobs to maximize weighted throughput. Oper. Res. Lett. **32**(3), 258–264 (2004)
4. Barringer, H.P.: Cost effective calibration intervals using Weibull analysis. In: Annual Quality Congress Proceedings-American Society For Quality Control, pp. 1026–1038 (1995)
5. Bender, M.A., Bunde, D.P., Leung, V.J., McCauley, S., Phillips, C.A.: Efficient scheduling to minimize calibrations. In: SPAA, pp. 280–287. ACM (2013)
6. Chau, V., Li, M., McCauley, S., Wang, K.: Minimizing total weighted flow time with calibrations. In: SPAA, pp. 67–76. ACM (2017)
7. Fineman, J.T., Sheridan, B.: Scheduling non-unit jobs to minimize calibrations. In: SPAA, pp. 161–170. ACM (2015)
8. Hochbaum, D.S.: Approximating covering and packing problems: set cover, vertex cover, independent set, and related problems. In: Approximation Algorithms for NP-Hard Problems, pp. 94–143. PWS Publishing Co., Boston (1997)
9. Lakin, J.R.: Establishing calibration intervals, how often should one calibrate? September 2014. http://www.inspec-inc.com/home/company/blog/inspec-insights/2014/09/30/establishing-calibration-intervals-how-often-should-one-calibrate. Accessed 30 Sept 2014
10. Lawler, E.L.: A dynamic programming algorithm for preemptive scheduling of a single machine to minimize the number of late jobs. Ann. Oper. Res. **26**(1), 125–133 (1990)
11. Lin, K.-H., Liu, B.-D.: A gray system modeling approach to the prediction of calibration intervals. IEEE Trans. Instrum. Measur. **54**(1), 297–304 (2005)
12. Mäcker, A., Malatyali, M., der Heide, F.M., Riechers, S.: Cost-efficient scheduling on machines from the cloud. In: Chan, T.-H.H., Li, M., Wang, L. (eds.) COCOA 2016. LNCS, vol. 10043, pp. 578–592. Springer, Cham (2016). https://doi.org/10.1007/978-3-319-48749-6_42
13. Nunzi, E., Panfilo, G., Tavella, P., Carbone, P., Petri, D.: Stochastic and reactive methods for the determination of optimal calibration intervals. IEEE Trans. Instrum. Measur. **54**(4), 1565–1569 (2005)

14. Postlethwaite, S.R., Ford, D.G., Morton, D.: Dynamic calibration of CNC machine tools. Int. J. Mach. Tools Manuf. **37**(3), 287–294 (1997)
15. Wilken, T., et al.: High-precision calibration of spectrographs. Mon. Not. R. Astron. Soc. Lett. **405**(1), L16–L20 (2010)
16. Wyatt, D.W., Castrup, H.T.: Managing calibration intervals. In: Annual Workshop and Symposium on National Conference of Standards Laboratories (NCSL) (1991)
17. Zhang, G., Hocken, R.: Improving the accuracy of angle measurement in machine calibration. CIRP Ann.-Manufact. Technol. **35**(1), 369–372 (1986)
18. Zhang, Z.: A flexible new technique for camera calibration. IEEE Trans. Pattern Anal. Mach. Intell. **22**(11), 1330–1334 (2000)

Maximizing Dominance in the Plane
and Its Applications

Jongmin Choi[1], Sergio Cabello[2,3], and Hee-Kap Ahn[1]($\boxtimes$)

[1] Department of Computer Science and Engineering,
Pohang University of Science and Technology, Pohang, South Korea
{icothos,heekap}@postech.ac.kr
[2] Department of Mathematics, IMFM, University of Ljubljana, Ljubljana, Slovenia
[3] Department of Mathematics, FMF, University of Ljubljana, Ljubljana, Slovenia
sergio.cabello@fmf.uni-lj.si

Abstract. Given a set P of n points with weights (possibly negative), a set Q of m points in the plane, and a positive integer k, we consider the optimization problem of finding a subset of Q with at most k points that dominates a subset of P with maximum total weight. We say a set of points Q' dominates p if some point q of Q' satisfies $x(p) \leqslant x(q)$ and $y(p) \leqslant y(q)$. We present an efficient algorithm solving this problem in $O(k(n + m) \log m)$ time and $O(n + m)$ space. Our result implies algorithms with better time bounds for related problems, including the disjoint union of cliques problem for interval graphs (equivalently, the hitting intervals problem) and the top-k representative skyline points problem in the plane.

Keywords: Dominance · Disjoint cliques · Hitting intervals

1 Introduction

We consider the following optimization problem in the plane. Given two sets P and Q of points in the plane, and a positive integer k, find a subset of Q with at most k points that dominates as many points of P as possible. We provide an efficient algorithm solving the problem in $O(k(n + m) \log m)$ time, where n and m are the numbers of points in P and Q, respectively. The algorithm is quite simple and can be easily implemented. In fact, we solve a weighted variant, where the points of P have weights, possibly negative, and we want to dominate a subset of P with maximum total weight.

The (unweighted) problem generalizes problems that have been considered before in the context of scheduling, optimization in graphs and databases, and

Work by Choi and Ahn was supported by the MSIT (Ministry of Science and ICT), Korea, under the SW Starlab support program (IITP-2017-0-00905) supervised by the IITP (Institute of Information & communications Technology Planning & Evaluation.). Work by Cabello was supported by the Slovenian Research Agency, program P1-0297 and projects J1-8130, J1-8155, J1-9109.

© Springer Nature Switzerland AG 2019
Z. Friggstad et al. (Eds.): WADS 2019, LNCS 11646, pp. 325–338, 2019.
https://doi.org/10.1007/978-3-030-24766-9_24

our algorithm directly improves the time bounds for the algorithms discussed on those contexts.

Problem Statement. For a point p in the plane, $x(p)$ and $y(p)$ denote its x and y-coordinate, respectively. A point q *dominates* a point p if $x(p) \leqslant x(q)$ and $y(p) \leqslant y(q)$. A set of points Q' dominates p if some point of Q' dominates p.

For each point $q \in Q$, let $\text{dom}(q)$ be the closed 2-sided range $(-\infty, x(q)] \times (-\infty, y(q)]$. Note that $\text{dom}(q)$ is the set of points in the plane dominated by q. For each subset $Q' \subseteq Q$, we define $\text{dom}(Q') = \bigcup_{q \in Q'} \text{dom}(q)$, that is, the set of points in the plane dominated by Q'.

We are given sets P and Q of points in the plane, and a function $w \colon P \to \mathbb{R}$ assigning a weight $w(p)$ to each point p of P. The weights may be negative. We extend the function w to regions of $\mathbb{R}^2$ by summing up the weights of the points of P contained in the region: $w(R) := \sum_{p \in P \cap R} w(p)$ for each $R \subseteq \mathbb{R}^2$. We refer to $w(R)$ as the weight of region R. If $R \cap P = \emptyset$, then $w(R) = 0$.

Given a positive integer k, we want to select a subset $Q' \subset Q$ with at most k points such that the total weight of the points of P dominated by Q' is maximized. We refer to this problem as maxDominance. Formally, we want to compute

$$\text{maxDominance}(P, Q, k) := \max \big\{ w(\text{dom}(Q')) \mid Q' \subset Q, |Q'| \leqslant k \big\}$$

and to obtain an optimal solution, that is, a subset $Q^* \subset Q$ satisfying $|Q^*| \leqslant k$ and $w(\text{dom}(Q^*)) = \text{maxDominance}(P, Q, k)$. It is important to ensure that each point $p \in \text{dom}(Q') \cap P$ contributes $w(p)$ to $w(\text{dom}(Q'))$ exactly once, even if p is covered multiple times by the dominance regions $\text{dom}(q)$, $q \in Q'$.

Previous Work. We are not aware of any previous work on the maxDominance problem in its full generality. A related problem is the *maximum volume subset selection* problem for anchored boxes, where we select k boxes among n given axis-parallel boxes, each box contained in the positive quadrant of $\mathbb{R}^d$ with a corner at the origin, that maximize the volume of the union of the selected boxes. The volume of the union is known as the *hypervolume indicator*. It can be solved in $O((n - k)k + n \log n)$ time [4,11] in $\mathbb{R}^2$, but it is NP-hard already in $\mathbb{R}^3$ [3]. This problem can be interpreted as a continuous version of the maxDominance problem. Placing a point in each cell of the arrangement defined by the anchored boxes with weight equal to be the volume of the cell, we reduce the maximum volume subset selection problem to the maxDominance problem. However, this reduction introduces $O(n^d)$ points, and thus not very useful.

Other related problems include the disjoint union of cliques problem for interval graphs, the hitting intervals problem, and the top-k representative skyline points problem in the plane. We will describe them in the next subsection.

1.1 Our Contribution

Henceforth, we use $n = |P|$ and $m = |Q|$ to bound the asymptotic time of the algorithms. We show that the maxDominance problem can be solved in

$O(k(n+m)\log m)$ time and $O(n+m)$ space. We employ a dynamic programming approach where we use a data structure based on segment trees to speed up the computation. Our result implies algorithms with better time bounds for problems considered by other authors in other contexts. We describe this next.

Disjoint Cliques in Interval Graphs and Hitting Intervals. In the disjoint union of cliques problem (DUC for short), we are given a graph $G = (V, E)$ and a positive integer k, and we are to find a set C of k disjoint cliques of G such that $\sum_{c \in C} |c|$ is maximized. Here $|c|$ is the number of vertices in a clique c, and a set of cliques is *disjoint* if each node of G belongs to at most one of the cliques.

For general graphs, the DUC problem is NP-hard and difficult in practice, as it includes the problem of finding a largest clique ($k = 1$) as a special case. The problem keeps getting attention because of its applications in analyzing data; see for example [8,15] for some recent accounts. Perhaps not so obvious, the problem encounters applications also in scheduling [5,10].

Algorithmically, research has focused on particular classes of graphs. The DUC problem remains hard even in split (and thus chordal) graphs [14], where finding a maximum clique is solvable in polynomial time. On the positive side, Gavril [9], Yannakis and Gavril [14] and Jansen, Scheffler and Woeginger [10] provide polynomial-time algorithms for interval graphs, comparability graphs, co-comparability graphs, directed path graphs, cographs, and for partial m-trees for a constant m.

We are interested in the DUC problem in *interval graphs*. It is well-known, and we discuss it in more detail in Sect. 4, that the DUC problem in interval graphs is equivalent to the HittingIntervals problem: given a set $\mathcal{I}$ of intervals on the real line and a value k, find k points on the real line such that the number of intervals of $\mathcal{I}$ hit by the points is maximized. Here, we say that a point $x \in \mathbb{R}$ hits an interval I when $x \in I$. This problem is sometimes also called as the maximum piercing problem.

Jansen, Scheffler and Woeginger [10] and Chrobak et al. [5] provide algorithms to solve the DUC problem in interval graphs in $O(k|V|^2)$ time. Damaschke [6] improves the running time to $O(k|E|)$ for connected interval graphs, which is relevant for graphs that are not very dense.

The *weighted* versions of the DUC problem in interval graphs and the HittingIntervals problem are also equivalent when the weights are positive. However, when negative weights are present, both problems are not equivalent anymore. Indeed, in the DUC problem we will never put a vertex with negative weight into the optimal solution, as we can remove it from the solution and increase the total value. In contrast, in the HittingIntervals problem we do not have any flexibility as to which intervals are counted. In any case, the DUC problems in interval graphs can be reduced to the HittingIntervals problem by first removing vertices with negative weights, and then solving the HittingIntervals problem in the remaining intervals. The reduction does not work in the other direction.

In this paper, we use a simple reduction to the maxDominance problem which, together with our new algorithm for the maxDominance problem, solves the (weighted version of the) HittingIntervals problem in $O(k|V|\log|V|)$ time. This also implies that the (weighted version of the) DUC problem in interval graphs can be solved also in $O(k|V|\log|V| + |E|)$ time. (The term $O(|E|)$ comes from the graph complexity $O(|V| + |E|)$.) This improves all previous results when $|E| = \omega(|V|\log|V|)$.

Representative Skyline Points. The skyline points of a set P of points are those that are not dominated by any other point of P. Lin et al. [12] introduce the *top-k representative skyline points* problem: given a set P of n points, select k of them maximizing the number of points from P they dominate. Among other results, they provide an algorithm to solve this problem in the plane in $O(km^2 + n\log m)$ time and $O(n + m^2)$ space, where m is the number of points in the skyline. (Additional data structures can be used to reduce the space bound to $O(n + km)$). In the worst case we have $m = \Theta(n)$, the running time is $O(kn^2)$, and the space is $O(n^2)$ (or $O(kn)$ using additional data structures).

Setting Q to be the points in the skyline and $w(p) = 1$ for all $p \in P$, this problem is a particular case of our problem. Thus, we can solve the top-k representative skyline points problem in $O(kn\log m)$ time and $O(n)$ space.

Lin et al. [12] also show that the problem is NP-hard in 3-dimensional space. Thus, our problem is NP-hard in 3 dimensions as well.

Several alternative measures have been introduced to choose the most representative points in the skyline, also in a variety of fields. See for example [1,4,13] for a small sample. It is beyond our possibilities to survey them.

2   Data Structure

We want to store n values $a_1, \ldots, a_n$, initially set to arbitrary values in $\mathbb{R} \cup \{-\infty\}$, in a data structure that supports the following operations:

- $\mathsf{max}(i, j)$ returns the pair (t, a) with $i \leqslant t \leqslant j$ and $a_t = a = \max\{a_i, \ldots, a_j\}$. As a special case, we can get a_i by querying $\mathsf{max}(i, i)$.
- $\mathsf{add}(i, j, c)$ adds the value $c \in \mathbb{R}$ to $a_i, \ldots, a_j$, where $i \leqslant j$.
- $\mathsf{set}(i, c)$ sets the value of a_i to $c \in \mathbb{R} \cup \{-\infty\}$.

Using an adaptation of segment trees [7, Section 10.3] we have the following.

Lemma 1. *There is a data structure to maintain $a_1, \ldots, a_n$ in $O(n)$ space that supports the operations **max**, **add** and **set** in $O(\log n)$ time per operation.*

Proof. We use a balanced binary search tree $\mathcal{T}$ with keys $\{1, \ldots, n\}$ at the leaf nodes. We use $\nu(i)$ to denote the leaf node of $\mathcal{T}$ that has key i and r for the root. For a nonleaf node u of $\mathcal{T}$, we use u_ℓ and u_r to denote its left and right child, respectively. For any two nodes u, v of $\mathcal{T}$, let $\pi(u, v)$ denote the path from u to v in $\mathcal{T}$ and let $p(u)$ denote the parent node of $u \neq r$.

Each node u of $\mathcal{T}$ represents a contiguous sequence of integers $I_u = \mathbb{Z} \cap [x, x']$ for two integers x and x' with $1 \leqslant x \leqslant x' \leqslant n$. More precisely, a leaf node u with key i represents $I_u = \{i\}$, and for a nonleaf node u we have $I_u = I_{u_\ell} \cup I_{u_r}$. A key property of this representation is that, for each contiguous sequence of integers $\mathbb{Z} \cap [x, x']$, there is a set $U(x, x')$ of $O(\log n)$ nodes of $\mathcal{T}$ such that $\mathbb{Z} \cap [x, x']$ is the disjoint union of I_u, $u \in U(x, x')$. In particular, for each $i \in \mathbb{Z} \cap [x, x']$ there is exactly one node $u \in U(x, x')$ that is on the path $\pi(r, \nu(i))$. Furthermore, for any given integers x and x' with $1 \leqslant x \leqslant x' \leqslant n$, the set $U(x, x')$ can be found in $O(\log n)$ time using the search paths in $\mathcal{T}$ from the root to x and to x', and all the nodes of $U(x, x')$ together have $O(\log n)$ ancestors in $\mathcal{T}$.

At each node u of $\mathcal{T}$ we store three values, $\alpha(u)$, $\beta(u)$ and $\gamma(u)$. For any two nodes u and v of $\mathcal{T}$, we denote by $\mathsf{sum}_\alpha(u, v)$ the sum of $\alpha(w)$ along all vertices w in $\pi(u, v)$. We maintain the following invariants:

- For each leaf node u of $\mathcal{T}$, we have $\alpha(u) \in \mathbb{R} \cup \{-\infty\}$. For each nonleaf node u of $\mathcal{T}$, $\alpha(u) \in \mathbb{R}$; thus $\alpha(u) \neq -\infty$.
- For each leaf node $\nu(i)$ of $\mathcal{T}$, $\mathsf{sum}_\alpha(r, \nu(i))$ is a_i.
- For each node u of $\mathcal{T}$, $\beta(u)$ is the maximum among $\mathsf{sum}_\alpha(u, v)$ over all leaf nodes v in the subtree rooted at u. An alternative, useful way to think of it is the following recursive formulation:

$$\beta(u) = \begin{cases} \alpha(u) & \text{if } u \text{ is a leaf node,} \\ \alpha(u) + \max\{\beta(u_\ell), \beta(u_r)\} & \text{if } u \text{ is a nonleaf node.} \end{cases} \tag{1}$$

- For each node u of $\mathcal{T}$, $\gamma(u)$ tells the index i of the leaf node that determines the value $\beta(u)$. Therefore,

$$\gamma(u) = \begin{cases} i & \text{if } u = \nu(i) \text{ is a leaf node,} \\ \gamma(u_\ell) & \text{if } u \text{ is a nonleaf node and } \beta(u_\ell) \geqslant \beta(u_r), \\ \gamma(u_r) & \text{if } u \text{ is a nonleaf node and } \beta(u_\ell) < \beta(u_r). \end{cases} \tag{2}$$

Note, for example, that $\beta(r)$ is $\max\{a_1, \ldots, a_n\} = a_{\gamma(r)}$. It is clear that $\mathcal{T}$ takes $O(n)$ space because we store only a constant amount of information per node. See Fig. 1 for an example.

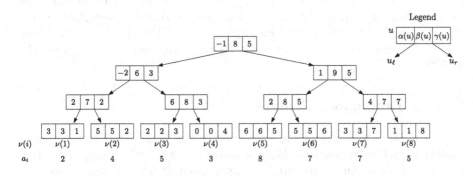

Fig. 1. An example of the tree used in the proof of Lemma 1.

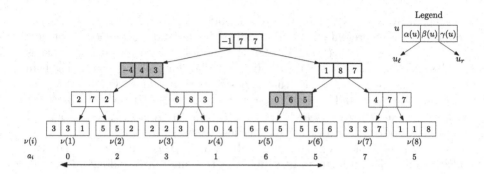

Fig. 2. The tree of Fig. 1 after the operation $\mathsf{add}(1, 6, -2)$. Grey nodes are nodes in $U(1, 6)$. Nodes with thicker, green boundary are those that have to be updated, as they are in $U(1, 6)$ or are their ancestors. (Color figure online)

We start with setting $\alpha(\nu(i)) = a_i$ and $\gamma(i) = i$ for all leaf nodes $\nu(i)$ of $\mathcal{T}$, and $\alpha(u) = 0$ for all nonleaf nodes u of $\mathcal{T}$. We then compute $\beta(u)$ and $\gamma(u)$ in $\mathcal{T}$ from the bottom up using Eqs. (1) and (2). With this, we establish the invariant before making any operations.

Consider now the operation $\mathsf{max}(i, j)$, where $i \leqslant j$. We find the set $U(i, j)$ of $O(\log n)$ nodes such that $\{i, \ldots, j\}$ is the disjoint union $\bigsqcup_{u \in U(i,j)} I_u$. Then $\max\{a_i, \ldots, a_j\}$ is the maximum of $\beta(u) + \mathsf{sum}_\alpha(r, p(u))$ over $u \in U(i, j)$. These values $\mathsf{sum}_\alpha(r, p(u))$ can be computed in $O(\log n)$ time in total because the nodes of $U(i, j)$ altogether have $O(\log n)$ ancestors. The index t where the maximum is attained can be computed similarly, but using the indices $\gamma(\ )$.

Consider now the operation $\mathsf{add}(i, j, c)$, which adds $c \in \mathbb{R}$ to each of the elements $a_i, \ldots, a_j$. Again, we start with finding the set $U(i, j)$ of $O(\log n)$ nodes. For each $u \in U(i, j)$, we set $\alpha(u) = \alpha(u) + c$. This settles the invariant for the values $\alpha(\ )$. For the ancestors v of the nodes of $U(i, j)$, traversing $\mathcal{T}$ in bottom-up manner, we update $\beta(v)$ and $\gamma(v)$ using Eqs. (1) and (2). Since the nodes of $U(i, j)$ altogether have $O(\log n)$ ancestors, we spend $O(\log n)$ time in total. See Fig. 2 for an example. Consider now the operation $\mathsf{set}(i, c)$, which sets a_i to $c \in \mathbb{R} \cup \{-\infty\}$. We compute the value $\mathsf{sum}_\alpha(p(\nu(i)), r)$ in $O(\log n)$ time by following $\pi(\nu(i), r)$ from $p(\nu(i))$, and at the leave $\nu(i)$ set $\alpha(\nu(i))$ to $c - \mathsf{sum}_\alpha(p(\nu(i)), r)$. With this we get the property $\mathsf{sum}_\alpha(\nu(i), r) = c$. Then we update $\beta(\ )$ and $\gamma(\ )$ for the nodes on $\pi(\nu(i), r)$ by following the path from $\nu(i)$ using Eqs. (1) and (2). □

3   Max Dominance Problem

We first consider the most general case, where the points of P have weights, possibly negative. We want to choose a subset Q' of $Q = \{q_1, \ldots, q_m\}$ with at most k points such that $w(\mathrm{dom}(Q'))$ is maximized. It is convenient to assume that the points of Q have different x- and y-coordinates. This can be enforced symbolically, for example replacing each point $q_i \in Q$ by $q_i + i \cdot (\varepsilon, \varepsilon)$ for an infinitesimal $\varepsilon > 0$. This can also be done explicitly in $O((n + m) \log m)$ time.

For ease of description, we add a point $q_{new} = (x_{max}+1, y_{min}-1)$ to Q, where $x_{max} = \max\{x(t) \mid t \in P \cup Q\}$ and $y_{min} = \min\{y(t) \mid t \in P \cup Q\}$. In particular, q_{new} does not dominate any other input point. Note that now we always have a solution $Q' = \{q_{new}\}$ of weight 0.

Let $q_1, \ldots, q_{m+1}$ be the points of Q sorted by decreasing y-coordinate, that is, $y(q_i) > y(q_{i+1})$ for each $i \in \{1, \ldots, m\}$. Clearly, $q_{m+1} = q_{new}$. For each $i \in \{1, \ldots, m+1\}$, let $P_i = \{p \in P \mid y(p) > y(q_i)\}$ and $Q_i^{\lrcorner} = \{q \in Q \mid x(q) \leqslant x(q_i), \ y(q) \geqslant y(q_i)\}$. Note that $q_i \in Q_i^{\lrcorner}$. See Fig. 4, left. We use $w_i(R)$ to denote the sum of the point weights in $R \cap P_i$ for any region R in the plane. Note that $w_i(\mathrm{dom}(q_i)) = 0$ for each i, $P_{m+1} = P$, $w_{m+1}() = w()$ and $Q_{m+1}^{\lrcorner} = Q$.

In general, we use the index ℓ to bound the number of points selected from Q, while the indices i and j encode a point of Q, so that we use $Q_j^{\lrcorner}$, $Q_i^{\lrcorner}$ or P_i. For each $\ell \in \{0, \ldots, k\}$ and $i \in \{1, \ldots, m+1\}$, let

$$T_\ell(i) = \max\left\{w_i(\mathrm{dom}(Q')) \mid Q' \subset Q_i^{\lrcorner}, |Q'| \leqslant \ell\right\}.$$

This means that we consider the weights of candidate solutions, restricted to P_i, with at most ℓ points selected among the points of $Q_i^{\lrcorner}$. Thus, $T_\ell(i) = \mathrm{maxDominance}(P_i, Q_i^{\lrcorner}, \ell)$, and we are interested in the value $T_k(m+1)$. We have $T_\ell(i) \geqslant 0$ because $Q' = \emptyset$ is a candidate solution. Obviously, $T_0(i) = 0$ for each $i \in \{1, \ldots, m+1\}$.

For each $\ell \in \{1, \ldots, k\}$, $i \in \{1, \ldots, m+1\}$ and $j \in \{1, \ldots, i\}$, let

$$S_\ell(i,j) = \max\left\{w_i(\mathrm{dom}(Q')) \mid Q' \subset Q_j^{\lrcorner}, |Q'| \leqslant \ell, q_j \in Q'\right\}. \tag{3}$$

This means that we consider the weights of nonempty candidate solutions, restricted to P_i, with at most ℓ points selected among the points of $Q_j^{\lrcorner}$ that must include the point q_j. We have $S_\ell(i,i) \geqslant 0$ because $Q' = \{q_i\}$ is a candidate solution considered in (3), and $w_i(\mathrm{dom}(q_i)) = 0$.

We will use dynamic programming to compute the values $S_\ell(i,j)$ and $T_\ell(i)$ for all ℓ, i and $j \leqslant i$. The following lemma provides the recursive formulas for the computation. See Fig. 3, right, for some intuition.

Lemma 2. *For each $\ell \in \{1, \ldots, k\}$, $i \in \{1, \ldots, m+1\}$ and $j \in \{1, \ldots, i\}$,*

$$T_\ell(i) = \max\{S_\ell(i,j) \mid q_j \in Q_i^{\lrcorner}\},$$
$$S_\ell(i,j) = T_{\ell-1}(j) + w_i(\mathrm{dom}(q_j)).$$

Proof. We first show that $T_\ell(i) \geqslant \max\{S_\ell(i,j) \mid q_j \in Q_i^{\lrcorner}\}$. Let j_{opt} be an index j that determines $\max\{S_\ell(i,j) \mid q_j \in Q_i^{\lrcorner}\}$ and let Q_{opt} be an optimal solution defining $S_\ell(i, j_{opt})$. This means that $|Q_{opt}| \leqslant \ell$ and $Q_{opt} \subset Q_{j_{opt}}^{\lrcorner} \subset Q_i^{\lrcorner}$ because $q_{j_{opt}} \in Q_i^{\lrcorner}$. Therefore Q_{opt} is considered as a candidate solution in the definition of $T_\ell(i)$, and thus $T_\ell(i) \geqslant w_i(\mathrm{dom}(Q_{opt})) = S_\ell(i, j_{opt}) = \max\{S_\ell(i,j) \mid q_j \in Q_i^{\lrcorner}\}$.

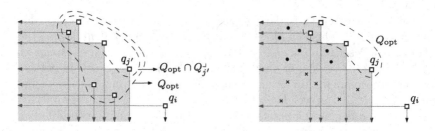

Fig. 3. Illustration for parts of the proof of Lemma 2. The gray region is $\mathrm{dom}(Q_{\mathrm{opt}})$. Right: the black dots contribute to $w_j(\mathrm{dom}(Q_{\mathrm{opt}} \setminus \{q_j\}))$, while the black crosses contribute to $w_i(\mathrm{dom}(q_j))$.

Now we show that $T_\ell(i) \leqslant \max\{S_\ell(i,j) \mid q_j \in Q_i^{\lrcorner}\}$. Let Q_{opt} be an optimal subset of $Q_i^{\lrcorner}$ defining $T_\ell(i)$. If $Q_{\mathrm{opt}} = \emptyset$, then $T_\ell(i) = 0 = w_i(\mathrm{dom}(q_i)) \leqslant S_\ell(i,i) \leqslant \max\{S_\ell(i,j) \mid q_j \in Q_i^{\lrcorner}\}$. For $Q_{\mathrm{opt}} \neq \emptyset$, let $q_{j'}$ be the rightmost point of Q_{opt}, that is, the one with largest x-coordinate. Note that $q_{j'} \in Q_i^{\lrcorner}$ because $Q_{\mathrm{opt}} \subset Q_i^{\lrcorner}$. For any point $q \in Q_{\mathrm{opt}}$ with $y(q) \leqslant y(q_{j'})$, we have $\mathrm{dom}(q) \subset \mathrm{dom}(q_{j'})$. Thus $\mathrm{dom}(Q_{\mathrm{opt}}) = \mathrm{dom}(Q_{\mathrm{opt}} \cap Q_{j'}^{\lrcorner})$. See Fig. 3, left. We then conclude that $T_\ell(i) = w_i(\mathrm{dom}(Q_{\mathrm{opt}})) = w_i(\mathrm{dom}(Q_{\mathrm{opt}} \cap Q_{j'}^{\lrcorner})) \leqslant S_\ell(i,j') \leqslant \max\{S_\ell(i,j) \mid q_j \in Q_i^{\lrcorner}\}$.

Now we show the second equality, for $S_\ell(i,j)$. First, we show that $S_\ell(i,j) \leqslant T_{\ell-1}(j) + w_i(\mathrm{dom}(q_j))$ for $\ell \geqslant 1$. Let Q_{opt} be an optimal solution defining $S_\ell(i,j)$. Since $q_j \in Q_{\mathrm{opt}}$ and $Q_{\mathrm{opt}} \subset Q_j^{\lrcorner}$, for $Q_{\mathrm{opt}}' = Q_{\mathrm{opt}} \setminus \{q_j\}$, we have (see Fig. 3, right)

$$
\begin{aligned}
w_i(\mathrm{dom}(Q_{\mathrm{opt}})) &= w_i(\mathrm{dom}(Q_{\mathrm{opt}}' \cup \{q_j\})) \\
&= w_i(\mathrm{dom}(Q_{\mathrm{opt}}')) - w_i(\mathrm{dom}(Q_{\mathrm{opt}}') \cap \mathrm{dom}(q_j)) + w_i(\mathrm{dom}(q_j)) \\
&= w_j(\mathrm{dom}(Q_{\mathrm{opt}}')) + w_i(\mathrm{dom}(q_j)).
\end{aligned}
$$

Since $|Q_{\mathrm{opt}} \setminus \{q_j\}| \leqslant \ell - 1$ and $Q_{\mathrm{opt}} \setminus \{q_j\} \subset Q_j^{\lrcorner}$, $S_\ell(i,j) = w_i(\mathrm{dom}(Q_{\mathrm{opt}})) = w_j(\mathrm{dom}(Q_{\mathrm{opt}} \setminus \{q_j\})) + w_i(\mathrm{dom}(q_j)) \leqslant T_{\ell-1}(j) + w_i(\mathrm{dom}(q_j))$.

Finally, we show that $S_\ell(i,j) \geqslant T_{\ell-1}(j) + w_i(\mathrm{dom}(q_j))$ for $\ell \geqslant 1$. Let Q_{opt} be an optimal solution defining $T_{\ell-1}(j)$. This means that $Q_{\mathrm{opt}} \subset Q_j^{\lrcorner}$, $|Q_{\mathrm{opt}}| \leqslant \ell - 1$ and $w_j(\mathrm{dom}(Q_{\mathrm{opt}})) = T_{\ell-1}(j)$. Therefore, $Q' = Q_{\mathrm{opt}} \cup \{q_j\}$ is a candidate solution considered in the definition of $S_\ell(i,j)$, and thus

$$
\begin{aligned}
S_\ell(i,j) &\geqslant w_i(\mathrm{dom}(Q_{\mathrm{opt}} \cup \{q_j\})) \\
&= w_j(\mathrm{dom}(Q_{\mathrm{opt}} \setminus \{q_j\})) + w_i(\mathrm{dom}(q_j)) \\
&= w_j(\mathrm{dom}(Q_{\mathrm{opt}})) + w_i(\mathrm{dom}(q_j)) \\
&= T_{\ell-1}(j) + w_i(\mathrm{dom}(q_j)).
\end{aligned}
$$

The result follows. $\square$

The values $T_0(i) = 0$ can be computed and stored in $O(m)$ time. For each $\ell \geqslant 1$, the straightforward computation of $T_\ell(i)$ using the formulas of Lemma 2 takes $\Theta(m^2)$ time, even when the values $w_i(\mathrm{dom}(q_j))$ are already available. Using the data structure of Lemma 1, we can do this step in $O((n+m)\log m)$ time.

Lemma 3. *Consider a fixed index $\ell \geqslant 1$ and assume that the values $T_{\ell-1}(i)$ are available for all i. We can compute $T_\ell(i)$ for all i in $O((n+m)\log m)$ time and $O(n+m)$ space.*

Proof. We use an iterative algorithm, with a for loop with $i = 1, \ldots, m+1$, where we maintain the values $S_\ell(i, j)$ in a data structure. First we analyze how we can transform $S_\ell(i-1, j)$ into $S_\ell(i, j)$.

Consider first the case $j < i$. Note that, because of Lemma 2,

$$
\begin{aligned}
S_\ell(i, j) &= T_{\ell-1}(j) + w_i\big(\mathrm{dom}(q_j)\big) \\
&= T_{\ell-1}(j) + w_{i-1}\big(\mathrm{dom}(q_j)\big) + w_i\big(\mathrm{dom}(q_j)\big) - w_{i-1}\big(\mathrm{dom}(q_j)\big) \\
&= S_\ell(i-1, j) + w_i\big(\mathrm{dom}(q_j)\big) - w_{i-1}\big(\mathrm{dom}(q_j)\big).
\end{aligned}
$$

Observe that the term $w_i(R) - w_{i-1}(R)$ is the sum of the weights of the points of $P_i \setminus P_{i-1}$ contained in a region R. This means that

$$
\forall j < i: \quad S_\ell(i, j) = S_\ell(i-1, j) + \sum_{\substack{p \in P_i \setminus P_{i-1} \\ x(p) \leqslant x(q_j)}} w(p). \tag{4}
$$

Our algorithm exploits the fact that each point $p \in P_i \setminus P_{i-1}$ contributes the same change, $+w(p)$, to all the indices j with $x(p) \leqslant x(q_j)$.

For $j = i$, Lemma 2 implies that

$$
S_\ell(i, i) = T_{\ell-1}(i) + w_i\big(\mathrm{dom}(q_i)\big) = T_{\ell-1}(i).
$$

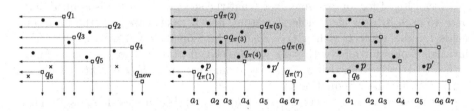

Fig. 4. Left: An example showing the notation. The black dots correspond to P_5 and the black crosses to $P \setminus P_5$. We have $Q_5^{\lrcorner} = \{q_1, q_3, q_5\}$ and $Q_7^{\lrcorner} = Q$. If $w(p) = 1$ for all $p \in P$, $w_5(\mathrm{dom}(q_3)) = 2$ and $w_6(\mathrm{dom}(q_4)) = 5$. Center: At the end of iteration $i = 5$, $a_1, a_7 = -\infty$ (because $\pi(1) = 6, \pi(7) = 7$) while $a_t = S_\ell(i, \pi(t))$ for $t \in \{2, \ldots, 6\}$. Right: During iteration $i = 6$, we handle $p, p' \in P_6 \setminus P_5$. We have $\tau(p) = 2, \tau(p') = 6$. Thus, we make $\mathsf{add}(2, 7, w(p))$, $\mathsf{add}(6, 7, w(p'))$ and $\mathsf{set}(1, T_{\ell-1}(6))$.

Now we explain how we maintain the values $S_\ell(i, j)$ through the algorithm. See Fig. 4, center and right, for an example illustrating the description. We want to consider the points of Q sorted by x-coordinate. Let $\pi: \{1, \ldots, m+1\} \to \{1, \ldots, m+1\}$ be a permutation that sorts (the indices of) the points of Q by non-decreasing x-coordinate. Since the points of Q have different x-coordinates, $x(q_{\pi(1)}) < x(q_{\pi(2)}) < \cdots < x(q_{\pi(m+1)})$. Note that $\pi(m+1) = m+1$ as $q_{m+1} = q_{\text{new}}$. We can compute π and its inverse π^{-1} in $O(m \log m)$ time.

We maintain values $a_1, \ldots, a_{m+1}$. It is convenient to interpret a_t as a value associated to the point $q_{\pi(t)}$, that is, the tth point of Q from left to right. Through the algorithm we maintain the invariant that, at the end of the ith iteration,

$$\forall t: \quad a_t = \begin{cases} -\infty & \text{if } \pi(t) > i, \\ S_\ell(i, \pi(t)) & \text{if } \pi(t) \leqslant i. \end{cases} \tag{5}$$

Note that $S_\ell(i, \pi(t))$ is not defined, when $\pi(t) > i$. Thus, we can interpret $-\infty$ as *undefined*. Before the for loop, we set $a_t = -\infty$ for all t and store the values $a_1, \ldots, a_{m+1}$ in the data structure of Lemma 1.

For each point $p \in P$, let $\tau(p)$ be the smallest integer such that $x(p) \leqslant x(q_{\pi(\tau(p))})$. This means that p is to the right of $q_{\pi(\tau(p)-1)}$ and to the left of or on the vertical line through $q_{\pi(\tau(p))}$. We can compute $\tau(p)$ for all $p \in P$ together in $O((n+m) \log m)$ time, as follows. First, we sort Q by x-coordinates in $O(m \log m)$. Then, for each $p \in P$ we perform a binary search on the x-coordinates of the points of Q. This takes $O(\log m)$ time per point of P. (Note that this computation can be reused for different values of ℓ.)

Computationally, at the ith iteration we maintain the invariant (5) making the following operations:

- For each point p in $P_i \setminus P_{i-1}$, apply $\mathsf{add}(\tau(p), m+1, w(p))$. This takes care that the value $w(p)$ has to be added to $S_\ell(i-1, j)$ to obtain $S_\ell(i, j)$, for all j with $\pi(j) \geqslant \pi(\tau(p))$. See equation (4).
- If $\pi(t) = i$, set a_t to $T_{\ell-1}(i)$ by applying $\mathsf{set}(t, T_{\ell-1}(i))$.

Recall Fig. 4 for an example. Note that the second operation is applied only once to each a_t over the whole algorithm, and it takes care that the invariant holds for the value t with $\pi(t) = i$. The first operation takes care that the invariant holds for all t with $\pi(t) \neq i$. Indeed, if $\pi(t) > i$, we add some finite values to $a_t = -\infty$, and if $\pi(t) < i$, we transform $a_t = S_\ell(i-1, \pi(t))$ to $a_t = S_\ell(i, \pi(t))$ because of equation (4).

Recall that $T_\ell(i) = \max\{S_\ell(i, j) \mid q_j \in Q_i^{\lrcorner}\}$ because of Lemma 2. This implies that, at the end of the ith iteration, we get the value $T_\ell(i)$ by querying $\max(1, \pi^{-1}(i))$. Indeed, if we denote by t the value such that $\pi(t) = i$, $\max(1, t)$ returns $\max\{a_1, \ldots, a_t\} = \max(\{-\infty\} \cup \{S(i, j) \mid j \leqslant i, x(q_j) \leqslant x(q_i)\})$.

Each point of P is considered only once over all iterations. Thus, we perform $O(n+m)$ operations in the representation of $a_1, \ldots, a_{m+1}$. Since each operation takes $O(\log m)$ time, the result follows.   ⌑

Making repeated used of Lemma 3, we can solve the problem maxDominance in $O(k(n+m) \log m)$ time. With some care we can use linear space.

Theorem 1. *The maxDominance problem in the plane for n weighted points in P and m points in Q using at most k points of Q can be solved in $O(k(n+m) \log m)$ time using $O(n+m)$ space.*

Proof. We use the notation $T_\ell(*)$ to denote the table $T_\ell(i)$ for all i. We can compute $T_0(*)$ in $O(m)$ time.

For each $\ell = 1, \ldots, k$, once we have $T_{\ell-1}(*)$ we can compute $T_\ell(*)$ in $O((n + m) \log m)$ time using $O(n + m)$ space because of Lemma 3. The running time over all applications of Lemma 3 together is $O(k(n + m) \log m)$. We reuse the space when computing each $T_\ell(*)$ so that we compute the final $T_k(*)$ in linear space. Then $T_k(m + 1)$ gives the optimal value for the problem maxDominance, but we *do not* find the optimal solution yet because we do not have access to $T_\ell(*)$ for $\ell \leqslant k - 2$ anymore.

We use a standard technique to construct an optimal solution without affecting the asymptotic running time or space. Unrolling the relations in Lemma 2,

$$T_\ell(i) \;=\; \max\{T_{\ell-1}(j) + w_i(\mathrm{dom}(q_j)) \mid q_j \in Q_i^{\lrcorner}\}.$$

During the computation of the table $T_\ell(*)$, we maintain a table of pointers $\pi_\ell(*)$ telling in which index the maximum was achieved. Thus, for each i we have

$$T_\ell(i) \;=\; T_{\ell-1}(\pi_\ell(i)) + w_i\big(\mathrm{dom}(q_{\pi_\ell(i)})\big).$$

Following the pointers $\pi_\ell(i)$, starting from $\pi_k(m+1)$, we can recover an optimal solution. We have $\pi_\ell(i) = i$ when the solution has fewer than k elements.

Set $k' = \lceil k/2 \rceil$. For each $\ell > k'$ and i, we want to store an index $\tau_\ell(i)$ in the table $T_{k'}(*)$ that leads to an optimal solution for $T_\ell(i)$. We achieve this with a small adaptation: for $\ell \leqslant k'$, we just compute the tables $T_\ell(*)$. For $\ell = k' + 1$, we store the pointer $\tau_\ell(i) = \pi_\ell(i)$. For $\ell > k' + 1$, we store the pointer $\tau_\ell(i) = \tau_{\ell-1}(\pi_\ell(i))$.

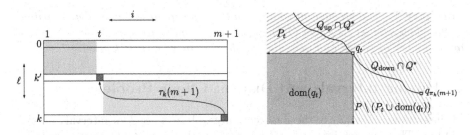

Fig. 5. Left: The two-dimensional table $T(\ell, i)$; the shaded regions are the ones computed during the recursion. Right: Geometric interpretation of the recursion.

Recall that $T_k(m + 1)$ is the optimal value. Set $t = \tau_k(m + 1)$. See Fig. 5. There exists an optimal solution Q^* such that $q_t \in Q^*$. Moreover, such an optimal solution Q^* contains at most $k' - 1 \leqslant k/2$ points from $Q_{\mathrm{up}} := Q_t^{\lrcorner} \setminus \{q_t\}$ and at most $k - k' \leqslant k/2$ points from $Q_{\mathrm{down}} := \{q \in Q \mid y(q) < y(q_t), \; x(q) > x(q_t)\}$. We can use the index t to split the problem because

$$\mathrm{maxDominance}(P, Q, k) \;:=\; w(\mathrm{dom}(q_t)) + \mathrm{maxDominance}(P_t, Q_{\mathrm{up}}, k' - 1)$$
$$+ \,\mathrm{maxDominance}(P \setminus (P_t \cup \mathrm{dom}(q_t)), Q_{\mathrm{down}}, k - k').$$

See Fig. 5, right. Note that each point of P and Q goes to at most one of the recursive problems. We solve the problems recursively. Let $A(k, m, n)$ denote the time needed to construct the optimal solution using this recursive approach. Since we need $O(k(n + m) \log m)$ time to compute the table $T_k(*)$ and $\tau_k(*)$, we have

$$A(k, m, n) \leqslant O(k(n + m) \log m) + A(k' - 1, t - 1, n_1) + A(k - k', m - t, n_2)$$
$$\leqslant O(k(n + m) \log m) + A(k/2, t - 1, n_1) + A(k/2, m - t, n_2),$$

where $n_1 + n_2 \leqslant n$. This solves to $A(k, m, n) = O(k(n + m) \log m)$. (Since $(k/2)(t - 1 + n_1) + (k/2)(m - t + n_2) \leqslant (k/2)(m + n)$, the size (number of cells) of the tables we have to compute at the same level of the recursion decreases geometrically through the recursive calls. See Fig. 5, left.)

At each level of the recursion we need $O(1)$ additional space to keep the indices defining the optimal solution. All other space can be reused. Thus, we spend $O(n + m + k) = O(n + m)$ space.   ▣

When the points have positive weights, we can make some simplifying step as in the following lemma. Its proof can be found in the full version.

Lemma 4. *Assume that the points of P have positive weights. After a preprocessing that takes $O((n + m) \log m)$ time and $O(n + m)$ space, we can assume that no point of Q is dominated by another point of Q and each point of P is dominated by some point of Q.*

As a consequence of Theorem 1 and Lemma 4, we improve the results by Lin et al. [12] mentioned in the introduction when $n = O(m^2 / \log m)$.

Corollary 1. *The problem of computing the top-k representative skyline points in the plane can be solved in $O(kn \log m)$ time and $O(n)$ space, where n is the total number of points and m is the number of points in the skyline.*

4   Hitting Intervals and Disjoint Clique Problem

An interval graph $G(\mathcal{I})$ is the intersection graph of a set $\mathcal{I}$ of intervals on the real line. That is, each interval of $\mathcal{I}$ is a node of the graph $G(\mathcal{I})$, and there is an edge between two nodes of $G(\mathcal{I})$ if and only if the two corresponding intervals intersect. The set of intervals defining the interval graph is the geometric representation or geometric model of the graph. Given an interval graph $G = (V, E)$, we can find a geometric representation in $O(|V| + |E|)$ time [2]. In the opposite direction, given a set $\mathcal{I}$ of n intervals, we can construct the interval graph $G(\mathcal{I})$ in $O(n \log n + |E(G(\mathcal{I}))|)$ time, where $E(G(\mathcal{I}))$ is the edge set of $G(\mathcal{I})$. (The term $O(n \log n)$ comes from sorting the endpoints of the intervals, and becomes $O(n)$ if the input is sorted.) Thus, up to sorting, there is no difference in assuming the input between an interval graph and its geometric model.

A set of intervals on the real line intersect pairwise if and only if they all have a common point. For an interval graph $G(\mathcal{I})$, a subset $\mathcal{I}' \subseteq \mathcal{I}$ of intervals defines a clique if and only if there is a point piercing all the intervals of $\mathcal{I}'$. Thus, the total number of the nodes in k disjoint cliques of an interval graph is the same as the total number of intervals hit (or pierced) by k points corresponding the k cliques. Here, intervals that are hit by more than one point get assigned arbitrarily to only one clique. The same equivalence holds for weighted variants of the problem, if the weights are positive. When there are negative weights, the equivalence does not hold because we may have intervals with negative weights that are hit by a point, but we do not include them into the clique.

Assume that we are given a set $\mathcal{I} = \{I_1, \ldots, I_n\}$ of intervals and we want to solve the HittingIntervals problem. For simplicity, we assume that each interval is closed and that the endpoints of the intervals are all distinct. The intervals $\mathcal{I}$ split the lines into cells (intervals).

Map each interval $I_i = [a_i, b_i]$ to the point $p_i = (a_i, b_i)$ in the xy-plane, and let $P = \{p_1, \ldots, p_n\}$. Since $a_i \leqslant b_i$, every interval is mapped to a point lying on or above the line $y = x$. A point $q \in \mathbb{R}$ hits an interval I_i if and only if p_i is on the top-left quadrant defined by the point (q, q), and denoted by $R_{(q,q)}$. By a counterclockwise rotation of the setting by $\pi/2$ around the origin, $R_{q,q}$ becomes exactly a dominance region, namely $\mathrm{dom}((-q, q))$.

Let X be a set of endpoints of all intervals in $\mathcal{I}$. For the HittingIntervals problem, it is sufficient to choose hitting points from X. Consider the set of $O(n)$ points $Q = \{(q, q) \mid q \in X\}$, which lies on the diagonal $y = x$. Then there is bijection between each candidate solution, restricted to X, for the HittingIntervals problem for $\mathcal{I}$ and each candidate solution for the maxDominance problem for P and Q (after a rotation). By Theorem 1 we conclude the following.

Theorem 2. *The* HittingIntervals *problem for n weighted intervals using at most k hitting points can be solved in $O(kn \log n)$ time and $O(n)$ space. The result also holds when there are negative weights.*

We can also solve the DUC problem for weighted interval graphs. We remove the intervals with negative weights because they will never be in an optimal solution. Then, the problem is equivalent to the weighted HittingIntervals problem.

Theorem 3. *The* DUC *problem for weighted interval graphs $G = (V, E)$ using at most k cliques can be solved in $O(k|V| \log |V| + |E|)$ time and $O(|V| + |E|)$ space.*

References

1. Alrifai, M., Skoutas, D., Risse, T.: Selecting skyline services for QoS-based web service composition. In: Proceedings of the 19th International Conference on World Wide Web, WWW 2010, pp. 11–20. ACM (2010)
2. Booth, K.S., Lueker, G.S.: Testing for the consecutive ones property, interval graphs, and graph planarity using PQ-tree algorithms. J. Comput. Syst. Sci. **13**(3), 335–379 (1976)

3. Bringmann, K., Cabello, S., Emmerich, M.T.M.: Maximum volume subset selection for anchored boxes. In: 33rd International Symposium on Computational Geometry (SoCG 2017), vol. 77. Leibniz International Proceedings in Informatics (LIPIcs), pp. 22:1–22:15. Schloss Dagstuhl-Leibniz-Zentrum fuer Informatik (2017)
4. Bringmann, K., Friedrich, T., Klitzke, P.: Two-dimensional subset selection for hypervolume and epsilon-indicator. In: GECCO, pp. 589–596. ACM (2014)
5. Chrobak, M., Golin, M., Lam, T.-W., Nogneng, D.: Scheduling with gaps: new models and algorithms. In: Paschos, V.T., Widmayer, P. (eds.) CIAC 2015. LNCS, vol. 9079, pp. 114–126. Springer, Cham (2015). https://doi.org/10.1007/978-3-319-18173-8_8
6. Damaschke, P.: Refined algorithms for hitting many intervals. Inf. Process. Lett. **118**, 117–122 (2017)
7. de Berg, M., Cheong, O., van Kreveld, M.J., Overmars, M.H.: Computational Geometry: Algorithms and Applications, 3rd edn. Springer, Heidelberg (2008). https://doi.org/10.1007/978-3-540-77974-2
8. Ertem, Z., Lykhovyd, E., Wang, Y., Butenko, S.: The maximum independent union of cliques problem: complexity and exact approaches. J. Glob. Optim. (2018)
9. Gavril, F.: Algorithms for maximum k-colorings and k-coverings of transitive graphs. Networks **17**(4), 465–470 (1987)
10. Jansen, K., Scheffler, P., Woeginger, G.: The disjoint cliques problem. RAIRO Recherhe Opérationnelle **31**, 45–66 (1997)
11. Kuhn, T., Fonseca, C.M., Paquete, L., Ruzika, S., Duarte, M.M., Figueira, J.R.: Hypervolume subset selection in two dimensions: formulations and algorithms. Evol. Comput. **24**(3), 411–425 (2016)
12. Lin, X., Yuan, Y., Zhang, Q., Zhang, Y.: Selecting stars: the k most representative skyline operator. In: Proceedings of the 23rd International Conference on Data Engineering, ICDE 2007, pp. 86–95 (2007)
13. Tao, Y., Ding, L., Lin, X., Pei, J.: Distance-based representative skyline. In: Proceedings of the 25th International Conference on Data Engineering, ICDE 2009, pp. 892–903. IEEE Computer Society (2009)
14. Yannakakis, M., Gavril, F.: The maximum k-colorable subgraph problem for chordal graphs. Inf. Process. Lett. **24**(2), 133–137 (1987)
15. Yuan, L., Qin, X.L., Chang, L., Zhang, W.: Diversified top-k clique search. VLDB J. **25**(2), 171–196 (2016)

Extending Upward Planar
Graph Drawings

Giordano Da Lozzo$^{(\boxtimes)}$, Giuseppe Di Battista, and Fabrizio Frati

Roma Tre University, Rome, Italy
{giordano.dalozzo,giuseppe.dibattista,fabrizio.frati}@uniroma3.it

Abstract. In this paper we study the computational complexity of the UPWARD PLANARITY EXTENSION problem, which takes as input an upward planar drawing Γ_H of a subgraph H of a directed graph G and asks whether Γ_H can be extended to an upward planar drawing of G.

We show that the UPWARD PLANARITY EXTENSION problem is NP-complete, even if G has a prescribed upward embedding, the vertex set of H coincides with the one of G, and H contains no edge. Conversely, we show that the UPWARD PLANARITY EXTENSION problem can be solved in $O(n \log n)$ time if G is an n-vertex upward planar st-graph. This result improves upon a known $O(n^2)$-time algorithm, which however applies to all n-vertex single-source upward planar graphs. We also show how to solve in polynomial time a surprisingly difficult version of the UPWARD PLANARITY EXTENSION problem, in which the underlying graph of G is a path or a cycle, G has a prescribed upward embedding, H contains no edges, and no two vertices share the same y-coordinate in Γ_H.

1 Introduction

The study of the extensibility of partial representations of graphs has recently become a mainstream in the graph drawing community; see, e.g., [5,12,14–16,23–27,29]. Major contributions in this scenario are the result of Angelini et al. [5], which states that the existence of a planar drawing of a graph G extending a given planar drawing of a subgraph of G can be tested in linear time, and the result of Brückner and Rutter [12], which states that the problem of testing the extensibility of a given partial level planar drawing of a level graph (where each vertex has a prescribed y-coordinate, called *level*) is NP-complete.

Upward planarity is the natural counterpart of planarity for directed graphs. In an upward planar drawing of a directed graph no two edges cross and an edge directed from a vertex u to a vertex v is represented by a curve monotonically increasing in the y-direction from u to v. The study of upward planar drawings is a most prolific topic in the theory of graph visualization [2–4,6–11,13,18,19,21,22,30]. Garg and Tamassia showed that deciding the existence of an upward

This research was supported in part by MIUR Project "MODE" under PRIN 20157EFM5C, by MIUR Project "AHeAD" under PRIN 20174LF3T8, by H2020-MSCA-RISE project 734922 – "CONNECT", and by MIUR-DAAD JMP N° 34120.

© Springer Nature Switzerland AG 2019
Z. Friggstad et al. (Eds.): WADS 2019, LNCS 11646, pp. 339–352, 2019.
https://doi.org/10.1007/978-3-030-24766-9_25

planar drawing is an NP-complete problem [22]. On the other hand, Bertolazzi et al. [7] showed that testing for the existence of an upward planar drawing belonging to a fixed isotopy class of planar embeddings can be done in polynomial time. Further, Di Battista et al. [19] proved that any upward planar graph is a subgraph of an upward planar st-graph and as such it admits a straight-line upward planar drawing.

In this paper, we consider the extensibility of upward planar drawings of directed graphs. Namely, we introduce and study the complexity of the UPWARD PLANARITY EXTENSION (for short, UPE) problem, which is defined as follows. The input is a triple $\langle G, H, \Gamma_H \rangle$, where Γ_H is an upward planar drawing of a subgraph H of a directed graph G; we call H and Γ_H the *partial graph* and the *partial drawing*, respectively. The UPE problem asks whether Γ_H can be extended to an upward planar drawing of G; or, equivalently, whether an upward planar drawing of G exists which coincides with Γ_H when restricted to the vertices and edges of H. We also study the UPWARD PLANARITY EXTENSION WITH FIXED UPWARD EMBEDDING (for short, UPE-FUE) problem, which is the UPE problem with the additional requirement that the drawing of G we seek has to respect a given *upward embedding*, i.e., a left-to-right order of the edges entering and exiting each vertex.

The NP-hardness of the UPWARD PLANARITY TESTING problem [22] directly implies the NP-hardness of the UPE problem, as the former coincides with the special case of the latter in which the partial graph is the empty graph. In the full version of the paper [17], we prove two stronger NP-hardness results. First, we show that the UPE problem is NP-hard even if the partial graph contains all the vertices and no edges, and no three vertices share the same y-coordinate in the partial drawing. This result is established by means of a simple reduction from the ORDERED LEVEL PLANARITY (OLP) problem, introduced and proved to be NP-complete by Klemz and Rote [28]. The input of the OLP problem is a partial drawing of a level graph containing all the vertices and no edges; the problem asks for the existence of a level planar drawing of the graph extending the partial one. Second, we show that the UPE-FUE problem is NP-hard even for connected instances whose partial graph contains all the vertices and no edges. This result is established by means of a non-trivial reduction from the already mentioned PARTIAL LEVEL PLANARITY (PLP) problem by Brückner and Rutter [12]. Our result is in contrast with several constrained embedding problems that are NP-hard when the graph has a variable embedding and efficiently solvable in the fixed embedding setting. Some examples are the UPWARD PLANARITY TESTING problem [7,22], the WINDROSE PLANARITY TESTING problem [1], and the notorious BEND MINIMIZATION IN PLANAR ORTHOGONAL DRAWINGS problem [22,31].

We now present an overview of our algorithmic results. First, we identify two main factors that contribute to the complexity of the UPE and UPE-FUE problems: (i) The presence of edges in the partial graph and (ii) the existence of vertices with the same y-coordinate in the partial drawing. These two properties are strictly tied together. Namely, any instance of the UPE or UPE-FUE problems can be efficiently transformed into an equivalent instance $\langle G, H, \Gamma_H \rangle$ of the same problem in which H contains no edges *or* no two vertices share the same y-coordinate in Γ_H (see Sect. 2). Hence, the NP-hardness results for the UPE and UPE-FUE problems discussed above carry over to such instances, even when $V(G) = V(H)$. When the partial graph contains no edges *and* no two vertices share the same y-coordinate in the partial drawing, then the UPE and UPE-FUE problems appear to be more tractable. Indeed, while we could not establish their computational complexity in general, we could solve them for instances $\langle G, H, \Gamma_H \rangle$ such that the underlying graph of G is a path or a cycle (see Sect. 4). In particular, in order to solve the UPE-FUE problem for paths, we employ a sophisticated dynamic programming approach.

Second, we look at the UPE and UPE-FUE problems for instances $\langle G, H, \Gamma_H \rangle$ such that G is an upward planar *st-graph* (see Sect. 3), i.e., it has a unique source s and a unique sink t. The upward planarity of an n-vertex st-graph is known to be decidable in $O(n)$ time [19,21]. We observe that a result of Brückner and Rutter [12] implies the existence of an $O(n^2)$-time algorithm to solve the UPE problem for upward planar st-graphs; their algorithm works more in general for upward planar single-source graphs. We present $O(n \log n)$-time algorithms for the UPE and UPE-FUE problems for upward planar st-graphs. Notably, these results assume neither that the edge set of H is empty, nor that any two vertices have distinct y-coordinates in Γ_H, nor that $V(G) = V(H)$.

Due to space limitations some theorems and proofs are omitted or sketched and can be found in the full version of the paper [17].

2 Preliminaries

In this section we give some preliminaries and definitions.

Let G be a directed graph. We denote by (u, v) an edge from a vertex u to a vertex v. A path $(u_1, \ldots, u_n)$ in G is *directed* if it consists of the edges (u_i, u_{i+1}), for $i = 1, \ldots, n - 1$. A vertex v is a *successor* (*predecessor*) of a vertex u if G contains a directed path from u to v (from v to u). We denote by $S_G(u)$ (by $P_G(u)$) the set of successors (predecessors) of a vertex u in G.

A drawing of a directed graph G is *upward* if each edge (u, v) is represented by a curve monotonically increasing in the y-direction from u to v. A graph is *upward planar* if it admits an upward planar drawing. Consider an upward planar drawing and a vertex v. The list $\mathcal{S}(v) = [w_1, \ldots, w_k]$ contains the adjacent successors of v in "left-to-right order". That is, consider a half-line ℓ starting at v and directed leftwards; rotate ℓ around v in clockwise direction and append a vertex w_i to $\mathcal{S}(v)$ when ℓ overlaps the tangent to the edge (v, w_i) at v. The list $\mathcal{P}(v) = [z_1, \ldots, z_l]$ of the adjacent predecessors of v is defined similarly. Then two upward planar drawings of a connected directed graph are *equivalent* if they have the same lists $\mathcal{S}(v)$ and $\mathcal{P}(v)$ for each vertex v. An *upward embedding* is an equivalence class of upward planar drawings. Given an upward planar graph G with a fixed upward embedding, and given a subgraph G' of G, we always implicitly assume that G' inherits the upward embedding from G.

We assume that any instance $\langle G, H, \Gamma_H \rangle$ of the UPE and UPE-FUE problems is such that Γ_H is a drawing in which the edges are represented as polygonal

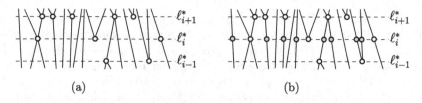

Fig. 1. The drawings Γ_H (a) and $\Gamma_{H'}$ (b) in the proximity of ℓ_i^*. The vertices that are inserted on ℓ_i^* are gray; those inserted on ℓ_{i-1}^* and ℓ_{i+1}^* are not shown.

lines. Then the *size* of $\langle G, H, \Gamma_H \rangle$ is $|\langle G, H, \Gamma_H \rangle| = |V(G)| + |E(G)| + s$, where s is the number of segments of the polygonal lines representing the edges in Γ_H.

Consider an upward planar st-graph G with a fixed upward embedding. In any upward planar drawing of G, every face f is delimited by two directed paths $(u_1, \ldots, u_k)$ and $(v_1, \ldots, v_l)$ connecting the same two vertices $u_1 = v_1$ and $u_k = v_l$. Assuming that $\mathcal{S}(u_1) = [\ldots, u_2, v_2, \ldots]$, we call $(u_1, \ldots, u_k)$ the *left boundary* of f and $(v_1, \ldots, v_l)$ the *right boundary* of f. For a vertex $v \neq t$, the *leftmost outgoing path* $\mathcal{L}_G^+(v) = (w_1, \ldots, w_m)$ of v is the directed path such that $w_1 = v$, $w_m = t$, and $\mathcal{S}(w_i) = [w_{i+1}, \ldots]$, for each $i = 1, \ldots, m - 1$. The *rightmost outgoing path* $\mathcal{R}_G^+(v)$, the *leftmost incoming path* $\mathcal{L}_G^-(v)$ and the *rightmost incoming path* $\mathcal{R}_G^-(v)$ are defined similarly. The paths $\mathcal{L}_G^+(s)$ and $\mathcal{R}_G^+(s)$ are also called *leftmost* and *rightmost path* of G, respectively. Note that these paths delimit the outer face of G. Consider a directed path $\mathcal{Q}$ from s to t. Let $\mathcal{Q}^*$ be obtained by extending $\mathcal{Q}$ with a y-monotone curve directed upwards from t to infinity and with a y-monotone curve directed downwards from s to infinity. Then a vertex u is *to the left* (*to the right*) of $\mathcal{Q}$ if it lies in the region to the left (resp. to the right) of $\mathcal{Q}^*$. In particular, u is *to the left* of a vertex v if it lies to the left of the directed path composed of $\mathcal{L}_G^+(v)$ and $\mathcal{L}_G^-(v)$. Similarly, u is *to the right* of v if it lies to the right of $\mathcal{R}_G^+(v) \cup \mathcal{R}_G^-(v)$. We denote by $L_G(v)$ ($R_G(v)$) the set of vertices that are to the left (resp. right) of a vertex v in G.

2.1   Simplifications

In this section we prove that it is not a loss of generality to restrict our attention to instances $\langle G, H, \Gamma_H \rangle$ of the UPE and UPE-FUE problems in which H contains no edges *or* no two vertices share the same y-coordinate in Γ_H.

Lemma 1. *Let $\langle G, H, \Gamma_H \rangle$ be an instance of the UPE or UPE-FUE problem and let $n = |\langle G, H, \Gamma_H \rangle|$. There exists an equivalent instance $\langle G', H', \Gamma_{H'} \rangle$ of the UPE or UPE-FUE problem, respectively, such that: (i) $E(H') = \emptyset$, (ii) if $V(H) = V(G)$, then $V(H') = V(G')$, and (iii) if G is an st-graph, then G' is an st-graph. Further, the instance $\langle G', H', \Gamma_{H'} \rangle$ has $O(n)$ size and can be constructed in $O(n \log n)$ time. The drawing $\Gamma_{H'}$ may contain vertices with the same y-coordinate even if Γ_H does not.*

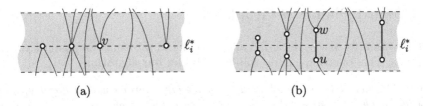

Fig. 2. The strip $\mathcal{S}_i^*$ in the construction of $\langle G', H', \Gamma_{H'} \rangle$.

Proof sketch. The graph G' is obtained from G by replacing the edges of H by directed paths, as described below. Property (iii) is then satisfied. The graph H' is composed of all the vertices of H plus all the internal vertices of the directed paths that are inserted in G' to replace the edges of H. Property (ii) is hence satisfied. Further, H' contains no edge, hence Property (i) is also satisfied. The drawing $\Gamma_{H'}$ coincides with Γ_H when restricted to the vertices in H. It remains to specify the lengths of the directed paths that are inserted in G' to replace the edges of H and to describe how to place their internal vertices in $\Gamma_{H'}$. This is done in the following.

We compute the increasing order $y_1^*, \ldots, y_m^*$ of the y-coordinates of the vertices of H in Γ_H. Let ℓ_i^* be the line with equation $y = y_i^*$. Refer to Fig. 1. We look at the left-to-right order X_i^* in which the vertices of H lying on ℓ_i^* and the edges of H crossing ℓ_i^* appear in Γ_H. We place a vertex v in $\Gamma_{H'}$ at the point in which an edge e of H crosses ℓ_i^* if: (i) e is preceded or followed by a vertex of H in X_i^*; or (ii) e has an end-vertex whose y-coordinate in Γ_H is y_{i-1}^* or y_{i+1}^*; in both such cases v is also a vertex that is internal to the directed path that is inserted in G' to replace e. This concludes the construction of $\langle G', H', \Gamma_{H'} \rangle$. The proof is completed by showing that $\langle G, H, \Gamma_H \rangle$ is a positive instance of the UPE or UPE-FUE problem if and only if $\langle G', H', \Gamma_{H'} \rangle$ is.   □

Lemma 2. *Let $\langle G, H, \Gamma_H \rangle$ be an instance of the UPE or UPE-FUE problem and let $n = |\langle G, H, \Gamma_H \rangle|$. There exists an equivalent instance $\langle G', H', \Gamma_{H'} \rangle$ of the UPE or UPE-FUE problem, respectively, such that:* (i) *no two vertices of H' share the same y-coordinate in $\Gamma_{H'}$ and* (ii) *if $V(H) = V(G)$, then $V(H') = V(G')$. Further, the instance $\langle G', H', \Gamma_{H'} \rangle$ has $O(n)$ size and can be constructed in $O(n \log n)$ time. The graph H' may contain edges even if H does not.*

Proof sketch. By Lemma 1 we can assume that H contains no edges. Let $y_1^*, \ldots, y_m^*$ be the y-coordinates of the vertices of H in Γ_H in increasing order. Let ℓ_i^* be the line with equation $y = y_i^*$. Let $\mathcal{S}_1^*, \ldots, \mathcal{S}_m^*$ be disjoint horizontal strips, where ℓ_i^* is in the interior of $\mathcal{S}_i^*$, for $i = 1, \ldots, m$. Refer to Fig. 2. We define $\langle G', H', \Gamma_{H'} \rangle$ by initializing $G' = G$ and by replacing each vertex $v \in V(H)$ with an edge (u, w), where u gets all the incoming edges of v, while w gets all the outgoing edges of v. The edge (u, w) belongs to H' and is represented in $\Gamma_{H'}$ by a vertical segment with its midpoint at v. Vertical segments corresponding to distinct vertices of H lying on ℓ_i^* have different lengths and lie inside $\mathcal{S}_i^*$.   □

3   Upward Planar st-Graphs

In this section we study the UPE and UPE-FUE problems for upward planar st-graphs. The following lemma will be useful for our algorithms.

Lemma 3. *Let G be an n-vertex upward planar st-graph with a given upward embedding. There exists a data structure to test in $O(1)$ time, for any two vertices u and v of G, whether $v \in S_G(u)$, $v \in P_G(u)$, $v \in L_G(u)$, or $v \in R_G(u)$. Further, such a data structure can be constructed in $O(n)$ time.*

Proof sketch. First, we construct the *transitive reduction* G^* of G, that is, the upward planar st-graph obtained from G by removing all its transitive edges. This can be done in $O(n)$ time. We then exploit the fact that, for each vertex v of G (and of G^*), it holds $S_{G^*}(v) = S_G(v)$, $P_{G^*}(v) = P_G(v)$, $L_{G^*}(v) = L_G(v)$, and $R_{G^*}(v) = R_G(v)$. We use the $O(n)$-time algorithm by Di Battista et al. [21] to construct a *dominance drawing* Γ^* of G^* such that: (i) $x(v) < x(u)$ if and only if $v \in P_G(u) \cup L_G(u)$; and (ii) $y(v) < y(u)$ if and only if $v \in P_G(u) \cup R_G(u)$. Hence, for a query $v \in P_G(u)$, we check whether $x(v) < x(u)$ and $y(v) < y(u)$ in Γ^*. The other queries can be similarly answered in $O(1)$ time.   □

We now present one of our main tools to deal with the UPE and UPE-FUE problems for upward planar st-graphs.

Lemma 4. *An instance $\langle G, H, \Gamma_H \rangle$ of the UPE-FUE problem such that G is an upward planar st-graph with a given upward embedding and such that H contains no edges is a positive instance if and only if:*

Condition 1: *For each vertex v of H, all its successors (predecessors) in G that belong to H have a y-coordinate in Γ_H that is larger (smaller) than $y(v)$; and*

Condition 2: *For each vertex v of H, all the vertices of H whose y-coordinate is the same as $y(v)$ and whose x-coordinate is larger (smaller) than $x(v)$ in Γ_H are to the right (to the left) of v in G.*

Proof sketch. Condition 1 is obviously necessary for the existence of an upward drawing of G extending Γ_H. Suppose that two vertices u, v exist in H such that (i) u is to the left and on the same horizontal line as v in Γ_H and (ii) u is to the right of v in G. Then any two minimal directed paths Q_{uw} and Q_{vw} from u and v to a common vertex w determine, in any upward planar drawing of G extending Γ_H, a list $\mathcal{P}(w)$ of adjacent predecessors of w that does not respect the upward embedding of G. This proves the necessity of Condition 2.

For the sufficiency, we construct an upward planar drawing Γ_G of G that extends Γ_H. First we draw every vertex of G not in H at an exclusive y-coordinate larger than those of its predecessors and smaller than those of its successors; then the instance still satisfies Conditions 1 and 2. Now we draw the edges of G "one face at a time". After each step we maintain the invariants that: (i) the subgraph of G currently drawn consists of an upward planar st-graph G' plus a set of isolated vertices; (ii) the current drawing of G' in Γ_G is upward planar; and (iii) the rightmost path of G' is represented by a y-monotone curve

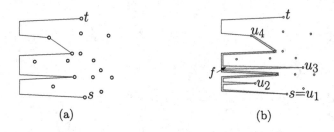

Fig. 3. (a) Drawing the leftmost path $\mathcal{L}_G^+(s)$ of G. (b) Drawing the right boundary $(u_1, \ldots, u_l)$ of a face f.

that has all the isolated vertices to its right. We first draw $\mathcal{L}_G^+(s)$ so to keep all the vertices not in it to its right; see Fig. 3a. Then, repeatedly, we consider a face f whose left boundary belongs to G' and whose right boundary $(u_1, \ldots, u_l)$ consists of edges not in G'; see Fig. 3b. Such a right boundary is a directed path which is drawn upward (this is possible by Condition 1), to the right of the left boundary of f and so close to it that no vertex which is still isolated in the drawing lies to the left of it (this is possible by Condition 2).   □

We can now prove the following algorithmic theorem.

Theorem 1. *The* UPE-FUE *problem can be solved in* $O(n \log n)$ *time for instances* $\langle G, H, \Gamma_H \rangle$ *with size* $n = |\langle G, H, \Gamma_H \rangle|$ *such that* G *is an upward planar* st-graph with a given upward embedding.

Proof sketch. We apply Lemma 1 in $O(n \log n)$ time to modify $\langle G, H, \Gamma_H \rangle$ so that H contains no edges while G remains an upward planar st-graph. Next, we test whether $\langle G, H, \Gamma_H \rangle$ satisfies Conditions 1 and 2 of Lemma 4 in $O(n \log n)$ time.

In order to test Condition 1, we construct an auxiliary graph $\mathcal{A}$, which we initialize to G. We construct in $O(n \log n)$ time a sequence $\mathcal{S}$ in which the vertices of H are ordered by increasing y-coordinates and, secondarily, by increasing x-coordinates in Γ_H. We partition $\mathcal{S}$ into maximal subsequences $\mathcal{S}_1, \ldots, \mathcal{S}_k$ such that all the vertices in $\mathcal{S}_i$ have the same y-coordinate. For every pair $\mathcal{S}_i, \mathcal{S}_{i+1}$ we add to $\mathcal{A}$ a vertex v_i and directed edges from every vertex in $\mathcal{S}_i$ to v_i and from v_i to every vertex in $\mathcal{S}_{i+1}$. Then $\langle G, H, \Gamma_H \rangle$ satisfies Condition 1 if and only if $\mathcal{A}$ is acyclic. The graph $\mathcal{A}$ can be constructed in $O(n \log n)$ time; further, it has $O(n)$ vertices and edges, hence it can be tested in $O(n)$ time whether it is acyclic.

In order to test Condition 2, we look at every pair u, v of consecutive vertices in each sequence $\mathcal{S}_i$ and test whether $u \in L_G(v)$. By Lemma 3, this can be done in $O(1)$ time per query, after an $O(n)$-time preprocessing.   □

Next, we deal with the UPE problem. Notice that an instance $\langle G, H, \Gamma_H \rangle$ of the UPE problem such that G is an upward planar st-graph can be easily transformed into an equivalent instance of the PLP problem. This is due to the fact that Condition 1 of Lemma 4 does not depend on the upward embedding of G and that we can assume: (i) the edge set of H to be empty, by Lemma 1;

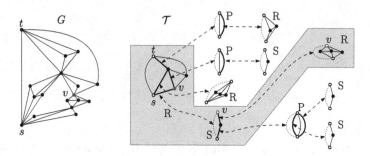

Fig. 4. (left) A biconnected upward planar st-graph G and (right) the SPQR-tree $\mathcal{T}$ of G. The skeletons of all the non-leaf nodes of $\mathcal{T}$ are depicted. The allocation nodes of a vertex v are in the yellow-shaded region.

and (ii) the partial drawing to contain all the vertices of G, by drawing each vertex in $V(G) \setminus V(H)$ as in the proof of Lemma 4 without violating neither Condition 1 nor Condition 2 of the lemma. Hence, the UPE problem for upward planar st-graphs can be solved in quadratic time, due to the results of Brückner and Rutter about the PLP problem for single-source graphs [12]. However, in the following theorem we show how to reduce the time bound to almost linear.

Theorem 2. *The* UPE *problem can be solved in $O(n \log n)$ time for instances $\langle G, H, \Gamma_H \rangle$ with size $n = |\langle G, H, \Gamma_H \rangle|$ such that G is an upward planar st-graph.*

Proof sketch. First, we test in $O(n \log n)$ time whether G satisfies Condition 1 of Lemma 4; this is done as in the proof of Theorem 1. If the test fails, we reject the instance, otherwise we apply Lemma 1 in order to modify $\langle G, H, \Gamma_H \rangle$ so that H contains no edges while G remains an upward planar st-graph.

In order to test whether G admits an upward embedding satisfying Condition 2 of Lemma 4 we proceed as follows. First, we add the edge (s, t) to G, so to ensure the biconnectivity of G. Second, we compute in $O(n \log n)$ time the order $\mathcal{O} = v_1, v_2, \ldots, v_h$ of the vertices in H by increasing y-coordinates and, secondarily, by increasing x-coordinates in Γ_H. Third, we compute in $O(n)$ time the SPQR-tree $\mathcal{T}$ of G (see [20] and Fig. 4). The tree $\mathcal{T}$ represents the recursive arrangement of the triconnected components of G. Roughly speaking, these components might be arranged in a cycle (this corresponds to an *S-node* in $\mathcal{T}$), or might share two vertices and be arranged in parallel (this corresponds to a *P-node* in $\mathcal{T}$), or might be arranged as in a triconnected graph (this corresponds to an *R-node* in $\mathcal{T}$). An auxiliary graph, called *skeleton* and denoted by $sk(\nu)$, is associated to each node ν of $\mathcal{T}$ and represents the corresponding arrangement. Each edge of $sk(\nu)$ corresponds to a subgraph of G, called *pertinent graph*.

Any upward embedding of G can be obtained by choosing a left-to-right order for the edges of the skeleton of each P-node of $\mathcal{T}$ and an upward embedding for the skeleton of each R-node of $\mathcal{T}$. We outline the approach for performing these choices so to satisfy Condition 2. First, we consider each R-node ν of $\mathcal{T}$ and we arbitrarily choose one of the two upward embeddings of $sk(\nu)$; we associate to ν two boolean variables $preserve(\nu)$ and $flip(\nu)$, that we both initially set to $\mathtt{false}$, respectively indicating whether the arbitrarily chosen embedding of $sk(\nu)$ has to be maintained or changed; finally, we set up in $O(|sk(\nu)|)$ time a data structure that, for a pair (x, y) of vertices or edges of $sk(\nu)$, determines in $O(1)$ time whether $x \in L_{sk(\nu)}(y)$, $x \in R_{sk(\nu)}(y)$, or none of the previous, in the chosen upward embedding of $sk(\nu)$; this can be done by Lemma 3.

We now consider any two vertices $u = v_i$ and $v = v_{i+1}$ with the same y-coordinate in Γ_H. Note that $x(u) < x(v)$ in Γ_H. Then $u \in L_G(v)$ in the upward embedding of G we look for. This imposes a constraint on the skeleton $sk(\nu)$ of the lowest common ancestor ν of the proper allocation nodes of u and v in $\mathcal{T}$. Specifically: **(1)** If ν is an **S-node**, then we reject the instance. **(2)** If ν is a **P-node**, then we constrain the edge of $sk(\nu)$ whose pertinent graph contains u to precede the edge of $sk(\nu)$ whose pertinent graph contains v. **(3)** If ν is an **R-node**, then let x_u be the *representative* of u in $sk(\nu)$, that is, if u is a vertex of $sk(\nu)$ then $x_u = u$, otherwise x_u is the edge of $sk(\nu)$ whose pertinent graph contains u. The representative x_v of v in $sk(\nu)$ is defined in the same way. We query in $O(1)$ time the data structure associated to $sk(\nu)$ to test whether $x_u \in L_{sk(\nu)}(x_v)$ (then we set $preserve(\nu) = \mathtt{true}$), or $x_u \in R_{sk(\nu)}(x_v)$ (then we set $flip(\nu) = \mathtt{true}$), or $x_u \notin L_{sk(\nu)}(x_v)$ and $x_u \notin R_{sk(\nu)}(x_v)$ (then we reject the instance).

Finally, for each P-node ν of $\mathcal{T}$, we test whether the precedence constraints imposed on the edges of $sk(\nu)$ induce an acyclic relationship. In case of a negative answer, we reject the instance. For each R-node ν of $\mathcal{T}$, we test whether $preserve(\nu) = \mathtt{false}$ or $flip(\nu) = \mathtt{false}$. In case of a negative answer, we reject the instance. Finally, if we did not reject the instance, then we accept it.   □

4   Paths and Cycles

In this section we deal with the UPE and UPE-FUE problems for instances $\langle G, H, \Gamma_H \rangle$ such that the underlying graph of G is a path or a cycle, H contains no edges, and no two vertices share the same y-coordinate in Γ_H. For the sake of readability, in the following we often just say "path" or "cycle" to address a directed graph whose underlying graph is a path or cycle, respectively.

It turns out that paths and cycles are easy to handle if they do not come with a prescribed upward embedding. Namely, as long as obvious conditions on the y-coordinates of the vertices in the partial drawing are satisfied, an upward planar drawing can be constructed one directed path at a time, so that every new directed path leaves to its left the already drawn directed paths. Hence, we immediately get the following.

Theorem 3. *The* UPE *problem can be solved in* $O(n)$ *time for instances* $\langle G, H, \Gamma_H \rangle$ *such that* G *is an* n-*vertex directed graph whose underlying graph is a path or a cycle,* H *contains no edges, and no two vertices share the same* y-*coordinate in* Γ_H.

Conversely, solving the UPE-FUE problem for paths and cycles, despite the simplicity of their structure, has proved to be challenging.

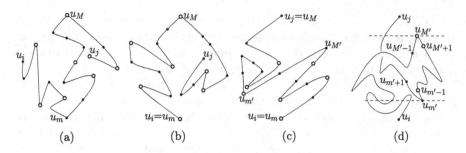

Fig. 5. (a)–(c) Three cases for the computation of the value of $t(u_i, u_j, u_m, u_M)$. (d) Illustration for the necessity of Condition (6).

Theorem 4. *The* UPE-FUE *problem can be solved in* $O(n^4)$ *time for instances* $\langle G, H, \Gamma_H \rangle$ *such that* G *is an n-vertex directed graph whose underlying graph is a path with a given upward embedding,* H *contains no edges, and no two vertices share the same y-coordinate in* Γ_H.

Proof sketch. Let $G = (u_1, \ldots, u_n)$. We show a decision algorithm for the UPE-FUE problem employing dynamic programming. Namely, we fill a table with entries $t(u_i, u_j, u_m, u_M)$, for all the indices $i, j, m, M \in \{1, \ldots, n\}$ with $i \leq m \leq j$, $i \leq M \leq j$, $i \neq j$, and $m \neq M$. Let $G_{i,j} = (u_i, \ldots, u_j)$ and let $\Gamma_{H,i,j}$ be the restriction of Γ_H to the vertices of $G_{i,j}$. Then $t(u_i, u_j, u_m, u_M) =$ TRUE if and only if there is an upward planar drawing $\Gamma_{G,i,j}$ of $G_{i,j}$ that extends $\Gamma_{H,i,j}$ in which u_m and u_M are the vertices with the smallest and largest y-coordinate, respectively. If such a drawing $\Gamma_{G,i,j}$ exists, then we say it is *valid* for $t(u_i, u_j, u_m, u_M)$. The UPE-FUE problem is positive if and only if $t(u_1, u_n, u_m, u_M) =$ TRUE, for some $1 \leq m \leq n$ and $1 \leq M \leq n$ with $m \neq M$.

We start by computing the entries $t(u_i, u_j, u_m, u_M)$ such that $G_{i,j}$ is a directed path. Assume that the edge between u_i and u_{i+1} is outgoing u_i, the other case is symmetric. Then $t(u_i, u_j, u_m, u_M) =$ TRUE if and only if the following conditions are satisfied: (1) $m = i$; (2) $M = j$; and (3) for any two indices i' and j' such that $i' < j'$ and such that $u_{i'}, u_{j'} \in V(H_{i,j})$, we have $y(u_{i'}) < y(u_{j'})$ in $\Gamma_{H,i,j}$.

Assume now that $G_{i,j}$ is not a directed path and that the values of all the entries $t(u_i, u_j, u_m, u_M)$ such that $1 \leq j - i \leq x$ have been computed, for some $x \in \{1, 2, \ldots\}$. After the computation of the entries $t(u_i, u_j, u_m, u_M)$ such that $G_{i,j}$ is a directed path, this is indeed the case with $x = 1$. We compute the values of the entries $t(u_i, u_j, u_m, u_M)$ such that $j - i = x + 1$. We distinguish three cases, based on how many of the equalities $i = m$, $i = M$, $j = m$, and $j = M$ are satisfied, that is, based on how many vertices among u_m and u_M are end-vertices of $G_{i,j}$. Refer to Figs. 5a to c.

In each case we characterize whether $t(u_i, u_j, u_m, u_M) =$ TRUE based on the values of already computed entries of the table and on the possibility of u_m and u_M to be the lowest and highest vertex in an upward planar drawing of $G_{i,j}$ extending $\Gamma_{H,i,j}$, respectively. In this proof sketch, we present such a characterization only for the (most difficult) case in which u_m and u_M are both end-vertices of $G_{i,j}$. We have that $t(u_i, u_j, u_i, u_j) = \text{TRUE}$ if and only if there exist indices $M' \in \{i+1, \ldots, j-2\}$ and $m' \in \{M'+1, \ldots, j-1\}$ such that:

(1) $t(u_i, u_{M'}, u_i, u_{M'}) = \text{TRUE}$;

(2) $t(u_{M'}, u_{m'}, u_{m'}, u_{M'}) = \text{TRUE}$;

(3) $t(u_{m'}, u_j, u_{m'}, u_j) = \text{TRUE}$;

(4) either u_i does not belong to H or u_i has the smallest y-coordinate among the vertices of $G_{i,j}$ in Γ_H;

(5) either u_j does not belong to H or u_j has the largest y-coordinate among the vertices of $G_{i,j}$ in Γ_H; and

(6) either $\mathcal{P}(u_{M'}) = [u_{M'-1}, u_{M'+1}]$ and $\mathcal{S}(u_{m'}) = [u_{m'-1}, u_{m'+1}]$, or $\mathcal{P}(u_{M'}) = [u_{M'+1}, u_{M'-1}]$ and $\mathcal{S}(u_{m'}) = [u_{m'+1}, u_{m'-1}]$.

For the necessity, consider a valid drawing $\Gamma_{G,i,j}$ for $t(u_i, u_j, u_i, u_j)$. Then define $u_{M'}$ as the internal sink of $G_{i,j}$ with the largest y-coordinate in $\Gamma_{G,i,j}$ and $u_{m'}$ as the internal source of $G_{M',j}$ with the smallest y-coordinate in $\Gamma_{G,i,j}$. Restricting $\Gamma_{G,i,j}$ to the vertices and edges of $G_{i,M'}$, $G_{M',m'}$, and $G_{m',j}$ yields valid drawings for $t(u_i, u_{M'}, u_i, u_{M'})$, $t(u_{M'}, u_{m'}, u_{m'}, u_{M'})$, and $t(u_{m'}, u_j, u_{m'}, u_j)$, respectively, which proves the necessity of Conditions (1)–(3). Conditions (4)–(5) hold true since u_i and u_j are the vertices with the smallest and largest y-coordinate in $\Gamma_{G,i,j}$, respectively. Finally, the necessity of Condition (6) is proved by observing that if, say, $\mathcal{P}(u_{M'}) = [u_{M'-1}, u_{M'+1}]$ and $\mathcal{S}(u_{m'}) = [u_{m'+1}, u_{m'-1}]$, then $\Gamma_{G,i,j}$ contains a crossing, as in Fig. 5d.

For the sufficiency, we start from valid drawings $\Gamma_{G,i,M'}$, $\Gamma_{G,M',m'}$, and $\Gamma_{G,m',j}$ for $t(u_i, u_{M'}, u_i, u_{M'})$, $t(u_{M'}, u_{m'}, u_{m'}, u_{M'})$, and $t(u_{m'}, u_j, u_{m'}, u_j)$. We modify $\Gamma_{G,i,M'}$, $\Gamma_{G,M',m'}$, and $\Gamma_{G,m',j}$ so that $u_{M'}$ is at the same point in $\Gamma_{G,i,M'}$ and $\Gamma_{G,M',m'}$, $u_{m'}$ is at the same point in $\Gamma_{G,M',m'}$ and $\Gamma_{G,m',j}$, and u_i (u_j) has the smallest (resp. largest) y-coordinate among all the vertices in $\Gamma_{G,i,M'}$, $\Gamma_{G,M',m'}$, and $\Gamma_{G,m',j}$. Satisfying these properties might require modifying the placement of u_i, $u_{M'}$, $u_{m'}$, and u_j, and scaling parts of $\Gamma_{G,i,M'}$, $\Gamma_{G,M',m'}$, and $\Gamma_{G,m',j}$. Gluing together these drawings results in an upward drawing $\Gamma_{G,i,j}$ of $G_{i,j}$ that extends $\Gamma_{H,i,j}$ and in which u_i and u_j are the vertices with the smallest and largest y-coordinate, respectively. However, $\Gamma_{G,i,j}$ might contain crossings and the left-to-right order of the edges incoming at $u_{M'}$ (outgoing from $u_{m'}$) in $\Gamma_{G,i,j}$ might not correspond to $\mathcal{P}(u_{M'})$ (resp. to $\mathcal{S}(u_{m'})$). We overcome these issues by redrawing the curves representing the edges of $G_{i,M'}$, $G_{M',m'}$, and $G_{m',j}$ in internally-disjoint regions of the plane, without changing the position of any vertex.

The quartic running time comes from the number of entries of the dynamic programming table, which is in fact $\Theta(n^4)$. $\qquad\square$

By exploiting arguments analogous to those in the proof of Theorem 4 we can extend our quartic-time algorithm to cycles.

Theorem 5. *The* UPE-FUE *problem can be solved in* $O(n^4)$ *time for instances* $\langle G, H, \Gamma_H \rangle$ *such that* G *is an n-vertex cycle with given upward embedding, H contains no edges, and no two vertices share the same y-coordinate in* Γ_H.

Proof. Suppose that an upward planar drawing Γ_G of $G = (u_1, \ldots, u_n)$ extending Γ_H exists. Since no two vertices share the same y-coordinate in Γ_H, we can assume w.l.o.g. that no two vertices share the same y-coordinate in Γ_G either. Our strategy is to test, for every possible pair of vertices (u_m, u_M) with $m, M \in \{1, \ldots, n\}$ and with $m \neq M$, whether there is an upward planar drawing Γ_G of G extending Γ_H in which the vertices with the smallest and largest y-coordinate are u_m and u_M, respectively. For any pair (u_m, u_M), the cycle G consists of two paths connecting u_m and u_M, call them $G_{m,M} = (u_m, u_{m+1}, \ldots, u_M)$ and $G_{M,m} = (u_M, u_{M+1}, \ldots, u_m)$, where indices are modulo n. Let $\Gamma_{H,m,M}$ and $\Gamma_{H,M,m}$ be the restrictions of Γ_H to the vertices that belong to $G_{m,M}$ and $G_{M,m}$, respectively. Then, G has an upward planar drawing extending Γ_H in which the vertices with the smallest and largest y-coordinate are u_m and u_M, respectively, if and only if: **(1)** $G_{m,M}$ has an upward planar drawing extending $\Gamma_{H,m,M}$ in which the vertices with the smallest and largest y-coordinate are u_m and u_M, respectively; **(2)** $G_{M,m}$ has an upward planar drawing extending $\Gamma_{H,M,m}$ in which the vertices with the smallest and largest y-coordinate are u_m and u_M, respectively; and **(3)** either $\mathcal{P}(u_M) = [u_{M-1}, u_{M+1}]$ and $\mathcal{S}(u_m) = [u_{m+1}, u_{m-1}]$, or $\mathcal{P}(u_M) = [u_{M+1}, u_{M-1}]$ and $\mathcal{S}(u_m) = [u_{m-1}, u_{m+1}]$.

From a computational point of view, we act as follows. First we compute, for every possible pair of vertices (u_m, u_M) with $m, M \in \{1, \ldots, n\}$ and with $m \neq M$, whether there are upward planar drawings of $G_{m,M}$ and $G_{M,m}$ extending $\Gamma_{H,m,M}$ and $\Gamma_{H,M,m}$, respectively, in which the vertex with the smallest y-coordinate is u_m and the vertex with the largest y-coordinate is u_M. This can be done by considering the $2n$-vertex path $(u_1, u_2, \ldots, u_n, u_{n+1} = u_1, u_{n+2} = u_2, \ldots, u_{2n} = u_n)$ and by setting up a dynamic programming table with entries $t(u_i, u_j, u_{m'}, u_{M'})$, for all the indices $i, j, m', M' \in \{1, \ldots, 2n\}$ such that $i \leq m' \leq j$ and $i \leq M' \leq j$, with $i \neq j$, $m' \neq M'$, and $j - i \leq n$. The values of the entries of this table can be computed in total $O(n^4)$ time as in the proof of Theorem 4.

Then, for each of the $O(n^2)$ pairs of vertices (u_m, u_M) with $m, M \in \{1, \ldots, n\}$ and with $m \neq M$, we query the table constructed in the step above to check whether G has an upward planar drawing extending Γ_H in which the vertices with the smallest and largest y-coordinate are u_m and u_M, respectively. Concerning Conditions (1) and (2), we check in $O(1)$ time whether $t(u_m, u_M, u_m, u_M) = \text{TRUE}$ and $t(u_M, u_{n+m}, u_{n+m}, u_M) = \text{TRUE}$ (if $m < M$) or whether $t(u_M, u_m, u_m, u_M) = \text{TRUE}$ and $t(u_m, u_{n+M}, u_m, u_{n+M}) = \text{TRUE}$ (if $m > M$). Condition (3) can also be trivially checked in $O(1)$ time. □

5   Conclusions and Open Problems

In this paper we introduced and studied the UPWARD PLANARITY EXTENSION (UPE) problem, which takes as input an upward planar drawing Γ_H of a subgraph H of a directed graph G and asks whether an upward planar drawing of G exists which coincides with Γ_H when restricted to the vertices and edges of H.

We proved that the UPE problem is NP-complete, even if G has a prescribed upward embedding and H contains all the vertices and no edges. Conversely, the problem can be solved efficiently for upward planar st-graphs.

Several questions are left open by our research. We cite our favorite two. First, is it possible to solve the UPE-FUE problem in polynomial time for instances $\langle G, H, \Gamma_H \rangle$ such that H contains no edges and no two vertices have the same y-coordinate in Γ_H? We proved that if any of the two conditions is dropped, then the UPE-FUE problem is NP-hard, however we can positively answer the above question only if G is a directed path or cycle. Second, are the UPE and UPE-FUE problems polynomial-time solvable for directed paths and cycles? Even when H contains no edges and no two vertices have the same y-coordinate in Γ_H, answering the above question in the affirmative was not a trivial task.

Acknowledgments. Lemma 4 comes from a research session the third author had with Ignaz Rutter, to which our thanks go.

References

1. Angelini, P., et al.: Windrose planarity: embedding graphs with direction-constrained edges. ACM Trans. Algorithms **14**(4), 54:1–54:24 (2018)
2. Angelini, P., Da Lozzo, G., Di Battista, G., Frati, F.: Strip planarity testing for embedded planar graphs. Algorithmica **77**(4), 1022–1059 (2017)
3. Angelini, P., Da Lozzo, G., Di Battista, G., Frati, F., Patrignani, M., Rutter, I.: Beyond level planarity. In: Hu, Y., Nöllenburg, M. (eds.) GD 2016. LNCS, vol. 9801, pp. 482–495. Springer, Cham (2016). https://doi.org/10.1007/978-3-319-50106-2_37
4. Angelini, P., Da Lozzo, G., Di Battista, G., Frati, F., Roselli, V.: The importance of being proper: (in clustered-level planarity and T-level planarity). Theor. Comput. Sci. **571**, 1–9 (2015)
5. Angelini, P., et al.: Testing planarity of partially embedded graphs. ACM Trans. Algorithms **11**(4), 32:1–32:42 (2015)
6. Bertolazzi, P., Di Battista, G., Didimo, W.: Quasi-upward planarity. Algorithmica **32**(3), 474–506 (2002)
7. Bertolazzi, P., Di Battista, G., Liotta, G., Mannino, C.: Upward drawings of tri-connected digraphs. Algorithmica **12**(6), 476–497 (1994)
8. Bertolazzi, P., Di Battista, G., Mannino, C., Tamassia, R.: Optimal upward planarity testing of single-source digraphs. SIAM J. Comput. **27**(1), 132–169 (1998)
9. Binucci, C., Didimo, W.: Computing quasi-upward planar drawings of mixed graphs. Comput. J. **59**(1), 133–150 (2016)
10. Binucci, C., Lozzo, G.D., Giacomo, E.D., Didimo, W., Mchedlidze, T., Patrignani, M.: Upward book embeddings of st-graphs. In: Barequet, G., Wang, Y. (eds.) 35th Symposium on Computational Geometry (SoCG 2019), LIPIcs (2019)
11. Brandenburg, F.: Upward planar drawings on the standing and the rolling cylinders. Comput. Geom. **47**(1), 25–41 (2014)
12. Brückner, G., Rutter, I.: Partial and constrained level planarity. In: Klein, P.N. (ed.) SODA 2017, pp. 2000–2011. SIAM (2017)
13. Chaplick, S., et al.: Planar L-drawings of directed graphs. In: Frati, F., Ma, K.-L. (eds.) GD 2017. LNCS, vol. 10692, pp. 465–478. Springer, Cham (2018). https://doi.org/10.1007/978-3-319-73915-1_36

14. Chaplick, S., Dorbec, P., Kratochvíl, J., Montassier, M., Stacho, J.: Contact representations of planar graphs: extending a partial representation is hard. In: Kratsch, D., Todinca, I. (eds.) WG 2014. LNCS, vol. 8747, pp. 139–151. Springer, Cham (2014). https://doi.org/10.1007/978-3-319-12340-0_12

15. Chaplick, S., Fulek, R., Klavík, P.: Extending partial representations of circle graphs. In: Wismath, S., Wolff, A. (eds.) GD 2013. LNCS, vol. 8242, pp. 131–142. Springer, Cham (2013). https://doi.org/10.1007/978-3-319-03841-4_12

16. Chaplick, S., Guspiel, G., Gutowski, G., Krawczyk, T., Liotta, G.: The partial visibility representation extension problem. Algorithmica 80(8), 2286–2323 (2018)

17. Da Lozzo, G., Di Battista, G., Frati, F.: Extending upward planar graph drawings. CoRR, abs/1902.06575 (2019)

18. Da Lozzo, G., Di Battista, G., Frati, F., Patrignani, M., Roselli, V.: Upward planar morphs. In: Biedl, T., Kerren, A. (eds.) GD 2018. LNCS, vol. 11282, pp. 92–105. Springer, Cham (2018). https://doi.org/10.1007/978-3-030-04414-5_7

19. Di Battista, G., Tamassia, R.: Algorithms for plane representations of acyclic digraphs. Theor. Comput. Sci. 61, 175–198 (1988)

20. Di Battista, G., Tamassia, R.: On-line planarity testing. SIAM J. Comput. 25(5), 956–997 (1996)

21. Di Battista, G., Tamassia, R., Tollis, I.G.: Area requirement and symmetry display of planar upward drawings. Discrete Comput. Geom. 7, 381–401 (1992)

22. Garg, A., Tamassia, R.: On the computational complexity of upward and rectilinear planarity testing. SIAM J. Comput. 31(2), 601–625 (2001)

23. Jelínek, V., Kratochvíl, J., Rutter, I.: A Kuratowski-type theorem for planarity of partially embedded graphs. Comput. Geom. 46(4), 466–492 (2013)

24. Klavík, P., Kratochvíl, J., Krawczyk, T., Walczak, B.: Extending partial representations of function graphs and permutation graphs. In: Epstein, L., Ferragina, P. (eds.) ESA 2012. LNCS, vol. 7501, pp. 671–682. Springer, Heidelberg (2012). https://doi.org/10.1007/978-3-642-33090-2_58

25. Klavík, P., et al.: Extending partial representations of proper and unit interval graphs. Algorithmica 77(4), 1071–1104 (2017)

26. Klavík, P., Kratochvíl, J., Otachi, Y., Saitoh, T.: Extending partial representations of subclasses of chordal graphs. Theor. Comput. Sci. 576, 85–101 (2015)

27. Klavík, P., Kratochvíl, J., Otachi, Y., Saitoh, T., Vyskocil, T.: Extending partial representations of interval graphs. Algorithmica 78(3), 945–967 (2017)

28. Klemz, B., Rote, G.: Ordered level planarity, geodesic planarity and bi-monotonicity. In: Frati, F., Ma, K.-L. (eds.) GD 2017. LNCS, vol. 10692, pp. 440–453. Springer, Cham (2018). https://doi.org/10.1007/978-3-319-73915-1_34

29. Patrignani, M.: On extending a partial straight-line drawing. Int. J. Found. Comput. Sci. 17(5), 1061–1070 (2006)

30. Rextin, A., Healy, P.: Dynamic upward planarity testing of single source embedded digraphs. Comput. J. 60(1), 45–59 (2017)

31. Tamassia, R.: On embedding a graph in the grid with the minimum number of bends. SIAM J. Comput. 16(3), 421–444 (1987)

Reconfiguring Undirected Paths

Erik D. Demaine[1], David Eppstein[2(✉)], Adam Hesterberg[3], Kshitij Jain[4],
Anna Lubiw[4], Ryuhei Uehara[5], and Yushi Uno[6]

[1] MIT Computer Science and Artificial Intelligence Laboratory,
32 Vassar Street, Cambridge, MA 02139, USA
edemaine@mit.edu
[2] Computer Science Department, University of California, Irvine, Irvine,
CA 92697, USA
eppstein@uci.edu
[3] MIT Mathematics Department, 77 Massachusetts Avenue, Cambridge,
MA 02139, USA
achesterberg@gmail.com
[4] University of Waterloo, Waterloo, Canada
{k22jain,alubiw}@uwaterloo.ca
[5] Japan Advanced Institute of Science and Technology, Nomi, Japan
uehara@jaist.ac.jp
[6] Graduate School of Engineering, Osaka Prefecture University, Sakai, Japan
uno@cs.osakafu-u.ac.jp

Abstract. We consider problems in which a simple path of fixed length, in an undirected graph, is to be shifted from a start position to a goal position by moves that add an edge to either end of the path and remove an edge from the other end. We show that this problem may be solved in linear time in trees, and is fixed-parameter tractable when parameterized either by the cyclomatic number of the input graph or by the length of the path. However, it is PSPACE-complete for paths of unbounded length in graphs of bounded bandwidth.

1 Introduction

In this paper, we consider the problem of sliding a fixed-length simple path within an undirected graph from a given starting position to a given goal position. The path may move in steps where we add an edge to either end of the path and simultaneously remove the edge from the opposite end, maintaining its length. Effectively, this can be thought of as sliding the path one step along its length in either direction. The allowed movements of the path are similar to those of trains in a switchyard, or of the model trains in any of several train shunting puzzles; the edges of the path can be thought of as the cars of a train. However, unlike train tracks, we do not constrain connections at junctions of track segments to be smooth: a path that enters a vertex along an incident edge can exit the vertex along any other incident edge. Additionally, we do not distinguish the two ends of the path from each other.

Supported in part by NSF grants CCF-1618301 and CCF-1616248.

© Springer Nature Switzerland AG 2019
Z. Friggstad et al. (Eds.): WADS 2019, LNCS 11646, pp. 353–365, 2019.
https://doi.org/10.1007/978-3-030-24766-9_26

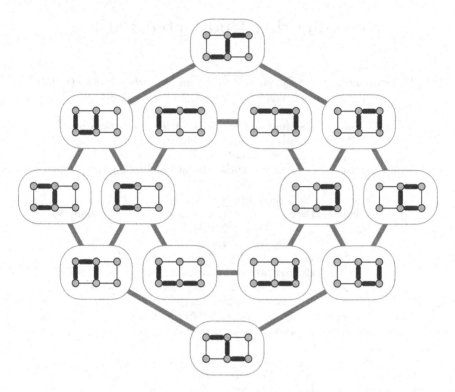

Fig. 1. State space of three-edge paths on a six-vertex graph

Our aim is to understand the computational complexity of two natural reconfiguration problems for such paths: the *decision problem*, of testing whether it is possible to reach the goal position from the start position, and the *optimization problem*, of reaching the goal from the start in as few moves as possible. One natural upper bound for the complexity of these problems is the size of the state space for the problem, a graph whose vertices are paths of equal length on the given graph and whose edges represent moves from one path to another (Fig. 1). If a given graph has N paths of the given length, and M moves from one path to another, we can solve either the decision problem or the optimization problem in time $O(M + N)$ (after constructing the state space) by a simple breadth-first search. As we will see, it is often possible to achieve significantly faster running times than this naive bound. On the other hand, the general problem is hard, even on some highly restricted classes of graphs.

Specifically, we prove the following results:

1. The decision problem for path reconfiguration is fixed-parameter tractable when parameterized by the length of the path. This stands in contrast to the size of the state space for the problem which (for paths of length k in n-vertex graphs) can have as many as $\Omega(n^{k+1})$ states.

2. For paths of unbounded length in graphs parameterized by the circuit rank, both the decision and the optimization problems can be solved in fixed-parameter tractable time by state space search. The same problem can be solved in polynomial (but not fixed-parameter tractable) time when parameterized by feedback vertex set number.
3. The optimization problem for path reconfiguration in trees can be solved in linear time, even though the state space for the problem has quadratic size.
4. The decision problem for path reconfiguration is PSPACE-complete for paths of unbounded length, even when restricted to graphs of bounded bandwidth. Therefore (unless P = PSPACE) path reconfiguration is not fixed-parameter tractable when parameterized by bandwidth, treewidth, or related graph parameters.

Because of limited space, the detailed versions of several of our results are deferred to the full version of this paper (arXiv:1905.00518).

1.1 Related Work

There has been much past research on reconfiguring structures in graphs, with motivations that include motion planning, understanding the mixing of Markov chains and bounding the computational complexity of popular games and puzzles. See, for instance, Ito et al. [1] for many early references, and Mouawad et al. [2] for more recent work on the parameterized complexity of these problems. Often, in these problems, one considers moves in which the structure changes by the removal of one element and the addition of an unrelated replacement element (token moving) or in which an element of the structure changes only locally, by moving along an edge of the graph (token sliding).

Several authors have considered problems of reconfiguring paths or shortest paths under token jumping or token sliding models of reconfiguration [3–5]. However, the path sliding moves that we consider are different. Token sliding moves only a single vertex or edge of a path along a graph edge, while we move the whole path. And although our path sliding moves can be seen as a special case of token jumping, because they remove one edge and add a different edge, token jumping in general would allow the replacement of edges or vertices in the middle of a path, while we allow changes only at the ends of the path.

The path reconfiguration problem that we study here is also closely related to a popular video game, Snake, which has a very similar motion to the path sliding moves that we consider. Our problem differs somewhat from Snake in that we consider bidirectional movement, while in Snake the motion must always be forwards. Snake is typically played on grid graphs, and it is known to be PSPACE-complete to determine whether the Snake can reach a specific goal state from a given start state on generalized grid graphs [6]. Independently of our work, Gupta et al. [7] have found that reconfiguring snakes (paths that can move only unidirectionally) is fixed-parameter tractable in the length of the path, analogously to our Theorem 1.

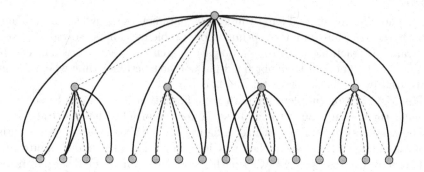

Fig. 2. A graph G of tree-depth 2 (solid black edges) and a tree T realizing this depth (dashed blue edges). (Color figure online)

2   Preliminaries

2.1   Reconfiguration Sequences and Time Reversal

Definition 1. *We define a* reconfiguration step *in a graph G to be a pair of edges (e, f), and a* reconfiguration sequence *to be a sequence σ of reconfiguration steps. We may apply a reconfiguration step to a path P by adding edge e to P and removing edge f, whenever f is one of the two edges at the ends of P, e is incident to the vertex at the other end, and the result of the application is another simple path. We may apply a reconfiguration sequence to a path by performing a sequence of applications of its reconfiguration steps. If applying reconfiguration sequence σ to path P produces another path Q we say that we can reconfigure P into Q or that σ takes P to Q.*

If (e, f) is a reconfiguration step, then we define its time reversal *to be the step (f, e). We define the time-reversal of a reconfiguration sequence σ to be the sequence of time reversals of the steps of σ, taken in the reverse order. If σ takes P to Q, then its time reversal takes Q to P. For this reason, when we seek the existence of a reconfiguration sequence (the path reconfiguration decision problem) or the shortest reconfiguration sequence (the path reconfiguration optimization problem), reconfiguring a path P to Q is equivalent under time reversal to reconfiguring Q to P. We call this equivalence* time-reversal symmetry.

We define the length *$|P|$ of a path P to be its number of edges, and the length $|\sigma|$ of a reconfiguration sequence to be its number of steps.*

2.2   Tree-Depth

Tree-depth is a graph parameter that can be defined in several equivalent ways [8], but the most relevant definition for us is that the tree-depth of a connected graph G is the minimum depth of a rooted tree T on the vertices of G such that each edge of G connects an ancestor-descendant pair of T (Fig. 2). Here, the depth of a tree is the length of the longest root-to-leaf path. Another way of expressing the connection between G and T is that T is a depth-first search tree for a supergraph of G. For disconnected graphs one can use a forest in place of a tree, but we will only consider tree-depth for connected graphs.

Tree-depth is a natural graph parameter to use for path configuration, because it is closely connected to the lengths of paths in graphs. If a graph G has maximum path-length ℓ, then clearly its tree-depth can be at most ℓ, because any depth-first search tree of G itself will achieve that depth. In the other direction, a graph with tree-depth d has maximum path-length at most $2^{d+1} - 2$, as can be proven inductively by splitting any given path at the vertex closest to the root of a tree T realizing the tree-depth. Therefore, the tree-depth and maximum path-length are equivalent for the purposes of determining fixed-parameter tractability. The parameterized complexity of reconfiguration problems on graphs of bounded tree-depth has been studied by Wrochna [9]. However, these graphs are highly constrained, so algorithms that are parameterized by tree-depth are not widely applicable.

We will prove as a lemma that path reconfiguration is fixed-parameter tractable for the graphs of bounded tree-depth. Because these graphs have bounded path lengths, this result will be subsumed in our theorem that path reconfiguration is fixed-parameter tractable when parameterized by path-length. However, we will use this lemma as a stepping-stone to the theorem, by proving that in arbitrary graphs we can either find a structure that allows us to solve the problem easily or restrict the input to a subgraph of bounded tree-depth.

3   Parameterized by Path Length

In this section we show that path reconfiguration is fixed-parameter tractable when parameterized by path length. As discussed above, our strategy is to find a structure (*loose paths*, defined below), whose existence allows us to solve the reconfiguration problem directly. When these structures do not exist or exist but cannot be used, we will instead restrict our attention to a subgraph of bounded tree-depth. We begin with the lemma that the problem is fixed-parameter tractable when parameterized by tree-depth instead of path length.

3.1   Tree-Depth

Our method for graphs of low tree-depth is based on the fact that, when these graphs are large, they contain a large amount of redundant structure: subgraphs that are all connected to the rest of the graph in the same way as each other. When this happens, we can eliminate some copies of the redundant structures and reduce the problem to a smaller instance size.

Definition 2. *Given a graph G and a vertex set S, we define an S-flap to be a subset X of the vertices of G such that X is disjoint from S and there are no edges from X to $G \setminus \{S \cup X\}$. We say that two S-flaps X and Y are equivalent when the induced subgraphs $G[S \cup X]$ and $G[S \cup Y]$ are isomorphic, by an isomorphism that reduces to the identity mapping on S (Fig. 3).*

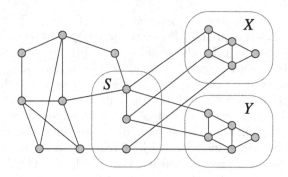

Fig. 3. Two equivalent S-flaps X and Y in a graph G

Observation 1. *For any graph G and any vertex set S, a path of length k can include vertices from at most $\lceil (k-1)/2 \rceil$ S-flaps of G.*

Proof. The path has $k+1$ vertices, and any two vertices in distinct flaps must be separated by at least one vertex of S. □

Lemma 1. *Suppose we are given an instance of path reconfiguration for paths of length k in a graph G, and that G contains a subset S that is disjoint from the start and goal positions of the path and has more than $\lceil (k+1)/2 \rceil$ pairwise equivalent S-flaps $X_1, X_2, \ldots$, all disjoint from the start and goal. Then we can construct an equivalent and smaller instance by removing all but $\lceil (k+1)/2 \rceil$ of these equivalent S-flaps.*

Proof. Any reconfiguration sequence in the original graph can be transformed into a reconfiguration sequence for the reduced graph by using one of the remaining S-flaps whenever the sequence for the original graph enters an S-flap. Because the S-flaps are equivalent, the moves within the flap can be mapped to each other by the isomorphism defining their equivalence, and by Observation 1 there will always be a free S-flap to use in the reduced graph. □

Lemma 2. *We can solve the decision or optimization problems for path reconfiguration in time that is fixed-parameter tractable in the tree-depth of the input graph.*

Proof. We provide a polynomial-time kernelization algorithm that uses Lemma 1 to reduce the instance to an equivalent instance whose size is a function only of the given tree-depth d. The problem can then be solved by a brute-force search on the resulting smaller instance. We assume without loss of generality that we already have a tree decomposition T of depth d, as it is fixed-parameter tractable to find such a decomposition when one is not already given [8, p. 138]. Recall that, for graphs of tree-depth d, the length k of the paths being reconfigured can be at most $2^{d+1} - 2$.

We apply Lemma 1 in a sequence of stages so that, after stage i, all vertices at height i in T have $O(1)$ children. As a base case, for stage 0, all vertices at height 0 in T automatically have 0 children, because they are the leaves of T. Therefore, suppose by induction on i that all vertices at height less than i in T have $O(1)$ children.

For a given vertex v at height i, let S_v be the set of ancestors of v in T (including v itself). Then, for each child w of v in T, let X_w be the set of descendants of w (including w itself). Then X_w is an S_v-flap, because S_v includes all of its ancestors in T and it can have no edges to vertices that are not ancestors in T. If we label each vertex in T by the set of heights of its adjacent ancestors, then the isomorphism type of $G[S_v \cup X_w]$ is determined by these labels, so two children u and w of T have equivalent S_v-flaps whenever they correspond to isomorphic labeled subtrees of W. Trees of constant size with a constant number of label values can have a constant number of isomorphism types, so there are a constant number of equivalence classes of S_v flaps among the sets W_x. Within each equivalence class, we apply Lemma 1 to reduce the number of flaps within that equivalence class to a constant. After doing so, we have caused the vertices of T at height i to have a constant number of children, completing the induction proof.

To implement this method in polynomial time, we can use any polynomial time algorithm for isomorphism of labeled trees [10]. The equivalence of subtrees of T by labeled isomorphism may be finer than the equivalence of the corresponding subgraphs of G by graph isomorphism (because two different labeled trees may correspond to isomorphic subgraphs) but using the finer equivalence relation nevertheless leaves us with a kernel of size depending only on d. The time for this algorithm can be bounded by a polynomial, independent of the parameter.   □

As the following observation shows, this result is nontrivial in the sense that its time bound is significantly smaller than the worst-case bound on the size of the state space for the problem.

Observation 2. *In graphs of tree-depth d, the number of paths of a given length can be $\Theta(n^{2^d})$.*

Proof. Let T be a tree realizing the depth of the given graph. To prove that the number of paths is $O(n^{2^d})$, consider the vertex v in any path that is highest in tree T, and apply the same bound inductively for the two parts of the path on either side of v, both of which must live in lower-depth subtrees. The total number of paths can be at most the product of the numbers of choices for these two smaller paths.

To prove that the number of paths can be $\Omega(n^{2^d})$, let T be a star as the base case for depth one (with $\Omega(n^2)$ paths of length two) and at each higher depth connect two inductively-constructed subtrees through a new root vertex v. Given a tree T constructed in this way, let G be the graph of all ancestor-descendant pairs in T (Fig. 4). Each two paths in the two subtrees can be connected to each other through v, so the number of paths in the whole graph is the product of the numbers of paths in the two subtrees.   □

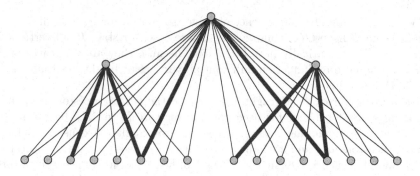

Fig. 4. One of $\Omega(n^4)$ paths of length 6 in a graph of tree-depth 2

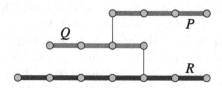

Fig. 5. A loose path R for start and goal paths P and Q

Therefore, an algorithm that searched the entire state space would only be in XP, not FPT.

3.2 Loose Paths

We have seen that graphs without long paths are easy for path reconfiguration. Next, we show that graphs with long paths are also easy. The following definition is central to this part of our results:

Definition 3. *Consider an instance of path reconfiguration consisting of a graph G, a start path P of length k, and a goal path Q of length k. We define a loose path to be a simple path R of length $2k$ in G, such that R is vertex-disjoint from both P and Q (Fig. 5).*

Lemma 3. *Let R be a loose path for an instance (G, P, Q) of path reconfiguration, such that it is possible to reconfigure path P into a path that uses at least one vertex of R. Then for every vertex v in R, it is possible to reconfigure path P into a sub-path of R for which v is an endpoint.*

Proof. Consider a sequence σ of reconfiguration steps starting from P that results in a path using at least one vertex of R and is as short as possible. Because σ is as short as possible and R is disjoint from P, the last move of σ must cause exactly one vertex u of R to be an endpoint of the reconfigured path. Because R has length $2k$, at least one endpoint of R is at distance k or more along R from u. By sliding the path along R towards this endpoint, we can reconfigure it so that it lies entirely along R. Again, because R has length $2k$, one of the two sub-paths of R ending at v has length at least k. By concatenating to σ an additional sequence of steps that slide the path along R (if necessary) we can reconfigure the starting path so that it lies within this sub-path and ends at v. □

We call a loose path R that meets the conditions of Lemma 3 a *reachable loose path*.

Lemma 4. *If an instance (G, P, Q) of path reconfiguration has a reachable loose path, and the graph G is connected, then all loose paths for that instance are reachable.*

Proof. Let R be a reachable loose path, and L be any other loose path. If R and L share a vertex v, then it is possible to slide any sub-path of R so that it includes this vertex, showing that L meets the conditions of Lemma 3. If R and L are disjoint, let T be a shortest path between them in G, and let v be the unique vertex of T that belongs to R. By Lemma 3, we can reconfigure the starting path so that it lies along R and ends at v. From there, we can slide the path along T until it reaches the other endpoint of T, a vertex of L. This shows that L meets the conditions of Lemma 3. □

It will be helpful to bound the tree-depth of graphs with no loose path.

Observation 3. *If an instance of path reconfiguration for paths of length k has no loose path, then its graph has tree-depth less than $4k$.*

Proof. Form a depth-first-search forest F of the subgraph formed by removing all vertices of the start and goal paths. Because there is no loose path, F has depth at most $2k - 1$. Form a single rooted path R of the vertices of the start and goal paths, in an arbitrary order. Connect R and F into a single tree T (not necessarily a subtree of the input graph) by making each root of F be a child of the leaf node of R. Then every edge in the given graph connects an ancestor–descendant pair in T, because either it connects two vertices in the depth-first-search forest or it has at least one endpoint on the ancestral path R. Thus, T meets the condition for trees realizing the tree-depth of a graph, and its depth is at most $4k - 1$, so the given graph has tree-depth at most $4k - 1$. □

3.3 Win-Win

We show now that we can either restrict our attention to a subgraph of bounded tree-depth or find a reachable loose path, in either case giving a structure that allows us to solve path reconfiguration.

Definition 4. *Given an instance (G, P, Q) of path reconfiguration, we say that S is a reachable set of vertices if, for every vertex v in S, there exists a sequence of reconfiguration steps that takes P into a path that uses vertex v. We say that S is an inescapable set of vertices if, for every vertex v that is not in S, there does not exist a sequence of reconfiguration steps that takes P into a path that uses vertex v.*

Lemma 5. *Given an instance (G, P, Q) of path reconfiguration, parameterized by the length k of the start and goal paths, we can in fixed-parameter-tractable time find either a reachable loose path, or a reachable and inescapable set S of vertices that induces a subgraph $G[S]$ of tree-depth at most $4k - 1$.*

Proof. We will maintain a vertex set S that is reachable and induces a subgraph of tree-depth less than $4k$ until either finding reachable loose path or finding that S is inescapable and has no path. Initially, S will consist of all vertices of the start path P; clearly, this satisfies the invariants that S is reachable and has tree-depth less than $4k$.

Then, while we have not terminated the algorithm, we perform the following steps:

- For each edge uv where $u \in S$ and $v \notin S$, use the algorithm of Lemma 2 to test whether P can be reconfigured within $S \cup \{v\}$ (a graph of tree-depth at most $4k$) into a path that uses vertex v. If we find any single edge uv for which this test succeeds, we go on to the next step. Otherwise, if no edge uv passes this test, S is inescapable and we terminate the algorithm.
- Test whether the graph $S \cup \{v\}$ contains a loose path. Finding a path of fixed length is fixed-parameter tractable for arbitrary graphs [11–13] and can be solved even more easily by standard dynamic programming techniques for graphs of bounded tree-depth. If this test succeeds, the loose path must contain v, as the remaining vertices have no loose path. In this case, we have found a reachable loose path (as v is reachable) and we terminate the algorithm.
- Add v to S and continue with the next iteration of the algorithm. Because (in this case) v is reachable but $S \cup \{v\}$ contains no loose path, it follows that including v in S maintains the invariants that S be reachable and induce a subgraph with tree-depth at most $4k - 1$.

Because each iteration adds a vertex to S, the loop must eventually terminate, either with a reachable inescapable subgraph of low tree-depth (from the first step) or with a reachable loose path (from the second step).   □

3.4   Fixed-Parameter Tractability

We are now ready to prove our main result:

Theorem 1. *The path reconfiguration decision problem is fixed-parameter tractable when parameterized by the length of the start and goal paths.*

Proof. Our algorithm for path reconfiguration begins by applying Lemma 5 to find either a reachable inescapable subgraph of low tree-depth or a reachable loose path. If we find a reachable inescapable subgraph that does not include all the goal path vertices, the reconfiguration problem has no solution. If we find a reachable inescapable subgraph that does include all the goal path vertices, we can solve the reconfiguration problem by applying Lemma 2.

If we find a reachable loose path R for the given instance (G, P, Q), we apply Lemma 5 a second time, to the equivalent reversed instance (G, Q, P). If we find a reachable inescapable subgraph that does not include all the vertices of the original start path P, the reconfiguration problem has no solution. If we find a reachable inescapable subgraph that does include all the vertices of P, we can solve the reconfiguration problem by applying Lemma 2.

If we find a second reachable loose path R', one that (by time-reversal symmetry) can reach the goal configuration, then the original reconfiguration problem has a positive solution. For, in this case, we can reconfigure P to a path that lies along R, then (by Lemma 4) to a path that lies along R', then (by the reverse of the reconfiguration sequence found by the second instance of Lemma 5) to Q. □

We leave as open the question of whether a similar result can be obtained for the optimization problem.

4   Tree-Like Graphs

In the full version of this paper we show that several special classes of graphs have polynomial algorithms for path reconfiguration regardless of path length. The prototypical example are the trees, for which the existence of a polynomial time algorithm follows immediately from the fact that any n-vertex tree has $O(n^2)$ distinct paths. In the full version, we refine this idea and provide a linear time algorithm for path reconfiguration in trees.

We also observe that the graphs of bounded circuit rank, and the graphs of bounded feedback vertex number, have polynomial algorithms for path reconfiguration, because in these graphs the size of the state space (the number of distinct paths in the graph) is bounded by a polynomial. For circuit rank the exponent of the polynomial is a constant, and we obtain a fixed-parameter tractable algorithm. For feedback vertex number, the exponent depends on the feedback vertex number. We defer the details to the full version of the paper.

5   Hardness

In the full version of this paper we describe a reduction from nondeterministic constraint logic showing that path reconfiguration (with unbounded path length) is PSPACE-complete even on graphs of bounded bandwidth. This result rules out the possibility (unless P = PSPACE) that our results on tree-like graph classes from Sect. 4 can be extended to another tree-like class of graphs, the graphs of bounded treewidth.

Theorem 2. *The path reconfiguration decision problem is* PSPACE-*complete, even for graphs of bounded bandwidth.*

An example of our reduction is depicted in Fig. 6.

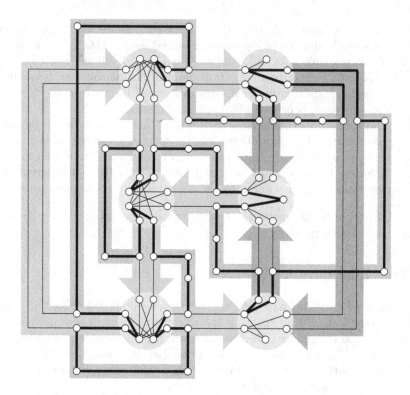

Fig. 6. Reduction from nondeterministic constraint logic to path reconfiguration. The underlying constraint logic instance has six vertices (yellow shaded circles) and nine edges (thick red and blue shaded arrows). Within each of the shaded circles is a vertex gadget of our reduction, and within each thick shaded arrow is an edge gadget of our reduction. The thin green shaded regions contain connection gadgets of our reduction, which the path that is undergoing reconfiguration uses to pass from one edge or vertex gadget to another. The heavy black edges depict one possible state of the path to be reconfigured. (Color figure online)

References

1. Ito, T., et al.: On the complexity of reconfiguration problems. Theoret. Comput. Sci. **412**(12–14), 1054–1065 (2011)
2. Mouawad, A.E., Nishimura, N., Raman, V., Simjour, N., Suzuki, A.: On the parameterized complexity of reconfiguration problems. Algorithmica **78**(1), 274–297 (2017)
3. Kamiński, M., Medvedev, P., Milanič, M.: Shortest paths between shortest paths. Theoret. Comput. Sci. **412**(39), 5205–5210 (2011)
4. Bonsma, P.: The complexity of rerouting shortest paths. Theoret. Comput. Sci. **510**, 1–12 (2013)
5. Hanaka, T., et al.: Reconfiguring spanning and induced subgraphs. In: Wang, L., Zhu, D. (eds.) COCOON 2018. LNCS, vol. 10976, pp. 428–440. Springer, Cham (2018). https://doi.org/10.1007/978-3-319-94776-1_36

6. De Biasi, M., Ophelders, T.: The complexity of Snake. In: Demaine, E.D., Grandoni, F. (eds.) 8th International Conference on Fun with Algorithms, FUN 2016, 8–10 June 2016, La Maddalena, Italy, vol. 49. Leibniz International Proceedings in Informatics (LIPIcs), Schloss Dagstuhl - Leibniz-Zentrum für Informatik, pp. 11:1–11:13 (2016)
7. Gupta, S., Sa'ar, G., Zehavi, M.: The parameterized complexity of motion planning for snake-like robots. Electronic preprint arxiv:1903.02445, March 2019
8. Nešetřil, J., de Mendez, P.O.: Bounded height trees and tree-depth. Sparsity. AC, vol. 28, pp. 115–144. Springer, Heidelberg (2012). https://doi.org/10.1007/978-3-642-27875-4_6
9. Wrochna, M.: Reconfiguration in bounded bandwidth and tree-depth. J. Comput. Syst. Sci. **93**, 1–10 (2018)
10. Hopcroft, J.E., Wong, J.K.: Linear time algorithm for isomorphism of planar graphs (preliminary report). In: Proceedings of the Sixth Annual ACM Symposium on Theory of Computing (STOC 1974), pp. 172–184 (1974)
11. Bodlaender, H.L.: On linear time minor tests with depth-first search. J. Algorithms **14**(1), 1–23 (1993)
12. Fellows, M.R., Langston, M.A.: On search, decision, and the efficiency of polynomial-time algorithms. J. Comput. Syst. Sci. **49**(3), 769–779 (1994)
13. Alon, N., Yuster, R., Zwick, U.: Color-coding. J. ACM **42**(4), 844–856 (1995)

Online Circle Packing

Sándor P. Fekete[1][ID], Sven von Höveling[2][ID], and Christian Scheffer[1(✉)][ID]

[1] Department of Computer Science, TU Braunschweig, Braunschweig, Germany
{s.fekete,c.scheffer}@u-bs.de
[2] Department of Computing Science, University of Oldenburg, Oldenburg, Germany
sven.von.hoeveling@uol.de

Abstract. We consider the online problem of packing circles into a square container. A sequence of circles has to be packed one at a time, without knowledge of the following incoming circles and without moving previously packed circles. We present an algorithm that packs any online sequence of circles with a combined area not larger than 0.350389 of the square's area, improving the previous best value of $\pi/10 \approx 0.31416$; even in an offline setting, there is an upper bound of $\pi/(3+2\sqrt{2}) \approx 0.5390$. If only circles with radii of at least 0.026622 are considered, our algorithm achieves the higher value 0.375898.

As a byproduct, we give an online algorithm for packing circles into a $1 \times b$ rectangle with $b \geq 1$. This algorithm is worst case-optimal for $b \geq 2.36$.

Keywords: Circle packing · Online algorithms · Packing density

1 Introduction

Packing a set of circles into a given container is a natural geometric optimization problem that has attracted considerable attention, both in theory and practice. Some of the many real-world applications are loading a shipping container with pipes of varying diameter [10], packing paper products like paper rolls into one or several containers [9], machine construction of electric wires [19], designing control panels [2], placing radio towers with a maximal coverage while minimizing interference [20], industrial cutting [20], and the study of macro-molecules or crystals [21]. See the survey paper of Castillo, Kampas, and Pintér [2] for an overview of other industrial problems. In many of these scenarios, the circles have to be packed *online*, i.e., one at a time, without the knowledge of further objects, e.g., when punching out a sequence of shapes from the raw material.

Even in an offline setting, deciding whether a given set of circles fits into a square container is known to be NP-hard [4], which is also known for packing squares into a square [11]. Furthermore, dealing with circles requires dealing with possibly complicated irrational numbers, incurring very serious additional geometric difficulties. This is underlined by the slow development of provably

A full version of this extended abstract is available at [8].

© Springer Nature Switzerland AG 2019
Z. Friggstad et al. (Eds.): WADS 2019, LNCS 11646, pp. 366–379, 2019.
https://doi.org/10.1007/978-3-030-24766-9_27

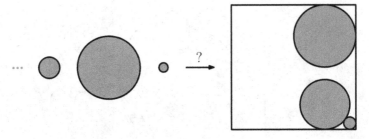

Fig. 1. Circles are arriving one at a time and have to be packed into the unit square. At this stage, the packed area is about 0.33. What is the largest $A \geq 0$ for which *any* sequence of total area A can be packed?

optimal packings of n identical circles into the smallest possible square. In 1965, Schaer [17] gave the optimal solution for $n = 7$ and $n = 8$ and Schaer and Meir [18] gave the optimal solution for $n = 9$. A quarter of a century later, Würtz et al. [3] provided optimal solutions for 10,11,12, and 13 equally sized circles. In 1998, Nurmela and Ostergård [15] provided optimal solutions for $n \leq 27$ circles by making use of computer-aided optimality proofs. Markót and Csendes [13] gave optimal solutions for $n = 28, 29, 30$ also by using computer-assisted proofs within tight tolerance values. Finally, in 2002 optimal solutions were provided for $n \leq 35$ by Locatelli and Raber [12]; at this point, this is still the largest n for which optimal packings are known. The extraordinary challenges of finding densest circle packings are also underlined by a long-standing open conjecture by Erdős and Oler from 1961 [16] regarding optimal packings of n unit circles into an equilateral triangle, which has only been proven up to $n = 15$.

These difficulties make it desirable to develop relatively simple criteria for the packability of circles. A natural bound arises from considering the *packing density*, i.e., the total area of objects compared to the size of the container; the *critical packing density* δ is the largest value for which any set of objects of total area at most δ can be packed into a unit square; see Fig. 1.

In an offline setting, two equally sized circles that fit exactly into the unit square show that $\delta \leq \delta^* = \pi/(3 + 2\sqrt{2}) \approx 0.5390$. This is indeed tight: Fekete, Morr and Scheffer [7] gave a worst-case optimal algorithm that packs any instance with combined area at most δ^*; see Fig. 2 (left). More recently, Fekete, Keldenich and Scheffer [6] established 0.5 as the critical packing density of circles in a circular container.

The difficulties of offline circle packing are compounded in an online setting. This is highlighted by the situation for packing squares into a square, which does not encounter the mentioned issues with irrational coordinates. It was shown in 1967 by Moon and Moser [14] that the critical offline density is 0.5: Refining an approach by Fekete and Hoffmann [5], Brubach [1] established the currently best lower bound for online packing density of 0.4. This yields the previous best bound for the online packing density of circles into a square: Inscribing circles into bounding boxes yields a value of $\pi/10 \approx 0.3142$.

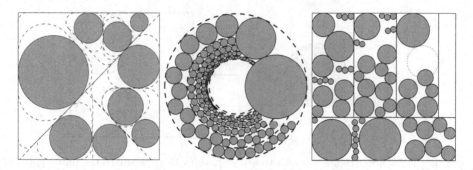

Fig. 2. Examples of algorithmic circle packings. (Left) The worst-case optimal offline algorithm of Fekete et al. [7] for packing circles into the unit square. (Center) The worst-case optimal offline algorithm of Fekete et al. [6] for packing circles into the unit circle. (Right) Our online algorithm for packing circles into the unit square.

1.1   Our Results

In this paper, we establish new lower bounds for the online packing density of circles into a square and into a rectangle. Note that in the online setting, a packing algorithm has to stop as soon as it cannot pack a circle. We provide three online circle packing results for which we provide constructive proofs, i.e., corresponding algorithms guaranteeing the claimed packing densities.

Theorem 1. *Let $b \geq 1$. Any online sequence of circles with a total area no larger than* $\min\left(0.528607 \cdot b - 0.457876, \frac{\pi}{4}\right)$ *can be packed into the $1 \times b$-rectangle R. This is worst-case optimal for $b \geq 2.36$.*

We use the approach of Theorem 1 as a subroutine and obtain the following:

Theorem 2. *Any online sequence of circles with a total area no larger than 0.350389 can be packed into the unit square.*

If the incoming circles' radii are lower bounded by 0.026623, the density guaranteed by the algorithm of Theorem 2 improves to 0.375898.

Theorem 3. *Any online sequence of circles with radii not smaller than 0.026623 and with a total area no larger than 0.375898 can be packed into the unit square.*

We describe the algorithm of Theorem 1 in Sect. 2 and the algorithm of the Theorems 2 and 3 in Sect. 3.

2   Packing into a Rectangle

In this section, we describe the algorithm, *Double-Sided Structured Lane Packing (DSLP)*, of Theorem 1. In particular, DSLP uses a packing strategy called *Structured Lane Packing (SLP)* and an extended version of SLP as subroutines.

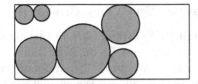 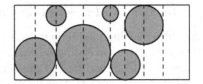

Fig. 3. Comparison of Tight Lane Packing (left) with Structured Lane Packing (right) for the same input. The former has a smaller packing length.

2.1 Preliminaries for the Algorithms

A *lane* $L \subset \mathbb{R}^2$ is an x- and y-axis-aligned rectangle. The *length* $\ell(L)$ and the *width* $w(L)$ of L are the dimensions of L such that $w(L) \leq \ell(L)$. L is *horizontal* if the length of L is given via the extension of L w.r.t. the x-axis. Otherwise, L is *vertical*. The *distance* between two circles packed into L is the distance between the orthogonal projections of the circles' midpoints onto the longer side of L. A lane is either *open* or *closed*. Initially, each lane is open.

Packing a circle C into a lane L means placing C inside L such that C does not intersect another circle that is already packed into L or into another lane. A *(packing) strategy* for a lane L is a set of rules that describe how a circle has to be packed into L. The *(packing) orientation* of a strategy for a horizontal lane is either *rightwards* or *leftwards* and the *(packing) orientation* of a strategy for a vertical lane is either *downwards* or *upwards*.

Let w be the width of L. Depending on the radius r of the current circle C, we say: C is *medium* (Class 1) if $w > r \geq \frac{w}{4}$, C is small if $\frac{w}{4} > r \geq 0.0841305w$ (Class 2), C is *tiny* (Class 3 or 4) if $0.0841305w > r \geq 0.023832125w$, and C is *very tiny* if $0.023832125w > r$ (Classes 5,6, ...). For a more refined classification of r, we refer to Sect. 2.3. The general idea is to reach a certain density within a lane by packing only relatively equally sized circles into a lane with SLP.

For the rest of Sect. 2, for $0 < w \leq b$, let L be a horizontal $w \times b$ lane.

2.2 Structured Lane Packing (SLP) – The Standard Version

Rightwards *Structured Lane Packing (SLP)* packs circles into L alternating touching the bottom and the top side of L from left to right, see Fig. 3 (right).

In particular, we pack a circle C into L as far as possible to the left while guaranteeing: (1) C does not overlap a vertical lane packed into L, see Sect. 2.3 for details[1]. (2) The distance between C and the circle C' packed last into L is at least $\min\{r, r'\}$ where r, r' are the radii of C, C', see Fig. 3 (right).

Leftwards *Structured Lane Packing* packs circles by alternatingly touching the bottom and the top side of L from right to left. Correspondingly, upwards and downwards *Structured Lane Packing* packs circles alternatingly touching the left and the right side of L from bottom to top and from top to bottom.

---

[1] Requiring that C does not overlap a vertical lane placed inside L is only important for the extension of SLP (see Sect. 2.3), because the standard version of SLP does not place vertical lanes inside L.

2.3   Extension of SLP – Filling Gaps by (Very) Tiny Circles

Now consider packing medium circles with SLP. We extend SLP for placing tiny and very tiny circles within the packing strategy, see Fig. 4. Note that small circles are not considered for the moment, such that (very) tiny circles are relatively small compared to the medium ones. In particular, if the current circle C is medium, we apply the standard version of SLP, as described in Sect. 2.2. If C is (very) tiny, we pack C into a vertical lane inside L, as described next.

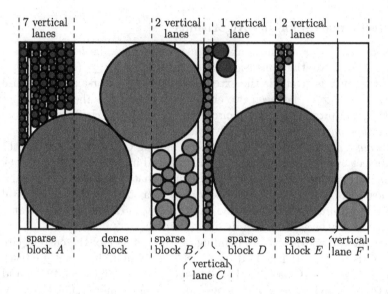

Fig. 4. A packing produced by extended SLP: 128 (3 medium, 17 tiny, and 108 very tiny) circles packed into 15 (1 medium, 4 tiny and 10 very tiny) lanes. A possible input order of the circles is 1 medium circle, 23 very tiny circles (filling the sparse block A), 1 medium circle, 13 tiny circles (filling the sparse block B), 24 very tiny circles (filling the vertical lane C), 1 medium circle, 2 tiny circles (filling the sparse block D), 11 very tiny circles (filling the sparse block E), and 2 tiny circles (filling the vertical lane F). (Color figure online)

We pack (very) tiny circles into vertical lanes inside L, see Fig. 4. The vertical lanes are placed inside *blocks* that are the rectangles induced by vertical lines touching medium circles already packed into L, see Fig. 3 (right).

Blocks that include two halves of medium circles are called *dense* blocks, while blocks that include one half of a medium circle are called *sparse* blocks. The area of L that is neither covered by a dense block or a sparse block is called a *free* block. *Packing* a vertical lane L' into a sparse block B means placing L' inside B as far as possible to the left, such that L' does not overlap another vertical lane already packed into B. Packing a vertical lane L' into L means placing L' inside L as far as possible to the left, such that L' does neither overlap another vertical lane packed into L, a dense block of L, or a sparse block of L.

We extend our classification of circles by defining classes i of lane widths w_i and *relative lower bounds* q_i for the circles' radii as described in Table 1. This means a circle with radius r belongs to class 1 if $0.5w \geq r > q_1 w_1$ and to class i if $q_{i-1} w_{i-1} \geq r > q_i w_i$, for $i \geq 2$. Only circles of class i are allowed to be packed into lanes of class i.

Table 1. Circles are classified into the listed classes. Note that the lower bounds to the circles' radii is relative to the lane width, e.g., the absolute lower bound for circles inside a small lane is $q_S w_2 = 0.168261 w_2$.

Class i	(Relative) lower bound q_i	Lane width w_i
1 (Medium)	$q_1 := q_M := 0.25$	$w_1 := w$
2 (Small)	$q_2 := q_S := 0.168261$	$w_2 := 2q_M w = 0.5 \cdot w$
3 (Tiny)	$q_3 := 0.371446$	$w_3 := 4q_M q_S w = 0.168261 \cdot w$
4 (Tiny)	$q_4 := 0.190657$	$w_4 := 8q_M q_S q_3 w \approx 0.125 \cdot w$
5 (Very tiny)	$q_5 := 0.175592$	$w_5 := 16q_M q_S q_3 q_4 w \approx 0.047664 \cdot w$
6 (Very tiny)	$q_6 := 0.170699$	$w_6 := 32q_M q_S q_3 q_4 q_5 w \approx 0.016739 \cdot w$
7 (Very tiny)	$q_7 := 0.169078$	$w_7 \approx 0.005715 \cdot w$
8 (Very tiny)	$q_8 := 0.168354$	$w_8 \approx 0.001932 \cdot w$
9 (Very tiny)	$q_9 := 0.168293$	$w_9 \approx 0.000651 \cdot w$
10 (Very tiny)	$q_{10} := 0.168272$	$w_{10} \approx 0.000219 \cdot w$
11 (Very tiny)	$q_{11} := 0.168265$	$w_{11} \approx 0.000074 \cdot w$
12 (Very tiny)	$q_{12} := 0.168263$	$w_{12} \approx 0.000025 \cdot w$
13 (Very tiny)	$q_{13} := 0.168262$	$w_{13} \approx 0.000008 \cdot w$
...	...	...
k (Very tiny)	$q_k := 0.168262$	$w_k := 2^{k-1} q_M q_S q_3 q_4 \cdot \ldots \cdot q_{k-1} w$

A sparse block is either *free*, *reserved* for class 3, *reserved* for class 4, *reserved* for all classes $i \geq 5$, or *closed*. Initially, each sparse block is free.

We use SLP in order to pack a circle C of class $i \geq 3$, into a vertical lane $L_i \subset L$ of class i and width w_i by applying the Steps 1–5 in increasing order as described below. When one of the five steps achieves that C is packed into a vertical lane L_i, the approach stops and returns successful.

- **Step (1):** If there is no open vertical lane $L_i \subset L$ of class i go to Step 2. Assume there is an open vertical lane L_i of class i. If C can be packed into L_i, we pack C into L_i. Else, we declare L_i to be closed.
- **Step (2):** We close all sparse blocks B that are free or reserved for class i in which a vertical lane of class i cannot be packed into B.
- **Step (3):** If there is an open sparse block B that is free or reserved for class i and a vertical lane of class i can be packed into B:

- **(3.1)**: We pack a vertical lane $L_i \subset L$ of class i into B. If the circle half that is included in B touches the bottom of L, we apply downwards SLP to L_i. Otherwise, we apply upwards SLP to L_i.
- **(3.2)**: If B is free and $i \in \{3,4\}$, we reserve B for class i. If B is free and $i \geq 5$, we reserve B for all classes $i \geq 5$.
- **(3.3)**: We pack C into L_i.
- **Step (4)**: If a vertical lane of class i can be packed into L:
  - **(4.1)**: We pack a vertical lane L_i of class i into L and apply upwards SLP to L_i.
  - **(4.2)**: We pack C into L_i
- **Step (5)**: We declare L to be closed and return failed.

2.4 Double-Sided Structured Lane Packing (DSLP)

We use SLP as a subroutine in order to define our packing strategy *Double-Sided Structured Lane Packing (DSLP)* of Theorem 1. In particular, additionally to L, we consider two small lanes L^1, L^2 that partition L, see Fig. 5.

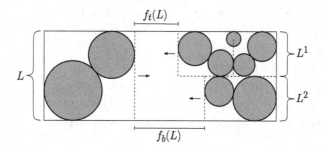

Fig. 5. A packing produced by DSLP: The medium lane is packed from left to right by medium circles. The two contained small lanes are packed simultaneously in parallel from right to left by small circles.

Rightwards *Double-Sided Structured Lane Packing (DSLP)* applies the extended version of rightwards SLP to L and leftwards SLP to L^1, L^2. If the current circle C is medium or (very) tiny, we pack C into L. If C is small, we pack C into that lane of L^1, L^2, resulting in a smaller packing length.

Leftwards DSLP is defined analogously, such that the extended version of leftwards SLP is applied to L and rightwards SLP to L^1, L^2. Correspondingly, upwards and downwards DSLP are defined for vertical lanes.

3   Packing into the Unit Square

We extend our circle classification by the class 0 of *large* circles and define a relative lower bound $q_0 := \frac{w}{2}$ and the lane width of corresponding *large* lanes as $w_0 := 1 - w$.

We set w to 0.288480 and 0.277927 for Theorem 2 respectively Theorem 3. In order to pack large circles, we use another packing strategy called *Tight Lane Packing (TLP)* defined as SLP, but without restrictions (1) and (2), see Fig. 3.

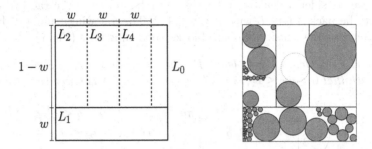

Fig. 6. Left: The unit square is divided into four lanes L_1, L_2, L_3, and L_4, into which medium, small, tiny, and very tiny circles are packed. Large circles are packed into a lane L_0 that overlaps L_1, L_2, and L_3. Right: An example packing. A medium circle (dotted) does not fit.

We cover the unit square by the union of one large lane L_0 and four medium lanes $L_1, \ldots, L_k$ for $k = 4$, see Fig. 6. We apply TLP to L_0 and DSLP to $L_1, \ldots, L_4$. The applied orientations for L_0, L_1, L_2 are leftwards, rightwards, and downwards. For $i = 3, 4$, the orientation for L_i is chosen such that the first circle packed into L_i is placed adjacent to the bottom side of L_i.

If the current circle to be packed is large, we pack C into the large lane L_0 and stop if C does not fit in L_0. Otherwise, in increasing order we try to pack C into $L_1, \ldots, L_4$.

4   Analysis of the Algorithms

In this section we sketch the analysis of our approaches and refer to the appendix for full details. First, we analyze the packing density guaranteed by DSLP. Based on that, we prove our main results Theorems 1, 2, and 3.

4.1   Analysis of SLP

In this section, we provide a framework for analyzing the packing density guaranteed by DSLP for a horizontal lane L of width w. It is important to note that this framework and its analysis in this subsection deals with the packing of only one class into a lane.

We introduce some definitions. The *packing length* $p(L)$ is the maximal difference of x-coordinates of points from circles packed into L. The *circle-free length* $f(L)$ of L is defined as $\ell(L) - p(L)$. We denote the *total area* of a region $R \subset \mathbb{R}^2$ by $area(R)$ and the area of an $a \times b$-rectangle by $\mathcal{R}(a, b)$. Furthermore, we denote the area of a semicircle for a given radius r by $\mathcal{H}(r) := \frac{\pi}{2}r^2$. The total area of the circles packed into R is called *occupied area* denoted by $occ(R)$. Finally, the density $den(R)$ is defined by $occ(R)/area(R)$.

In order to apply our analysis for different classes of lanes, i.e., different lower bounds, we consider a general (relative) lower bound q for the radii of circles allowed to be packed into L with $0 < q \leq 1/2$. The following lemma deduces a lower bound for the density of dense blocks depending on q (Fig. 7).

Lemma 1. *Consider a dense block D containing two semicircles of radii r_1 and r_2 such that $0 < qw \leq r_1, r_2 \leq 1/2w$. Then $den(D)$ is lower-bounded by*

$$\delta : \left(0, \frac{1}{2}\right] \to \mathbb{R} \; with \; q \mapsto \begin{cases} \pi q & 0 < q < \frac{1}{3\sqrt{3}} \\ \frac{\pi}{3\sqrt{3}} \approx 0.6046 & \frac{1}{3\sqrt{3}} \leq q \leq \frac{1}{3} \\ \frac{\pi q^2}{\sqrt{4q-1}} & \frac{1}{3} < q \leq \frac{1}{2}. \end{cases}$$

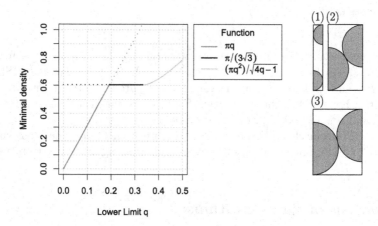

Fig. 7. (Left): A plot of $\delta(q)$ for its complete range. It provides the minimal density of a dense block whose two semicircles have a radius of at least $q \cdot w$. A lower bound of $q = 1/2$ leads to a minimal density of $\pi/4$ which is the ratio of a circle to its minimal bounding square. (Right): (1): $\delta(0.15) \approx 0.47123$, (2): $\delta(\frac{1}{3\sqrt{3}}) \approx 0.6046$, and (3): $\delta(0.4) \approx 0.6489$.

We continue with the analysis of sparse blocks. Sparse blocks have a minimum length qw. Lemma 2 states a lower bound for the occupied area of sparse blocks. This lower bound consists of a constant summand and a summand that is linear with respect to the actual length.

Lemma 2. *Given a density bound $\delta_{min} \leq \delta(q)$ for dense blocks. Let S be a sparse block and z be the lower bound for $\ell(S)$ with $\ell(S) \geq z \geq qw$. Then $occ(S) \geq \mathcal{R}(\ell(S) - z, w) \cdot \delta_{min} + \mathcal{H}(z)$.*

The occupied area of a sparse block is at least a semicircle of a smallest possible circle plus the remaining length multiplied by the lane width and by the minimal density δ_{min} of dense blocks. This composition is shown in Fig. 8.

Next, we combine the results of Lemmas 1 and 2. We define the term $min_{SLP}\big(p, w, z, \delta_{min}\big) := \mathcal{R}(p - 2z, w) \cdot \delta_{min} + 2 \cdot \mathcal{H}(z)$ for some $p, w, z, \delta_{min} > 0$ and state the following (see also Fig. 8 (4)).

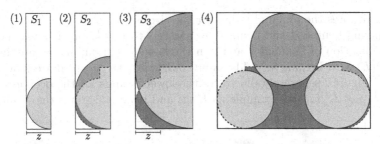

Fig. 8. (1)+(2)+(3) Three sparse blocks S_1, S_2, S_3 in order of ascending length. The grey coloured area (light and dark unified) represents the occupied area. The dashed area shows the lower bound of Lemma 2, which is composed of the smallest possible semicircle plus a linear part. The dark grey parts symbolize the area that exceeds the bound, whereas the red parts symbolize the area missing to the bound. Block S_1 has the minimal length p so that the occupied area and the bound are equal. For blocks of larger length, represented by S_2 and S_3, the dark grey area is larger than the red area. (4) A packing produced by SLP and the lower bound of Lemma 3.

Lemma 3. *Given a lane L packed by SLP, a lower bound q, and a density bound $\delta_{min} \leq \delta(q)$ for dense blocks. Let w be the width of L. The occupied area in L is lower-bounded by* $min_{SLP}\big(p(L), w, qw, \delta_{min}\big)$.

4.2   Analysis of DSLP

Let L be a horizontal lane packed by DSLP. We define $p_t(L)$ $(p_b(L))$ as the sum of the packing lengths of the packing inside L and the length of the packing inside the top (bottom) small lane inside L. Furthermore, we define $f_t(L) := \ell(L) - p_t(L)$ and $f_b(L) := \ell(L) - p_b(L)$, see Fig. 5.

By construction, vertical dense blocks packed into L have a density of at least $\delta(q_2)$. In fact, the definitions of circle sizes for all classes $i \geq 2$ ensure the common density bound $\hat{\delta} := \delta(q_2)$ for dense blocks.

We consider mixed dense blocks, that were defined as sparse blocks of L in which vertical lanes are packed, also as dense blocks and extend the lower bound $den(D) \geq \hat{\delta}$ to all kinds of dense blocks by the following Lemma.

Lemma 4. *Let D be a dense block of L. Assume all vertical lanes packed into D to be closed. Then $den(D) \geq \hat{\delta}$.*

As some vertical lanes may not be closed, we upper bound the error $\mathcal{O}(L)$ that we make by assuming that all vertical lanes $L_1, \ldots, L_n \subset L$ are closed.

Lemma 5. $\mathcal{O}(L_1 \cup \ldots \cup L_n) < 0.213297 \cdot w^2$.

We lower bound the occupied area inside L by using the following term:

$$min_{DSLP}(p_t(L), p_b(L), w, z, \delta_{min}) := \mathcal{R}\big(p_t(L) + p_b(L) - w - 4z, w_2\big) \cdot \delta_{min}$$
$$+ 2 \cdot \mathcal{H}\Big(\frac{w}{4}\Big) + 4\mathcal{H}(z),$$

where z denotes the minimal radius for the circles, i.e., $z := q_2 w_2$.

Applying Lemma 4 and Lemma 2 separately to L, L^1, and L^2, analogous to the combination of Lemmas 1 and 2 in the last subsection, and estimating the error $\mathcal{O}(L)$ with Lemma 5, yields lower bounds for the occupied areas of L, L^1, and L^2. Figure 9 separately illustrates the lower bounds for the occupied areas of L, L^1, and L^2 for two example packings and Lemma 6 states the result.

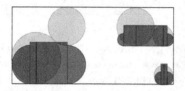 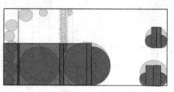

Fig. 9. Two example packings and the compositions of our lower bounds (red) for the occupied area implied by Lemma 6. Note that $\mathcal{O}(L)$ is not visualized. (Color figure online)

Lemma 6. $min_{DSLP}\big(p_t(L), p_b(L), w, z, \hat{\delta}\big) - \mathcal{O}(L) \geq occ(L)$.

4.3 Analysis of Packing Circles into a Rectangle

Given a $1 \times b$ rectangle R, we apply DSLP for packing the input circles into R.

Theorem 1. *Let $b \geq 1$. Any online sequence of circles with a total area no larger than* $\min\Big(0.528607 \cdot b - 0.457876, \frac{\pi}{4}\Big)$ *can be packed into the $1 \times b$-rectangle R. This is worst-case optimal for $b \geq 2.36$.*

The lower bound for the occupied area implied by Lemma 6 is equal to $\big(b - \frac{3}{4} - q_2\big) \cdot \hat{\delta} + \frac{\pi}{16} + \frac{\pi}{2}(q_2)^2 - 0.213297$. This is lower bounded by $\frac{\pi}{4}$ for $b \geq 2.36$. Hence, the online sequence consisting of one circle with a radius of $\frac{1}{2} + \epsilon$ and resulting total area of $\frac{\pi}{4} + \epsilon$ is a worst case online sequence for $b \geq 2.36$, see Fig. 10. This concludes the proof of Theorem 1.

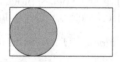

Fig. 10. A worst case for packing circles into an $1 \times b$-rectangle with $b \geq 2.36$ consists of one circle with radius $\frac{1}{2} + \epsilon$. The shown circle with radius $\frac{1}{2}$ just fits.

4.4   Analysis of Packing Circles into the Unit Square

In this section, we analyze the packing density of our overall approach for online packing circles into the unit square. In order to prove Theorem 3, i.e., a lower bound for the achieved packing density, we show that if there is an overlap or if there is no space in the last lane, then the occupied area must be at least this lower bound. Our analysis distinguishes six different higher-level cases of where and how the overlap can happen, see Fig. 11 left.

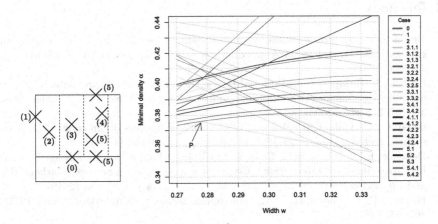

Fig. 11. (Left): Different cases for an overlap. Case 0: A single circle is too large for L_0. Case 1: L_0 exceeded. Case 2: Overlap in L_2. Case 3: Overlap in L_3. Case 4: Overlap in L_4 with large circle being involved. Case 5: Overlap in L_4 with no large circle being involved. (Right): A plot of 24 terms for corresponding 24 (sub-)cases for $q_S = 0.191578$. The point $P = (0.277927, 0.375898)$ is the highest point of the 0-level. Its y-value is the highest guaranteed packing density for circles with minimal radii of $0.191578 \cdot 0.277927/2 < 0.0266223$.

For each of the six cases and its subcases, we explicitly give a density function, providing the guaranteed packing density depending on the choice of w, see Fig. 11 right. The shown functions are constructed for the case of no (very) tiny circles with an alternative $q_2 = 0.191578$, which was chosen numerically in order to find a high provable density. The w with the highest guaranteed packing density of 0.375898 is $w = 0.277927$. This concludes the proof of Theorem 3.

Theorem 3. *Any online sequence of circles with radii not smaller than 0.026623 and with a total area no larger than 0.375898 can be packed into the unit square.*

With the same idea but with $w = 0.288480$ and all circles classes, especially with classes $i \geq 2$ as defined in Table 1, we prove Theorem 2.

Theorem 2. *Any online sequence of circles with a total area no larger than 0.350389 can be packed into the unit square.*

5 Conclusion

We provided online algorithms for packing circles into a square and a rectangle. For the case of a rectangular container, we guarantee a packing density which is worst-case optimal for rectangles with a skew of at least 2.36. For the case of a square container, we provide a packing density of 0.350389 which we improved to 0.375898 if the radii of incoming circles are lower-bounded by 0.026622.

References

1. Brubach, B.: Improved online square-into-square packing. In: Proceedings 12th International Workshop on Approximation and Online Algorithms (WAOA), pp. 47–58 (2014)
2. Castillo, I., Kampas, F.J., Pintér, J.D.: Solving circle packing problems by global optimization: numerical results and industrial applications. Eur. J. Oper. Res. **191**(3), 786–802 (2008)
3. de Groot, C., Peikert, R., Würtz, D.: The optimal packing of ten equal circles in a square. Technical report 90–12, ETH Zurich, Switzerland (1990)
4. Demaine, E.D., Fekete, S.P., Lang, R.J.: Circle packing for origami design is hard. In: Proceedings 5th International Conference on Origami in Science, Mathematics and Education (Origami[5]), pp. 609–626. A. K. Peters/CRC Press (2010)
5. Fekete, S.P., Hoffmann, H.: Online square-into-square packing. Algorithmica **77**(3), 867–901 (2017)
6. Fekete, S.P., Keldenich, P., Scheffer, C.: Packing disks into disks with optimal worst-case density. In: Proceedings 35th Symposium on Computational Geometry (SoCG) (2019, to appear)
7. Fekete, S.P., Morr, S., Scheffer, C.: Split packing: algorithms for packing circles with optimal worst-case density. Discrete Comput. Geom. (2018). https://doi.org/10.1007/s00454-018-0020-2
8. Fekete, S.P., von Hoeveling, S., Scheffer, C.: Online circle packing, CoRR (2019). http://arXiv.org/abs/1905.00612
9. Fraser, H.J., George, J.A.: Integrated container loading software for pulp and paper industry. Eur. J. Oper. Res. **77**(3), 466–474 (1994)
10. George, J.A., George, J.M., Lamar, B.W.: Packing different-sized circles into a rectangular container. Eur. J. Oper. Res. **84**(3), 693–712 (1995)
11. Leung, J.Y.T., Tam, T.W., Wong, C.S., Young, G.H., Chin, F.Y.: Packing squares into a square. J. Parallel Distrib. Comput. **10**(3), 271–275 (1990)
12. Locatelli, M., Raber, U.: Packing equal circles in a square: a deterministic global optimization approach. Discrete Appl. Math. **122**(1), 139–166 (2002)
13. Markót, M.C., Csendes, T.: A new verified optimization technique for the "packing circles in a unit square" problems. SIAM J. Optim. **16**(1), 193–219 (2005)
14. Moon, J.W., Moser, L.: Some packing and covering theorems. Colloquium Math. **17**(1), 103–110 (1967)
15. Nurmela, J.K., Ostergård, J.P.R.: More optimal packings of equal circles in a square. Discrete Comput. Geom. **22**(3), 439–457 (1998)
16. Oler, N.: A finite packing problem. Canad. Math. Bull. **4**, 153–155 (1961)
17. Schaer, J.: The densest packing of nine circles in a square. Canad. Math. Bull. **8**, 273–277 (1965)

18. Schaer, J., Meir, A.: On a geometric extremum problem. Canad. Math. Bull. **8**(1) (1965)
19. Sugihara, K., Sawai, M., Sano, H., Kim, D.S., Kim, D.: Disk packing for the estimation of the size of a wire bundle. Jpn. J. Ind. Appl. Math. **21**(3), 259–278 (2004)
20. Szabó, P.G., Markót, M.C., Csendes, T., Specht, E., Casado, L.G., García, I.: New Approaches to Circle Packing in a Square: With Program Codes. Springer Optimization and Its Applications, vol. 6. Springer, Boston (2007)
21. Würtz, D., Monagan, M., Peikert, R.: The history of packing circles in a square. Maple Tech. Newsl., 35–42 (1994)

Guess Free Maximization of Submodular and Linear Sums

Moran Feldman[(⊠)]

The Open University of Israel, Ra'anana, Israel
moranfe@openu.ac.il

Abstract. We consider the problem of maximizing the sum of a monotone submodular function and a linear function subject to a general solvable polytope constraint. Recently, Sviridenko et al. [16] described an algorithm for this problem whose approximation guarantee is optimal in some intuitive and formal senses. Unfortunately, this algorithm involves a guessing step which makes it less clean and significantly affects its time complexity. In this work we describe a clean alternative algorithm that uses a novel weighting technique in order to avoid the problematic guessing step while keeping the same approximation guarantee as the algorithm of [16].

Keywords: Submodular maximization · Continuous greedy · Curvature

1 Introduction

The last decade has seen a surge of work on submodular maximization problems. Arguably, the main factor that allowed this surge was the invention of the multilinear relaxation for submodular maximization problems as well as algorithms for (approximately) solving this relaxation [4,6,8,9,11]. The invention of the multilinear relaxation was so influential because it allowed algorithms for submodular maximization to use the technique of first solving a relaxed version of the problem, and then rounding the fractional solution obtained. This technique is well-known, and it is often used in the design of algorithms for other kinds of problems; but, prior to the invention of the multilinear relaxation, it was not known how to apply it to submodular maximization problems.

An algorithm based on the above mentioned technique usually has two main components: a solver that (approximately) solves the relaxation and a rounding procedure. Historically, the first solver described for multilinear relaxations was the Continuous Greedy algorithm that solves such relaxations up to an approximation ratio of $1 - 1/e$ when the objective function is non-negative and monotone

This research was partially supported by Israel Science Foundation grant number 1357/16.

© Springer Nature Switzerland AG 2019
Z. Friggstad et al. (Eds.): WADS 2019, LNCS 11646, pp. 380–394, 2019.
https://doi.org/10.1007/978-3-030-24766-9_28

(in addition to being submodular) [6].[1] While the invention of continuous greedy was very significant, one can note that unlike standard solvers for more familiar relaxations such as LPs and SDPs, the approximation ratio of continuous greedy is quite far from 1. Unfortunately, a hardness result due to [14] implies that its approximation ratio cannot be improved in general.

This situation motivates the question of how well can one approximate multilinear relaxations whose objective includes both monotone submodular and linear components. Specifically, it is interesting to know whether the approximation ratio that can be achieved in such cases improves gradually as the linear component of the objective becomes more prominent. Recently, Sviridenko et al. [16] answered this question in the affirmative. More formally, they considered the following problem. Given a non-negative monotone submodular function $g: 2^{\mathcal{N}} \to \mathbb{R}_{\geq 0}$, a linear function ℓ and a solvable polytope $P \subseteq [0,1]^{\mathcal{N}}$,[2] find a point $x \in P$ that approximately maximizes $G(x) + \ell(x)$, where G is the multilinear extension of g (see Sect. 2 for a definition). Sviridenko et al. [16] described a variant of continuous greedy that, given an instance of this problem, outputs a vector x obeying the inequality $G(x) + \ell(x) \geq (1 - e^{-1}) \cdot g(OPT) + \ell(OPT)$ up to a small error term, where OPT is an optimal integral solution for the problem.[3]

Intuitively, the result of Sviridenko et al. [16] is tight since it approximates the submodular component of the objective up an approximation ratio of $1 - 1/e$ and the linear component up to a ratio of 1. More formally, Sviridenko et al. [16] showed that their result is tight since it yields optimal approximation ratios for two problems of interest: maximizing a non-negative monotone submodular function with a bounded curvature subject to a matroid constraint, and minimizing a non-negative non-increasing supermodular function with a bounded curvature subject to the same kind of constraint.

1.1   Our Result

Despite being optimal in terms of its approximation guarantee, in the senses described above, the algorithm of Sviridenko et al. [16] suffers from a significant drawback. Namely, it is based on guessing the contribution of the linear component of the objective to the optimal solution, and this guessing step is quite problematic for the following reasons.

– The guessing is done by enumerating $\Theta(n\varepsilon^{-1} \log n)$ different possible values, and thus, increases the time complexity of the algorithm by this factor. Moreover, to guarantee that one of the enumerated values is a good enough

---

[1] A set function $f: 2^{\mathcal{N}} \to \mathbb{R}$ is *monotone* if $f(S) \leq f(T)$ for every two sets $S \subseteq T \subseteq \mathcal{N}$ and *submodular* if $f(S \cup \{u\}) - f(S) \geq f(T \cup \{u\}) - f(T)$ for every two such sets and element $u \in \mathcal{N} \setminus T$.

[2] A polytope P is solvable if one can optimize linear functions subject to it.

[3] Technically, Sviridenko et al. [16] considered only the special case of the problem in which P is a matroid polytope, and designed two algorithms for this case. However, one of these algorithms (the continuous greedy based one) trivially extends to arbitrary solvable polytopes.

guess, the set of values tried is constructed in a non-trivial way, which is then reflected in the complexity of the algorithm's analysis.

- The original continuous greedy algorithm repeatedly maximizes linear functions subject to the constraint polytope P. While this computational step is quite slow in general, for many cases of interest (for example, when P is a matroid polytope) it can be implemented very efficiently. In contrast, the algorithm of [16] maximizes linear functions subject to the intersection of P with a polytope defined by the guessed value, which might be a very slow operation even when optimizing linear functions subject to P itself is fast. Moreover, various techniques have been described to speed up continuous greedy when it is applied to a matroid polytope [3,5], and these techniques fail to apply to the algorithm of [16] because it considers the intersection of P with another polytope rather than P itself.

In this work we present a clean alternative algorithm that has the same approximation guarantee as the algorithm of Sviridenko et al. [16], but avoids the guessing step and all the problems resulting from it. In a nut shell, our algorithm is a modification of continuous greedy in which the weight assigned to each component of the objective function varies over time. At the beginning of an execution of the algorithm, the linear component has much more weight than the monotone submodular component, and over time their weights become equal. Intuitively, this kind of weighting makes sense because the standard analysis of continuous greedy for submodular functions uses a lower bound on the gain of the algorithm in its later steps which decreases if the algorithm has already made a significant gain in earlier steps. Thus, any gain from the submodular component of the objective that is obtained early in the algorithm's execution is partially cancelled by the resulting decrease in the gain guaranteed in later steps of the execution. In contrast, gain obtained from the linear component of the objective in the same early steps of the execution does not suffer from such partial cancellation, and thus, should get more weight.

It should be mentioned that in addition to the algorithm of Sviridenko et al. [16] discussed above, Sviridenko et al. [16] also presented another algorithm for the same problem which is based on local search techniques. This other algorithm has the advantage that, like our algorithm, it does not need to resort to guessing. However, unlike our algorithm, this other algorithm of [16] applies only to matroid polytopes, and moreover, it has a very large time complexity of $\tilde{O}(\varepsilon^{-3} n^8)$.

1.2 Additional Related Work

When the linear function is non-negative, its sum with the monotone submodular function is still monotone and submodular. Thus, in this case the work of Sviridenko et al. [16] can be viewed as improving over the guarantee of continuous greedy in a special case. More recently, Soma and Yoshida [15] used an algorithm based on a similar technique to improve over the guarantee of continuous greedy in the more general case in which the monotone submodular objective can be decomposed into a monotone submodular component and a significant $M^\natural$-concave component. In an earlier work, Feldman et al. [11] took the complementing approach of using properties of the constraint polytope, rather than the objective function, to improve over the guarantee of continuous greedy. Specifically, they described a variant of continuous greedy, named Measured Continuous Greedy, which achieves an improved approximation ratio when the constraint is dense (in some sense).

As mentioned above, an algorithm that works by solving a relaxation and then rounding the solution has two main components: a relaxation solver and a rounding procedure. All the discussion up to this point was devoted to relaxation solvers because the current work is about such a solver and also because, unlike solvers, rounding procedures tend to be very problem specific. Nevertheless, there are a few more noticeable such procedures. A large portion of the work done so far on submodular maximization has been in the context of matroid constraints, for which there are two known rounding procedures that do not lose anything in the objective: Pipage Rounding [6] and Swap Rounding [7]. In another line of work, Chekuri et al. [8] designed a framework called "contention resolution schemes" which yields a rounding procedure for every constraint that can be presented as the intersection of few simple constraints. Later works extended the contention resolution schemes framework into online and stochastic settings [1,12,13].

2   Preliminaries

In this section we describe the notation that we use and give a few relevant definitions. Using these definitions we then formally describe the guarantee of the algorithm we analyze.

Given a set S and an element u, we use $S + u$ and $S - u$ as shorthands for the union $S \cup \{u\}$ and the expression $S \setminus \{u\}$, respectively. If we are also given a set function f, then the marginal contribution of u to S with respect to f is denoted by $f(u \mid S) \triangleq f(S + u) - f(S)$. Notice that using this notation we get that a function $f \colon 2^{\mathcal{N}} \to \mathbb{R}$ is submodular if and only if for every two sets $S \subseteq T \subseteq \mathcal{N}$ and element $u \in \mathcal{N} \setminus T$ it holds that $f(u \mid S) \geq f(u \mid T)$. Occasionally, we are also interested in the marginal contribution of a set T to a set S with respect to f, which we denote by $f(T \mid S) \triangleq f(S \cup T) - f(S)$.

The *multilinear* extension of a set function $f \colon 2^{\mathcal{N}} \to \mathbb{R}$ is a function $F \colon [0,1]^{\mathcal{N}} \to \mathbb{R}$ whose value for a vector $x \in [0,1]^{\mathcal{N}}$ is defined as

$$F(x) = \mathbb{E}[f(\mathtt{R}(x))] = \sum_{S \subseteq \mathcal{N}} \left(\prod_{u \in S} x_u \right) \cdot \left(\prod_{u \in \mathcal{N} \setminus S} (1 - x_u) \right) \cdot f(S) \ ,$$

where $\mathtt{R}(x)$ is a random set containing every element $u \in \mathcal{N}$ with probability x_u, independently. One can observe that F is an extension of f in the sense that for every set $S \subseteq \mathcal{N}$, if we denote by $\mathbf{1}_S$ the characteristic vector of S, then it holds that $f(S) = F(\mathbf{1}_S)$. Additionally, observe that the rightmost side of F's definition implies that F is indeed a multilinear function, as suggested by its name. The multilinearity of F implies that for every vector $y \in [0,1]^{\mathcal{N}}$ and element $u \in \mathcal{N}$ the partial derivative of F at y with respect to u is given by

$$\frac{\partial F(x)}{\partial x_u}\bigg|_{x=y} = F(y \vee \mathbf{1}_{\{u\}}) - F(y \wedge \mathbf{1}_{\mathcal{N}-u}) = \mathbb{E}[f(u \mid \mathsf{R}(y) - u)] \ ,$$

where the vector operations $\vee$ and $\wedge$ represent coordinate-wise maximum and minimum, respectively.

A linear function is defined by a vector $\ell \in \mathbb{R}^{\mathcal{N}}$. We abuse notation and identify the vector ℓ with the linear function it defines. Accordingly, we denote the value of the function for a vector $x \in [0,1]^{\mathcal{N}}$ by $\ell(x) \triangleq \langle \ell, x \rangle$—where the notation $\langle \cdot, \cdot \rangle$ denotes the dot product. In a further abuse of notation, given a set $S \subseteq \mathcal{N}$, we use $\ell(S)$ as a shorthand for $\ell(\mathbf{1}_S) = \sum_{u \in S} \ell_u$.

An instance of the problem we consider in this work consists of a non-negative monotone submodular function $g \colon 2^{\mathcal{N}} \to \mathbb{R}_{\geq 0}$, a linear function $\ell \colon 2^{\mathcal{N}} \to \mathbb{R}$ and a solvable polytope $P \subseteq [0,1]^{\mathcal{N}}$. We make the standard assumption that the submodular function g, whose description might be exponential in terms of the size n of $\mathcal{N}$, is accessible to the algorithm through a value oracle that given a set $S \subseteq \mathcal{N}$ returns $g(S)$. The objective of the problem is to find a vector $x \in P$ maximizing $G(x) + \ell(x)$, where G is the multilinear extension of g. The result that we prove for this problem is given by the next theorem. Let OPT be the set corresponding to an optimal integral solution for the problem, $i.e.$, $OPT = \arg\max_{S \subseteq 2^{\mathcal{N}}, \mathbf{1}_S \in P}\{g(S) + \ell(S)\}$, and let $m = \max_{u \in \mathcal{N}}\{g(u \mid \varnothing)\}$.

Theorem 1. *There exists a polynomial time algorithm for the above problem that given a value $\varepsilon \in (0,1)$ outputs a vector $x \in P$ such that with high probability $G(x) + \ell(x) \geq (1 - e^{-1}) \cdot g(OPT) + \ell(OPT) - O(\varepsilon) \cdot m$.*

Theorem 1 is very similar to the corresponding result of Sviridenko et al. [16]. However, there are two differences between the two. First, the error term of [16] depends also on $\max_{u \in \mathcal{N}} |\ell_u|$, which is unnecessary for the analysis of our cleaner algorithm. Second, Sviridenko et al. [16] considered only matroid polytopes, for which there are known lossless rounding methods [6,7], and thus, their result is stated in terms of sets rather than vectors.

It should also be noted that Theorem 1 guarantees an algorithm that outputs a fractional vector x which has a certain approximation guarantee with respect to the discrete optimal solution OPT. One might argue that, since x is fractional, it more natural to have the approximation guarantee of x with respect to the optimal fractional solution $\mathbf{opt} = \arg\max_{x \in P}\{G(x) + \ell(x)\}$. Interestingly, this can be done. In particular, by making only a few technical changes in the proof of Theorem 1, one can show that the vector x produced by the algorithm used to prove Theorem 1 in fact has a value of at least $(1 - e^{-1}) \cdot G(\mathbf{opt}) + \ell(\mathbf{opt})$. Nevertheless, we chose to prove the stated version of Theorem 1 in this paper because its proof is slightly cleaner, and most of the theorem's applications require a guarantee only with respect to the integral optimum anyhow.

3 Continuous Time Algorithm

In this section we give a non-formal proof of Theorem 1. This proof demonstates our new ideas, but uses some non-formal simplifications such as allowing a direct oracle access to the multilinear extension G of g and giving the algorithm in the form of a continuous time algorithm (which cannot be implemented on a discrete computer). There are known techniques for getting rid of these simplifications (see, *e.g.*, [6]), and Sect. 4 includes a formal proof of Theorem 1 based on these techniques.

The algorithm we use for the non-formal proof of Theorem 1 is given as Algorithm 1. Like the original continuous greedy algorithm of [6], this algorithm grows a solution $y(t)$ over time. The solution starts as $\mathbf{1}_\varnothing$ at time $t = 0$, and the output of the algorithm is the solution at time $t = 1$. Our algorithm differs, however, from the original continuous greedy algorithm in the method used to determine the direction in which the solution is grown at every given time point. Specifically, our algorithm defines a weight vector $w(t)$ for every time $t \in [0, 1)$ based on the derivatives of the multilinear extension G. It then looks for a vector $z(t)$ in P maximizing a *weighted* combination of $w(t)$ with the linear function ℓ, and this vector $z(t)$ determines the direction in which the solution $y(t)$ is grown.

Algorithm 1. Distorted Continuous Greedy(g, ℓ, P)

---

1 Let $y(0) \leftarrow \mathbf{1}_\varnothing$.

2 **foreach** *time $t \in [0, 1)$* **do**

3 $\quad$ For each $u \in \mathcal{N}$, let $w_u(t) \leftarrow \left.\frac{\partial G(x)}{\partial x_u}\right|_{x=y(t)}$.

4 $\quad$ Let $z(t)$ be the vector in P maximizing $\langle z(t), e^{t-1} \cdot w(t) + \ell \rangle$.

5 $\quad$ Increase $y(t)$ at a rate of $\frac{dy(t)}{dt} = z(t)$.

6 **return** $y(1)$.

We begin the analysis of Algorithm 1 by observing that its output is a vector in P.

Observation 1. $y(1) \in P$.

Proof. By definition, $z(t)$ is a vector in P for every time $t \in [0, 1)$. Hence, $y(1) = \int_0^1 z(t)dt$ is a convex combination of vectors in P, and thus, belongs to P by the convexity of P. $\qquad\square$

Let us consider now the function $\Phi(t) \triangleq e^{t-1} \cdot G(y(t)) + \ell(y(t))$. This function is a central component in our analysis of the approximation ratio of Algorithm 1. The following technical lemma gives an expression for the derivative of this important function.

Lemma 1.

$$\frac{d\Phi(t)}{dt} = e^{t-1} \cdot G(y(t)) + \langle z(t), e^{t-1} \cdot w(t) + \ell \rangle \ .$$

Proof. By the chain rule,

$$\frac{d\Phi(t)}{dt} = e^{t-1} \cdot G(y(t)) + e^{t-1} \cdot \frac{dG(y(t))}{dt} + \frac{d\ell(y(t))}{dt}$$

$$= e^{t-1} \cdot G(y(t)) + e^{t-1} \cdot \sum_{u \in N} \frac{dy_u(t)}{dt} \cdot \frac{\partial G(x)}{\partial x_u}\bigg|_{x=y(t)} + \sum_{u \in N} \frac{dy_u(t)}{dt} \cdot \frac{\partial \ell(x)}{\partial x_u}\bigg|_{x=y(t)}$$

$$= e^{t-1} \cdot G(y(t)) + e^{t-1} \cdot \sum_{u \in N} z_u(t) \cdot w_u(t) + \sum_{u \in N} z_u(t) \cdot \ell_u$$

$$= e^{t-1} \cdot G(y(t)) + \langle z(t), e^{t-1} \cdot w(t) + \ell \rangle \ . \qquad \square$$

The next lemma lower bounds the expression given by the last lemma for the derivative of $\Phi(t)$.

Lemma 2. *For every* $t \in [0, 1)$,

$$e^{t-1} \cdot G(y(t)) + \langle z(t), e^{t-1} \cdot w(t) + \ell \rangle \geq e^{t-1} \cdot g(OPT) + \ell(OPT) \ .$$

Proof. Recall that $z(t)$ is chosen by Algorithm 1 as the vector in P maximizing $\langle z(t), e^{t-1} \cdot w(t) + \ell \rangle$. Since $1_{OPT} \in P$, we get

$$\langle z(t), e^{t-1} \cdot w(t) + \ell \rangle \geq \langle 1_{OPT}, e^{t-1} \cdot w(t) + \ell \rangle = e^{t-1} \cdot \sum_{u \in OPT} w_u(t) + \ell(OPT) \ .$$

Note now that, by the multilinearity of G,

$$\sum_{u \in OPT} w_u(t) = \sum_{u \in OPT} \frac{\partial G(x)}{\partial x_u}\bigg|_{x=y(t)} = \sum_{u \in OPT} [G(y(t) \vee 1_{\{u\}}) - G(y(t) \wedge 1_{N-u})]$$

$$\geq \sum_{u \in OPT} [G(y(t) \vee 1_{\{u\}}) - G(y(t))] \geq G(y(t) \vee 1_{OPT}) - G(y(t))$$

$$\geq g(OPT) - G(y(t)) \ ,$$

where the first and last inequalities follow from the monotonicity of g, and the remaining inequality follows from its submodularity.

Combining the two above inequalities yields

$$\langle z(t), e^{t-1} \cdot w(t) + \ell \rangle \geq e^{t-1} \cdot [g(OPT) - G(y(t))] + \ell(OPT) \ ,$$

and the lemma now follows by adding $e^{t-1} \cdot G(y(t))$ to both sides of this inequality. $\qquad \square$

We are now ready to prove Theorem 1.

Proof (Proof of Theorem 1). Observation 1 shows that $y(1) \in P$. Thus, to prove the theorem it only remains to prove $G(y(1)) + \ell(y(1)) \geq (1 - e^{-1}) \cdot g(OPT) + \ell(OPT)$.

Lemmata 1 and 2 prove together that

$$\frac{d\Phi(t)}{dt} = e^{t-1} \cdot G(y(t)) + \langle z(t), e^{t-1} \cdot w(t) + \ell \rangle \geq e^{t-1} \cdot g(OPT) + \ell(OPT) \ .$$

Integrating both sides of this inequality from $t = 0$ to $t = 1$, we get

$$\Phi(1) - \Phi(0) \geq (1 - e^{-1}) \cdot g(OPT) + \ell(OPT) ,$$

and the theorem now follows by noticing that

$$\Phi(1) = G(y(1)) + \ell(y(1)) \qquad \text{and} \qquad \Phi(0) = e^{-1} \cdot G(y(0)) + \ell(y(0)) \geq 0$$

(the last inequality holds since g is non-negative and $\ell(y(0)) = \ell(1_\varnothing) = 0$).   □

4 Discrete Algorithm

In this section we give a formal proof of Theorem 1. The algorithm that we use for this proof is given as Algorithm 2. Notice that this algorithm considers only discrete times that are integer multiples of a value δ. This value δ is chosen in a way that guarantees $\delta \leq \varepsilon n^{-2}/2 \leq 1/2$ and also ensures that 1 is an integer multiple of δ.

Algorithm 2. Distorted Continuous Greedy – Formal$(g, \ell, P, \varepsilon)$

---
1 Let $y(0) \leftarrow 1_\varnothing$, $t \leftarrow 0$ and $\delta \leftarrow \lceil 2n^2/\varepsilon \rceil^{-1}$.
2 **while** $t < 1$ **do**
3   For each $u \in \mathcal{N}$, let $w_u(t)$ be an estimate for $\mathbb{E}[g(u \mid R(y(t)) - u)]$ obtained
        by averaging the value of the expression within this expectation for
        $r = \lceil -2n^2\varepsilon^{-2} \ln(\delta/n^2) \rceil$ independent samples of $R(y(t))$.
4   Let $z(t)$ be the vector in P maximizing $\langle z(t), (1 + \delta)^{(t-1)/\delta} \cdot w(t) + \ell \rangle$.
5   Let $y(t + \delta) \leftarrow y(t) + \delta \cdot z(t)$.
6   Update $t \leftarrow t + \delta$.
7 **return** $y(1)$.

Let T be the set of times considered by Algorithm 2, i.e., $T = \{i\delta \mid i \in \mathbb{Z}, 0 \leq i < \delta^{-1}\}$. The following observation, which corresponds to Observation 1, shows that the output of Algorithm 2 is feasible.

Observation 2. $y(1) \in P$.

Proof. By definition, $z(t)$ is a vector in P for every time $t \in T$. Observe also that $|T| = \delta^{-1}$. Hence, $y(1) = \sum_{t \in T} \delta \cdot z(t)$ is a convex combination of vectors in P, and thus, belongs to P by the convexity of P.   □

Next, we need to lower bound the probability that any of the estimates made by Algorithm 2 has a significant error. This is done by Lemma 4, whose proof is based on the following known lemma.

Lemma 3 (The symmetric version of Theorem A.1.16 in [2]). *Let X_i, $1 \leq i \leq k$, be mutually independent with all $\mathbb{E}[X_i] = 0$ and all $|X_i| \leq 1$. Set $S = X_1 + \cdots + X_k$. Then, $\Pr[|S| > a] \leq 2e^{-a^2/2k}$.*

Let $\mathcal{E}$ be the event that $|w_u(t) - \mathbb{E}[g(u \mid \mathrm{R}(y(t)) - u)]| \leq \varepsilon m / n$ for every element $u \in \mathcal{N}$ and time $t \in T$.

Lemma 4. $\Pr[\mathcal{E}] \geq 1 - 2n^{-1}$, and hence, $\mathcal{E}$ is a high probability event.

Proof. Consider an arbitrary element $u \in \mathcal{N}$ and time $t \in T$, and let us denote by R_i the i-th independent sample of $\mathrm{R}(y(t))$ used for calculating $w_u(t)$. We now define for every $1 \leq i \leq r$

$$X_i = \frac{g(u \mid R_i - u) - \mathbb{E}[g(u \mid \mathrm{R}(y(t)) - u)]}{m} .$$

Clearly $\mathbb{E}[X_i] = 0$ due to the linearity of the expectation. Additionally, note that $X_i \in [-1, 1]$ because the monotonicity of g guarantees that $g(u \mid R_i - u)$ and $\mathbb{E}[g(u \mid \mathrm{R}(y(t)) - u)]$ are both non-negative, and the submodularity of g guarantees that these expressions are upper bounded by $g(u \mid \varnothing)$—and therefore, also by m because m is defined as $\max_{u \in \mathcal{N}} \{g(u \mid \varnothing)\}$. Hence, by Lemma 3,

$$\Pr\left[|w_u(t) - \mathbb{E}[g(u \mid \mathrm{R}(y(t)) - u)]| > \frac{\varepsilon m}{n}\right] = \Pr\left[\frac{m}{r} \cdot \left|\sum_{i=1}^r X_i\right| > \frac{\varepsilon m}{n}\right]$$

$$= \Pr\left[\left|\sum_{i=1}^r X_i\right| > \frac{r\varepsilon}{n}\right] \leq 2e^{-(r\varepsilon n^{-1})^2/2r}$$

$$= 2e^{-r\varepsilon^2/(2n^2)} \leq 2e^{\ln(\delta/n^2)} = \frac{2\delta}{n^2} .$$

Using the union bound, we now get that the probability that there is any pair of element $u \in \mathcal{N}$ and time $t \in T$ for which

$$|w_u(t) - \mathbb{E}[g(u \mid \mathrm{R}(y(t)) - u)]| > \frac{\varepsilon m}{n}$$

is at most $|\mathcal{N}| \cdot |T| \cdot (2\delta/n^2) = 2/n$. □

Let us define now $\Phi(t) \triangleq (1 + \delta)^{(t-1)/\delta} \cdot G(y(t)) + \ell(y(t))$. Lemma 6 bounds the rate in which this expression increases as a function of t (and thus, can be viewed as a counterpart of Lemma 1). The following technical lemma is used in the proof of Lemma 6. Since similar lemmata have been proved in other places (see, for example, [10,11]), we defer the proof of this lemma to Appendix A.

Lemma 5. *Given two vectors* $y, y' \in [0, 1]^{\mathcal{N}}$ *such that* $0 \leq y'_u - y_u \leq \delta \leq 1$ *and a non-negative monotone submodular function* $f \colon 2^{\mathcal{N}} \to \mathbb{R}_{\geq 0}$ *whose multilinear extension is* F,

$$F(y') - F(y) \geq \sum_{u \in \mathcal{N}} (y'_u - y_u) \cdot \left.\frac{\partial F(x)}{\partial x_u}\right|_{x=y} - n^2 \delta^2 \cdot \max_{u \in \mathcal{N}} f(u \mid \varnothing) .$$

Lemma 6. *If the event $\mathcal{E}$ happens, then, for every time $t \in T$,*

$$\frac{\Phi(t+\delta) - \Phi(t)}{\delta} \geq (1+\delta)^{(t-1)/\delta} \cdot G(y(t)) + \langle z(t), (1+\delta)^{(t-1)/\delta} \cdot w(t) + \ell \rangle - 2\varepsilon m \ .$$

Proof. Since $y(t+\delta) - y(t) = \delta z(t)$, and every coordinate of $z(t)$ is between 0 and 1, we get by Lemma 5 that

$$G(y(t+\delta)) - G(y(t)) \geq \sum_{u \in \mathcal{N}} (y(t+\delta) - y(t)) \cdot \left.\frac{\partial G(x)}{\partial x_u}\right|_{x=y(t)} - n^2 \delta^2 \cdot \max_{u \in \mathcal{N}} g(u \mid \varnothing)$$

$$\geq \sum_{u \in \mathcal{N}} \delta z_u(t) \cdot \mathbb{E}[g(u \mid \mathrm{R}(y(t))) - u)] - \varepsilon \delta m \qquad (1)$$

$$\geq \sum_{u \in \mathcal{N}} \delta z_u(t) \cdot [w_u(t) - \varepsilon m/n] - \varepsilon \delta m \geq \langle \delta z(t), w(t) \rangle - 2\varepsilon \delta m \ ,$$

where the second inequality hold since $\delta \leq \varepsilon n^{-2}$ by definition, and the third inequality holds since we assume that the event $\mathcal{E}$ happened.

Using the linearity of ℓ and the definition of Φ, we now get

$$\frac{\Phi(t+\delta) - \Phi(t)}{\delta}$$

$$= \frac{(1+\delta)^{(t+\delta-1)/\delta} \cdot G(y(t+\delta)) - (1+\delta)^{(t-1)/\delta} \cdot G(y(t))}{\delta}$$

$$+ \frac{\ell(y(t+\delta)) - \ell(y(t))}{\delta}$$

$$= \frac{(1+\delta)^{(t-1)/\delta} \cdot [G(y(t+\delta)) - G(y(t))]}{\delta} + (1+\delta)^{(t-1)/\delta} \cdot G(y(t+\delta))$$

$$+ \langle \ell, z(t) \rangle \ .$$

Plugging Inequality (1) and the inequality $G(y(t+\delta)) \geq G(y(t))$ (which holds due to monotonicity) into the last equality, we get

$$\frac{\Phi(t+\delta) - \Phi(t)}{\delta} \geq (1+\delta)^{(t-1)/\delta} \cdot G(y(t))$$

$$+ \langle z(t), (1+\delta)^{(t-1)/\delta} \cdot w(t) + \ell \rangle - 2\varepsilon m \cdot (1+\delta)^{(t-1)/\delta} \ .$$

The lemma now follows from the last inequality by observing that $(1+\delta)^{(t-1)/\delta} \leq 1$ since $(t-1)/\delta \leq 0$. □

The last lemma gives a lower bound on the increase in $\Phi(t)$ as a function of t. Unfortunately, this lower bound depends on a lot of entities (such as $z(t)$ and $w(t)$), and thus, it is difficult to use it. The following lemma allows us to simplify the lower bound.

Lemma 7. *If the event $\mathcal{E}$ happens, then $\langle z(t), (1 + \delta)^{(t-1)/\delta} \cdot w(t) + \ell \rangle \geq (1 + \delta)^{(t-1)/\delta} \cdot [g(OPT) - G(y(t))] + \ell(OPT) - \varepsilon m$.*

Proof. Recall that $z(t)$ is the vector in the polytope P maximizing the dot product $\langle z(t), (1+\delta)^{(t-1)/\delta} \cdot w(t) + \ell \rangle$. Since $\mathbf{1}_{OPT} \in P$, we get

$$
\begin{aligned}
\langle z(t), (1+\delta)^{(t-1)/\delta} \cdot w(t) + \ell \rangle &\geq \langle \mathbf{1}_{OPT}, (1+\delta)^{(t-1)/\delta} \cdot w(t) + \ell \rangle \\
&= (1+\delta)^{(t-1)/\delta} \cdot \sum_{u \in OPT} w(u) + \ell(OPT) \\
&\geq (1+\delta)^{(t-1)/\delta} \cdot \sum_{u \in OPT} \{\mathbb{E}[g(u \mid \mathsf{R}(y(t))) - u)] - \varepsilon m/n\} + \ell(OPT) \\
&\geq (1+\delta)^{(t-1)/\delta} \cdot \sum_{u \in OPT} \mathbb{E}[g(u \mid \mathsf{R}(y(t))) - u)] + \ell(OPT) - \varepsilon m \;,
\end{aligned}
$$

where the second inequality holds since we assume that the event $\mathcal{E}$ happened. Observe now that the submodularity and monotonicity of f yield

$$
\sum_{u \in OPT} \mathbb{E}[g(u \mid \mathsf{R}(y(t))) - u)] \geq \sum_{u \in OPT} \mathbb{E}[g(u \mid \mathsf{R}(y(t)))] \geq \mathbb{E}[g(OPT \mid \mathsf{R}(y(t)))]
$$

$$
= \mathbb{E}[g(OPT \cup \mathsf{R}(y(t))) - g(\mathsf{R}(y(t)))] \geq g(OPT) - G(y(t)) \;.
$$

The lemma now follows by combining the two above inequalities. □

Combining the last two lemmata, we immediately get the following corollary, which is the promised simplified lower bound on the increase in $\Phi(t)$ as a function of t.

Corollary 1.

$$
\frac{\Phi(t+\delta) - \Phi(t)}{\delta} \geq (1+\delta)^{(t-1)/\delta} \cdot g(OPT) + \ell(OPT) - 3\varepsilon m \;.
$$

Using the last corollary, we can now get a lower bound on the value of $G(y(1)) + \ell(y(t))$ conditioned on the event $\mathcal{E}$.

Lemma 8. *If the event $\mathcal{E}$ happens, then $G(y(1)) + \ell(y(1)) \geq (1 - e^{-1}) \cdot g(OPT) + \ell(OPT) - 4\varepsilon m$.*

Proof. Observe that

$$
G(y(1)) + \ell(y(1)) = \Phi(1) = \Phi(0) + \sum_{t \in T} [\Phi(t+\delta) - \Phi(t)] \tag{2}
$$

$$
\geq \Phi(0) + \sum_{t \in T} \left[\delta(1+\delta)^{(t-1)/\delta} \cdot g(OPT) + \delta \cdot \ell(OPT) - 3\varepsilon\delta m \right]
$$

$$
\geq e^{-1} \cdot g(\varnothing) + g(OPT) \cdot \sum_{t \in T} \delta(1+\delta)^{(t-1)/\delta} + \ell(OPT) - 3\varepsilon m \;,
$$

where the first inequality holds due to Corollary 1 and the second inequality holds since $|T| = \delta^{-1}$ and $\Phi(0) = (1+\delta)^{-1/\delta} \cdot g(\varnothing) + \ell(\varnothing) = (1+\delta)^{-1/\delta} \cdot g(\varnothing) \geq e^{-1} \cdot g(\varnothing)$ because ℓ is linear and g is non-negative.

We now need to lower bound the sum on the rightmost side of the last inequality. Notice that this sum can be presented as the sum of a geometrical series as follows.

$$\sum_{t \in T} \delta(1+\delta)^{(t-1)/\delta} = \sum_{i=0}^{\delta^{-1}-1} \delta(1+\delta)^{(i\delta-1)/\delta} = \delta(1+\delta)^{-\delta^{-1}} \cdot \sum_{i=0}^{\delta^{-1}-1} (1+\delta)^i$$

$$= \delta(1+\delta)^{-\delta^{-1}} \cdot \frac{1-(1+\delta)^{\delta^{-1}}}{1-(1+\delta)} = 1 - (1+\delta)^{-\delta^{-1}}$$

$$\geq 1 - e^{-1}(1-\delta)^{-1} \geq 1 - e^{-1}(1+2\delta) \geq 1 - e^{-1} - \varepsilon e^{-1}/n \ ,$$

where the first inequality holds since it is known that $(1+1/a)^a \geq e(1-1/a)$ for every $a \geq 1$ (and in particular for $a = \delta^{-1}$), the second inequality holds since $(1-a)^{-1} \leq 1+2a$ for every $a \leq 1/2$, and the last inequality holds since $\delta \leq \varepsilon n^{-2}/2$. Plugging the last inequality into Inequality (2) (and using the non-negativity of g), we get

$$G(y(1)) + \ell(y(1)) \geq e^{-1} \cdot g(\varnothing) + (1 - e^{-1} - \varepsilon e^{-1}/n) \cdot g(OPT) + \ell(OPT) - 3\varepsilon m$$

$$\geq e^{-1} \cdot g(\varnothing) + (1 - e^{-1}) \cdot g(OPT) - (\varepsilon e^{-1}/n) \cdot [g(\varnothing) + mn] + \ell(OPT) - 3\varepsilon m$$

$$\geq (1 - e^{-1}) \cdot g(OPT) + e^{-1} \cdot (1 - \varepsilon/n) \cdot g(\varnothing) + \ell(OPT) - 4\varepsilon m$$

$$\geq (1 - e^{-1}) \cdot g(OPT) + \ell(OPT) - 4\varepsilon m \ ,$$

where the second inequality holds since the submodularity of g guarantees

$$g(OPT) \leq g(\varnothing) + \sum_{u \in OPT} g(u \mid \varnothing) \leq g(\varnothing) + mn \ . \qquad \qquad \square$$

We are now ready to prove Theorem 1.

Proof (Proof of Theorem 1). Observation 2 shows that $y(1) \in P$. Additionally, Lemmata 4 and 8 show together that with high probability

$$G(y(1)) + \ell(y(1)) \geq (1 - e^{-1}) \cdot g(OPT) + \ell(OPT) - 4\varepsilon m$$

$$= (1 - e^{-1}) \cdot g(OPT) + \ell(OPT) - O(\varepsilon) \cdot m \ . \qquad \qquad \square$$

A   Proof of Lemma 5

In this section we prove Lemma 5. Let us begin by recalling the lemma itself.

Lemma 5. *Given two vectors $y, y' \in [0,1]^{\mathcal{N}}$ such that $0 \leq y'_u - y_u \leq \delta \leq 1$ and a non-negative monotone submodular function $f \colon 2^{\mathcal{N}} \to \mathbb{R}_{\geq 0}$ whose multilinear extension is F,*

$$F(y') - F(y) \geq \sum_{u \in \mathcal{N}} (y'_u - y_u) \cdot \left. \frac{\partial F(x)}{\partial x_u} \right|_{x=y} - n^2 \delta^2 \cdot \max_{u \in \mathcal{N}} f(u \mid \varnothing) \ .$$

Let us denote the elements of $\mathcal{N}$ by $u_1, u_2, \ldots, u_n$ in an arbitrary order. We define $y^{(i)}$ for every integer $0 \leq i \leq n$ as the vector in $[0,1]^{\mathcal{N}}$ that agrees with

y' on the coordinates 1 to i and with y or the remaining coordinates. Note that this definition implies, in particular, $y^{(0)} = y$ and $y^{(n)} = y'$. The next lemma bounds the amount by which the partial derivative $\frac{\partial F(x)}{\partial x_u}$ can differ between the points $x = y$ and $x = y'$.

Lemma 9. *For every integer $0 \leq i \leq n$ and element $u \in \mathcal{N}$,*

$$\left. \frac{\partial F(x)}{\partial x_u} \right|_{x=y^{(i)}} \geq \left. \frac{\partial F(x)}{\partial x_u} \right|_{x=y} - n\delta \cdot f(u \mid \varnothing) \ .$$

Proof. For the sake of the proof, we assume that $R(y^{(i)})$ is formed from $R(y)$ using the following process. Every element of $\mathcal{N} \setminus R(y)$ is added to a set D with probability of $1 - (1 - y_u^{(i)})/(1 - y_u)$. Then, $R(y^{(i)})$ is chosen as $R(y) \cup D$. Observe that every element $u \in \mathcal{N}$ gets into D with probability $y_u^{(i)} - y_u \leq \delta$, independently, and thus, $R(y) \cup D$ indeed has the distribution that $R(y^{(i)})$ should have.

Using the above definitions, we get

$$\left. \frac{\partial F(x)}{\partial x_u} \right|_{x=y^{(i)}} = \mathbb{E}[f(u \mid R(y^{(i)}) - u)] = \mathbb{E}[f(u \mid R(y) \cup D - u)]$$

$$\geq \Pr[D = \varnothing] \cdot \mathbb{E}[f(u \mid R(y) - u) \mid D = \varnothing] \ ,$$

where the inequality follows from the law of total expectation and the monotonicity of f. Additionally, by the submodularity of f we also get

$$f(u \mid R(y) - u) \leq f(u \mid \varnothing) \ .$$

Combining this inequality with the previous one yields

$$\left. \frac{\partial F(x)}{\partial x_u} \right|_{x=y^{(i)}} + \Pr[D \neq \varnothing] \cdot f(u \mid \varnothing)$$

$$\geq \Pr[D = \varnothing] \cdot \mathbb{E}[f(u \mid R(y) - u) \mid D = \varnothing]$$

$$+ \Pr[D \neq \varnothing] \cdot \mathbb{E}[f(u \mid R(y) - u) \mid D \neq \varnothing]$$

$$= \mathbb{E}[f(u \mid R(y) - u)] = \left. \frac{\partial F(x)}{\partial x_u} \right|_{x=y} \ .$$

One can verify that the last inequality will imply the lemma if we have an upper bound of $n\delta$ on $\Pr[D \neq \varnothing]$. Thus, all we are left to do is to prove this upper bound. Since elements belong to D with probability at most δ and independently,

$$\Pr[D \neq \varnothing] = 1 - \prod_{u \in \mathcal{N}} \Pr[u \notin D] \leq 1 - \prod_{u \in \mathcal{N}} (1 - \delta) = 1 - (1 - \delta)^n \leq n\delta \ . \qquad \square$$

We are now ready to prove Lemma 5.

Proof (Proof of Lemma 5). Observe that for every integer $1 \leq i \leq n$ the vectors $y^{(i-1)}$ and $y^{(i)}$ differ only in coordinate i (in which they differ by $y'_u - y_u$). Recalling that $y^{(0)} = y$, $y^{(n)} = y'$ and F is multilinear, this observation yields

$$F(y') - F(y) = \sum_{i=1}^{n}(y'_{u_i} - y_{u_i}) \cdot \frac{\partial F(x)}{\partial x_{u_i}}\bigg|_{x=y^{(i-1)}}$$

$$\geq \sum_{u \in \mathcal{N}}(y'_u - y_u) \cdot \left[\frac{\partial F(x)}{\partial x_u}\bigg|_{x=y} - n\delta \cdot f(u \mid \varnothing) \right]$$

$$\geq \sum_{u \in \mathcal{N}}(y'_u - y_u) \cdot \frac{\partial F(x)}{\partial x_u}\bigg|_{x=y} - n^2\delta^2 \cdot \max_{u \in \mathcal{N}} f(u \mid \varnothing) \ ,$$

where the first inequality follows from Lemma 9 and the second inequality holds by the monotonicity of f and the fact that $y'_u - y_u \leq \delta$ for every $u \in \mathcal{N}$. □

References

1. Adamczyk, M., Wlodarczyk, M.: Random order contention resolution schemes. In: FOCS, pp. 790–801 (2018)
2. Alon, N., Spencer, J.H.: The Probabilistic Method, 2nd edn. Wiley Interscience, New York (2000)
3. Badanidiyuru, A., Vondrák, J.: Fast algorithms for maximizing submodular functions. In: SODA, pp. 1497–1514 (2014)
4. Buchbinder, N., Feldman, M.: Constrained submodular maximization via a non-symmetric technique. Math. Oper. Res (2019)
5. Buchbinder, N., Feldman, M., Schwartz, R.: Comparing apples and oranges: query trade-off in submodular maximization. Math. Oper. Res. **42**(2), 308–329 (2017)
6. Călinescu, G., Chekuri, C., Pál, M., Vondrák, J.: Maximizing a monotone submodular function subject to a matroid constraint. SIAM J. Comput. **40**(6), 1740–1766 (2011)
7. Chekuri, C., Vondrák, J., Zenklusen, R.: Dependent randomized rounding via exchange properties of combinatorial structures. In: FOCS, pp. 575–584 (2010)
8. Chekuri, C., Vondrák, J., Zenklusen, R.: Submodular function maximization via the multilinear relaxation and contention resolution schemes. SIAM J. Comput. **43**(6), 1831–1879 (2014)
9. Ene, A., Nguyen, H.L.: Constrained submodular maximization: Beyond 1/e. In: FOCS, pp. 248–257 (2016)
10. Feldman, M.: Maximizing symmetric submodular functions. ACM Trans. Algorithms **13**(3), 39:1–39:36 (2017)
11. Feldman, M., Naor, J., Schwartz, R.: A unified continuous greedy algorithm for submodular maximization. In: FOCS, pp. 570–579 (2011)
12. Feldman, M., Svensson, O., Zenklusen, R.: Online contention resolution schemes. In: SODA, pp. 1014–1033 (2016)
13. Gupta, A., Nagarajan, V.: A stochastic probing problem with applications. In: Goemans, M., Correa, J. (eds.) IPCO 2013. LNCS, vol. 7801, pp. 205–216. Springer, Heidelberg (2013). https://doi.org/10.1007/978-3-642-36694-9_18
14. Nemhauser, G.L., Wolsey, L.A.: Best algorithms for approximating the maximum of a submodular set function. Math. Oper. Res. **3**(3), 177–188 (1978)

15. Soma, T., Yoshida, Y.: A new approximation guarantee for monotone submodular function maximization via discrete convexity. In: ICALP, pp. 99:1–99:14 (2018)
16. Sviridenko, M., Vondrák, J., Ward, J.: Optimal approximation for submodular and supermodular optimization with bounded curvature. Math. Oper. Res. **42**(4) (2017)

Efficient Second-Order
Shape-Constrained Function Fitting

David Durfee[1], Yu Gao[1], Anup B. Rao[2], and Sebastian Wild[3(✉)] (iD)

[1] Georgia Institute of Technology, Atlanta, Georgia
{ddurfee,ygao380}@gatech.edu
[2] Adobe Research, San Jose, USA
anuprao@adobe.com
[3] University of Waterloo, Waterloo, Canada
wild@uwaterloo.ca

Abstract. We give an algorithm to compute a one-dimensional shape-constrained function that best fits given data in weighted-L_∞ norm. We give a *single* algorithm that works for a variety of commonly studied shape constraints including monotonicity, Lipschitz-continuity and convexity, and more generally, any shape constraint expressible by bounds on first- and/or second-order differences. Our algorithm computes an approximation with additive error ϵ in $O\left(n \log \frac{U}{\epsilon}\right)$ time, where U captures the range of input values. We also give a simple greedy algorithm that runs in $O(n)$ time for the special case of unweighted L_∞ convex regression. These are the first (near-)linear-time algorithms for second-order-constrained function fitting. To achieve these results, we use a novel geometric interpretation of the underlying dynamic programming problem. We further show that a generalization of the corresponding problems to directed acyclic graphs (DAGs) is as difficult as linear programming.

1 Introduction

We consider the fundamental problem of finding a function f that approximates a given set of data points $(x_1, y_1), \ldots, (x_n, y_n)$ in the plane with smallest possible error, i.e., $f(x_i)$ shall be close to y_i (formalized below), subject to shape constraints on the allowable functions f, such as being increasing and/or concave. More specifically, we present a new algorithm that can handle arbitrary constraints on the (discrete) first- and second-order derivatives of f.

When we only require f to be weakly increasing, the problem is known as isotonic regression, a classic problem in statistics; (see, e.g., [13] for history and applications). It has more recently also found uses in machine learning [12,15,16].

In certain applications, further shape restrictions are integral part of the model: For example, microeconomic theory suggests that production functions

---

An extended online version with full proofs is available at arxiv.org/abs/1905.02149.
D. Durfee—Supported in part by National Science Foundation Grant 1718533.
S. Wild—Supported by the Natural Sciences and Engineering Research Council of Canada and the Canada Research Chairs Programme.

© Springer Nature Switzerland AG 2019
Z. Friggstad et al. (Eds.): WADS 2019, LNCS 11646, pp. 395–408, 2019.
https://doi.org/10.1007/978-3-030-24766-9_29

are weakly increasing and concave (modeling diminishing marginal returns); similar reasoning applies to utility functions. Restricting f to functions with bounded derivative (Lipschitz-continuous functions) is desirable to avoid over-fitting [15]. All these shape restrictions can be expressed by inequalities for first and second derivatives of f; their discretized equivalents are hence amenable to our new method. Shape restrictions that we cannot directly handle are studied in [26] (f is piecewise constant and the number of breakpoints is to be minimized) and [24] (unimodal f). For a more comprehensive survey of shape-constrained function-fitting problems and their applications, see [14, §1]. Motivated by these applications, the problems have been studied in statistics (as a form of nonparametric regression), investigating, e.g., their consistency as estimators and their rate of convergence [4,13,14].

While fast algorithms for isotonic-regression variants have been designed [25], both [20] and [3] list shape constraints beyond monotonicity as important challenges. For example, fitting (multidimensional) convex functions is mostly done via quadratic or linear programming solvers [22]. In his PhD thesis, Balázs writes that current "methods are computationally too expensive for practical use, [so] their analysis is used for the design of a heuristic training algorithm which is empirically evaluated" [4, p. 1].

This lack of efficient algorithms motivated the present work. Despite a few limitations discussed below (implying that we do not yet solve Balázs' problem), we give the first *near-linear-time* algorithms for any function-fitting problem with second-order shape constraints (such as convexity). We use dynamic programming (DP) with a novel geometric encoding for the "states". Simpler versions of such geometric DP variants were used for isotonic regression [23] and are well-known in the competitive programming community; incorporating second-order constraints efficiently is our main innovation.

Problem Definition. Given the vectors $\boldsymbol{x} = (x_1, \ldots, x_n) \in \mathbb{R}^n$ and $\boldsymbol{y} \in \mathbb{R}^n$, an error norm d and shape constraints (formalized below), compute $\boldsymbol{f} = (f_1, \ldots, f_n)$ satisfying the shape constraints with minimal $d(\boldsymbol{f}, \boldsymbol{y})$, i.e., we represent f via its values $f_i = f(x_i)$ at the given points. d is usually an L_p norm, $d(\boldsymbol{x}, \boldsymbol{y}) = (\sum_i |x_i - y_i|^p)^{1/p}$; least squares ($p = 2$) dominate in statistics, but more general error functions have been studied for isotonic regression [3,18,20,21]. We will consider the *weighted* L_∞ norm, i.e., $d(\boldsymbol{f}, \boldsymbol{y}) = \max_{i \in [n]} w_i |f_i - y_i|$, where $[n] = \{1, \ldots, n\}$ and $\boldsymbol{w} \in \mathbb{R}_{\geq 0}^n$ is a given vector of weights.

Since we are dealing with discretized functions (a vector $\boldsymbol{f}$), restrictions for derivatives f' and f'' have to be discretized, as well. We define local slope and curvature as

$$f_i' = \frac{f_i - f_{i-1}}{x_i - x_{i-1}}, \quad (i \in [2..n]), \quad \text{and} \quad f_i'' = \frac{f_i' - f_{i-1}'}{x_i - x_{i-1}}, \quad (i \in [3..n]);$$

the shape constraints are then given in the form of vectors $\boldsymbol{f}'^-, \boldsymbol{f}'^+, \boldsymbol{f}''^-, \boldsymbol{f}''^+$ of bounds for the first- and second-order differences, i.e., we define the set of feasible answers as $F = \{\boldsymbol{f} \in \mathbb{R}^n \mid \boldsymbol{f}'^- \leq \boldsymbol{f}' \leq \boldsymbol{f}'^+ \wedge \boldsymbol{f}''^- \leq \boldsymbol{f}'' \leq \boldsymbol{f}''^+\}$ where inequalities on vectors mean the inequality on all components. The *weighted-L_∞ function-fitting problem with second-order shape constraints* is then to find

$$f^* = \underset{f \in F}{\arg\min} \left(\max_i w_i \cdot |f_i - y_i| \right). \tag{1}$$

Often, we only need a lower resp. upper bound; we can achieve that by allowing $-\infty$ and $+\infty$ entries in $f_i'^{\pm}$ and $f_i''^{\pm}$. For example, setting $f''^- = 0$, $f'^- = f''^- = +\infty$ and $f'- = -\infty$, we can enforce a convex function/vector. We also consider the decision-version of the problem: given a bound L, decide if there is an $f \in F$ with $\max_i w_i |f_i - y_i| \leq L$, and if so, report one.

Contributions. Our main result is a *single* $O(n)$-time algorithm for the decision problem of function fitting with second-order constraints; see Theorem 1.2 for the precise statement. With binary search, this readily yields an additive ϵ-approximation for (1), and thus weighted L_∞ isotonic regression, convex regression and Lipschitz convex regression, in $O\left(n \log \frac{U}{\epsilon}\right)$ time (Theorem 1.4), where $U = (\max_i w_i) \cdot (\max_i y_i - \min_i y_i)$. In the extended online version of this article (arxiv.org/abs/1905.02149), we give a simple greedy algorithm for *unweighted* ($w = 1$) L_∞ convex regression that runs in $O(n)$ time. Finally, we show that a generalization of the problem to DAGs (where the applied first- and second-order difference constraints are restricted by the graph), is as hard as linear programming.

Related Work. Stout [25] surveys algorithms for various versions of isotonic regression; they achieve near-linear or even linear time for many error metrics. He also considers the generalization to any partial order (instead of the total order corresponding to weakly increasing functions). A related task is to fit a piecewise-constant function (with a prescribed number of jumps) to given data. [9,10] solve this problem for L_∞ in optimal $O(n \log n)$ time. Since the geometric constraints are much easier than in our case, a simple greedy algorithm suffices to solve the decision version.

For more restricted shapes, less is known. [24] gives a $O(n \log n)$ solution for unimodal regression. [1] gives an $O(n \log n)$ algorithm for unweighted L_2 Lipschitz isotonic regression and a $O(n \operatorname{poly}(\log n))$ time algorithm for Lipschitz unimodal regression. [22] describes (multidimensional) L_2 convex regression algorithms based quadratic programming. Fefferman [8] studied a closely related problem of smooth interpolation of data in Euclidean space minimizing a certain norm defined on the derivatives of the function. His setup is much more general, but his algorithm cannot find arbitrarily good interpolations (ϵ is fixed for the algorithm). All fast algorithms above consider classes defined by constraints on the *first* derivative only, not the second derivative as needed for convexity. To our knowledge, the fastest prior solution for any convex regression problem is solving a linear program, which will imply super-linear time.

We use a geometric interpretation of dynamic-programming states and represent them implicitly. The work closest in spirit to ours is a recent article by Rote [23]; establishing the transformation of states is much more complicated in the presently studied problem, though. Implicitly representing a series of more complicated objects using data structures has been used in geometric and graph algorithms, such as multiple-source shortest paths [17] and shortest paths in polygons [5,7,19]. The only other work (we know of) that interprets dynamic programming geometrically is [26].

There is a rich literature on methods for speeding up dynamic programming [6,11,27,28]. They involve a variety of powerful techniques such as monotonicity of transition points, quadrangle inequalities, and Monge matrix searching [2], many of which have found applications in other settings. The focus of these methods is to reduce the (average) number of transitions that a state is involved in, often from $O(n)$ to $O(1)$. Therefore, their running times are lower bounded by the number of states in the dynamic programs.

1.1 Results

We formally state our theorem for the decision problem here; results for shape-constrained function fitting are obtained as corollaries. For our algorithm, the discrete derivatives (as defined above) are inconvenient because they involve the x-distance between points. We therefore *normalize* all x-distances to 1 (s.t. $x_i = i$); for the second-order constraints, this normalization makes the introduction of an additional parameter necessary, the scaling factors α_i (see below).

Definition 1.1 (1st/2nd-diff-constrained vectors). *Let n-dimensional vectors $\boldsymbol{x}^- \leq \boldsymbol{x}^+$ (value bounds), $\boldsymbol{y}^- \leq \boldsymbol{y}^+$ (difference bounds), $\boldsymbol{z}^- \leq \boldsymbol{z}^+$ (second-order difference bounds), and $\boldsymbol{\alpha} > 0$ be given. We define $\mathcal{S} \subset \mathbb{R}^n$ to be the set of all $\boldsymbol{b} \in \mathbb{R}^n$ that satisfy the following constraints:*

$$\forall i \in [1..n] \quad x_i^- \leq b_i \leq x_i^+ \qquad\qquad \text{(value constraints)}$$

$$\forall i \in [2..n] \quad y_i^- \leq b_i - b_{i-1} \leq y_i^+ \qquad\qquad \text{(first-order constr.)}$$

$$\forall i \in [3..n] \quad z_i^- \leq (b_i - b_{i-1}) - \alpha_i(b_{i-1} - b_{i-2}) \leq z_i^+ \quad \text{(second-order constr.)}$$

Moreover, we consider the "truncated problems" $\mathcal{S}_k$, where $\mathcal{S}_k$ is the set of all $\boldsymbol{b} \in \mathbb{R}^n$ that satisfy the constraints up to k (instead of n).

A visualization of an example is shown in Fig. 1. We can encode an instance $(\boldsymbol{x}, \boldsymbol{y}, \boldsymbol{f}'^\pm, \boldsymbol{f}''^\pm)$ of the decision version of the weighted-L_∞ function-fitting problem with second-order constraints as 1st/2nd-diff-constrained vectors by setting

$$x_i^\pm = y_i \pm L/w_i, \qquad\qquad y_i^\pm = f'^\pm \cdot (x_i - x_{i-1}),$$

$$z_i^\pm = f''^\pm \cdot (x_i - x_{i-1})^2, \qquad\qquad \alpha_i = \frac{x_i - x_{i-1}}{x_{i-1} - x_{i-2}}.$$

So, our goal is to efficiently compute some $\boldsymbol{b} \in \mathcal{S}$ or determine that $\mathcal{S} = \emptyset$. Our core technical result is a linear-time algorithm for this task:

Theorem 1.2 (1st/2nd-diff-constrained decision). *With the notation of Definition 1.1, in $O(n)$ time, we can compute $\boldsymbol{b} \in \mathcal{S}$ or determine that $\mathcal{S} = \emptyset$.*

Section 2 will be devoted to the proof. To simplify the presentation, we will assume throughout that $\boldsymbol{x}^+, \boldsymbol{x}^-, \boldsymbol{y}^+, \boldsymbol{y}^-, \boldsymbol{z}^+, \boldsymbol{z}^-$ are bounded.[1]For the optimization version of the problem, Eq. (1), we consider approximate solutions in the following sense.

---

[1] Some problems are stated with $\pm\infty$ values, but we can always replace unbounded values in the algorithms with an (input-specific) sufficiently large finite number.

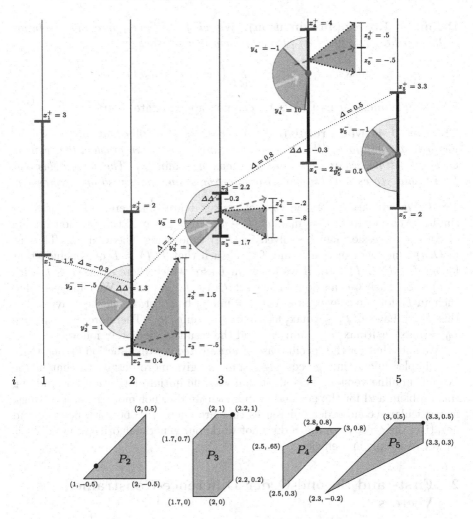

Fig. 1. Exemplary input for the 1st/2nd-diff-constrained decision problem (with $\alpha = 1$). Value constraints are illustrated as blue bars. First-order constraints are shown as green circles, indicating the allowable incoming angles/slopes; the green dot and the circle can be moved up and down within the blue range. Finally, second-order constraints are given as red triangles, in which the minimal and maximal allowable change in slope is shown (dotted red), based off an exemplary incoming slope (dashed red). The thin dotted line shows $\boldsymbol{b} = (1.7, 1.2, 2.2, 2.8, 3.3) \in \mathcal{S}$.

Below the visualization of the instance, we show the set of pairs $(b_i, b_i - b_{i-1})$ for $\boldsymbol{b} \in \mathcal{S}_i$, i.e., allowable combinations of value and slope at i for the truncated problems; our specific solution is shown as a black dot. These sets are the *feasibility polygons* P_i (defined in Sect. 2.1) that play a vital role in our algorithm. Given all P_i, one can easily construct a solution backwards, starting from any point in P_5. (Color figure online)

Definition 1.3 (ϵ-approximation). *We call $\boldsymbol{f} \in F$ an ϵ-approximate solution to the weighted L_∞ function-fitting problem if it satisfies*

$$\max_i w_i |f_i - y_i| \ \leq \ \min_{g \in F} \left(\max_i w_i |g_i - y_i| \right) + \epsilon.$$

By a simple binary search on L, we can find approximate solutions.

Theorem 1.4 (Main result). *There exists an algorithm that computes an ϵ-approximate solution to the weighted-L_∞ convex regression problem that runs in $O(n \log \frac{U}{\epsilon})$ time, where $U = (\max_i w_i)(\max_i y_i - \min_i y_i)$. The same holds true for isotonic regression, Lipschitz isotonic regression, convex isotonic regression.*

Proof. We will argue for the case of convex regression here, other cases are similar. Abbreviate $L(\boldsymbol{f}) = \max_i w_i |f_i - y_i|$. For a given L, the decision version of convex regression can be solved in $O(n)$ time using Theorem 1.2. That is, in $O(n)$ time, we can either find $\boldsymbol{f} \in F$ such that $L(\boldsymbol{f}) \leq L$ or conclude that for all $\boldsymbol{f} \in F$, $L(\boldsymbol{f}) > L$. If we know an L_0 for which there exists $\boldsymbol{f} \in F$ with $L(\boldsymbol{f}) \leq L_0$, then we can do a binary search for L_c in $[0, L_0]$. We can easily find such an L_0 for the convex case: Let $\boldsymbol{f} = \min y_j$ be constant (hence convex). For this $\boldsymbol{f}$, we have $L(\boldsymbol{f}) \leq (\max_j w_j)(\max_j y_j - \min_j y_j)$. Therefore, we can take $L_0 = (\max_j w_j)(\max_j y_j - \min_j y_j)$ and the result immediately follows. $\square$

We note that for the specific case of *unweighted* convex function fitting, there is a simpler linear-time greedy algorithm; we give more details on that in the extended online version. This algorithm was the initial motivation for studying this problem and for the geometric approach we use. For more general settings, in particular second-order differences that are allowed to be both positive and negative, the greedy approach does not work; our generic algorithm, however, is almost as simple and efficient.

2   First- and Second-Order Difference-Constrained Vectors

In this section, we present our main algorithm and prove Theorem 1.2. In Sect. 2.1, we give an overview and introduce the feasibility polygons P_i. Section 2.2 shows how P_i can be inductively computed from P_{i-1} via a geometric transformation. We finally show how this transformation can be computed efficiently, culminating in the proof of Theorem 1.2, in Sect. 2.3. Two proofs are deferred to the extended online version of this article.

2.1   Overview of the Algorithm

Recall that the problem we want to solve, in order to prove Theorem 1.2, is finding a feasible point $\boldsymbol{b}$ in $\mathcal{S}$ from Definition 1.1. Our algorithm will use dynamic programming (DP) where each state is associated with the feasible b_i in the truncated problem. We will iteratively determine all b_i such that b_i is the ith entry of some $\boldsymbol{b} \in \mathcal{S}_i$.

Feasible b_i have to respect the first- and second-order difference constraints. To check those, we also need to know the possible pairs (b_{i-1}, b_{i-2}) of $(i-1)$th and $(i-2)$th entries for some $\boldsymbol{b} \in \mathcal{S}_{i-1}$, so the states have to maintain more information than the b_i alone. It will be instrumental to *rewrite* this pair as $(b_{i-1}, b_{i-1} - b_{i-2})$, the combination of valid values b_{i-1} and valid *slopes* at which we entered b_{i-1} for a solution in $\mathcal{S}_{i-1}$. From that, we can determine the valid slopes at which we can *leave* b_{i-1} using our shape constraints. We thus define the *feasibility polygons*

$$P_i \;=\; \big\{(x,y) \;\big|\; \exists \boldsymbol{b} \in \mathcal{S}_i : x = b_i \wedge y = b_i - b_{i-1}\big\} \tag{2}$$

for $i = 2, \ldots, n$. See Fig. 1 for an example. We view each point in P_i as a "state" in our DP algorithm, and our goal becomes to efficiently compute P_i from P_{i-1}. The key observation is that each P_i is indeed an $O(n)$-vertex convex polygon, and we only need an efficient way to compute the *vertices* of P_i from those of P_{i-1}. This needs a clever representation, though, since all vertices can change when going from P_{i-1} to P_i. A closer look reveals that we can represent the vertex transformations *implicitly*, without actually updating each vertex, and we can combine subsequent transformations into a single one. More specifically, if we consider the boundary of P_{i-1}, the transformation to P_i consists of two steps: (1) a linear transformation for the upper and lower hull of P_{i-1}, and (2) a truncation of the resulting polygon by vertical and horizontal lines (i.e., an intersection of the polygon and a half-plane).

The first step requires a more involved proof and uses that all line segments of P_i have weakly positive slope ("+SLOPED", formally defined below). Implicitly computing the first transformation as we move between P_i is straightforward, only requiring a composition of linear operations (a different one, though, for upper and lower hull). We can apply the cumulative transformation whenever we need to access a vertex.

The second step is conceptually simpler, but more difficult to implement efficiently, as we have to determine where a line cuts the polygon in amortized constant time. For this operation, we separately store the vertices of the upper and lower hull of P_i in two arrays, sorted by increasing x-coordinate; since P_i is +SLOPED, y-values are also increasing. A linear search for intersections has overall $O(n)$ cost since we can charge individual searches to deleted vertices.

Finally, if $P_n \neq \emptyset$, we compute a feasible vector $\boldsymbol{b}$ backwards, starting from any point in P_n. Since we do not explicitly store the P_i, this requires successively "undoing" all operations (going from P_i back to P_{i-1}); see the extended version of this paper for details.

2.2   Transformation from State P_{i-1} to P_i

We first define the structural property "+SLOPED" that our method relies on.

Definition 2.1 (+SLOPED). *We say a polygon $P \subseteq \mathbb{R}^2$ with vertices $v_1, \ldots, v_k$ is +SLOPED if* $\mathrm{slope}(v_i, v_j) \geq 0$ *for all edges (v_i, v_j) of P. Here, the slope between two points $v_1 = (x_1, y_1)$, $v_2 = (x_2, y_2) \in \mathbb{R}^2$ is defined as* $\mathrm{slope}(v_1, v_2) = \frac{y_2 - y_1}{x_2 - x_1}$, *when $x_1 \neq x_2$, and $\mathrm{slope}(v_1, v_2) = \infty$, otherwise.*

We will now show that P_i can be computed by applying a simple geometric transformation to P_{i-1}. In passing, we will prove (by induction on i) that all P_i are +SLOPED. For the base case, note that $P_2 = \{(b_2, b_2 - b_1) \mid x_1^- \leq b_1 \leq x_1^+ \wedge x_2^- \leq b_2 \leq x_2^+ \wedge y_2^- \leq b_2 - b_1 \leq y_2^+\}$, which is an intersection of 6 half-planes. The slopes of the defining inequalities are all non-negative or infinite, so P_2 is +SLOPED.

Let us now assume that P_{i-1}, $i \geq 3$, is +SLOPED; we will consider the transformation from P_{i-1} to P_i and show that it preserves this property. We begin by separating the transformation from P_{i-1} to P_i into two main steps.

Step 1: Second-order constraint only. For the first step, we ignore the value and first-order constraints at index i. This will yield a convex polygon, $P_i^{(z)}$, that contains P_i; in Step 2, we will add the other constraints at i to obtain P_i itself.

Definition 2.2 ($P_i^{(z)}$: 2nd-order-only polygons). *For a fixed i, consider the modified problem with $x_i^-, y_i^- = -\infty$ and $x_i^+, y_i^+ = \infty$. Define the second-order-only polygon, $P_i^{(z)}$, as the polygon P_i of this modified problem (considering only the z_i constraints at i).*

The statement of the following lemma is very simple observation, but allows us to compute $P_i^{(z)}$ from P_{i-1} with an explicit geometric construction, (whereas such seemed not obvious for the original feasibility polygons).

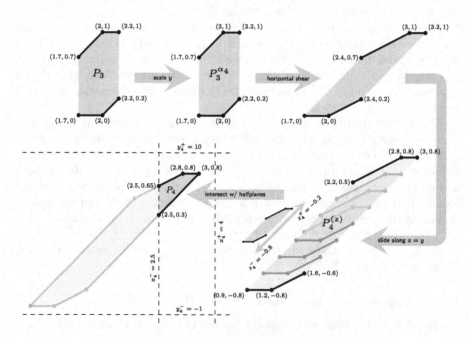

Fig. 2. The transformation from P_3 to P_4 for the example instance of Fig. 1. Upper and lower hulls are shown separately in green resp. red. (Color figure online)

Lemma 2.3 ($P_i^{(z)}$: scaled, sheared and shifted P_{i-1}). $P_i^{(z)} = \{(x + \alpha_i y + z, \alpha_i y + z) \mid (x,y) \in P_{i-1}, z \in [z_i^-, z_i^+]\}$.

Proof. The only constraint at i is $z_i^- \le (b_i - b_{i-1}) - \alpha_i(b_{i-1} - b_{i-2}) \le z_i^+$. We rewrite this as (a) a constraint for $b_i - b_{i-1}$, using that $b_{i-1} - b_{i-2}$ is the y-coordinate in P_{i-1}, and (b) a constraint for b_i, using that, additionally, b_{i-1} is the x-coordinate in P_{i-1}. □

Once we have computed this polygon $P_i^{(z)}$, computing P_i is easy: adding the constraints $x_i^- \le x \le x_i^+$ and $y_i^- \le y \le y_i^+$ requires only cutting $P_i^{(z)}$ with two horizontal and vertical lines. We give a visual representation of the mapping on an example in Fig. 2. We break the above mapping into two simpler stages:

Corollary 2.4 ($P_i^{(z)}$: sheared and shifted $P_{i-1}^{\alpha_i}$).
Setting $P_{i-1}^{\alpha_i} = \{(x, \alpha_i y) \mid (x,y) \in P_{i-1}\}$, we have
$P_i^{(z)} = \{(x + y + z, y + z) \mid (x,y) \in P_{i-1}^{\alpha_i}, z \in [z_i^-, z_i^+]\}$.

We note that scaling the y-coordinate by α_i preserves the +SLOPED-property:

Lemma 2.5. *Let $\alpha \ge 0$. If P is +SLOPED, so is $P^\alpha = \{(x, \alpha y) \mid (x, y \in P)\}$.*

Proof. Scaling the y-coordinates will preserve all of the vertices of P, and also scale the slope of each vertex pair by $\alpha \ge 0$. So, P^α is +SLOPED. □

That leaves us with the core of the transformation, from $P_{i-1}^{\alpha_i}$ to $P_i^{(z)}$. Intuitively, it can be viewed as sliding $P_{i-1}^{\alpha_i}$ along the line $x = y$ by any amount $z \in [z_i^-, z_i^+]$ and taking the union thereof, (see Fig. 2). To compute the result of this operation, we split the boundary into upper and lower hull.

Definition 2.6 (Upper/lower hull). *Let P be a convex polygon with vertex set V. We define the upper hull (vertices) resp. lower hull (vertices) of P as*

$$\text{U-HULL}(P) = \{u_i = (x_i, y_i) \in V \mid \nexists(x_i, y) \in P : y > y_i\}$$
$$\text{L-HULL}(P) = \{u_i = (x_i, y_i) \in V \mid \nexists(x_i, y) \in P : y < y_i\}$$

Unless specified otherwise, hull vertices are ordered by increasing x-coordinate.

Note that a vertex can be in both hulls. Moreover, the leftmost vertices in U-HULL(P) and L-HULL(P) always have the same x-coordinate, similarly for the rightmost vertices. As proved in Lemma 2.3, each point in $P_{i-1}^{\alpha_i}$ is mapped to a line-segment with slope 1; we give this mapping a name.

Definition 2.7 (2nd-order P transform). *Let $f_i((x,y))$ be the line-segment $\{(x + y + z, y + z) \mid z \in [z_i^-, z_i^+]\}$ and denote by $f_i^-((x,y)) = (x + y + z_i^-, y + z_i^-)$ and $f_i^+((x,y)) = (x + y + z_i^+, y + z_i^+)$ the two endpoints of $f_i((x,y))$.*
We write $f(S) = \bigcup_{(x,y) \in S} f((x,y))$ for the element-wise application of f to a set S of points.

The vertices of $P_i^{(z)}$ result from transforming the upper hull of $P_{i-1}^{\alpha_i}$ by f_i^+ and the lower hull by f_i^-. The next lemma formally establishes that applying f_i^+ resp. f_i^- to the hulls of $P_{i-1}^{\alpha_i}$ correctly computes $P_i^{(z)}$, (again, compare Fig. 2).

Lemma 2.8 (From $P_{i-1}^{\alpha_i}$ to $P_i^{(z)}$ via hulls). *If $P_{i-1}^{\alpha_i}$ is +SLOPED, then $P_i^{(z)}$ is +SLOPED and* U-HULL$(P_i^{(z)}) = \{f_i^-(v_{ll})\} \cup f_i^+($U-HULL$(P_{i-1}^{\alpha_i}))$ *and* L-HULL$(P_i^{(z)})$ $= f_i^-($L-HULL$(P_{i-1})) \cup \{f_i^+(v_{ur})\}$, *where v_{ll} (lower-left) and v_{ur} (upper-right) are the first vertex of* L-HULL$(P_{i-1}^{\alpha_i})$ *and the last vertex of* U-HULL$(P_{i-1}^{\alpha_i})$, *respectively.*

A formal proof is given in the extended version. Intuitively, since each point in $P_{i-1}^{\alpha_i}$ is mapped to a line-segment with slope 1 in $P_i^{(z)}$, $P_i^{(z)}$ is obtained by sliding $P_{i-1}^{\alpha_i}$ along the line $x = y$. The full transformation from P_{i-1} to $P_i^{(z)}$ can now be stated as:

Lemma 2.9 (P_{i-1} to $P_i^{(z)}$). *Let f_i^{*,α_i} be the function $f_i^{*,\alpha_i}(x,y) = (x + \alpha_i y + z_i^*, \alpha_i y + z_i^*)$ for $* \in \{-,+\}$. If P_{i-1} is +SLOPED, then $P_i^{(z)}$ is +SLOPED with*

$$\text{U-HULL}(P_i^{(z)}) = \{f_i^{-,\alpha_i}(v_{ll})\} \cup f_i^{+,\alpha_i}(\text{U-HULL}(P_{i-1}))$$
$$\text{L-HULL}(P_i^{(z)}) = f_i^{-,\alpha_i}(\text{L-HULL}(P_{i-1})) \cup \{f_i^{+,\alpha_i}(v_{ur})\}$$

with v_{ll} and v_{ur} the lower-left resp. upper-right vertex of P_{i-1}.

Proof. This follows immediately from Corollary 2.4 and Lemmas 2.5 and 2.8. □

Step 2: Truncating by value and slope. To complete the transformation, we need to add the constraints $x_i^- \leq b_i \leq x_i^+$ and $y_i^- \leq b_i - b_{i-1} \leq y_i^+$ to $P_i^{(z)}$. This is equivalent to cutting our polygon with two vertical and horizontal planes. The following lemma shows that this preserves the +SLOPED-property.

Lemma 2.10 (# new vertices). *If P_{i-1} is +SLOPED with k vertices, then P_i is either empty or +SLOPED with at most $k + 6$ vertices.*

It follows that over the course of the algorithm, only $O(n)$ vertices are added in total. This will be instrumental for analyzing the running time.

Proof. We know that $P_i^{(z)}$ is +SLOPED, and it follows easily from the definition that cutting by horizontal and vertical planes will preserve this property. Furthermore, note that cutting a convex polygon will increase the total number of vertices by at most one. We added at most 2 vertices to P_{i-1} to obtain $P_i^{(z)}$. We then cut $P_i^{(z)}$ by the inequalities $x \leq x_i^+$, $x \geq x_i^-$, $y \leq y_i^-$, and $y \geq y_i^+$, i.e., two horizontal and vertical planes. Each adds at most one vertex, giving the desired upper bound.   □

2.3 Algorithm

A direct implementation of the transformation of Lemma 2.9 yields a "brute force" algorithm that maintains all vertices of P_i and checks if P_n is empty; (the running time would be quadratic). It works as follows:

1. *[Init]*: Compute the vertices of P_2.
2. *[Compute P_i]*: For $i = 3, \ldots, n$, do the following:
   2.1. At step i, scale the y-coordinate of each vertex by α_i.
   2.2. Apply f_i^+ resp. f_i^- to each vertex, depending on which hull it is in.
   2.3. Add the new vertex to U-HULL and L-HULL, as per Lemma 2.9.
   2.4. Delete all the vertices outside $[x_i^-, x_i^+] \times [y_i^-, y_i^+]$ and
        add the vertices created by intersecting with $[x_i^-, x_i^+] \times [y_i^-, y_i^+]$.
3. *[Compute b]* : If $P_n \neq \emptyset$, compute $(b_1, \ldots, b_n)$ by backtracing.

Observe that Lemma 2.9 applies the *same* linear function (multiplication of y-coordinate by α_i and f_i^+ or f_i^-) to *all* vertices in U-HULL resp. L-HULL. So, we do not need to modify every vertex each time; instead, we can store – separately for U-HULL and L-HULL – the *composition* of the linear transformations as a matrix. Whenever we access a vertex, we take the unmodified vertex and apply the cumulative transformation in $O(1)$ time.

At each step, after applying the linear transformations, by Lemma 2.9 we also need to copy the leftmost vertex of L-HULL, add it to the left of U-HULL and copy the rightmost vertex of U-HULL and add it to the right of L-HULL. To add these vertices, we simply apply the inverse of each respective cumulative transformation such that all stored vertices require the same transformation. This will also take $O(1)$ time.

Since all the slopes of $P_i^{(z)}$ are non-negative (+SLOPED) and we keep vertices sorted by x-coordinate, the truncation by a horizontal or vertical plane can only remove a prefix or suffix from U-HULL and L-HULL of $P_i^{(z)}$. Depending on the constraint we are adding, ($x \leq x_i^+$, $x \geq x_i^-$, $y \leq y_i^-$, or $y \geq y_i^+$), we start at the rightmost or leftmost vertex of the U-HULL and L-HULL, and continue until we find the intersection with the cutting plane. We remove all visited vertices.

This could take $O(n)$ time in any single iteration, but the total cost over all iterations is $O(n)$ since we start with $O(1)$ vertices and add $O(n)$ vertices throughout the entire procedure (by Lemma 2.10). This allows us to use two deques (double-ended queues), represented as arrays, to store the vertices of U-HULL and L-HULL. Putting this all together gives the linear time algorithm for the decision problem "$\mathcal{S} = \emptyset$?".

To compute an actual solution when $\mathcal{S} \neq \emptyset$, we compute $b_n, \ldots, b_1$, in this order. From the last P_n, we can find a feasible b_n (the x-coordinate of any point in P_n). Then, we retrace the steps of our algorithm through specific points in each P_i. Since intermediate P_i were only implicitly represented, we have to recover P_i by "undoing" the algorithm's operations in reverse order; this is possible in overall time $O(n)$ by remembering the operations from the forward phase. The details on the backtracing step are presented in the extended version of this article.

3   Conclusion

In this article, we presented a linear-time dynamic-programming algorithm to decide whether there is a vector b that lies (componentwise) between given upper and lower bounds and additionally satisfies inequalities on its first- and second-order (successive) differences. This method can be used to approximate weighted-L_∞ shape-restricted function-fitting problems, where the shape restrictions are given as bounds on first- and/or second-order differences (local slope and curvature).

This is a first step towards much sought-after efficient methods for more general convex regression tasks. A main limitation of our approach is the restriction to one-dimensional problems. We show in the extended version of this article (arxiv.org/abs/1905.02149) that a natural extension of the problem studied here to directed acyclic graphs is already as hard as linear programming, leaving little hope for an efficient generic solution. This is in sharp contrast to isotonic regression, where similar extensions to arbitrary partial orders do have efficient algorithms (for L_∞) [25]. This might also be bad news for multidimensional regression with second-order constraints, since higher dimensions entail, among other complications, a non-total order over the inputs.

A second limitation is the L_∞ error metric, which might not be adequate for all applications. We leave the question whether similarly efficient methods are also possible for other metrics for future work. A further extension to study is convex *unimodal* regression; here, finding the maximum is part of the fitting problem, and so not directly possible with our presented method.

Acknowledgments. We thank Richard Peng, Sushant Sachdeva, and Danny Sleator for insightful discussions, and our anonymous referees for further relevant references and insightful comments that significantly improved the presentation.

References

1. Agarwal, P.K., Phillips, J.M., Sadri, B.: Lipschitz unimodal and isotonic regression on paths and trees. In: López-Ortiz, A. (ed.) LATIN 2010. LNCS, vol. 6034, pp. 384–396. Springer, Heidelberg (2010). https://doi.org/10.1007/978-3-642-12200-2_34
2. Aggarwal, A., Klawe, M.M., Moran, S., Shor, P., Wilber, R.: Geometric applications of a matrix-searching algorithm. Algorithmica **2**(1–4), 195–208 (1987). https://doi.org/10.1007/bf01840359
3. Bach, F.: Efficient algorithms for non-convex isotonic regression through submodular optimization. In: Bengio, S., Wallach, H., Larochelle, H., Grauman, K., Cesa-Bianchi, N., Garnett, R. (eds.) Advances in Neural Information Processing Systems, vol. 31, pp. 1–10. Curran Associates, Inc. (2018)
4. Balázs, G.: Convex Regression: Theory, Practice, and Applications. Ph.D. thesis (2016). https://doi.org/10.7939/R3T43J98B
5. Chazelle, B.: A theorem on polygon cutting with applications. In: Symposium on Foundations of Computer Science (SFCS), pp. 339–349. IEEE (1982). https://doi.org/10.1109/SFCS.1982.58
6. Eppstein, D., Galil, Z., Giancarlo, R.: Speeding up dynamic programming. In: Symposium on Foundations of Computer Science (SFCS), IEEE (1988). https://doi.org/10.1109/sfcs.1988.21965

7. Erickson, J.: Shortest homotopic paths (2009). http://jeffe.cs.illinois.edu/teaching/comptop/2009/notes/shortest-homotopic-paths.pdf. lecture notes for computational topology
8. Fefferman, C.: Smooth interpolation of data by efficient algorithms. In: Excursions in Harmonic Analysis, vol. 1, pp. 71–84. Birkhäuser, Boston, November 2012. https://doi.org/10.1007/978-0-8176-8376-4_4
9. Fournier, H., Vigneron, A.: Fitting a step function to a point set. Algorithmica **60**(1), 95–109 (2009). https://doi.org/10.1007/s00453-009-9342-z
10. Fournier, H., Vigneron, A.: A deterministic algorithm for fitting a step function to a weighted point-set. Inf. Process. Lett. **113**(3), 51–54 (2013). https://doi.org/10.1016/j.ipl.2012.11.003
11. Galil, Z., Giancarlo, R.: Speeding up dynamic programming with applications to molecular biology. Theor. Comput. Sci. **64**(1), 107–118 (1989). https://doi.org/10.1016/0304-3975(89)90101-1
12. Ganti, R.S., Balzano, L., Willett, R.: Matrix completion under monotonic single index models. In: Cortes, C., Lawrence, N.D., Lee, D.D., Sugiyama, M., Garnett, R. (eds.) Advances in Neural Information Processing Systems, vol. 28, pp. 1873–1881. Curran Associates, Inc. (2015)
13. Groeneboom, P., Jongbloed, G.: Nonparametric Estimation Under Shape Constraints, vol. 38. Cambridge University Press (2014)
14. Guntuboyina, A., Sen, B.: Nonparametric shape-restricted regression. Stat. Sci. **33**(4), 568–594 (2018). https://doi.org/10.1214/18-sts665
15. Kakade, S.M., Kanade, V., Shamir, O., Kalai, A.: Efficient learning of generalized linear and single index models with isotonic regression. In: Shawe-Taylor, J., Zemel, R.S., Bartlett, P.L., Pereira, F., Weinberger, K.Q. (eds.) Advances in Neural Information Processing Systems, vol. 24, pp. 927–935. Curran Associates, Inc. (2011)
16. Kalai, A.T., Sastry, R.: The isotron algorithm: High-dimensional isotonic regression. In: Annual Conference on Learning Theory (COLT) (2009)
17. Klein, P.N.: Multiple-source shortest paths in planar graphs. In: Symposium on Discrete Algorithms (SODA), pp. 146–155. SIAM (2005)
18. Kyng, R., Rao, A., Sachdeva, S.: Fast, provable algorithms for isotonic regression in all l_p-norms. In: Cortes, C., Lawrence, N.D., Lee, D.D., Sugiyama, M., Garnett, R. (eds.) Advances in Neural Information Processing Systems, vol. 28, pp. 2719–2727. Curran Associates, Inc. (2015)
19. Lee, D.T., Preparata, F.P.: Euclidean shortest paths in the presence of rectilinear barriers. Networks **14**(3), 393–410 (1984). https://doi.org/10.1002/net.3230140304
20. Lim, C.H.: An efficient pruning algorithm for robust isotonic regression. In: Bengio, S., Wallach, H., Larochelle, H., Grauman, K., Cesa-Bianchi, N., Garnett, R. (eds.) Advances in Neural Information Processing Systems, vol. 31, pp. 219–229. Curran Associates, Inc. (2018)
21. Luss, R., Rosset, S.: Generalized isotonic regression. J. Comput. Graph. Stat. **23**(1), 192–210 (2014). https://doi.org/10.1080/10618600.2012.741550
22. Mazumder, R., Choudhury, A., Iyengar, G., Sen, B.: A computational framework for multivariate convex regression and its variants. J. Am. Stat. Assoc. 1–14 (2018). https://doi.org/10.1080/01621459.2017.1407771
23. Rote, G.: Isotonic regression by dynamic programming. In: Fineman, J.T., Mitzenmacher, M. (eds.) Symposium on Simplicity in Algorithms (SOSA 2019), OASIcs, vol. 69, pp. 1:1–1:18. Schloss Dagstuhl-Leibniz-Zentrum fuer Informatik (2018). https://doi.org/10.4230/OASIcs.SOSA.2019.1

24. Stout, Q.F.: Unimodal regression via prefix isotonic regression. Comput. Stat. Data Anal. **53**(2), 289–297 (2008). https://doi.org/10.1016/j.csda.2008.08.005
25. Stout, Q.F.: Fastest isotonic regression algorithms (2014). http://web.eecs.umich.edu/~qstout/IsoRegAlg.pdf
26. Tsourakakis, C.E., Peng, R., Tsiarli, M.A., Miller, G.L., Schwartz, R.: Approximation algorithms for speeding up dynamic programming and denoising aCGH data. J. Exp. Algorithmics **16**, 1 (2011). https://doi.org/10.1145/1963190.2063517
27. Yao, F.F.: Efficient dynamic programming using quadrangle inequalities. In: Symposium on Theory of Computing (STOC). ACM Press (1980). https://doi.org/10.1145/800141.804691
28. Yao, F.F.: Speed-up in dynamic programming. SIAM J. Algebraic Discrete Methods **3**(4), 532–540 (1982). https://doi.org/10.1137/0603055

Dynamic Dictionary Matching
in the Online Model

Shay Golan, Tomasz Kociumaka, Tsvi Kopelowitz$^{(\boxtimes)}$, and Ely Porat

Bar-Ilan University, Ramat-Gan, Israel
{golansh1,porately}@cs.biu.ac.il, kociumaka@mimuw.edu.pl,
kopelot@gmail.com

Abstract. In the classic *dictionary matching problem*, the input is a dictionary of patterns $\mathcal{D} = \{P_1, P_2, \ldots, P_k\}$ and a text T, and the goal is to report all the occurrences in T of every pattern from $\mathcal{D}$. In the *dynamic* version of the dictionary matching problem, patterns may be either added or removed from $\mathcal{D}$. In the *online* version of the dictionary matching problem, the characters of T arrive online, one at a time, and the goal is to establish, immediately after every new character arrival, which of the patterns in $\mathcal{D}$ are a suffix of the current text.

In this paper, we consider the *dynamic* version of the *online dictionary matching problem*. For the case where all the patterns have the same length m, we design an algorithm that adds or removes a pattern in $\mathcal{O}(m \log \log \|\mathcal{D}\|)$ time and processes a text character in $O(\log \log \|\mathcal{D}\|)$ time, where $\|\mathcal{D}\| = \sum_{P \in \mathcal{D}} |P|$. For the general case where patterns may have different lengths, the cost of adding or removing a pattern P is $\mathcal{O}(|P| \log \log \|\mathcal{D}\| + \log d / \log \log d)$ while the cost per text character is $\mathcal{O}(\log \log \|\mathcal{D}\| + (1 + occ) \log d / \log \log d)$, where $d = |\mathcal{D}|$ is the number of patterns in $\mathcal{D}$ and occ is the size of the output. These bounds improve on the state of the art for dynamic dictionary matching, while also providing online features. All our algorithms are Las-Vegas randomized and the time costs are in the worst-case with high probability. A by-product of our work is a solution for the *fringed colored ancestor problem*, resolving an open question of Breslauer and Italiano [J. Discrete Algorithms, 2013].

1 Introduction

In the classic *dictionary matching problem* [2–8,11,13,15,17–20,22], the input is a dictionary of patterns $\mathcal{D} = \{P_1, P_2, \ldots, P_{|\mathcal{D}|}\}$ and a text T, both over alphabet Σ, and the goal is to report all the occurrences of patterns from $\mathcal{D}$ in T. The dictionary matching problem is one of the most fundamental pattern matching problems and was already considered in the 70s [1]. For example, one of the crucial components of Network Intrusion Detection Systems (NIDS) is the ability

This research is supported by ISF grants no. 824/17 and 1278/16 and by an ERC grant MPM under the EU's Horizon 2020 Research and Innovation Programme (grant no. 683064). The authors thank Tatiana Starikovskaya for early discussions on the online dynamic dictionary problem.

© Springer Nature Switzerland AG 2019
Z. Friggstad et al. (Eds.): WADS 2019, LNCS 11646, pp. 409–422, 2019.
https://doi.org/10.1007/978-3-030-24766-9_30

to detect the presence of viruses and malware in streaming data. This task is typically executed by searching for occurrences of special digital signatures which indicate the presence of harmful intent. However, while searching for one such signature is a relatively simple task, NIDS have to deal with the task of searching for many signatures in parallel, which is exactly the dictionary matching problem where the patterns are the special digital signatures. Indeed, the task of finding these signatures dominates the performance of such security tools [25], and several practical approaches have been suggested [9,24].

A common objective in many dictionary matching applications is to support *online* arrivals of the text characters. In NIDS, for example, the typical goal is to report a special digital signature as soon as it appears, before the next packet of data arrives. To this end, in the *online* version of the dictionary matching problem, the text T arrives one character at a time, and the goal is to report all of the occurrences of patterns from $\mathcal{D}$ in T as soon as they appear, before the next character arrives. Thus, if an occurrence of a pattern $P \in \mathcal{D}$ ends at the ith character of T, then P must be reported before the $(i + 1)$th character arrives.

Online Dynamic Dictionary Matching Problem. Another common task in applications that utilize dictionary matching is to support changes to $\mathcal{D}$. In NIDS, for example, one might introduce new digital signatures on the fly or remove from consideration digital signatures creating too many false positives. To this end, this paper focuses on the *online dynamic dictionary matching problem*, where the goal is to solve the online dictionary problem so that patterns may be added or removed from $\mathcal{D}$ between the arrivals of text characters. In particular, the requirement now is that when the ith character $T[i]$ arrives, the algorithm must report all patterns in the *current* $\mathcal{D}$ with an occurrence that ends at the ith character of T, and this reporting must take place before $T[i + 1]$ arrives.

In this work, we investigate the online dynamic dictionary matching problem with a focus on randomized Las-Vegas algorithms, with worst-case high probability[1] guarantees on the runtime. A summary of our results follows.

1.1   Previous Work and New Results

For (offline) dictionary matching the Aho–Corasick (AC) data structure [1] provides an optimal linear time solution. However the AC data structure was not initially designed for being neither dynamic nor online.

Let $\|\mathcal{D}\| = \sum_{P \in \mathcal{D}} |P|$ be the total length of all patterns in $\mathcal{D}$, let $d = |\mathcal{D}|$ be the number of patterns in $\mathcal{D}$, and let occ denote the size of the output.

Dynamic Dictionary Matching. In the following we provide an overview of previous work on dynamic dictionary matching that does not support online reporting. Amir et al. [3] improved on Amir et al. [2] by designing a dynamic dictionary matching data structure that allows for adding/deleting a pattern P in $O(|P| \log \|\mathcal{D}\| / \log \log \|\mathcal{D}\|)$ time, and processing a text T costs $O((|T| +$

---

[1] Throughout this paper, an event happens *with high probability* (whp) if the probability of the event not happening is polynomially small in the size of the input.

occ) log $\|\mathcal{D}\|$/ log log $\|\mathcal{D}\|$) time. Sahinalp and Vishkin [23] provide an optimal solution for a restricted family of dictionaries[2]. Feigenblat et al. [14] sacrifice the update time in order to obtain a faster text processing time. Remarkably, there have been no significant improvements to date over these time bounds.

Online Dictionary Matching. The AC data structure is an online dictionary matching, but the time guarantee per text character is in an amortized sense. Kopelowitz et al. [22] provide an online data structure that is based on the AC data structure, and has a worst-case time of $O(\log \log |\Sigma|)$ per text character.

In recent years there has been a renewed interest in dictionary matching in the streaming model which inherently provide online solutions. Let m be the length of the longest pattern in $\mathcal{D}$. Clifford et al. [11] introduced an algorithm that costs $O(\log \log m)$ time per character, which was improved by Golan and Porat [20] to an algorithm that costs $O(\log \log |\Sigma|)$ time per character.

Our Results: Online Dynamic Dictionary Matching. While there has been a large body of work on either online or dynamic dictionary matching, not much is known about the combination, beyond trivial solutions. In order to simplify the description of our results, we first focus on the special case in which all of the patterns in $\mathcal{D}$ have the same length.

Theorem 1. *Let $m > 0$ be a positive integer and let $\mathcal{D}$ be a dictionary where each $P \in \mathcal{D}$ has $|P| = m$. Then there exists a linear-space data structure for online dynamic dictionary matching that supports the following operations on $\mathcal{D}$ (the worst-case time costs are given in parentheses):*

1. insert(*P*) *(assumes $|P| = m$): Insert a pattern P into $\mathcal{D}$ ($O(m \log \log \|\mathcal{D}\|)$ with high probability).*
2. delete(*P*): *Delete a pattern $P \in \mathcal{D}$ ($O(m \log \log \|\mathcal{D}\|)$ with high probability).*
3. online_search(*T*): *Report all of the occurrences of patterns from P in $\mathcal{D}$ in an online fashion, processing one character at a time ($O(\log \log \|\mathcal{D}\|)$ time per character).*

Moreover, the updates to $\mathcal{D}$ can take place during an online_search *operation. If such an update takes place during an* online_search *operation, then from that time onwards the algorithm applies the update to the output.*

Notice that the time costs of our algorithm are only a log log $\|\mathcal{D}\|$ factor away from optimal, which is an exponential improvement over [3] and [2]. Moreover, our algorithm is online and has worst-case guarantees per text character.

For the case of a general dictionary $\mathcal{D}$, we present the following algorithm.

Theorem 2. *There exists a linear-space data structure for online dynamic dictionary matching that supports the following operations on a dictionary $\mathcal{D}$ (the worst-case time costs are given in parentheses):*

---

[2] Sahinalp and Vishkin [23] claim that their results can be extended to general dictionaries, but the details were left for a full version that was never made available. The missing details are unclear and do not seem to be straightforward.

1. insert(P): *Insert a new pattern P into $\mathcal{D}$* ($O(|P|\log\log\|\mathcal{D}\| + \log d/\log\log d)$
   with high probability).
2. delete(P): *Delete a pattern $P \in \mathcal{D}$* ($O(|P|\log\log\|\mathcal{D}\| + \log d/\log\log d)$ *with*
   high probability).
3. online_search(T): *Report all of the occurrences of patterns from P in $\mathcal{D}$ in*
   an online fashion, processing one character at a time ($O(\log\log\|\mathcal{D}\| + (1 +$
   $occ)\log d/\log\log d)$ *time per character*).

Moreover, the updates to $\mathcal{D}$ can take place during an online_search *operation. If*
such an update takes place during an online_search *operation, then from that time*
onwards the algorithm applies the update to the output.

The main difference between Theorem 1 and Theorem 2 is that when the
lengths of patterns are all the same then there can be at most one pattern whose
occurrence ends at a given text location, while if the lengths of the patterns are
different then there could be several patterns whose occurrence ends at a given
location. This difference ends up costing a factor of $\log d/\log\log d$ per output
element and per text location in the general case. Nevertheless, this overhead is a
significant improvement compared to the $\log\|\mathcal{D}\|/\log\log\|\mathcal{D}\|$ overhead from [3].

Fringe Colored Ancestors. As a consequence of our data structure, we also
address an open problem described in [10] which is known as the *fringe colored*
ancestors problem. The problem definition and solution are given in Sect. 3.

2   Algorithmic Preliminaries

2.1   The Aho-Corasick Data Structure

The following provides an overview of the AC algorithm [1], which is helpful
for gaining intuition for our new algorithm. The AC algorithm stores a trie
of all of the patterns in $\mathcal{D}$ where each edge is marked with the corresponding
character from Σ. Each node u in the trie corresponds to a string $S(u)$ that is
a prefix of some pattern (possibly more than one) from $\mathcal{D}$ and is comprised of
the concatenation of the characters on the unique simple path from the root to
u. Thus, if r is the root, then $S(r)$ is the empty string. Moreover, each node u
in the trie has at most $|\Sigma|$ edges to children of u, each edge corresponding to
a different character from Σ, which are called *forward-links*. In addition, each
non-root node u has a *failure-link* which points to a node u' if and only if $S(u')$
is the longest prefix of some pattern from $\mathcal{D}$ that is also a proper suffix of $S(u)$.
Notice that a failure-link exists for every non-root node in the trie since the
string corresponding to the root is the empty string. Finally, each node u stores
a *reporting-link* to the node $\hat{u}$, if such a node exists, where $S(\hat{u})$ is the longest
proper suffix of $S(u)$ such that $S(\hat{u}) \in \mathcal{D}$ (that is, $\hat{u}$ represents a pattern in $\mathcal{D}$).

In order to solve the dictionary matching problem, the AC algorithm starts
from the root of the trie and scans T. Suppose that after scanning $i-1$ characters
of T, the algorithm is at node u and the ith character is c. If there exists a child x
of u where the edge (u, x) is marked with c, then the algorithm transitions to x.

Otherwise, if u is the root then the algorithm continues to the next text character, and if u is not the root then the algorithm follows the failure-link (u, u'), sets u to be u', and attempts to transition from the new u with c (this last step is recursive which may entail following additional failure-links if needed). Once a transition is successful, then the algorithm uses reporting-links to report all of the occurrences of patterns in T that end at the ith location.

Using standard amortization arguments together with hashing, it is straightforward to show that the total time for scanning a text T is $O(|T| + occ)$ time, where occ is the size of the output.

Worst-Case Versions of the AC Data Structure. The main downside of the AC data structure regarding the time cost is that the *worst-case* (as opposed to amortized) time for treating a single character from T could be $\Omega(\max_{P \in \mathcal{D}}\{|P|\})$. One straightforward way of reducing this worst-case time is to store a *direct failure-link* for each pair of a node u and a character c for which there is no forward-link for u that has character c. The direct failure-link for u and c points to a node v such that $S(v)$ is the longest suffix of $S(u)c$. The direct failure-links allow for a constant worst-case time per character, but the space usage becomes $O(n|\Sigma|)$.

Challenges with the Dynamic Setting. When considering the dynamic dictionary case, one of the challenges in using an AC-like data structure is that a single update to $\mathcal{D}$ has the potential of updating a large number of failure-links (whether we are using direct failure-links or not).

2.2 Generalized Suffix Tree

The backbone of the new data structure is the *generalized suffix tree* (GST) $\mathsf{T}(\mathcal{D})$ of the patterns in $\mathcal{D}$, which is also the main component of the solution by Amir et al. [2] and its subsequent improved variants [3]. The GST $\mathsf{T}(\mathcal{D})$ is a compacted trie of the suffixes of the patterns $P \in \mathcal{D}$. That is, we construct the trie containing the suffixes of the patterns $P \in \mathcal{D}$ and compress every non-branching path with no terminal inner nodes (a path whose internal nodes have a single child and do not represent suffixes of the patterns) into a single edge. Since edges in the compressed trie may correspond to multiple characters (which happens when the edge represents a non-branching path in the uncompacted trie), the string corresponding to a given edge is stored implicitly as a pointer to a fragment of one of the patterns $P \in \mathcal{D}$. By standard arguments the size of $\mathsf{T}(\mathcal{D})$ is $O(\|\mathcal{D}\|)$.

Notice that the nodes of $\mathsf{T}(\mathcal{D})$ form a subset of the nodes of the underlying uncompacted trie. A node u of the uncompacted trie is *explicit* if u is in $\mathsf{T}(\mathcal{D})$, and otherwise u is *implicit*. As in the AC data structure, for a node u (implicit or explicit), let $S(u)$ denote the unique string corresponding to u, which is obtained by concatenating the labels of edges on the path from the root to u.

For each substring S of a pattern in $\mathcal{D}$, there exists either an implicit or explicit node u such that $S(u) = S$; we call u the *locus* of S. Since the locus might be implicit, u is represented by a pair (v, β), where v is the nearest explicit *descendant* of u and β is the *offset*—the length of the path from u to v in the uncompacted trie. Notice that if u is explicit, then u is represented as $(u, 0)$.

Suffix Links. A crucial and extremely helpful feature of suffix trees is that *suffix-links* are stored in explicit nodes. The *suffix-link* from a non-root node u is a pointer to the node v such that $S(u) = cS(v)$ for some $c \in \Sigma$. The suffix-links play a role similar to that of failure-links in the AC data structure: $S(v)$ is the longest *substring* of some pattern in $\mathcal{D}$ that is also a proper suffix of $S(u)$[3]. By standard suffix-tree arguments, if u is explicit, then v is also explicit.

The suffix-links of $\mathsf{T}(\mathcal{D})$ span a tree $\mathsf{SLT}(\mathcal{D})$ on the set of implicit and explicit nodes of $\mathsf{T}(\mathcal{D})$. The *suffix-link path* from a node u (either implicit or explicit) in $\mathsf{SLT}(\mathcal{D})$, denoted by π_u, is the path in $\mathsf{SLT}(\mathcal{D})$ from u to the root of $\mathsf{SLT}(\mathcal{D})$. Notice that π_u may contain implicit nodes (but only if u is implicit).

GST Construction. Amir et al. [2] showed that a GST (and its suffix-links) can be maintained efficiently while undergoing insertions and deletions of strings. The implementation of [2] was designed for constant-size alphabets, but the only issue with supporting large integer alphabets is the following primitive: given a node u and a character c, retrieve the outgoing edge from u whose label starts with c. Such a primitive can be implemented in $O(1)$ worst-case time using dynamic hash tables [16], which also support updates in $O(1)$ time whp.

Lemma 3 (Dynamic GST; Amir et al. [2]). *A generalized suffix tree* $\mathsf{T}(\mathcal{D})$, *including the suffix-links for explicit nodes, can be maintained in linear space so that inserting a string to $\mathcal{D}$ takes $O(|P|)$ time with high probability.*

The Extended GST of the Reversed Dictionary. The *reversed dictionary* of $\mathcal{D}$ is the dictionary $\mathcal{D}^R = \{P^R : P \in \mathcal{D}\}$, where P^R is the string obtained by reversing P. Notice that $\mathsf{SLT}(\mathcal{D})$ is isomorphic to the generalized suffix *trie* of $\mathcal{D}^R$, which is the uncompacted version of $\mathsf{T}(\mathcal{D}^R)$. Specifically, a node u that is either explicit or implicit in $\mathsf{T}(\mathcal{D})$ corresponds to a unique node v (either implicit or implicit) in $\mathsf{T}(\mathcal{D}^R)$ such that $S(v) = (S(u))^R$. Due to the isomorphism, we denote $v = u^R$. In this work, $\mathsf{T}(\mathcal{D}^R)$ is represented using an extension $\mathsf{E}(\mathcal{D}^R)$ (the E stands for "extension") obtained by explicitly adding all implicit nodes of $\mathsf{T}(\mathcal{D}^R)$ which correspond to explicit nodes of $\mathsf{T}(\mathcal{D})$. We emphasize that while this is not the first time that the generalized suffix trie of $\mathcal{D}^R$ [14,22] is used for solving dictionary matching, as far as we know this is the first time that an algorithm uses $\mathsf{E}(\mathcal{D}^R)$.

2.3   Traversing the Generalized Suffix Tree

As text characters arrive, the GST enables maintaining the *main pointer*—the locus of the longest suffix of T that is a substring of some pattern in $\mathcal{D}$[4]. In an online solution, the locus must be updated before the next character arrives.

---

[3] This is in contrast to the AC failure-links, which lead to the locus of the longest *prefix* of some pattern in $\mathcal{D}$ that is also a proper suffix of $S(u)$.

[4] That is, for each *prefix* of the text, the algorithm finds the longest *suffix* that is a substring of some pattern in $\mathcal{D}$. This is in contrast to [2] where the goal is to find, for each *suffix* of T, the longest *prefix* of the suffix that is a substring of some pattern in $\mathcal{D}$. Notice that in general the sets of these strings is not necessarily the same.

Suppose that the main pointer is at a node u. When the next character c arrives, the algorithm must find the node v representing the longest suffix of Tc. Notice that v is the locus of the longest suffix of $S(u)c$ that is also a substring of a pattern in $\mathcal{D}$. If u has a forward-link ℓ labeled with c, then v is the other endpoint of ℓ, and finding v costs $O(1)$ time using standard suffix tree traversal methods. Otherwise, a naïve way of finding v is to follow the path of suffix-links, starting from u, until reaching a node u' with a forward-link ℓ' labeled with c, and then to traverse ℓ'. If such a node does not exist (which is only possible if c does not occur in any $P \in \mathcal{D}$), then v is the root.

If an amortized bound for processing a character of T is sufficient, then one can traverse the suffix-link path from u (while suffix-links are not stored for implicit nodes, they can be simulated in constant amortized time). Our goal is to obtain efficient worst-case running time, so a different solution is needed. The main idea is to implement *direct failure-links*, which allow to directly reach v from u and c. Formally, given a node u and a character $c \in \Sigma$, the c-labeled direct failure-link leads from u to the locus of the longest suffix of $S(u)c$ occurring as a substring of a pattern $P \in \mathcal{D}$.

As observed above, the task of simulating a direct failure-link reduces to finding the lowest ancestor u' of u in $\mathsf{SLT}(\mathcal{D})$ that has a c-labeled forward-link. If u' exists, then, by standard suffix-tree arguments, all ancestors u'' of u' in $\mathsf{SLT}(\mathcal{D})$ must also have a c-labeled forward-link. In other words, the nodes with a c-labeled forward-link satisfy the so-called *fringe property* [10] on $\mathsf{SLT}(\mathcal{D})$: a set M of nodes in a rooted tree T satisfies the fringe property if $v \in M$ implies that either v is the root of T or the parent of v is also in M. Moreover, the ancestors of u in $\mathsf{SLT}(\mathcal{D})$ define π_u, and π_u corresponds to the path in $\mathsf{T}(\mathcal{D}^R)$ from u^R to the root. Thus, the task of finding u' can be expressed in terms of *fringe colored ancestor* queries (see Sect. 3 for a formal definition) in $\mathsf{T}(\mathcal{D}^R)$. Moreover, the algorithm uses $\mathsf{E}(\mathcal{D}^R)$, instead of $\mathsf{T}(\mathcal{D}^R)$, since if u does not have a c-labeled forward-link, then u' must be explicit in $\mathsf{T}(\mathcal{D})$ (see Claim 8), and therefore u'^R is explicit in $\mathsf{E}(\mathcal{D}^R)$.

To conclude, traversing the GST reduces to simulating direct failure-links from explicit and implicit nodes of $\mathsf{T}(\mathcal{D})$. A subroutine designed for this task is among our main technical contributions and is described in Sect. 4.

2.4 Reporting the Patterns

As in previous amortized procedures for scanning the text T [2,3], the most expensive phase is reporting the patterns. If the main pointer is to a node u, then the loci of the patterns to be reported lie on π_u. The algorithm maintains a mark on each node v in $\mathsf{SLT}(\mathcal{D})$ where $S(v) \in \mathcal{D}$, and so the reporting procedure reduces to the task of repeatedly finding lowest marked ancestors in $\mathsf{SLT}(\mathcal{D})$, starting from u. However, finding the marked ancestors is performed on $\mathsf{E}(\mathcal{D}^R)$, since all of the terminal nodes of $\mathsf{T}(\mathcal{D})$ correspond to explicit nodes in $\mathsf{E}(\mathcal{D}^R)$.

While any algorithm for reporting marked ancestors could be used, in order to derive our main results, we tweak existing marked ancestor algorithms. The proof of Theorem 1 exploits the fact that a locus of a pattern P is marked only when P is inserted, and so when marking the locus of P, one can spend time proportional to $|P|$ which is also the depth of the locus. Furthermore, due to uniform pattern lengths, there is at most one ancestor to be reported. The proof of Theorem 2 uses a new algorithm for reporting marked ancestors whose cost depends on the number of marked nodes d rather than on the tree size $\|\mathcal{D}\|$.

In Sect. 5 we provide more details on how to report the patterns.

2.5   Updates to the Dictionary

Insertions. When a pattern P is inserted into $\mathcal{D}$, the GST, the data structure for implementing direct failure-links, and the reporting data structure need to be updated accordingly. In addition, the main pointer may need to be updated (because a longer suffix of the current T may now occur as a substring of P). If the node u represented by the main pointer is at depth at least $|P|$, then the main pointer does not need to change. However, if the depth of u is less than $|P|$, then right after P is inserted into the GST, the algorithm traverses from the root of the GST with the last $|P|$ characters of T (the time cost of the traversal is amortized over the time cost of inserting P).

Deletions. The only immediate effect of deleting a pattern P is that P should not be reported anymore, but the suffixes of P may remain in the GST for some time. Thus, when deleting P, the only immediate consequence is that the algorithm removes P from the reporting data structure. Since we are interested in a linear space solution, the algorithm employs a lazy approach for updating the GST and the data structure for implementing direct failure-links by using the standard doubling technique: if the GST becomes too large, the algorithm initiates a background process of constructing a new GST (including the auxiliary data structures) on the non-deleted patterns. By using the doubling technique, the algorithm avoids the technical details of explicitly supporting pattern deletions.

3   Fringe Colored Ancestors

In this section, we describe a solution for the *fringe colored ancestor* problem. We begin by recalling two results from the literature that we shall use as subroutines.

Lemma 4 (Kopelowitz [21, Theorem 5.1]). *There exists a linear-space data structure maintaining an ordered list L and disjoint subsets $S_1, \ldots, S_k \subseteq L$ that supports the following operations (the runtimes are given in parentheses[5]):*

1. insert(u, v): *Insert a new item u after a given item $v \in L$ ($O(1)$ time whp).*
2. set_insert(v, i) *(assumes that $v \notin \bigcup_{j=1}^{k} S_j$): Insert an element $v \in L$ into S_i ($O(\log \log |L|)$ time whp).*
3. pred(v, i): *Locate the predecessor of $v \in L$ in S_i ($O(\log \log |L|)$ time whp).*

---

[5] Theorem 5.1 in [21] is expressed in terms of expected runtimes. However, the only randomization is due to hashing, and the same time bounds hold whp.

Lemma 5 (Cole and Hariharan [12, Theorem 8.1]). *There exists a linear-space data structure maintaining a dynamic rooted tree T subject to the following operations, supported in $O(1)$ time:*

1. insert_leaf(u, v): *Insert a new leaf u as a child of a node v.*
2. insert(u, v): *Insert a new node u by subdividing the edge from a non-root node v to the parent of v.*
3. LCA(u, v): *Return the lowest common ancestor of two nodes u and v.*

We now describe the new component for fringe colored ancestors.

Lemma 6. *Let T be a rooted tree such that each node in T is colored using 0 or more colors from a set Δ. If a node u in T has 0 colors, then u is said to be uncolored. For each color $\delta \in \Delta$, let M_δ be the nodes of T colored with color δ. Let $N = |T| + \sum_{\delta \in \Delta} |M_\delta|$. Suppose that for each color $\delta \in \Delta$, the set M_δ satisfies the fringe property. Then there exists an $O(N)$-space data structure that supports each of the following operations in $O(\log \log N)$ time with high probability:*

1. insert_leaf(u, v): *Insert a new leaf u as a child of node v.*
2. insert(u, v) *(assumes that v is a non-root uncolored node): Insert a new node u by subdividing the edge connecting node v and the parent of v.*
3. color(v, δ) *(assumes that v is either the root or the parent of v is already colored with δ): Color node v with color δ, i.e., add v to M_δ.*
4. colored_ancestor(v, δ): *Return the lowest ancestor of node v that is colored with color δ.*

Proof. The data structure uses Lemma 5 on T in order to support LCA queries. In addition, the data structure uses Lemma 4 to store the parenthesis representation of T, defined recursively as follows. Let r be the root of T and let $v_1, \dots, v_k$ be the children of r (in an arbitrary order). The parenthesis representation of the tree is the concatenation of the representations of the subtrees rooted at $v_1, v_2, \dots v_k$, wrapped with (at the beginning and) at the end. The two enclosing parentheses represent the root r.

We extend the parenthesis representation to represent T together with the colors of nodes in T, and denote this new representation by L. The representation L is defined as follows. For a node v with $c(v) > 0$ colors, instead of creating just one pair of wrapping parentheses representing v, the parenthesis representation uses $c(v) + 1$ nested pairs of parentheses, one additional pair for each color. Thus, each extra pair of parentheses is assigned a unique color from Δ, representing one of the colors of v. The algorithm maintains the invariant that the *innermost* parentheses for a node v are the uncolored parentheses. Each node $v \in T$ maintains pointers to the corresponding uncolored parentheses, and each (colored or uncolored) parenthesis stores a pointer to the corresponding node.

For each $\delta \in \Delta$, let $S_\delta \subseteq L$ be the set of parentheses in L corresponding to color δ. Notice that for different $\delta, \delta' \in \Delta$, the sets S_δ and $S_{\delta'}$ are disjoint, and thus the algorithm uses Lemma 4 to store the sets S_δ. The total space usage of all of the data structures is $O(|T|)$ machine words.

Supporting Operations. To insert a new leaf u with parent v, the algorithm locates the closing uncolored parenthesis $)_v$ corresponding to v and adds a new pair of matching uncolored parentheses $(_u$ and $)_u$ immediately prior to $)_v$.

To subdivide the edge from v to its parent, the algorithm locates the uncolored parentheses corresponding to v and inserts an opening (closing) uncolored parenthesis $(_u ()_u)$ immediately before (after) $(_v$.

To add a color δ to a node v, the algorithm locates the uncolored parentheses $(_v$ and $)_v$ representing v, inserts a pair of parentheses $(_{v,\delta}$ and $)_{v,\delta}$ immediately before $(_v$ and after $)_v$, respectively, and adds both $(_{v,\delta}$ and $)_{v,\delta}$ to S_δ.

To answer a lowest colored ancestor query, the algorithm finds the node $x \in T$ corresponding to the predecessor in S_δ of the opening uncolored parenthesis $(_v$ representing v, and then returns $y = \mathsf{LCA}(x, v)$. The time cost is $O(\log \log N)$ time using the data structures of Lemmas 4 and 5.

The correctness of the query algorithm follows from the fringe property: Since x is colored with δ, then y is both an ancestor of v and colored with δ. Suppose by contradiction that there exists a lower ancestor u of v that is colored with δ. Then the opening δ-colored parenthesis of u appears between the parenthesis that defines x and $(_v$, contradicting x being the node corresponding to the predecessor of $(_v$ in S_δ. Thus, y is the lowest ancestor of v colored with δ.   □

4   Simulating Direct Failure Links

We begin by proving two claims that are used in the proof of Lemma 9.

Claim 7. Let (x^R, y^R) be an edge in $\mathsf{T}(\mathcal{D}^R)$ with subsequent internal implicit nodes $z_1^R, \ldots, z_k^R$. Then $z_1^R, \ldots, z_k^R$ are either all explicit or all implicit in $\mathsf{E}(\mathcal{D}^R)$.

Proof. Denote $z_0^R = x^R$ and $z_{k+1}^R = y^R$. If a node z_i^R, for $1 \le i \le k$, is explicit in $\mathsf{E}(\mathcal{D}^R)$, then z_i is explicit in $\mathsf{T}(\mathcal{D})$, and the occurrences of $S(z_i)$ in $\mathcal{D}$ are followed by at least two distinct characters a, b. On the other hand, as z_i^R is implicit in $\mathsf{T}(\mathcal{D}^R)$, the occurrences of $S(z_i)$ in $\mathcal{D}$ are all preceded by the same character c. Hence, both a and b follow some occurrences of $S(z_{i+1}) = cS(z_i)$ in $\mathcal{D}$, so z_{i+1} is explicit in $\mathsf{T}(\mathcal{D})$ and z_{i+1}^R is explicit in $\mathsf{E}(\mathcal{D}^R)$. Furthermore, z_{i-1} is the target of the suffix-link from z_i, so z_{i-1} is explicit in $\mathsf{T}(\mathcal{D})$ and z_{i-1}^R is explicit in $\mathsf{E}(\mathcal{D}^R)$. By induction, $z_1^R, \ldots, z_k^R$ are all explicit in $\mathsf{E}(\mathcal{D}^R)$ if at least one of them is.   □

Claim 8. Let u be a node of $\mathsf{T}(\mathcal{D})$ with no c-labeled forward-link. If u' is the lowest ancestor of u in $\mathsf{SLT}(\mathcal{D})$ with a c-labeled forward-link, then u' is explicit in $\mathsf{T}(\mathcal{D})$.

Proof. If u is explicit in $\mathsf{T}(\mathcal{D})$, then all of the nodes in π_u are also explicit, and in particular u' is explicit. Otherwise, all of the nodes in π_u (including u) have a forward-link with another label $d \ne c$. Consequently, u' has at least two forward-links and thus u' is explicit in $\mathsf{T}(\mathcal{D})$.   □

Lemma 9. *There exists a data structure that augments* $\mathsf{T}(\mathcal{D})$ *with* $O(\|\mathcal{D}\|)$ *space, and supports inserting a string* P *into* $\mathcal{D}$ *in* $O(|P| \log \log \|\mathcal{D}\|)$ *time with high probability, and simulating direct failure-links in* $O(\log \log \|\mathcal{D}\|)$ *time.*

Proof. The proof begins by listing the components augmenting $\mathsf{T}(\mathcal{D})$.

Components. The first augmenting component is $\mathsf{E}(\mathcal{D}^R)$. Additionally, for every explicit node u of $\mathsf{T}(\mathcal{D})$, the data structure stores a bidirectional pointer between u and the corresponding node u^R in $\mathsf{E}(\mathcal{D}^R)$. If u has a c-labeled forward link, then u^R is colored with color c. Notice that every node on π_u is then explicit and has a c-labeled forward-link. Moreover, the nodes on π_u correspond to the ancestors of u^R in $\mathsf{E}(\mathcal{D}^R)$, and so all of the ancestors of u^R in $\mathsf{E}(\mathcal{D}^R)$ are colored with color c. Therefore, the set M_c of nodes in $\mathsf{E}(\mathcal{D}^R)$ with color c satisfies the fringe property, and so the algorithm augments $\mathsf{E}(\mathcal{D}^R)$ with the data structure of Lemma 6 to maintain the sets M_c, while supporting colored ancestor queries.

In order to support updating $\mathsf{E}(\mathcal{D}^R)$, the algorithm also maintains $\mathsf{T}(\mathcal{D}^R)$ and, for every explicit node of $\mathsf{T}(\mathcal{D}^R)$, a bidirectional pointer between the node and its counterpart in $\mathsf{E}(\mathcal{D}^R)$. Claim 7 classifies edges of $\mathsf{T}(\mathcal{D}^R)$ into two types depending on whether the internal nodes are all explicit or implicit in $\mathsf{E}(\mathcal{D}^R)$. If the internal nodes of an edge e from $\mathsf{T}(\mathcal{D}^R)$ are explicit in $\mathsf{E}(\mathcal{D}^R)$, then the data structure stores an array A_e with pointers to these explicit nodes in $\mathsf{E}(\mathcal{D}^R)$, and e maintains a pointer to A_e.

For each explicit node u of $\mathsf{T}(\mathcal{D})$, the algorithm maintains a pointer to a fragment[6] $P[\ell..r]$ of a pattern $P \in \mathcal{D}$ such that $P[\ell..r] = S(u)$. Finally, for each pattern $P \in \mathcal{D}$, and each $r \in \{1, \ldots, |P|\}$, the data structure stores a pointer $\lambda_{P,r}$ to the locus of $(P[1..r])^R$ in $\mathsf{E}(\mathcal{D}^R)$ (which is a terminal node).

The only non-trivial aspect of the space complexity is the size of the component of Lemma 6, which is $O(\|\mathcal{D}\|)$ since the number of colorings $\sum_{c \in \Sigma} |M_c|$ is bounded by the number of edges in the GST.

Implementing Direct Failure-Links. Given a node u in $\mathsf{T}(\mathcal{D})$ and a character $c \in \Sigma$, let v be the target of the c-labeled direct failure-link from u. The algorithm locates v as follows. Recall from Sect. 2 that, if u' is the lowest ancestor of u in $\mathsf{SLT}(\mathcal{D})$ that has a c-labeled forward-link, then v is the target of the c-labeled forward-link from u'. Thus, the goal of the algorithm is to locate u' and then return v using the hash table stored at u'. However, u' may not exist (in the case c does not occur in any pattern of $\mathcal{D}$). For simplicity, the algorithm description first assumes that u' exists and later shows how this assumption is supported.

If u has a c-labeled forward-link, then $u' = u$. Otherwise, notice that π_u contains u' and corresponds to a path in $\mathsf{E}(\mathcal{D}^R)$ from u^R to the root. Node u' can be either explicit or implicit. In either case, if the algorithm locates an *explicit* node w^R in $\mathsf{E}(\mathcal{D}^R)$ such that u' is the lowest ancestor of the (possibly implicit) corresponding node w in $\mathsf{SLT}(\mathcal{D})$ that has a c-labeled forward-link, then u'^R is the lowest ancestor of w^R colored with c (since u' is explicit in $\mathsf{T}(\mathcal{D})$ by Claim 8), and u' is located using a single colored ancestor query on w^R and c.

If u is explicit, then u^R satisfies the requirements for w^R. If u is implicit, let x be the highest explicit descendant of u in $\mathsf{T}(\mathcal{D})$. Recall that the main pointer provides direct access to x. Through x, the algorithm gains access to the fragment $P[\ell..r]$ of a pattern $P \in \mathcal{D}$ such that $P[\ell..r] = S(u)$. Let z be the

---

[6] Recall that the pointers to fragments are one of the standard ways of storing edge labels in a GST.

locus of $P[1..r]$ in $\mathsf{T}(\mathcal{D})$. Notice that π_u is a sub-path of π_z and the sub-path of π_z from z to u does not contain any explicit nodes. Hence, u' is the lowest ancestor of z in $\mathsf{SLT}(\mathcal{D})$ that has a c-labeled forward-link, and so z satisfies the requirements for w. Therefore, w^R is the locus of $(P[1..r])^R$ in $\mathsf{E}(\mathcal{D}^R)$, which is accessible through $\lambda_{P,r}$.

Finally, if u' does not exist, then w'^R does not exist either. Thus, the algorithm identifies this case when executing the colored ancestor query. Such a case implies that c does not appear in any pattern from $\mathcal{D}$ (since the root of $\mathsf{E}(\mathcal{D}^R)$ is colored by all characters that appear in $\mathcal{D}$), and so v is the root of $\mathsf{T}(\mathcal{D})$.

Updates. Upon insertion of a pattern P, both $\mathsf{T}(\mathcal{D})$ and $\mathsf{T}(\mathcal{D}^R)$ are updated in $O(|P|)$ time using Lemma 3. Updates to $\mathsf{T}(\mathcal{D}^R)$ are reiterated in $\mathsf{E}(\mathcal{D}^R)$ as follows. If a new node u is inserted by subdividing an edge e, and the implicit nodes on e are implicit in $\mathsf{E}(\mathcal{D}^R)$ too, then the algorithm executes the same insertion in $\mathsf{E}(\mathcal{D}^R)$. Otherwise, u is explicit in $\mathsf{E}(\mathcal{D}^R)$, and so the array A_e already stores a pointer to u in $\mathsf{E}(\mathcal{D}^R)$. Note that the task of maintaining array A_e subject to splits is a straightforward task. The insertion of a leaf u is also straightforward since parent of u already knows its counterpart in $\mathsf{E}(\mathcal{D}^R)$.

The second phase of updates to $\mathsf{E}(\mathcal{D}^R)$ is to introduce an explicit node u^R for each new explicit node u of $\mathsf{T}(\mathcal{D})$. The new explicit nodes of $\mathsf{T}(\mathcal{D})$ are processed by non-decreasing depths, and so when processing a node u, the target v of the suffix-link from u is guaranteed to have already been processed; hence, v already has a pointer to v^R in $\mathsf{E}(\mathcal{D}^R)$. Notice that $S(u) = cS(v)$ for a character $c \in \Sigma$ and u^R is the target of the c-labeled forward-link from v^R (in $\mathsf{E}(\mathcal{D}^R)$). The algorithm retrieves u^R using the hash table stored at v^R. If u^R is an implicit node on an edge e in $\mathsf{E}(\mathcal{D}^R)$, then the algorithm converts all implicit nodes on e into explicit nodes so that Claim 7 holds at all times (including during the update process). The number of implicit nodes converted to explicit nodes is proportional to the length of e. Nevertheless, by Claim 7, the total number of nodes converted during an update is still $O(|P|)$.

However, every update to $\mathsf{E}(\mathcal{D}^R)$ imposes an update to the data structure of Lemma 6. Moreover, for every new edge (u,v) added to $\mathsf{T}(\mathcal{D})$, the algorithm retrieves the first character c of the edge label and colors u^R with color c. Thus, the total update time is $O(|P| \log\log \|\mathcal{D}\|)$ whp. $\square$

5   Reporting Patterns

Having updated the location u of the main pointer (i.e., the locus of the longest substring of some $P \in \mathcal{D}$ which occurs as a suffix of T), the algorithm reports all patterns which occur as suffixes of T. Notice that π_u contains the locus of every suffix of $S(u)$, so the task is reduced to reporting all patterns whose loci lie on π_u. The terminal nodes are explicit in $\mathsf{T}(\mathcal{D})$, so the corresponding nodes are explicit in $\mathsf{E}(\mathcal{D}^R)$, and the algorithm *marks* them.

If u is explicit, then the task reduces to reporting all marked ancestors of the corresponding node u^R of $\mathsf{E}(\mathcal{D}^R)$. Otherwise, via the same arguments as in

Sect. 4, the locus of $(P[1..r])^R$, where $P[\ell..r] = S(u)$ for $P \in \mathcal{D}$, is used instead of u^R. In either case, the task reduces to reporting marked ancestors of a node w^R in $\mathsf{E}(\mathcal{D}^R)$.

If all the patterns are of the same length, the algorithm maintains marks in $\mathsf{E}(\mathcal{D}^R)$ for all ancestors of the loci of P^R for $P \in \mathcal{D}$, and so the marked nodes satisfy the fringe property. In this case, if w^R has an ancestor v^R representing a pattern, then v^R is the nearest marked ancestor of w^R, and v^R is found in $O(\log \log \|\mathcal{D}\|)$ time using a fringe marked ancestor data structure [10]. Upon inserting a pattern P, at most $O(|P|)$ nodes may need to be marked, which costs $O(|P| \log \log \|\mathcal{D}\|)$ time. This completes the proof of Theorem 1.

The proof of Theorem 2 is deferred to the full version of the paper.

References

1. Aho, A.V., Corasick, M.J.: Efficient string matching: an aid to bibliographic search. Commun. ACM **18**(6), 333–340 (1975)
2. Amir, A., Farach, M., Galil, Z., Giancarlo, R., Park, K.: Dynamic dictionary matching. J. Comput. Syst. Sci. **49**(2), 208–222 (1994)
3. Amir, A., Farach, M., Idury, R.M., Poutré, J.A.L., Schäffer, A.A.: Improved dynamic dictionary matching. Inf. Comput. **119**(2), 258–282 (1995)
4. Amir, A., Kopelowitz, T., Levy, A., Pettie, S., Porat, E., Shalom, B.R.: Mind the gap! Online dictionary matching with one gap. Algorithmica **81**(6), 2123–2157 (2019)
5. Amir, A., Levy, A., Porat, E., Shalom, B.R.: Dictionary matching with one gap. In: Kulikov, A.S., Kuznetsov, S.O., Pevzner, P. (eds.) CPM 2014. LNCS, vol. 8486, pp. 11–20. Springer, Cham (2014). https://doi.org/10.1007/978-3-319-07566-2_2
6. Amir, A., Levy, A., Porat, E., Shalom, B.R.: Dictionary matching with a few gaps. Theor. Comput. Sci. **589**, 34–46 (2015)
7. Athar, T., et al.: Fast circular dictionary-matching algorithm. Math. Struct. Comput. Sci. **27**(2), 143–156 (2017)
8. Belazzougui, D.: Succinct dictionary matching with no slowdown. In: Amir, A., Parida, L. (eds.) CPM 2010. LNCS, vol. 6129, pp. 88–100. Springer, Heidelberg (2010). https://doi.org/10.1007/978-3-642-13509-5_9
9. Bremler-Barr, A., Hay, D., Koral, Y.: Compactdfa: scalable pattern matching using longest prefix match solutions. IEEE/ACM Trans. Netw. **22**(2), 415–428 (2014)
10. Breslauer, D., Italiano, G.F.: Near real-time suffix tree construction via the fringe marked ancestor problem. J. Discrete Algorithms **18**, 32–48 (2013)
11. Clifford, R., Fontaine, A., Porat, E., Sach, B., Starikovskaya, T.: Dictionary matching in a stream. In: Bansal, N., Finocchi, I. (eds.) ESA 2015. LNCS, vol. 9294, pp. 361–372. Springer, Heidelberg (2015). https://doi.org/10.1007/978-3-662-48350-3_31
12. Cole, R., Hariharan, R.: Dynamic LCA queries on trees. SIAM J. Comput. **34**(4), 894–923 (2005)
13. Feigenblat, G., Porat, E., Shiftan, A.: Linear time succinct indexable dictionary construction with applications. In: 2016 Data Compression Conference, DCC, pp. 13–22 (2016)
14. Feigenblat, G., Porat, E., Shiftan, A.: A grouping approach for succinct dynamic dictionary matching. Algorithmica **77**(1), 134–150 (2017)

15. Fischer, J., Gagie, T., Gawrychowski, P., Kociumaka, T.: Approximating LZ77 via small-space multiple-pattern matching. In: Bansal, N., Finocchi, I. (eds.) ESA 2015. LNCS, vol. 9294, pp. 533–544. Springer, Heidelberg (2015). https://doi.org/10.1007/978-3-662-48350-3_45

16. Fredman, M.L., Komlós, J., Szemerédi, E.: Storing a sparse table with $O(1)$ worst case access time. J. ACM **31**(3), 538–544 (1984)

17. Ganguly, A., Hon, W., Sadakane, K., Shah, R., Thankachan, S.V., Yang, Y.: Space-efficient dictionaries for parameterized and order-preserving pattern matching. In: 27th Annual Symposium on Combinatorial Pattern Matching, CPM, LIPIcs, vol. 54, pp. 2:1–2:12 (2016)

18. Ganguly, A., Hon, W., Shah, R.: A framework for dynamic parameterized dictionary matching. In: 15th Scandinavian Symposium and Workshops on Algorithm Theory, SWAT, LIPIcs, vol. 53, pp. 10:1–10:14 (2016)

19. Golan, S., Kopelowitz, T., Porat, E.: Towards optimal approximate streaming pattern matching by matching multiple patterns in multiple streams. In: 45th International Colloquium on Automata, Languages, and Programming, ICALP, LIPIcs, vol. 107, pp. 65:1–65:16 (2018)

20. Golan, S., Porat, E.: Real-time streaming multi-pattern search for constant alphabet. In: 25th Annual European Symposium on Algorithms, ESA, LIPIcs, vol. 87, pp. 41:1–41:15 (2017)

21. Kopelowitz, T.: On-line indexing for general alphabets via predecessor queries on subsets of an ordered list. In: 53rd Annual IEEE Symposium on Foundations of Computer Science, FOCS, pp. 283–292 (2012)

22. Kopelowitz, T., Porat, E., Rozen, Y.: Succinct online dictionary matching with improved worst-case guarantees. In: 27th Annual Symposium on Combinatorial Pattern Matching, CPM, LIPIcs, vol. 54, pp. 6:1–6:13 (2016)

23. Sahinalp, S.C., Vishkin, U.: Efficient approximate and dynamic matching of patterns using a labeling paradigm. In: 37th Annual Symposium on Foundations of Computer Science, FOCS, pp. 320–328 (1996)

24. Tan, L., Sherwood, T.: A high throughput string matching architecture for intrusion detection and prevention. In: 32st International Symposium on Computer Architecture, ISCA, pp. 112–122 (2005)

25. Tuck, N., Sherwood, T., Calder, B., Varghese, G.: Deterministic memory-efficient string matching algorithms for intrusion detection. In: 23rd IEEE International Conference on Computer Communications, INFCOM, pp. 2628–2639 (2004)

Balanced Stable Marriage:
How Close Is Close Enough?

Sushmita Gupta[1], Sanjukta Roy[2($\boxtimes$)], Saket Saurabh[2,3], and Meirav Zehavi[4]

[1] National Institute of Science Education and Research, Bhubaneswar, India
sushmitagupta@niser.ac.in
[2] The Institute of Mathematical Sciences, HBNI, Chennai, India
{sanjukta,saket}@imsc.res.in
[3] University of Bergen, Bergen, Norway
[4] Ben-Gurion University, Beersheba, Israel
meiravze@bgu.ac.il

Abstract. BALANCED STABLE MARRIAGE (BSM) is a central optimization version of the classic STABLE MARRIAGE (SM) problem. We study BSM from the viewpoint of Parameterized Complexity. Informally, the input of BSM consists of n men, n women, and an integer k. Each person a has a (sub)set of acceptable partners, $\mathcal{A}(a)$, who a ranks strictly; we use $p_a(b)$ to denote the position of $b \in \mathcal{A}(a)$ in a's preference list. The objective is to decide whether there exists a stable matching μ such that $\mathsf{balance}(\mu) \triangleq \max\{\sum_{(m,w)\in\mu} p_m(w), \sum_{(m,w)\in\mu} p_w(m)\} \leq k$. In SM, all stable matchings match the same set of agents, $A^\star$ which can be computed in polynomial time. As $\mathsf{balance}(\mu) \geq \frac{|A^\star|}{2}$ for any stable matching μ, BSM is trivially fixed-parameter tractable (FPT) with respect to k. Thus, a natural question is whether BSM is FPT with respect to $k - \frac{|A^\star|}{2}$. With this viewpoint in mind, we draw a line between tractability and intractability in relation to the target value. This line separates additional natural parameterizations higher/lower than ours (e.g., we automatically resolve the parameterization $k - \frac{|A^\star|}{2}$). The two extreme stable matchings are the man-optimal μ_M and the woman-optimal μ_W. Let $O_M = \sum_{(m,w)\in\mu_M} p_m(w)$, and $O_W = \sum_{(m,w)\in\mu_W} p_w(m)$. In this work, we prove that
- BSM parameterized by $t = k - \min\{O_M, O_W\}$ admits (1) a kernel where the number of people is linear in t, and (2) a parameterized algorithm whose running time is single exponential in t.
- BSM parameterized by $t = k - \max\{O_M, O_W\}$ is W[1]-hard.

Keywords: Balanced Stable Marriage · Parameterized Complexity · Kernelization

This work is supported by the European Research Council (ERC) via grant LOPPRE, reference no. 819416.
Meirav Zehavi was supported by ISF (1176/18).

© Springer Nature Switzerland AG 2019
Z. Friggstad et al. (Eds.): WADS 2019, LNCS 11646, pp. 423–437, 2019.
https://doi.org/10.1007/978-3-030-24766-9_31

1   Introduction

Over the last two decades, Parameterized Complexity has evolved to be a central field of research in theoretical computer science. However, the scope of this field was essentially limited to applications to NP-hard optimization problems *on graphs*. There is no inherent reason why this should be the case. Indeed, the main idea of Parameterized Complexity is very general—measure running time in terms of both input size and a parameter that captures structural properties of the input instance. The idea of a multivariate analysis of algorithms holds the potential to address the need for a framework for refined algorithm analysis for all kinds of problems across all domains and subfields of computer science. Recently, techniques in Parameterized Complexity were successfully applied in the area of Computational Social Choice Theory. In particular, Voting has become a subject of intensive study from the viewpoint of Parameterized Complexity (for a few examples, see [1, Chapter 10, 11] and [2]; for more information on the current state-of-the-art, we refer to excellent surveys such as [3–5,11]. However, Voting is only one topic under the rich umbrella of Computational Social Choice. Parameterized analysis of other topics has been few and far between. In the past few months, a collective effort to study Matching under Preferences through the lens of Parameterized Complexity was initiated [6,17,18,21,22,24,25,28–31]. Our work is the first to introduce *kernelization* to this topic. We note that recently Chen et al. [6] have presented results on this topic (for different problems) that include a kernelization algorithm. However, it is fair to say that our work is the first to introduce "above guarantee parameterization" to this topic.

In Matching under Preferences, a matching is an allocation (or assignment) of agents to resources that satisfies some predefined criterion of compatibility/acceptability. Here, (arguably) the best known model is the *two-sided model*, where the agents on one side are referred to as *men*, and the agents on the other side are referred to as *women*. A few illustrative examples of real life situations where this model is relevant include matching hospitals to residents, students to colleges, kidney patients to donors and users to servers in a distributed Internet service. At the heart of all of these applications lies the fundamental STABLE MARRIAGE problem. In particular, the Nobel Prize in Economics was awarded to Shapley and Roth in 2012 "for the theory of stable allocations and the practice of market design." Moreover, several books have been dedicated to the study of STABLE MARRIAGE as well as optimization variants of this classical problem such as the EGALITARIAN STABLE MARRIAGE, MINIMUM REGRET STABLE MARRIAGE, SEX-EQUAL STABLE MARRIAGE and BALANCED STABLE MARRIAGE problems [19,20,23,27].

The input of STABLE MARRIAGE consists of a set of men, M, and a set of women, W. Each person a has a set of *acceptable partners*, $\mathcal{A}(a)$, who are ranked by a in a strict order. Consequently, each person a has a so-called *preference list*, where $p_a(b)$ is the position of $b \in \mathcal{A}(a)$ in a's preference list. Without loss of generality, it is assumed that if a person a ranks a person b, then the person b ranks the person a as well (if not then b can be deleted from a's preference list, since they can not be matched to each other). The sets of preference lists of the men and the women are denoted by $\mathcal{L}_M$ and $\mathcal{L}_W$, respectively. In this context, we say that a pair of a man and a woman, (m, w), is an *acceptable pair* if both $m \in \mathcal{A}(w)$ and $w \in \mathcal{A}(m)$. Accordingly, the notion of a *matching* refers to a matching between men and women, where two people matched to one another form an acceptable pair. Roughly speaking, the goal of the STABLE MARRIAGE problem is to *find* a matching that is *stable* in the following sense: there should not exist two people who prefer being matched to one another over their current "status". More precisely, a matching μ is said to be stable if it does not have a *blocking pair*, which is an acceptable pair (m, w) such that **(i)** either m is unmatched by μ or $p_m(w) < p_m(\mu(m))$, and **(ii)** either w is unmatched by μ or $p_w(m) < p_w(\mu(w))$. Here, the notation $\mu(a)$ indicates the person to whom μ matches the person a. Note that a person always prefers being matched to an acceptable partner over being unmatched.

The seminal paper [13] by Gale and Shapely on stable matchings shows that given an instance of STABLE MARRIAGE, a stable matching necessarily *exists*, but it is not necessarily unique. In fact, for a given instance of STABLE MARRIAGE, there can be an *exponential* number of stable matchings, and they should be viewed as a *spectrum* where the two extremes are known as the *man-optimal stable matching* and the *woman-optimal stable matching*. Formally, the man-optimal stable matching, denoted by μ_M, is a stable matching such that every stable matching μ satisfies the following condition: every man m is either unmatched by both μ_M and μ or $p_m(\mu_M(m)) \leq p_m(\mu(m))$. The woman-optimal stable matching, denoted by μ_W, is defined analogously. These two extremes, which give the best possible solution for one party at the expense of the other party, always exist and can be computed in polynomial time [13]. Naturally, it is desirable to analyze matchings that lie somewhere in the middle. Here, the quantity $p_a(\mu(a))$ is the "satisfaction" of a in a matching μ, where a smaller value signifies a greater amount of satisfaction. The most well-known measures are as follows:

- μ is *globally desirable* if it minimizes $\sum_{(m,w) \in \mu}(p_m(\mu(m)) + p_w(\mu(w)))$, called the *egalitarian stable matching*;
- μ is *minimum regret* if it minimizes $\max_{(m,w) \in \mu}\{\max\{p_m(\mu(m)), p_w(\mu(w))\}\}$, called the *minimum regret stable matching*;
- μ is *fair towards both sides* if it minimizes $|\sum_{(m,w) \in \mu} p_m(\mu(m)) - \sum_{(m,w) \in \mu} p_w(\mu(w))|$, called the *sex-equal stable matching*;
- μ is *desirable by both sides* if it minimizes $\max\{\sum_{(m,w) \in \mu} p_m(w), \sum_{(m,w) \in \mu} p_w(m)\}$, called the *balanced stable matching*.

Each notion above leads to a natural, *different* well-studied optimization problem (see Related Work). We focus on the NP-hard BALANCED STABLE MARRIAGE (BSM) problem, where the objective is to find a stable matching μ that minimizes

$$\mathsf{balance}(\mu) = \max\{\sum_{(m,w) \in \mu} p_m(w), \sum_{(m,w) \in \mu} p_w(m)\}.$$

This problem was introduced in the influential work of Feder [12] on stable matchings, and was shown to be NP-hard and admitting a 2-approximation algorithm. We refer reader to [16] for the full version of this paper.

Our Contribution. Above guarantee parameterization is a topic of extensive study in Parameterized Complexity [7]. We introduce two "above-guarantee parameterizations" of BSM. Consider the minimum value O_M (O_W) of the total dissatisfaction of men (women) realizable by a stable matching. Formally, $O_M = \sum_{(m,w) \in \mu_M} p_m(w)$, and $O_W = \sum_{(m,w) \in \mu_W} p_w(m)$, where μ_M (μ_W) is the man-optimal (woman-optimal) stable matching. Denote $\mathsf{Bal} = \min_{\mu \in \mathrm{SM}} \mathsf{balance}(\mu)$, where SM is the set of all stable matchings. An input integer k would indicate that our objective is to decide whether $\mathsf{Bal} \leq k$. In our first parameterization, the parameter is $k - \min\{O_M, O_W\}$, and in the second one, it is $k - \max\{O_M, O_W\}$.

ABOVE-MIN BALANCED STABLE MARRIAGE (ABOVE-MIN BSM)
Input: An instance $(M, W, \mathcal{L}_M, \mathcal{L}_W)$ of BALANCED STABLE MARRIAGE, and a non-negative integer k.
Question: Is $\mathsf{Bal} \leq k$?
Parameter: $t = k - \min\{O_M, O_W\}$

ABOVE-MAX BALANCED STABLE MARRIAGE (ABOVE-MAX BSM)
Input: An instance $(M, W, \mathcal{L}_M, \mathcal{L}_W)$ of BALANCED STABLE MARRIAGE, and a non-negative integer k.
Question: Is $\mathsf{Bal} \leq k$?
Parameter: $t = k - \max\{O_M, O_W\}$

Choice of Parameters: Note that the minimum dissatisfaction the party of men can hope for (call it *optimum satisfaction*) is O_M, and the minimum dissatisfaction the party of women can hope for (also call it optimum satisfaction) is O_W. First, consider the parameter $t = k - \min\{O_M, O_W\}$. Whenever we have a solution such that the amounts of satisfaction of *both* parties are *close enough* to the optimum, this parameter is small. (When the parameter is small, we cannot simply pick μ_M or μ_W since $\mathsf{balance}(\mu_M)$ and $\mathsf{balance}(\mu_W)$ can be *arbitrarily larger* than $\min\{O_M, O_W\}$.) Indeed, the closer the satisfaction of both parties to the optimum (which is exactly the case where both parties would find the solution desirable), the smaller the parameter is, and the smaller the parameter is, the faster a parameterized algorithm is. In this above guarantee parameterization, the guarantee is already quite high—for example, our parameter is significantly stronger than $k - n'$, where n' is the number of men (or women) matched by a stable matching, since $k - \min\{O_M, O_W\}$ is (i) never larger than $k - n'$, and (ii) can be *arbitrarily smaller* than $k - n'$, e.g. $k - n'$ can be of the magnitude of $\mathcal{O}(n)$ while our parameter is 0.

Since we are taking the *minimum* of $\{O_M, O_W\}$, we need the satisfaction of *both* parties to be close to optimal in order to have a small parameter. As we are able to show that BSM is FPT with respect to this parameter, it is very natural to next examine the case where we take the *max* of $\{O_M, O_W\}$. In this case, the closer the satisfaction of *at least one* party to the optimum, the smaller the parameter is. In other words, now to have a small parameter, the demand from a solution is weaker than before. In the vocabulary of Parameterized Complexity, it is said that the parameterization by $t = k - \max\{O_M, O_W\}$ is "above a higher guarantee" than the parameterization by $t = k - \min\{O_M, O_W\}$, since it is *always* the case that $\max\{O_M, O_W\} \geq \min\{O_M, O_W\}$. Interestingly, as we show in this paper, the parameterization by $k - \max\{O_M, O_W\}$ results in a problem that is W[1]-hard. Hence, the complexities of the two parameterizations behave very differently. We remark that in Parameterized Complexity, it is *not at all* the rule that when one takes an "above a higher guarantee" parameterization, the problem would become W[1]-hard, as can be evidenced by the VERTEX COVER problem, the classical above guarantee parameterizations in this field, for which three distinct above guarantee parameterizations yielded FPT algorithms [8,15,26,32]. Overall, our results draw a nontrivial line between tractability and intractability of above guarantee parameterization of BSM.

Finally, the three main theorems that we establish in this study are as follows.

Theorem 1. ABOVE-MAX BSM *is* W[1]-hard.

Theorem 2. ABOVE-MIN BSM *admits a kernel that has at most 3t men, at most 3t women, and such that each person has at most $2t+1$ acceptable partners.*

Note that Theorem 2 implies that ABOVE-MIN BSM has an FPT algorithm with running time $2^{\mathcal{O}(t \log t)}$. However, we present an algorithm whose running time is single exponential in t utilizing the method of bounded search trees on top of our kernel (See [16]).

Theorem 3. ABOVE-MIN BSM *can be solved in time* $\mathcal{O}^*(8^t)^1$.

Our Techniques: The reduction we develop to prove Theorem 1 is quite technically intricate; the overview followed by a detailed reduction is presented in Sect. 2. The proof of Theorem 2 is based on the introduction and analysis of a "functional" variant of ABOVE-MIN BSM. Our kernelization algorithm consists of several phases, each simplifying a different aspect of ABOVE-MIN BSM, and shedding light on structural properties of the YES-instances of this problem. We stress that the design and order of our reduction rules are very carefully tailored to ensure correctness. For example, it is tempting to execute Step 2 before Step 1 (see the outline in Sect. 3.1), but this is simply incorrect as it alters the set of stable matchings. The reduction rules are easy to express for this functional variant of ABOVE-MIN BSM. Hence, we choose to define the reduction rules for this functional variant of ABOVE-MIN BSM instead of ABOVE-MIN BSM directly.

Preliminaries. Let f be a function $f : A \to B$. For a subset $A' \subseteq A$, we let $f|_{A'}$ denote the restriction of f to A'. That is, $f|_{A'} : A' \to B$, and $f|_{A'}(a) = f(a)$ for all $a \in A'$. For basic notions in Parameterized Complexity, see preliminaries in [16]. We refer the reader to the books [7,10] for more information on Parameterized Complexity.

---

[1] $\mathcal{O}^*$ hides terms that are polynomial in the input size.

Throughout the paper, whenever the instance $\mathcal{I}$ of BSM under discussion is not clear from context or we would like to put emphasis on $\mathcal{I}$, we add "$(\mathcal{I})$" to the appropriate notation. For example, we use the notation $t(\mathcal{I})$ rather than t. When we would like to refer to the balance of a stable matching μ in a specific instance $\mathcal{I}$, we would use the notation $\mathsf{balance}_{\mathcal{I}}(\mu)$. A matching is called a *perfect matching* if it matches every person (to some other person).

A Functional Variant of STABLE MARRIAGE. To obtain our kernelization algorithm for ABOVE-MIN BSM, it will be convenient to work with a "functional" definition of preferences, resulting in a "functional" variant of this problem which we call ABOVE-MIN FBSM. Here, instead of the sets of preferences lists $\mathcal{L}_M$ and $\mathcal{L}_W$, the input consists of sets of preference *functions* $\mathcal{F}_M$ and $\mathcal{F}_W$, where $\mathcal{F}_M$ replaces $\mathcal{L}_M$ and $\mathcal{F}_W$ replaces $\mathcal{L}_W$. Specifically, every person $a \in M \cup W$ has an injective (one-to-one) function $f_a : \mathcal{A}(a) \to \mathbb{N}$, called a *preference function*. Intuitively, a lower function value corresponds to a higher preference. Since every preference function is injective, it defines a total order over a set of acceptable partners. Note that all of the definitions presented in the introduction extend to our functional variant in the natural way (the required adaptations can be found in [16]. Clearly, it is straightforward to turn an instance of ABOVE-MIN BSM into an equivalent instance of ABOVE-MIN FBSM as stated in the following observation. We list some classical results, which were originally presented in the context of STABLE MARRIAGE. These results also hold in the context of FUNCTIONAL STABLE MARRIAGE (see [16]).

Observation 1. *Let* $\mathcal{I} = (M, W, \mathcal{L}_M, \mathcal{L}_W, k)$ *be an instance of* ABOVE-MIN BSM. *For each* $a \in M \cup W$, *define* $f_a : \mathcal{A}(a) \to \mathbb{N}$ *by setting* $f_a(b) = p_a(b)$ *for all* $b \in \mathcal{A}(a)$. *Then,* $\mathcal{I}$ *is a* YES-*instance of* ABOVE-MIN BSM *if and only if* $(M, W, \mathcal{F}_M = \{f_m\}_{m \in M}, \mathcal{F}_W = \{f_w\}_{w \in W}, k)$ *is a* YES-*instance of* ABOVE-MIN FBSM.

Proposition 1 ([13]). *For any instance of* STABLE MARRIAGE *(or* FUNCTIONAL STABLE MARRIAGE*), there exist a man-optimal stable matching,* μ_M, *and a woman-optimal stable matching,* μ_W, *and both* μ_M *and* μ_W *can be computed in polynomial time.*

Proposition 2 ([14]). *Given an instance of* STABLE MARRIAGE *(or* ABOVE-MIN FBSM*), the set of men and women that are matched is the same for all stable matchings.*

Proposition 3 ([20]). *For any instance of* STABLE MARRIAGE *(or* FUNCTIONAL STABLE MARRIAGE*), every stable matching* μ *satisfies the following conditions: every woman* w *is either unmatched by both* μ_M *and* μ *or* $p_w(\mu_M(w)) \geq p_w(\mu(w))$, *and every man* m *is either unmatched by both* μ_W *and* μ *or* $p_m(\mu_W(m)) \geq p_m(\mu(m))$.

2   Hardness

In this section, we prove Theorem 1. For this purpose, we give a reduction from a W[1]-hard problem, CLIQUE [9]. Thus, to prove Theorem 1, it is sufficient to prove the following result.

Lemma 1. *Given an instance* $\mathcal{I} = (G = (V, E), k)$ *of* CLIQUE, *an equivalent instance* $\widehat{\mathcal{I}} = (M, W, \mathcal{L}_M, \mathcal{L}_W, \widehat{k})$ *of* ABOVE-MAX BSM *such that* $t = 6k + 3k(k-1)$ *can be constructed in time* $\mathcal{O}(f(k) \cdot |\mathcal{I}|^{\mathcal{O}(1)})$ *for some function* f.

The goal is to construct (in "FPT time") an instance $\widehat{\mathcal{I}} = (M, W, \mathcal{L}_M, \mathcal{L}_W, \widehat{k})$ of ABOVE-MAX BSM. The following subsection contains an informal explanation of the intuition. A formal description of the gadget construction and the analysis are presented in [16].

Reduction. Let $\mathcal{I} = (G = (V, E), k)$ be some instance of CLIQUE. We select arbitrary orders on V and E, and accordingly we denote $V = \{v_1, v_2, \ldots, v_{|V|}\}$ and $E = \{e_1, e_2, \ldots, e_{|E|}\}$.

First, to construct the sets M and W, we define three pairwise-disjoint subsets of M, called M_V, M_E and $\widetilde{M}$, and three pairwise-disjoint subsets of W, called W_V, W_E and $\widetilde{W}$. Then, we set $M = M_V \cup M_E \cup \widetilde{M} \uplus \{m^*\}$ and $W = W_V \cup W_E \cup \widetilde{W} \uplus \{w^*\}$, where m^* and w^* denote a new man and a new woman, respectively.

- $M_V = \{m_v^i : v \in V, i \in \{1, 2\}\}$; $W_V = \{w_v^i : v \in V, i \in \{1, 2\}\}$.
- $M_E = \{m_e^i : e \in E, i \in \{1, 2\}\}$; $W_E = \{w_e^i : e \in E, i \in \{1, 2\}\}$.
- Let $\delta = 2(|V| + |E| + |V||E| + |V||E|^2) - k(4 + 4k + 2|E| + (k-1)|V||E|)$.
  Then, $\widetilde{M} = \{\widetilde{m}^i : i \in \{1, 2, \ldots, \delta\}\}$ and $\widetilde{W} = \{\widetilde{w}^i : i \in \{1, 2, \ldots, \delta\}\}$.

Note that $|M| = |W|$. We remark that in what follows, we assume w.l.o.g. that $\delta \geq 0$ and $|V| > k + k(k-1)/2$, else the size of the input instance $\mathcal{I}$ of CLIQUE is bounded by a function of k and can therefore, by using brute-force, be solved in FPT time.

Roughly speaking, each pair of men, m_v^1 and m_v^2, represents a vertex, and we aim to ensure that either both men will be matched to their best partners (in the man-optimal stable matching) or both men will be matched to other partners (where there would be only *one* choice that ensures stability). Accordingly, we will guarantee that the choice of matching these two men to their best partners translates to *not* choosing the vertex they represent into the clique, and the other choice translates to choosing this vertex into the clique.

Now, having just the set M_V, we can encode selection of vertices into the clique, but we cannot ensure that the vertices we select indeed form a clique. For this purpose, we also have the set M_E which, in a manner similar to M_V, encodes selection of edges into the clique. By designing the instance in a way that the situation of the men in the man-optimal stable matching is significantly worse than that of the women in the women-optimal stable matching, we are able to ensure that at most $2(k + k(k-1)/2)$ men in $M_V \cup M_E$ will not be assigned their best partners (here, we exploit the condition that $\mathsf{balance}(\mu) \leq \widehat{k}$ for a solution μ). Here the man m^* plays a crucial role—by using dummy men and women (in the sets $\widetilde{M}$ and $\widetilde{W}$) that prefer each other over all other people, we ensure that the situation of m^* is always "extremely bad" (from his viewpoint), while the situation of his partner, w^*, is always "excellent" (from her viewpoint).

At this point, we first need to ensure that the edges that we select indeed connect the vertices that we select. For this purpose, we carefully design our reduction so that when a pair of men representing some edge e obtain partners worse than those they have in the man-optimal stable matching, it must be that the men representing the endpoints of e have also obtained partners worse than those they have in the man-optimal stable matching, else stability will not be preserved—the partners of the men representing the endpoints of e will form blocking pairs together with the men representing e.

Finally we observe that we still need to ensure that among our $2\,(k + k(k-1)/2)$ distinguished men in $M_V \cup M_E$, which are associated with $k + k(k-1)/2$ selected elements (vertices and edges), there will be exactly $2k$ distinguished men from M_V and exactly $k(k-1)$ distinguished men from M_E, which would mean we have chosen k vertices and $k(k-1)/2$ edges. For this purpose, we construct an instance where for the women, it is only somewhat "beneficial" that the men in M_V will not be matched to their best partners, but it is extremely beneficial that the men in M_E will not be matched to their best partners. This objective is achieved by carefully placing dummy men (from $\widetilde{M}$) in the preference lists of women in W_E. By again exploiting the condition that $\mathsf{balance}(\mu) \leq \widehat{k}$ for a solution μ, we are able to ensure that there would be at least $k(k-1)$ distinguished men from M_E.

3   Kernel

In this section, we design a kernelization algorithm for ABOVE-MIN BSM. More precisely, we prove Theorem 2.

3.1   Functional Balanced Stable Marriage

To prove Theorem 2, we first prove the following result for the ABOVE-MIN FBSM problem. Proofs of result marked with (*) can be found in [16].

Lemma 2 (*). ABOVE-MIN FBSM *admits a kernel with at most $2t$ men, at most $2t$ women, and such that the image of the preference function of each person is a subset of $\{1, 2, \ldots, t + 1\}$.*

To obtain the desired kernelization algorithm, we execute the following plan.

1. **Cleaning Prefixes and Suffixes.** Simplify the preference functions by "cleaning" suffixes and thereby also "cleaning" prefixes.
2. **Perfect Matching.** Zoom into the set of people matched by every stable matching.

3. **Overcoming Sadness.** Bound the number of "sad" people. Roughly speaking, a "sad" person a is one whose best attainable partner, b, does not reciprocate by considering a as the best attainable partner.
4. **Marrying Happy People.** Remove "happy" people from the instance.
5. **Truncating High-Values.** Obtain "compact" preference functions by truncating "high-values".
6. **Shrinking Gaps.** Shrink some of the gaps created by previous steps.

Each of the following subsections captures one of the steps above. In what follows, we let $\mathcal{I}$ denote our current instance of ABOVE-MIN FBSM. Initially, this instance is the input instance, but as the execution of our algorithm progresses, the instance is modified. The reduction rules that we present are applied *exhaustively* in the order of their presentation. In other words, at each point of time, the first rule whose condition is true is the one that we apply next. In particular, the execution terminates once the value of t drops below 0, as implied by the following rule.

Reduction Rule 1. *If $k < \max\{O_M, O_W\}$, then return* NO.

Lemma 3 (*). *Reduction Rule 1 is safe.*

Note that if $t < 0$, then $k < \min\{O_M, O_W\} \le \max\{O_M, O_W\}$. From Proposition 1 we can infer that each of our reduction rules can indeed be implemented in polynomial time.

Cleaning Prefixes and Suffixes. We begin by modifying the images of the preference functions. We remark that it is *necessary* to perform this step first as otherwise the following steps would not be correct. To clean prefixes while ensuring *both* safeness and that the parameter t does not increase, we would actually need to clean *suffixes* first. Formally, we define suffixes as follows.

Definition 1. *Let (m, w) denote an acceptable pair. If m is matched by μ_W and $f_m(w) > f_m(\mu_W(m))$, then we say that w belongs to the suffix of m. Similarly, if w is matched by μ_M and $f_w(m) > f_w(\mu_M(w))$, then we say that m belongs to the suffix of w.*

By Proposition 3, we have the following observation.

Observation 2. *Let (m, w) denote an acceptable pair such that one of its members belongs to the suffix of the other member. Then, there is no $\mu \in$ SM$(\mathcal{I})$ that matches m with w.*

For every person a, let worst(a) be the person in $\mathcal{A}(a)$ to whom f_a assigns its worst preference value. More precisely, worst$(a) = \text{argmax}_{b \in \mathcal{A}(a)} f_a(b)$. We will now clean suffixes.

Reduction Rule 2. *If there exists a person a such that worst(a) belongs to the suffix of a, then define the preference functions as follows.*

- $f'_a = f_a|_{\mathcal{A}(a)\setminus\{\text{worst}(a)\}}$ and $f'_{\text{worst}(a)} = f_{\text{worst}(a)}|_{\mathcal{A}(\text{worst}(a))\setminus\{a\}}$.
- For all $b \in M \cup W \setminus \{a, \text{worst}(a)\}$: $f'_b = f_b$

The new instance is $\mathcal{J} = (M, W, \{f'_{m'}\}_{m' \in M}, \{f'_{w'}\}_{w' \in W}, k)$.

Lemma 4 (*). *Reduction Rule 2 is safe, and $t(\mathcal{I}) = t(\mathcal{J})$.*

By cleaning suffixes, we actually also accomplish the objective of cleaning prefixes, which are defined as follows.

Definition 2. *Let (m, w) denote an acceptable pair. If m is matched by μ_M and $f_m(w) < f_m(\mu_M(m))$, then we say that w belongs to the prefix of m. Similarly, if w is matched by μ_W and $f_w(m) < f_w(\mu_W(w))$, then we say that m belongs to the prefix of w.*

Lemma 5 (*). *Let $\mathcal{I}$ be an instance of* ABOVE-MIN *FBSM on which Reduction Rules 1 to 2 have been exhaustively applied. Then, there does not exist an acceptable pair (m, w) such that one of its members belongs to the prefix of the other one.*

Corollary 1. *Let $\mathcal{I}$ be an instance of* ABOVE-MIN *FBSM on which Reduction Rules 1 to 2 have been exhaustively applied. Then, for every acceptable pair (m, w) in $\mathcal{I}$ where m and w are matched (not necessarily to each other) by both μ_M and μ_W, it holds that $f_m(\mu_M(m)) \leq f_m(w) \leq f_m(\mu_W(m))$ and $f_w(\mu_W(w)) \leq f_w(m) \leq f_w(\mu_M(w))$.*

Perfect Matching. Having Corollary 1 at hand, we are able to provide a simple rule that allows us to assume that every solution matches all people.

Reduction Rule 3. *If there exists a person unmatched by μ_M, then let M' and W' denote the subsets of men and women, respectively, who are matched by μ_M. For each $a \in M' \cup W'$, denote $\mathcal{A}'(a) = \mathcal{A}(a) \cap (M' \cup W')$, and define $f'_a = f_a|_{\mathcal{A}'(a)}$. The new instance is $\mathcal{J} = (M', W', \{f'_m\}_{m \in M'}, \{f'_w\}_{w \in W'}, k)$.*

To prove the safeness of this rule, we first prove the following lemma.

Lemma 6 (*). *Let $\mathcal{I}$ be an instance of* ABOVE-MIN *FBSM on which Reduction Rules 1 to 2 have been exhaustively applied. Then, for every person a not matched by μ_M, it holds that $\mathcal{A}(a) = \emptyset$.*

Lemma 7 (*). *Reduction Rule 3 is safe, and $t(\mathcal{I}) = t(\mathcal{J})$.*

By Proposition 2, from now onwards, we have that for the given instance, any stable matching is a perfect matching. Due to this observation, we can denote $n = |M| = |W|$, and for any stable matching μ, we have the following equalities.

$$\sum_{(m,w)\in\mu} f_m(w) = \sum_{m\in M} f_m(\mu(m)); \qquad \sum_{(m,w)\in\mu} f_w(m) = \sum_{w\in W} f_w(\mu(w)). \quad (I)$$

Overcoming Sadness. As every stable matching is a perfect matching, every person is matched by every stable matching, including the man-optimal and woman-optimal stable matchings. Thus, it is well defined to classify the people who do not have the same partner in the man-optimal and woman-optimal stable matchings as "sad". That is,

Definition 3. *A person* $a \in M \cup W$ *is* sad *if* $\mu_M(a) \neq \mu_W(a)$.

We let M_S and W_S denote the sets of sad men and sad women, respectively. People who are not sad are termed *happy*. Accordingly, we let M_H and W_H denote the sets of happy men and happy women, respectively. Note that $M_S = \emptyset$ if and only if $W_S = \emptyset$. Note that by the definition, a happy person a and $\mu_M(a) = \mu_W(a)$ are matched to each other in every stable matching. Next, we bound the number of sad people in a YES-instance.

Reduction Rule 4. *If* $|M_S| > 2t$ *or* $|W_S| > 2t$, *then return* NO.

Lemma 8 (*). *Reduction Rule 4 is safe.*

Marrying Happy People. Towards the removal of happy people, we first need to handle the special case where there are no sad people. In this case, there is exactly one stable matching, which is the man-optimal stable matching (that is equal, in this case, to the woman-optimal stable matching). This immediately implies the safeness of the following rule.

Reduction Rule 5. *If* $M_S = W_S = \emptyset$, *then return* YES *if* balance$(\mu_M) \leq k$ *and* NO *otherwise.*

Observation 3. *Reduction Rule 5 is safe.*

Next, we turn to removing happy people. For this, we need to ensure that the balance of the instance is preserved. This is because we do not know which side (men or women) attains the Bal$(\mathcal{I})$ value, hence we cannot reduce the quantity k by the dissatisfaction of the happy people on that side. Consequently, we need to ensure that Bal$(\mathcal{I})$ = Bal$(\mathcal{J})$, where $\mathcal{J}$ denotes the new instance resulting from the removal of some happy people. Let (m_h, w_h) denote a *happy pair*, a pair of a happy man and a happy woman who are matched to each other in every stable matching.[2] Then, we redefine the preference functions in a manner that allows us to transfer the "contributions" of m_h and w_h from Bal$(\mathcal{I})$ to Bal$(\mathcal{J})$ via some sad man and woman. Note that these sad people exist because Reduction Rule 5 does not apply.

Reduction Rule 6. *If there exists a happy pair* (m_h, w_h), *then proceed as follows. Select an arbitrary sad man* m_s *and an arbitrary sad woman* w_s. *Denote* $M' = M \setminus \{m_h\}$ *and* $W' = W \setminus \{w_h\}$. *For each person* $a \in M' \cup W'$, *the new preference function* $f'_a : \mathcal{A}(a) \setminus \{m_h, w_h\} \to \mathbb{N}$ *is defined as follows.*

---

[2] Such a pair is often referred to as fixed pair in literature.

- *For each $w \in \mathcal{A}(m_s) \setminus \{w_h\}$: $f'_{m_s}(w) = f_{m_s}(w) + f_{m_h}(w_h)$.*
- *For each $m \in \mathcal{A}(w_s) \setminus \{m_h\}$: $f'_{w_s}(m) = f_{w_s}(m) + f_{w_h}(m_h)$.*
- *For each $w \in W' \setminus \{w_s\}$: $f'_w = f_w|_{M'}$, and for each $m \in M' \setminus \{m_s\}$: $f'_m = f_m|_{W'}$.*

The new instance is $\mathcal{J} = (M', W', \{f'_m\}_{m \in M'}, \{f'_w\}_{w \in W'}, k)$.

The following lemma proves one direction of the safeness of Reduction Rule 6.

Lemma 9 (*). *Let $\mu \in \mathrm{SM}(\mathcal{I})$. Then, $\mu' = \mu \setminus \{(m_h, w_h)\}$ is a stable matching in $\mathcal{J}$ such that $\mathsf{balance}_{\mathcal{J}}(\mu') = \mathsf{balance}_{\mathcal{I}}(\mu)$.*

The following observation helps to prove the other direction of the safeness of Reduction Rule 6.

Observation 4 (*). *Let $\mathcal{I}$ be an instance of ABOVE-MIN FBSM on which Reduction Rules 1 to 2 have been exhaustively applied. Then, for every happy pair (m_h, w_h), it holds that $\mathcal{A}(m_h) = \{w_h\}$ and $\mathcal{A}(w_h) = \{m_h\}$.*

Lemma 10 (*). *Let $\mu' \in \mathrm{SM}(\mathcal{J})$. Then, $\mu = \mu' \cup \{(m_h, w_h)\}$ is a stable matching in $\mathcal{I}$ such that $\mathsf{balance}_{\mathcal{I}}(\mu) = \mathsf{balance}_{\mathcal{J}}(\mu')$.*

Lemma 11 (*). *Reduction Rule 6 is safe, and $t(\mathcal{I}) = t(\mathcal{J})$.*

Before we examine the preference functions closely, we prove the following.

Lemma 12 (*). *Given an instance $\mathcal{I}$ of ABOVE-MIN FBSM, one can exhaustively apply Reduction Rules 1 to 6 in polynomial time to obtain an instance $\mathcal{J}$ such that $t(\mathcal{J}) \leq t(\mathcal{I})$. All people in $\mathcal{J}$ are sad and matched by every stable matching, and there exist at most $2t$ men and at most $2t$ women.*

Truncating High-Values. So far we have bounded the number of people. However, the images of the preference functions can contain integers that are not bounded by a function polynomial in the parameter. Thus, even though the number of people is upper bounded by $4t$, the total size of the instance can be huge. Hence, we need to process the images of the preference functions. Formally, we have the following rule.

Reduction Rule 7. *If there exists an acceptable pair (m, w) such that $f_m(w) > (k - O_M) + f_m(\mu_M(m))$ or $f_w(m) > (k - O_W) + f_w(\mu_W(w))$, then define the preference functions as follows:*

- *$f'_m = f_m|_{\mathcal{A}(m) \setminus \{w\}}$ and $f'_w = f_w|_{\mathcal{A}(w) \setminus \{m\}}$.*
- *For all $a \in M \cup W \setminus \{m, w\}$: $f'_a = f_a$.*

The new instance is $\mathcal{J} = (M, W, \{f'_{m'}\}_{m' \in M}, \{f'_{w'}\}_{w' \in W}, k)$.

Lemma 13 (*). *Reduction Rule 7 is safe, and $t(\mathcal{I}) \geq t(\mathcal{J})$.*

Shrinking Gaps. Currently, there might still exist a man m or a woman w such that $f_m(\mu_M(m)) > 1$ or $f_w(\mu_W(w)) > 1$, respectively. In the following rule, we would like to decrease some values assigned by the preference functions of such men and women in a manner that preserves equivalence.

Reduction Rule 8. *If there exist $m \in M$ and $w \in W$ such that $f_m(\mu_M(m)) > 1$ and $f_w(\mu_W(w)) > 1$, then define the preference functions as follows.*

- *For all $w' \in \mathcal{A}(m)$: $f'_m(w') = f_m(w') - 1$ and for all $m' \in \mathcal{A}(w)$: $f'_w(m') = f_w(m') - 1$.*
- *For all $a \in M \cup W \setminus \{m, w\}$: $f'_a = f_a$.*

The new instance is $\mathcal{J} = (M, W, \{f'_{m'}\}_{m' \in M}, \{f'_{w'}\}_{w' \in W}, k - 1)$.

Lemma 14 (*). *Reduction Rule 8 is safe, and $t(\mathcal{I}) = t(\mathcal{J})$.*

After the exhaustive application of Reduction Rule 8, at least one of the two parties does not have any member without a person assigned 1 by his/her preference function. Thus,

Observation 5. *Let $\mathcal{I}$ be an instance of* ABOVE-MIN *FBSM that is reduced with respect to Reduction Rules 1 to 8. Then, either* (i) *for every $m \in M$, we have that $f_m(\mu_M(m)) = 1$, or* (ii) *for every $w \in W$, we have that $f_w(\mu_W(w)) = 1$. In particular, either* (i) *$O_M = |M|$ or* (ii) *$O_W = |W|$.*

This concludes the description of our reduction rules yielding Lemma 2.

3.2 Balanced Stable Marriage

We would like to use the kernel for ABOVE-MIN FBSM, given by Lemma 2, to design a kernel for ABOVE-MIN BSM. In order to view the preference functions as preference lists, we need to remove the "gaps" from the preference functions. The details of the construction can be found in the full version [16]. We conclude by stating the following result which completes the proof of Theorem 2.

Lemma 15. ABOVE-MIN *BSM admits a kernel that has at most $3t$ men among whom at most $2t$ are sad, at most $3t$ women among whom at most $2t$ are sad, and such that each happy person has at most $2t + 1$ acceptable partners and each sad person has at most $t + 1$ acceptable partners. Moreover, every stable matching in the kernel is a perfect matching.*

References

1. Trends in Computational Social Choice. AI Access (2017)
2. Betzler, N.: A Multivariate Complexity Analysis of Voting Problems. Ph.D. thesis, Friedrich-Schiller-Universität Jena (2010)

3. Betzler, N., Bredereck, R., Chen, J., Niedermeier, R.: Studies in computational aspects of voting. In: Bodlaender, H.L., Downey, R., Fomin, F.V., Marx, D. (eds.) The Multivariate Algorithmic Revolution and Beyond. LNCS, vol. 7370, pp. 318–363. Springer, Heidelberg (2012). https://doi.org/10.1007/978-3-642-30891-8_16
4. Brandt, F., Conitzer, V., Endriss, U., Lang, J., Procaccia, A.: Handbook of Computational Social Choice. Cambridge University Press, London (2016)
5. Bredereck, R., Chen, J., Faliszewski, P., Guo, J., Niedermeier, R., Woeginger, G.: Parameterized algorithmics for computational social choice: nine research challenges. Tsinghua Sci. Technol. 19(4), 358 (2014)
6. Chen, J., Hermelin, D., Sorge, M., Yedidsion, H.: How hard is it to satisfy (almost) all roommates. In: 45th International Colloquium on Automata, Languages, and Programming (ICALP), pp. 35:1–35:15 (2018)
7. Cygan, M., et al.: Parameterized Algorithms. Springer, Cham (2015). https://doi.org/10.1007/978-3-319-21275-3
8. Cygan, M., Pilipczuk, M., Pilipczuk, M., Wojtaszczyk, J.O.: On multiway cut parameterized above lower bounds. ACM Trans. Comput. Theory 5(1), 3:1–3:11 (2013)
9. Downey, R.G., Fellows, M.R.: Fixed-parameter tractability and completeness II: On completeness for W[1]. Theor. Comput. Sci. 141(1&2), 109–131 (1995)
10. Downey, R.G., Fellows, M.R.: Fundamentals of Parameterized Complexity. Springer, London (2013). https://doi.org/10.1007/978-1-4471-5559-1
11. Faliszewski, P., Niedermeier, R.: Parameterization in computational social choice. In: Encyclopedia of Algorithms, pp. 1516–1520 (2016)
12. Feder, T.: Stable networks and product graphs. Ph.D. thesis, Stanford University (1990)
13. Gale, D., Shapley, L.S.: College admissions and the stability of marriage. Am. Math. Monthly 69, 9–15 (1962)
14. Gale, D., Sotomayor, M.: Some remarks on the stable matching problem. Discrete Appl. Math. 11(4), 223–232 (1985)
15. Garg, S., Philip, G.: Raising the bar for vertex cover: Fixed-parameter tractability above a higher guarantee. In: Proceedings of the 27th Annual ACM-SIAM Symposium on Discrete Algorithms (SODA), pp. 1152–1166 (2016)
16. Gupta, S., Roy, S., Saurabh, S., Zehavi, M.: Balanced stable marriage: how close is close enough? CoRR abs/1707.09545 (2017)
17. Gupta, S., Roy, S., Saurabh, S., Zehavi, M.: Group activity selection on graphs: Parameterized analysis. In: Proceedings of 10th International Symposium Algorithmic Game Theory (SAGT), pp. 106–118 (2017)
18. Gupta, S., Saurabh, S., Zehavi, M.: On treewidth and stable marriage. CoRR abs/1707.05404 (2017)
19. Gusfield, D.: Three fast algorithms for four problems in stable marriage. SIAM J. Comput. 16(1), 111–128 (1987)
20. Gusfield, D., Irving, R.W.: The Stable Marriage Problem - Structure and Algorithms. Foundations of Computing Series. MIT Press (1989)
21. Igarashi, A., Bredereck, R., Elkind, E.: On parameterized complexity of group activity selection problems on social networks. In: Proceedings of the 16th Conference on Autonomous Agents and MultiAgent Systems (AAMAS), pp. 1575–1577 (2017)
22. Igarashi, A., Peters, D., Elkind, E.: Group activity selection on social networks. In: Proceedings of the 31st Conference on Artificial Intelligence (AAAI), pp. 565–571 (2017)

23. Knuth, D.E.: Stable Marriage and Its Relation to Other Combinatorial Problems: An Introduction to the Mathematical Analysis of Algorithms. CRM Proceedings & Lecture notes, American Mathematical Society (1997)

24. Lee, H., Williams, V.: Complexity of the stable invitations problem. In: Proceedings of the 31st Conference on Artificial Intelligence (AAAI), pp. 579–585 (2017)

25. Lee, H., Williams, V.: Parameterized complexity of group activity selection. In: Proceedings of the 16th Conference on Autonomous Agents and MultiAgent Systems (AAMAS), pp. 353–361 (2017)

26. Lokshtanov, D., Narayanaswamy, N.S., Raman, V., Ramanujan, M.S., Saurabh, S.: Faster parameterized algorithms using linear programming. ACM Trans. Algorithms $11(2)$, 15:1–15:31 (2014)

27. Manlove, D.F.: Algorithmics of Matching Under Preferences, Series on Theoretical Computer Science, vol. 2. WorldScientific (2013)

28. Marx, D., Schlotter, I.: Parameterized complexity and local search approaches for the stable marriage problem with ties. Algorithmica $58(1)$, 170–187 (2010)

29. Marx, D., Schlotter, I.: Stable assignment with couples: parameterized complexity and local search. Discrete Optim. $8(1)$, 25–40 (2011)

30. Meeks, K., Rastegari, B.: Solving hard stable matching problems involving groups of similar agents. CoRR **abs/1708.04109** (2017)

31. Mnich, M., Schlotter, I.: Stable marriage with covering constraints-a complete computational trichotomy. In: Proceedings of the 10th International Symposium of Algorithmic Game Theory (SAGT), pp. 320–332 (2017)

32. Raman, V., Ramanujan, M.S., Saurabh, S.: Paths, flowers and vertex cover. In: Proceedings of the 19th Annual European Symposium on Algorithms (ESA), pp. 382–393 (2011)

Improved Streaming Algorithms for Maximizing Monotone Submodular Functions Under a Knapsack Constraint

Chien-Chung Huang[1] and Naonori Kakimura[2(✉)]

[1] CNRS, École Normale Supérieure, Paris, France
villars@gmail.com
[2] Keio University, Yokohama, Japan
kakimura@math.keio.ac.jp

Abstract. In this paper, we consider the problem of maximizing a monotone submodular function subject to a knapsack constraint in the streaming setting. In particular, the elements arrive sequentially and at any point of time, the algorithm has access only to a small fraction of the data stored in primary memory. For this problem, we propose a $(0.4 - \varepsilon)$-approximation algorithm requiring only a single pass through the data. This improves on the currently best $(0.363 - \varepsilon)$-approximation algorithm. The required memory space depends only on the size of the knapsack capacity and ε.

Keywords: Submodular functions · Streaming algorithm · Approximation algorithm

1 Introduction

A set function $f : 2^E \to \mathbb{R}_+$ on a ground set E is *submodular* if it satisfies the *diminishing marginal return property*, i.e., for any subsets $S \subseteq T \subsetneq E$ and $e \in E \setminus T$,

$$f(S \cup \{e\}) - f(S) \geq f(T \cup \{e\}) - f(T).$$

A set function is *monotone* if $f(S) \leq f(T)$ for any $S \subseteq T$. Submodular functions play a fundamental role in combinatorial optimization, as they capture rank functions of matroids, edge cuts of graphs, and set coverage, just to name a few examples. Besides their theoretical interests, submodular functions have attracted much attention from the machine learning community because they can model various practical problems such as online advertising [1,24,35], sensor location [25], text summarization [30,31], and maximum entropy sampling [28].

Many of the aforementioned applications can be formulated as the maximization of a monotone submodular function under a knapsack constraint. In this problem, we are given a monotone submodular function $f : 2^E \to \mathbb{R}_+$, a size

---

Supported by JSPS KAKENHI Grant Numbers JP17K00028 and JP18H05291.

© Springer Nature Switzerland AG 2019
Z. Friggstad et al. (Eds.): WADS 2019, LNCS 11646, pp. 438–451, 2019.
https://doi.org/10.1007/978-3-030-24766-9_32

function $c : E \to \mathbb{N}$, and an integer $K \in \mathbb{N}$, where $\mathbb{N}$ denotes the set of positive integers. The problem is defined as

$$\text{maximize } f(S) \quad \text{subject to } c(S) \leq K, \quad S \subseteq E, \tag{1}$$

where we denote $c(S) = \sum_{e \in S} c(e)$ for a subset $S \subseteq E$. Note that, when $c(e) = 1$ for every item $e \in E$, the constraint coincides with a cardinality constraint. Throughout this paper, we assume that every item $e \in E$ satisfies $c(e) \leq K$ as otherwise we can simply discard it.

The problem of maximizing a monotone submodular function under a knapsack or a cardinality constraint is classical and well-studied [20,37]. The problem is known to be NP-hard but can be approximated within the factor of $1 - e^{-1}$ (or $1 - e^{-1} - \varepsilon$); see e.g., [3,15,21,26,36,38].

In some applications, the amount of input data is much larger than the main memory capacity of individual computers. In such a case, we need to process data in a *streaming* fashion (see e.g., [32]). That is, we consider the situation where each item in the ground set E arrives sequentially, and we are allowed to keep only a small number of the items in memory at any point. This setting effectively rules out most of the techniques in the literature, as they typically require random access to the data. In this work, we assume that the function oracle of f is available at any point of the process. Such an assumption is standard in the submodular function literature and in the context of streaming setting [2,13,39].

Our main contribution is to propose a single-pass $(2/5 - \varepsilon)$-approximation algorithm for the problem (1), which improves on the previous work [23,39] (see Table 1). The space complexity is independent of the number of items in E.

Table 1. The knapsack-constrained problem. The algorithms [16,36] are not for the streaming setting. See also [15,26].

	Approx. ratio	#passes	Space	Running time
Ours	$2/5 - \varepsilon$	1	$O\left(K\varepsilon^{-4}\log^4 K\right)$	$O\left(n\varepsilon^{-4}\log^4 K\right)$
Huang et al. [23]	$4/11 - \varepsilon$	1	$O\left(K\varepsilon^{-4}\log^4 K\right)$	$O\left(n\varepsilon^{-4}\log^4 K\right)$
Yu et al. [39]	$1/3 - \varepsilon$	1	$O\left(K\varepsilon^{-1}\log K\right)$	$O\left(n\varepsilon^{-1}\log K\right)$
Huang et al. [23]	$2/5 - \varepsilon$	3	$O\left(K\varepsilon^{-4}\log^4 K\right)$	$O\left(n\varepsilon^{-4}\log^4 K\right)$
Huang-Kakimura [22]	$1/2 - \varepsilon$	$O\left(\varepsilon^{-1}\right)$	$O\left(K\varepsilon^{-7}\log^2 K\right)$	$O\left(n\varepsilon^{-8}\log^2 K\right)$
Ene and Nguyễn [16]	$1 - e^{-1} - \varepsilon$	—	—	$O\left((1/\varepsilon)^{O(1/\varepsilon^4)}n\log n\right)$
Sviridenko [36]	$1 - e^{-1}$	—	—	$O\left(Kn^4\right)$

Theorem 1. *There exists a single-pass streaming $(2/5 - \varepsilon)$-approximation algorithm for the problem (1) requiring $O\left(K\varepsilon^{-4}\log^4 K\right)$ space.*

Our Technique. Let us first describe approximation algorithms for the knapsack-constrained problem (1) in the offline setting. The simplest algorithm is a greedy algorithm, that repeatedly takes an item with maximum marginal return. The greedy algorithm admits a $(1 - 1/\sqrt{e})$-approximation, together with taking one item with the maximum return, although it requires to read all the items K times. Sviridenko [36] showed that, by applying the greedy algorithm from each set of three items, we can find a $(1 - 1/e)$-approximate solution. Recently, such partial enumeration is replaced by a more sophisticated multi-stage guessing strategies (where fractional items are added based on the technique of multilinear extension) to improve the running time in nearly linear time [16]. However, all of them require large space and/or a large number of passes to implement.

In the streaming setting, Badanidiyuru *et al.* [2] proposed a single-pass thresholding algorithm that achieves a $(0.5 - \varepsilon)$-approximation for the cardinality-constrained problem. The algorithm just takes an arriving item e when the marginal return exceeds a threshold and the feasibility is maintained. However, this strategy gives us only a $(1/3 - \varepsilon)$-approximation for the knapsack-constrained problem. This drop in approximation ratio comes from the fact that, while we can freely add an item as long as our current set is of size less than K for the cardinality constraint, we cannot take a new item if its addition exceeds the capacity of the knapsack.

To overcome this issue, in [23] a branching technique is introduced, where one stops at some point of the thresholding algorithm and use a different strategy to collect subsequent items. The ratio of this branching algorithm depends on the size of the largest item o_1 in the optimal solution; when o_1 is overly large, other strategies must be employed. Overall, the proposed approach of [23] gives a $(4/11 - \varepsilon)$-approximation.

How does one improve the ratio further when $c(o_1)$ is large? One can certainly guess the size $c(o_1)$ and the f-value $f(\{o_1\})$ beforehand and in the stream pick the item of similar size and f-value. The difficulty lies in how to pick such an item that, *together with the rest of the optimal solution (excluding o_1)*, guarantees a decent f-value. Namely, we need a good substitute of o_1. In [23], a single-pass procedure, called PickOneItem, is designed to find such an item (see Sect. 2 for details). Once equipped with such an item, it is not difficult to collect other items so as to improve the approximation ratio to $2/5 - \varepsilon$. The down-side of this approach is that one needs multiple passes.

In this paper, we introduce new techniques to achieve the same ratio *without* the need to waste a pass to collect a good substitute of o_1. Depending on the relative size of o_1 and o_2 (second largest item in the optimal solution), we combine PickOneItem with the thresholding algorithm in two different ways. The first one is to perform both of them *dynamically*, that is, each time we find a candidate e for an approximation of o_1, we perform the thresholding algorithm starting from e with the current set. In contrast, the other runs both of them in a *parallel* way; we perform the thresholding algorithm and PickOneItem independently for some subset of items, and combine their results in the end. The details of these algorithms are described in Sects. 3.2 and 3.3, respectively.

Related Work. Maximizing a monotone submodular function subject to various constraints is a subject that has been extensively studied in the literature. We do not attempt to give a complete survey here and just highlight the most relevant results. Besides a knapsack constraint or a cardinality constraint mentioned above, the problem has also been studied under (multiple) matroid constraint(s), p-system constraint, multiple knapsack constraints. See [9,11,12,15,19,26,29] and the rences therein.

In the streaming setting, single-pass algorithms have been proposed for the problem with matroid constraints [10,18] and knapsack constraint [23,39], and without monotonicity [13,34]. *Multi-pass streaming algorithms*, where we are allowed to read a stream of the input multiple times, have also been studied [3,10,22,23]. In particular, Chakrabarti and Kale [10] gave an $O(\varepsilon^{-3})$-pass streaming algorithms for a generalization of the maximum matching problem and the submodular maximization problem with cardinality constraint. Huang and Kakimura [22] designed an $O(\varepsilon^{-1})$-pass streaming algorithm with approximation guarantee $1/2 - \varepsilon$ for the knapsack-constrained problem. Other than the streaming setting, recent applications of submodular function maximization to large data sets have motivated new directions of research on other computational models including parallel computation model such as the MapReduce model [6,7,27] and the adaptivity analysis [4,5,14,17].

The maximum coverage problem is a special case of monotone submodular maximization under a cardinality constraint where the function is a set-covering function. For the special case, McGregor and Vu [33] and Batani *et al.* [8] gave a $(1 - e^{-1} - \varepsilon)$-approximation algorithm in the multi-pass streaming setting.

2   Preliminaries

For a subset $S \subseteq E$ and an element $e \in E$, we use the shorthand $S + e$ and $S - e$ to stand for $S \cup \{e\}$ and $S \setminus \{e\}$, respectively. For a function $f : 2^E \to \mathbb{R}_+$, we also use the shorthand $f(e)$ to stand for $f(\{e\})$. The *marginal return* of adding $e \in E$ with respect to $S \subseteq E$ is defined as $f(e \mid S) = f(S + e) - f(S)$. Thus the submodularity means that $f(e \mid S) \geq f(e \mid T)$ for any subsets $S \subseteq T \subsetneq E$ and $e \in E \setminus T$.

In the rest of the paper, let $\mathcal{I} = (f, c, K, E)$ be an input instance of the problem (1). Let $\text{OPT} = \{o_1, \ldots, o_\ell\}$ denote an optimal solution with $c(o_1) \geq c(o_2) \geq \cdots \geq c(o_\ell)$. We denote $r_i = c(o_i)/K$ for $i = 1, 2, \ldots, \ell$. Let v be an approximated value of $f(\text{OPT})$ such that $v \leq f(\text{OPT}) \leq (1 + \varepsilon)v$.

In the following sections, we review the previous results: the thresholding algorithm and the procedure PickOneItem.

2.1   Thresholding Algorithms

In this section, we present a thresholding algorithm with a single pass [2,23,39]. The algorithm just takes an arriving item e when the marginal return exceeds a threshold. That is, when a new item e arrives, we decide to add e to our current set S if $c(S + e) \leq K$ and $f(e \mid S) \geq \alpha \frac{c(e)}{K} v$, where α is a parameter. The performance is due to the following fact, which follows from submodularity.

Lemma 1. *Let $S = \{e_1, e_2, \ldots, e_s\}$. Suppose that $f(e_i \mid \{e_1, e_2, \ldots, e_{i-1}\}) \geq \alpha \frac{c(e_i)}{K} v$ for each $i = 1, \ldots, s$. Then it holds that $f(S) \geq \alpha \frac{c(S)}{K} v$.*

By setting $\alpha = 1/2$, the algorithm finds a set S such that $f(S) \geq v/2$ with a single pass for the cardinality-constrained problem [2]. Setting $\alpha = 2/3$, together with taking a singleton with maximum return in parallel, we can find a set S such that $f(S) \geq v/3$ with a single pass [23].

2.2   Guessing the Large Item

We here consider a procedure to approximate the largest item o_1 in OPT. It is difficult to correctly identify o_1 among the items in E, but we can nonetheless find a reasonable approximation of it by a single pass. This procedure is used to design multi-pass streaming algorithms [22,23]. Recall that we are given an approximated value v of $f(\mathrm{OPT})$ such that $v \leq f(\mathrm{OPT}) \leq (1 + \varepsilon)v$.

We first present the following fact.

Lemma 2 ([23]). *Let $E_1 \subseteq E$ such that $e^* \in E_1 \cap \mathrm{OPT}$. Suppose that θ satisfies $\theta v/(1 + \varepsilon) \leq f(e^*) \leq \theta v$. For a number t with $t > \frac{1}{\theta} - 2$, define*

$$\lambda = 2 \left(\frac{\theta}{t + 1} - \frac{1}{(t + 1)(t + 2)} \right). \tag{2}$$

Suppose that a set $X = \{e_1, e_2, \ldots, e_x\} \subseteq E_1$ satisfies that $f(e_i \mid \{e_1, e_2, \ldots, e_{i-1}\}) \geq (\theta - \lambda(i - 1))v$ for each $i = 1, \ldots, x$. Then the following holds:

(i) If $x = t+1$, then at least one item $e \in X$ guarantees that $f(\mathrm{OPT} - e^* + e) \geq \Gamma(\theta)v - O(\varepsilon)v$.

(ii) If $x < t+1$ and $f(e^* \mid X) < (\theta - \lambda x)v$, then at least one item $e \in X$ satisfies $f(\mathrm{OPT} - e^* + e) \geq \Gamma(\theta)v - O(\varepsilon)v$.

Here Γ is the function defined by

$$\Gamma(\theta) = \frac{t(t + 3)}{(t + 1)(t + 2)} - \frac{t - 1}{t + 1} \theta. \tag{3}$$

This lemma suggests the following procedure to approximate o_1, which we call PickOneItem. Suppose that we are given approximations $\underline{r}_1, \overline{r}_1$ of r_1 such that $\underline{r}_1 \leq r_1 \leq \overline{r}_1$ and $\overline{r}_1 \leq (1 + \varepsilon)\underline{r}_1$. Define $E_1 = \{e \in E \mid \underline{r}_1 K \leq c(e) \leq \overline{r}_1 K, \theta v/(1+\varepsilon) \leq f(e) \leq \theta v\}$. Then we see that $o_1 \in E_1$. In a single pass, starting from $X = \emptyset$, we decide to add an item $e \in E_1$ to X if $f(e \mid X) \geq (\theta - \lambda|X|)v$. We stop this decision when $|X| = t + 1$. Then, in each step, X always satisfies the assumption in Lemma 2, that is, $X = \{e_1, e_2, \ldots, e_x\} \subseteq E_1$ satisfies that $f(e_i \mid \{e_1, e_2, \ldots, e_{i-1}\}) \geq (\theta - \lambda(i - 1))v$ for each $i = 1, \ldots, x$.

We claim that the output X contains an item $e \in X$ such that $f(\text{OPT} - o_1 + e) \geq \Gamma(\theta)v - O(\varepsilon)v$. Indeed, we consider the situation just before o_1 arrives. If the current set X has size $t+1$, then Lemma 2 (i) implies that there exists $e \in X$ such that $f(\text{OPT} - o_1 + e) \geq \Gamma(\theta)v - O(\varepsilon)v$. If X has size less than $t + 1$, then either o_1 is put in X, or there exists $e \in X$ such that $f(\text{OPT} - o_1 + e) \geq \Gamma(\theta)v - O(\varepsilon)v$ by Lemma 2 (ii). Hence, in any case, at least one item $e \in X$ guarantees that $f(\text{OPT} - o_1 + e) \geq \Gamma(\theta)v - O(\varepsilon)v$.

By choosing an optimal value t for a given θ, we can obtain $\Gamma(\theta) \geq 2/3$. More specifically, we have the following theorem.

Theorem 2 ([23]). *Let $E_1 \subseteq E$ such that $e^* \in E_1 \cap \text{OPT}$. Suppose that θ satisfies $\theta v/(1 + \varepsilon) \leq f(e^*) \leq \theta v$. Define t to be*

$$t = \begin{cases} 1 & \text{if } \theta \geq \frac{1}{2} \\ 2 & \text{if } \frac{1}{2} \geq \theta \geq \frac{2}{5} \\ 3 & \text{if } \frac{2}{5} \geq \theta \geq 0. \end{cases} \tag{4}$$

Then, with a single pass and $O(1)$ space, we can find a set $X \subseteq E_1$ such that there exists $e \in X$ such that $f(\text{OPT} - e^* + e) \geq \Gamma(\theta)v - O(\varepsilon)v$, where

$$\Gamma(\theta) \geq \begin{cases} \frac{2}{3} & \text{if } \theta \geq \frac{1}{2} \\ \frac{5}{6} - \frac{\theta}{3} & \text{if } \frac{1}{2} \geq \theta \geq \frac{2}{5} \\ \frac{9}{10} - \frac{\theta}{2} & \text{if } \frac{2}{5} \geq \theta \geq 0. \end{cases}$$

3   Single-Pass $(2/5 - \varepsilon)$-Approximation Algorithm

In this section, we present a single-pass $(2/5 - \varepsilon)$-approximation algorithm for the problem (1). We first show that, if $c(o_1)$ is at most $K/2$ or more than $2/3K$, then the algorithm in [23] can be used. So we focus on the case when $c(o_1)$ is in $[K/2, 2/3K]$. For this case, we develop two algorithms by combining the technique in Sect. 2.2 into the thresholding algorithm in Sect. 2.1. The first one is useful when $c(o_2)$ is at most $K/3$, while the other is applied when $c(o_2)$ is more than $K/3$. Some proofs are omitted due to the page limitation.

In what follows, we often assume that we know in advance approximations of r_1 and r_2. That is, we are given $\underline{r}_\ell, \overline{r}_\ell$ such that $\underline{r}_\ell \leq r_\ell \leq \overline{r}_\ell$ and $\overline{r}_\ell \leq (1+\varepsilon)\underline{r}_\ell$ for $\ell \in \{1, 2\}$. These values can be guessed from a geometric series of some interval.

3.1   Algorithm When $c(o_1)$ is Small

It is known that when $c(o_1) \leq K/2$, we can improve the thresholding algorithm so that we can find a $(2/5 - \varepsilon)$-approximate solution in $O(K\varepsilon^{-4} \log^4 K)$ space with a single pass.

Theorem 3 ([23]). *Suppose that $c(o_1) \leq K/2$. We can find a $(2/5 - \varepsilon)$-approximate solution with a single pass for the problem (1). The space complexity of the algorithm is $O(K\varepsilon^{-4} \log^4 K)$.*

The above theorem can be extended to the case where we aim to find a set S of items that maximizes $f(S)$ subject to the relaxed constraint that the total size is at most pK, for a given number $p \geq 1$. We say that a set S of items is a (p, α)-approximate solution if $c(S) \leq pK$ and $f(S) \geq \alpha f(\text{OPT})$, where OPT is an optimal solution of the original instance.

Theorem 4 ([23])**.** *For a constant number $p \geq 2r_1$, there is a $\left(p, \frac{2p}{2p+3} - \varepsilon\right)$-approximation streaming algorithm with a single pass. The space complexity of the algorithm is $O(K\varepsilon^{-3} \log^3 K)$.*

With the aid of this algorithm, we can find a $(2/5 - \varepsilon)$-approximate solution for some special cases even when $c(o_1) \geq K/2$.

Corollary 1. *If $c(o_1) > 2/3K$, then we can find a $(2/5-\varepsilon)$-approximate solution with a single pass. The space complexity of the algorithm is $O(K\varepsilon^{-3} \log^3 K)$.*

Corollary 2. *Suppose that $c(o_1) > K/2$. If $f(o_1) \leq 3/10f(\text{OPT})$ then we can find a $(2/5 - \varepsilon)$-approximate solution with a single pass. The space complexity of the algorithm is $O(K\varepsilon^{-3} \log^3 K)$.*

3.2   Algorithm for Small $c(o_2)$

By Theorem 3 and Corollary 1, we may assume that $c(o_1)$ is in $[K/2, 2/3K]$. In this section, we describe a single-pass algorithm that works well when $c(o_2)$ is small.

Suppose that we know in advance the approximate value v of $f(\text{OPT})$, i.e., $v \leq f(\text{OPT}) \leq (1+\varepsilon)v$. This assumption can be removed with dynamic update technique using $O(\varepsilon^{-1} \log K)$ additional space in a similar way to [2,23,39].

In addition, we suppose that we are given θ_1 such that $\theta_1 v/(1 + \varepsilon) \leq f(o_1) \leq \theta_1 v$, which is an approximation of $f(o_1)$. Define $E_1 = \{e \in E \mid c(e) \in [r_1K, \overline{r}_1K], f(e) \in [\theta_1 v/(1+\varepsilon), \theta_1 v]\}$. We can assume that E is the disjoint union of E_1 and $\overline{E_1} = \{e \mid c(e) \leq \overline{r}_2K\}$, as we can discard the other items. We note that $o_1 \in E_1$ and $\text{OPT} - o_1 \subseteq \overline{E_1}$.

We propose a single-pass streaming algorithm, where the target approximation ratio is $\beta = 2/5$. The algorithm description is given in Algorithm 1.

In the algorithm, we basically run the thresholding algorithm for $\overline{E_1}$ to collect a set S of items. In the same pass in parallel, we try to find a subset $X \subseteq E_1$ that contains a good approximation of o_1, based on Lemma 2. That is, when an item e in E_1 arrives, we add e to X if $|X| < t+1$ and $f(e \mid X) \geq (\theta_1 - \lambda|X|)v$. Each time an item e is added to X, since e may be a good approximation of o_1, we create a new feasible set $S + e$, and start to run the thresholding algorithm to $S + e$ in parallel. Thus a feasible set is generated for each e in X, and the family of these feasible sets is maintained as $\mathcal{T}$ in the algorithm. We remark that, to guarantee the approximation ratio of the algorithm starting from $S + e$, we need to satisfy the thresholding condition for e in X as well (Line 7): $f(e \mid S) \geq \alpha \frac{c(e)}{K} v$ for the current set S. Thus the above algorithm performs *dynamically* the thresholding algorithm to $\overline{E_1}$ and $\overline{E_1} + e$ for each $e \in X$.

Algorithm 1.

1: **procedure** Dynamic(v)
2:     $S := \emptyset; \mathcal{T} := \emptyset; X := \emptyset.$
3:     $\alpha := \beta \frac{1}{1-\bar{r}_2}.$
4:     Define t and λ from θ_1 by (4) and (2).
5:     **while** item e is arriving **do**   ▷ First phase
6:         **if** $e \in E_1$ **then**
7:             **if** $|X| < t+1$ and $f(e \mid X) \geq (\theta_1 - \lambda|X|)v$ and $f(e \mid S) \geq \alpha \frac{c(e)}{K}v$ **then**
8:                 $\mathcal{T} := \mathcal{T} \cup \{S+e\}$ and $X := X + e.$
9:         **else**
10:             **if** $f(e \mid S) \geq \alpha \frac{c(e)}{K}v$ and $c(S+e) \leq K$ **then** $S := S + e.$
11:             **for** each $T \in \mathcal{T}$ **do**
12:                 **if** $f(e \mid T) \geq \alpha \frac{c(e)}{K}v$ and $c(T+e) \leq K$ **then** $\mathcal{T} := \mathcal{T} \setminus \{T\} \cup \{T+e\}.$
13:             **if** $c(S) \geq (1 - \bar{r}_1 - \bar{r}_2)K$ **then** $S'_0 := S$ **and break.**
14:     $S' := S'_0.$
15:     **while** item e is arriving **do**   ▷ Second phase
16:         **if** $e \in E_1$ **then**
17:             **if** $f(S') < f(S'_0 + e)$ **then** $S' := S'_0 + e.$
18:         **else**
19:             **if** $f(e \mid S) \geq \alpha \frac{c(e)}{K}v$ and $c(S+e) \leq K$ **then** $S := S + e.$
20:             **for** any $T \in \mathcal{T}$ **do**
21:                 **if** $f(e \mid T) \geq \alpha \frac{c(e)}{K}v$ and $c(T+e) \leq K$ **then** $\mathcal{T} := \mathcal{T} \setminus \{T\} \cup \{T+e\}.$
    return the best one among $\{S, S'\} \cup \mathcal{T}.$

However, the above strategy does not work when the size of S becomes large. Indeed, as we perform the thresholding algorithm to $S + e$ for each $e \in X$, it is necessary that $S+e$ is feasible, that is, $c(S) \leq K - c(e)$ when e arrives. Moreover, since we have the additional condition $f(e \mid S) \geq \alpha \frac{c(e)}{K}v$ to pick an item to X, we may throw away an approximation of o_1 when $f(e \mid S)$ is small (even if Lemma 2 is applicable). To avoid them, we adopt another strategy when $c(S)$ becomes large. Let S'_0 be the set we have the first time when $c(S)$ is at least $(1 - \bar{r}_1 - \bar{r}_2)K$. It follows from Lemma 1 that $f(S'_0)$ is relatively large as (6) below. Moreover, since $c(S)$ is at most $(1 - \bar{r}_1)K$, we can add any item in E_1 to S'_0. In the rest of a stream, we just take one item $e \in E_1$ that maximizes $f(S'_0 + e)$. At the same time, we continue to run the thresholding algorithm to S and every set in $\mathcal{T}$. In the end, the algorithm returns the best one among all the candidates.

Theorem 5. *Suppose that $v \leq f(\text{OPT}) \leq (1+\varepsilon)v$. Then Algorithm* Dynamic(v) *returns a set S such that $c(S) \leq K$ and*

$$f(S) \geq \min\left\{\beta, 1 - \beta\frac{1}{1-\bar{r}_2}, \Gamma(\theta) - \beta\frac{1-r_1}{1-\bar{r}_2}\right\}v - O(\varepsilon)v. \qquad (5)$$

The space complexity is $O(K)$.

Let $\tilde{S}$ be the final set of S, and $\tilde{S}'$ be the output obtained by adding one item in E_1 to S'_0 (Line 17). Let $\tilde{T}_e$ be the final set in $\mathcal{T}$ containing an item $e \in X$.

We remark that the sets $\tilde{S}$ and $\tilde{T}_e$ are obtained by adding an item satisfying the thresholding condition repeatedly. Also, $\tilde{S}'$ and $\tilde{T}_e$ contain exactly one item e in E_1.

It is not difficult to see that all the obtained sets are of size at most K. We note that $c(S_0') < (1 - \bar{r}_1)K$, since S_0' is the set the first time the size exceeds $(1 - \bar{r}_1 - \bar{r}_2)K$ by adding an item of size at most $\bar{r}_2 K$.

In the rest of this subsection, we will show (5). We first claim that, by Lemma 1, we have

$$f(S_0') \geq \alpha(1 - \bar{r}_1 - \bar{r}_2)v, \tag{6}$$

since $c(S_0') \geq (1 - \bar{r}_1 - \bar{r}_2)K$. Let $\tilde{X}$ be the final set of X.

Lemma 3. *At the end of the algorithm, one of the following holds.*

- *There exists an item $e \in \tilde{X}$ such that $f(\mathrm{OPT} - o_1 + e) \geq \Gamma(\theta)v - O(\varepsilon)v$.*
- *It holds that $f(o_1 \mid \tilde{S}) < \alpha \bar{r}_1 v$.*
- *It holds that $f(\tilde{S}') \geq \beta v$.*

Proof. Suppose that o_1 arrives during the first while-loop. Let $X = \{e_1, e_2, \ldots, e_x\}$ be the set just before o_1 arrives such that items are sorted in the ordering of the addition. Then X satisfies that $f(e_i \mid \{e_1, e_2, \ldots, e_{i-1}\}) \geq (\theta - \lambda(i - 1))v$ for each $i = 1, \ldots, x$. Note that, when o_1 will be contained in X, clearly the first statement holds. Thus we may assume that o_1 does not satisfy the condition in Line 7, that is, one of the following three conditions holds: $|X| = t + 1$, $f(o_1 \mid X) < (\theta - \lambda|X|)v$, and $f(o_1 \mid S) < \alpha c(o_1)v \leq \alpha \bar{r}_1 v$. It follows from Lemma 2 that, if one of the first two conditions holds, then at least one item $\bar{e} \in X$ satisfies $f(\mathrm{OPT} - o_1 + \bar{e}) \geq \Gamma(\theta)v - O(\varepsilon)v$. If $f(o_1 \mid S) < \alpha \bar{r}_1 v$, then $f(o_1 \mid \tilde{S}) \leq f(o_1 \mid S) < \alpha \bar{r}_1 v$ by submodularity. Thus one of the first two statements of Lemma 3 is satisfied since $X \subseteq \tilde{X}$.

Next suppose that o_1 arrives after constructing S_0', and suppose that $f(\tilde{S}') < \beta v$. From Line 17, we see that $f(S_0' + o_1) \leq f(\tilde{S}') < \beta v$. Hence we have

$$f(o_1 \mid S_0') = f(S_0' + o_1) - f(S_0') < \beta v - f(S_0').$$

By (6), it holds that

$$f(o_1 \mid S_0') < \beta v - \alpha(1 - \bar{r}_1 - \bar{r}_2)v \leq \alpha \bar{r}_1 v = \beta \frac{\bar{r}_1}{1 - \bar{r}_2}v,$$

where we recall $\beta = \alpha(1 - \bar{r}_2)$. Therefore, by submodularity, we obtain $f(o_1 \mid \tilde{S}) \leq f(o_1 \mid S_0') \leq \alpha \bar{r}_1 v$. Thus the lemma follows. □

We then show that, for each case of Lemma 3, the approximation ratio can be bounded as below. Combining all the cases, we can prove Theorem 5.

Lemma 4. *Suppose that there exists $e \in \tilde{X}$ such that $f(\mathrm{OPT} - o_1 + e) \geq \Gamma(\theta)v - O(\varepsilon)v$. Then it holds that*

$$f(\tilde{T}_e) \geq \min\left\{\beta, \Gamma(\theta) - \beta \frac{1 - r_1}{1 - \bar{r}_2}\right\}v - O(\varepsilon)v.$$

Lemma 5. *If $f(o_1 \mid \tilde{S}) < \alpha \bar{r}_1 v$, then we have $f(\tilde{S}) \geq \min\{\beta, 1 - \alpha\}v - O(\varepsilon)v$.*

It turns out from Theorem 5 that the algorithm works well when $c(o_2)$ is small or $f(o_2)$ is small.

Corollary 3. *Suppose that v satisfies $v \leq f(\text{OPT}) \leq (1 + \varepsilon)v$. If $c(o_1) \geq K/2$ and $c(o_2) \leq K/3$, then we find a set S such that $c(S) \leq K$ and $f(S) \geq (2/5 - O(\varepsilon))v$. The space complexity is $O(K\varepsilon^{-3} \log K)$.*

Corollary 4. *Suppose that $2K/3 \geq c(o_1) > K/2$ and $c(o_2) > K/3$. If $f(o_2) < 1/5f(\text{OPT})$, then we can find a set S such that $c(S) \leq K$ and $f(S) \geq (2/5 - O(\varepsilon))f(\text{OPT})$. The space complexity of the algorithm is $O(K\varepsilon^{-4} \log^3 K)$.*

In summary, we have the following theorem, together with the dynamic update technique to guess the approximate value v of $f(\text{OPT})$.

Theorem 6. *If $c(o_1) \geq K/2$ and $c(o_2) \leq K/3$, then we can find a $(2/5 - \varepsilon)$-approximate solution with a single pass. The space complexity is $O(K\varepsilon^{-4} \log^2 K)$.*

3.3   Algorithm for Large $c(o_2)$

In this section, we propose our second algorithm that is efficient when $c(o_2)$ is large. Since $o_1, o_2 \in \text{OPT}$, it is clear that $c(o_1) + c(o_2) \leq K$, and hence $r_2 \leq 1 - r_1$. We here assume that $r_2 \leq 1 - r_1 - \varepsilon$, where the other case when r_2 is too large is easier as in the following lemma. Note that the assumption implies that $\bar{r}_1 + \bar{r}_2 \leq 1$.

Lemma 6. *If $c(o_1) + c(o_2) \geq (1-\varepsilon)K$, then we can find a $(2/5-\varepsilon)$-approximate solution with a single pass using $O(K\varepsilon^{-3} \log^3 K)$ space.*

Similarly to the previous section, we assume that we know in advance the approximate value v of $f(\text{OPT})$, i.e., $v \leq f(\text{OPT}) \leq (1 + \varepsilon)v$. This assumption can be removed using $O(\varepsilon^{-1} \log K)$ additional space. We also assume that we are given θ_ℓ such that $\theta_\ell v/(1 + \varepsilon) \leq f(o_\ell) \leq \theta_\ell v$ for $\ell \in \{1, 2\}$. Define $E_\ell = \{e \in E \mid c(e) \in [\underline{r}_\ell K, \bar{r}_\ell K], f(e) \in [\theta_\ell v/(1+\varepsilon), \theta_\ell v]\}$ for $\ell \in \{1, 2\}$. Then $o_\ell \in E_\ell$ holds. We can assume that E is the union of E_1, E_2 and $\overline{E} = \{e \mid c(e) \leq \bar{r}_2 K\}$, as we can discard the other items.

In the algorithm, we perform the thresholding algorithm and the procedure PickOneItem in Sect. 2.2 *in parallel*. We apply PickOneItem to both E_1 and E_2 to obtain approximations of o_1 and o_2. Then X_ℓ includes an approximation of o_ℓ for $\ell = 1, 2$. While finding X_1 and X_2, we check in Line 11 whether there exists a pair of items each from X_1 and X_2, respectively, whose f-value is more than βv. In parallel, we run the thresholding algorithm with α_ℓ to $\overline{E}$ to obtain a set S_ℓ, where $\alpha_\ell := \frac{\beta}{1 - \bar{r}_\ell}$, for $\ell = 1, 2$. If the output S_ℓ has large size, then Lemma 1 guarantees that $f(S_\ell)$ is large. Otherwise, $c(S_\ell)$ is small, meaning that there is a room for adding an item from X_ℓ. The algorithm returns the set that maximizes $f(S_\ell + e)$ for $e \in X_\ell$ and $\ell = 1, 2$. Intuitively, the algorithm partitions the ground set E into three parts E_1, E_2 and $\overline{E}$, and then it returns the best set that can be obtained from two of the three parts.

Algorithm 2.

1: **procedure** Parallel(v)
2:      $S_\ell := \emptyset; X_\ell := \emptyset$ for $\ell = 1, 2$.
3:      $\alpha_\ell := \frac{\beta}{1-\overline{r}_\ell}$ for $\ell = 1, 2$.
4:      Define t_ℓ and λ_ℓ from θ_ℓ by (4) and (2) for $\ell = 1, 2$.
5:      **while** item e is arriving **do**
6:          **for** each $\ell \in \{1, 2\}$ **do**
7:              **if** $e \in E_\ell$ **then**
8:                  **if** $|X_\ell| < t_\ell + 1$ and $f(e \mid X_\ell) \geq (\theta_\ell - \lambda_\ell|X_\ell|)v$ **then**
9:                      $X_\ell := X_\ell + e$.
10:             **else if** $e \in E_1 \cup E_2 \setminus E_\ell$ **then**
11:                 **if** there exists an item $\bar{e} \in X_\ell$ such that $f(\{\bar{e}, e\}) \geq \beta v$ **then return**
     $\{\bar{e}, e\}$.
12:             **else**
13:                 **if** $f(e \mid S_\ell) \geq \alpha_\ell \frac{c(e)}{K} v$ and $c(S_\ell + e) \leq K$ **then** $S_\ell := S_\ell + e$.
14:         **if** $c(S_\ell) \geq (1 - \overline{r}_\ell)K$ for some $\ell \in \{1, 2\}$ **then return** S_ℓ.
15:         **else** **return** the set that achieves $\max_{\ell \in \{1,2\}, e \in X_\ell} f(S_\ell + e)$.

Theorem 7. *Suppose that $v \leq f(\mathrm{OPT}) \leq (1 + \varepsilon)v$. If $r_1 + r_2 \leq 1 - \varepsilon$, then Algorithm Parallelv returns a set S such that $c(S) \leq K$ and $f(S) \geq \gamma v - O(\varepsilon)v$, where*

$$\gamma = \min \left\{ \beta, \Gamma(\theta_2) + \theta_2 - \beta \frac{2 - 2r_2 - r_1}{1 - \overline{r}_2}, \Gamma(\theta_1) + \theta_1 - \beta \frac{2 - 2r_1 - r_2}{1 - \overline{r}_1} \right\}. \quad (7)$$

The space complexity is $O(K)$.

Let $\tilde{S}_\ell$ ($\ell = 1, 2$) be the final set of S_ℓ in the algorithm. We also denote by $\tilde{X}_\ell$ the final set of X_ℓ. Let $\tilde{S}'_\ell$ be the set that achieves $\max_{e \in \tilde{X}_\ell} f(\tilde{S}_\ell + e)$ for $\ell = 1, 2$. The set $\tilde{S}_\ell$ is obtained by adding an item based on the thresholding condition $f(e \mid S_\ell) \geq \alpha_\ell \frac{c(e)}{K} v$.

In the algorithm, each item in $\overline{E}$ is added to S_1 or S_2 only when it does not exceed the knapsack capacity. Hence $c(\tilde{S}_\ell) \leq K$ for $\ell = 1, 2$. Also clearly $c(\tilde{S}'_\ell) \leq K$ for $\ell = 1, 2$ if $c(\tilde{S}_\ell) \leq (1 - \overline{r}_\ell)K$. On the other hand, if the algorithm terminates in Line 11, then the output has only two items each from E_1 and E_2, and hence the size is at most K since $\overline{r}_1 + \overline{r}_2 \leq 1$ by the assumption. Thus the output of the algorithm is of size at most K.

From now on, we will prove (7). We consider the following two cases separately: the case when o_2 arrives before o_1 and when o_1 arrives before o_2.

Case 1: Suppose that o_2 Arrives Before o_1. We consider the case when $\ell = 2$. We may assume that the algorithm terminates in the end (not in Line 11). Moreover, if $c(\tilde{S}_2) \geq (1 - \overline{r}_2)K$, then $f(\tilde{S}_2) \geq \beta v$ by Lemma 1. Thus we may assume that $c(\tilde{S}_2) < (1 - \overline{r}_2)K$.

Let $X_2 = \{e_1, e_2, \ldots, e_x\}$ be the set collected just before o_2 arrives. Then X_2 satisfies that $f(e_j \mid \{e_1, e_2, \ldots, e_{j-1}\}) \geq (\theta_2 - \lambda_2(j - 1))$ for each $j = 1, \ldots, x$.

When o_1 arrives, we return the set $\{e, o_1\}$ for some $e \in X_2$ if $f(\{o_1, e\}) \geq \beta v$ at Line 11. Thus we may assume that $f(\{o_1, e\}) < \beta v$ for any $e \in X_2$.

Lemma 7. *Suppose that $c(\tilde{S}_2) < (1 - \bar{r}_2)K$ and that, for any $e \in X_2$, we have $f(\{o_1, e\}) < \beta v$. There exists an item $e \in X_2$ such that*

$$f(\tilde{S}_2 + e) \geq \Gamma(\theta_2)v + \theta_2 v - \beta \frac{2 - 2r_2 - r_1}{1 - \bar{r}_2} v.$$

Proof. By Lemma 2, we have $e \in X_2$ such that $f(\text{OPT} - o_2 + e) \geq \Gamma(\theta_2)v - O(\varepsilon)v$. Since $f(\{o_1, e\}) < \beta v$ and $f(e) \geq \theta_2 v$, we see that $f(o_1 \mid e) = f(\{o_1, e\}) - f(e) < (\beta - \theta_2)v$. It then holds by submodularity that

$$f(\text{OPT} - o_2 + e) \leq f(e) + f(o_1 \mid e) + f(\text{OPT} - o_1 - o_2 \mid e)$$
$$\leq (\beta - \theta_2)v + f(\text{OPT} - o_1 - o_2 + e).$$

Hence, since $f(\text{OPT} - o_2 + e) \geq \Gamma(\theta_2)v - O(\varepsilon)v$,

$$\Gamma(\theta_2)v - (\beta - \theta_2)v - O(\varepsilon)v \leq f(\text{OPT} - o_1 - o_2 + e).$$

On the other hand, it follows from submodularity that

$$f(\text{OPT} - o_1 - o_2 + e) \leq f(\tilde{S}_2 + e) + f(\text{OPT} - o_1 - o_2 - \tilde{S}_2 \mid \tilde{S}_2 + e)$$
$$\leq f(\tilde{S}_2 + e) + f(\text{OPT} - o_1 - o_2 - \tilde{S}_2 \mid \tilde{S}_2)$$
$$\leq f(\tilde{S}_2 + e) + \alpha_2(1 - r_1 - r_2)v,$$

where the last inequality follows from the fact that, since $c(\tilde{S}_2) \leq (1 - c(o_2))K$, any item $o \in \text{OPT} - o_1 - o_2 - \tilde{S}_2$ is discarded due to the thresholding condition, implying $f(o \mid \tilde{S}_2) \leq \alpha_2 c(o)v$. Combining them, we obtain

$$f(\tilde{S}_2 + e) \geq (\Gamma(\theta_2) - (\beta - \theta_2) - \alpha(1 - r_1 - r_2)) v - O(\varepsilon)v$$
$$\geq \left(\Gamma(\theta_2) + \theta_2 - \beta \frac{2 - 2r_2 - r_1}{1 - \bar{r}_2} \right) v - O(\varepsilon)v.$$

$\square$

Case 2: Suppose that o_1 Arrives Before o_2. Let X_1 be the set just before o_1 arrives. We can use the symmetrical argument to Case 1. We omit the proof.

Lemma 8. *Suppose that $c(\tilde{S}_1) < (1 - \bar{r}_1)K$ and that, for any $e \in X_1$, we have $f(\{o_2, e\}) < \beta v$. There exists an item $e \in X_1$ such that*

$$f(\tilde{S}_1 + e) \geq \Gamma(\theta_1)v + \theta_1 v - \beta \frac{2 - 2r_1 - r_2}{1 - \bar{r}_1} v.$$

Combining the above two lemmas, we have Theorem 7. It follows from Theorem 7 that the algorithm admits a $(2/5 - \varepsilon)$-approximation when $r_2 \geq 1/3$.

Corollary 5. *Suppose that $v \leq f(\mathrm{OPT}) \leq (1 + \varepsilon)v$. Consider the case when $K/2 < c(o_1) \leq 2/3K$ and $K/3 \leq c(o_2) < (1 - r_1 - \varepsilon)K$. We can find a set S such that $c(S) \leq K$ and $f(S) \geq (2/5 - O(\varepsilon))v$. The space complexity of the algorithm is $O(K\varepsilon^{-2})$.*

Theorem 8. *If $c(o_1) \geq K/2$ and $K/3 \leq c(o_2) < (1 - r_1 - \varepsilon)K$, then we can find a $(2/5 - \varepsilon)$-approximate solution with a single pass using $O(K\varepsilon^{-3} \log K)$ space.*

Theorem 1 follows from Theorem 6, Lemma 6, and Theorem 8.

References

1. Alon, N., Gamzu, I., Tennenholtz, M.: Optimizing budget allocation among channels and influencers. In: WWW, pp. 381–388 (2012)
2. Badanidiyuru, A., Mirzasoleiman, B., Karbasi, A., Krause, A.: Streaming submodular maximization: massive data summarization on the fly. In: KDD, pp. 671–680 (2014)
3. Badanidiyuru, A., Vondrák, J.: Fast algorithms for maximizing submodular functions. In: SODA, pp. 1497–1514 (2013)
4. Balkanski, E., Rubinstein, A., Singer, Y.: An exponential speedup in parallel running time for submodular maximization without loss in approximation. In: SODA, pp. 283–302 (2019)
5. Balkanski, E., Singer, Y.: The adaptive complexity of maximizing a submodular function. In: STOC, STOC 2018, pp. 1138–1151 (2018)
6. Barbosa, R.D.P., Ene, A., Nguyen, H.L., Ward, J.: A new framework for distributed submodular maximization. In: FOCS, pp. 645–654 (2016)
7. Barbosa, R., Ene, A., Le Nguyen, H., Ward, J.: The power of randomization: Distributed submodular maximization on massive datasets. In: ICML, ICML 2015, pp. 1236–1244 (2015). JMLR.org
8. Bateni, M., Esfandiari, H., Mirrokni, V.: Almost optimal streaming algorithms for coverage problems. In: SPAA, SPAA 2017, pp. 13–23. ACM, New York (2017)
9. Calinescu, G., Chekuri, C., Pál, M., Vondrák, J.: Maximizing a monotone submodular function subject to a matroid constraint. SIAM J. Comput. **40**(6), 1740–1766 (2011)
10. Chakrabarti, A., Kale, S.: Submodular maximization meets streaming: matchings, matroids, and more. Math. Program. **154**(1–2), 225–247 (2015)
11. Chan, T.H.H., Huang, Z., Jiang, S.H.C., Kang, N., Tang, Z.G.: Online submodular maximization with free disposal: Randomization beats for partition matroids online. In: SODA, pp. 1204–1223 (2017)
12. Chan, T.H.H., Jiang, S.H.C., Tang, Z.G., Wu, X.: Online submodular maximization problem with vector packing constraint. In: ESA (2017)
13. Chekuri, C., Gupta, S., Quanrud, K.: Streaming algorithms for submodular function maximization. In: Halldórsson, M.M., Iwama, K., Kobayashi, N., Speckmann, B. (eds.) ICALP 2015. LNCS, vol. 9134, pp. 318–330. Springer, Heidelberg (2015). https://doi.org/10.1007/978-3-662-47672-7_26
14. Chekuri, C., Quanrud, K.: Submodular function maximization in parallel via the multilinear relaxation. In: SODA, pp. 303–322 (2019)
15. Chekuri, C., Vondrák, J., Zenklusen, R.: Submodular function maximization via the multilinear relaxation and contention resolution schemes. SIAM J. Comput. **43**(6), 1831–1879 (2014)

16. Ene, A., Nguyễn, H.L.: A nearly-linear time algorithm for submodular maximization with a knapsack constraint. In: ICALP (2019)
17. Ene, A., Nguyễn, H.L.: Submodular maximization with nearly-optimal approximation and adaptivity in nearly-linear time. In: SODA, pp. 274–282 (2019)
18. Feldman, M., Karbasi, A., Kazemi, E.: Do less, get more: streaming submodular maximization with subsampling. In: NeurIPS 2018, pp. 730–740 (2018)
19. Filmus, Y., Ward, J.: A tight combinatorial algorithm for submodular maximization subject to a matroid constraint. SIAM J. Comput. **43**(2), 514–542 (2014)
20. Fisher, M.L., Nemhauser, G.L., Wolsey, L.A.: An analysis of approximations for maximizing submodular set functions I. Math. Program. **14**, 265–294 (1978)
21. Fisher, M.L., Nemhauser, G.L., Wolsey, L.A.: An analysis of approximations for maximizing submodular set functions II. Math. Program. Study **8**, 73–87 (1978)
22. Huang, C., Kakimura, N.: Multi-pass streaming algorithms for monotone submodular function maximization (2018). arXiv http://arxiv.org/abs/1802.06212
23. Huang, C.C., Kakimura, N., Yoshida, Y.: Streaming algorithms for maximizing monotone submodular functions under a knapsack constraint. In: APPROX (2017)
24. Kempe, D., Kleinberg, J., Tardos, É.: Maximizing the spread of influence through a social network. In: KDD, pp. 137–146 (2003)
25. Krause, A., Singh, A.P., Guestrin, C.: Near-optimal sensor placements in gaussian processes: theory, efficient algorithms and empirical studies. J. Mach. Learn. Res. **9**, 235–284 (2008)
26. Kulik, A., Shachnai, H., Tamir, T.: Maximizing submodular set functions subject to multiple linear constraints. In: SODA, pp. 545–554 (2013)
27. Kumar, R., Moseley, B., Vassilvitskii, S., Vattani, A.: Fast greedy algorithms in mapreduce and streaming. ACM Trans. Parallel Comput. **2**(3), 14:1–14:22 (2015)
28. Lee, J.: Maximum Entropy Sampling, Encyclopedia of Environmetrics, vol. 3, pp. 1229–1234. Wiley (2006)
29. Lee, J., Sviridenko, M., Vondrák, J.: Submodular maximization over multiple matroids via generalized exchange properties. Math. Oper. Res. **35**(4), 795–806 (2010)
30. Lin, H., Bilmes, J.: Multi-document summarization via budgeted maximization of submodular functions. In: NAACL-HLT, pp. 912–920 (2010)
31. Lin, H., Bilmes, J.: A class of submodular functions for document summarization. In: ACL-HLT, pp. 510–520 (2011)
32. McGregor, A.: Graph stream algorithms: a survey. SIGMOD Rec. **43**(1), 9–20 (2014)
33. McGregor, A., Vu, H.T.: Better streaming algorithms for the maximum coverage problem. In: ICDT (2017)
34. Mirzasoleiman, B., Jegelka, S., Krause, A.: Streaming non-monotone submodular maximization: personalized video summarization on the fly. In: AAAI (2018)
35. Soma, T., Kakimura, N., Inaba, K., Kawarabayashi, K.: Optimal budget allocation: theoretical guarantee and efficient algorithm. In: ICML, pp. 351–359 (2014)
36. Sviridenko, M.: A note on maximizing a submodular set function subject to a knapsack constraint. Oper. Res. Lett. **32**(1), 41–43 (2004)
37. Wolsey, L.: Maximising real-valued submodular functions: primal and dual heuristics for location problems. Math. Oper. Res. **7**, 410–425 (1982)
38. Yoshida, Y.: Maximizing a monotone submodular function with a bounded curvature under a knapsack constraint (2016). https://arxiv.org/abs/1607.04527
39. Yu, Q., Xu, E.L., Cui, S.: Streaming algorithms for news and scientific literature recommendation: submodular maximization with a d-knapsack constraint. In: IEEE Global Conference on Signal and Information Processing (2016)

Inventory Routing Problem with Facility Location

Yang Jiao[✉][iD] and R. Ravi[iD]

Tepper School of Business, Carnegie Mellon University,
5000 Forbes Ave, Pittsburgh, PA 15213, USA
{yangjiao,ravi}@andrew.cmu.edu

Abstract. We study problems that integrate depot location decisions
along with the inventory routing problem of serving clients from these
depots over time balancing the costs of routing vehicles from the depots
with the holding costs of demand delivered before they are due. Since
the inventory routing problem is already complex, we study the version that assumes that the daily vehicle routes are direct connections
from the depot thus forming stars as solutions, and call this problem
the Star Inventory Routing Problem with Facility Location (SIRPFL).
As a stepping stone to solving SIRPFL, we first study the Inventory
Access Problem (IAP), which is the single depot, single client special
case of IRP. The Uncapacitated IAP is known to have a polynomial time
dynamic program. We provide an NP-hardness reduction for Capacitated IAP where each demand cannot be split among different trips. We
give a 3-approximation for the case when demands can be split and a
6-approximation for the unsplittable case. For Uncapacitated SIRPFL,
we provide a 12-approximation by rounding an LP relaxation. Combining the ideas from Capacitated IAP and Uncapacitated SIRPFL, we
obtain a 24-approximation for Capacitated Splittable SIRPFL and a 48-
approximation for the most general version, the Capacitated Unsplittable
SIRPFL.

Keywords: Inventory routing problem · Facility Location ·
Approximation algorithms

1 Introduction

We initiate the integrated study of facility opening and inventory routing problems. Facility location has many applications such as the placement of factories,
warehouses, service centers, etc. The facility location problem involves selecting
a subset of locations to open facilities to serve demands minimizing the facility
opening costs plus connection costs between demands and the opened locations.

This material is based upon research supported in part by the U. S. Office of Naval
Research under award number N00014-18-1-2099, and the U. S. National Science Foundation under award number CCF-1527032.

© Springer Nature Switzerland AG 2019
Z. Friggstad et al. (Eds.): WADS 2019, LNCS 11646, pp. 452–465, 2019.
https://doi.org/10.1007/978-3-030-24766-9_33

Inventory routing arises from Vendor Managed Inventory systems in which a product supplier and its retailers cooperate in the inventory planning. First, the retailers share with the supplier the demand patterns for its product and the storage costs for keeping early deliveries per retailer location. Then the supplier is responsible for planning a delivery schedule that serves all the demands on time. The inventory routing problem (IRP) trades off visits from fixed depots over a planning horizon to satisfy deterministic daily demands at clients to minimize routing costs plus the holding costs of demand delivered before they are due at clients. We integrate the decision of which depots to open in the problem and study the joint problem of opening depots (given opening costs), and using these depots to minimize the total inventory routing costs, i.e. the sum of the routing costs from these depots and the holding costs at clients.

The IRP has been challenging to study by itself from an approximation perspective: constant-factor approximations are known only in very structured metrics like trees [3] or when the routes are periodic [5]. Hence we simplify the routing considerably to gain a better understanding of the integrated problem. In particular, we assume that the visits from each client go to the closest opened depots via a direct edge: the routing solution is thus a collection of stars rather than the Steiner trees or tours considered in the original IRP. We call this simplified variant of IRP the Star IRP or SIRP for short.

1.1 Problem Definitions

The *Star Inventory Routing Problem with Facility Location (SIRPFL)* is inventory routing with the extra choice to build depots at a subset of the locations for additional costs before the first day, which then can be used to route deliveries throughout the entire time horizon. Formally, we are given an undirected graph $G = (V, E)$ with edge weights w_e, a time horizon $1, \ldots, T$, a set D of demand points (v, t) with d_t^v units of demand due by day t, facility opening costs f_v for vertex v, holding costs $h_{s,t}^v$ per unit of demand delivered on day s serving (v, t). The objective is to open a set $F \in V$ of facilities that can be used throughout the entire time horizon, determine the set of demands to serve/visit per day, and connect any visited clients to opened facilities per day so that the total cost from facility openings, client-facility connections, and storage costs for early deliveries is minimized. Three natural variants of the problem arise based on whether the delivery vehicles are uncapacitated, and if not, whether or not any single day's demand can be split among different visits. We call the first variant the *Uncapacitated* version. For the *Capacitated* version, we assume all vehicles have a fixed capacity U and arrive at two variants: *Unsplittable* where every daily demand is satisfied wholly in one visit and the *Splittable* where it can be split across multiple visits (even across multiple days). We assume that any single demand never exceeds the capacity of the vehicle so that the splittable problem is always feasible.

Once the facility decisions are fixed, the resulting SIRP instances can be decomposed across the clients due to the assumption that the visits are direct edges from the client to an open facility. Thus, for each client, the routing solution is the direct edge to the closest open facility, and the only decisions are the delivery days and in the capacitated case, the number of trips on such days. We call the single-depot single-client problem the *Inventory Access Problem (IAP)*. Even the simple IAP has the three variants alluded to above.

1.2 Contributions

1. We initiate the study of inventory routing problems integrated with Facility Location (IRPFL) and supply the first complexity and approximation results.
2. For the simpler Inventory Access Problem, we show that the unsplittable capacitated case is already weakly NP-hard. The uncapacitated problem is a single-item lot-sizing problem, for which a polynomial time exact solution exists [20]. For the latter and its splittable counterpart, we give constant approximation algorithms using LP rounding.
3. For the Star versions of the IRPFL we consider, we give constant-factor approximation for all three versions by deterministically rounding new linear programming relaxations for the problems. The table below summarizes the approximation guarantees.

	IAP	SIRPFL
Uncapacitated	polynomial time [20]	12-approx
Capacitated Splittable	3-approx	24-approx
Capacitated Unsplittable	NP-hard, 6-approx	48-approx

4. Our algorithms need to modify and adapt current facility location LP rounding methods, since none of the variables in the objective function can directly be used for the rounding methods. These methods may be useful in future work involving time-indexed formulations integrating network design and facility location.

We review related work in the next section. We then present a complete description of our 12-approximation for the Uncapacitated SIRPFL in Sect. 3. To handle the capacitated versions, we need to strengthen the LP relaxation to better bound multiple visits per day: we illustrate this using the simpler example of the Capacitated Splittable IAP by providing a 3-approximation in Sect. 4. Building on the 3-approximation, we show a 6-approximation for the Capacitated Unsplittable IAP. Finally, Sect. 5 summarizes the results and open problems. Details of the other results are in the full version of this paper [11].

2 Related Work

UFL: The first constant approximation for Uncapacitated Facility Location was a 3.16-approximation by Shmoys et el. [19] using the filtering method of Lin and Vitter [15]. Various LP-based methods made further improvements [4,9,10]. More recently, Li gave a 1.488-approximation [14].

IRP: Without facility opening decisions, IRP itself on general metrics has an $O(\frac{\log T}{\log \log T})$-approximation by Nagarajan and Shi [16] and an $O(\log N)$-approximation by Fukunaga et al. [5]. For variants of periodic IRP, Fukunaga et al. [5] provide constant approximations. IRP on tree metrics have a constant approximation [3]. Another special case of IRP is the joint replenishment problem (JRP), which has also been extensively studied [1,2,12,13,17].

TreeIRPFL: Another related problem is the *Tree IRPFL*, which has the same requirements except that the connected components for the daily visits are trees (instead of tours in the regular IRP, or stars in the Star IRP version we study). Tree IRPFL differs from Star IRPFL by allowing savings in connection costs by connecting clients through various other clients who are connected to an opened facility.

Single-day variants of Tree IRPFL have been studied extensively. In these problems, there is no holding cost component and thus they trade off the facility location placements with the routing costs from these facilities. We use ρ_Π to denote the best existing approximation ratio for problem Π. For uncapacitated single-day Tree IRPFL, the problem can directly be modeled as a single Steiner tree problem: attach a new root node with edges to each facility of cost equal to its opening cost; finding a Steiner tree from this root to all the clients gives the required solution. Thus, this problem has a ρ_{ST}-approximation algorithm. If clients are given in groups such that only one client per group needs to be served, Glicksman and Penn [6] generalize the Steiner tree approximation method of Goemans and Williamson [7] to $(2 - \frac{1}{|V|-1})L$-approximation, where L is the largest size of a group. For the capacitated single-day case of Tree IRPFL, Harks et al. [8] provide a 4.38-approximation. They also give constant approximations for the prize collecting variant and a cross-docking variant. For the group version of the problem, Harks and König show a $4.38L$-approximation.

Integrated Logistics: Ravi and Sinha [18] originated the study of more general integrated logistics problems, and give a $(\rho_{ST} + \rho_{UFL})$-approximation for a generalization of the capacitated single-day Tree IRPFL called Capacitated-Cable Facility Location (CCFL). Here, ST stands for Steiner Tree and UFL stands for Uncapacitated Facility Location. In CCFL, the amount of demand delivered through each edge must be supported by building enough copies of cables on the edge. They give a bicriteria $(\rho_{k-\text{MEDIAN}} + 2)$-approximation opening $2k$ depots for the k-median version of the CCFL, which allows k depots to be located at no cost.

3   Uncapacitated SIRPFL

In this section, we give a constant approximation for Uncapacitated SIRPFL. First, we state the LP formulation for Uncapacitated SIRPFL. Let z_v indicate whether a facility at v is opened, y_s^{uv} indicate whether edge uv is built on day s, y_{st}^{uv} indicate whether to deliver the demand of (v,t) on day s from facility u, and $x_{s,t}^v$ indicate whether demand point (v,t) is served on day s. Then Uncapacitated SIRPFL has the following LP relaxation. To simplify notation, define $H_{s,t}^v = d_t^v h_{s,t}^v$, i.e., $H_{s,t}^v$ is holding cost of storing all of the demand for demand point (v,t) from day s to day t.

$$\min \quad \sum_{v \in V} f_v z_v + \sum_{s \leq T} \sum_{e \in E} w_e y_s^e + \sum_{(v,t) \in D} \sum_{s \leq t} H_{s,t}^v x_{s,t}^v$$

$$\text{s.t.} \quad \sum_{s \leq t} x_{s,t}^v \geq 1 \qquad \forall (v,t) \in D \tag{1}$$

$$\sum_{u \in V} y_{st}^{uv} \geq x_{s,t}^v \qquad \forall (v,t) \in D, s \leq t \tag{2}$$

$$z_u \geq \sum_{s=1}^{T} y_{st}^{uv} \qquad \forall (v,t) \in D, u \in V \tag{3}$$

$$y_s^{uv} \geq y_{st}^{uv} \qquad \forall (v,t) \in D, u \in V, s \leq t \tag{4}$$

$$z_u \geq y_s^{uv} \qquad \forall u, v \in V, s \leq T \tag{5}$$

$$\sum_{u \in V} \sum_{s=s'}^{t_2} y_{st_2}^{uv} \geq \sum_{u \in V} \sum_{s=s'}^{t_2} y_{st_1}^{uv} \qquad \forall v \in V, t_2 > t_1 \geq s' \tag{6}$$

$$z_u, y_r^e, y_{l,m}^a x_{s,t}^v \geq 0 \qquad \forall u, v \in V, e, a \in E, r, m, t \leq T, l \leq m, s \leq t. \tag{7}$$

Constraint 1 requires that every demand point is served by its deadline. Constraint 2 enforces that v gets connected to some facility on day s if (v,t) is served on day s. Constraint 3 ensures that facility u is open if u is assigned to any demand point over the time horizon. Constraint 4 ensures that whenever (v,t) is served on day s from u, an edge between u and v must be built on day s. Constraint 5 ensures that whenever some client v is connected to u on some day s, a facility must be built at u. Constraint 6 is valid for optimal solutions since for any v, if there is a service to (v,t_1) within $[s',t_1]$ and $t_1 < t_2$, then the service to t_2 is either on the same day or later, i.e., there must be a service to (v,t_2) within $[s',t_2]$. Here we are using the property that in an optimal solution the demands from a client over time are served in order without loss of generality, which is a consequence of the monotonicity of the unit holding costs at any location.

Using the above LP formulation, we provide an LP rounding algorithm. Before stating the algorithm, we define the necessary notation. First, let (x,y,z) be an optimal LP solution. Let $f(x,y,z)$, $r(x,y,z)$, and $h(x,y,z)$ denote the facility cost, routing cost, and holding cost of (x,y,z) respectively. Define $s_{v,t}$ to be the latest day s^* such that $\sum_{u \in V} \sum_{s=s^*}^{t} y_{st}^{uv} \geq \frac{1}{2}$.

The key idea is to apportion the visit variable y_{uv}^s at day s to different demand days t that it serves using the additional variable y_{uv}^s. The latter variables for any demand at node v on day t provide a stronger lower bound, via Constraint 3, on how much facility must be installed at node u than any lower bound from

y_s^{uv} alone. Constraint 3 is a crucial component in the proof of Lemma 1, which ultimately allows us to bound the facility cost.

Ideally, we would like to use $s_{v,t}$ to bound the holding cost incurred when serving (v,t) on day $s_{v,t}$. However, to avoid high routing costs, not all demands will get to be served by the desired $s_{v,t}$. Instead, for each client v, an appropriately chosen subset of $\{s_{v,t} : t \leq T\}$ will be selected to be the days that have service to v. To determine facility openings and client-facility connections, the idea is to pick "balls" that gather enough density of z_u values so that the cheapest facility within it can be paid for by the facility cost part of the LP objective. To be able to bound the routing cost, we would like to pick the radii of the balls based on the amount of y_s^{uv} values available from the LP solution. However, y_s^{uv} by itself does not give a good enough lower bound for z_u. So we will carefully assign disjoint portions of y_s^{uv} to y_{st}^{uv} for different t's. In this way, we use y_{st}^{uv} to bound the facility cost, and the disjoint portions of y^{uv} to pay for the routing cost. With these goals in mind, we now formally define the visit days and the radius for each client.

Fix a client v. The set A_v of demand days t that v gets visited on their $s_{v,t}$ will be assigned based on collecting enough y_{st}^{uv} over u and s. We call the days in A_v *anchors* of v. Denote by t_{L_v} the latest day that has positive demand at v. We use S_v to keep track of the service days for the anchors.

Algorithm 1. Visits for v

1: Initialize $A_v \leftarrow \{t_{L_v}\}$.
2: Initialize $S_v \leftarrow \{s_{v,t_{L_v}}\}$.
3: Denote by $\tilde{t}$ the earliest anchor in A_v.
4: **while** there is a positive unserved demand at v on some day before $\tilde{t}$ **do**
5:     Denote by t the latest day before $\tilde{t}$ with positive demand at (v,t).
6:     **if** $t \geq s_{v,\tilde{t}}$ **then**
7:         Serve (v,t) on day $s_{v,\tilde{t}}$.
8:     **else**
9:         Update $A_v \leftarrow A_v \cup \{t\}$.
10:        Update $S_v \leftarrow S_v \cup \{s_{v,t}\}$.
11:        Update $\tilde{t} \leftarrow t$.
12:    **end if**
13: **end while**
14: Output the visit set S_v for v.

Define $W_{v,t} = \sum_{u \in V} \sum_{s=s_{v,t}}^{t} w_{uv} y_{st}^{uv}$. Let $W_v = \min_{t \in A_v} W_{v,t}$. Finally, define $B_v = \{u \in V : w_{uv} \leq 4W_v\}$, which is a ball of radius $4W_v$ centered at v. For ball B_v, let $F_v = \arg\min_{q \in B_v} f_q$. Simply, F_v is a location in B_v with the lowest facility cost. Now we are ready to state the algorithm for opening facilities in Algorithm 2.

Denote by $B_{v_1}, \ldots, B_{v_l}$ the balls picked into $\mathcal{B}$ by Algorithm 2.

Proposition 1. *The holding cost of the solution from the algorithm is at most* $2h(x,y,z)$.

Algorithm 2. 12-approximation for Uncapacitated SIRPFL

1: $\mathcal{B} \leftarrow \emptyset$
2: **while** there is any ball B_v disjoint from all balls in $\mathcal{B}$ **do**
3:     Add to $\mathcal{B}$ the ball B_{v_i} of smallest radius
4: **end while**
5: Within each ball B_{v_i}, open a facility at F_{v_i}.
6: Assign each client v to the closest opened facility $u(v)$.
7: For each v, serve it on all days in S_v by building an edge from facility $u(v)$ to v per day $s \in S_v$.

Proof. For each demand point (v, t), we will charge a disjoint part of twice the x values in the LP solution to pay for the holding cost. In particular, to pay for the holding cost incurred by (v, t), we charge $\sum_{s=1}^{s_{v,t}} H_{s,t}^v x_{s,t}^v$ part of the LP solution. We consider two cases: $t \in A_v$ and $t \notin A_v$.

1. In this case, assume that $t \in A_v$. Then (v, t) is served on day $s_{v,t}$, and incurs a holding cost of $H_{s_{v,t},t}^v$. By definition of $s_{v,t}$, we have $\sum_{u \in V} \sum_{s=s_{v,t}+1}^{t} y_{st}^{uv} < \frac{1}{2}$. Then

$$\sum_{s=1}^{s_{v,t}} x_{s,t}^v \geq 1 - \sum_{s=s_{v,t}+1}^{t} x_{s,t}^v \geq 1 - \sum_{s=s_{v,t}+1}^{t} \sum_{u \in V} y_{st}^{uv} > 1 - \frac{1}{2} = \frac{1}{2}.$$

So our budget of $\sum_{s=1}^{s_{v,t}} H_{s,t}^v x_{s,t}^v$ is at least $H_{s_{v,t},t}^v \sum_{s=1}^{s_{v,t}} x_{s,t}^v \geq \frac{H_{s_{v,t},t}^v}{2}$.

2. In this case, assume that $t \notin A_v$. Let $\tilde{t}$ be the earliest anchor after t. Since t is not an anchor, $[s_{v,t}, t]$ must have overlapped $[s_{v,\tilde{t}}, \tilde{t}]$. So $s_{v,\tilde{t}} \leq t$. So (v, t) is served on $s_{v,\tilde{t}}$. By constraint 6, we have $s_{v,t} \leq s_{v,\tilde{t}}$. By monotonicity of holding cost, the holding cost incurred by serving (v, t) on $s_{v,\tilde{t}}$ is at most $H_{s_{v,t},t}^v \leq 2\sum_{s=1}^{s_{v,t}} H_{s,t}^v x_{s,t}^v$.

Proposition 2. *The routing cost of the solution from the algorithm is at most* $12r(x, y, z)$.

Proof. We will charge a disjoint portion of 12 times the y values in the LP solution to pay for the routing cost. Note that only anchors cause new visit days to be created in the algorithm. So consider a demand point (v, t) such that t is an anchor for v.

1. First, consider the case that $v \in \{v_1, \ldots, v_l\}$, the set of vertices for whose balls were picked in $\mathcal{B}$ in Algorithm 2. Then the routing cost to connect (v, t) to the nearest opened facility is

$$w_{F_v,v} \leq 4W_v \leq 4W_{v,t} (\text{by definition of } W_v)$$

$$\leq 4 \sum_{u \in V} \sum_{s=s_{v,t}}^{t} w_{uv} y_s^{uv} (\text{by constraint 4}).$$

Since it is within 4 times the LP budget, the desired claim holds.

2. Now, assume that $v \notin \{v_1, \ldots, v_l\}$. Then B_v overlaps $B_{v'}$ for some v' of smaller radius than B_v (otherwise B_v would have been chosen into $\mathcal{B}$ instead of the larger balls that overlap B_v). Then the edge built to serve (v, t) connects $F_{v'}$ to v. So the routing cost to serve (v, t) is

$$w_{F_{v'},v} \leq W_{v,v'} + W_{v',F_{v'}}$$

$$\leq 2 \cdot 4W_v + 4W_v \text{ (since radius of } B_v \text{ is at least radius of } B_{v'})$$

$$\leq 12W_v \leq 12W_{v,t} \leq 12 \sum_{u \in V} \sum_{s=s_{v,t}}^{t} w_{uv} y_s^{uv} \text{ (by constraint 4)}.$$

Observe that for every v and any two anchors t_1, t_2 for v, we have $[s_{v,t_1}, t_1] \cap [s_{v,t_2}, t_2] = \emptyset$ by the construction of anchors in Algorithm 1. So each y_s^{uv} is charged at most once among all demands whose deadline correspond to anchors.

Before bounding the facility costs, we show a Lemma that will help prove the desired bound.

Lemma 1. *For all* $i \in \{1, \ldots, l\}$, *we have* $\sum_{v \in B_{v_i}} z_v \geq \frac{1}{4}$.

Proof. Suppose there is some $i \in \{1, \ldots, l\}$ such that $\sum_{u \in B_{v_i}} z_u < \frac{1}{4}$. Let $\hat{t} = \arg\min_t W_{v_i,t}$. Then

$$W_{v_i} = W_{v,\hat{t}} = \sum_{u \in V} \sum_{s=s_{v,t}}^{\hat{t}} w_{uv} y_{s\hat{t}}^{uv} \geq \sum_{u \notin B_{v_i}} \sum_{s=s_{v,t}}^{\hat{t}} w_{uv} y_{s\hat{t}}^{uv}$$

$$\geq 4W_v \sum_{u \notin B_{v_i}} \sum_{s=s_{v,t}}^{\hat{t}} y_{s\hat{t}}^{uv} \text{ (since } u \notin B_{v_i})$$

$$\geq 4W_v \left(\sum_{u \in V} \sum_{s=s_{v,t}}^{\hat{t}} y_{s\hat{t}}^{uv} - \sum_{u \in B_{v_i}} \sum_{s=s_{v,t}}^{\hat{t}} y_{s\hat{t}}^{uv} \right)$$

$$\geq 4W_v \left(\frac{1}{2} - \sum_{u \in B_{v_i}} \sum_{s=s_{v,\hat{t}}}^{\hat{t}} y_{s\hat{t}}^{uv} \right) \text{ (by definition of } s_{v,t})$$

$$\geq 4W_v \left(\frac{1}{2} - \sum_{u \in B_{v_i}} z_u \right) \text{ (by constraint 3)}$$

$$> W_v \text{ (by the supposition } \sum_{u \in B_{v_i}} z_u < \frac{1}{4}, \text{ which leads to a contradiction)}.$$

Proposition 3. *The facility cost of the algorithm's solution is at most* $4f(x, y, z)$.

Proof. We will charge four times the z values of the LP solution to pay for the facilities opened by the algorithm. Since the balls picked by Algorithm 2 are disjoint, we can pay for each facility opened using the LP value in its ball. Consider ball B_{v_i} picked by the algorithm and its cheapest facility F_{v_i}. Then the cost of opening F_{v_i} is at most f_v for all $v \in B_{v_i}$. So the facility cost for F_{v_i} is

$$f_{F_{v_i}} \leq 4 \sum_{v \in B_{v_i}} z_v f_{F_{v_i}} \leq 4 \sum_{v \in B_{v_i}} z_v f_v.$$

The first inequality follows from Lemma 1. The second inequality is due to F_{v_i} being the cheapest facility in the ball.

Since facility, holding and routing costs are bounded within 12 times their respective optimal values, we have the following result.

Theorem 1. *Algorithm 2 is a 12-approximation for Uncapacitated SIRPFL.*

4   Capacitated IAP

Recall that the *Inventory Access Problem (IAP)* is the single client case of the Inventory Routing Problem. The only decision needed is to determine on each day whether to visit the client and how much supply to drop off. In SIRPFL, if we know where to build the facilities, then the best way to connect clients would be to the closest opened facility. So once facility openings are determined, the remaining problem decomposes into solving IAP for every client.

4.1   A 3-Approximation for Capacitated Splittable IAP

Here, we consider *Capacitated Splittable IAP*, in which a single demand is allowed to be served in parts over multiple days. Let W be the distance between the depot and the client. Denote by $h_{s,t}$ the holding cost to store one unit of demand from s to deadline t. The demand with deadline t is denoted by d_t. Recall that U denotes the capacity of the vehicle. We model Capacitated Splittable IAP by the following LP relaxation.

$$\min \quad \sum_{s \leq T} W y_s + \sum_{t \in D} \sum_{s \leq t} h_{s,t} d_t x_{s,t}$$

$$\text{s.t.} \quad \sum_{s \leq t} x_{s,t} \geq 1 \qquad\qquad \forall t \in D \tag{8}$$

$$y_s \geq \sum_{t=s}^{T} \frac{x_{s,t} d_t}{U} \qquad \forall s \leq T \tag{9}$$

$$y_s \geq x_{s,t} \qquad\qquad \forall t \leq T, s \leq t \tag{10}$$

$$x_{s,t} \geq 0 \qquad\qquad \forall t \in D, s \leq t \tag{11}$$

$$y_s \geq 0 \qquad\qquad \forall s \leq T \tag{12}$$

The variable y_s indicates the number of trips on day s. Variable $x_{s,t}$ indicates the fraction of d_t to deliver on day s. Note that the objective only counts the cost of the visit to the client as a single copy of the trip variable y_s reflecting the star constraint (if a return trip needs to be accounted for, we can multiply this term by 2 and all our results generalize easily). Constraint 8 requires that each demand becomes entirely delivered by the due date (possibly split over multiple days). Constraint 9 ensures that the total demand that day s serves do not exceed the total capacity among all trips on day s. Constraint 10 ensures that there is a trip whenever some delivery is made on day s.

Let (x, y) be an optimal LP solution. For convenience of the analysis, let $r(x, y) = \sum_{s \leq T} W y_s$ and $h(x, y) = \sum_{t \in D} \sum_{s \leq t} h_{s,t} d_t x_{s,t}$ denote the routing and holding cost of the solution respectively. We will use the LP values $x_{s,t}$ to determine when to visit the client and which demands to drop per visit. For each $t \in D$, let s_t be the latest day for which $\sum_{s=s_t}^{t} x_{s,t} \geq \frac{1}{2}$. We will keep track of a visit set S of days when visits are scheduled along with an anchor set A consisting of demand days that caused the creation of new visits.

Algorithm 3. Visit Rule for Capacitated Splittable IAP

1: Initialize $A \leftarrow \emptyset$.
2: Initialize $S \leftarrow \emptyset$.
3: **while** there is any unsatisfied demand **do**
4:    Denote by t the unsatisfied demand day with the latest s_t
5:    $A \leftarrow A \cup \{t\}$.
6:    $S \leftarrow S \cup \{s_t\}$.
7:    Satisfy t by dropping off d_t on day s_t.
8:    **for** unsatisfied demand day $\hat{t} \geq s_t$ **do**
9:        satisfy $\hat{t}$ by dropping off $d_{\hat{t}}$ on day s_t.
10:    **end for**
11: **end while**
12: Output the visit set S.

For the analysis, denote by T_s the set of all demand days t such that t was satisfied by s in Algorithm 3.

Proposition 4. *The holding cost of the solution from Algorithm 3 is at most* $2h(x, y)$.

Proof. 1. Assume that $t \in A$. Then t was served on day s_t, i.e., incurs holding cost $h_{s_t, t} d_t$. To pay for the holding cost, we use the following part of the LP.

$$\sum_{s=1}^{s_t} h_{s,t} d_t x_{s,t} \geq h_{s_t, t} d_t \sum_{s=1}^{s_t} x_{s,t} \geq \frac{h_{s_t, t} d_t}{2}.$$

2. Assume that $t \notin A$. Let $\tilde{s}$ be the latest day in S such that $\tilde{s} \leq t$. Then the holding cost incurred by the demand on day t is $h_{\tilde{s}, t} d_t$. By definition of the

chosen visit days S, t was not chosen as anchor because s_t was earlier than $\tilde{s}$. So we pay for the holding cost using

$$\sum_{s=1}^{\tilde{s}} h_{s,t} d_t x_{s,t} \geq \sum_{s=1}^{s_t} h_{s,t} d_t x_{s,t} \ (\text{by} s_t \leq \tilde{s})$$

$$\geq \frac{h_{s_t,t} d_t}{2} \geq \frac{h_{\tilde{s},t} d_t}{2} \ (\text{by monotonicity of holding costs}).$$

Proposition 5. *The routing cost of the solution from Algorithm 3 is at most* $3r(x,y)$.

Proof. For each visit day $s_{\tilde{t}} \in S$, the number of trips made is $\left\lceil \dfrac{\sum_{t \in T_{s_{\tilde{t}}}} d_t}{U} \right\rceil \leq \dfrac{\sum_{t \in T_{s_{\tilde{t}}}} d_t}{U} + 1$. So the total number of trips made is at most $\sum_{\tilde{t} \in A} \left(\dfrac{\sum_{t \in T_{s_{\tilde{t}}}} d_t}{U} + 1 \right) \leq \left(\sum_{\tilde{t} \in A} \dfrac{\sum_{t \in T_{s_{\tilde{t}}}} d_t}{U} \right) + |A|$. We will use 3 copies of $\sum_{s=1}^{T} y_s$ to pay for the routing cost–1 copy to pay for the first term and 2 copies to pay for the second term. The total LP budget for the number of trips is

$$\sum_{s=1}^{T} y_s \geq \frac{\sum_{s=1}^{T} \sum_{t=s}^{T} x_{s,t} d_t}{U} \ (\text{by constraint 9})$$

$$\geq \frac{\sum_{t=1}^{T} \sum_{s=1}^{t} x_{s,t} d_t}{U} \geq \sum_{t=1}^{T} \frac{d_t}{U} \geq \sum_{\tilde{t} \in A} \frac{\sum_{t \in T_{s_{\tilde{t}}}} d_t}{U}.$$

So we can pay for the first term using one copy of the LP budget from all the y variables.

To pay for the second term, we will use constraint 10 instead so that we can use disjoint intervals of y for different anchors. In particular, for anchor $\tilde{t}$, we will charge

$$2 \sum_{s=s_{\tilde{t}}}^{\tilde{t}} y_s \geq \sum_{s=s_{\tilde{t}}}^{\tilde{t}} x_{s,\tilde{t}} \geq 2 \cdot \frac{1}{2} \ (\text{by definition of } s_{\tilde{t}}).$$

By the construction of A, for any $t_1, t_2 \in A$, we have $[s_{t_1}, t_1] \cap [s_{t_2}, t_2] = \emptyset$. So the payment for different anchors use disjoint portions of y. Hence the second term can be paid for within 2 copies of the budget provided by y.

Since both holding and routing costs are bounded within 3 times their respective optimal values, we have the following result.

Theorem 2. *Algorithm 3 is a 3-approximation for the Capacitated Splittable Inventory Access Problem.*

4.2   A 6-Approximation for Capacitated Unsplittable IAP

Here, we show that Capacitated Unsplittable IAP has a $2\alpha_{CSIAP}$-approximation, where α_{CSIAP} is the best approximation factor for Capacitated Splittable IAP.

Proposition 6. *There is a $2\alpha_{CSIAP}$-approximation for Capacitated Unsplittable IAP.*

Proof. Given a Capacitated Unsplittable IAP instance, solve the corresponding Capacitated Splittable IAP instance obtaining a solution (x, y) with approximation factor α_{CSIAP}. To obtain a solution that does not split the demands, we will repack the demands per visit day of (x, y). For each visit day s of the solution (x, y), let D^s be the set of demands assigned to be served on day s by (x, y). Let $D^s_{\leq 1/2} = \{t \in D^s : d_t \leq U/2\}$ and $D^s_{>1/2} = D^s \setminus D^s_{\leq 1/2}$. Denote by $n(s)$ the number of trips on day s in the splittable solution. Note that $n(s) \geq \lceil \frac{\sum_{t \in D^s} d_t}{U} \rceil$.

For each trip, for each demand in $D^s_{>1/2}$, give each demand its own trip. Then, fill all demands of $D^s_{\leq 1/2}$ (without splitting) greedily into the previous trips and new ones as long as the capacity is not exceeded. This means that all trips involving demands in $D^s_{\leq 1/2}$, except for possibly one trip, will be filled to strictly more than half the capacity. Let $n'(s)$ be the number of trips in the unsplittable solution thus obtained. If there are no trips of more than half the capacity, then $n'(s) = 1 = n(s)$. Otherwise, the total sum of demands across the trips is strictly more than $(n'(s) - 1) \cdot \frac{U}{2}$. Since $n(s) \geq \lceil \frac{\sum_{t \in D^s} d_t}{U} \rceil$, we get $n(s) > \frac{n'(s)-1}{2}$, i.e., $n'(s) < 2n(s) + 1$, which implies that $n'(s) \leq 2n(s)$ since $n'(s)$ is an integer. Since we kept all deliveries to the days they occurred in (x, y), the holding cost does not change. Hence, the unsplittable solution has cost at most 2 times the splittable solution.

Applying Proposition 6 with the 2-approximation for Capacitated Splittable IAP, we obtain the following result.

Theorem 3. *Capacitated Unsplittable IAP has a 6-approximation.*

In the full paper [11], we show weak NP-hardness for the Capacitated Unsplittable IAP.

5   Conclusion

We studied the Uncapacitated, Capacitated Unsplittable, and Capacitated Splittable variants of IAP and SIRPFL. For the Uncapacitated IAP, a polynomial time dynamic program is known [20]. For the Capacitated Splittable IAP, we proved a 3-approximation by rounding the LP. For the Capacitated Unsplittable IAP, we gave an NP-hardness reduction from Number Partition and a 6-approximation. For the more general Uncapacitated Star Inventory Routing Problem with Facility Location (Uncapacitated SIRPFL), we gave a 12-approximation by combining rounding ideas from Facility Location and the visitation ideas from our 3-approximation for Capacitated Splittable IAP. For Capacitated Splittable SIRPFL, we provided at 24-approximation. Following that, we have a 48-approximation for Capacitated Unsplittable SIRPFL. It remains open whether Capacitated Splittable IAP is NP-hard. Since we tried to keep the proofs simple and did not optimize for the approximation factors, it may not be difficult to improve the factors.

References

1. Arkin, E., Joneja, D., Roundy, R.: Computational complexity of uncapacitated multi-echelon production planning problems. Oper. Res. Lett. **8**, 61–66 (1989)
2. Bienkowski, M., Byrka, J., Chrobak, M., Jeż, L., Nogneng, D., Sgall, J.: Better approximation bounds for the joint replenishment problem. In: Proceedings of the 25th Annual ACM-SIAM Symposium on Discrete Algorithms, pp. 42–54 (2014)
3. Cheung, M., Elmachtoub, A., Levi, R., Shmoys, D.: The submodular joint replenishment problem. Math. Program. **158**(1–2), 207–233 (2016)
4. Chudak, F.A., Shmoys, D.B.: Improved approximation algorithms for the uncapacitated facility location problem. SIAM J. Comput. **33**(1), 1–25 (2004)
5. Fukunaga, T., Nikzad, A., Ravi, R.: Deliver or hold: approxmation algorithms for the periodic inventory routing problem. In: Proceedings of the 17th International Workshop on Approximation Algorithms for Combinatorial Optimization Problems, pp. 209–225 (2014)
6. Glicksman, H., Penn, M.: Approximation algorithms for group prize-collecting and location-routing problems. Discrete Appl. Math. **156**(17), 3238–3247 (2008)
7. Goemans, M., Williamson, D.: A general approximation technique for constrained forest problems. SIAM J. Comput. **24**(2), 296–317 (1995)
8. Harks, T., König, F., Matuschke, J.: Approximation algorithms for capacitated location routing. Transp. Sci. **47**(1), 3–21 (2013)
9. Jain, K., Mahdian, M., Markakis, E., Saberi, A., Vazirani, V.V.: Greedy facility location algorithms analyzed using dual fitting with factor-revealing LP. J. ACM **50**, 795–824 (2003)
10. Jain, K., Vazirani, V.V.: Approximation algorithms for metric facility location and k-median problems using the primal-dual schema and lagrangian relaxation. J. ACM **48**(2), 274–296 (2001)
11. Jiao, Y., Ravi, R.: Inventory routing problem with facility location. CoRR abs/1905.00148 (2019). https://arxiv.org/abs/1905.00148
12. Levi, R., Roundy, R., Shmoys, D.: Primal-dual algorithms for deterministic inventory problems. Math. Oper. Res. **31**(2), 267–284 (2006)
13. Levi, R., Roundy, R., Shmoys, D., Sviridenko, M.: First constant approximation algorithm for the one-warehouse multi-retailer problem. Manage. Sci. **54**(4), 763–776 (2008)
14. Li, S.: A 1.488 approximation algorithm for the uncapacitated facility location problem. Inf. Comput. **222**, 45–58 (2013)
15. Lin, J., Vitter, J.S.: Approximation algorithms for geometric median problems. Inf. Process. Lett. **44**, 245–249 (1992)
16. Nagarajan, V., Shi, C.: Approximation algorithms for inventory problems with submodular or routing costs. Math. Program. **160**, 1–20 (2016)
17. Nonner, T., Souza, A.: Approximating the joint replenishment problem with deadlines. Discrete Math. Alg. Appl. **1**(2), 153–174 (2009)

18. Ravi, R., Sinha, A.: Approximation algorithms for problems combining facility location and network design. Oper. Res. **54**(1), 73–81 (2006)
19. Shmoys, D.B., Tardos, E., Aardal, K.: Approximation algorithms for facility location problems (extended abstract). In: Proceedings of the Twenty-Ninth Annual ACM Symposium on Theory of Computing, STOC 1997, pp. 265–274. ACM (1997)
20. Wagner, H.M., Whitin, T.M.: Dynamic version of the economic lot sizing model. Manage. Sci. **5**, 89–96 (1958)

A Linear-Time Algorithm
for Radius-Optimally Augmenting Paths
in a Metric Space

Christopher Johnson and Haitao Wang$^{(\boxtimes)}$

Department of Computer Science, Utah State University, Logan, UT 84322, USA
christopherajohnson42@gmail.com, haitao.wang@usu.edu

Abstract. Let P be a path graph of n vertices embedded in a metric space. We consider the problem of adding a new edge to P to minimize the radius of the resulting graph. Previously, a similar problem for minimizing the diameter of the graph was solved in $O(n \log n)$ time. To the best of our knowledge, the problem of minimizing the radius has not been studied before. In this paper, we present an $O(n)$ time algorithm for the problem, which is optimal.

1 Introduction

In this paper, we consider the problem of augmenting a path graph embedded in a metric space by adding a new edge so that the radius of the new graph is minimized.

Let P be a path graph of n vertices, $v_1, v_2, \ldots, v_n$, ordered from one end to the other. Let $e(v_i, v_{i+1})$ denote the edge connecting v_i and v_{i+1} for $i \in [1, n-1]$. Let V be the set of all vertices of P. We assume that P is embedded in a metric space, i.e., $(V, |\cdot|)$ is a metric space and $|v_i v_j|$ is the distance of any two vertices v_i and v_j of V. Specifically, the following properties hold: (1) the triangle inequality: $|v_i v_k| + |v_k v_j| \geq |v_i v_j|$; (2) $|v_i v_j| = |v_j v_i| \geq 0$; (3) $|v_i v_j| = 0$ iff $i = j$. For each edge $e(v_i, v_{i+1})$ of P, its *length* is equal to $|v_i v_{i+1}|$.

Suppose we add a new edge e connecting two vertices v_i and v_j of P, and let $P \cup \{e\}$ denote the resulting graph. Note that a *point* of P can be either a vertex of P or in the interior of an edge. A point c on $P \cup \{e\}$ is called a *center* if it minimizes the largest shortest path length from c to all vertices of P, and the largest shortest path length from the center to all vertices is called the *radius*. Our problem is to add a new edge e to connect two vertices of P such that the radius of $P \cup \{e\}$ is minimized. We refer to the problem as *the radius-optimally augmenting path problem*, or ROAP for short.

To the best of our knowledge, the problem has not been studied before. In this paper, we present an $O(n)$ time algorithm. We assume that the distance $|v_i v_j|$ can be obtained in $O(1)$ time for any two vertices v_i and v_j of P.

A full version of this paper is available at https://arxiv.org/abs/1904.12061.

© Springer Nature Switzerland AG 2019
Z. Friggstad et al. (Eds.): WADS 2019, LNCS 11646, pp. 466–480, 2019.
https://doi.org/10.1007/978-3-030-24766-9_34

1.1   Related Work

A similar problem for minimizing the diameter of the augmenting graph was studied before. Große et al. [9] first gave an $O(n \log^3 n)$ time algorithm, and later Wang [15] solved the problem in $O(n \log n)$ time.

Some variations of the diameter problem have also been considered in the literature. If the path P is in the Euclidean space $\mathbb{R}^d$ for a constant d, then Große et al. [9] gave an $O(n + 1/\epsilon^3)$ time algorithm that can find a $(1 + \epsilon)$-approximate solution for the diameter problem, for any $\epsilon > 0$. If P is in the Euclidean plane $\mathbb{R}^2$, De Carufel et al. [5] gave a linear time algorithm for adding a new edge to P to minimize the *continuous diameter* (i.e., the diameter is defined with respect to all points of P, not only vertices). For a geometric tree T of n vertices embedded in the Euclidean plane, De Carufel et al. [6] gave an $O(n \log n)$ time algorithm for adding a new edge to T to minimize the *continuous* diameter. For the discrete diameter problem where T is embedded in a metric space, Große et al. [10] first proposed an $O(n^2 \log n)$ time algorithm and later Bilò [3] solved the problem in $O(n \log n)$ time. Oh and Ahn [13] studied the problem on a general tree (i.e., the tree is not embedded in a metric space) and gave $O(n^2 \log^3 n)$ time algorithms for both the discrete and continuous versions of the diameter problem, and later Bilò [3] gave an optimal algorithm of $O(n^2)$ time for the discrete case.

The more general problem of adding k edges to a graph G so that the diameter of the resulting graph is minimized has also been considered before. The problem is NP-hard [14] and some other variants are even W[2]-hard [7,8]. Approximation algorithms have been proposed [4,7,12]. The upper and lower bounds on the diameters of the augmented graphs were also studied, e.g., [1,11]. Bae *et al.* [2] considered the problem of adding k shortcuts to a circle in the plane to minimize the diameter of the resulting graph.

Like the diameter, the radius is a critical metric of network performance, which measures the worst-case cost between a "center" and all other nodes. Therefore, our problem of augmenting graphs to minimize the radius potentially has many applications. As an example, suppose there is a highway that connects several cities and we want to build a facility along the highway to provide certain service for all these cities. To reduce the transportation time, we plan to build a new highway connecting two cities such that the radius (i.e., the maximum distance from the cities to the facility located at the center) is as small as possible.

1.2   Our Approach

Note that in general the radius of $P \cup \{e\}$ is not equal to the diameter divided by two. For example, suppose e connects v_1 and v_n (i.e., $P \cup \{e\}$ is a cycle). Assume that the edges of the cycle have the same length and n is even. Suppose the total length of the cycle is 1. Then, the diameter of the cycle is $1/2$ while the radius is $(1 - 1/n)/2$, very close to the diameter.

An easy way to solve ROAP is to try all edges e connecting v_i and v_j for all $i, j \in [1, n]$, which would take $\Omega(n^2)$ time. We instead use the following approach. Suppose an optimal edge e connecting two vertices v_{i^*} and v_{j^*}. Depending on

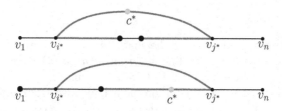

Fig. 1. Illustrating two configurations for the optimal solution, where c^* is the center and the thick (blue) paths are shortest paths from c^* to its two farthest vertices, depicted by larger points. In the top configuration, c^* is on the new edge e and both farthest vertices are on the sub-path of P between v_{i^*} and v_{j^*}. In the bottom configuration, c^* is on the sub-path of P between v_{i^*} and v_{j^*}; v_1 is a farthest vertex of c^* and the other one is on the sub-path of P between v_{i^*} and v_{j^*}. There are also other configurations, e.g., c^* is on the sub-path of P between v_1 and v_{i^*}. (Color figure online)

the locations of the center c^* and its two farthest vertices in $P \cup \{e\}$, there are several possible *configurations* for the optimal solution (e.g., see Fig. 1). For each such configuration, we compute the best solution for it in linear time, such that if there is an optimal solution conforming with the configuration, our solution is also optimal. The efficiency of our approach relies on many observations and properties, which help us avoid the brute-force method. In fact, our algorithm, which involves several kinds of linear scans, is relatively simple. The challenge, however, is on discovering and proving these observations and properties.

The remaining paper is organized as follows. In Sect. 2, we introduce some notation. In Sect. 3, we present our linear time algorithm for ROAP. Due to the space limit, most proofs are omitted but can be found in the full paper.

2 Preliminaries

Denote by $e(v_i, v_j)$ the edge connecting two vertices v_i and v_j for any $i, j \in [1, n]$. The length of $e(v_i, v_j)$ is $|v_i v_j|$. This implies that for any two points p and q on $e(v_i, v_j)$, the length of the portion of $e(v_i, v_j)$ between p and q is $|pq|$. Later we will use this property directly without further explanations.

For any two points p and q on P, we use $P(p, q)$ to denote the subpath of P between p and q. Unless otherwise stated, we assume $i \leq j$ for each index pair (i, j) discussed in the paper. For any pair (i, j), we use $G(i, j)$ to denote the new graph $P \cup \{e(v_i, v_j)\}$ and use $C(i, j)$ to denote the cycle $P(v_i, v_j) \cup e(v_i, v_j)$. Note that if $j \leq i + 1$, then $G(i, j) = P$ and $C(i, j) = P(v_i, v_j)$.

For any graph G, we use $d_G(p, q)$ to denote the length of the shortest path between two points p and q in G, and we also call $d_G(p, q)$ the *distance* between p and q in G. In our paper, G is usually a subgraph of $G(i, j)$, e.g., P or $C(i, j)$. For example, $d_P(p, q)$ denotes the length of $P(p, q)$. We perform a linear time preprocessing so that $d_P(v_i, v_j)$ can be computed in $O(1)$ time for any pair of (i, j). Recall that $|v_i v_j| > 0$ unless $i = j$, and thus, $d_P(v_i, v_j) > 0$ unless $i = j$.

A *center* of $G(i,j)$ is defined as a point (either a vertex or in the interior of an edge) that minimizes the maximum distance from it to all vertices in $G(i,j)^1$, and the maximum distance is called the *radius* of $G(i,j)$. Hence, the problem ROAP is to find a pair of indices (i,j) such that the radius of $G(i,j)$ is minimized.

We assume that P from v_1 to v_n is oriented from left to right, so that we can talk about the relative positions of the points of P (i.e., a point p is to the left of another point q on P if p is closer to v_1 than q is). Similarly, each edge $e(v_i, v_j)$ with $i < j$ from v_i to v_j is oriented from left to right.

3   Our Algorithm for ROAP

In this section, we present our algorithm for solving the problem ROAP. Let (i^*, j^*) be an optimal solution with $i^* \leq j^*$ and c^* be a center of $G(i^*, j^*)$. Let r^* denote the radius of $G(i^*, j^*)$. We begin with the following observation.

Observation 1. *In $G(i^*, j^*)$, there are two vertices v_{a^*} and v_{b^*} such that the following are true.*

1. $d_{G(i^*,j^*)}(c^*, v_{a^*}) = d_{G(i^*,j^*)}(c^*, v_{b^*}) = r^*$.
2. *There is a shortest path from c^* to v_{a^*}, denoted by π_{a^*}, and a shortest path from c^* to v_{b^*}, denoted by π_{b^*}, such that c^* is at the middle of $\pi_{a^*} \cup \pi_{b^*}$ (i.e., the concatenation of the two paths).*

Proof. If this were not true, then we could slightly move c^* so that the maximum distance from the new position of c^* to all vertices in $G(i^*, j^*)$ becomes smaller than r^*, which contradicts with the definition of r^*.   □

Let a^* and b^* be the indices of the two vertices v_{a^*} and v_{b^*}, and π^* be the union of the two paths π_{a^*} and π_{b^*} stated in Observation 1.

Without loss of generality, we assume $a^* < b^*$. Depending on the locations of c^*, a^*, b^*, as well as whether $e(v_{i^*}, v_{j^*}) \in \pi^*$, there are several possible configurations. For each configuration, we will give a linear time algorithm to compute a candidate solution, i.e., a pair (i,j) (along with a radius r and a center c), so that if there is an optimal solution conforming with the configuration then (i,j) is also an optimal solution with c as the center and $r = r^*$. On the other hand, each such solution is *feasible* in the sense that the distances from c to all vertices in $G(i,j)$ is at most r. There are a constant number of configurations. Since we do not know which configuration has an optimal solution, we will compute a candidate solution for each configuration, and among all candidate solutions we return the one with the smallest radius. The runtime of the algorithm is $O(n)$.

For example, one configuration is that $a^* = 1$ and c^* is on $P(v_1, v_{i^*})$. In this case, r^* is equal to $d_P(v_1, c^*)$ and also equal to $d_P(c^*, v_{i^*})$ plus the distance from v_{i^*} to its farthest vertex v_k for all $k \in [i^*, n]$, i.e., $\max_{k \in [i^*, n]} d_{G(i^*,j^*)}(v_{i^*}, v_k)$.

---

[1] The concept of center is defined with respect to the graph instead of to the metric space.

In other words, r^* is equal to half of $d_P(v_1, v_{i^*}) + \max_{k \in [i^*, n]} d_{G(i^*, j^*)}(v_{i^*}, v_k)$. Further, it can be verified that j^* must be the index j that minimizes the value $\max_{k \in [i^*, n]} d_{G(i^*, j)}(v_{i^*}, v_k)$ among all $j \in [i^*, n]$. Therefore, r^* is equal to half of $d_P(v_1, v_{i^*}) + \min_{j \in [i^*, n]} \max_{k \in [i^*, n]} d_{G(i^*, j)}(v_{i^*}, v_k)$. Also, since $c^* \in P(v_1, v_{i^*})$, $d_P(v_1, v_{i^*}) \geq \max_{k \in [i^*, n]} d_{G(i^*, j^*)}(v_{i^*}, v_k)$. Correspondingly, we can compute a candidate solution as follows.

For any $i \in [1, n]$, define $\lambda_i = \min_{j \in [i, n]} \max_{k \in [i, n]} d_{G(i, j)}(v_i, v_k)$, and let $j(i)$ denote the index $j \in [i, n]$ that achieves λ_i. Suppose λ_i and $j(i)$ for all $i \in [1, n]$ are known (which will be computed below). Then, in $O(n)$ time we can find the index i that minimizes the value $d_P(v_1, v_i) + \lambda_i$ among all $i \in [1, n]$ with $d_P(v_1, v_i) \geq \lambda_i$. We return the pair $(i, j(i))$ (with radius $r = (d_P(v_1, v_i) + \lambda_i)/2$ and center c as the point on $P(v_1, v_i)$ such that $d_P(v_1, c) = r$) as the candidate solution for the configuration. It is not difficult to see that if the configuration has an optimal solution, then $(i, j(i))$ is an optimal solution with the center at c and $r^* = r$. Further, by our definition of λ_i, the distance from c to every vertex in $G(i, j(i))$ is at most r, and thus our candidate solution is feasible.

It remains to compute λ_i and $j(i)$ for all $i \in [1, n]$, which is done in the following lemma, with the algorithm given in the next subsection.

Lemma 1. λ_i *and* $j(i)$ *for all* $i \in [1, n]$ *can be computed in* $O(n)$ *time.*

3.1   The Algorithm for Lemma 1

The success of our approach hinges on several monotonicity properties that we shall prove first in the following. Consider any $i \in [1, n]$. For any $j \in [i, n]$, define $\alpha(i, j) = \max_{k \in [i, n]} d_{G(i, j)}(v_i, v_k)$, $\beta(i, j) = \max_{k \in [i, j]} d_{G(i, j)}(v_i, v_k)$, and $\gamma(i, j) = \max_{k \in [j+1, n]} d_{G(i, j)}(v_i, v_k)$ if $j < n$ and $\gamma(i, j) = 0$ otherwise. Clearly, $\alpha(i, j) = \max\{\beta(i, j), \gamma(i, j)\}$ and $\lambda_i = \min_{j \in [i, n]} \alpha(i, j)$.

Note that for any $k \in [i, j]$, the shortest path from v_i to v_k in $G(i, j)$ must be in the cycle $C(i, j)$. Hence, $\beta(i, j) = \max_{k \in [i, j]} d_{C(i, j)}(v_i, v_k)$. Also, it is not difficult to see that $\gamma(i, j) = d_{G(i, j)}(v_i, v_n)$. For $d_{G(i, j)}(v_i, v_n)$, there are two paths from v_i to v_n in $G(i, j)$: $P(v_i, v_n)$ and $e(v_i, v_j) \cup P(v_j, v_n)$. The length of the latter path is $|v_i v_j| + d_P(v_j, v_n)$. Due to the triangle inequality in the metric space, it holds that $|v_i v_j| \leq d_P(v_i, v_j)$. Hence, $\gamma(i, j) = |v_i v_j| + d_P(v_j, v_n)$. Our first monotonicity property is given in Lemma 2.

Lemma 2. $\gamma(i, j) \geq \gamma(i, j + 1)$ *for all* $j \in [i, n - 1]$.

Proof. Since $\gamma(i, j) = |v_i v_j| + d_P(v_j, v_n)$ and $\gamma(i, j + 1) = |v_i v_{j+1}| + d_P(v_{j+1}, v_n)$, we have $\gamma(i, j) - \gamma(i, j + 1) = |v_i v_j| + d_P(v_j, v_{j+1}) - |v_i v_{j+1}| = |v_i v_j| + |v_j v_{j+1}| - |v_i v_{j+1}| \geq 0$. The last inequality is due to the triangle inequality. □

Let $I(i, j)$ be the index k in $[i, j]$ such that $\beta(i, j) = d_{C(i, j)}(v_i, v_k)$ (if there more than one such k, then we let $I(i, j)$ refer to the smallest one). In our algorithm given later, we will need to compute $\beta(i, j)$ for some pairs (i, j). For each $k \in [i, j]$, observe that $d_{C(i, j)}(v_i, v_k) = \min\{d_P(v_i, v_k), |v_i v_j| + d_P(v_k, v_j)\}$. Hence, if $I(i, j)$ is known, then $\beta(i, j)$ can be computed in constant time due to our preprocessing in Sect. 2. In order to determine $I(i,j)$, we introduce a new notation. Define $I'(i,j)$ to be the smallest index $k \in [i,j]$ such that $d_P(v_i, v_k) \geq |v_i v_j| + d_P(v_k, v_j)$. Note that such k must exist since $d_P(v_i, v_j) \geq |v_i v_j|$.

Observation 2. $I(i,j)$ is either $I'(i,j)$ or $I'(i,j) - 1$.

Proof. Let $h = I'(i,j)$. We first assume that $h > i$. By the definition of $I'(i,j)$, $d_P(v_i, v_h) \geq |v_i v_j| + d_P(v_j, v_h)$ and $d_P(v_i, v_{h-1}) < |v_i v_j| + d_P(v_j, v_{h-1})$. Thus, $d_{C(i,j)}(v_i, v_h) = |v_i v_j| + d_P(v_j, v_h)$ and $d_{C(i,j)}(v_i, v_{h-1}) = d_P(v_i, v_{h-1})$.

Consider any $k \in [i,j]$. If $k > h$, then $d_{C(i,j)}(v_i, v_h) = |v_i v_j| + d_P(v_j, v_h) \geq |v_i v_j| + d_P(v_j, v_k) \geq d_{C(i,j)}(v_i, v_k)$. If $k < h - 1$, then we have $d_{C(i,j)}(v_i, v_{h-1}) = d_P(v_i, v_{h-1}) > d_P(v_i, v_k) \geq d_{C(i,j)}(v_i, v_k)$. Note that $d_P(v_i, v_{h-1}) > d_P(v_i, v_k)$ holds because $d_P(v_i, v_{h-1}) = d_P(v_i, v_k) + d_P(v_k, v_{h-1})$ and $d_P(v_k, v_{h-1}) > 0$. Therefore, one of h and $h - 1$ must be $I(i,j)$.

If $h = i$, since $d_P(v_i, v_h) = 0$, by the definition of h, it must be the case that $i = j$. Thus, $I(i,j) = i = h$. □

Observation 2 tells that if we know $I'(i,j)$, then $\beta(i,j)$ is equal to the minimum of $d_{C(i,j)}(v_i, v_{I'(i,j)})$ and $d_{C(i,j)}(v_i, v_{I'(i,j)-1})$, which can be computed in constant time. Hence, to compute $\beta(i,j)$, it is sufficient to determine $I'(i,j)$. To efficiently compute $I'(i,j)$ during our algorithm, the following monotonicity properties on $I'(i,j)$ will be quite helpful.

Lemma 3. *1.* $I'(i,j) \leq I'(i,j+1)$ *for all* $j \in [i, n-1]$.
2. $I'(i,j) \leq I'(i+1,j)$ *for all* $i \in [1, j-1]$.

The following lemma characterizes a monotonicity property of the β values.

Lemma 4. $\beta(i,j) \leq \beta(i,j+1)$ *for all* $j \in [i, n-1]$.

Proof. Let $h = I(i,j)$. Then, $\beta(i,j) = d_{C(i,j)}(v_i, v_h) = \min\{d_P(v_i, v_h), |v_i v_j| + d_P(v_j, v_h)\}$. By the triangle inequality, $|v_i v_j| + d_P(v_j, v_h) \leq |v_i v_{j+1}| + d_P(v_{j+1}, v_h)$. Since $\beta(i, j + 1) = \max_{k \in [i, j+1]} \min\{d_P(v_i, v_k), |v_i v_{j+1}| + d_P(v_{j+1}, v_k)\}$, we obtain

$$\beta(i, j+1) \geq \min\{d_P(v_i, v_h), |v_i v_{j+1}| + d_P(v_{j+1}, v_h)\}$$
$$\geq \min\{d_P(v_i, v_h), |v_i v_j| + d_P(v_j, v_h)\} = \beta(i,j).$$

□

Consider any $i \in [1, n]$. Recall that $j(i)$ is the index j that minimizes the value $\alpha(i,j)$ for all $j \in [i, n]$, and $\alpha(i,j) = \max\{\beta(i,j), \gamma(i,j)\}$. If we consider $\alpha(i,j)$, $\beta(i,j)$, and $\gamma(i,j)$ as functions of $j \in [i, n]$, then by Lemmas 2 and 4, $\alpha(i,j)$ is a unimodal function (first decreases and then increases; e.g., see Fig. 2). In order to compute $j(i)$ and thus λ_i during our algorithm, we define $j'(i)$ to be the smallest index $j \in [i, n]$ such that $\gamma(i,j) \leq \beta(i,j)$. Note that such j must exist because $\gamma(i,n) \leq \beta(i,n)$. We have the following observation.

Observation 3. $j(i)$ is either $j'(i) - 1$ or $j'(i)$.

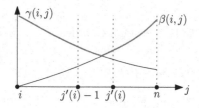

Fig. 2. Illustrating the two functions $\beta(i,j)$ and $\gamma(i,j)$ for $j \in [i,n]$. The function $\alpha(i,j)$, depicted by the thick (red) curve, is the pointwise maximum of them. The index $j'(i)$ is also shown. (Color figure online)

Lemma 5 gives our last monotonicity, which will help us to determine $j'(i)$.

Lemma 5. $j'(i) \le j'(i+1)$ *for all* $i \in [n-1]$.

Based on the above several monotonicity properties, we present our linear time algorithm for computing λ_i and $j(i)$ for all $i \in [1,n]$, as follows. Recall that we have done preprocessing so that $d_P(v_i, v_j)$ can be computed in $O(1)$ time for any pair (i,j).

Starting with $i = 1$ and $j = 1$, we increment j from 1 to n. For each j, we maintain the four values $\gamma(i, j-1)$, $\gamma(i,j)$, $\beta(i, j-1)$, and $\beta(i,j)$. So $\alpha(i, j-1)$ and $\alpha(i,j)$ can be obtained in $O(1)$ time. Since $\gamma(i,j) = |v_i v_j| + d_P(v_j, v_n)$, $\gamma(i,j)$ can be computed in $O(1)$ time, and the same applies to $\gamma(i, j-1)$. We will explain how to compute the β values later. During the increasing of j, if the first time we find $\gamma(i,j) \le \beta(i,j)$, then $j'(i) = j$. By Observation 3, $\lambda_i = \min\{\alpha(i, j-1), \alpha(i,j)\}$.

Then, we increase i by one (for differentiation, we use $i+1$ to denote the increased i). By Lemma 5, to determine $j'(i+1)$, we only need to start j from $j = j'(i)$. Following the similar procedure as above, we increase j and maintain $\gamma(i+1, j-1)$, $\gamma(i+1, j)$, $\beta(i+1, j-1)$, and $\beta(i+1, j)$. Initially when $j = j'(i)$, $\gamma(i+1, j-1)$ and $\gamma(i+1, j)$ can be computed in $O(1)$ time as discussed before; for $\beta(i+1, j-1)$ and $\beta(i+1, j)$, we will show later that they can be computed in $O(1)$ amortized time. In this way, the total time for computing λ_i and $j(i)$ for all $i \in [1,n]$ is $O(n)$.

It remains to describe how to compute the values $\beta(i,j)$. As discussed before, by Observation 2, it is sufficient to determine $I'(i,j)$, after which $\beta(i,j)$ can be computed in $O(1)$ time.

Our algorithm relies on the monotonicity properties of Lemma 3. Initially, when $i = j = 1$, we let $k = 1$. As j increases, we also increase k. We can compute both $d_P(v_i, v_k)$ and $|v_i v_j| + d_P(v_k, v_j)$ in constant time for each triple (i, k, j). During the increasing of k, if we find $d_P(v_i, v_k) \ge |v_i v_j| + d_P(v_k, v_j)$ for the first time, then $I'(i,j)$ is k. After j is increased, we need to compute $\beta(i, j+1)$, i.e., determine $I'(i, j+1)$ (for differentiation, we use $j+1$ to refer to the increased j). To this end, by Lemma 3, we have k start from $I'(i,j)$, which is the exactly current value of k. Similarly, when i is increased and we need to determine $I'(i+1, j)$, we also have k start from $I'(i, j)$, i.e., the current value of k. Thus, in the entire algorithm, the index k continuously increases from 1 to n.

In summary, in the overall algorithm i and j simultaneously increase from 1 to n with $i \le j$. Hence, the number of $\beta(i, j)$ values computed in the entire algorithm is at most $2n$. Further, the procedure for computing all β values increases k from 1 to n. Thus, the total time for computing all β values in the algorithm is $O(n)$, and the amortized time for computing each β value is $O(1)$.

3.2   The Configurations and Our Algorithm

In this section, we present our algorithm for computing an optimal solution. As discussed before, we will consider all possible configurations for the optimal solution and compute a candidate solution for each such configuration.

Recall the definitions of c^*, a^*, b^*, r^*, and π^* in the beginning of Sect. 3. We already discussed one configuration above, i.e., c^* is on $P(v_1, v_{i^*})$. With the help of Lemma 1, we gave a linear time algorithm for it. Another configuration, which is symmetric, is that c^* is on $P(v_{j^*}, v_n)$. Correspondingly, we can use an analogous algorithm (e.g., reverse the indices of P and then apply the same algorithm) to compute a candidate solution in linear time. We omit the details. For the reference purpose, we consider the above two configurations as Case 0.

It remains to consider the configuration $c^* \in C(i^*, j^*) \setminus \{v_{i^*}, v_{j^*}\}$. Here, c^* is in the interior of either $e(v_{i^*}, v_{j^*})$ or $P(v_{i^*}, v_{j^*})$. It is not difficult to see that v_{a^*} is either v_1 or a vertex in $P(v_{i^*}, v_{j^*})$, i.e., $a^* = 1$ or $a^* \in [i^*, j^*]$. Similarly, $b^* = n$ or $b^* \in [i^*, j^*]$. Depending on whether $a^* = 1$, $b^* = n$, or both a^* and b^* are in $[i^*, j^*]$, there are three main cases.

Case 1. $a^* = 1$. In this case, depending on whether $b^* = n$ or $b^* \in [i^*, j^*]$, there are two cases.

Case 1.1. $b^* = n$. In this case, if π^* does not contain $e(i^*, j^*)$, then π^* is the path P, and we keep a candidate solution with $d_P(v_1, v_n)/2$ as the radius. Below, we focus on the case where π^* contains $e(i^*, j^*)$. As $c^* \in C(v_{i^*}, v_{j^*}) \setminus \{v_{i^*}, v_{j^*}\}$ and $c^* \in \pi^*$, c^* must be in the interior of $e(v_{i^*}, v_{j^*})$ (e.g., see Fig. 3).

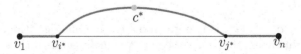

Fig. 3. Illustrating the configuration for Case 1.1, where $c^* \in e(v_{i^*}, v_{j^*})$, $a^* = 1$, and $b^* = n$. The thick (blue) path is π^*. (Color figure online)

We make an assumption on j^* that no index $j > j^*$ exists such that (i^*, j) is also an optimal solution with the same configuration as (i^*, j^*), since otherwise we could instead consider (i^*, j) as (i^*, j^*). We also assume that none of the previously discussed configurations has an optimal solution since otherwise our previously obtained candidate solutions already have an optimal one. With these assumptions, we have the following key lemma for our algorithm.

Lemma 6. *Let k^* be the smallest index such that $d_P(v_{i^*}, v_{k^*}) > d_P(v_1, v_{i^*})$. Such an index k^* must exist in $[i^*, j^*]$. Further, j^* is the largest index $j \in [k^*, n]$ such that $d_P(v_{k^*}, v_j) \leq d_P(v_j, v_n)$.*

Based on Lemma 6, our algorithm for this case works as follows. For each index $i \in [1, n]$, define $k(i)$ as the smallest index $k \in [i, n]$ such that $d_P(v_i, v_k) > d_P(v_1, v_i)$ (let $k(i) = n + 1$ if no such index exists), and if $k(i) \leq n$, define $j(i)^2$ as the largest index $j \in [k(i), n]$ such that $d_P(v_{k(i)}, v_j) \leq d_P(v_j, v_n)$ (let $j(i) = n + 1$ if $k(i) = n + 1$). The following observation is self-evident.

Observation 4. *For each $i \in [1, n-1]$, $k(i) \leq k(i+1)$ and $j(i) \leq j(i+1)$.*

By the above observation, we can easily compute $k(i)$ and $j(i)$ for all $i \in [1, n]$ in $O(n)$ time by a linear scan on P. We omit the details.

For each i, if $j(i) \leq n$, let $r(i) = (d_P(v_1, v_i) + |v_i v_{j(i)}| + d_P(v_{j(i)}, v_n))/2$, and if $d_P(v_1, v_i) < r(i)$ and $d_P(v_{j(i)}, v_n) < r(i)$ (implies that the center is in the interior of $e(v_i, v_{j(i)})$), then we keep $(i, j(i))$ as a candidate solution with $r(i)$ as the radius (and the center is a point c in $e(v_i, v_{j(i)})$ with $d_P(v_1, v_i) + |v_i c| = r(i)$). Note that due to our definitions of $k(i)$ and $j(i)$, the solution is feasible, i.e., the distances from c to all vertices in the graph $G(i, j(i))$ are no more than $r(i)$. The above computes at most n candidate solutions, and among them, we keep the one with the smallest $r(i)$ value as our candidate solution for this configuration. Based on our discussions, if this configuration has an optimal solution, then our solution is also optimal. The running time of the algorithm is $O(n)$.

Case 1.2: $b^* \in [i^*, j^*]$. Note that π^* either contains $e(i^*, j^*)$ or does not contain any interior point of the edge. Depending on whether π^* contains $e(i^*, j^*)$, there are two cases.

Case 1.2.1: π^* contains $e(i^*, j^*)$. Recall that c^* is in the interior of either $e(v_{i^*}, v_{j^*})$ or $P(v_{i^*}, v_{j^*})$. We discuss the two cases below.

Case 1.2.1.1: $c^* \in e(i^*, j^*)$, e.g., see Fig. 4. Our algorithm for this case is somewhat similar to that for Case 1.1. We make an assumption on j^* that no index $j < j^*$ exists such that (i^*, j) is also an optimal solution with the same configuration as (i^*, j^*) since otherwise we could instead consider (i^*, j) as (i^*, j^*). We also assume that none of the previously discussed configurations has an optimal solution. We have the following lemma.

---

^{2} This notation was used differently before. As we have several cases to consider, to save notation, we may use the same notation as long as the context is clear.

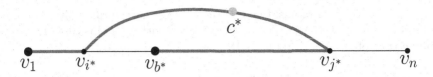

Fig. 4. Illustrating the configuration for Case 1.2.1.1, where $c^* \in e(v_{i^*}, v_{j^*})$, $a^* = 1$, and $b^* \in [i^*, j^*]$. The thick (blue) path is π^*. (Color figure online)

Lemma 7. *1. Let k^* be the smallest index such that $d_P(v_{i^*}, v_{k^*}) > d_P(v_1, v_{i^*})$. Such an index k^* must exist in $[i^*, j^*]$.*
2. $b^* = k^*$.
3. j^* is the smallest index $j \in [k^*, n]$ such that $d_P(v_{k^*}, v_j) > d_P(v_j, v_n)$.

Based on Lemma 7, our algorithm for this case works as follows. For each index $i \in [1, n]$, define $k(i)$ as the smallest $k \in [i, n]$ such that $d_P(v_i, v_k) > d_P(v_1, v_i)$ (let $k(i) = n + 1$ if no such index exists), and if $k(i) \leq n$, define $j(i)$ as the smallest index $j \in [k(i), n]$ such that $d_P(v_{k(i)}, v_j) > d_P(v_j, v_n)$ (let $j(i) = n + 1$ if no such index exists or if $k(i) = n + 1$). It is easy to see that for each $i \in [1, n-1]$, $k(i) \leq k(i+1)$ and $j(i) \leq j(i+1)$. The indices $k(i)$ and $j(i)$ can be computed in $O(n)$ time by a linear scan on P. We omit the details.

For each i, if $j(i) \leq n$, then let $r(i) = (d_P(v_1, v_i) + |v_i v_{j(i)}| + d_P(v_{j(i)}, v_{k(i)}))/2$, and if $d_P(v_1, v_i) < r(i)$ and $d_P(v_{j(i)}, v_{k(i)}) < r(i)$ (this implies that the center is in the interior of $e(v_i, v_{j(i)})$), then we have a candidate solution $(i, j(i))$ with $r(i)$ as the radius. By our definitions of $k(i)$ and $j(i)$, the solution is feasible. Finally, among the at most n candidate solutions, we keep the one with the smallest $r(i)$ as our solution for this configuration. The running time of the algorithm is $O(n)$.

Case 1.2.1.2: $c^* \in P(i^*, j^*)$. Since π^* contains $e(v_{i^*}, v_{j^*})$, c^* must be to the right of v_{b^*} (e.g., see the bottom example in Fig. 1). Further, $d_P(v_1, v_{c^*}) = d_P(v_1, v_{i^*}) + |v_{i^*} v_{j^*}| + d_P(c^*, v_{j^*}) = r^*$. We make an assumption on j^* that no index $j < j^*$ exists such that (i^*, j) is also an optimal solution with the same configuration as (i^*, j^*). We also assume that none of the previously discussed configurations has an optimal solution. The following lemma is literally the same as Lemma 7, although the proof is different.

Lemma 8. *1. Let k^* be the smallest index such that $d_P(v_{i^*}, v_{k^*}) > d_P(v_1, v_{i^*})$. Such an index k^* must exist in $[i^*, j^*]$.*
2. $b^* = k^*$.
3. j^* is the smallest index $j \in [k^*, n]$ such that $d_P(v_{k^*}, v_j) > d_P(v_j, v_n)$.

Based on Lemma 8, our algorithm for this case works as follows. For each index $i \in [1, n]$, define $k(i)$ and $j(i)$ in the same way as in the above Case 1.2.1.1. We also compute them in $O(n)$ time. For each i, if $j(i) \leq n$, then let $r(i) = (d_P(v_1, v_i) + |v_i v_{j(i)}| + d_P(v_{j(i)}, v_{k(i)}))/2$, and if $d_P(v_1, v_i) + |v_i v_{j(i)}| < r(i)$ (implies that the center is on $P(v_{k(i)}, v_{j(i)})$), then we have a candidate solution

$(i, j(i))$ with $r(i)$ as the radius. Finally, among the at most n candidate solutions, we keep the one with the smallest radius as our solution for this case. The total running time of the algorithm is $O(n)$.

Case 1.2.2: π^* does not contain $e(i^*, j^*)$. In this case, the shortest path from c^* to v_1 in $G(i^*, j^*)$ is $P(v_1, c^*)$ and the shortest path from c^* to v_{b^*} is $P(c^*, v_{b^*})$. Since c^* is in the middle of π^*, π^* is $P(v_1, v_{b^*})$ (e.g., see Fig. 5). Further, it is not difficult to see that for any $j \in [b^* + 1, n]$, the shortest path from c^* to v_j in $G(i^*, j^*)$ is $P(c^*, v_{i^*}) \cup e(v_{i^*}, v_{j^*}) \cup P(v_{j^*}, v_j)$. Also note that $b^* < j^*$, since otherwise (i.e., $b^* = j^*$, which is smaller than n as $b^* \neq n$) $d_{G(i^*, j^*)}(c^*, v_n) = d_{G(i^*, j^*)}(c^*, v_{b^*}) + d_P(v_{b^*}, v_n) > d_{G(i^*, j^*)}(c^*, v_{b^*}) = r^*$, a contradiction. We make an assumption on j^* that no index $j < j^*$ exists such that (i^*, j) is also an optimal solution with the same configuration as (i^*, j^*). We also assume that none of the previously discussed configurations has an optimal solution. We begin with the following observation.

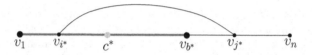

Fig. 5. Illustrating the configuration for Case 1.2.2, where $a^* = 1$, $b^* \in [i^*, j^*]$, and $c^* \in P(v_{i^*}, v_{j^*})$. The thick (blue) path is π^*. (Color figure online)

Observation 5. *For any $i \in [1, n]$, the value $|v_i v_j| + d_P(v_j, v_n)$ is monotonically decreasing as j increases from i to n.*

Lemma 9. *1. Let k^* be the largest index in $[i^*, j^*]$ such that $d_P(v_1, v_{i^*}) < |v_{i^*} v_{j^*}| + d_P(v_{j^*}, v_{k^*})$. Such an index k^* must exist.*
2. $b^* = k^*$.
3. j^* must be the smallest index $j \in [i^*, n]$ such that $d_P(v_1, v_{i^*}) \geq |v_{i^*} v_j| + d_P(v_j, v_n)$.
4. $d_P(v_1, v_{i^*}) < d_P(v_{i^*}, v_n)$.

Based on Lemma 9, our algorithm works as follows. Let i_1 be the largest index i in $[1, n]$ such that $d_P(v_1, v_i) < d_P(v_i, v_n)$. Let i_2 be the smallest index i in $[1, n]$ such that $d_P(v_1, v_i) \geq |v_i v_n|$. By Lemma 9 and Observation 5, if $i_2 \leq i_1$[3], we only need to consider the indices in $[i_2, i_1]$ as the candidates for i^*. For each $i \in [i_2, i_1]$, define $j(i)$ as the smallest index $j \in [i, n]$ such that $d_P(v_1, v_i) \geq |v_i v_j| + d_P(v_j, v_n)$[4], and define $k(i)$ as the largest $k \in [i, j(i)]$ such that $d_P(v_1, v_i) < |v_i v_{j(i)}| + d_P(v_{j(i)}, v_k)$ (let $k(i) = 0$ if no such index k exists).

The monotonicity properties of $j(i)$ and $k(i)$ in the following lemma will lead to an efficient algorithm to compute them.

---

[3] Note that $i_2 \leq i_1 + 1$ always holds because $d_P(v_1, v_{i_1+1}) \geq d_P(v_{i_1+1}, v_n) \geq |v_{i_1+1} v_n|$.
[4] The index j must exist because $d_P(v_1, v_i) \geq |v_i v_n|$ due to the definition of i_2.

Lemma 10. *For any $i \in [i_2, i_1 - 1]$, $j(i + 1) \leq j(i)$ and $k(i + 1) \leq k(i)$.*

Our algorithm for this configuration works as follows. We first compute the two indices i_1 and i_2. If $i_2 > i_1$, then we do not keep any solution for this case. Otherwise, by the monotonicity properties of $j(i)$ and $k(i)$ in Lemma 10, we can compute $j(i)$ and $k(i)$ for all $i \in [i_2, i_1]$ in $O(n)$ time by a linear scan on P. The details are omitted. Then, for each $i \in [i_2, i_1]$, if $k(i) \neq 0$ and $d_P(v_1, v_i) < d_P(v_i, v_{k(i)})$ (this makes sure that the center is in $P(v_i, v_{k(i)})$), then we have a candidate solution $(i, j(i))$ with radius $r(i) = d_P(v_1, v_{k(i)})/2$. By our definition of $j(i)$ and $k(i)$, the solution is feasible. Finally, among all the at most n candidate solutions, we keep the one with the smallest radius as our solution for this case. The algorithm runs in $O(n)$ time.

Case 2: $b^* = n$. This case is symmetric to Case 1 ($a^* = 1$), so we omit the details.

Case 3: Both a^* and b^* are in $[i^*, j^*]$. Since $a^* \neq 1$ and $b^* \neq n$, a^* cannot be i^* and b^* cannot be j^*. Hence, both a^* and b^* are in $[i^* + 1, j^* - 1]$. As in Case 1, depending on whether c^* is in $e(i^*, j^*)$ or $P(v_{i^*}, v_{j^*})$, there are two subcases.

Case 3.1. $c^* \in e(i^*, j^*)$. More precisely, c^* is in the interior of $e(i^*, j^*)$, which implies that $e(i^*, j^*)$ is in π^*. It is not difficult to see that $b^* = a^* + 1$ (e.g., see Fig. 6). We make an assumption on $[i^*, j^*]$ that there is no smaller interval $[i, j] \subset [i^*, j^*]$ such that (i, j) is also an optimal solution with the same configuration as (i^*, j^*) (since otherwise we could instead consider (i, j) as (i^*, j^*)).

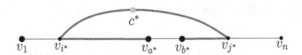

Fig. 6. Illustrating the configuration for Case 3.1, where $a^*, b^* \in [i^*, j^*]$ and $c^* \in e(v_{i^*}, v_{j^*})$. The thick (blue) path is π^*. (Color figure online)

We also assume that none of the previously discussed cases happens. This implies that neither v_1 nor v_n is a farthest vertex of c^* in $G(i^*, j^*)$. To see this, suppose to the contrary that v_1 is also a farthest vertex. Then, if we consider v_1 and v_{b^*} as two farthest vertices stated in Observation 1, then the configuration becomes Case 1.2.1.1 (the bottom example in Fig. 1), which incurs contradiction. Similarly, v_n is not a farthest vertex. Since neither v_1 nor v_n is a farthest vertex of c^*, $d_P(v_1, v_{i^*}) < d_P(v_{i^*}, v_{a^*})$ and $d_P(v_{j^*}, v_n) < d_P(v_{b^*}, v_{j^*})$.

Lemma 11. *i^* is the largest index $i \in [1, a^*]$ such that $d_P(v_1, v_i) < d_P(v_i, v_{a^*})$. j^* is the smallest index $j \in [b^*, n]$ such that $d_P(v_j, v_n) < d_P(v_{b^*}, v_j)$.*

Based on Lemma 11, our algorithm works as follows. For each interval $[k, k+1]$ with $k \in [2, n-2]$ (since $a^* > 1$ and $b^* < n$, we do not need to consider the case where $k = 1$ or $k+1 = n$), define $i(k)$ as the largest index $i \in [1, k]$ such that $d_P(v_1, v_i) < d_P(v_i, v_k)$, and define $j(k)$ as the smallest index $j \in [k+1, n]$ such that $d_P(v_j, v_n) < d_P(v_{k+1}, v_j)$. It can be verified that for any $k \in [2, n-3]$, $i(k) \leq i(k+1)$ and $j(k) \leq j(k+1)$. Thus, we can easily compute $i(k)$ and $j(k)$ for all $k \in [2, n-2]$ in $O(n)$ time. Then, for each $k \in [2, n-2]$, let $r(i) = (d_P(v_{i(k)}, v_k) + |v_{i(k)}v_{j(k)}| + d_P(v_{k+1}, v_{j(k)}))/2$, and if $r(i) > d_P(v_{i(k)}, v_k)$ and $r(i) > d_P(v_{k+1}, v_{j(k)})$ (this makes sure that the center is on the edge $e(v_{i(k)}, v_{j(k)})$), then we have a candidate solution $(i(k), j(k+1))$ with $r(i)$ as the radius. By the definitions of $i(k)$ and $j(k+1)$, the solution is feasible. Finally, among the at most n candidate solutions, we keep the one with the smallest radius as our solution for this case. The algorithm runs in $O(n)$ time.

Case 3.2. $c^* \in P(v_{i^*}, v_{j^*})$. More precisely, c^* is in the interior of $P(v_{i^*}, v_{j^*})$. We first have the following observation.

Lemma 12. π^* must contain $e(v_{i^*}, v_{j^*})$; $b^* = a^* + 1$; v_{a^*} and v_{b^*} are on the same side of c^* (e.g., see Fig. 7).

v_1 $v_{i^*}\ v_{a^*}$ v_{b^*} $c^*\ v_{j^*}$ v_n

Fig. 7. Illustrating the configuration for Case 3.2, where $a^*, b^* \in [i^*, j^*]$ and $c^* \in P(v_{i^*}, v_{j^*})$. The thick (blue) path is π^*. (Color figure online)

In the following, we only discuss the case where c^* is to the right of v_{a^*} and v_{b^*} (e.g., see Fig. 7), and the algorithm for the other case is symmetric. We make an assumption on $[i^*, j^*]$ such that there is no smaller interval $[i, j] \subset [i^*, j^*]$ such that (i, j) is also an optimal solution with the same configuration as (i^*, j^*). We again assume that none of the previously discussed cases happens. The following lemma is literally the same as Lemma 11, although the proof is different.

Lemma 13. i^* is the largest index $i \in [1, a^*]$ such that $d_P(v_1, v_i) < d_P(v_i, v_{a^*})$. j^* is the smallest index $j \in [b^*, n]$ such that $d_P(v_j, v_n) < d_P(v_j, v_{b^*})$.

Based on Lemma 13, our algorithm for this configuration works as follows. We define $i(k)$ and $j(k)$ for each $k \in [2, n-2]$ in the same way as those for Case 3.1, and their values have already been computed in Case 3.1. Then, for each $k \in [2, n-2]$, let $r(i) = (d_P(v_k, v_{i(k)}) + |v_{i(k)}v_{j(k)}| + d_P(j(k), v_{k+1}))/2$, and if $r(i) < d_P(v_{k+1}, v_{j(k)})$ (this makes sure that the center is on $P(v_{k+1}, v_{j(k)})$), then we keep $(i(k), j(k))$ as a candidate solution with $r(i)$ as the radius.

The definitions of $i(k)$ and $j(k)$ guarantee that it is a feasible solution. Finally, among the at most n candidate solutions, we keep the one with the smallest radius as the solution for this configuration. The total time of the algorithm is $O(n)$.

Remark. The above gives the algorithm for Case 3.2 when c^* is to the right of v_{b^*}. If c^* is to the left of v_{a^*}, then we also use the above same values $i(k)$, $j(k)$, and $r(i)$. We keep the candidate solution only if $r(i) < d_P(v_{i(k)}, v_k)$ (this makes sure that the center is on $P(v_{i(k)}, v_k)$. In fact, the can unify our algorithms for Case 3.1 and Case 3.2 to obtain an algorithm for Case 3, as follows. We compute the same values $i(k)$, $j(k)$, and $r(i)$ as before. Then, for each $k \in [2, n-2]$, we keep $(i(k), j(k))$ as a candidate solution with $r(i)$ as the radius. Finally, among the at most n candidate solutions, we keep the one with the smallest radius.

Theorem 1. *The ROAP problem is solvable in linear time.*

Proof. The above provides a linear time algorithm for computing at most one candidate solution for each configuration. As there are $O(1)$ configurations, the total time is $O(n)$. $\square$

References

1. Alon, N., Gyárfás, A., Ruszinkó, M.: Decreasing the diameter of bounded degree graphs. J. Graph Theory **35**, 161–172 (2000)
2. Bae, S., de Berg, M., Cheong, O., Gudmundsson, J., Levcopoulos, C.: Shortcuts for the circle. In: Proceedings of the 28th International Symposium on Algorithms and Computation (ISAAC), pp. 9:1–9:13 (2017)
3. Bilò, D.: Almost optimal algorithms for diameter-optimally augmenting trees. In: Proceedings of the 29th International Symposium on Algorithms and Computation (ISAAC), pp. 40:1–40:13 (2018)
4. Bilò, D., Gualà, L., Proietti, G.: Improved approximability and non-approximability results for graph diameter decreasing problems. Theor. Comput. Sci. **417**, 12–22 (2012)
5. Carufel, J.L.D., Grimm, C., Maheshwari, A., Smid, M.: Minimizing the continuous diameter when augmenting paths and cycles with shortcuts. In: Proceedings of the 15th Scandinavian Workshop on Algorithm Theory (SWAT), pp. 27:1–27:14 (2016)
6. Carufel, J.L.D., Grimm, C., Schirra, S., Smid, M.: Minimizing the continuous diameter when augmenting a tree with a shortcut. In: Proceedings of the 15th Algorithms and Data Structures Symposium (WADS), pp. 301–312 (2017)
7. Frati, F., Gaspers, S., Gudmundsson, J., Mathieson, L.: Augmenting graphs to minimize the diameter. Algorithmica **72**, 995–1010 (2015)
8. Gao, Y., Hare, D., Nastos, J.: The parametric complexity of graph diameter augmentation. Discrete Appl. Math. **161**, 1626–1631 (2013)
9. Große, U., Gudmundsson, J., Knauer, C., Smid, M., Stehn, F.: Fast algorithms for diameter-optimally augmenting paths. In: Proceedings of the 42nd International Colloquium on Automata, Languages and Programming (ICALP), pp. 678–688 (2015)
10. Große, U., Gudmundsson, J., Knauer, C., Smid, M., Stehn, F.: Fast algorithms for diameter-optimally augmenting paths and trees. arXiv:1607.05547 (2016)

11. Ishii, T.: Augmenting outerplanar graphs to meet diameter requirements. J. Graph Theory **74**, 392–416 (2013)
12. Li, C.L., McCormick, S., Simchi-Levi, D.: On the minimum-cardinality-bounded-diameter and the bounded-cardinality-minimum-diameter edge addition problems. Oper. Res. Lett. **11**, 303–308 (1992)
13. Oh, E., Ahn, H.K.: A near-optimal algorithm for finding an optimal shortcut of a tree. In: Proceedings of the 27th International Symposium on Algorithms and Computation (ISAAC), pp. 59:1–59:12 (2016)
14. Schoone, A., Bodlaender, H., Leeuwen, J.V.: Diameter increase caused by edge deletion. J. Graph Theory **11**, 409–427 (1997)
15. Wang, H.: An improved algorithm for diameter-optimally augmenting paths in a metric space. Comput. Geometry Theory Appl. **75**, 11–21 (2018)

Geometric Firefighting in the Half-Plane

Sang-Sub Kim, Rolf Klein, David Kübel, Elmar Langetepe,
and Barbara Schwarzwald[✉]

Department of Computer Science, University of Bonn, 53115 Bonn, Germany
{sang-sub,rolf.klein,dkuebel,schwarzwald}@uni-bonn.de
elmar.langetepe@cs.uni-bonn.de

Abstract. In 2006, Alberto Bressan [3] suggested the following problem. Suppose a circular fire spreads in the Euclidean plane at unit speed. The task is to build, in real time, barrier curves to contain the fire. At each time t the total length of all barriers built so far must not exceed $t \cdot v$, where v is a speed constant. How large a speed v is needed? He proved that speed $v > 2$ is sufficient, and that $v > 1$ is necessary. This gap of $(1, 2]$ is still open. The crucial question seems to be the following. *When trying to contain a fire, should one build, at maximum speed, the enclosing barrier, or does it make sense to spend some time on placing extra delaying barriers in the fire's way?* We study the situation where the fire must be contained in the upper L_1 half-plane by an infinite horizontal barrier to which vertical line segments may be attached as delaying barriers. Surprisingly, such delaying barriers are helpful when properly placed. We prove that speed $v = 1.8772$ is sufficient, while $v > 1.66$ is necessary.

Keywords: Barrier · Firefighting · Geodesic circle

1 Introduction and Problem Statement

Fighting wildfires is a difficult problem, involving many parameters one can neither foresee nor control. But there seem to be two main techniques firefighters employ, namely to extinguish the fire by dropping water or chemicals from aircraft, and to prevent the fire from spreading further by firebreaks. In 2006, Alberto Bressan [3] developed a rather general model for containing a fire by means of barrier curves that must be built in real time, subject to velocity constraints. Barriers are impenetrable by fire, they do not burn and cannot be moved once built.

In addition to general optimality results [5–7], in [3] Bressan proposed the following problem. Suppose a circular fire spreads in the plane at unit speed. In real time, barrier curves must be built to contain it. At each time t, the total length of barriers built so far must not exceed t times v, for some velocity constant v. The question is how large a velocity is needed to contain the fire.

This work has been supported by DFG grant Kl 655/19 as part of a DACH project.

© Springer Nature Switzerland AG 2019
Z. Friggstad et al. (Eds.): WADS 2019, LNCS 11646, pp. 481–494, 2019.
https://doi.org/10.1007/978-3-030-24766-9_35

Bressan showed that $v > 1$ is necessary and that $v > 2$ is sufficient; see also [15] for short proofs. He conjectured that speed $v = 2$ is necessary. But the gap $(1, 2]$ is still open, even though a 500 USD reward has been offered [4] in 2011.

It seems that the difficulty lies with the following question. *To contain a fire, should one build an enclosing barrier at maximum speed, or is it better to invest some time in building extra delaying barriers that will not be part of the final enclosure but can slow the fire down during construction?* If delaying barriers could be shown to be useless, Bressan's proof of the lower bound 1 could be easily extended to prove his conjecture, the lower bound of 2. In fact they consider a special variant in [6], where the fire spreads in a half plane. In that case they can construct an optimal strategy without delaying barriers, that encloses the fire between the boundary of the half plane and the barrier curve.

To study the effectiveness of delaying barriers we study a different setting where an infinite horizontal barrier has to be built to contain the fire in the upper half-plane, instead of the interior of a closed barrier curve. To this horizontal barrier, vertical line segments may be attached as delaying barriers. Without vertical barriers speed $v = 2$ is necessary and sufficient to build the horizontal barrier. While it takes extra time to build vertical barriers, they offer some respite because the expanding fire has to overcome them before it reaches the horizontal barrier again. To simplify matters further we are working in the L_1 norm, so that distances are free of square roots. Also, all intersections of the fire's boundary with the barriers advance at unit speed.

Our main result is the following. In our setting, speed $v > 1.66$ is necessary, and, with a careful placement of delaying barriers, speed $v = 1.8772$ is sufficient. While this result does not disprove Bressan's conjecture it casts a new light on the problem by showing that building delaying barriers can be helpful. Also, the gap we leave open is smaller than the one for the original containment problem.

Previous, but weaker results have been presented at EuroCG'18 [14].

1.1 Related Work

Among theoretical work on *extinguishing* a fire, the "lion and man" problem stands out [1,2,8,13]. Here, r fighters are tasked with quenching a fire in an $n \times n$ grid. In every step, fighters and fire move simultaneously to adjacent cells, subject to certain rules. While $r = n$ fighters can easily extinguish the fire, $\lfloor n/2 \rfloor$ fighters are not enough. The gap in between is still open, despite serious efforts.

How to *contain* a fire has received a lot of attention in graph theory, see, e.g., [9–11]. In quite a few examples, in each round, a stationary guard can be placed in a vertex not on fire, then the fire spreads to all unguarded adjacent vertices. This continues until the fire cannot spread any further. The problem to determine the maximum number of vertices that can be protected is NP-hard, even in trees of degree 3.

Similar in spirit is a geometric firefighting problem in simple polygons [18], where barriers must be chosen from a set of pairwise disjoint diagonals, to save an area of maximum size. Even for convex polygons, the problem is NP-hard, but a 0.086 approximation algorithm exists.

It is interesting to see what happens when building a barrier along the boundary of an expanding circular fire [5,6,16,17]. A spiraling curve results that closes on itself, and thus contains the fire, if the speed of building is larger than 2.6144. Then the number of rounds to completion can be determined by residue calculus. Below this threshold, the curve keeps winding forever.

The rest of this paper is organized as follows. Section 2 formally introduces the problem as well as terms and definitions required for the analysis. In Sect. 4 we develop a lower bound of $v > 1.66$. In Sect. 5 we show that $v = {}^{17}/_9 = 1.\overline{8}$ is sufficient and discuss how this value can even be reduced to $v = 1.8772$.

2   Model

In our model, the fire spreads from the origin and continuously expands over time with speed 1 according to the L_1 metric. To prevent the fire from immediately spreading into the lower half-plane, we allow an arbitrarily small head-start of barrier of length s into both directions along the x-axis.

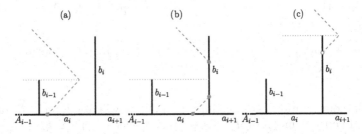

Fig. 1. Fire spreading along delaying barriers. The dashed line shows the fire front at different times t, solid points represent consumption points, while empty points represent places, where the fire burns along the back of already consumed parts of the barrier b_i. In (a) there is one consumption point, so there is a 1-interval in the right direction. In (b) there are three consumption points and in (c) there is a 0-interval in the right direction as there are no consumption points.

Assume that a system of barriers has been built. The barrier system consists of a horizontal barrier containing the fire in the upper half-plane and several vertical delaying barriers attached to it.

To describe a barrier system, we denote the i-th delaying barrier to the right by b_i. The part of the horizontal barrier between b_{i-1} and b_i is denoted by a_i. For simplicity, we also refer to their length by a_i and b_i. For the other direction, we use c_i and d_i respectively. For convenience, $A_i := \sum_{j=1}^{i} a_j$ will denote the total length of horizontal barriers in the right direction up till and including a_i and $B_i := \sum_{j=1}^{i} b_j$ will denote the total length of vertical barriers in the right direction up till and including b_i. Equivalently for the left direction we define C_i and D_i.

As the fire spreads over the barrier system, it represents a geodesic L_1 circle, which consumes the barriers when burning along them. The fire-front is the set of all points in the plane, which shortest non-barrier-crossing path to the fire origin has length t. We consider a point x on a barrier as *consumed* at time t if the fire has reached this point at time t. That means there exists a non-barrier-crossing path of length at most t from the fire origin to the point x. Hence, any piece of the barrier is not consumed all at once, but as the fire burns along it. We call a point on a barrier, which shortest non-barrier-crossing path to the fire has exactly length t a consumption point at time t, so the consumption points are a subset of the fire front. We call the number of consumption points at time t the *current consumption* and a time interval with constant k consumption points at all times a *k-interval*.

The fire front, consumption points and the effect of vertical delaying barriers are illustrated in Fig. 1. As one can see, after the fire reaches a delaying barrier for the first time, it may burn along multiple barriers at multiple points. However, after reaching both ends and passing the top of a barrier there might be no consumption for a while as the delaying barrier has already been burned along from the other side.

We define the *total consumption* $\mathcal{C}$ and *consumption-ratio* $\mathcal{Q}$ for a time interval $[t_1, t_2]$ in a barrier system:

$$\mathcal{C}(t_1, t_2) := \text{length of barrier pieces consumed by the fire between } t_1 \text{ and } t_2$$

$$\mathcal{Q}(t_1, t_2) := \frac{\mathcal{C}(t_1, t_2)}{t_2 - t_1}.$$

For the consumption in a time interval $[0, t]$, we will also write $\mathcal{C}(t)$ and $\mathcal{Q}(t)$ for short. In our setting, if $[t_0, t_1]$ is a k-interval, then $\mathcal{C}(t_1) = \mathcal{C}(t_0) + (t_1 - t_0) \cdot k$.

Note that all these definitions can easily be applied to either side of the barrier system, denoted by $\mathcal{Q}^l(t)$, $\mathcal{Q}^r(t)$ and $\mathcal{C}^l(t)$, $\mathcal{C}^r(t)$ equivalently. Obviously, $\mathcal{Q}(t) = \mathcal{Q}^l(t) + \mathcal{Q}^r(t)$ and $\mathcal{C}(t) = \mathcal{C}^l(t) + \mathcal{C}^r(t)$.

It is clear that when building a barrier system simultaneously to the fire spreading, then every piece of barrier should be build before the fire reaches it. For a limited build speed v, it is necessary and sufficient to have $\mathcal{C}(t) \leq v \cdot t$ for all times t, which means $v \geq \sup_t \mathcal{Q}(t)$. The question then obviously is: What is the minimum speed v for which such a barrier system exists?

3   Prerequisites

Observe that a vertical barrier which is shorter than the predecessor in the same direction does not delay the fire. Hence, we can assume that vertical barriers in one direction increase strictly in length, so $b_i > b_{i-1}$ and $d_i > d_{i-1}$ for all $i > 1$. But we can show an even stronger bound on the growth of successive vertical barriers.

Lemma 1. *If there exists a barrier system with $\mathcal{C}(t) \leq v \cdot t$ at all times t, then there also exists such a barrier system in which any vertical barrier b_i (or d_i) is more than twice as long as the previous barrier b_{i-1} (or d_{i-1}) in the same direction.*

This can be proven constructively by transforming any barrier system S into a slightly different barrier system S' with $C_{S'}(t) \leq C_S(t)$ at all times fulfilling both $b_i > 2b_{i-1}$ and $d_i > 2d_{i-1}$ for all $i > 1$. The details of this can be found in [12].

This means that when given an arbitrary barrier system, we can assume $b_i > 2b_{i-1}$ and $d_i > 2d_{i-1}$ for all $i > 1$. From this we can derive a helpful observation about the order of consumption of vertical and horizontal barriers in a barrier system: when the fire reaches the top of a vertical barrier b_i at some time t (compare Fig. 2), every barrier a_k and b_k with $k \leq i$ has been completely consumed, as for every point on a_k or b_k the shortest non-barrier-crossing path has length smaller than $A_i + b_i = t$. Hence, a 0-interval in the right direction will begin at such times t and $C^r(t) = A_i + B_i - s$, where s denotes the length of the head-start not contributing to the consumption. This observation holds equivalently for both directions.

4   A Lower Bound of $v > 1.66$

Assume there exists a barrier system S consisting of horizontal barriers along the x-axis and vertical barriers attached to it. Further assume for S that $C(t) \leq v \cdot t$ at all times t for some $v = (1 + V)$ with $V \leq \frac{2}{3}$. For this we will construct a contradiction by identifying a specific time t_S, for which $C(t_S) > (1 + V) \cdot t_S$.

By Lemma 1, we can assume $b_i > 2b_{i-1}$ and $d_i > 2d_{i-1}$ for all $i > 1$ in S.

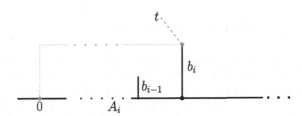

Fig. 2. At some time $t = A_i + b_i$ the fire will reach the top of a vertical barrier b_i.

As without vertical delaying barriers, the consumption-ratio just goes towards 2, S has an unbounded number of vertical barriers in at least one direction. W. l. o. g. assume this is the right direction. Consider a moment when the fire reaches the end of some barrier b_i as illustrated in Fig. 2. As explained in Sect. 3, this happens at time $t = b_i + A_i$ and Lemma 1 implies we have $C^r(t) = A_i + B_i - s$.

$$C^r(t) = A_i + B_i - s = A_i + b_i + B_{i-1} - s \qquad | \; B_{i-1} > 2s \text{ for } i \text{ large enough}$$
$$> A_i + b_i + s > t + s > t \qquad\qquad\qquad\qquad (1)$$

Hence for t large enough, $Q^r(t) > 1$ at times t, when the fire reaches the top of a vertical barrier. Therefore, S has repeated 0-intervals in the left direction as well, or else $Q^l(t)$ would go towards 1 and $Q(t) > 2$ at such times t.

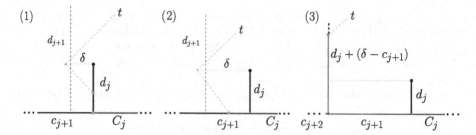

Fig. 3. All three possible situations for the left side to be in at time t. Note that in case (1) and (2) the fire might have reached d_{j+1}, which does not affect our considerations.

We now consider the situation in the left direction at time $t = b_i + A_i$. Let d_j denote the last vertical barrier, whose upper end was reached by the fire, so $t = d_j + C_j + \delta$ with $0 \leq \delta < c_{j+1} + d_{j+1} - d_j$. W.l.o.g. we assume that $b_{i+1} + A_{i+1} \geq d_{j+1} + C_{j+1}$. Otherwise, there must be multiple vertical barriers in the right direction whose upper ends are reached by the fire after it reaches the upper end of d_j and before it reaches the upper end of d_{j+1}. In that case, we can assume that b_i is the last among those, such that $b_{i+1} + A_{i+1} \geq d_{j+1} + C_{j+1}$ holds.

We split our consideration in three cases, which are all illustrated in Fig. 3:

1. $0 \leq \delta < d_j$
2. $d_j \leq \delta < d_j + c_{j+1}$
3. $d_j + c_{j+1} \leq \delta < c_{j+1} + d_{j+1} - d_j$

In the first case, the fire has not reached the horizontal barrier c_{j+1} yet after passing over d_j; in the second case, it has reached c_{j+1}, but not its end; in the third case the fire has completely consumed c_{j+1}.

In Case 3, $\delta = d_j + c_{j+1} + \epsilon$ and then $\mathcal{C}^l(t) \geq C_{j+1} + D_j + 2d_j + \epsilon - s > (d_j + C_j) + (d_j + c_{j+1}) + \epsilon = t$, which together with Inequality (1) already gives $\mathcal{C}(t) > 2t > (1 + V) \cdot t$ which is a contradiction.

For both remaining cases, we will derive a lower bound for d_j. We will then consider the moment $t_1 = 2d_j + C_{j+1}$, when the fire reaches the end of the horizontal barrier c_{j+1}. Using the lower bound on d_j, we will prove $\mathcal{C}(t_1) > (1 + V) \cdot t_1$.

4.1   Case 1: $0 \leq \delta < d_j$

In Case 1, $\mathcal{C}^l(t) > C_j + D_j - s = C_j + d_j + D_{j-1} - s > C_j + d_j$, since $D_{j-1} > s$ for j large enough. Now at time t, it must hold:

$$\mathcal{C}(t) = \mathcal{C}^r(t) + \mathcal{C}^l(t) < (1 + V) \cdot t \qquad | \text{ Inequality (1)}$$
$$\Rightarrow \qquad C_j + d_j < V(d_j + C_j + \delta)$$
$$\Rightarrow \qquad (V - 1)C_j > (1 - V)d_j - V\delta \qquad | (V < 1)$$
$$\Leftrightarrow \qquad C_j < V/(1-V) \cdot \delta - d_j \qquad (2)$$

$V \leq \frac{2}{3}$ implies $V/(1-V) \leq 2$ by direct calculation, which gives bounds for C_j, d_j:

$$C_j < 2\delta - d_j < d_j \qquad | \ \delta < d_j \text{ in Case 1}$$
$$\Rightarrow \quad 2d_j > C_j + \delta$$
$$\Leftrightarrow \quad d_j > \frac{1}{2}(C_j + \delta) \tag{3}$$

4.2   Case 2: $d_j \leq \delta < d_j + c_{j+1}$

In Case 2 a part of c_{j+1} of length $(\delta - d_j)$ has already been consumed, so $C^l(t) \geq D_j + C_j + (\delta - d_j) - s > d_j + C_j + (\delta - d_j) = C_j + \delta$, as $D_{j-1} > s$ for j large enough. Now at time t it must hold

$$\mathcal{C}(t) = \mathcal{C}^r(t) + \mathcal{C}^l(t) < (1 + V) \cdot t \qquad | \text{ Inequality (1)}$$
$$\Rightarrow \qquad C_j + \delta < V(d_j + C_j + \delta)$$
$$\Rightarrow \qquad (1 - V)(C_j + \delta) < V d_j$$
$$\Rightarrow \qquad d_j > (1-V)/V(C_j + \delta) \tag{4}$$

$V \leq \frac{2}{3}$ implies $(1-V)/V \geq \frac{1}{2}$ by direct calculation, which gives the bound:

$$d_j > \frac{1}{2}(C_j + \delta) \tag{5}$$

This is the same bound as found for Case 1 in Inequality (3).

4.3   Deriving the Contradiction $\mathcal{C}(t_1) > (1 + V) \cdot t_1$

Now we consider time $t_1 = C_{j+1} + 2d_j > t$, when the fire reaches the end of the horizontal barrier c_{j+1}; see Fig. 4. As for any time, at time t_1, it must hold

$$\mathcal{C}(t_1) = \mathcal{C}^r(t_1) + \mathcal{C}^l(t_1) \leq (1 + V) \cdot t_1$$
$$\Leftrightarrow \quad \mathcal{C}^l(t_1) \leq (1 + V) \cdot t_1 - \mathcal{C}^r(t_1)$$
$$= (1 + V) \cdot t + (1 + V)(t_1 - t) - (\mathcal{C}^r(t) + \mathcal{C}^r(t, t_1))$$
$$\leq Vt + (1 + V)(t_1 - t) + t - \mathcal{C}^r(t) \quad | \text{ Ineq. (1)} \tag{6}$$
$$\Rightarrow \quad \mathcal{C}^l(t_1) + s < Vt + (1 + V)(t_1 - t) \tag{7}$$

By construction, $t_1 = C_{j+1} + 2d_j$. As $t = d_j + C_j + \delta$, this means $t_1 = t + (d_j + c_{j+1} - \delta)$. Due to Lemma 1, we know that the fire has not reached the end of d_{j+1} yet, hence $\mathcal{C}^l(t_1) \geq 3d_j + C_{j+1} - s$. Hence, we arrive at the following inequalities:

$$3d_j + C_{j+1} < V(d_j + C_j + \delta) + (1 + V)(d_j + c_{j+1} - \delta)$$
$$\Leftrightarrow \quad -V(c_{j+1} - \delta) < (V - 1)\delta + (V - 1)C_j + (2V - 2)d_j \qquad | \ (1 > V)$$
$$\Leftrightarrow \quad c_{j+1} - \delta > \frac{1 - V}{V}\delta + \frac{1 - V}{V}C_j + 2\frac{1 - V}{V}d_j. \tag{8}$$

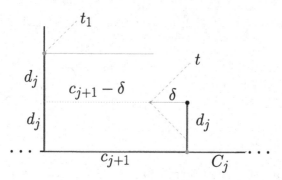

Fig. 4. After $d_j + c_{j+1} - \delta$ additional time after t, the fire has reached the end of c_{j+1} and has also consumed a piece of length $2d_j$ of the next vertical barrier.

$V \leq \frac{2}{3}$ implies $(1-V)/V \geq \frac{1}{2}$ by direct calculation, which gives the bound:

$$c_{j+1} - \delta > \frac{1}{2}\delta + \frac{1}{2}C_j + d_j$$

$$\Leftrightarrow \quad d_j + c_{j+1} - \delta > \frac{1}{2}\delta + \frac{1}{2}C_j + 2d_j \tag{9}$$

Now in both cases we got $d_j > 1/2(C_j + \delta)$ (Inequalities (3) and (5)), so we can apply that and conclude:

$$t_1 - t = d_j + c_{j+1} - \delta > C_j + d_j + \delta = t = A_i + b_i \tag{10}$$

So we know, that in both cases $t_1 - t > b_i + A_i$. Now consider the situation in the right direction again (compare Fig. 2). At $t + b_i$ the fire reaches the horizontal barrier a_{i+1} behind b_i. Additionally, by assumption $b_{i+1} + A_{i+1} \geq d_{j+1} + C_{j+1}$, the fire has not reached the top of the next barrier b_{i+1} at t_1. This means, that between $t + b_i$ and t_1, there is always at least consumption 1 in the right direction, which means the fire has consumed barriers of length at least A_i, hence $\mathcal{C}^r(t, t_1) \geq A_i$.

As our whole consideration is based on inequalities, we will consider an edge case with a contradiction that can be extended to our given barrier system $\mathcal{S}$. More precisely, assume, that Inequality (6) is tight for some t_1^*, so:

$$\mathcal{C}^l(t_1^*) = Vt + (1 + V)(t_1^* - t) + t - \mathcal{C}^r(t)$$

$$\Leftrightarrow \quad \mathcal{C}^r(t) + \mathcal{C}^l(t_1^*) = (1 + V)t_1^*$$

By our arguments above, $\mathcal{C}^r(t, t_1^*) \geq A_i$ and hence $\mathcal{C}(t_1^*) = \mathcal{C}^r(t, t_1^*) + \mathcal{C}^r(t) + \mathcal{C}^l(t_1^*) \geq (1+V)t_1^* + A_i > (1+V)t_1^*$, which is a contradiction for this edge case.

Now in our given barrier system $\mathcal{S}$ it holds $t_1 = t_1^* + x$ for some $x > 0$. As everything except c_{j+1} is fixed at t, this additional time results in additional consumption of at least horizontal barriers of length x in both directions in comparison to the edge case. Hence we can extend the contradiction:

$$\begin{aligned}
\mathcal{C}(t_1) &= \mathcal{C}^l(t_1) + \mathcal{C}^r(t) + \mathcal{C}^r(t, t_1) \\
&= \mathcal{C}^l(t_1^*) + \mathcal{C}^r(t) + \mathcal{C}^r(t, t_1^*) + 2x \\
&= (1 + V)t_1^* + 2x + A_i > (1 + V)(t_1^* + x) = (1 + V)t_1.
\end{aligned}$$

Theorem 1. *The fire can not be contained in the upper half-plane with speed $v \leq 1.66$ by a barrier system consisting of a horizontal barrier along the x-axis and vertical barriers attached to it.*

5   Upper Bounds

We prove the upper bound by defining a barrier system with bounded consumption-ratio. Before we present the construction, we give some intuition. We choose the following conditions:

$$\begin{aligned}
&a_{i+1} \geq b_i \text{ and } b_{i+1} \geq 2b_i \ \forall i \geq 1, \\
&\text{similarly } c_{i+1} \geq d_i \text{ and } d_{i+1} \geq 2d_i \ \forall i \geq 1.
\end{aligned} \quad (11)$$

This forces the 0-intervals generated by b_i to be of length of b_i. For a single direction this results in a repeating sequence of k-intervals of specific lengths and k as shown in Fig. 5.

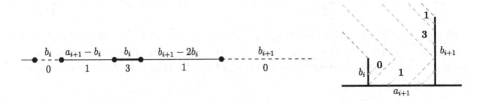

Fig. 5. A sequence of k-intervals to the right of $(0,0)$. The length is given above each interval and the current consumption below.

The idea is to construct the barrier system in such a way that the 0-intervals always appear in an alternating fashion, so the local maxima in the consumption-ratio of one direction can be countered by the 0-intervals of the other direction.

To show that this idea can be realized, we consider the periodic interlacing of time intervals as illustrated in Fig. 6. There, the ends of the 0-intervals in one direction coincide with the ends of the 3-intervals in the other direction, that is, at t_3 and t_6.

The current consumption is always greater than 1, since the 0-intervals do not overlap. Also, the combined consumption-ratio $\mathcal{Q}(t)$ must be smaller than 2 at all times. This also implies that t_3 is no local maximum and the consumption-ratio grows towards 2 between t_3 and t_4. Hence, by setting $d_i > 2b_i$ we make t_1, t_4, t_7 the local maxima and t_2, t_5 the local minima of $\mathcal{Q}(t)$.

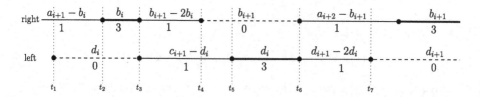

Fig. 6. The periodic interlacing of time intervals.

Let us now consider the consumption-ratio $\mathcal{Q}(t_1, t_4)$ of the cycle from t_1 to t_4. There are two 1-intervals involved in this cycle in the right direction. The first one, where the fire burns along a_{i+1}, is of length $a_{i+1} - b_i$ and lies partially in this cycle. The second one, where the fire crawls up along b_{i+1}, is of length $b_{i+1} - 2b_i$ and lies completely in this cycle. As the beginning of this cycle is given by the start of the 0-interval on one side and the end is given by the end of the second 1-interval on the other side, we know that the length of this cycle is $d_i + (b_{i+1} - 2b_i)$. The total consumption in this cycle is $1 \cdot d_i + 2 \cdot b_i + 2(b_{i+1} - 2b_i)$. Now we define $d_i = \beta \cdot b_i$, $b_{i+1} = \beta \cdot d_i$, and $d_i = \alpha + 2b_i$ for some $\alpha, \beta \in \mathbb{R}_{>0}$. Note that this choice satisfies all our conditions, including $d_i > 2b_i$, and that $\alpha = (\beta - 2)b_i$ and $b_{i+1} = \beta^2 b_i$. Then the consumption-ratio $\mathcal{Q}(t_1, t_4)$ of the cycle is given by

$$\frac{\mathcal{C}(t_1, t_4)}{t_4 - t_1} = \frac{(\alpha + 2b_i) + 2b_i + 2(b_{i+1} - 2b_i)}{(\alpha + 2b_i) + b_{i+1} - 2b_i} = \frac{\alpha + 2b_{i+1}}{\alpha + b_{i+1}} = \frac{(\beta - 2) + 2\beta^2}{(\beta - 2) + \beta^2}$$

and attains a minimal value of $17/9$ for $\beta = 4$. Note that by design, $\mathcal{Q}(t_1, t_2)$ and $\mathcal{Q}(t_1, t_3)$ stay below $17/9$, as well. Moreover, if the consumption-ratio has a maximum of $17/9$ at the beginning of the cycle at t_1, this will also be the case at the end at t_4 as

$$\mathcal{Q}(t_4) = \frac{\mathcal{C}(t_1) + \mathcal{C}(t_1, t_4)}{t_4} = \frac{t_1}{t_4} \cdot \frac{\mathcal{C}(t_1)}{t_1} + \frac{t_4 - t_1}{t_4} \cdot \frac{\mathcal{C}(t_1, t_4)}{t_4 - t_1} \leq \frac{17}{9}.$$

Since the cycles change their roles at t_4 such that the 0-interval occurs on the right side of $(0, 0)$, the same argument can be used to bound the local consumption-ratio in the following interval and for all subsequent cycles, recursively. Note that by looking at the time interval from t_3 to t_6, we can derive a closed form for c_{i+1}. Similarly we proceed for a_{i+1}.

To prove the final theorem, it remains to find initial values to get the interlacing started, while maintaining $\mathcal{Q}(t) \leq 17/9$. Suitable values are

$$a_1 := s \quad b_1 := 17s \quad a_2 := 34s \quad a_{i+1} := 7.5b_i \quad b_{i+1} := 4d_i$$
$$c_1 := s \quad d_1 := 34s \quad c_2 := 238s \quad c_{i+1} := 7.5d_i \quad d_{i+1} := 4b_{i+1},$$

which results in the starting intervals given in Fig. 7. The local maxima at t_1 and t_4 then have consumption-ratio exactly $17/9$. The interval between t_2 and t_3 is set up equivalent to the one between t_3 and t_6 in Fig. 6, which means the interlacing construction can be applied to all intervals beyond. Note that all barriers scale with s. An example of this construction for $s = 1$ is given in Fig. 8.

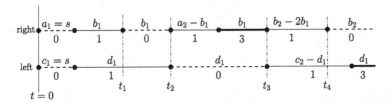

Fig. 7. Illustration of time intervals at the start. Due to their growth, the sizes of the intervals are not true to scale.

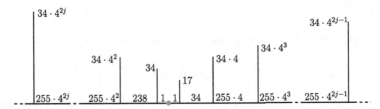

Fig. 8. Example for the final barrier system for $s = 1$, also not true to scale.

Theorem 2. *The fire can be contained in the upper half-plane with speed* $v = \frac{17}{9} = 1.\overline{8}$

5.1 Improving the Upper Bound

It is possible to reduce the upper bound of $v = 1.\overline{8}$ slightly. As shown in Fig. 6, the end of the 3-interval in one direction coincides with the end of the 0-interval in the other direction, which makes t_4 the only local maximum of the interval $[t_1, t_4]$. We introduce a regular shift by a factor of δ, see Fig. 9. This allows the 3-interval in one direction to lie completely inside the 0-interval of the other direction, as shown in Fig. 9. Then, there are two local maxima in the equivalent interval $[t_1, t_5]$, namely at t_3 and t_5. We force both maxima to attain the same value to minimize both at the same time. Again, we set $d_i = \beta \cdot b_i$ and $b_{i+1} = \beta \cdot d_i$, for some $\beta \geq 1$ determined below. Then the value of the first local maximum can be expressed as

$$\mathcal{Q}(t_1, t_3) = \frac{\mathcal{C}(t_1, t_3)}{t_3 - t_1} = \frac{1 \cdot (\delta \cdot b_i) + 3 \cdot b_i}{\delta \cdot b_i + b_i} = \frac{\delta + 3}{\delta + 1} = 1 + \frac{2}{\delta + 1}.$$

Considering the cycle from t_1 to t_5 in Fig. 9, we can conclude that $c_{i+1} = b_{i+1} - b_i + \delta b_i + \delta d_i$. Similarly, we can proceed on the interval from t_5 to t_9 to express a_{i+1} in terms of β, δ and b_i.

Using these identities, we obtain for the second local maximum

$$\begin{aligned}
\mathcal{Q}(t_1, t_5) &= \frac{\mathcal{C}(t_1, t_5)}{t_5 - t_1} = \frac{1 \cdot (\delta \cdot b_i) + 3 b_i + 1 \cdot (b_{i+1} - 2 b_i) + 1 \cdot (c_{i+1} - d_i - \delta \cdot d_i)}{d_i + (c_{i+1} - d_i) - \delta \cdot d_i} \\
&= \frac{c_{i+1} - \delta \cdot d_i}{c_{i+1} - \delta \cdot d_i} + \frac{b_{i+1} + \delta \cdot b_i + b_i - d_i}{c_{i+1} - \delta \cdot d_i} = 1 + \frac{b_{i+1} - b_i + \delta \cdot b_i + 2 b_i - d_i}{b_{i+1} - b_i + \delta b_i} \\
&= 2 + \frac{2 b_i - d_i}{b_{i+1} - b_i + \delta b_i} = 2 + \frac{2 - \beta}{\beta^2 - 1 + \delta}.
\end{aligned}$$

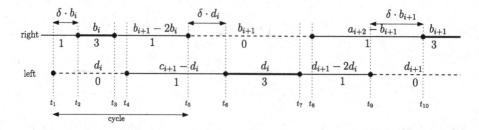

Fig. 9. A general periodic interlacing of time intervals.

As mentioned above, we set both local maxima to be equal, solve for δ and obtain

$$\delta = \frac{1}{2}\left(\beta - \beta^2 + \sqrt{-12 + 4\beta + 5\beta^2 - 2\beta^3 + \beta^4}\right).$$

Plugging this into either one of the two local maxima and minimizing the resulting function for $\beta \geq 1$, we obtain

$$\beta = \frac{3}{2} + \frac{1}{6}\left(513 - 114\sqrt{6}\right)^{1/3} + \frac{\left(19(9 + 2\sqrt{6})\right)^{1/3}}{2 \cdot 3^{2/3}} \approx 4.06887$$

for the optimal value of β, $\delta \approx 1.2802$ and

$$v = \frac{1}{6}\left(10 - \frac{19^{2/3}}{\sqrt[3]{2(4 + 3\sqrt{6})}} + \frac{\sqrt[3]{19(4 + 3\sqrt{6})}}{2^{2/3}}\right) \approx 1.8771$$

as the minimum speed.

Note that the optimal value for β satisfies our conditions given in Eq. 11, so that the barrier system can in fact be realized. Finally, we give suitable values to get the interlacing started:

$$b_1 := 1 \qquad\qquad d_1 := 2b_1$$
$$s := \frac{(4\beta + 2\delta + 1) - v(2\beta + \delta + 1)}{v} \cdot b_1 \qquad a_1 := c_1 := s$$

$$a_2 := (\delta + 1) \cdot b_1 \qquad\qquad c_2 := (2\beta + 3\delta - 1) \cdot b_1$$
$$a_{i+1} := (\delta - 1)d_i + (\beta + \delta)b_{i+1} \qquad b_{i+1} := \beta \cdot d_i$$
$$c_{i+1} := (\delta - 1)b_i + (\beta + \delta)d_i \qquad d_{i+1} := \beta \cdot b_{i+1}.$$

To keep the expression simple, we fixed the value of b_1 and scaled the value of s as listed above. These values can be rescaled to work for any given s.

Theorem 3. *The fire can be contained in the upper half-plane with speed* $v = 1.8772$.

6   Conclusion

We have shown non-trivial bounds for the problem of protecting the lower half-plane from fire with an infinite horizontal barrier. Our results show that delaying barriers – in this case vertical segments attached to the horizontal barrier– can help to break the obvious upper bound of 2 for the building speed. More complex delaying barriers, e. g., free-floating ones, were not analysed specifically, however it is hard to imagine a way for those to have improving effects. It will be interesting to see if such an effect can also be achieved for the problem of containing the fire by a closed barrier curve, i. e., for Bressan's original problem. As a intermediate result in that direction, one ought to extend these results to the Euclidean metric first, where the effect of delaying barriers is less pronounced and harder to analyse.

Acknowledgements. We thank the anonymous referees for their valuable input.

References

1. Berger, F., Gilbers, A., Grüne, A., Klein, R.: How many lions are needed to clear a grid? Algorithms **2**(3), 1069–1086 (2009)
2. Brass, P., Kim, K.D., Na, H.S., Shin, C.S.: Escaping offline searchers and isoperimetric theorems. Comput. Geom. **42**(2), 119–126 (2009)
3. Bressan, A.: Differential inclusions and the control of forest fires. J. Diff. Eqn. **243**(2), 179–207 (2007)
4. Bressan, A.: Price offered for a dynamic blocking problem (2011). http://personal. psu.edu/axb62/PSPDF/prize2.pdf
5. Bressan, A., Burago, M., Friend, A., Jou, J.: Blocking strategies for a fire control problem. Anal. Appl. **6**(3), 229–246 (2008)
6. Bressan, A., Wang, T.: The minimum speed for a blocking problem on the half plane. J. Math. Anal. Appl. **356**(1), 133–144 (2009)
7. Bressan, A., Wang, T.: On the optimal strategy for an isotropic blocking problem. Calc. Var. PDE **45**, 125–145 (2012)
8. Dumitrescu, A., Suzuki, I., Żyliński, P.: Offline variants of the "lion and man" problem. Theor. Comput. Sci. **399**(3), 220–235 (2008)
9. Finbow, S., King, A., MacGillivray, G., Rizzi, R.: The firefighter problem for graphs of maximum degree three. Discrete Math. **307**(16), 2094–2105 (2007)
10. Finbow, S., MacGillivray, G.: The firefighter problem: a survey of results, directions and questions. Technical report (2007)
11. Fomin, F.V., Heggernes, P., van Leeuwen, E.J.: The firefighter problem on graph classes. Theor. Comput. Sci. **613**(C), 38–50 (2016)
12. Kim, S.S., Klein, R., Kübel, D., Langetepe, E., Schwarzwald, B.: Geometric firefighting in the half-plane. CoRR abs/1905.02067 (2019). https://arxiv.org/abs/ 1905.02067
13. Klein, R.: Reversibility properties of the fire-fighting problem in graphs. Comput. Geom. **67**, 38–41 (2018)
14. Klein, R., Kübel, D., Langetepe, E., Schwarzwald, B.: Protecting a highway from fire. In: Abstracts EuroCG 2018 (2018)

15. Klein, R., Langetepe, E.: Computational geometry column 63. SIGACT News **47**(2), 34–39 (2016)
16. Klein, R., Langetepe, E., Levcopoulos, C.: A fire-fighter's problem. In: Proceedings 31st Symposium on Computational Geometry (SoCG 2015) (2015)
17. Klein, R., Langetepe, E., Levcopoulos, C., Lingas, A., Schwarzwald, B.: On a fire fighter's problem. Int. J, Found. Comput. Sci. (2018, to appear)
18. Klein, R., Levcopoulos, C., Lingas, A.: Approximation algorithms for the geometric firefighter and budget fence problem. Algorithms **11**, 45 (2018)

Most Vital Segment Barriers

Irina Kostitsyna[1], Maarten Löffler[2], Valentin Polishchuk[3],
and Frank Staals[2(✉)]

[1] Department of Mathematics and Computer Science,
TU Eindhoven, Eindhoven, The Netherlands
i.kostitsyna@tue.nl
[2] Department of Information and Computing Sciences,
Utrecht University, Utrecht, The Netherlands
{m.loffler,f.staals}@uu.nl
[3] Communications and Transport Systems, ITN,
Linköping University, Linköping, Sweden
polishchuk@liu.se

Abstract. We study continuous analogues of "vitality" for discrete network flows/paths, and consider problems related to placing segment barriers that have highest impact on a flow/path in a polygonal domain. This extends the graph-theoretic notion of "most vital arcs" for flows/paths to geometric environments. We give hardness results and efficient algorithms for various versions of the problem, (almost) completely separating hard and polynomially-solvable cases.

Keywords: Simple polygon · Geodesic distance · Flows and paths

1 Introduction

This paper addresses the following kind of questions:

> *Given a polygonal domain with an "entry" and an "exit", where should one place a given set of "barriers" so as to decrease the maximum entry-exit flow as much as possible ("flow" version), or to increase the length of the shortest entry-exit path as much as possible ("path" version)?*

Figure 1 illustrates these questions in their simplest form (placing a single barrier in a simple polygon). We call the solutions to the problems *most vital* segment barriers for the flow and the path resp. The name derives from the notion of *most vital arcs* in a network – those whose deletion decreases the flow or increases the length of the shortest path as much as possible. While the graph problems are well studied [1–4,16,18,22,27], to our knowledge, geometric versions of locating "most vital" facilities have not been explored. Throughout the paper, the segment barriers will be called simply barriers. When several segments are aligned to form a longer barrier, we call this longer segment a *super-barrier*. We focus only on segment barriers because already with segments there are a number

© Springer Nature Switzerland AG 2019
Z. Friggstad et al. (Eds.): WADS 2019, LNCS 11646, pp. 495–509, 2019.
https://doi.org/10.1007/978-3-030-24766-9_36

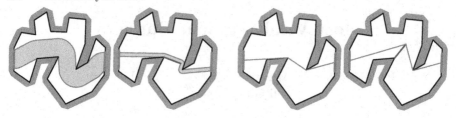

Fig. 1. A polygon in which a single barrier is placed to minimize the flow between two edges of the polygon (left) or lengthen the shortest path between two points (right).

of interesting problem versions, and in principle, any polygonal barrier may be created from sufficiently many segments; however, our results imply that the optimal blocking is always attained by gluing the barriers into super-barriers (no other configuration of segments is most vital).

Determining the most vital barriers is related to resilience and critical infrastructure protection, as it identifies the most vulnerable spots ("bottlenecks, weakest links") in the environment by quantifying how fragile or robust the flow/path is, how much it can be hurt, in the worst case, due to an adversarial act. It is thus an example of optimizing from an *adversarial* point of view: do as much harm as possible using available budget. In practice, the abstract "bad" and "good" may swap places, e.g., when the "good guys" build a defense wall, under constrained resources, to make the "evil" (epidemics, enemy, predator, flood) reach a treasure as late as possible (for the path version) or in a small amount (for the flow version). Our problem may also be viewed as a Stackelberg game (in networks/graphs parlance aka *interdiction problems* [8,11,13,28,30], extensively studied due to its relation to security) where the leader places the blockers and the follower computes the maximum flow or the shortest path around them.

Our paper also contributes to the plethora of work on uncertain environments [7,17,24]. Motion planning under uncertainty is important, e.g., in computing aircraft paths: locations of hazardous storm systems and other no-fly zones are not known precisely in advance, and it is of interest to understand how much the path or the whole traffic flow may be hurt, in the worst case, if new obstacles pop up (of course, there are many other ways to model weather uncertainty).

Finally, similar types of problems arise when barriers are installed for managing the queue to an airline check-in desk or controlling the flow of spectators to an event entrance.

Taxonomy. Since our input consists of the domain and the barriers, several problem versions may be defined:

H/h The domain may have an arbitrary number of holes (such versions will be denoted by H) or a constant number of holes (denoted by h)

B/b There may be arbitrarily many barriers (denoted B) or $O(1)$ barriers in the input (denoted b)

D/1 The barriers may have different lengths (denoted D) or all have the same length – w.l.o.g. unit (denoted 1)

Overall, for each of the two problems—flow blocking and path blocking—we have 8 versions (HBD, HB1, HbD, Hb1, hBD, hB1, hbD, hb1); e.g., flow-hBD is the problem of blocking the flow in a polygonal domain with $O(1)$ holes using arbitrarily many barriers of different lengths, etc. We allow barriers to intersect the holes. Depending on the nature of the barriers and the environment, in some of the envisioned applications these may be impractical (e.g., if a hole is pillar in the building, a barrier cannot run through it) while in others the assumptions are natural (e.g., if a hole is a pond near the entrance to an event). From the theoretical point of view, in most of our problems these assumptions are w.l.o.g. because in the optimal solution the barriers just touch the holes, not "wasting" their length inside a hole (one exception is HBD in which the solution may change if the barriers must avoid the holes).

Overview of the Results. Section 3 describes our main technical contribution: a linear-time algorithm for the fundamental problem of finding *one* most vital barrier for the shortest s-t path in a *simple* polygon. The algorithm is based on observing that the barrier must be "rooted" at a vertex of the polygon. The main challenge is thus to trace the locations of the barrier's "free" endpoint (the one not touching the polygon boundary) through the overlay of shortest path maps from s and t. The overlay has quadratic complexity, so instead of building it, we show that only a linear number of the maps' cells can be intersected and work out an efficient way to go through all the cells. Furthermore, we prove that when placing multiple barriers they can be lined up into a single super-barrier; this reduces the problem to that of placing one barrier. In the remainder of the paper we consider polygons with holes. Section 4 shows hardness of the most general problems flow-HBD and path-HBD, i.e., blocking with multiple different-length barriers in polygons with (a large number of) holes. We also prove weak hardness of the versions with small number of holes (flow-hBD and path-hBD). Finally, we argue that path blocking is weakly hard if the barriers have the same length (path-HB1). Section 5 presents polynomial-time algorithms for path blocking with few barriers (path-HbD), implying that path-hbD, path-Hb1 and path-hb1 are also polynomial. The section then describes polynomial-time algorithms for the remaining versions of flow blocking. We first show that the problem is pseudopolynomial if the barriers have the same length (flow-HB1). We then prove that blocking with few barriers (flow-HbD) is strongly polynomial, implying that flow-hbD, flow-Hb1 and flow-hb1 are also polynomial. Finally, we show polynomiality of the version with constant number of holes (flow-hB1). Table 1 summarizes the hardness and polynomiality of our results.

Table 1. When the number of holes and barriers exceeds 1, the problem may become (weakly or strongly) NP-hard. This table shows which combinations of parameters lead to polynomial or hard problems. The results for *Hb1*, *hbD* and *hb1* follow directly from the result for *HbD*.

	HBD	HB1	HbD	Hb1	hBD	hB1	hbD	hb1
Path	NP-hard	weakly NP-hard	poly	poly	weakly NP-hard	?	poly	poly
Flow	NP-hard	pseudo-poly	poly	poly	weakly NP-hard	poly	poly	poly

2 Preliminaries

Let P be a polygonal domain with n vertices, and let the *source* S and the *sink* T be two given edges on the outer boundary of P (Fig. 2). A *flow* in P is a vector field $F : P \to \mathbb{R}^2$ with the following properties: $\operatorname{div} F(p) = 0 \ \forall p \in P$ (there are no source/sinks inside the domain), $F(p) \cdot \mathbf{n}(p) = 0 \ \forall p \in \partial P \setminus \{S \cup T\}$ where $\mathbf{n}(p)$ is the unit normal to the boundary of P at point p (the flow enters/exits P only through the source/sink), and $|F(p)| \le 1 \ \forall p \in P$ (the permeability of any point is 1, i.e., not more than a unit of flow can be pushed through any point—the flow respects the capacity constraint). Similarly to the discrete network flow, the *value* of a continuous flow F is the total flow coming in from the source ($\int_S F \cdot \mathbf{n} \ ds$) – since in the interior of P the flow is divergence-free (flow conserves inside P), by the divergence theorem, the value is equal to the total flow out of the sink ($-\int_T F \cdot \mathbf{n} \ dt$). A *cut* is a partition of P into 2 parts with S, T in different parts (analogous to a cut in a network); the *capacity* of the cut is the length of the boundary between the parts. Finally, the source and the sink split the outer boundary of P into two parts called the *bottom* $\mathcal{B}$ and the *top* $\mathcal{T}$, and the *critical graph* of the domain [10] is the complete graph on the domain's holes, $\mathcal{B}$ and $\mathcal{T}$, whose edge lengths equal to the distances between their endpoints (we assume that the edges are embedded to connect the closest points on the corresponding holes, $\mathcal{B}$ or $\mathcal{T}$). The celebrated Flow Decomposition and MaxFlow/MinCut theorems for network flows have continuous counterparts: (the support of) a flow decomposes into (thick) paths [25], and the maximum value of the S-T flow is equal to the capacity of the minimum cut [29]; moreover, the mincut is defined by the shortest $\mathcal{B}$-$\mathcal{T}$ path in the critical graph [21].

For shortest path blocking, the setup is a bit more elaborated. Let s be a point on the outer boundary of P, and let S^* be the edge containing s. We assume that s is actually an infinitesimally small gap $s^- s^+$ in the boundary of P (with s^- below s and s^+ above), and that the union of the barriers and the holes is not allowed to contain a path that starts on S^* below s^- and ends on

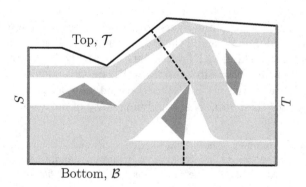

Fig. 2. Flow setup. An S-T flow decomposed into 3 thick paths (blue); two edges of the critical graph, defining a cut (dashed). (Color figure online)

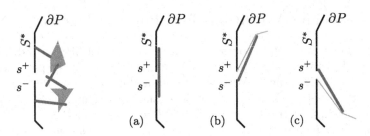

Fig. 3. Path setup; barriers are red and s-t path is blue. Surrounding s^-s^+ (left) is forbidden, even if no barrier touches the gap. Completely "shutting the door" s^-s^+ with one barrier (right (a)) is not allowed: if a barrier is at s, it must touch at most one of s^-, s^+ (right (b,c)). (Color figure online)

S^* above s^+, completely cutting out s (Fig. 3).[1] W.l.o.g. we treat s^- and s^+ as vertices of P. Similarly, we are given a point t, modeled as a gap t^+t^- in another edge T^* on the outer boundary of P.

Let $\mathsf{SP}(p,q)$ denote a shortest path (a geodesic) between points p and q in P. Where it creates no confusion, we will identify a path with its length; in particular, for two points p, q, we will use pq to denote both the segment pq and its length. The *shortest path map* from s, denoted $\mathsf{SPM}(s)$, is the decomposition of P into cells such that shortest paths $\mathsf{SP}(s,p)$ from s to all points p within a cell visit the same sequence of vertices of P; the last vertex in this sequence is called the *root* of the cell and is denoted by $r_s(p)$. The shortest path map from t ($\mathsf{SPM}(t)$) and the roots of its cells ($r_t(p)$) are defined analogously. The maps have linear complexity and can be built in $O(n \log n)$ time (in $O(n)$ time if P is simple) [20]. Our algorithm for path blocking in a simple polygon uses:

Lemma 1. *[26, Lemma 1] Let p, q, and r be three points in a simple polygon P. The geodesic distance from p to a point $x \in \mathsf{SP}(q,r)$ is a convex function of x.*

Finally, let $E(u,v,p)$ denote the ellipse with foci u and v, going through the point p. It is well known that the sum of distances to the foci is constant along the ellipse; for the points outside (resp. inside) the ellipse, the sum is larger (resp. smaller) than $up + pv$. It is also well known that the tangent to the ellipse at p is perpendicular to the bisector of the angle upv (the light from u reaches v after reflecting from the ellipse at p).

---

[1] Other modeling choices could have been made; e.g, another way to avoid complete blockage could be to introduce a "protected zone" around s à la in works on *geographic mincut* [23]. Also a more generic view, outside our scope, could be to combine the flow and path problems into considering *minimum-cost* flows [9,25] (the shortest path is the mincost flow of value 0) and explore how the barriers could influence both the capacity of the domain and the cost of the flow.

3   Linear-Time Algorithms for Simple Polygons

In this section P is a simple polygon. For a set $X \subset P$, let $\mathsf{SP}_X(p,q)$ denote the shortest path between points p, q in $P \setminus X$ (and the length of the path), i.e., the shortest p-q path avoiding X. We first consider finding the most vital unit barrier for the shortest path, i.e., finding the unit segment ab maximizing $\mathsf{SP}_{ab}(s,t)$. For the path blocking, we (re)define the bottom $\mathcal{B}$ and top $\mathcal{T}$ of P as the $t^- $-$s^-$ and s^+-t^+ parts of ∂P resp. (which mimics the flow setup, replacing the entrance S and exit T with $s^- s^+$ and $t^- t^+$). We will treat s^-, s^+, t^-, and t^+ as vertices of P. We then prove that a most vital barrier is placed at a vertex of P (Sect. 3.1). We focus on placing the barrier at (a vertex of) $\mathcal{B}$; placing at $\mathcal{T}$ is symmetric. In Sect. 3.2 we test whether it is possible for any unit barrier ab touching $\mathcal{B}$ to also touch $\mathcal{T}$ (while not lying on S^* or T^*): if this is possible, the barrier separates s from y completely and $\mathsf{SP}_{ab}(s,t) = \infty$. We test this by computing the Minkowski sum of $\mathcal{B}$ with a unit disk and intersecting the resulting shape with $\mathcal{T}$, taking special care around s and t (to disallow having $ab \subset S^*$). In Sect. 3.3 we then proceed to our main technical contribution: showing how to optimally place a barrier touching (a vertex of) $\mathcal{B}$ given that no such barrier can simultaneously touch $\mathcal{T}$. For this, we compute the shortest s-t path H around the Minkowski sum of $\mathcal{B}$ with the unit disk and argue that an optimal barrier will have one endpoint on (a vertex of) $\mathcal{B}$ and the other endpoint on H. Furthermore, we show that this path H intersects edges of the shortest path maps $\mathsf{SPM}(s)$ and $\mathsf{SPM}(t)$ only linearly many times. We subdivide H at these intersection points, and show that for each edge e of H we can then calculate the optimal placement of a point on e maximizing the sum of distances to s and t. This gives us a linear-time algorithm for finding a single most vital barrier. In Sect. 3.4 we then show that even if we have multiple barriers, it is best to glue the barriers together into a single super-barrier.

3.1   A Most Vital Barrier is "Rooted" at a Vertex of P

We start by establishing the following lemma. It's complete proof can be found in the full version of this paper [15].

Lemma 2. *There exists a most vital barrier ab in which one endpoint, say b, lies on a vertex of P.*

Proof sketch. The main idea is to first show that there is a most vital barrier that touches ∂P, then that we can shift it to touch ∂P in an endpoint, and finally that we can shift it along ∂P until it's endpoint coincides with a vertex.   □

3.2   Blocking the Path from s to t Completely

We now argue that we can check in linear time whether it is possible to completely block passage from s to t, by placing a barrier that connects $\mathcal{B}$ to $\mathcal{T}$ (without placing the barrier along S^* or T^*, which is forbidden by our model; see Sect. 2).

Observation 1. *Let u and v be two vertices of* $\mathsf{SP}(s,t)$ *in* $\mathcal{B}$*. The geodesic makes a right turn at u if and only if it makes a right turn at v. Let u' and v' be two vertices of* $\mathsf{SP}(s,t)$ *in* $\mathcal{T}$*. The geodesic makes a left turn at u' if and only if it makes a left turn at v'. Moreover, if* $\mathsf{SP}(s,t)$ *makes a right turn in u then it makes a left turn in u'.*

Assume w.l.o.g. that $\mathsf{SP}(s,t)$ makes a right turn at a vertex $u \in \mathcal{B}$. By Observation 1 it thus makes right turns at all vertices of $\mathsf{SP}(s,t) \cap \mathcal{B}$, and left turns at all vertices of $\mathsf{SP}(s,t) \cap \mathcal{T}$.

Observation 2. *If* $\mathsf{SP}(s,t)$ *makes a right turn at $u \in \mathcal{B}$, and we place a barrier ur at u, then* $\mathsf{SP}_{ur}(s,t)$ *makes a right turn at r.*

For every point p on $\mathcal{B}$, consider placing a barrier pq of length at most one, with one endpoint on p. The possible placements D_p of the other endpoint, q, form a subset of the unit disk centered at p. Let $\mathcal{D} = \bigcup_{p \in \mathcal{B}} D_p$ denote the union of all these regions (see Fig. 4).

Observation 3. *There is a barrier that separates s from t if and only if s and t are in different components of $P \setminus \mathcal{D}$.*

We now observe that $\mathcal{D}$ is essentially the Minkowski sum of $\mathcal{B}$ with a unit disk D. More specifically, let $A \oplus B = \{a + b \mid a \in A \wedge b \in B\}$ denote the Minkowski sum of A and B, let $S_{\mathcal{B}}^* = S^* \cap \mathcal{B}$ denote the part of S^* in $\mathcal{B}$, let $S_{\mathcal{T}}^*$, $T_{\mathcal{B}}^*$, and $T_{\mathcal{T}}^*$ be defined analogously, and let $\mathcal{B}' = \mathcal{B} \setminus (S_{\mathcal{B}}^* \cup T_{\mathcal{B}}^*)$.

Lemma 3. *We have that $\mathcal{D} = \mathcal{D}' \cup X_S \cup X_T$, where $\mathcal{D}' = \mathcal{B}' \oplus D$, $X_A = (A_{\mathcal{B}}^* \oplus D) \setminus A_{\mathcal{T}}^*$, and D is the unit disk centered at the origin. Moreover, $\mathcal{D}$ can be computed in $O(n)$ time.*

Proof. The equality follows directly from the definition of $\mathcal{D}$ and the Minkowski sum. It then also follows $\mathcal{D}$ has linear complexity. So we focus on computing $\mathcal{D}$. To this end we separately compute $\mathcal{D}'$, X_S, and X_T, and take their union. More specifically, we construct the Voronoi diagram of $\mathcal{B}'$ using the algorithm of Chin, Snoeyink, and Wang [6], and use it to compute $\mathcal{B}' \oplus D$ [14]. Both of these steps can be done in linear time. Since S^*, T^*, and D have constant complexity, we can compute X_S and X_T in constant time. The resulting sets still have constant complexity, so unioning them with $\mathcal{B}' \oplus D$ takes linear time. □

Lemma 4. *We can test if s and t lie in the same component C of $P \setminus \mathcal{D}$, and compute C if it exists, in $O(n)$ time.*

Proof. Using Lemma 3 we compute $\mathcal{D}$ in linear time. If s or t lies inside $\mathcal{D}$, which we can test in linear time, then C does not exist. Otherwise, by definition of X_S and X_T, s and t must lie on the boundary of $\mathcal{D}$. We then extract the curve σ connecting s to t along the boundary of $\mathcal{D}$, and test if σ intersects the top of the polygon $\mathcal{T}$. If (and only if) σ and $\mathcal{T}$ do not intersect, their concatenation delineates a single component C' of $P \setminus \mathcal{D}$. Since C' contains both s and t we have $C = C'$. So, all that is left is to test if σ and $\mathcal{T}$ intersect. This can be done in linear time by explicitly constructing C' and testing if it is simple [5]. □

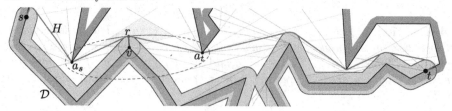

Fig. 4. Our algorithm constructs the region $\mathcal{D}$ describing possible placements of a barrier incident to $\mathcal{B}$, and the shortest path H around $\mathcal{D}$. An optimal barrier incident to $\mathcal{B}$ has one endpoint on H.

Theorem 4. *Given a simple polygon P with n vertices and two points s and t on the boundary of P, we can test whether there exists a placement of a unit length barrier that disconnects s from t in $O(n)$ time.*

3.3 Maximizing the Length from s to t with a Single Barrier

In the remainder of the section we assume that we cannot place a barrier on (a vertex of) $\mathcal{B}$ that completely separates s from t. Fix a distance d, and consider all points $p \in P$ such that $\mathsf{SP}(s,p) + \mathsf{SP}(p,t) = d$. Let C_d denote this set of points, and define $C_{\leq d} = \bigcup_{d' \leq d} C_{d'}$.

Observe that an optimal barrier will have one of its endpoints on the boundary of $\mathcal{D}$. Let $H = \mathsf{SP}_{\mathcal{D}}(s,t)$ be the shortest path from s to t avoiding $\mathcal{D}$. We will actually show that there is an optimal barrier V^* whose endpoint a lies on H, and that H has low complexity. This then gives us an efficient algorithm to compute an optimal barrier. To show that a lies on H we use that if V^* realizes detour d^* (i.e., $\mathsf{SP}_{V^*}(s,t) = d^*$), the endpoint a also lies on C_{d^*}. First, we prove some properties of C_{d^*} towards this end.

Observation 5. *Let Δ_s be a cell in $\mathsf{SPM}(s)$ with root a_s, and Δ_t be a cell in $\mathsf{SPM}(t)$ with root a_t. We have that $C_d \cap \Delta_s \cap \Delta_t$ consists of a constant number of intervals along the boundary of the ellipse with foci a_s and a_t.*

Proof. A point $p \in C_d$ satisfies $\mathsf{SP}(s,p) + \mathsf{SP}(p,t) = d$. For $p \in \Delta_s \cap \Delta_t$ we thus have $\mathsf{SP}(s,a_s) + \|a_s p\| + \|p a_t\| + \mathsf{SP}(a_t,t) = d$. Since d, $\mathsf{SP}(s,a_s)$, and $\mathsf{SP}(a_t,t)$ are constant, this equation describes an ellipse with foci a_s and a_t. Since Δ_s and Δ_t have constant complexity the lemma follows. □

Lemma 5. *$C_{\leq d}$ is a geodesically convex set (it contains shortest paths between its points).*

Proof. Let p and q be two points on C_d, and assume, by contradiction, that there is a point r on $\mathsf{SP}(p,q)$ outside of $C_{\leq d}$. By Lemma 1 the geodesic distance from s to $\mathsf{SP}(p,q)$ is a convex function. Similarly, the distance from t to $\mathsf{SP}(p,q)$ is convex. It then follows that the function $f(x) = \mathsf{SP}(s,x) + \mathsf{SP}(x,t)$, for x on $\mathsf{SP}(p,q)$ is also convex, and thus has its local maxima at p and/or q. Contradiction. □

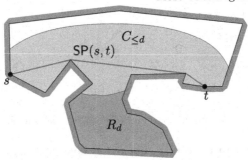

Fig. 5. A sketch of the regions $C_{\leq d}$ (purple) and R_d. Observe that R_d cannot contain any vertices of $\mathcal{T}$, otherwise $\mathcal{T}$ would have to pierce $\mathsf{SP}(s,t)$ and thus $C_{\leq d}$. (Color figure online)

Lemma 6. *If there is an optimal barrier ua incident to a vertex u of $\mathcal{B}$, then the ray ρ from u through a intersects H.*

Proof. The ray ρ splits P into two subpolygons P_1 and P_2. Since $\mathsf{SP}_{ua}(s,t)$ makes a right bend at a (Observation 2 and our assumption that $\mathsf{SP}(s,t)$ makes a right turn at u) it intersects both subpolygons P_1 and P_2. It is easy to show that therefore s and t must be in different subpolygons (otherwise the geodesic crosses ρ a second time, and we could shortcut the path along ρ). Since H connects s to t it must thus also intersect ρ. $\qquad\square$

Next, we define the region R_d "below" $C_{\leq d}$. More formally, let R' be the region enclosed by $\mathcal{B}$ and $\mathsf{SP}(s,t)$, let $d \geq \mathsf{SP}(s,t)$, and let $R_d = R' \setminus C_{\leq d}$. See Fig. 5. We then argue that it is separated from the top part of our polygon $\mathcal{T}$, which allows us to prove that there is an optimal barrier with an endpoint on H.

Observation 6. *Region R_d contains no vertices of $\mathcal{T}$.*

Proof. Assume, by contradiction that there is a vertex of $\mathcal{T}$ in R_d. Observe that this disconnects $C_{\leq d}$. However, since $C_{\leq d}$ is geodesically convex (Lemma 5) and non-empty it is a connected set. Contradiction. $\qquad\square$

Lemma 7. *If there is an optimal barrier ua where u is a vertex of $\mathcal{B}$, then there is an optimal barrier ur where r is a point on $D_u \cap H$ (recall that D_u is the unit disk centered at u).*

Proof. Assume, by contradiction, that there is no optimal barrier incident to u that has its other endpoint on H. Consider the ray from u in the direction of a. By Lemma 6, the ray hits H in a point r' (Fig. 6(a)). Because a lies on C_{d^*} and $C_{\leq d^*}$ is geodesically convex (Lemma 5), r' lies outside $C_{\leq d^*}$. Let $H[p,q] = \mathsf{SP}_\mathcal{D}(p,q)$ be the maximal (open ended) subpath of H that contains r' and lies outside of $C_{\leq d^*}$. We then distinguish two cases, depending on whether or not $H[p,q]$ intersects (touches) $\mathcal{D}$: $\qquad\square$

$H[p,q]$ **does not intersect (touch)** $\mathcal{D}$. It follows that $H[p,q]$ is a geodesic in P as well, i.e. $H[p,q] = \mathsf{SP}(p,q)$. Since $p,q \in C_{\leq d^*}$, and $C_{\leq d^*}$ is geodesically convex (Lemma 5) we then have that $H[p,q] \subseteq C_{\leq d^*}$. Contradiction.

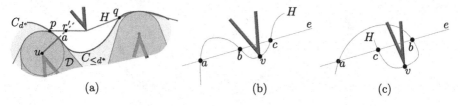

Fig. 6. (a) Illustration of Lemma 7. (b) and (c) The two cases in the proof of Lemma 8.

$H[p, q]$ **intersects** $\mathcal{D}$ **in a point** z. Let $v \in B$ be a point such that $z \in D_v$. We distinguish two subcases, depending on whether z lies in the region R_{d^*}.

- $z \in R_{d^*}$. In this case z lies "below" $C_{\leq d^*}$. From $z \in H$ it follows that $H[p, q] \subset R_{d^*}$. However, as C_{d^*} is geodesically convex, this must mean that $H[p, q]$ has a vertex w in R_{d^*} at which it makes a left turn. This implies that w is a vertex of $\mathcal{T}$. By Observation 6 there are no vertices of $\mathcal{T}$ in R_{d^*}. Contradiction.
- $z \notin R_{d^*}$. Observe that vz is a valid candidate barrier. Since $z \notin C_{\leq d^*}$, the point z actually lies above (i.e. to the left of) $\mathsf{SP}(s, t)$, and thus $\overline{\mathsf{SP}}_{vz}(s, t)$ makes a right turn at z. Using that $z \notin C_{\leq d^*}$ it follows that $\mathsf{SP}_{vz}(s, t) > d^*$. This contradicts that d^* is the maximal detour we can achieve.

Since all cases end in a contradiction this concludes the proof.   □

We now know there exists an optimal barrier with an endpoint on H. Next, we focus on the complexity of H.

Observation 7. *Let b and c be two points on H, such that H makes a left turn in between b and c (i.e. the subcurve $H[b, c]$ of H between b and c intersects the half-plane right of the supporting line of bc). Then $H[b, c]$ contains a vertex of $\mathcal{T}$.*

Lemma 8. *The curve H intersects an edge e of $SPM(z)$, with $z \in \{s, t\}$, at most twice. Hence, H intersects $\mathsf{SPM}(z)$ at most $O(n)$ times.*

Proof. If e is a polygon edge, then H cannot intersect e at all, so consider the case when e is interior to P. Assume, by contradiction, that H intersects e at least three times, in points a, b, and c, in that order along H (Fig. 6(b) and (c)).

If the intersections a, b, and c, are also consecutive along e, then H makes both a left and right turn in between a and c. It is easy to see that since H can bend to the left only at vertices of $\mathcal{T}$ (Observation 7), the region (or one of the two regions) enclosed by H and ac must contain a polygon vertex. Since both e and $H[a, c]$ lie inside P, this means that P has a hole. Contradiction.

If the intersections are not consecutive, (say a, c, b), then again there is a region enclosed by $H[a, c]$ and ab, containing a polygon vertex. Since both $H[b, c]$ and cb lie inside P, this vertex must lie on a hole. Contradiction.   □

Algorithm. We compute intersections of H with the shortest path maps $\mathsf{SPM}(s)$ and $\mathsf{SPM}(t)$, and subdivide H at each intersection point. By Lemma 8, the resulting curve H' still has only linear complexity. Consider the edges of H' in which H' follows the boundary of D_v, for the vertices v of $\mathcal{B}$. By Lemma 7 for some $v \in \mathcal{B}$ there is an optimal barrier that has one endpoint on such an edge of H' and the other at v. Since H' has only $O(n)$ edges we simply try each edge e of H'. For all points $r \in e$, the geodesics $\mathsf{SP}(s, r)$ and $\mathsf{SP}(t, r)$ have the same combinatorial structure, i.e., the roots $a_s = \mathsf{r}_s(r), a_t = \mathsf{r}_t(r)$ stay the same. It follows that we have a constant-size subproblem in which we can compute an optimal barrier in constant time. Specifically, we compute the smallest ellipse E with foci a_s and a_t that contains e and goes through the point r in which E and $e = D_v$ have a common tangent (if such a point exists). See Fig. 4. For that point r, we then also know the length of the shortest path $\mathsf{SP}_{vr}(s, t) = \mathsf{SP}(s, r) + \mathsf{SP}(r, t)$, assuming that we place the barrier vr. We then report the point r that maximizes this length over all edges of H'.

Constructing the connected component P' of $P \backslash \mathcal{D}$ that contains s and t takes linear time (Lemma 4). This component P' is a simple splinegon, in which we can compute the shortest path H connecting s to t in $O(n)$ time [19]. Computing $\mathsf{SPM}(s)$ and $\mathsf{SPM}(t)$ also requires linear time [12]. We can then walk along H, keeping track of the cells of $\mathsf{SPM}(s)$ and $\mathsf{SPM}(t)$ containing the current point on H. Computing the ellipse, the point p on the current edge e, and the length of the geodesic takes constant time. It follows that we can compute an optimal barrier incident to $\mathcal{B}$ in linear time. We use the same procedure to compute an optimal barrier incident to $\mathcal{T}$. We thus obtain the following result.

Theorem 8. *Given a simple polygon P with n vertices and two points s and t on ∂P, we can compute a unit length barrier that maximizes the length of the shortest path between s and t in $O(n)$ time.*

3.4 Using Multiple Vital Barriers

We prove a structural property that even when we are given many barriers, there always exists an optimal solution in which they glued into a single super-barrier. This implies that our linear-time algorithm from the previous section can still be used to solve the problem.

Clearly, any solution distributes the barriers over some (unknown) number of super-barriers. First observe that, similarly to Sect. 3.1, any super-barrier must have a vertex at a vertex of P. We only need to argue that it is suboptimal to have more than one such super-barrier. Let $a_1 b_1$ and $a_2 b_2$ be two segments inside P, and let $m_1 \in a_1 b_1, m_2 \in a_2 b_2$ be two points that divide the segments in the same proportion, that is $\boldsymbol{m_1} = \gamma \boldsymbol{a_1} + (1 - \gamma) \boldsymbol{b_1}, \boldsymbol{m_2} = \gamma \boldsymbol{a_2} + (1 - \gamma) \boldsymbol{b_2}$ for some $\gamma \in [0, 1]$. Then we may argue, similar to Lemma 1, that $f(\gamma) = \mathsf{SP}(m_1, m_2)$ is convex for $\gamma \in [0, 1]$. The result then follows from multiple application of the triangle inequality, using that $\mathsf{SP}(m_1, m_2) \leq \gamma \mathsf{SP}(a_1, a_2) + (1 - \gamma) \mathsf{SP}(b_1, b_2)$. The complete argument can be found in the full version of this paper [15].

Theorem 9. *Given a simple polygon P with n vertices, two points s and t on ∂P, and k unit-length barriers, the optimal placement of the barriers which maximizes the length of the shortest path between s and t consists of a single super-barrier.*

3.5   Most Vital Barriers for the Flow

In simple polygons the critical graph has only two vertices – $\mathcal{B}$ and $\mathcal{T}$ (which, for the flow blocking, are the T-S and S-T parts of ∂P; refer to Fig. 2). Flow blocking thus boils down to finding the shortest $\mathcal{B}$-$\mathcal{T}$ connection (then all the barriers will be placed along the connecting segment) – a problem that was solved in linear time in [21].

4   Hardness Results

In the remainder of the paper P is a polygonal domain with holes (as defined in Sect. 2). We show that when there are many barriers, it is hard to decide whether full blockage can be achieved, by reduction from Partition which reduces to deciding whether equal-width channels between S and T can be blocked (the reduction for path-HB1 is more involved, as it is not based on deciding full blockage); the details are in full version [15]. We summarize our results in the following two theorems.

Theorem 10. *Flow-HBD and path-HBD are NP-hard.*

Theorem 11. *Path-HB1, Flow-HBD, and path-hBD are weakly NP-hard.*

Membership in NP for our problems is open, since verifying solutions involve summing square roots.

5   Polynomial-Time Algorithms

For path blocking, we show that $O(1)$ barriers have only a polynomial number of "combinatorially different" placements and for each placement the different-homotopy path lengths are given by fixed functions of the barriers' locations. Flow blocking is reduced to shortening the $\mathcal{B}$-$\mathcal{T}$ path in the critical graph. The details are in the full version. We summarize our results in the following theorems.

Theorem 12. *Path-HbD, and hence path-hbD, path-Hb1 and path-hb1, are polynomial.*

Theorem 13. *Flow-HB1 can be solved in pseudopolynomial time.*

Theorem 14. *Flow-HbD, and hence flow-hbD, flow-Hb1 and flow-hb1, are polynomial.*

Theorem 15. *Flow-hB1 is polynomial.*

6   Conclusion

We introduced geometric versions of the graph-theoretic most vital arcs problem. We presented efficient solutions for simple polygons, and gave hardness results and algorithms for various versions of the problem. The most intriguing open problem is the hardness of path-hB1 (path blocking with few holes, our only unresolved version); we conjecture that it is polynomial, as still only a constant-number of super-barriers may be needed. Another interesting question is whether the flow and the path blocking have fundamentally different complexities: we proved that the complexities are the same for all versions except HB1 – for path-HB1 we showed weak hardness but lack a pseudopolynomial-time algorithm, while for flow-HB1 we have a pseudopolynomial-time algorithm but no (weak) hardness proof. More generally, various other setups may be considered. For instance, one may be given a budget on the total length of the barriers– the problem then is how to split the budget between the barriers and where to locate them. For minimizing the maximum flow this version is easy: just place the barriers along the shortest $\mathcal{B}$-$\mathcal{T}$ path in the critical graph of the domain. For maximizing the shortest path in a simple polygon the solution is trivial: make a single barrier of the full length (and use our algorithm to find the optimal barrier location). Blocking shortest paths in polygons with holes in this setting is an open problem.

Acknowledgements. M.L. and F.S. were partially supported by the Netherlands Organisation for Scientific Research (NWO) through project no 614.001.504 and 612.001.651, respectively.

References

1. Ahuja, R.K., Magnanti, T.L., Orlin, J.B.: Network Flows: Theory, Algorithms, and Applications. Prentice Hall, Upper Saddle River (1993)
2. Alderson, D.L., Brown, G.G., Carlyle, W.M., Cox Jr., L.A.: Sometimes there is no most-vital arc: assessing and improving the operational resilience of systems. Technical report, Naval Postgraduate School Monterey CA (2013)
3. Ball, M.O., Golden, B.L., Vohra, R.V.: Finding the most vital arcs in a network. Oper. Res. Lett. **8**(2), 73–76 (1989)
4. Bazgan, C., Nichterlein, A., Niedermeier, R.: A refined complexity analysis of finding the most vital edges for undirected shortest paths. In: Paschos, V.T., Widmayer, P. (eds.) CIAC 2015. LNCS, vol. 9079, pp. 47–60. Springer, Cham (2015). https://doi.org/10.1007/978-3-319-18173-8_3
5. Chazelle, B.: Triangulating a simple polygon in linear time. Discrete Comput. Geom. **6**(5), 485–524 (1991)
6. Chin, F., Snoeyink, J., Wang, C.A.: Finding the medial axis of a simple polygon in linear time. Discrete Comput. Geom. **21**(3), 405–420 (1999)
7. Citovsky, G., Mayer, T., Mitchell, J.S.B.: TSP with locational uncertainty: the adversarial model. In: 33rd International Symposium on Computational Geometry. Leibniz International Proceedings in Informatics (LIPIcs), vol. 77, pp. 32:1–32:16. Schloss Dagstuhl-Leibniz-Zentrum fuer Informatik (2017)

8. Collado, R.A., Papp, D.: Network interdiction-models, applications, unexplored directions. Rutcor Research Report, RRR4, Rutgers University, New Brunswick, NJ (2012)
9. Eriksson-Bique, S., Polishchuk, V., Sysikaski, M.: Optimal geometric flows via dual programs. In: Proceedings of the Thirtieth Annual Symposium on Computational Geometry, p. 100. ACM (2014)
10. Gewali, L., Meng, A., Mitchell, J.S.B., Ntafos, S.: Path planning in $0/1/\infty$ weighted regions with applications. ORSA J. Comput. **2**(3), 253–272 (1990)
11. Golden, B.: A problem in network interdiction. Naval Research Logistics (NRL) **25**(4), 711–713 (1978)
12. Guibas, L.J., Hershberger, J., Leven, D., Sharir, M., Tarjan, R.E.: Linear-time algorithms for visibility and shortest path problems inside triangulated simple polygons. Algorithmica **2**, 209–233 (1987)
13. Guo, Q., An, B., Tran-Thanh, L.: Playing repeated network interdiction games with semi-bandit feedback. In: Proceedings of the 26th International Joint Conference on Artificial Intelligence, IJCAI 2017, pp. 3682–3690. AAAI Press (2017)
14. Kim, D.S.: Polygon offsetting using a voronoi diagram and two stacks. Comput. Aided Des. **30**(14), 1069–1076 (1998)
15. Kostitsyna, I., Löffler, M., Staals, F., Polishchuk, V.: Most vital segment barriers. CoRR abs/1905.01185 (2019)
16. Lin, K.C., Chern, M.S.: Finding the most vital arc in the shortest path problem with fuzzy arc lengths. In: Tzeng, G.H., Wang, H.F., Wen, U.P., Yu, P.L. (eds.) Multiple Criteria Decision Making, pp. 159–168. Springer, New York (1994). https://doi.org/10.1007/978-1-4612-2666-6_17
17. Löffler, M.: Existence and computation of tours through imprecise points. Int. J. Comput. Geom. Appl. **21**(1), 1–24 (2011)
18. Lubore, S.H., Ratliff, H., Sicilia, G.: Determining the most vital link in a flow network. Naval Research Logistics (NRL) **18**(4), 497–502 (1971)
19. Melissaratos, E.A., Souvaine, D.L.: Shortest paths help solve geometric optimization problems in planar regions. SIAM J. Comput. **21**(4), 601–638 (1992)
20. Mitchell, J.S.B.: Geometric shortest paths and network optimization. In: Sack, J.R., Urrutia, J. (eds.) Handbook of Computational Geometry, pp. 633–701. Elsevier, Amsterdam (2000)
21. Mitchell, J.S.: On maximum flows in polyhedral domains. J. Comput. Syst. Sci. **40**(1), 88–123 (1990)
22. Nardelli, E., Proietti, G., Widmayer, P.: A faster computation of the most vital edge of a shortest path. Inf. Process. Lett. **79**(2), 81–85 (2001)
23. Neumayer, S., Efrat, A., Modiano, E.: Geographic max-flow and min-cut under a circular disk failure model. Comput. Netw. **77**, 117–127 (2015)
24. Papadimitriou, C.H., Yannakakis, M.: Shortest paths without a map. In: Ausiello, G., Dezani-Ciancaglini, M., Della Rocca, S.R. (eds.) ICALP 1989. LNCS, vol. 372, pp. 610–620. Springer, Heidelberg (1989). https://doi.org/10.1007/BFb0035787
25. Polishchuk, V., Mitchell, J.S.: Thick non-crossing paths and minimum-cost flows in polygonal domains. In: Proceedings of the Twenty-Third Annual Symposium on Computational Geometry, SCG 2007, pp. 56–65. ACM (2007)
26. Pollack, R., Sharir, M., Rote, G.: Computing the geodesic center of a simple polygon. Discrete Comput. Geom. **4**(6), 611–626 (1989)
27. Ratliff, H.D., Sicilia, G.T., Lubore, S.: Finding the n most vital links in flow networks. Manage. Sci. **21**(5), 531–539 (1975)

28. Smith, J.C., Prince, M., Geunes, J.: Modern network interdiction problems and algorithms. In: Pardalos, P.M., Du, D.-Z., Graham, R.L. (eds.) Handbook of Combinatorial Optimization, pp. 1949–1987. Springer, New York (2013). https://doi.org/10.1007/978-1-4419-7997-1_61
29. Strang, G.: Maximal flow through a domain. Math. Program. **26**, 123–143 (1983)
30. Zhang, P., Fan, N.: Analysis of budget for interdiction on multicommodity network flows. J. Global Optim. **67**(3), 495–525 (2017)

Splaying Preorders and Postorders

Caleb C. Levy[1,2]([⊠]) and Robert E. Tarjan[1,2]

[1] Princeton University, Princeton, NJ, USA
cclevy@princeton.edu, ret@cs.princeton.edu
[2] Intertrust Technologies, Sunnyvale, CA, USA

Abstract. Let T be a binary search tree of n nodes with root r, left subtree $L = \text{left}(r)$, and right subtree $R = \text{right}(r)$. The *preorder* and *postorder* of T are defined as follows: the preorder and postorder of the empty tree is the empty sequence, and

$$\text{preorder}(T) = (r) \oplus \text{preorder}(L) \oplus \text{preorder}(R)$$
$$\text{postorder}(T) = \text{postorder}(L) \oplus \text{postorder}(R) \oplus (r),$$

where $\oplus$ denotes sequence concatenation. (We will refer to any such sequence as *a preorder* or *a postorder*). We prove the following results about the behavior of splaying [21] preorders and postorders:

1. Inserting the nodes of preorder(T) into an empty tree via splaying costs $O(n)$. (Theorem 2.)
2. Inserting the nodes of postorder(T) into an empty tree via splaying costs $O(n)$. (Theorem 3.)
3. If T' has the same keys as T and T is *weight-balanced* [18] then splaying either preorder(T) or postorder(T) starting from T' costs $O(n)$. (Theorem 4.)

For 1 and 2, we use the fact that preorders and postorders are *pattern-avoiding*: i.e. they contain no subsequences that are order-isomorphic to $(2,3,1)$ and $(3,1,2)$, respectively. Pattern-avoidance implies certain constraints on the manner in which items are inserted. We exploit this structure with a simple potential function that counts inserted nodes lying on access paths to uninserted nodes. Our methods can likely be extended to permutations that avoid more general patterns. The proof of 3 uses the fact that preorders and postorders of balanced search trees do not contain many large "jumps" in symmetric order, and exploits this fact using the dynamic finger theorem [5,6]. Items 2 and 3 are both novel. Item 1 was originally proved by Chaudhuri and Höft [4]; our proof simplifies theirs. These results provide further evidence in favor of the elusive *dynamic optimality conjecture* [21].

Keywords: Binary search trees · Pattern avoidance · Amortized time

Research at Princeton University partially supported by an innovation research grant from Princeton and a gift from Microsoft.

© Springer Nature Switzerland AG 2019
Z. Friggstad et al. (Eds.): WADS 2019, LNCS 11646, pp. 510–522, 2019.
https://doi.org/10.1007/978-3-030-24766-9_37

Outline. Section 1 discusses the mathematical preliminaries, historical background, and context for this investigation, and Sect. 2 samples some related work. Familiar readers may skip directly to the main results and their proofs, in Sects. 3 and 4. Section 3 proves that inserting both preorders and postorders via splaying takes linear time. Section 4 establishes that splaying preorders and postorders of *balanced* search trees [18] takes linear time, regardless of starting tree. Section 5 provides our thoughts on how to analyze insertion splaying permutations that avoid more general patterns, particularly the class of "*k*-increasing" sequences [3].

1 Preliminaries

Binary Search Trees. A *binary tree T* contains of a finite set of *nodes*, with one node designated to be the *root*. All nodes have a *left* and a *right child* pointer, each leading to a different node. Either or both children may be *missing*, and we denote a missing child by null. Every node in T, save for the root, has a single *parent* node of which it is a child. (The root has no parent.) The *size* of T is the number of nodes it contains, and is denoted $|T|$.

There is a unique path from root(T) to every other node x in T, called the *access path for x in T*. If x is on the access path for y then we say x is an *ancestor* of y, and y is a *descendent* of x. We refer to the subtree comprising x and all of its descendants as the subtree rooted at x. Nodes thus have *left* and *right subtrees* rooted respectively at their left and right children. (Subtrees are *empty* for null children.) The *depth* of the node x, denoted $d_T(x)$, is the number of edges on its access path. Its *right-depth* is the number of right pointers followed, and its *left-depth* is the number of left pointers followed.

In a *binary search tree*, every node has a unique *key*, and the tree satisfies the *symmetric order* condition: every node's key greater than those in its left subtree and smaller than those in its right subtree. The binary search tree derives its name from how its structure enables finding keys. To find a key k initialize the current node to be the root. While the current node is not null and does not contain the given key, replace the current node by its left or right child depending on whether k is smaller or larger than the key in the current node, respectively. The search returns the last current node, which contains k if k is in the tree and otherwise null.

The *lowest common ancestor* of x and y in T, denoted $\mathrm{lca}_T(x,y)$, is the deepest node shared by the access paths of both x and y. Since the root is a common ancestor of any pair of nodes in T and T is finite, $\mathrm{lca}_T(x,y)$ exists and is well defined. Furthermore $\min\{x,y\} \leq \mathrm{lca}_T(x,y) \leq \max\{x,y\}$.

To *insert* a new key k into a binary search tree T, we first do a search for k in T. When the search reaches a missing node, we replace this node with a node containing the key k. (Inserting into an empty tree makes k the root key.)

Rotation. Binary search trees are the canonical data structure for maintaining an ordered set of elements, and are building blocks in countless algorithms.

Perhaps the most attractive feature of binary search trees is that the number of comparisons required to find an item in an n-node binary search tree is $O(\log n)$, provided that the tree is properly arranged, which is good in theory and practice. However, without exercising care when inserting nodes, a binary search tree can easily become unbalanced (for example when inserting $1, 2, \ldots, n$ in order), leading to search costs as high as $\Omega(n)$. Thus, binary search trees require some form of maintenance and restructuring for good performance.

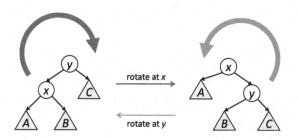

Fig. 1. Rotation at node x with parent y, and reversing the effect by rotating at y.

We will employ a restructuring primitive called *rotation*. A rotation at left child x with parent y makes y the right child of x while preserving symmetric order. A rotation at a right child is symmetric, and rotation at the root is undefined. (See Fig. 1). A rotation changes three child pointers in the tree.

Rotations were first employed in "balanced" search trees, which include AVL trees [1], Red-Black trees [10], weight-balanced trees [18], and more recently weak AVL trees [11]. These trees augment nodes with bits that provide rough information about how "balanced" each node's subtree is. Whenever an item is inserted or deleted, rotations are performed to restore invariants on the balance bits that ensure all search paths have $O(\log n)$ nodes. While balanced searched trees are not the focus of this work, they were progenitors for the main algorithm of interest.

Splaying. The Splay algorithm [21] eschews keeping track of balance information, replacing it with an intriguing notion: instead of adjusting the search tree only after insertion and deletion, Splay modifies the tree after *every* search.

The algorithm begins with a binary search for a key in the tree. Let x be the node returned by this search. If x is not `null` then the algorithm repeatedly applies a "splay step" until x becomes the root. A splay step applies a certain series of rotations based on the relationship between x, its parent, and its grandparent, as follows. If x has no grandparent (i.e. x's parent is the root), then rotate at x (this case is always terminal). Otherwise, if x is a *left* child and its parent is a *right* child, or vice-versa, rotate at x twice. Otherwise, rotate at x's parent, and then rotate at x. Sleator and Tarjan [21] assigned the respective names *zig*, *zig-zag* and *zig-zig* to these three cases. The series of splay steps that

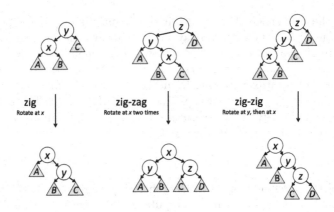

Fig. 2. A splaying step at node x. Symmetric variants not shown. Triangles denote subtrees.

bring x to the root are collectively called to as *splaying at* x, or simply *splaying* x. The three cases are depicted in Fig. 2.

The *cost* of splaying a single item x in T is defined to be $d_T(x) + 1$.[1] If $X = (x_1, \ldots, x_m)$ is a sequence of requested keys in T then the cost of splaying X starting from T is defined as $m + \sum_{i=1}^{m} d_{T_{i-1}}(x_i)$, where $T_0 = T$, and for $1 \leq i \leq m$, we form T_i by splaying x_i in T_{i-1}. To perform *insertion splaying*, insert a key into the tree and then splay the newly created node. The cost of an insertion splay is the cost splaying the new node.

While splaying individual items can cost $\Omega(n)$, the *total* cost of splaying m requested items in a tree of size $n > 0$ is $O((m+n) \log n)$. Hence, the worst case cost of a splay operation, *amortized* over all the requests, is the same as any balanced binary search tree. This is perhaps surprising for an algorithm that keeps no record of balance information.

What makes Splay truly remarkable is how it takes advantage of "latent structure" in the request sequence, and provides more than simple "worst-case" guarantees. As just one example, if $t_X(i)$ is the number of different items accessed before access i since the last access to item x_i (or since the beginning of the sequence if i is the first access to x_i), then the cost to splay X starting from T is $O(n \log n + \sum_{j=1}^{m} \log(t_X(j) + 1))$ [21].[2] (This is called the "working set" property.) Thus, Splay exploits "temporal locality" in the access pattern.

Splay simultaneously exploits "spatial" locality, as shown by the following theorem (originally conjectured in [21]) that we will use later on:

Theorem 1 (Dynamic Finger [5,6]). *Let the rank of* x *in* T, *denoted* $r_T(x)$, *be the number of nodes in* T *whose keys are less than or equal the key in* x. *The cost of splaying* $X = (x_1, \ldots, x_m)$ *starting from* T *is* $O(|T| + m + \mathrm{DF}_T(X))$, *where* $\mathrm{DF}_T(X) \equiv \sum_{i=2}^{m} \log_2(|r_T(x_i) - r_T(x_{i-1})| + 1)$.

---

[1] We absorb the search cost into the rotations.

[2] Note that $O(\log n)$ amortized cost per splay is a corollary of this.

In fact, the properties of Splay inspired the authors of [21] to speculate on a much stronger possibility: that Splay's cost is always within a constant factor of the "optimal" way of executing the requests. Formally, an *execution E* for (X, T) comprises the following. Let $T_0 = T$, and for $1 \leq i \leq m$, we perform some number $e_i \geq 0$ of rotations starting from T_{i-1} to form T_i, followed by a search for x_i. The *cost* of this execution is $\sum_{i=1}^{m}(1 + e_i + d_{T_{i-1}}(x_i))$. The *optimal cost* $\mathrm{OPT}(X, T) \equiv \min\{\mathrm{cost}(E) \mid E \text{ executes } (X, T)\}$. The following conjecture has spawned a great deal of related research (see Sect. 2):

Conjecture 1 (Dynamic Optimality [21]). $\mathrm{cost}_{\mathrm{splay}}(X, T) = O(\mathrm{OPT}(X, T))$.

The conjecture remains open. In fact, there is no sub-exponential time algorithm *whatsoever* that is known to compute, even to within a constant factor, the cost of an optimum binary search tree execution for an instance. There are several known lower bounds [7,25], none known to be tight (though some conjectured to be).

Pattern-Avoidance. For simplicity, we restrict subsequent discussion to *permutation* request sequences (i.e. no key is requested twice). By [3], any algorithm that achieves optimal cost on all permutations can be extended to an algorithm that is optimal for all request sequences.

An auxiliary question to determining if Splay (or any other algorithm) is dynamically optimal is: "what class(es) of permutations have optimum executions with 'low' cost?" This issue is not a mere curiosity, as almost every permutation of length n has optimal execution cost $\Theta(n \log n)$ [14], a bound achieved by any balanced search tree. Thus, in the absence of insertions or deletions, adjusting the tree after every access only gives an advantage on a small subset of "structured" request sequences. In addition, these structured request sequences provide candidate counter-examples to dynamic optimality. In this work, we focus on certain *pattern-avoiding* permutations: those that do not contain any subsequences of a specified type. More formally:[3]

Two permutations $\alpha = (a_1, \ldots, a_n)$ and $\beta = (b_1, \ldots, b_n)$ of the same length are *order-isomorphic* if their entries have the same relative order, i.e. $a_i < a_j \iff b_i < b_j$. For example, $(5, 8, 1)$ is order-isomorphic to $(2, 3, 1)$. A sequence π *avoids* a sequence α (or is called α-*avoiding*) if it has no subsequence that is order-isomorphic with α. If π is α-avoiding then all subsequences of π are α-avoiding. We use $\pi \backslash \alpha$ as shorthand for "an (arbitrary) permutation π that avoids α." Both preorders and postorders may be characterized as pattern-avoiding permutations:

Lemma 1 (Lemma 1.4 from [13]). *For any permutation π:*

(a) $\pi = \mathrm{preorder}(T)$ for some binary search tree $T \iff \pi$ avoids $(2, 3, 1)$.
(b) $\pi = \mathrm{postorder}(T)$ for some binary search tree $T \iff \pi$ avoids $(3, 1, 2)$.

---

[3] The following definitions and theorems are taken from [13, Chapter 1.3], almost verbatim.

Proof (Sketch). For preorders, Kozma builds a bijection between binary search trees and $(2, 3, 1)$-avoiding sequences, and uses a simple argument by contradiction to show preorders avoid $(2, 3, 1)$ [13]. The proof for postorders is a nearly symmetric variation of this argument.   □

2 Related Work

The first result about Splay's behavior on pattern-avoiding request sequences was the *sequential access* theorem [24]: the cost of splaying the nodes of T in order is $O(|T|)$. This is a special case of a corollary[4] of dynamic optimality:

Conjecture 2 (Traversal [21]). There exists $c > 0$ for which the cost of splaying preorder(T) starting from T' is at most $c|T|$ for all pairs of binary search trees T, T' with the same keys.

Theorem 2 and [4] is another special case, when $T = T'$. In Sect. 3 we prove a new special case: when T is α-weight balanced.

Interest in the behavior of binary search tree algorithms on "structured" request sequences was revived by Seth Pettie's analysis of the performance of Splay-based deque data structures using Davenport-Schinzel sequences [19], and his later reproof of the sequential access theorem via the theory of forbidden submatrices [20].

This analysis was later adapted to and greatly extended for another binary search tree algorithm, colloquially known as "Greedy," that was first proposed as an *off*-line algorithm independently by Lucas [16] and Munro [17]. Greedy is widely conjectured to be dynamically optimal, and is known to have many of the same properties of Splay, including the working set [8] and dynamic finger [12] bounds.

Greedy was later recast as an on-line algorithm in a "geometric" view of binary search trees [7]. This geometric view of Greedy is especially amenable to forbidden submatrix analysis. In [3], Chalermsook et. al. show that Greedy has nearly-optimal run-time on a broad class of pattern-avoiding permutations. Moreover, they demonstrate that if Greedy is optimal on a certain class of "non-decomposable" permutations then it is dynamically optimal. Chalermsook et al.'s analysis was later simplified in [9].

3 Insertion Splaying

If $\pi = (p_1, \ldots, p_n)$ is a permutation then the *insertion tree for* π, denoted $\text{BST}(\pi)$, is the binary search tree obtained by starting from an empty tree and inserting keys in order of their first appearance in π.

Lemma 2. *If x is a proper ancestor of y in $\text{BST}(\pi)$ then x precedes y in π.*

---

[4] A priori, the traversal conjecture follows from dynamic optimality *conditioned* on Splay being optimal with low "additive overhead." The authors recently proved that this corollary is actually *unconditional* [15].

Proof. Let $\pi_{\prec y}$ denote the prefix of π containing the elements preceding y. By construction, y is inserted as a child of some node z in $\mathrm{BST}(\pi_{\prec y})$. Every proper ancestor of y is an ancestor of z, thus $x \in \mathrm{BST}(\pi_{\prec y})$. Hence, x precedes y.   □

Insertion splaying π has the same cost as splaying π starting from $\mathrm{BST}(\pi)$.[5] For the purposes of analysis we will assume that, initially, every node in $\mathrm{BST}(\pi)$ is marked as *untouched*. An insertion splay marks the node as *touched*, and then splays the node. The touched nodes form a connected subtree containing the root, called the *touched subtree*. The untouched nodes form subtrees each of which contains no touched node. Call an untouched node with a touched parent a *sub-root*. The subtrees rooted at sub-roots have identical structure in both the splayed tree and $\mathrm{BST}(\pi)$. By Lemma 2, the next node to be touched is always a sub-root.

For $1 \leq i \leq n$, form T_i by touching and then splaying p_i in T_{i-1}, where $T_0 = \mathrm{BST}(\pi)$ starts with all nodes untouched. At any time we define the *potential* to be the twice the number of touched nodes that are ancestors of sub-roots, and we define Φ_i to be the potential of T_i. The *amortized* cost of splaying p_i in T_{i-1} is defined as $c_i = t_i + \Phi_i - \Phi_{i-1}$, where t_i denotes the actual cost. By a standard telescoping sum argument, the cost of insertion splaying π is $\sum_{i=1}^n t_i = \sum_{i=1}^n c_i + \Phi_0 - \Phi_n$ [23]. Since $\Phi_0 = \Phi_n = 0$, an upper bound on amortized cost provides an upper bound on the actual cost.

Pattern-avoidance provides certain information about both $\mathrm{BST}(\pi)$ and about which sub-root can be touched next. We exploit this information in the next two sections.

Insertion Splaying Preorders. There are no restrictions on the possible structure of preorder insertion trees as $\mathrm{BST}(\mathrm{preorder}(T)) = T$.[6] However, the manner in which sub-roots are chosen is particularly simple.

Lemma 3. *If $\pi \backslash (2,3,1) = (p_1,\ldots,p_n)$ is a preorder then, for $1 \leq i \leq n$, p_i is the smallest sub-root of T_{i-1}, where all nodes begin untouched in $T_0 = \mathrm{BST}(\pi)$ and T_i is formed by touching and splaying p_i in T_{i-1}.*

Proof. The statement is vacuously true for $i = 1$. We prove for $i > 1$ by contradiction, as follows. Suppose T_{i-1} has some sub-root q that is smaller than p_i. Since q and p_i are *both* sub-roots in T_{i-1}, they are both children of respective (though not necessarily distinct) nodes a and b in T_{i-1}. Let $r = \mathrm{lca}_{T_{i-1}}(a,b)$. Since $q \neq a$ and $p_i \neq b$, all of p_i, q and r are distinct nodes in T_{i-1}, and furthermore $q < r < p_i$. By Lemma 2, r precedes both q and p_i in π, and by construction p_i precedes q. We thus have (r,p_i,q) is a subsequence of π. But (r,p_i,q) is order-isomorphic with $(2,3,1)$, contradicting $\pi \backslash (2,3,1)$.   □

---

[5] This is because the manner in which Splay restructures the access path is independent of nodes outside the path.

[6] In fact, this property is shared by *any* permutation π for which every node in T appears in π before those in its left and right subtrees.

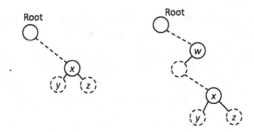

Fig. 3. Possible locations for the next sub-root x to be insertion splayed in $\pi\backslash(2,3,1)$. The case on the left occurs when the next splayed node has left-depth 0, and the case on the right occurs when it has left-depth 1. Dashed nodes may or may not be present, and any number of nodes may lie on the paths denoted by dashed lines.

Theorem 2. *Insertion splaying* preorder(T) *keeps each sub-root at left-depth at most 1 and takes $O(1)$ amortized time per splay operation.*

Proof. The theorem is trivial for the first insertion splay. The inductive hypothesis is that every sub-root has left depth 0 or 1. Let x be the next sub-root to be splayed, and let y and z (either or both of which can be missing) be its left and right children. Touching x makes y and z into sub-roots.

Suppose x has left depth 0 before it is touched. Converting x from untouched to touched (without splaying it) increases the potential by at most 2 and gives the new sub-roots y and z left depths of 1 and 0, respectively. (In this case they are the only two sub-roots.) Each splay step, except possibly the last, is a zig-zig in which x starts as a left child with parent p and grandparent g. After completing the zig-zig, g is no longer an ancestor of any untouched node, which decreases the potential by 2. The zig-zig also preserves the left depths of y and z. (y becomes the right child of p.) No other sub-roots can increase left-depth, as x is the smallest sub-root. If the last splay step is a zig, the potential does not change (although the length of the path to y increases by 1).

More complicated is the case in which x has left depth 1. Converting x from untouched to touched (without splaying it) makes y a sub-root of left depth 2 and z a sub-root of left depth 1. Let w be the parent of the ancestor of x that is a left child. All other sub-roots are in the right subtree of w, which is unaffected by splaying x. The splay of x consists of 0 or more left zig-zigs, followed by a zig-zag (which can either be left-right or right-left), followed by zero of more left zig-zigs, followed possibly by a zig. Each zig-zig reduces the potential by 2 and preserves the left depths of all sub-roots. The zig-zag does not increase the potential, reduces the left depth of y from 2 to 1, and that of x from 1 to 0, and preserves the left depth of z. Now x has left depth 0, and the argument above applies to the remaining splay steps.

By Lemma 3, the next node to be splayed will be y if present, otherwise z if present, otherwise w if present. All three of these items have left-depth 0 or 1, hence an identical form to Fig. 3. Thus the hypothesis holds.

To obtain the constant factor, we observe that converting x from untouched to touched increases the potential by 2. Each zig-zig step pays for itself: it requires 2 rotations, paid for by the potential decreasing by at least 2. The zig-zag requires 2 rotations, and the zig requires 1 rotation. If the cost of a splay is the number of nodes on the splay path, equal to the number of rotations plus 1, we have an amortized cost of 6 per splay.   □

Insertion Splaying Postorders. Postorder insertion trees are more restricted. A binary search tree C is a *(left-toothed) comb* if the access path for $x \in C$ always comprises some number $j \geq 0$ of right children followed by some number $k \geq 0$ of left children. The nodes of C are partitioned into *teeth*, where every node in the ith tooth has right-depth $i - 1$. The shallowest node in a tooth is called the *head*. The insertion trees of postorders are combs:

Lemma 4. *If π is a postorder then no left child in $\mathrm{BST}(\pi)$ has a right child.*

Proof. By contradiction. Let y be a left child in $\mathrm{BST}(\pi)$ with right child z, and let $x = \mathrm{parent}(y)$. As z is y's right child, $y < z$. Similarly, as both y and z are in x's left subtree, $y < z < x$. By Lemma 2, y can be an ancestor of z only if y precedes z in π, and similarly x must precede y. Thus, (x, y, z) is a subsequence of π that is order-isomorphic to $(3, 1, 2)$. By Lemma 1(b), π is not a postorder.   □

While postorder insertion trees are less varied than for preorders, there may be many postorders with a given insertion tree. This affords some amount of freedom for choosing different sub-roots.

Lemma 5. *Let $\pi \backslash (3, 1, 2) = (p_1, \ldots, p_n)$ be a postorder with insertion tree sequence $T_0, T_1, \ldots, T_n$. For $1 \leq i \leq n$, p_i is either:*

(a) *The single sub-root greater than $\max\{T_{i-1}\}$ (if present), or*
(b) *The largest sub-root smaller than $\max\{T_{i-1}\}$ (if present).*

Proof. The result is vacuous for $i = 1, 2$. If p_i is case (a), we merely note that if p_i is a new maximum then it must be the right child of the largest node in $\max\{T_{i-1}\}$. There can be at most one sub-root in this position. Hence, p_i is unique.

For the sake of contradiction, suppose p_i is not of the form in case (a) or (b), and let q be the largest sub-root smaller than $\max\{T_{i-1}\}$. By Lemma 2, the items of each tooth are added in decreasing order. As q is not the head of its tooth, its successor r must be in T_{i-1}, and furthermore r precedes both p_i and q in π. By construction, (r, p_i, q) is a subsequence of π. Yet this subsequence is isomorphic to $(3, 1, 2)$ since $p_i < q < r$, contradicting Lemma 1(b).   □

Theorem 3. *Insertion splaying postorders maintains the following invariants:*

1. *After each insertion splay, the path to every sub-root comprises $j \geq 0$ left pointers followed by $k \geq 0$ right pointers. (Furthermore, after the first insertion, $k \geq 1$.)*

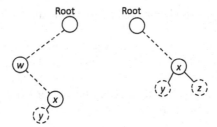

Fig. 4. Possible locations for the next sub-root x to be insertion splayed in $\pi\backslash(3,1,2)$. The case on the left occurs when the next splayed node is less than the root, and the case on the right occurs when the next sub-root is the new tree maximum. Dashed nodes may or may not be present, and any number of nodes may lie on the paths denoted by dashed lines.

2. *The left-depth of every sub-root decreases from smallest to largest.*[7]
3. *The splay operation takes constant amortized time.*

Proof. The base case is trivial. Lemma 5 dictates that the next splayed sub-root is either greater than all marked items, or is the largest sub-root smaller than the tree root. Let x be the next node to be insertion splayed, y its left child, and z its right child (either or both children may be missing).

Suppose x is greater than the current tree root. Marking x increases the potential by 2 and makes y and z new sub-roots. The splay operation brings x to the root by a sequence of left zig-zigs followed possibly by a left zig (depending on whether the length of the access path is odd or even). After each one of these zigs or zig-zigs, y's left-depth remains 1, and z's left depth remains 0. Let v be the root prior to the splay operation. If the last splay step is a zig then the last splay operation increases the left depth of v and everything in its left subtree by either 1 or 2. Since the left-depth of x was 0 and x was the largest sub-root, the inductive hypothesis ensures that all sub-roots had left-depth at least 1 before the splay operation, and therefore at least 2 afterward. Thus, when x becomes the root, the left-depths of each sub-root decrease from left to right.

Otherwise, x is the largest sub-root less than the root. Marking x again increases the potential by at most 2. By Lemma 4, x has no right child (see Fig. 4), so we only need to worry about its left child y. Let w be the last ancestor of x that is a left child. Each left zig-zig prior to the splay step involving w maintains the left-depth of y to be one greater than the left-depth of x. The splay step involving w will either be a left zig-zig or a left-right zig-zag, depending on the length of the original path connecting w to x. Regardless, immediately after the splay step involving w, the ancestor of y that is the left child of x is either the left child of w or the left child of w's parent. Since all the sub-roots less than y are in the left subtree of w, and thus have left-depth greater than the left-depth

---

[7] The first two invariants dictate that the ancestors of sub-roots form a *right*-toothed comb.

of y, the invariant is restored, and remains true after each right zig-zig or zig that brings x to the root.

All that remains is showing constant amortized time. As noted before, marking x costs 2. If x is greater than the root then each left zig-zig, except possibly the last, pays for itself, giving amortized cost of 4. In the other case, all splay steps except for the one involving w and the one making x the root pay for themselves, giving amortized cost at most 6.   □

4   Balanced Trees

Let $|x|$ denote the size of the subtree rooted at x. Following [18], we say T is α *weight balanced* for $\alpha \in (0, 1/2]$ if $\min\{|\,\text{left}(x)|, |\,\text{right}(x)|\} + 1 \geq \alpha \cdot (|x| + 1)$ for all $x \in T$, and write $T \in \text{BB}[\alpha]$.

Theorem 4. *For any* (fixed) $0 < \alpha \leq 1/2$, *if* $S \in \text{BB}[\alpha]$ *and* T *has the same keys as* S, *then the cost of splaying* $\text{preorder}(S)$ *or* $\text{postorder}(S)$ *starting from* T *is* $O(|T|)$.

Proof. By Theorem 1, it suffices to show that $\text{DF}_T(\text{preorder}(S)) = O(|T|)$. Let

$$A_\alpha(n) \equiv \max\{\text{DF}_T(\text{preorder}(S)) \mid S \in \text{BB}[\alpha] \text{ and } |T| = n\}.$$

Recall that $\text{preorder}(S) = (\text{root}(S)) \oplus \text{preorder}(L) \oplus \text{preorder}(R)$, where L and R are the left and right subtrees of the root of S, respectively. Notice that the rank differences between $\text{root}(S)$ and the first item in $\text{preorder}(L)$, and between the last item in $\text{preorder}(L)$ and the first item in $\text{preorder}(R)$, are at most $|T|$ by definition. Hence,

$$\text{DF}_T(\text{preorder}(S)) \leq \text{DF}_T(\text{preorder}(L)) + \text{DF}_T(\text{preorder}(R)) + 2\log_2(|T| + 1).$$

Observe that $(|L|+1)/(|S|+1) \in [\alpha, 1-\alpha]$ since $S \in \text{BB}[\alpha]$, and by definition $|R| < |S| - |L|$. Hence[8],

$$A_\alpha(n) = \max_{\alpha \leq \beta \leq 1/2}\{A_\alpha(\beta \cdot n) + A_\alpha((1-\beta) \cdot n)\} + O(\log n).$$

Akra-Bazzi's result [2] suffices to show $A_\alpha(n) = O(n)$ for fixed α. The proof for postorders is identical.   □

Remark 1. In actuality, $A_\alpha(n) = O(f(\alpha) \cdot n)$ for some function f of α. Unfortunately, the computation appears to be messy. We have declined to do the necessary footwork, as we strongly suspect that, regardless, $A_\alpha(n)$ does not tightly bound the cost of splaying these sequences.

Remark 2. This result extends to any binary search tree algorithm that satisfies the dynamic finger bound. Iacono and Langerman proved Greedy also has the dynamic finger property [12]; their analysis does not consider initial trees, however.

---

[8] Technically, since $|L|/|S| < (|L| + 1)/(|S| + 1)$, we need to pick S sufficiently large for a given alpha, and offset the recurrence term by a corresponding constant. This does not asymptotically affect the result.

5   Remarks

Patterns that avoid $(2, 1, 3)$ are "symmetric" to those that avoid $(2, 3, 1)$: if $\pi \backslash (2, 1, 3)$ then π is the preorder of the mirror image of $\mathrm{BST}(\pi)$. Similarly, patterns that avoid $(1, 3, 2)$ are symmetric to patterns that avoid $(3, 1, 2)$. Thus, insertion splaying $\pi \backslash (2, 1, 3)$ and $\pi \backslash (1, 3, 2)$ takes linear time.

The only other patterns of length three are $(3, 2, 1)$ and its symmetric counterpart $(1, 2, 3)$. The pattern $(3, 2, 1)$ was explored in [3], where it was shown that Greedy executes $(3, 2, 1)$-avoiding permutations in linear time starting from an *arbitrary* tree. In fact, they showed that executing $\pi \backslash (k, \ldots, 2, 1)$ takes time proportional to $n \cdot 2^{O(k^2)}$; this is linear in n for fixed k. These permutations are called *k-increasing* because they can be partitioned into $k - 1$ disjoint monotonically increasing subsequences [3]. They form the natural generalization of *sequential access*, which is the (unique) permutation of the tree nodes that avoids $(2, 1)$.

More general invariants can be derived about insertion tree structure and subroot insertion order based on pattern-avoidance. As one particularly interesting example:

Theorem 5. *If* $\pi \backslash (k, \ldots, 2, 1)$ *then no node in* $\mathrm{BST}(\pi)$ *has left-depth more than* $k - 2$, *and the next sub-root inserted (without splaying) is always the smallest sub-root with its given left-depth.*

The proof is similar to Lemmas 4 and 5. In particular, the insertion trees of $(3, 2, 1)$-avoiding permutations look like the combs of postorder insertion trees, except the teeth are rightward, instead of leftward paths.

For k-increasing sequences, the potential used for Theorems 2 and 3 needs modifications. The main issue is that in both of these cases, the zig-zigs paid for themselves because the nodes knocked off the access path did not have sub-root descendants. This structure no longer holds for $(3, 2, 1)$-avoiding sequences, since we must splay the nodes of the teeth in increasing order. The proof seems to require a generalization of the sequential access theorem. It is possible that the notion of *kernel trees* used by Sundar in [22] for a potential-based proof of the sequential access theorem could be useful.

References

1. Adel'son-Vel'skii, G., Landis, E.: An algorithm for the organization of information. Sov. Math. Dokl. **3**, 1259–1262 (1962)
2. Akra, M., Bazzi, L.: On the solution of linear recurrence equations. Comput. Optim. Appl. **10**(2), 195–210 (1998)
3. Chalermsook, P., Goswami, M., Kozma, L., Mehlhorn, K., Saranurak, T.: Pattern-avoiding access in binary search trees. In: FOCS, pp. 410–423 (2015)
4. Chaudhuri, R., Höft, H.: Splaying a search tree in preorder takes linear time. ACM SIGACT News **24**(2), 88–93 (1993)
5. Cole, R.: On the dynamic finger conjecture for splay trees. Part II: the proof. SICOMP **30**(1), 44–85 (2000)

6. Cole, R., Mishra, B., Schmidt, J., Siegel, A.: On the dynamic finger conjecture for splay trees. Part I: splay sorting $\log n$-block sequences. SICOMP **30**(1), 1–43 (2000)
7. Demaine, E., Harmon, D., Iacono, J., Kane, D., Pătraşcu, M.: The geometry of binary search trees. In: SODA, pp. 496–505 (2009)
8. Fox, K.: Upper bounds for maximally greedy binary search trees. In: WADS, pp. 411–422 (2011)
9. Goyal, N., Gupta, M.: Better analysis of binary search tree on decomposable sequences. Theor. Comput. Sci. **776**, 19–24 (2019)
10. Guibas, L., Sedgewick, R.: A dichromatic framework for balanced trees. In: FOCS, pp. 8–21 (1978)
11. Haeupler, B., Sen, S., Tarjan, R.: Rank-balanced trees. TALG **11**(4), 30 (2015)
12. Iacono, J., Langerman, S.: Weighted dynamic finger in binary search trees. In: SODA, pp. 672–691 (2016)
13. Kozma, L.: Binary search trees, rectangles and patterns. Ph.D. thesis, Saarland University (2016)
14. Kujala, J., Elomaa, T.: The cost of offline binary search tree algorithms and the complexity of the request sequence. TCS **393**, 231–239 (2008)
15. Levy, C., Tarjan, R.: A new path from splay to dynamic optimality. In: SODA, pp. 1311–1330 (2019)
16. Lucas, J.: Canonical forms for competitive binary search tree algorithms. Technical report, Rutgers University (1988)
17. Munro, I.: On the competitiveness of linear search. In: ESA, pp. 338–345 (2000)
18. Nievergelt, J., Reingold, E.: Binary search trees of bounded balance. SICOMP **2**(1), 33–43 (1973)
19. Pettie, S.: Splay trees, davenport-schinzel sequences, and the deque conjecture. In: SODA, pp. 1115–1124 (2008)
20. Pettie, S.: Applications of forbidden 0–1 matrices to search tree and path compression-based data structures. In: SODA, pp. 1457–1467 (2010)
21. Sleator, D., Tarjan, R.: Self-adjusting binary search trees. J. ACM **32**(3), 652–686 (1985)
22. Sundar, R.: On the deque conjecture for the splay algorithm. Combinatorica **12**(1), 95–124 (1992)
23. Tarjan, R.: Amortized computational complexity. SIAM J. Algebraic Discrete Meth. **6**(2), 306–318 (1985)
24. Tarjan, R.: Sequential access in splay trees takes linear time. Combinatorica **5**(4), 367–378 (1985)
25. Wilber, R.: Lower bounds for accessing binary search trees with rotations. SICOMP **18**(1), 56–67 (1989)

Wannabe Bounded Treewidth Graphs Admit a Polynomial Kernel for DFVS

Daniel Lokshtanov[1], M. S. Ramanujan[2], Saket Saurabh[3], Roohani Sharma[3(✉)], and Meirav Zehavi[4]

[1] University of California, Santa Barbara, USA
daniello@ucsb.edu
[2] University of Warwick, Coventry, UK
R.Maadapuzhi-Sridharan@warwick.ac.uk
[3] Institute of Mathematical Sciences, HBNI, Chennai, India
{saket,roohani}@imsc.res.in
[4] Ben-Gurion University of the Negev, Beersheba, Israel
meiravze@bgu.ac.il

Abstract. In the DIRECTED FEEDBACK VERTEX SET (DFVS) problem, given a digraph D and $k \in \mathbb{N}$, the goal is to check if there exists a set of at most k vertices whose deletion from D leaves a directed acyclic graph. Resolving the existence of a polynomial kernel for DFVS parameterized by the solution size k is a central open problem in Kernelization. In this paper, we give a polynomial kernel for DFVS parameterized by k *plus* the size of a treewidth-η modulator. Our choice of parameter strictly encompasses previous positive kernelization results on DFVS. Our main result is based on a novel application of the tool of *important separators* embedded in state-of-the-art machinery such as protrusion decompositions.

Keywords: DFVS · Kernel · Important separator · Treewidth

1 Introduction

FEEDBACK SET problems are fundamental combinatorial optimization problems. Typically, in these problems, we are given a (directed or undirected) graph G and an integer k, and the objective is to select at most k vertices, edges or arcs to hit all cycles of the input graph. FEEDBACK SET problems are among Karp's 21 NP-complete problems and have been a subject of active research from algorithmic [3,5,6,9–11,13,15,18,19,25,31,32,34,36,44,48] as well as structural point of view [24,33,35,43,45–47]. In particular, such problems constitute one of the most important topics of research in parameterized algorithms [9,11,13,15,18,19,32,34,36,44,48], spearheading the development of several new techniques.

In this paper, we study the DIRECTED FEEDBACK VERTEX SET (DFVS) problem, whose input consists of a digraph D on n vertices and m arcs, and

Appeared as a Brief Announcement at ICALP 2018.

© Springer Nature Switzerland AG 2019
Z. Friggstad et al. (Eds.): WADS 2019, LNCS 11646, pp. 523–537, 2019.
https://doi.org/10.1007/978-3-030-24766-9_38

an integer k that is the parameter. The goal is to check whether there exists a vertex subset of size at most k that intersects every directed cycle in D. In other words, we ask whether there exists a set of vertices S of size at most k such that $F = D - S$ is a directed acyclic graph (DAG). For over a decade, resolving the parameterized complexity of DFVS was considered one of the most important open problems in parameterized complexity. In fact, this question was posed as an open problem in the first few papers on fixed-parameter tractability (FPT) [21,22]. In a breakthrough paper, DFVS was shown to be fixed-parameter tractable by Chen et al. [13] in 2008, who gave an algorithm that runs in time $\mathcal{O}(4^k \cdot k! \cdot k^4 \cdot n^4)$. Subsequently, it was observed that, in fact, the running time of this algorithm is $\mathcal{O}(4^k \cdot k! \cdot k^4 \cdot nm)$ (see, e.g., [16]). Since this breakthrough, the techniques used to solve DFVS have found numerous applications. However, apart from the design of a parameterized algorithm for DFVS with a linear dependency on $m + n$ [40], the question of the existence of a polynomial kernel for the same problem has seen close to no progress. To be specific, the following fundamental question about the problem remains open:

Does DFVS admit a polynomial kernel?

That is, does there exist a polynomial-time algorithm (called a *kernelization algorithm*) that, given an instance (D, k) of DFVS, returns an equivalent instance (D', k') (called a *kernel*) of DFVS whose size is bounded by a polynomial function of k? We refer the reader to the surveys [29,30,37,39], as well as the books [16,23,26,28,42], for a detailed treatment of the area of kernelization.

The lack of progress on the kernelization complexity of DFVS has led to the study of this problem on restrictive input instances. In particular, we know of polynomial kernels for DFVS when the input digraph is a tournament or even various other generalizations of it [1,4,20,38]. However, the existence of a polynomial kernel for DFVS is open even when the input digraph is a planar digraph. Recently, in order to shed some light on the kernelization complexity of DFVS, the following two directions have been proposed.

1. Study the kernelization complexity of DFVS where, in addition to the solution size, we parameterize by a structural parameter (such as the size of a modulator to a graph of constant treewidth). Throughout the paper, by treewidth of a digraph we refer to the *treewidth of its underlying undirected graph*.
2. Study the kernelization complexity of DFVS with an additional restriction on the resulting DAG $F = D - S$.

In this paper, we aim to significantly broaden the scope of both directions as much as possible—our efforts are mostly aimed at the first approach, but we also deal with the second one. Towards our contribution for the first direction, we give a polynomial kernel for DFVS parameterized by the solution size plus the size of a treewidth-η modulator. Let $\mathcal{F}_\eta$ be the family of digraphs of treewidth at most η. For a directed graph D, a subset $M \subseteq V(D)$ is called a *treewidth-η-modulator* if $D - M \in \mathcal{F}_\eta$. We consider the following parameterized problem parameterized by $k + \ell$.

DFVS/DFVS+TREEWIDTH-η MODULATOR (TW-η MOD)

Input: A digraph D, $k \in \mathbb{N}$, $M \subseteq V(D)$ where $|M| = \ell$ and $D - M \in \mathcal{F}_\eta$.

Output: Is there $S \subseteq V(D)$ where $|S| \le k$ and $D - S$ is a DAG?

Observe that DFVS/DFVS+TW-η MOD is the same problem as DFVS with just a different parameter. Our main contribution is the following theorem.

Theorem 1.1. DFVS/DFVS+TW-η MOD *admits a kernel of size* $(k \cdot \ell)^{\mathcal{O}(\eta^2)}$.

Notably, our result can be viewed as a proof that DFVS parameterized *only* by k, admits a polynomial kernel on the class of *all* graphs whose treewidth can be made constant by the removal of $k^{\mathcal{O}(1)}$ vertices. Yet another justification for our choice of parameter is the following. Parameterized by k alone, the problem has been open for a very long time. On the other hand, parameterized by ℓ alone, it can be easily seen that the problem does not exhibit a polynomial kernel (by a reduction from Vertex Cover parameterized by the size of a treewidth-2 modulator) unless NP $\subseteq$ coNP/poly [17]. Thus, $k + \ell$ is a natural parameter to explore.

We also remark that the proof of Theorem 1.1 required the development of a novel use of important separators, among other ideas for finding protrusions and using the state-of-the-art protrusion machinery. Thus, as a side reward, the ideas developed in this article may be insightful, helping to design reduction rules for a polynomial kernel of DFVS. Lastly, our result encompasses the recent result of Bergougnoux et al. [7], where they studied DFVS parameterized by the feedback vertex set number of the underlying undirected graph, and gave a polynomial kernel for this problem. Specifically, they gave a kernel of size $\mathcal{O}(\mathbf{fvs}^4)$, where **fvs** is the feedback vertex set number of the underlying undirected graph of D. Note that our parameter $k + \ell$ is not only upper bounded by $\mathcal{O}(\mathbf{fvs})$, but it can be arbitrarily smaller than **fvs**. We also remark that DFVS has already been parameterized by treewidth in the literature (not for kernelization purposes)— recently, Bonamy et al. [8] showed that DFVS parameterized by the treewidth of the input graph, t, can be solved in time $2^{\mathcal{O}(t \log t)} n^{\mathcal{O}(1)}$, and that unless the Exponential Time Hypothesis fails, it cannot be solved in time $2^{o(t \log t)} n^{\mathcal{O}(1)}$.

We now consider our contribution towards the second direction viz. understanding the kernelization complexity of DFVS with an additional restriction on the resulting DAG $F = D - S$. This direction was proposed by Mnich and van Leeuwen [41]. Essentially, the basic philosophy of their program is the following: What happens to the kernelization complexity of DFVS when we consider deletion to subclasses of DAGs? Specifically, Mnich and van Leeuwen [41] obtained polynomial kernels for the classes of out-forests, out-trees and directed pumpkins. Note that for all these families, the treewidth of the graph obtained after deleting the solution is constant. In a follow-up paper [2], the kernel sizes given by Mnich and van Leeuwen [41] were reduced. We study the problem named $\mathcal{F}_\eta$-VERTEX DELETION SET which is defined as follows. Given a digraph D and an integer k, determine whether there exists a set S of size at most k such that $D - S$ is a DAG in $\mathcal{F}_\eta$. Observe that this problem is different from DFVS as the deletion set is required to bring in more structure to the resulting graph. Towards resolving the existence of a polynomial kernel for this problem (parameterized by k), we observe that the existing machinery can already be harnessed to resolve this question affirmatively. We omit the proof of this result (Theorem 1.2) in the short version of the paper.

Theorem 1.2. $\mathcal{F}_\eta$ -VERTEX DELETION SET *admits a polynomial kernel parameterized by* k.

Given a choice of which amongst the two directions may bring us closer to the resolution of the kernelization complexity of DFVS, we believe that studying DFVS parameterized by a non-trivial structural parameter larger than k has a major advantage over studying the DFVS problem by restricting the resulting DAG—the study of a larger parameter does not alter the problem at hand, that is, the focus is still aimed at DFVS itself rather than at a variant of it. In fact, the second approach may derail us from the track of resolving the kernelization complexity of DFVS as each restriction of the output DAG results in its own definition of a variant of DFVS that may have its own properties. Thus, if the ultimate goal is to design a polynomial kernel for DFVS itself (or prove that such a kernel does not exist), we find the first approach more suitable. Nevertheless, it is also important to note that the questions raised by the second approach, namely, the study of the variants of DFVS, may be interesting in their own right.

Proof Idea of Theorem 1.1. Our kernelization algorithm can be divided into three main phases. We give a brief summary of each phase here.

1. Computing a Zone Decomposition of the Directed Graph: We first compute a decomposition of D into three components: the vertex set M (modulator), a collection of $\mathcal{O}(k\ell^2)$ vertex sets $\mathcal{Z}$ (zones), and a vertex set R (remainder) of size $\mathcal{O}(k\ell^2)$. All of these sets are pairwise disjoint and form a partition of $V(D)$. The aim of this decomposition is to achieve a few properties with respect to each zone $Z \in \mathcal{Z}$, which we will later exploit to design reduction rules to bound the size of each zone. Since the number of zones in the decomposition and the size of R is $\mathcal{O}(k\ell^2)$, in order to get the desired kernel, it would be enough to bound the size of each zone $Z \in \mathcal{Z}$ by $k\ell^{\mathcal{O}(1)}$, after such a decomposition is constructed.

Let us mention three important properties of a zone $Z \in \mathcal{Z}$ that this decomposition achieves, and which play a critical role in helping us bound the size of Z. The first property is that if D has a directed feedback vertex set of size at most k, then there exists a directed feedback vertex set, say S, in D of size at most k, whose intersection with Z is of constant size. The second property is that the neighborhood of Z is entirely contained in $M \cup R$ and the size of the neighborhood of Z in R is bounded by some constant. Finally, for any two vertices in the neighborhood of Z in M, the maximum value of a directed flow from one to the other is either extremely high or zero. We will exploit the first property to mark a "small" set of vertices in Z that in some sense "represents" all partial solutions in Z. Such a set, which is called Γ_{DFVS}, is then used to design reduction rules that eliminate arcs between $Z \setminus \Gamma_{\text{DFVS}}$ and M. The third property is critically used in these reduction rules. Then, from the second property, all the vertices in $Z \setminus \Gamma_{\mathrm{DFVS}}$ have a constant-sized neighborhood outside Z. Having this information at hand, we further partition Z into small slices, each of which is then replaced by constant sized sets by using protrusion machinery.

Though at first glance this decomposition of the graph may look very similar to the near-protrusion decomposition of [27], it is not a near-protrusion decomposition. In fact, for this problem we probably cannot find a near-protrusion decomposition.

2. Computing a k-DFVS Representative Set in Z: A k-DFVS representative in a zone Z is a subset of vertices of Z (say, Γ_{DFVS}) with the following property: If D has a directed feedback vertex set of size at most k, then there is a directed feedback vertex set S in D of size at most k and $S \cap Z \subseteq \Gamma_{\mathrm{DFVS}}$. We aim to compute such a set whose size is bounded by some polynomial in k and ℓ.

For this purpose, we first revisit the relation between our problem and SKEW MULTICUT. In particular, we see that for any directed feedback vertex set S in D, $S \cap Z$ is a solution to an "appropriate" instance of the SKEW MULTICUT problem. Thus, if we can compute solutions to all possible appropriate instances of SKEW MULTICUT, then we can set Γ_{DFVS} to be the union of all these solutions. In this overview, we prefer to keep the notion of an appropriate instance abstract.

Unfortunately, a single instance of our problem gives rise to a huge number of appropriate instances. In particular, if we naively construct Γ_{DFVS} by *individually* computing a solution for each possible choice for an appropriate instance, we do not obtain a set whose size is bounded by a polynomial function in k and ℓ. So, in the second step, we invest significant efforts to construct a set Γ_{DFVS} of the desired small size, which contains a solution for each possible choice of an appropriate instance. To this end, we observe that if such a set of the desired size exists, then solutions of "many" possible appropriate instances intersect a lot. Very roughly speaking, we aim to identify a small set of vertices that is guaranteed to be contained in solutions of "many" instances. If we can identify such a set, then we delete it from all appropriate instances in which it is guaranteed to be present in some solution, and recurse on the resulting instances. (Here, only one recursive call is performed.) From the properties of a zone decomposition, we are able to derive that there is a solution to our original problem whose intersection with Z is small, which in turn leads us to the observation that we can only focus on small solutions for each appropriate instance. Hence, we can bound the depth of the recursion. Though this description roughly conveys the broad picture, the implementation of these ideas is significantly more complex. For example, we are unable to find a small set of vertices that is contained in some solution for "many" instances. Instead, we find a *collection* of small sets such that at least one among them is the set that we want, though we do not know which one.

These abstract ideas are materialized with the help of important separators (defined in Sect. 2), the Pushing Lemma for SKEW MULTICUT and a new (simple) lemma, which we call the *Important Separator Preservation (via Small Sink Set)*

Lemma. This lemma says that if S is an important (X, Y)-separator of size α in some digraph, then S is also an important (X, Y')-separator for some subset Y' of Y of size at most $\alpha + 1$, where $Y \setminus Y'$ is removed from that digraph. We mainly use this lemma in situations (that arise when we try to compute the collection of sets mentioned above) that require guessing the set Y when X is given, so that we can compute important (X, Y)-separators. In these situations, since it is enough to guess Y' to compute all important (X, Y)-separators (from the Important Separator Preservation Lemma), the fact that Y' is small significantly reduces the search space for Y' compared to that for Y.

3. Reduction Rules for Bounding the Size of Each Zone Z**:** After computing a small set that is a k-DFVS representative in a zone in our zone decomposition, our final objective is to bound its size. To this end, we design reduction rules that decompose a zone Z into a "small" number of *protrusions*. (Roughly speaking, a protrusion in a graph G is an induced subgraph $G[U]$ of G for a subset $U \subseteq V(G)$ that has constant treewidth and only a constant number of vertices with neighbors in $G - U$.) More precisely, we first design a set of reduction rules that only bounds the size of the neighborhood of every zone Z (with Γ_{DFVS} removed) outside Z by a constant. Then, by computing a nice tree decomposition of $D[Z]$ and relying on properties of an LCA-closure in that tree, we decompose the set Z as $Z = \widetilde{\Gamma}_{\text{DFVS}} \uplus \biguplus_{U \in \mathcal{U}} U$ such that $\Gamma_{\text{DFVS}} \subseteq \widetilde{\Gamma}_{\text{DFVS}}$, the size of $\mathcal{U}$ is "small", and each set $U \in \mathcal{U}$ induces a protrusion. We then replace each protrusion $D[U]$ by a "small" digraph such that the resulting digraph is a "minor" of the original digraph, and the input modulator M is also a treewidth-η modulator in the resulting digraph. This concludes our kernelization algorithm.

We remark that the operations of the last step of our kernelization algorithm ensure that the input modulator M remains a modulator in the final returned kernel. If we allow the modulator in the returned instance to be of size larger than $|M|$, then we can bypass all of these reduction rules and the protrusion machinery, and directly create the kernelized instance by taking the torso of the set S that is the union of M, R and a k-DFVS representative set in each Z. Here, by torso we mean that for every two vertices $u, v \in S$ with a directed path from u to v whose internal vertices do not belong to S, we add an arc from u to v. Since the set S is of small size (polynomial in k and ℓ), we directly obtain a kernel by omitting the vertices outside S. However, when we perform the torso operation, we lose the property that M is a modulator for the final instance, which means that the parameter can increase to be of the magnitude of the entire kernel.

Road-map. In this extended abstract, we only present a high-level overview of our approach. In particular, our focus is to convey the main ideas of Step 2 above. In Sect. 3, we present a short description of our zone decomposition (Step 1). In Sect. 4, we describe the main difficulties, and the insights incorporated to overcome them, with respect to the design a procedure to find a k-DFVS representative set for any single zone (Step 2). In Sect. 5, we recall our final objective, that is, to bound the size of each zone (Step 3).

2 Preliminaries

Throughout the paper, parenthesis (resp. braces) notation denote ordered (resp. unordered) sets. For a digraph D and subsets $X, Y \subseteq V(D)$, an (X, Y) -*separator* in D is a set $S \subseteq V(D) \setminus (X \cup Y)$ such that there is no path from any vertex in X to any vertex in Y in $D - S$. When we consider a topological ordering of a DAG, suppose that no arc is directed from a vertex v to a vertex u that occurs *before* v in the ordering. Given a topological ordering π of a DAG D and $X \subseteq V(D)$, we say that π_X is *induced by* π if the vertices of X appear in the same order in π_X and in π. By the treewidth $\mathbf{tw}(D)$ of a digraph D and a (nice) tree decomposition of D, we refer to the treewidth and a (nice) tree decomposition of the underlying undirected graph of D, respectively. For any $X \subseteq V(D)$, we say that X is a η-*treewidth modulator* in D if $\mathbf{tw}(D - X) \leq \eta$.

Definition 2.1 (Important Separators). *Let D be a digraph and $X, Y \subseteq V(D)$. Let $S \subseteq V(D) \setminus (X \cup Y)$ be an (X, Y)-separator and let R be the set of vertices reachable from X in $D - S$. We say that S is an important (X, Y)-separator if it is inclusion-wise minimal and there is no (X, Y)-separator $S' \subseteq V(D) \setminus (X \cup Y)$ such that $|S'| \leq |S|$ and $R \subsetneq R'$, where R' is the set of vertices reachable from X in $D - S'$.*

Proposition 2.1 ([12,14]). *Let D be a digraph, $X, Y \subseteq V(D)$ and $k \in \mathbb{N} \cup \{0\}$. Then, D has at most 4^k important (X, Y)-separators of size at most k, and the set of all of them can be constructed in time $\mathcal{O}(4^k \cdot k^2 \cdot (n + m))$.*

3 Decomposing the Graph

Given an instance (D, k, M) of DFVS/DFVS+Tw-η MOD, the goal of this section is to compute a decomposition of D consisting of three components: the vertex set M (modulator), a collection of vertex sets $\mathcal{Z}$ (zones), and a vertex set R (remainder). All of these sets would be pairwise disjoint. The crux is to "divide-and-conquer" D so that each zone—that is, a set $Z \in \mathcal{Z}$—would correspond to a subproblem that is easier to solve than (D, k, M) because (1) the intersection of a minimum solution with Z would be necessarily small, and (2) the interaction of Z is "well-structured" with respect to M, "limited" with respect to R, and "non-existent" with respect to any other zone. Towards the computation of R, we compute three sets: (i) a solution S in $D - M$; (ii) a set F to separate vertices in M that have low-flow; (iii) an LCA-closure of bags derived from $S \cup F$. The arguments given on the way to construct these sets will only partially prove that we have derived the desired decomposition. At the end, we complete the proof by focusing on the property regarding the intersection of a minimum solution with each zone. The details are deferred to the full version of the paper.

Definition 3.1. *Let (D, k, M) be an instance of DFVS/DFVS+Tw-η MOD. A partition $V(D) = M \uplus R \uplus (\uplus_{Z \in \mathcal{Z}} Z)$ is a zone-decomposition if:*

1. $D - (M \cup R)$ is a DAG.

2. For all $Z \in \mathcal{Z}$, we have $N(Z) \subseteq M \cup R$, and $|N(Z) \cap R| \leq 2(\eta + 1)$.
3. For all $(u, v) \in M \times M \setminus E(D)$, either there is no path from u to v in the digraph $D - ((M \cup R) \setminus \{u, v\})$, or there are at least $k + 1$ internally vertex-disjoint paths from u to v in D. (For $u = v$, having no path refers to having no path on at least two vertices.)

Lemma 3.1. There is a polynomial-time algorithm that, given an instance (D, k, M) of DFVS/DFVS+Tw-η MOD, either correctly decides that (D, k, M) is a NO-instance, or constructs a zone-decomposition $V(D) = M \uplus R \uplus (\uplus_{Z \in \mathcal{Z}} Z)$ with $|\mathcal{Z}| \leq 6k(\ell^2 + 1)$ and $|R| \leq 2(\eta + 1)k(\ell^2 + 1)$.

We now argue that if (D, k, M) is a YES-instance, then the size of the intersection of each minimum(-size) solution with each zone is only a constant.

Lemma 3.2. Let (D, k, M) be a YES-instance of DFVS/DFVS+Tw-η MOD with zone-decomposition $V(D) = M \uplus R \uplus (\uplus_{Z \in \mathcal{Z}} Z)$. For any minimum-sized directed feedback vertex set S of D, we have $|S \cap Z| \leq |N(Z) \cap R| \leq 2(\eta+1)$ for all $Z \in \mathcal{Z}$.

4 Reducing Each Part: k-DFVS Representative Marking

For an instance (D, k, M) of DFVS/DFVS+Tw-η MOD, the kernelization algorithm starts by applying Lemma 3.1 and either concludes that (D, k, M) is a NO-instance, or obtains a zone decomposition $V(D) = M \uplus R \uplus (\uplus_{Z \in \mathcal{Z}} Z)$ with the properties in Lemma 3.1. In this section, we fix an arbitrary zone $Z \in \mathcal{Z}$ and give a polynomial time algorithm (Lemma 4.1) to mark a small set of vertices in Z which in some sense "represents" all partial solutions in Z. Such a set will then be used to design reduction rules that bound the degree of the vertices in Z that are not in the representative, which will further be useful to decompose Z into a small number of protrusions. We now formally define the desired set.

Definition 4.1 (k-DFVS Representative in Z). For a digraph D, $Z \subseteq V(D)$ and an integer k, we say that $\Gamma_{\mathrm{DFVS}} \subseteq Z$ is a k-DFVS representative in Z if the following holds. If D has a directed feedback vertex set of size at most k, then it also have a directed feedback vertex set S of size at most k where $S \cap Z \subseteq \Gamma_{\mathrm{DFVS}}$.

Lemma 4.1 (k-DFVS Representative Marking Lemma). There is an algorithm that given a digraph D, $Z \subseteq V(D)$ and an integer k, runs in time $2^{\mathcal{O}(\eta^2)} \cdot (k\ell)^{\mathcal{O}(\eta^2)} \cdot (n + m)$, and returns a set $\Gamma_{\mathrm{DFVS}} \subseteq Z$ of size $(k\ell)^{\mathcal{O}(\eta^2)}$ such that Γ_{DFVS} is a k-DFVS representative in Z.[1]

We prove Lemma 4.1 in two parts. In Sect. 4.1, we revisit the relation between DFVS and SKEW MULTICUT (defined later). Using this relation, we conclude

---

[1] Throughout the paper, we do not hide constants that depend on η in the $\mathcal{O}$ notation.

that Γ_{DFVS} can be computed by taking the union of skew multicuts of "appropriate" instances of SKEW MULTICUT. The problem at this stage stems from the fact that the set of appropriate instances that we need to consider is not polynomially bounded, and hence a naive approach of finding a solution to each of these instances and taking their union does not work. This issue is tackled in Sect. 4.2.

4.1   Revisiting the Relation with the Skew Multicut Problem

Towards the definition of SKEW MULTICUT, we first define a skew multicut in a digraph. Let D be a digraph, and $\mathcal{P} = ((s_1, t_1), \ldots, (s_p, t_p))$ be an *ordered* set of pairs of vertices (called *terminals*) of D. A *skew multicut* in D with respect to $\mathcal{P}$ is a set S of non-terminal vertices of D such that for all $i, j \in \{1, \ldots, p\}$ with $i \leq j$, there is no path from s_i to t_j in $D - S$. In SKEW MULTICUT (SMC), the input is a digraph D, an ordered set $\mathcal{P} = ((s_1, t_1), \ldots, (s_p, t_p))$ and an integer k. The goal is to decide whether D has a skew multicut of size at most k with respect to $\mathcal{P}$. The relation between DFVS and SKEW MULTICUT was established in [13]. We restate the relation here in the format that will be useful later. The proofs are deferred to the full version of the paper.

Definition 4.2. *Let D be a digraph and $B \subseteq V(D)$. The digraph $D_{\dagger B}$ is obtained from D as follows. Replace each vertex $v \in B$ by two new vertices v^{out} and v^{in}, add the arc (v^{in}, v^{out}) and replace each arc $(u, v) \in E(D)$ by (u, v^{in}) and each arc $(v, u) \in E(D)$ by (v^{out}, u).*

Lemmas 4.2 and 4.3 show that any DFVS solution restricted to Z is a skew multicut solution to an appropriate instance of SKEW MULTICUT and vice-versa.

Lemma 4.2. *Let D be a digraph, $Z \subseteq V(D)$ and $k \in \mathbb{Z}$. Let S be a directed feedback vertex set in D of size at most k. Let $B = N(Z) \setminus S$, and let $\pi_B = (v_1, \ldots, v_b)$ be an ordering of B induced by a topological ordering π of $D - S$. Denote $D' = D[Z \cup B]_{\dagger B}$, $\mathcal{P} = ((v_1^{out}, v_1^{in}), \ldots, (v_b^{out}, v_b^{in}))$ and $k' = |S \cap Z|$. Then, there is a skew multicut in D' with respect to $\mathcal{P}$ of size at most k', that is, $(D', \mathcal{P}, k')$ is a YES-instance of SKEW MULTICUT.*

Lemma 4.3. *Let D be a digraph, $Z \subseteq V(D)$ and $k \in \mathbb{Z}$. Let S be a directed feedback vertex set in D of size at most k. Let $B = N(Z) \setminus S$ and let $\pi_B = (v_1, \ldots, v_b)$ be an ordering of the vertices of B induced by a topological ordering π of $D - S$. Denote $D' = D[Z \cup B]_{\dagger B}$, and $\mathcal{P} = ((v_1^{out}, v_1^{in}), \ldots, (v_b^{out}, v_b^{in}))$. Let S' be any skew multicut in D' with respect to $\mathcal{P}$. Then, $S^* = (S \setminus Z) \cup S'$ is a directed feedback vertex set in D.*

The number of guesses for the SKEW MULTICUT instance for which the intersection of a potential DFVS solution with Z is a skew multicut solution is $2^{|N(Z)|} \cdot |N(Z)|^{|N(Z)|}$. This is not polynomially bounded in k and ℓ. In the next section, we see how to compute a set containing some skew multicut solution to each of these instances without having to go over the instances individually.

4.2   Computing Solutions for All Instances of Skew Multicut

We formalize the notion of "all possible choices for the appropriate instance of SKEW MULTICUT", by defining a family of instances of SKEW MULTICUT denoted by $\mathcal{F}_{\text{SMC}}$. To simplify notation, for a digraph D and a (not necessarily ordered) set $\mathcal{P}$ of terminal pairs, let $D - \mathcal{P}$ be the digraph obtained from D by deleting all terminals in $\mathcal{P}$. Similarly, for a subset $X \subseteq V(D)$, let $X - \mathcal{P}$ be the set of vertices obtained from X by deleting all terminals in $\mathcal{P}$. To improve readability, *unordered* sets of terminal pairs will be denoted by $\mathcal{Q}$ rather than $\mathcal{P}$. We also stress that in what follows, k should be thought of as a small constant, because here it does not refer to the original k in the input instance of DFVS, but to the parameter set up when we construct an instance of SKEW MULTICUT.

Definition 4.3. *Given a digraph D, an* unordered *set $\mathcal{Q} = \{(s_i, t_i) : i \in \{1, \ldots, p\}, s_i, t_i \in V(D)\}$ and an integer k, $\mathcal{F}_{\text{SMC}}(D, \mathcal{Q}, k)$ is a family of instances of* SKEW MULTICUT *such that for each $P^* \subseteq \{1, \ldots, p\}$, for each ordering π of P^* and for each $k' \leq k$, the instance $(D - (\mathcal{Q} \setminus \mathcal{P}^*), \mathcal{P}^*, k')$ belongs to $\mathcal{F}_{\text{SMC}}(D, \mathcal{Q}, k)$ where $\mathcal{P}^* = ((s_{\pi(i)}, t_{\pi(i)}) : i \in P^*)$.*

We clarify that the above notation $\mathcal{P}^* = ((s_{\pi(i)}, t_{\pi(i)}) : i \in P^*)$ means that for all $i, j \in P^*$, we have that $(s_{\pi(i)}, t_{\pi(i)})$ is ordered before $(s_{\pi(j)}, t_{\pi(j)})$ if and only if $i < j$. Similar to the notion of a k-DFVS representative of Z, we first define the notion of a k-SMC representative. The construction of a set that, for any instance in $\mathcal{F}_{\text{SMC}}$, contains some solution for that instance, is captured by the Lemma 4.4.

Definition 4.4 (k-SMC Representative). *Given a digraph D, a set $\mathcal{Q} = \{(s_i, t_i) : i \in \{1, \ldots, p\}, s_i, t_i \in V(D)\}$ and an integer k, a k-SMC representative in D with respect to $\mathcal{Q}$ is a subset $\Gamma_{\text{SMC}} \subseteq V(D)$ such that each* YES*-instance in the family $\mathcal{F}_{\text{SMC}}(D, \mathcal{Q}, k)$ has a solution that belongs to Γ_{SMC}.*

Lemma 4.4 (k-SMC Representative Marking Lemma). *There is an algorithm that, given a digraph D $p, k \in \mathbb{N}$ and $\mathcal{Q} = \{(s_i, t_i) : i \in \{1, \ldots, p\}, s_i, t_i \in V(D)\}$, runs in time $p^{\mathcal{O}(k^2)} \cdot (n + m)$ time and outputs a k-SMC representative in D with respect to $\mathcal{Q}$ of size at most $k^2(k + 1)^k \cdot p^{k(k+2)} \cdot 4^{k^2}$.*

In the rest of the section, we give an intuitive explanation of the algorithm of Lemma 4.4. The details can be found in the full version of the paper. Since $|\mathcal{F}_{\text{SMC}}(D, \mathcal{Q}, k)|$ is exponential in p, if a k-SMC representative in D with respect to $\mathcal{Q}$, say Γ_{SMC}, of the desired size exists, then the solutions of "many" instances in $\mathcal{F}_{\text{SMC}}(D, \mathcal{Q}, k)$ intersect "a lot". This is exactly what we want to exploit. Roughly speaking, we want to recursively apply the following step. In each recursive call, partition the instances of $\mathcal{F}_{\text{SMC}}(D, \mathcal{Q}, k)$ into $p^{\mathcal{O}(k)}$ classes, and for each class find a set that is guaranteed to be contained in some solution for each of the instances in the class. Delete this set from the instances in the class, and recurs. Note that we keep track of deleted vertices, since they are precisely the vertices that will form Γ_{SMC}. Since we are looking for a k-sized solution in each instance in $\mathcal{F}_{\text{SMC}}(D, \mathcal{P}, k)$, the depth of the recursion is at most k and hence, we can form the set Γ_{SMC} of the desired size (i.e. $\mathcal{O}(p^{g(k)})$ for some function g of k).

Now, we try to formalize this approach; depending on the obstacles faced, we add layers and machinery to the outline above. First consider the step of partitioning the instances of $\mathcal{F}_{\text{SMC}}(D, \mathcal{Q}, k)$ into some $p^{\mathcal{O}(k)}$ classes, such that for each class there is a set guaranteed to be contained in some solution for each of the instances in the class. To this end, we first try the power of the Pushing Lemma for SKEW MULTICUT, defined below. Roughly speaking, this lemma states that any YES-instance $(D, ((s_i, t_i) : i \in \{1, \dots, a\}), k)$ of SKEW MULTICUT (for instances in $\mathcal{F}_{\text{SMC}}(D, \mathcal{Q}, k)$, we have $a \leq p$) has a solution of size at most k that contains an important $(\{s_1\}, \{t_1, \dots, t_a\})$-separator of size at most k.

Proposition 4.1 (Pushing Lemma for SKEW MULTICUT, [13]). *For a* YES-*instance* $(D, \mathcal{P} = ((s_1, t_1), \dots, (s_p, t_p)), k)$ *of* SKEW MULTICUT, *there is a solution* S^* *containing an important* $(\{s_1\}, \{t_1, \dots, t_p\})$-*separator in* D *of size at most* k.

Now, consider any $P^* \subseteq \{1, \dots, p\}$. Assume w.l.o.g. that $P^* = \{1, \dots, p^*\}$. Consider all YES-instances in $\mathcal{F}_{\text{SMC}}(D, \mathcal{P}, k)$ where the first terminal pair is (s_1, t_1) and the other terminal pairs are $\{(s_2, t_2), \dots, (s_{p^*}, t_{p^*})\}$ in some order. Then, by Proposition 4.1, for each of these instances there exists a solution containing some important $(\{s_1\}, \{t_1, \dots, t_{p^*}\})$-separator of size at most k. This is not exactly what we wanted (since we do not obtain a *single* set that is contained in some solution for each of the instances), but we can still work with this as the number of important separators of size at most k is at most 4^k (from Proposition 2.1). Then, we can branch on which important separator to add to Γ_{SMC}. Thus, Proposition 4.1 seems to give a way to go about constructing Γ_{SMC}.

However, we are not done yet because if we naively utilize the Pushing Lemma approach, we need to partition $\mathcal{F}_{\text{SMC}}(D, \mathcal{P}, k)$ into $2^p \cdot p$ classes. Indeed, we have 2^p possibilities to choose a subset $P^* \subseteq \{1, \dots, p\}$ (which captures the indices of the terminals pairs in $\mathcal{P}$ that should not be deleted), and $p^* \leq p$ choices for which is the index in P^* of the first terminal pair (from which we push our solution as described above). For us, $2^p \cdot p$ is a huge number. To handle this issue, we introduce another tool, called the *Important Separator Preservation (via Small Sink Set) Lemma* (formally defined later). Intuitively, this lemma says that if I is an important (X, Y)-separator, then I is also an important (X, Y')-separator for some $Y' \subseteq Y$ where the size of Y' is at most the size of I plus one.

Lemma 4.5 (Important Separator Preservation (via Small Sink Set) Lemma). *Let* D *be a digraph with* $X, Y \subseteq V(D)$ *and an important* (X, Y)-*separator* $S \subseteq V(D)$ *of size* α. *There is* $Y' \subseteq Y$ *of size* $\alpha + 1$ *such that* S *is an important* (X, Y')-*separator in* $D - (Y \setminus Y')$.

The observation that we can exploit this lemma in our setting is a crucial insight in the design of our kernel. Recall that by Proposition 4.1, we can conclude that for some class of instances, the following property holds: There exists a pair (X, Y), where $X = \{s_i\}$ and Y is some set of terminals t_j, such that there is an important (X, Y)-separator of size at most k that is contained in some solution for each of the instances in the class. Since the number of important (X, Y)-separators of size at most k is small, we could branch on them. Basically, we combine Lemma 4.5 with Proposition 4.1 to add another layer of branching. Below, we briefly discuss the meaning of this extra layer.

Here, we partition our instances into p classes: All instances that have (s_i, t_i) as the first terminal pair (recall that the set of terminal pairs in SKEW MULTICUT is ordered) belong to the same class. While before we had a refined partition with $2^p \cdot p$ classes, here we only have p classes, but which *at first glance* seem non-informative. However, we show that (by Lemma 4.5) not much additional information is needed. More precisely, we argue that for all YES-instances in the same class of our rough partition, there exists some $p^{\mathcal{O}(k)}$ sized collection of pairs $\{(X_i, Y_i) : i \in p^{\mathcal{O}(k)}\}$ with the following property: For any instance in the class, there exists a pair in this collection, say (X_i, Y_i), such that there exists an important (X_i, Y_i)-separator of size at most k that is contained in some solution of that instance. Since the size of the collection is $p^{\mathcal{O}(k)}$, and for each pair in it there are at most 4^k important separators of size at most k, we branch into at most $p^{\mathcal{O}(k)} \cdot 4^k$ branches for each class. Since $p^{\mathcal{O}(k)} \cdot 4^k$ is small enough to obtain a kernel—recall that in SKEW MULTICUT, k is small (constant) but p is large—let us move ahead to see how we obtain the collection $\{(X_i, Y_i) : i \in p^{\mathcal{O}(k)}\}$.

We claim that for any class, whose first terminal pair is some (s_i, t_i), the collection $\{(s_i, T) : T \subseteq \{t_1, \ldots, t_p\}, |T| \leq k+1\}$ is precisely that collection that we want. To see this, consider any YES-instance $(D, \mathcal{P}^*, k)$ whose first terminal pair is (s_i, t_i). Let P^* denote the set of indices of the pairs in $\mathcal{P}^*$. By the Pushing Lemma for SKEW MULTICUT, there exists a solution to this instance that contains some important $(\{s_i\}, \{t_j \mid j \in P^*\})$-separator of size at most k. In turn, by the Important Separator Preservation Lemma, there exists $T \subseteq \{t_j \mid j \in P^*\}$ of size at most $k+1$, such that any important $(\{s_i\}, \{t_j \mid j \in P^*\})$-separator is also an important $(\{s_i\}, T)$-separator! Thus, we conclude that for each YES-instance in $\mathcal{F}_{\mathrm{SMC}}(D, \mathcal{P}, k)$ whose first terminal pair is (s_i, t_i), there exists a pair in the collection $\{(s_i, T) : T \subseteq \{t_1, \ldots, t_p\}\}$ such that one of the important separators of size at most k of this pair is contained in some solution for this instance.

5   Reduction Rules

In this section we give reduction rules to reduce the size of each "zone". More precisely, we first apply the algorithm of Lemma 3.1 which either correctly decides that (D, k, M) is a NO-instance, or constructs a zone-decomposition $V(D) = M \uplus R \uplus (\uplus_{Z \in \mathcal{Z}} Z)$ with $|\mathcal{Z}| \leq 6k(\ell^2 + 1)$ and $|R| \leq 2(\eta + 1)k(\ell^2 + 1)$. For a fixed zone $Z \in \mathcal{Z}$, we concentrate on reducing the size of Z. Once we are able to bound the size of each zone by a polynomial function of k and ℓ, we obtain a polynomial kernel for our problem. Thus, from now onwards we concentrate on bounding the size of a single zone Z. Let Γ_{DFVS} be a k-DFVS representative in Z, computed using the algorithm of Lemma 4.1. We bound the size of Z in a two step procedure. In the first step, we design reduction rules that remove all the arcs between M and $Z \setminus \Gamma_{\text{DFVS}}$ (at the cost of adding arcs between M and Γ_{DFVS}). Once this is done, we have that Z interacts with the "outside world" in a limited fashion via Γ_{DFVS} alone. After we have achieved this, in the second step, we will be able to partition $Z \setminus \Gamma_{\text{DFVS}}$ into a "small" number of slices such that each slice has treewidth at most η and has at most $\mathcal{O}(\eta)$ neighbors outside (that is, the slice is an $\mathcal{O}(\eta)$-protrusion). Each such slice can then be replaced by a constant size equivalent slice using the protrusion replacement machinery. The details of this can be found in the full version of the paper.

References

1. Abu-Khzam, F.: A kernelization algorithm for d-HS. JCSS **76**(7), 524–531 (2010)
2. Agrawal, A., Saurabh, S., Sharma, R., Zehavi, M.: Kernels for deletion to classes of acyclic digraphs. JCSS **92**, 9–21 (2018)
3. Bafna, V., Berman, P., Fujito, T.: A 2-approximation algorithm for the undirected feedback vertex set problem. SIDMA **12**(3), 289–297 (1999)
4. Bang-Jensen, J., Maddaloni, A., Saurabh, S.: Algorithms and kernels for feedback set problems in generalizations of tournaments. Algorithmica **76**, 1–24 (2015)
5. Bar-Yehuda, R., Geiger, D., Naor, J., Roth, R.M.: Approximation algorithms for the feedback vertex set problem with applications to constraint satisfaction and bayesian inference. SIAM J. Comput. **27**(4), 942–959 (1998)
6. Becker, A., Geiger, D.: Optimization of pearl's method of conditioning and greedy-like approximation algorithms for the vertex feedback set problem. AI **83**, 167–188 (1996)
7. Bergougnoux, B., Eiben, E., Ganian, R., Ordyniak, S., Ramanujan, M.: Towards a polynomial kernel for directed feedback vertex set. In: MFCS, vol. 83 (2017)
8. Bonamy, M., Kowalik, Ł., Nederlof, J., Pilipczuk, M., Socała, A., Wrochna, M.: On directed feedback vertex set parameterized by treewidth. In: WG, pp. 65–78 (2018)
9. Cao, Y., Chen, J., Liu, Y.: On feedback vertex set: new measure and new structures. Algorithmica **73**(1), 63–86 (2015)
10. Chekuri, C., Madan, V.: Constant factor approximation for subset feedback set problems via a new LP relaxation. In: SODA, pp. 808–820 (2016)
11. Chen, J., Fomin, F.V., Liu, Y., Lu, S., Villanger, Y.: Improved algorithms for feedback vertex set problems. JCSS **74**(7), 1188–1198 (2008)
12. Chen, J., Liu, Y., Lu, S.: An improved parameterized algorithm for the minimum node multiway cut problem. Algorithmica **55**(1), 1–13 (2009)
13. Chen, J., Liu, Y., Lu, S., O'Sullivan, B., Razgon, I.: A fixed-parameter algorithm for the directed feedback vertex set problem. J. ACM **55**(5), 21 (2008)
14. Chitnis, R., Hajiaghayi, M., Marx, D.: Fixed-parameter tractability of directed multiway cut parameterized by the size of the cutset. SIAM J. Comput. **42**(4), 1674–1696 (2013)
15. Chitnis, R.H., Cygan, M., Hajiaghayi, M.T., Marx, D.: Directed subset feedback vertex set is fixed-parameter tractable. ACM TALG **11**(4), 28 (2015)
16. Cygan, M., et al.: Parameterized Algorithms. Springer, Cham (2015). https://doi.org/10.1007/978-3-319-21275-3
17. Cygan, M., Lokshtanov, D., Pilipczuk, M., Pilipczuk, M., Saurabh, S.: On the hardness of losing width. TOCS **54**(1), 73–82 (2014)

18. Cygan, M., Nederlof, J., Pilipczuk, M., Pilipczuk, M., van Rooij, J.M.M., Wojtaszczyk, J.O.: Solving connectivity problems parameterized by treewidth in single exponential time. In: FOCS, pp. 150–159 (2011)
19. Cygan, M., Pilipczuk, M., Pilipczuk, M., Wojtaszczyk, J.O.: Subset feedback vertex set is fixed-parameter tractable. SIDMA 27(1), 290–309 (2013)
20. Dom, M., Guo, J., Hüffner, F., Niedermeier, R., Truß, A.: Fixed-parameter tractability results for feedback set problems in tournaments. JDA 8(1), 76–86 (2010)
21. Downey, R.G., Fellows, M.R.: Fixed-parameter intractability. In: CCC, pp. 36–49 (1992)
22. Downey, R.G., Fellows, M.R.: Fixed-parameter tractability and completeness I: basic results. SIAM J. Comput. 24(4), 873–921 (1995)
23. Downey, R.G., Fellows, M.R.: Fundamentals of Parameterized Complexity. TCS. Springer, London (2013). https://doi.org/10.1007/978-1-4471-5559-1
24. Erdős, P., Pósa, L.: On independent circuits contained in a graph. Can. J. Math. 17, 347–352 (1965)
25. Even, G., Naor, J., Schieber, B., Sudan, M.: Approximating minimum feedback sets and multicuts in directed graphs. Algorithmica 20(2), 151–174 (1998)
26. Flum, J., Grohe, M.: Parameterized Complexity Theory. TTCSAES. Springer, Heidelberg (2006). https://doi.org/10.1007/3-540-29953-X
27. Fomin, F.V., Lokshtanov, D., Misra, N., Saurabh, S.: Planar F-deletion: approximation, kernelization and optimal FPT algorithms. In: FOCS, pp. 470–479 (2012). http://www.ii.uib.no/~daniello/papers/PFDFullV1.pdf
28. Fomin, F.V., Lokshtanov, D., Saurabh, S., Zehavi, M.: Kernelization: Theory of Parameterized Preprocessing. Cambridge University Press, Cambridge (2018)
29. Fomin, F.V., Saurabh, S.: Kernelization methods for fixed-parameter tractability. In: Bordeaux, L., Hamadi, Y., Kohli, P. (eds.) Tractability, pp. 260–282. Cambridge University Press, Cambridge (2014)
30. Guo, J., Niedermeier, R.: Invitation to data reduction and problem kernelization. SIGACT News 38(1), 31–45 (2007)
31. Guruswami, V., Lee, E.: Inapproximability of H-transversal/packing. In: APPROX/RANDOM. LIPIcs, vol. 40, pp. 284–304 (2015)
32. Kakimura, N., Kawarabayashi, K., Kobayashi, Y.: Erdös-Pósa property and its algorithmic applications: parity constraints, subset feedback set, and subset packing. In: SODA, pp. 1726–1736 (2012)
33. Kakimura, N., ichi Kawarabayashi, K., Marx, D.: Packing cycles through prescribed vertices. JCTB 101(5), 378–381 (2011)
34. Kawarabayashi, K., Kobayashi, Y.: Fixed-parameter tractability for the subset feedback fet problem and the S-cycle packing problem. JCTB 102, 1020–1034 (2012)
35. Kawarabayashi, K., Král, D., Krcál, M., Kreutzer, S.: Packing directed cycles through a specified vertex set. In: SODA, pp. 365–377 (2013)
36. Kociumaka, T., Pilipczuk, M.: Faster deterministic feedback vertex set. IPL 114(10), 556–560 (2014)
37. Kratsch, S.: Recent developments in kernelization. Bull. EATCS 113, 58–97 (2014)
38. Le, T., Lokshtanov, D., Saurabh, S., Thomassé, S., Zehavi, M.: Subquadratic kernels for implicit 3-hitting set and 3-set packing problems. In: SODA, pp. 331–342 (2018)

39. Lokshtanov, D., Misra, N., Saurabh, S.: Kernelization – preprocessing with a guarantee. In: Bodlaender, H.L., Downey, R., Fomin, F.V., Marx, D. (eds.) The Multivariate Algorithmic Revolution and Beyond. LNCS, vol. 7370, pp. 129–161. Springer, Heidelberg (2012). https://doi.org/10.1007/978-3-642-30891-8_10
40. Lokshtanov, D., Ramanujan, M.S., Saurabh, S.: When recursion is better than iteration. In: SODA, pp. 1916–1933 (2018)
41. Mnich, M., van Leeuwen, E.J.: Polynomial kernels for deletion to classes of acyclic digraphs. Discrete Optim. **25**, 48–76 (2017)
42. Niedermeier, R.: Invitation to Fixed-Parameter Algorithms. Oxford University Press, Oxford (2006)
43. Pontecorvi, M., Wollan, P.: Disjoint cycles intersecting a set of vertices. JCTB **102**(5), 1134–1141 (2012)
44. Raman, V., Saurabh, S., Subramanian, C.R.: Faster fixed parameter tractable algorithms for finding feedback vertex sets. ACM TALG **2**(3), 403–415 (2006)
45. Reed, B.A., Robertson, N., Seymour, P., Thomas, R.: Packing directed circuits. Combinatorica **16**(4), 535–554 (1996)
46. Seymour, P.: Packing directed circuits fractionally. Combinatorica **15**(2), 281–288 (1995)
47. Seymour, P.: Packing circuits in eulerian digraphs. Combinatorica **16**(2), 223–231 (1996)
48. Wahlström, M.: Half-integrality, LP-branching and FPT algorithms. In: SODA, pp. 1762–1781 (2014)

Discrete Morse Theory for Computing Zigzag Persistence

Clément Maria[1]([⊠]) and Hannah Schreiber[2]

[1] Inria Sophia Antipolis-Méditerranée, Valbonne, France
clement.maria@inria.fr
[2] University of Technology of Graz, Graz, Austria
hschreiber@tugraz.at

Abstract. We introduce a theoretical and computational framework to use discrete Morse theory as an efficient preprocessing in order to compute zigzag persistent homology. From a zigzag filtration of complexes, we introduce a *zigzag Morse filtration* whose complexes are Morse reductions of the original complexes, and we prove that they both have same persistent homology. The key point of our construction is that it does not require any knowledge of past and future maps of the input filtration. We deduce an algorithm to compute the zigzag persistence of a filtration that depends mostly on the number of critical cells of the complexes, and show experimentally that it performs better in practice.

1 Introduction

Persistent homology is an algebraic method that permits to characterize the evolution of the topology of a growing sequences of spaces $X_1 \subseteq \ldots \subseteq X_n$, called a filtration. The theory has found many applications, especially in data analysis where it has been successfully applied to material science [25], shape classification [6,10], or clustering [9,11]. Its success relies on sound theoretical foundations [19,37], favorable stability properties [3,12,14], and fast algorithms for computation [1,2,4,13].

Another approach to fast computation consists of preprocessing the input filtration in order to drastically reduce the size of the domains X_i, while preserving persistence [5,18,30,34]. This has the double advantage of reducing both time and memory complexity. This goal has in particular been successfully reached by the use of *discrete Morse theory* [21], which led to the implementation of efficient software, such as Perseus [32] and Diamorse [17]. Additionally, noticeable successes, at the crossroad of persistence and discrete Morse theory, have been reached in the study of 3D images [34], allowing drastic improvements in memory and time performance, as well as the study of data ranging from medical imaging to material science [15,16,22].

H. Schreiber—Funded by the Austrian Science Fund (FWF) grant number P 29984-N35.

© Springer Nature Switzerland AG 2019
Z. Friggstad et al. (Eds.): WADS 2019, LNCS 11646, pp. 538–552, 2019.
https://doi.org/10.1007/978-3-030-24766-9_39

Zigzag persistent homology is a generalization of persistent homology that allows the measurement and tracking of the topology of sequences of spaces that both grow and shrink, known as a *zigzag filtrations* $X_1 \subseteq X_2 \supseteq X_3 \subseteq \cdots$. Zigzag persistence was introduced in [7], and theoretical [29] and practical [8,27] algorithms have been designed to compute it. They perform however much slower than standard persistence algorithms, due to the maintenance of heavier data structures to manage both insertions and deletions of faces. In particular, the optimizations of standard persistence do not apply to zigzag persistence.

Motivation and Applications for Zigzag Persistence. We give two important applications of zigzag persistence on which we test the experimental performance of our method.

(1) *Topology inference from data points P.* A standard approach [19] consists of computing the persistent homology of the Rips complex $\mathcal{R}^\rho(P)$ on the set of points P, for an increasing threshold $\rho \geq 0$. We compute instead the zigzag persistence of oscillating Rips zigzag filtrations [33]. These filtrations add data points progressively while reducing the scale of reconstruction in order to adapt to a more and more dense set of points. Specifically,

$$\cdots \longleftarrow \mathcal{R}^{\mu\varepsilon_i}(P_i) \overset{\subseteq}{\longrightarrow} \mathcal{R}^{\nu\varepsilon_i}(P_i \cup \{p_{i+1}\}) \overset{\supseteq}{\longleftarrow} \mathcal{R}^{\mu\varepsilon_i}(P_i \cup \{p_{i+1}\}) \longrightarrow \cdots , \quad (1)$$

where $\mathcal{R}^\alpha(P)$ is the Rips complex of threshold α on points P, and ε_i a measure of the "sparsity" of the set of points $P_i := \{p_1, \ldots, p_i\}$ that decreases when points are added. Finally, $0 < \mu \leq \nu$ are parameters. This filtration is known to furnish provably correct persistence diagrams, with much less noise than standard persistence [33], while naturally maintaining much smaller complexes during computation. This application is of importance in data analysis [9,11].

(2) *Levelset persistence of images.* Given a function $f \colon X \to \mathbb{R}$ on a domain X, classical persistence studies the persistent homology of sublevel sets $f^{-1}(-\infty, \rho]$ for an increasing ρ. Levelset persistence [8] studies instead the zigzag persistence of of the pre-images of intervals, for appropriate $s_1 \leq s_2 \leq \ldots$,

$$\cdots \longleftarrow f^{-1}[s_{i-1}, s_i] \overset{\subseteq}{\longrightarrow} f^{-1}[s_{i-1}, s_{i+1}] \overset{\supseteq}{\longleftarrow} f^{-1}[s_i, s_{i+1}] \longrightarrow \cdots . \quad (2)$$

From the levelset persistence, one can recover the sublevel set persistence [8], while maintaining again much smaller structures. This application is of particular importance for medical imaging and material science [15,16,22].

Streaming Model and Memory Efficiency. A main advantage of zigzag persistence is to consequently maintain much smaller complexes over the computation. To formalize this notion, we adopt a streaming model for the computation of zigzag persistence. The input is given by a stream of insertions and deletions of faces, with no knowledge of the entire zigzag filtration, and zigzag persistence is computed "on the fly". In particular, the memory complexity of our algorithms, depends solely on the maximal size of any complex in the filtration, $\max_i |X_i|$, as opposed to the entire number of insertions and deletions of faces, which is generally much larger.

Contributions and Existing Results. In the spirit of [30], we introduce a pre-processing reduction of a zigzag filtration based on discrete Morse theory [21]. After introducing some background in Sect. 2, we introduce in Sect. 3 a *zigzag Morse filtration* that generalizes the filtered Morse complex [30] of standard persistence, and we prove that it has same persistent homology as the input zigzag filtration. Because of removal of cells not agreeing with the Morse decomposition, the zigzag Morse filtration contains chain maps that are not inclusions. We study the effect of those maps on the boundary operator of the Morse complex in Sect. 4, and design a persistence algorithm for zigzag Morse complexes in Sect. 5. Finally, we report on the experimental performance of the zigzag persistence algorithm for Morse complexes in Sect. 6.

Note that a similar approach to adapt discrete Morse theory to zigzag persistence was followed by Escolar and Hiraoka [20]. Adapting [30], they define a *global* zigzag filtered Morse complex for a zigzag filtration, and study its interval decomposition. The main limitation of their approach is that the user must know the entirety of the input zigzag filtration to compute the Morse pairing, canceling the benefit of using "small complexes" in zigzag persistence. On the contrary, our approach requires no other than local knowledge of the input zigzag filtration, and all computation are done "on the fly" in the streaming model.

2   Background

Due to space constraints, we assume the reader has knowledge of the notions of *general abstract complexes* and *homology* [26], *persistent homology* [19], and *discrete Morse theory* [21]. Otherwise, we refer the reader to a longer arxiv version [28] of this paper for more detailed definitions and explanations.

Algebraic Topology. We fix some notations. The incidence function of a complex X is denoted by $[\cdot : \cdot]^X : X \times X \to \mathbf{D}$, for a fixed PID $\mathbf{D}$ (usually $\mathbb{Z}$, or a field $\mathbb{F}$), its boundary operator by ∂^X, and its homology groups by $H_d(X)$ when we assume the coefficients are in a fixed field $\mathbb{F}$.

A Morse matching on a set of cells X is denoted by $(\mathcal{A}, \mathcal{Q}, \mathcal{K}, \omega)$ where $\mathcal{A}$ are the critical cells, and $\omega \colon \mathcal{Q} \to \mathcal{K}$ is the bijection forming the Morse pairs $(\tau, \omega(\tau))$ of the Morse matching. A Morse matching induces an orientation of the Hasse diagram of X, denoted by $\mathcal{H}$. For critical cells σ of dimension $d + 1$, and τ of dimension d, the set of all gradient paths from σ to τ is denoted $\Gamma(\sigma, \tau)$.

Persistent Homology and Discrete Morse Theory. We refer the reader to [30] for the study of the (standard) persistence of discrete Morse complexes.

Persistent homology is the study of persistent modules induced by filtrations. Let $X_1 \subseteq \ldots \subseteq X_n$ be a filtration of complexes. A *standard Morse filtration* (called *filtered Morse complex* in [30]) for this filtration is a collection of Morse matchings $(\mathcal{A}_i, \mathcal{Q}_i, \mathcal{K}_i, \omega_i)_{i=1...n}$ for each X_i, with Morse complex $(\mathcal{A}_i, \partial^{\mathcal{A}_i})$ on the critical cells, and Morse pairs $\omega_i \colon \mathcal{K}_i \; -\text{bij.}\!\!\rightarrow \mathcal{Q}_i$, satisfying:

$$\mathcal{A}_i \subseteq \mathcal{A}_{i+1}, \quad \mathcal{Q}_i \subseteq \mathcal{Q}_{i+1}, \quad \mathcal{K}_i \subseteq \mathcal{K}_{i+1}, \quad \omega_{i+1}\big|_{\mathcal{Q}_i} = \omega_i, \quad \partial^{\mathcal{A}_{i+1}}\big|_{\mathcal{A}_i} = \partial^{\mathcal{A}_i}. \quad (3)$$

A filtered Morse complex consequently forms a filtration $\mathcal{A}_1 \subseteq \ldots \subseteq \mathcal{A}_n$ of Morse complexes connected by inclusions, which induces a persistence module.

Theorem 1 (Forman [21] **, Mischaikow and Nanda** [30]**).** *Let $(\mathcal{A}_i, \mathcal{Q}_i, \mathcal{K}_i, \omega_i)_{i=1...n}$ be a standard Morse filtration for a filtration $X_1 \subseteq \ldots \subseteq X_n$. There exist collections of chain maps $(\psi_i : C(X_i) \to C(\mathcal{A}_i))_{i=1...n}$ and $(\varphi_i : C(\mathcal{A}_i) \to C(X_i))_{i=1...n}$ for which the following diagrams commute for every i*

$$
\begin{array}{ccc}
C(X_i) & \overset{\subseteq}{\longrightarrow} & C(X_{i+1}) \\
\psi_i \downarrow & & \downarrow \psi_{i+1} \\
C(\mathcal{A}_i) & \overset{\subseteq}{\longrightarrow} & C(\mathcal{A}_{i+1})
\end{array}
\qquad\qquad
\begin{array}{ccc}
C(X_i) & \overset{\subseteq}{\longrightarrow} & C(X_{i+1}) \\
\varphi_i \uparrow & & \uparrow \varphi_{i+1} \\
C(\mathcal{A}_i) & \overset{\subseteq}{\longrightarrow} & C(\mathcal{A}_{i+1})
\end{array}
$$

and φ_i and ψ_i induce isomorphisms at the homology level, that are inverses of each other. They induce an isomorphism of persistence modules.

For a fixed ordering of the Morse pairs reducing X_i into $\mathcal{A}_i$, the map ψ_i (resp. φ_i) can be expressed as the composition of maps $\psi_{\tau,\sigma} \colon X' \to X' \setminus \{\tau, \sigma\}$, over Morse pairs (τ, σ), (resp. $\varphi_{\tau,\sigma} \colon X' \setminus \{\tau, \sigma\} \to X'$) between partially reduced complexes differing by one Morse pair. Maps $\psi_{\tau,\sigma}$ and $\varphi_{\tau,\sigma}$ induce isomorphisms in homology, inverse of each other (see [30] for explicit formulas).

3 Zigzag Morse Filtration and Persistence

For a zigzag filtration of complexes $\mathcal{F}$, we introduce in this article a canonical zigzag filtration $\mathcal{M}$ of Morse complexes admitting the same persistent homology.

3.1 Zigzag Morse Filtration

Without loss of generality, consider the zigzag filtration

$$\overline{\mathcal{F}} := \overline{X}_1 \overset{\Sigma_1}{\longleftrightarrow} \overline{X}_2 \overset{\Sigma_2}{\longleftrightarrow} \cdots \overset{\Sigma_{2k-1}}{\longleftrightarrow} \overline{X}_{2k-1} \overset{\Sigma_{2k}}{\longleftrightarrow} \overline{X}_{2k} , \quad (4)$$

where the $\overline{X}_i$ are complexes, $\overline{X}_1 = \overline{X}_{2k} = \emptyset$, and the ith arrow is an inclusion, either forward (i odd) or backward (i even), where complexes $\overline{X}_i$ and $\overline{X}_{i+1}$ differ by a set of cells Σ_i (possibly empty). We now further decompose $\overline{\mathcal{F}}$.

Atomic Operations. For each forward arrow $\bullet_i \longrightarrow \bullet_{i+1}$, i odd, let $(\hat{\mathcal{A}}_i, \hat{\mathcal{Q}}_i, \hat{\mathcal{K}}_i, \hat{\omega}_i)$ be a Morse matching of the set of cells Σ_i.

Because Morse matchings are acyclic, there exists a total ordering of the cells of Σ_i, compatible with the face partial ordering of Σ_i, such that paired cells in $(\hat{\mathcal{A}}_i, \hat{\mathcal{Q}}_i, \hat{\mathcal{K}}_i, \hat{\omega}_i)$ are consecutive with regard to that order. We can consequently decompose a forward inclusion $\overline{X}_i \subseteq \overline{X}_{i+1}$ into a sequence of inclusions of a single critical cell $\sigma \in \hat{\mathcal{A}}_i$, and of inclusions of a single Morse pair of cells $(\tau, \sigma) \in \hat{\mathcal{Q}}_i \times \hat{\mathcal{K}}_i$, with $\sigma = \hat{\omega}_i(\tau)$.

For every backward arrow $\bullet_i \longleftarrow \bullet_{i+1}$, i even, the Morse matchings $(\hat{\mathcal{A}}_j, \hat{\mathcal{Q}}_j, \hat{\mathcal{K}}_j, \hat{\omega}_j)$, for smaller odd indices $j < i$, induce a Morse matching on the cells of X_i. To avoid ambiguity, if a cell is reinserted in the filtration after being removed it is considered as a different element. By restriction, they consequently induce a valid Morse matching on all cells of Σ_i, except on those cells $\sigma \in \Sigma_i$ that form a Morse pair (τ, σ), with $\tau \notin \Sigma_i$. We decompose backward arrows into a sequence of removals of a single critical cell, of removals of a single Morse pair of cells, and of removals of a non-critical cell σ, without its paired cell $\tau \notin \Sigma_i$.

In summary, given an input filtration $\overline{\mathcal{F}}$ as above, and the Morse matchings $(\hat{\mathcal{A}}_i, \hat{\mathcal{Q}}_i, \hat{\mathcal{K}}_i, \hat{\omega}_i)$, we defined an *atomic* zigzag filtration

$$\mathcal{F} := \quad (\emptyset =) X_1 \longleftrightarrow X_2 \longleftrightarrow \cdots \longleftrightarrow X_{n-1} \longleftrightarrow X_n \, (= \emptyset) \, ,$$

where all arrows are of the following three types:

$$X \xleftarrow{\ \sigma\ } X' \tag{5}$$

$$X \xleftarrow{\ \{\tau, \sigma\}\ } X' \tag{6}$$

$$X \xrightarrow{\ \mathbb{1}\ } X \xleftarrow{\ \sigma\ } X \setminus \{\sigma\} \tag{7}$$

where σ is in each case a maximal cell in X, Diagrams (5) and (6) are forward or backward insertions of a critical cell or a Morse pair (τ, σ) of cells, respectively, and Diagram (7) is the removal of the cell σ from a Morse pair (τ, σ), where the cell τ is not removed. The identity arrow in this last diagram is a technicality that is clarified later. Naturally, one can recover the persistent homology of the zigzag filtration $\overline{\mathcal{F}}$ from the one of $\mathcal{F}$. We work with $\mathcal{F}$ for the rest of the article.

Morse Filtration. Given a zigzag filtration $\overline{\mathcal{F}}$, Morse matchings $(\mathcal{A}_i, \mathcal{Q}_i, \mathcal{K}_i, \omega_i)$, and an associated atomic filtration $\mathcal{F}$ as above, we define a *zigzag Morse filtration*

$$\mathcal{M} := \quad (\emptyset =) \mathcal{A}_1 \longleftrightarrow \mathcal{A}_2 \longleftrightarrow \cdots \longleftrightarrow \mathcal{A}_{n-1} \longleftrightarrow \mathcal{A}_n \, (= \emptyset) \, ,$$

of Morse complexes $(\mathcal{A}_i, \partial^{\mathcal{A}_i})$ of the complexes (X_i, ∂^{X_i}) of $\mathcal{F}$ inductively. Note that the maps of the zigzag Morse filtration are not all inclusions. Specifically, for a critical cell σ in both X_i and X_{i+1}, in general $\partial^{\mathcal{A}_i}(\sigma) \neq \partial^{\mathcal{A}_{i+1}}(\sigma)$.

All $X_1, X_n, \mathcal{A}_1$ and $\mathcal{A}_n$ are empty complexes. The zigzag Morse filtration is constructed inductively for the insertion of a critical cell (Diagram (5)) and the insertion of a Morse pair (Diagram (6)) as for standard Morse filtrations [30]:

$$
\begin{array}{ccc}
C(X) \xhookrightarrow{\ \sigma'\ } C(X \cup \{\sigma'\}) & \qquad & C(X) \xhookrightarrow{\ \{\tau, \sigma\}\ } C(X \cup \{\tau, \sigma\}) \\
\psi \downarrow \qquad\qquad \downarrow \psi & \qquad & \psi \downarrow \qquad\qquad\qquad \downarrow \psi_{\tau, \sigma} \circ \psi \\
C(\mathcal{A}) \xhookrightarrow{\ \sigma'\ } C(\mathcal{A} \cup \{\sigma'\}) & \qquad & C(\mathcal{A}) \xrightarrow{\ \mathbb{1}\ } C(\mathcal{A}) \, ,
\end{array}
\tag{8}
$$

where all horizontal arrows are inclusions of complexes, and in particular the boundary maps of $\mathcal{A}$ and $\mathcal{A} \cup \{\sigma'\}$ are equal when restricted to the cells of $\mathcal{A}$. The removal of critical cells and Morse pairs is symmetrical. The chain maps ψ and $\psi_{\tau,\sigma}$ are the ones of Theorem 1, and are used later.

For the removal of a non-critical cell σ without its paired cell τ (Diagram (7)), which is specific to zigzag persistence, the Morse filtration is constructed with:

$$
\begin{array}{ccc}
C(X) & \xrightarrow{\ 1\ } C(X) \xleftarrow{\ \ \sigma\ \ } C(X \setminus \{\sigma\}) \\
{\scriptstyle \psi_{\tau,\sigma} \circ \psi}\Big\downarrow & \Big\downarrow{\scriptstyle \psi} \qquad\qquad \Big\downarrow{\scriptstyle \psi} \\
C(\mathcal{A},\partial) & \xrightarrow{\ \varphi_{\tau,\sigma}\ } C(\mathcal{A}\cup\{\tau,\sigma\},\partial') \xleftarrow{\ \sigma\ } C(\mathcal{A}\cup\{\tau\},\partial'') \ .
\end{array}
\tag{9}
$$

The main technicality is that the boundary maps ∂ and ∂' differ in a non trivial way, that we study in Sect. 4. The map ∂'' is equal to the restriction of ∂' to the critical cells $\mathcal{A} \cup \{\tau\}$ (the right arrow is a backward inclusion of complexes). The chain maps $\psi_{\tau,\sigma}$ and $\varphi_{\tau,\sigma}$ are the ones from Theorem 1, and ψ is the compositions of all maps $\psi_{\mu,\omega(\mu)}$ over the Morse pairs $(\mu, \omega(\mu))$ of the Morse matching of X, except the pair (τ, σ). We give an example of zigzag Morse filtration in Fig. 1.

Diagrams (8) are studied in [30]. We now focus on the study of Diagram (9).

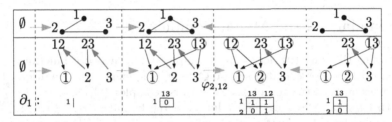

Fig. 1. Zigzag filtration (top) and its Morse filtration (bottom), given by Hasse diagrams and (Morse) boundary maps. Upward arrows in Hasse diagrams represent Morse matchings, critical faces are circled. Note that the rightmost operation illustrates Diagram (9), with a non trivial modification of $\partial_1(\{1,3\})$.

3.2 Isomorphism of Zigzag Modules

Theorem 1 implies that the atomic operations of Diagrams (8) induce commuting diagrams in homology, with vertical maps being isomorphisms. We prove the following lemma, which is specific to our zigzag Morse filtration.

Lemma 1. *Let X be a complex and $(\mathcal{A}, \mathcal{Q}, \mathcal{K}, \omega)$ a Morse complex obtained from X. Let σ be a maximal cell of X not in $\mathcal{A}$, which therefore forms a Morse pair with a cell τ, $[\sigma : \tau]^X \neq 0$. There exist isomorphisms ψ_*, $(\psi_{\tau,\sigma})_*$, and $(\varphi_{\tau,\sigma})_*$ such that the following diagram commutes:*

$$
\begin{array}{ccc}
H(X) & \xrightarrow{\ 1\ } H(X) \xleftarrow{\ \sigma_*\ } H(X \setminus \{\sigma\}) \\
{\scriptstyle (\psi_{\tau,\sigma})_* \circ \psi_*}\Big\downarrow & \Big\downarrow{\scriptstyle \psi_*} \qquad\qquad \Big\downarrow{\scriptstyle \psi_*} \\
H(\mathcal{A}) & \xrightarrow{(\varphi_{\tau,\sigma})_*} H(\mathcal{A}\cup\{\tau,\sigma\}) \xleftarrow{\ \sigma_*\ } H(\mathcal{A}\cup\{\tau\})
\end{array}
$$

Proof. Apply the homology functor to Diagram (9). The right square commutes, being induced by horizontal inclusions. Because the maps induced at homology level by $\psi_{\tau,\sigma}$ and $\varphi_{\tau,\sigma}$ are isomorphisms, inverse of each other (see Theorem 1), we get $(\varphi_{\tau,\sigma})_* \circ (\psi_{\tau,\sigma})_* \circ \psi_* = \psi_*$ and the left square commutes. By Theorem 1, the maps ψ, $\psi_{\tau,\sigma}$, and $\varphi_{\tau,\sigma}$ are isomorphisms.

Theorem 2. *The zigzag filtrations $\mathcal{F}$ and $\mathcal{M}$ have same persistent homology.*

Proof. Applying the homology functor to $\mathcal{F}$ and $\mathcal{M}$, we get the zigzag modules

$$
\begin{array}{ccccccc}
H(\mathcal{F}): & H(X_0) \longleftrightarrow H(X_1) \longleftrightarrow \cdots \longleftrightarrow H(X_m) \\
& \downarrow \psi_*^0 \qquad\quad \downarrow \psi_*^1 \qquad\qquad\qquad \downarrow \psi_*^m \\
H(\mathcal{M}): & H(\mathcal{A}_0) \longleftrightarrow H(\mathcal{A}_1) \longleftrightarrow \cdots \longleftrightarrow H(\mathcal{A}_m)
\end{array}
$$

where, by construction, every $\mathcal{A}_i$ is a Morse complex of X_i, and the ψ_*^i are the isomorphisms induced by the chain maps $\psi^i: C(X_i) \to C(\mathcal{A}_i)$, connecting a complex and its Morse reduction (Theorem 1). By Theorem 1 and Lemma 1, all squares commute and are compatible with each other, and the (ψ_*^i) define an isomorphism of zigzag modules.

4   Boundary of the Morse Complex

Referring to Diagram (9), let X be a complex with incidence function $[\cdot : \cdot]^X$, together with a Morse matching $(\mathcal{A}, \mathcal{Q}, \mathcal{K}, \omega)$, inducing an orientation of the Hasse diagram $\mathcal{H}$ of the complex, and a Morse complex $(\mathcal{A}, \partial)$.

In this section, we track the evolution of the boundary operators in Morse complexes under the evaluation of the map $\varphi_{\tau,\sigma}: (\mathcal{A}, \partial) \to (\mathcal{A} \cup \{\tau, \sigma\}, \partial')$ from Diagram (9). Both complexes are Morse complexes of the same X, whose matchings differ by exactly one pair (τ, σ), i.e., the Morse partition of complex $\mathcal{A} \cup \{\tau, \sigma\}$ is $(\mathcal{A} \cup \{\tau, \sigma\}) \sqcup (\mathcal{Q} \setminus \{\tau\}) \sqcup (\mathcal{K} \setminus \{\sigma\})$. We denote this last complex by $(\mathcal{A}', \partial')$, with incidence function $[\cdot : \cdot]^{\mathcal{A}'}$ in the following. We prove:

Lemma 2. *Let ν be a cell of the complex $(\mathcal{A}, \partial)$. Then, in the complex $(\mathcal{A}', \partial')$,*

$$\partial'(\nu) = \partial(\nu) + \left([\sigma : \tau]^X\right)^{-1} [\nu : \tau]^{\mathcal{A}'} \cdot \partial'\sigma. \tag{10}$$

Proof. First, note that σ is maximal in X, and so it is maximal in $\mathcal{A} \cup \{\tau, \sigma\}$.

Let $\mathcal{H}$ and $\mathcal{H}'$ be the Hasse diagrams of X induced by the Morse matchings of $\mathcal{A}$ and $\mathcal{A}'$, respectively. Because the matchings differ by a single Morse pair (τ, σ), $\mathcal{H}$ and $\mathcal{H}'$ only differ by the orientation of the edge $\tau \leftrightarrow \sigma$.

For a critical cell $\nu \in \mathcal{A}$, we have:

$$\partial\nu = \underbrace{\sum_{\substack{\mu \in \mathcal{A} \\ \gamma \in \Gamma(\nu,\mu)}} m(\gamma) \cdot \mu}_{} = \underbrace{\sum_{\mu \in \mathcal{A},\ \gamma \in \Gamma_{\tau \to \sigma}(\nu,\mu)} m(\gamma) \cdot \mu}_{(\star)} + \underbrace{\sum_{\mu \in \mathcal{A},\ \gamma \in \Gamma_{\tau \not\to \sigma}(\nu,\mu)} m(\gamma) \cdot \mu}_{\partial'\nu - [\nu:\tau]^{\mathcal{A}'} \cdot \tau},$$

where $\Gamma_{\tau\to\sigma}(\nu,\mu)$ are the gradient paths from ν to μ in $\mathcal{H}$ containing the upward arrow $\tau \to \sigma$, and $\Gamma_{\tau\nrightarrow\sigma}(\nu,\mu)$ are the ones not containing it. Moreover, $m(\gamma)$ is defined as a product of weights on the edges of the path γ (see [28] for details); in particular $m(\gamma_2 \circ \gamma_1) = m(\gamma_1) \cdot m(\gamma_2)$. Also, assume τ is of dimension d, and σ of dimension $d+1$.

Because σ is critical in $\mathcal{A}'$, it has no ingoing arrow from cells of dimension d in $\mathcal{H}'$. Consequently, $\Gamma_{\tau\nrightarrow\sigma}(\nu,\mu)$ contains exactly all gradient paths from ν to $\mu \neq \tau$ in $\mathcal{H}'$. Hence, the sum over $\Gamma_{\tau\nrightarrow\sigma}(\nu,\mu)$, for $\mu \in \mathcal{A}$, gives $\partial'\nu - [\nu : \tau]^{\mathcal{A}'}\tau$. Note that σ cannot appear in $\partial'\nu$ because σ is maximal by hypothesis.

Now, studying the left term $(\star)$, and splitting gradient paths passing through edge (τ,σ), then factorizing, we get

$$(\star) = \sum_{\mu\in\mathcal{A},\ \gamma_1\in\Gamma(\nu,\tau),\ \gamma_2\in\Gamma(\sigma,\mu)} m(\gamma_1) \cdot \left(-[\sigma:\tau]^X\right)^{-1} m(\gamma_2) \cdot \mu$$

$$= -\left([\sigma:\tau]^X\right)^{-1} \sum_{\mu\in\mathcal{A}} \underbrace{\left(\sum_{\gamma_2\in\Gamma(\sigma,\mu)} m(\gamma_2) \cdot \mu\right)}_{(\star_2)\ =\ \partial'\sigma - [\sigma:\tau]\cdot\tau} \cdot \underbrace{\left(\sum_{\gamma_1\in\Gamma(\nu,\tau)} m(\gamma_1)\right)}_{(\star_1)}.$$

The sum $(\star_1)$ over $\Gamma(\nu,\tau)$ is independent of μ, and equal to $[\nu:\tau]^{\mathcal{A}'}$ by definition.

Because τ is critical in $\mathcal{A}'$, it has no outgoing arrow towards cells of dimension $d+1$ in $\mathcal{H}'$. Consequently, $\Gamma(\sigma,\mu)$ contains exactly all gradient paths from σ to μ in $\mathcal{H}'$, where $\mu \neq \tau$. Hence, the sum $(\star_2)$ over $\Gamma(\sigma,\mu)$ gives $\partial'\sigma - [\sigma:\tau]^X \cdot \tau$.

Finally, putting terms together, the following allows us to conclude:

$$\partial\nu = \left(\partial'\nu - [\nu:\tau]^{\mathcal{A}'} \cdot \tau\right) - \frac{[\nu:\tau]^{\mathcal{A}'}}{[\sigma:\tau]^X}\left(\partial'\sigma - [\sigma:\tau]^X \cdot \tau\right).$$

5   Persistence Algorithm for Zigzag Morse Complexes

In this section, we describe an algorithm to perform the update of Diagram (9) to compute the zigzag persistent homology of the Morse filtration. Let $\mathcal{F}$ be an *atomic* zigzag filtration of complexes where all maps are forward or backward inclusions of a single cell: $\mathcal{F} : X_1 \longleftrightarrow X_2 \longleftrightarrow \cdots \longleftrightarrow X_n$.

Compatible Homology Matrix. Following [27,35] it is sufficient to maintain, at step j, a *homology matrix* for X_j that is *compatible* with the persistence module $H(\mathcal{F}_j) : H(X_1) \longleftrightarrow \cdots \longleftrightarrow H(X_j)$ made of the j first homology groups.

Definition 1 ([35]). Let X be a cell complex of size m and $\mathcal{B} = \{c_0,\ldots,c_{m-1}\}$ be a collection of m chains of $C(X)$. We say that $\mathcal{B}$ is a *homology matrix* at X if there exists an ordering $\sigma_0,\ldots,\sigma_{m-1}$ of the m cells of X such that:

(0) for all $0 \leq r < m$, the restriction $\{\sigma_0,\ldots,\sigma_r\} \subset X$ is a subcomplex of X,

(1) for all $0 \le r < m$, the leading term of c_r is σ_r for the chosen ordering, i.e., $c_r = \varepsilon_0\sigma_0 + \ldots + \varepsilon_{r-1}\sigma_{r-1} + \sigma_r$, for some $\varepsilon_i \in \mathbb{F}$,

and there exists a partition $\{0, \ldots, m-1\} = F \sqcup G \sqcup H$, and a bijective pairing $G \leftrightarrow H$, satisfying:

(2) for all indices $f \in F$, $\partial^{X_j}c_f = 0$,
(3) for all pairs $g \leftrightarrow h$ of $G \times H$, $\partial^{X_j}c_h = c_g$.

Additionally, the homology matrix at X_j is *compatible* if it agrees with the interval decomposition of the persistence module $H(\mathcal{F}_j)$; see [28] for details. The data of a compatible homology matrix is enough to compute the persistence diagram of a zigzag filtration [27,35].

The Morse theory algorithm for persistent homology of [30] can be applied to maintain a compatible homology matrix for a Morse filtration under the operations pictured in Diagrams (8). We design the update for the new operation of Diagram (9). Consider:

$$\mathcal{M}_j: \quad \mathcal{A}_1 \longleftrightarrow \cdots \longleftrightarrow \mathcal{A}_j \quad \text{and} \quad \mathcal{F}_j: \quad X_1 \longleftrightarrow \cdots \longleftrightarrow X_j \, ,$$

such that $\mathcal{M}_j$ is a zigzag Morse filtration for $\mathcal{F}_j$. Assume $\mathcal{A}_j$ has m cells, and let $\mathcal{B} = \{c_0, \ldots, c_{m-1}\}$ be a homology matrix at $\mathcal{A}_j$ compatible with $H(\mathcal{M}_j)$. Following Diagram (9), consider:

$$\overline{\mathcal{M}}_j: \mathcal{A}_1 \longleftrightarrow \cdots \longleftrightarrow \mathcal{A}_j \longrightarrow \mathcal{A}_j \cup \{\tau, \sigma\} \quad \text{and} \quad \overline{\mathcal{F}}_j: X_1 \longleftrightarrow \cdots \longleftrightarrow X_j \xrightarrow{1} X_j$$

such that $\overline{\mathcal{M}}_j$ is a zigzag Morse filtration for $\overline{\mathcal{F}}_j$. From $\mathcal{B}$, we define a homology matrix $\overline{\mathcal{B}} := \{c'_0, \ldots, c'_{m-1}, c_\tau, c_\sigma\}$ at $\mathcal{A}_j \cup \{\tau, \sigma\}$ that is compatible with $H(\overline{\mathcal{M}}_j)$.

Denote the two last complexes and their boundary maps in $\overline{\mathcal{M}}_j$ by $(\mathcal{A}_j, \partial)$ and $(\mathcal{A}'_j, \partial')$, with $\mathcal{A}'_j := \mathcal{A}_j \cup \{\tau, \sigma\}$. Then:

– for all indices $i \in F \sqcup H$, define

$$c'_i := c_i - \left([\sigma : \tau]^{X_j}\right)^{-1}\left(\sum_{\nu \in c_i}[\nu : \tau]^{\mathcal{A}'}\right) \cdot \sigma,$$

where the sum is taken over all cells ν in the support of chain c_i,
– define $c_\tau := \partial'\sigma$, and $c_\sigma := \sigma$, and put the index of c_τ in G, the index of c_σ in H, and pair them together,
– the pairing $G \leftrightarrow H$ inherited from $\mathcal{B}$ remains unchanged, and so does F.

Lemma 3. *The collection $\overline{\mathcal{B}}$ is a homology matrix at $\mathcal{A}_j \cup \{\tau, \sigma\}$.*

Proof. We prove that $\overline{\mathcal{B}}$ satisfies the three conditions of Definition 1.

(0) Because a Morse matching induces an acyclic Hasse Diagram, there exists r such that $\sigma_0, \ldots, \sigma_r, \tau, \sigma, \sigma_{r+1}, \ldots, \sigma_{m-1}$ is an ordering of the cells of $\mathcal{A}_j \cup \{\tau, \sigma\}$ such that the first k cells form a subcomplex, for any k, as in Definition 1.

(1) **Case** c_τ, c_σ. The leading term of c_σ is σ. We prove that the leading term of c_τ is τ in the ordering defined above. Let $\mathcal{H}$ be the oriented Hasse diagram of X_j for the Morse matching where (τ, σ) forms a Morse pair (complex $\mathcal{A}_j$), and $\mathcal{H}'$ for the matching where τ and σ are critical (complex $\mathcal{A}_j \cup \{\tau, \sigma\}$); they differ by the orientation of arrow $\sigma \leftrightarrow \tau$. First, $\langle \partial' \sigma, \tau \rangle^{\mathcal{A}_j \cup \{\tau, \sigma\}} \neq 0$ because there exists a unique gradient path from critical cell σ to critical cell τ in $\mathcal{A}_j \cup \{\tau, \sigma\}$, which is the one edge path $\gamma = (\tau, \sigma)$. The path γ exists because τ is a facet of σ in X_j. If there were another distinct gradient path from σ to τ in $\mathcal{H}'$, not containing the edge $\sigma \to \tau$, this path would exist in $\mathcal{H}$ and form a cycle with edge $\tau \to \sigma$ in $\mathcal{H}$; a contradiction with the definition of Morse matchings. Second, if $\mu \in \mathcal{A}_j \cup \{\tau, \sigma\}$, is critical such that $[\sigma : \mu]^{\mathcal{A}_j \cup \{\tau, \sigma\}} \neq 0$, then μ appears before σ (and τ) in the ordering. Indeed, there exists a gradient path $\gamma = (\sigma, \mu_1, \omega(\mu_1), \dots, \omega(\mu_{r-1}), \mu_r = \mu)$ from σ to μ in $\mathcal{H}'$. The cells $(\mu_i, \omega(\mu_i))$ of a pair are inserted consecutively by construction, and, for all i, μ_i is inserted before $\omega(\mu_{i-1})$ because it is a facet in X_j. By transitivity, μ is inserted before σ.

Case c_i'. The leading term of c_i' is σ_i. If $c_i' = c_i$, it is direct. Otherwise, by construction, $c_i' = c_i + \alpha \cdot \sigma$, $\alpha \neq 0$, and the chain c_i contains cells ν in its support such that $[\nu : \tau]^{\mathcal{A}_j \cup \{\tau, \sigma\}} \neq 0$, i.e., cofacets of τ in $\mathcal{A}_j \cup \{\tau, \sigma\}$. With a similar transitivity argument as above, τ (and σ) must consequently appear before such ν in the ordering of cells defined. The leading term of c_i' is then unchanged.

(2) Let c_i be a chain such that $i \in F \sqcup H$. By Lemma 2, it is a direct calculation from the definition of c_i' that $\partial' c_i' = \partial c_i$. Consequently, Conditions (2) and (3) of Definition 1 are satisfied for those chains. The pairing $G \leftrightarrow H$ remains valid, because $\partial' c_h' = \partial c_h = c_g = c_g'$ for $g \leftrightarrow h$, $(g, h) \in G \times H$.

(3) By definition, $\partial' c_\sigma = c_\tau$, their indices are in $H \times G$ and paired together.

Additionally, this new homology is compatible with the appropriate filtration; see [28] for more details. For the reproducibility of the experiments (Sect. 6), we give a detailed description of the use of the zigzag persistence algorithm [27] adapted to our Morse framework.

Implementation and Complexity. We represent $\mathcal{B} = \{c_0, \dots, c_{m-1}\}$ by an $(m \times m)$-sparse matrix data structure $M_\mathcal{B}$. Assume computing boundaries and coboundaries in a Morse complex of size m is given by an oracle of complexity $\mathcal{C}(m)$. We implement the transformation $\mathcal{B} = \{c_0, \dots, c_{m-1}\} \to \overline{\mathcal{B}} = \{c_0', \dots, c_{m-1}', c_\tau, c_\sigma\}$ presented above by:

- computing the boundary $\partial \sigma$ of σ in $\mathcal{A}_j \cup \{\tau, \sigma\}$, and the coboundary $\{\nu : [\nu : \tau]^{\mathcal{A}_j \cup \{\tau, \sigma\}} \neq 0\}$ of τ, in $O(\mathcal{C}(m))$ operations,
- adding columns c_τ and c_σ to the matrix in $O(m)$ operations,
- computing c_i' for all i, in $O(m^2)$. We can restrict the transformation to those c_i containing a cell of the coboundary of τ.

Consequently, we can perform the transformation above in $O(m^2 + \mathcal{C}(m))$ operations on a $(m \times m)$-matrix. The zigzag persistence algorithm of [8,27] deals with forward and backward insertions of a single cell in $O(m^2)$ operations.

In conclusion, let $\overline{\mathcal{F}} = (\, \overline{X}_i \leftarrow \Sigma_i \rightarrow \overline{X}_{i+1} \,)_{i=1\ldots 2k}$ be a general zigzag filtration (Diagram (4)), and let $\mathcal{M}$ be a zigzag Morse filtration as defined in Sect. 3, for a collection of Morse matchings $(\mathcal{A}_i, \mathcal{Q}_i, \mathcal{K}_i, \omega_i)$ on Σ_i, i odd. And:

- denote by n the total number of insertions and deletions critical cells in $\mathcal{M}$, and by $|\mathcal{A}_m|$ the maximal number of critical cells of a complex in $\mathcal{M}$,
- denote by N the total number of insertion and deletion of cells in $\overline{\mathcal{F}}$, and by $|X_m|$ the maximal number of cells of a complex in $\overline{\mathcal{F}}$.

Table 1. Experimental results for the oscillating Rips zigzag filtrations. For each experiment, the maximal dimension is 10, $\mu = 4$, $\nu = 6$, except for Sph3, where $\nu = 7$. The number of vertices is 2000.

	Without Morse reduction				With Morse reduction			
	$N \times 10^6$	$|X_m|$	time (s) cpx + pers	mem. peak (GB)	$n \times 10^6$	$|\mathcal{A}_m|$	time (s) cpx + pers	mem. peak (GB)
K1Bt5	63.3	187096	403 + 2912	4.7	4.9	11272	394 + 448	1.1
Spi3	66.1	47296	435 + 4438	5.2	3.8	12810	382 + 343	1.1
MoCh	75.7	37709	460 + 4680	5.8	4.1	11975	450 + 318	1.1
Sph3	99.4	66848	430 + 3498	7.5	4.2	13432	665 + 853	1.3
To3	32.8	32903	117 + 847	2.4	1.6	7570	173 + 79	0.47
By	30.5	18764	153 + 951	2.3	5.2	8677	165 + 287	0.96

Additionally, we compute Morse matchings using the fast coreduction algorithm of Mrozek and Batko [31]. Even if computing optimal Morse matchings is hard in general [24], this heuristic gives experimentally very small Morse complexes, with constant amortized cost per cell considered. We compute boundaries and coboundaries in a Morse complex $\mathcal{A}$ of a complex X by a linear traversal of the Hasse diagram of X. We store in memory the homology matrix of the Morse complex and the complex X. Consequently, the total cost of the algorithm is:

Theorem 3. *The persistent homology of $\overline{\mathcal{F}}$ can be computed in*

- *time:* $O(n \cdot |\mathcal{A}_m|^2 + n \cdot |X_m| + N)$,
- *memory:* $O(|\mathcal{A}_m|^2 + |X_m|)$.

In comparison, running the (practical) zigzag persistence algorithms [7,8,27] require $O(N \cdot |X_m|^2)$ operation and memory $O(|X_m|^2)$.

6   Experiments

In this section, we report on the performance of the zigzag persistence algorithm [27] with and without Morse reduction. The corresponding code will be available in a future release of the open source library GUDHI [36].

The following tests are made on a 64-bit Linux (Ubuntu) HP machine with a 3.50 GHz Intel processor and 63 GB RAM. The programs are all implemented in C++ and compiled with optimization level -O2 and gcc-8. Memory peaks are obtained via the /usr/bin/time -f Linux command, and timings are measured via the C++ std::chrono::system_clock::now() method. The timings for File IO are not included in any process time.

Table 2. Experimental results for the level set zigzag filtrations. For each experiment, the function $f : [0; 1]^3 \to [-14, 21]$ is applied to $129^3 = 2\,146\,689$ cells and the persistence is computed for maximal dimension 3. The interval size is denoted by ϵ. The infinity symbol ∞ corresponds to more than 12 h computing time.

		Without Morse reduction				With Morse reduction			
ϵ	max. noise	$N \times 10^6$	$\|X_m\|$	time (s) cpx + pers	mem. peak (GB)	$n \times 10^6$	$\|\mathcal{A}_m\|$	time (s) cpx + pers	mem. peak (GB)
0.1	0	34	286780	563 + 1725	3.9	6.3	48578	224 + 29	2.7
0.15	0	-	-	∞	-	9.3	115558	756 + 44	3.6
0.15	0.5	36.5	315305	417 + 3248	4.2	4.7	36144	221 + 59	2.8
0.2	0	-	-	∞	-	15.5	245360	2097 + 68	4.7
0.2	0.5	-	-	∞	-	5.6	56500	392 + 47	3.4

We run two types of experiments: homology inference from point clouds, using oscillating Rips zigzag filtrations, and levelset persistence of 3D-images. Both applications are described in the introduction.

For homology inference, we use both synthetic and real data points. The point clouds KlBt5, Spi3, Sph3, and To3 are synthetic samples of respectively the 5-dimensional Klein bottle, a 3-dimensional spiral wrapped around a torus, the 3-dimensional sphere, and the 3-dimensional torus. The point cloud MoCh and By are 3-dimensional measured samples of surface models: the MotherChild model, and the Stanford bunny model from the Stanford Computer Graphics Laboratory. The results with corresponding parameters are presented in Table 1.

Levelset persistence is computed for a function $f : [0; 1]^3 \to \mathbb{R}$, were f is a Fourier sum with random coefficients, as proposed in the DIPHA library[1] as representative of smooth data. The cube $[0; 1]^3$ and function f are discretized into equal size voxels. For some tests, we also added random noise to the values of f. The values of $s_1 \leq s_2 \leq \ldots$ are spaced out equally such that $s_{i+1} - s_i = \epsilon$ for all i. The results with corresponding parameters are presented in Table 2.

---

[1] github.com/DIPHA/dipha/blob/master/matlab/create_smooth_image_data.m.

In all experiments, timings are decomposed into 'cpx' for computation dedicated to the complex (construction, computation of (co)boundaries and of Morse matchings) and 'pers' for the computation of zigzag persistence.

Analysis of the Results. The results show a significant improvement when using Morse reduction. For homology inference (Table 1), the total running time is between 2.5 and 6.7 times faster when using Morse reduction. Moreover, most of the computation is transferred onto the computation of the Morse complex, which opens new roads to improvement in future implementation, such as parallelization of the Morse reduction [23] (note that parallelization of the computation of zigzag persistence is not possible in the streaming model). In particular, the computation of zigzag persistence is from 3.3 to 14.7 times faster. The better performance is due to filtrations being from 5.8 to 23.5 times shorter than the original ones (quantities n vs N in the complexity analysis) and smaller complexes, from 2.2 to 16.6 times smaller with the Morse reduction (quantities $|\mathcal{A}_m|$ and $|X_m|$ in the complexity analysis). Note that the memory consumption with Morse reduction is from 2.4 and up to 5.6 times smaller, which is critical on complex examples in practice.

For levelset persistence (Table 2), the total running time is at least 9 times faster, and the computation of zigzag persistence alone is itself approximatively 55 times faster, when the computation without Morse reduction finished. On those cases that finish, the filtration size is from 5.5 to 7.7 times shorter with Morse reduction, the maximal size of the complexes between 5.9 and 8.7 times smaller, and the memory consumption around 50% more efficient.

Additionally, using Morse reduction allows to handle cases where the standard zigzag algorithm never finishes (more than 12 h). On these examples, the Morse algorithm does not take more than 36 min. for the entire computation.

These results agree with the complexity analysis (Sect. 5) where terms $O(|\mathcal{A}_m|^2)$ and $O(|X_m|^2)$ dominate both time and memory complexities.

References

1. Bauer, U., Kerber, M., Reininghaus, J.: Clear and compress: computing persistent homology in chunks. In: Bremer, P.-T., Hotz, I., Pascucci, V., Peikert, R. (eds.) Topological Methods in Data Analysis and Visualization III. MV, pp. 103–117. Springer, Cham (2014). https://doi.org/10.1007/978-3-319-04099-8_7
2. Bauer, U., Kerber, M., Reininghaus, J.: Distributed computation of persistent homology. In: ALENEX, pp. 31–38 (2014)
3. Bauer, U., Lesnick, M.: Induced matchings and the algebraic stability of persistence barcodes. JoCG **6**(2), 162–191 (2015)
4. Boissonnat, J.D., Dey, T.K., Maria, C.: The compressed annotation matrix: An efficient data structure for computing persistent cohomology. Algorithmica (2014)
5. Boissonnat, J., Pritam, S., Pareek, D.: Strong collapse for persistence. In: ESA 2018, pp. 67:1–67:13 (2018)
6. Carlsson, G., Zomorodian, A., Collins, A., Guibas, L.J.: Persistence barcodes for shapes. Int. J. Shape Model. **11**(2), 149–187 (2005)

7. Carlsson, G.E., de Silva, V.: Zigzag persistence. Found. Comput. Math. **10**(4), 367–405 (2010)
8. Carlsson, G.E., de Silva, V., Morozov, D.: Zigzag persistent homology and real-valued functions. In: Symposium on Computational Geometry, pp. 247–256 (2009)
9. Chang, H.W., Bacallado, S., Pande, V.S., Carlsson, G.E.: Persistent topology and metastable state in conformational dynamics. PLoS ONE **8**, e58699 (2013)
10. Chazal, F., Cohen-Steiner, D., Guibas, L.J., Mémoli, F., Oudot, S.Y.: Gromov-Hausdorff stable signatures for shapes using persistence. In: Proceedings of SGP (2009)
11. Chazal, F., Guibas, L.J., Oudot, S., Skraba, P.: Persistence-based clustering in Riemannian manifolds. J. ACM **60**(6), 41:1–41:38 (2013)
12. Chazal, F., de Silva, V., Glisse, M., Oudot, S.Y.: The Structure and Stability of Persistence Modules. Springer Briefs in Mathematics. Springer (2016)
13. Chen, C., Kerber, M.: Persistent homology computation with a twist. In: Proceedings 27th European Workshop on Computational Geometry (2011)
14. Cohen-Steiner, D., Edelsbrunner, H., Harer, J.: Stability of persistence diagrams. Discrete Comput. Geom. **37**(1), 103–120 (2007)
15. Delgado-Friedrichs, O., Robins, V., Sheppard, A.: Morse theory and persistent homology for topological analysis of 3D images of complex materials. In: 2014 IEEE International Conference on Image Processing (ICIP), pp. 4872–4876 (2014)
16. Delgado-Friedrichs, O., Robins, V., Sheppard, A.: Skeletonization and partitioning of digital images using discrete Morse theory. IEEE Trans. Pattern Anal. Mach. Intell. **37**(3), 654–666 (2015)
17. Delgado-Friedrichs, O., Robins, V.: Diamorse. https://github.com/AppliedMathe maticsANU/diamorse
18. Dlotko, P., Wagner, H.: Computing homology and persistent homology using iterated Morse decomposition. CoRR abs/1210.1429 (2012)
19. Edelsbrunner, H., Harer, J.: Computational Topology - An Introduction. American Mathematical Society, USA (2010)
20. Escolar, E., Hiraoka, Y.: Morse reduction for zigzag persistence. J. Indonesian Math. Soc. **20**(1), 47–75 (2014)
21. Forman, R.: Morse theory for cell complexes. Adv. Math. **134**, 90–145 (1998)
22. Gunther, D., Reininghaus, J., Hotz, I., Wagner, H.: Memory-efficient computation of persistent homology for 3D images using discrete Morse theory. In: 2011 24th SIBGRAPI Conference on Graphics, Patterns and Images, pp. 25–32 (2011)
23. Gyulassy, A., Bremer, P., Pascucci, V.: Shared-memory parallel computation of morse-smale complexes with improved accuracy. IEEE Trans. Visual Comput. Graphics **25**(1), 1183–1192 (2019)
24. Joswig, M., Pfetsch, M.E.: Computing optimal Morse matchings. SIAM J. Discrete Math. **20**(1), 11–25 (2006)
25. Lee, Y., Barthel, S., Dlotko, P., Moosavi, S., Hess, K., Smit, B.: Quantifying similarity of pore-geometry in nanoporous materials. Nature Commun. **8** (2017)
26. Lefschetz, S.: Algebraic Topology. AMS books online, AMS (1942)
27. Maria, C., Oudot, S.Y.: Zigzag persistence via reflections and transpositions. In: Proceedings of SODA 2015, pp. 181–199 (2015)
28. Maria, C., Schreiber, H.: Discrete Morse theory for computing zigzag persistence. CoRR abs/1807.05172 (2018)
29. Milosavljevic, N., Morozov, D., Skraba, P.: Zigzag persistent homology in matrix multiplication time. In: Symposium on Computational Geometry (2011)
30. Mischaikow, K., Nanda, V.: Morse theory for filtrations and efficient computation of persistent homology. Discrete Comput. Geom. **50**(2), 330–353 (2013)

31. Mrozek, M., Batko, B.: Coreduction homology algorithm. Discrete Comput. Geom. **41**(1), 96–118 (2009)
32. Nanda, V.: Perseus. http://www.sas.upenn.edu/~vnanda/perseus
33. Oudot, S.Y., Sheehy, D.R.: Zigzag Zoology: rips zigzags for homology inference. Found. Comput. Math. **15**(5), 1151–1186 (2015)
34. Robins, V., Wood, P.J., Sheppard, A.P.: Theory and algorithms for constructing discrete Morse complexes from grayscale digital images. IEEE Trans. Pattern Anal. Mach. Intell. **33**(8), 1646–1658 (2011)
35. de Silva, V., Morozov, D., Vejdemo-Johansson, M.: Dualities in persistent (co)homology. CoRR abs/1107.5665 (2011)
36. The GUDHI Project: GUDHI (2015). http://gudhi.gforge.inria.fr
37. Zomorodian, A., Carlsson, G.E.: Computing persistent homology. Discrete Comput. Geom. **33**(2), 249–274 (2005)

Optimal Offline Dynamic
2, 3-Edge/Vertex Connectivity

Richard Peng[1], Bryce Sandlund[2(✉)], and Daniel D. Sleator[3]

[1] School of Computer Science, Georgia Institute of Technology, Atlanta, GA, USA
rpeng@cc.gatech.edu
[2] Cheriton School of Computer Science, University of Waterloo,
Waterloo, ON, Canada
bcsandlund@uwaterloo.ca
[3] Department of Computer Science, Carnegie Mellon University,
Pittsburgh, PA, USA
sleator@cs.cmu.edu

Abstract. We give offline algorithms for processing a sequence of 2-
and 3-edge and vertex connectivity queries in a fully-dynamic undirected
graph. While the current best fully-dynamic online data structures for
3-edge and 3-vertex connectivity require $O(n^{2/3})$ and $O(n)$ time per
update, respectively, our per-operation cost is only $O(\log n)$, optimal
due to the dynamic connectivity lower bound of Patrascu and Demaine.
Our approach utilizes a divide and conquer scheme that transforms a
graph into smaller equivalents that preserve connectivity information.
This construction of equivalents is closely-related to the development of
vertex sparsifiers, and shares important connections to several upcoming
results in dynamic graph data structures, including online models.

1 Introduction

Dynamic graph data structures seek to answer queries on a graph as it undergoes
edge insertions and deletions. Perhaps the simplest and most fundamental query
to consider is connectivity. A connectivity query asks for the existence of a path
connecting two vertices u and v in the current graph. As the insertion or deletion
of a single edge may have large consequences to connectivity across the entire
graph, constructing an efficient dynamic data structure to answer connectivity
queries has been a challenge to the data structure community. A number of
solutions have been developed, achieving a wide variety of runtime tradeoffs in
a number of different models [10–12, 18, 19, 22–24, 30, 34, 36]

The models typically addressed are *online*: each query must be answered
before the next query or update is given. A less demanding variant is the *offline*
setting, where the entire sequence of updates and queries is provided as input to
the algorithm. While an online data structure is more general, there are many
scenarios in which the entire sequence of operations is known in advance. This is
often the case when a data structure is used in a subroutine of an algorithm [6,
28], one specific example being the use of dynamic trees in the near-linear time
minimum cut algorithm of Karger [25].

© Springer Nature Switzerland AG 2019
Z. Friggstad et al. (Eds.): WADS 2019, LNCS 11646, pp. 553–565, 2019.
https://doi.org/10.1007/978-3-030-24766-9_40

In exchange for the loss of flexibility, one can hope to obtain faster and simpler algorithms in the offline setting. This has been shown to be the case in the dynamic minimum spanning tree problem. While an online fully-dynamic minimum spanning tree data structure requires about $O(\log^4 n)$ time per update [20], the offline algorithm of Eppstein requires only $O(\log n)$ time per update [10].

In this paper, we show similar, but stronger, performance gains for higher versions of connectivity. In particular, we consider the problems of 2, 3-edge/vertex connectivity on a fully-dynamic undirected graph. An extension of connectivity, c-edge connectivity asks for the existence of c edge-disjoint paths between two vertices u and v in the current graph. Vertex connectivity requires vertex-disjoint paths instead of edge-disjoint paths. Current online fully-dynamic 2-edge/vertex connectivity data structures require update time $\tilde{O}(\log^2 n)$[1] [19] and $\tilde{O}(\log^3 n)$[2] [34], respectively, and current online fully-dynamic 3-edge/vertex connectivity data structures require update time $O(n^{2/3})$ and $O(n)$, respectively [11]. In contrast, our offline algorithms for 2, 3-edge/vertex connectivity require only $O(\log n)$ time per operation. As the lower bound on dynamic connectivity [32], as well as most lower bounds in general [1–3,7], also apply in the offline model, our algorithms are optimal up to a constant factor. This paper further shows that any lower bound attempting to show hardness stronger than $\Omega(\log n)$ time per operation for online fully-dynamic 2, 3-edge/vertex connectivity must make use of the online model.

As a straightforward application of our results, one can consider the use of our algorithms when data regarding a dynamic network is collected, but not analyzed, until a later point in time. For example, to diagnose an issue of network latency across key routing hubs, or determine viability of a dynamically-changing network of roads, our algorithms can answer a batch of queries in time $O(t \log n)$, where t is the total number of updates and queries. Since online fully-dynamic algorithms for higher versions of connectivity are significantly slower, namely, $O(n^{2/3})$ and $O(n)$ time update for 3-edge and 3-vertex connectivity, respectively, our offline data structure makes these computations practical for large data sets when they would otherwise be prohibitively expensive.

Related to our work are papers by Łącki and Sankowski [30] and Karczmarz and Łącki [24], which also apply to the above applications but for lower versions of connectivity. Their work considers a fixed sequence of graph updates, given in advance, and is then able to answer connectivity queries regarding intervals of this sequence, online. This is more general than the model we consider because the queries need not be supplied in advance and data regarding an interval of time is richer than information from a specific point in time. For connectivity and 2-edge connectivity, Karczmarz and Łącki achieve $O(\log n)$ time per operation [24]. Both 2-vertex connectivity and 3-edge/vertex connectivity queries are not supported.

---

[1] The $\tilde{O}(\cdot)$ notation hides $\log \log n$ factors.

[2] This complexity is claimed in Thorup's STOC 2000 [34] result. As noted by Huang et al. [22], the paper provides few details, deferring to a journal version that has since not appeared. The best complexity for online fully-dynamic biconnectivity prior to this claim was $O(\log^5 n)$ by Holm and Thorup [18].

In competitive programming, the idea of using divide and conquer as an offline algorithm for connectivity is known. The authors are aware of several contest problems[3] that are solved with similar techniques to Eppstein's minimum spanning tree algorithm [10], as we do here. The master's thesis of Sergey Kopeliovich, a member of the competitive programming community, describes such an offline algorithm for fully-dynamic 2-edge connectivity, also achieving about $O(\log n)$ time per operation [26]. Unfortunately, the thesis only appears in Russian, but we speculate that the ideas used are similar.

The techniques developed in this paper may be of independent interest. Our work has close connections with recent developments in vertex sparsification, particularly vertex sparsification-based dynamic graph data structures [4,5,8,9,12–17,27]. In particular, the equivalent graphs at the core of our algorithms are akin to vertex sparsifiers, with the main difference that 2- and 3-connectivity require preserving far less information than the more general definitions of vertex sparsifiers [15,27]. A promising step in this direction is very recent work of Goranci et al. [17], which suggests the notion of a *local sparsifier*. This is a generalization of the sparsifier that we consider here, and leads to efficient incremental algorithms in the *online* setting.

Indeed, offline algorithms haven proven useful for the development of online counterparts in the past. One such example is recent development in the maintenance of dynamic effective resistance. Recent work in fully-dynamic data structures for maintaining effective resistances online [8] relied heavily upon ideas from earlier data structures for maintaining effective resistances in offline [9,29] or offline-online hybrid [9,28] settings.

The results of this paper were previously published online in the open access journal arXiv [33] and have recently been extended to offline 4- and 5-edge connectivity [31]. This new work achieves about $O(\sqrt{n})$ time complexity per operation.

The rest of this paper will be dedicated to proving the following theorem:

Theorem 1. *Given a sequence of t updates/queries on a graph of the form:*

- *Insert edge (u, v),*
- *Delete edge (u, v),*
- *Query if a pair of vertices u and v are 2-edge connected/3-edge connected/biconnected/tri-connected in the current graph,*

there exists an algorithm that answers all queries in $O(t \log n)$ time.

For simplicity, we will assume the graph is empty at the start and end of the sequence, but the results discussed are easily modified to start with an initial graph G, at the cost of an additive $O(m)$ term in the running time, where m is

---

[3] See https://codeforces.com/blog/entry/15296 and https://codeforces.com/gym/100551/problem/A, for example.

the number of edges of G. Further, we assume a fixed vertex set of size n. Any update sequence with arbitrary vertex endpoints can be modified to one on a fixed set of vertices, where the size of the fixed set is equal to the largest number of non-isolated vertices in any graph achieved in the given update sequence. Finally, we consider the graph G to be a multigraph, since at times during our constructions and definitions, we will need to work with multigraphs.

The paper is organized as follows. We describe our offline framework for reducing graphs to smaller equivalents in Sect. 2. We show how simple techniques can be used to create such equivalents for 2-edge connectivity and bi-connectivity in Sect. 3. In Sect. 4 we extend these constructions to 3-edge connectivity. Our most technical section is 3-vertex connectivity, where constructing equivalent graphs requires careful manipulation of SPQR trees. Unfortunately, due to page limits, this will only appear in the full version of this paper.

2   Offline Framework

The main idea of our offline framework is to perform divide and conquer on the input sequence, similar to what is done in Eppstein's offline minimum spanning tree algorithm [10]. Consider the full sequence of updates and queries $x_1, \ldots, x_t$, where each x_i is either an edge insertion, edge deletion, or query. Call each x_i an event.

Assume each inserted edge has a unique identity. Then for each inserted edge e, we may associate an interval $[I(e), D(e)]$, indicating that edge e was inserted at time $I(e)$ and removed at time $D(e)$. Plotting time along the x-axis and edges on the y-axis as in Fig. 1 gives a convenient way to view the sequence of events.

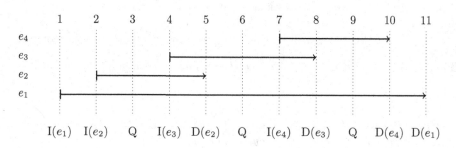

Fig. 1. A timeline diagram of four edge insertions(I)/deletions(D) and three queries(Q), with time on the x-axis and edges on the y-axis.

Fix some subinterval $[l, r]$ of the sequence of events. Let us classify all edges present at any point of time in the sequence $x_l, \ldots, x_r$ as one of two types.

1. Edges present throughout the duration ($i_e \leq l \leq r \leq d_e$), we call *permanent* edges.
2. Edges affected by an event in this range (one or both of $I(e), D(e)$ is in (l, r)), we call *non-permanent* edges.

While there may be a large number of permanent edges, the number of non-permanent edges is limited by the number of time steps, $r - l + 1$. Therefore, the graph can be viewed as a large static graph on which a smaller number of events take place.

Our goal will be to reduce this graph of permanent edges to one whose size is a small function of the number of events in the subinterval. If we may do so without affecting the answers to the queries, we can recursively apply the technique to achieve an efficient divide and conquer algorithm for the original dynamic c-connectivity sequence.

We will work in the dual-view, considering cuts instead of edge-disjoint or vertex-disjoint paths. Two vertices u and v are c-edge connected if there does not exist a cut of $c - 1$ edges separating them; further, u and v are c-vertex connected if there does not exist a cut of $c - 1$ vertices that separates them.

We need the following definition.

Definition 1. *Given a graph $G = (V_G, E_G)$ with vertex subset $W \subseteq V_G$ and a graph $H = (V_H, E_H)$ with $W \subseteq V_H$, we say that H and G are c-edge equivalent if, for any partition (A, B) of W, the size of a minimum cut separating A and B is the same in G and H whenever either of these sizes is less than c. Similarly, we say H and G are c-vertex equivalent if, for any partition (A, B, C) of W with $|C| < c$, the size of a minimum vertex cut D separating A and B such that $C \subseteq D$ and $D \cap A = \emptyset$, $D \cap B = \emptyset$, is the same in G and H whenever either of these sizes is less than c.*

This gives the following.

Lemma 1. *Suppose $G = (V_G, E_G)$ and $H = (V_H, E_H)$ are c-edge/c-vertex equivalent on vertex set W. Let E_W denote any set of edges between vertices of W. Then $H' = (V_H, E_H \cup E_W)$[4] and $G' = (V_G, E_G \cup E_W)$ are c-edge/c-vertex equivalent.*

Proof. We first show c-edge equivalence. Let (A, B) be any partition of W and consider the minimum cuts separating A and B in G' and H'. Since the edges in E_W are between vertices of W, they must cross the separation (A, B) in the same way. Therefore, if the minimum cut separating A and B had size less than c in either G or H, the minimum cuts separating A and B will have equivalent size in G' and H'. Further, if the minimum cuts separating A and B had size larger or equal to c in both G and H, the minimum cuts separating A and B will also have size larger or equal to c in G' and H', since we only add edges to G' and H'. Thus G' and H' are c-edge equivalent.

We now consider c-vertex equivalence. Consider a partition (A, B, C) of W. As with edge connectivity, if no vertex subset D exists satisfying the conditions of Definition 1, the introduction of additional edges between any vertices of W will not change the existence of such a set D in G' or H'. Furthermore, if an

---

[4] We take $\cup$ here to be in the multigraph sense; an edge $uv \in E_W$ is added regardless if there is already a uv edge in E_H or E_G.

edge of E_W connects a vertex of A to a vertex of B, no vertex cut separates A and B in G' and H'. Now suppose none of these cases is true, and there exists a vertex set D satisfying the conditions of Definition 1 such that the removal of D disconnects A and B in G and H and further that no edge of E_W connects a vertex of A to a vertex of B. Then the removal of vertex set D still disconnects A and B in G' and H'. Thus c-vertex equivalence of G' and H' follows from c-vertex equivalence of G and H.

Now consider the graph G of permanent edges for the subinterval $x_l, \ldots, x_r$ of events. Let W be the set of vertices involved in any event in the subinterval (that is, W is the set of endpoints of all non-permanent edges in the subinterval, as well as vertices involved in a query). We will refer to these vertices as *active* vertices, and all other vertices of G not in W as *inactive* vertices. Lemma 1 says that if we reduce G to a c-edge/ c-vertex equivalent graph H on set W, the result of all queries in $x_l, \ldots, x_r$ on H will be the same as on G. This is because all cuts in H and G that affect the queries (therefore of size less than c) are of equivalent size, even after the addition of non-permanent edges in H and G.

This idea can lead to a divide and conquer algorithm if we can produce such equivalent graphs H of small size efficiently. Specifically:

Lemma 2. *Given a graph G with m edges and vertex set W of size k, if there is an $O(m)$ time algorithm that produces a graph H of size $O(k)$ that is c-edge/c-vertex equivalent to G on W, then there is an algorithm that can answer all c-edge/c-vertex connectivity queries in a sequence of events $x_1, \ldots, x_t$ in $O(t \log n)$ time.*

Proof. We perform divide and conquer on the sequence of events. We take the sequence of events $x_1, \ldots, x_t$ and divide it in half. Over each half, we will take the graph of permanent edges, which we denote G, and reduce it to a c-edge/ c-vertex equivalent graph H. We repeat the scheme recursively. As the subintervals get smaller, non-permanent edges become permanent and are absorbed by the production of equivalent graphs. Eventually, we reduce to subintervals with a constant number of events, which can be answered by any algorithm of our choice on a graph of constant size.

Consider the sizes of the graphs in each step of recursion. The graph G is the graph produced in the previous level plus the edges that became permanent in this interval. The graph produced at the previous level has size linear in the number of events at the current level, and since we reduce the number of events by a factor 2 in each step of recursion, the number of edges that become permanent is also linear. It follows that the divide and conquer satisfies the recurrence $T(t) = 2T(t/2) + O(t)$, which solves to $T(t) = O(t \log t)$. If t is polynomially-bounded by n, $T(t) = O(t \log n)$. If not, we may first break the sequence of events $x_1, \ldots x_t$ into blocks of size, say, n^2. Since the size of the graph G cannot be more than $O(n^2)$ in any subinterval, we can therefore handle each block separately and answer all queries in $O(t \log n)$ time. This proves the lemma.

The remainder of the paper will show the construction of 2-edge, 2-vertex, 3-edge, and 3-vertex equivalent graphs.

3 Equivalent Graphs for 2-Edge Connectivity and Bi-connectivity

We now show offline algorithms for dynamic 2-edge connectivity and bi-connectivity by constructing 2-edge and 2-vertex equivalents needed by Lemma 2. These two properties ask for the existence of a single edge/vertex whose removal separates query vertices u and v. Since these cuts can affect at most one connected component, it suffices to handle each component separately.

The underlying structure for 2-edge connectivity and bi-connectivity is tree-like. This is perhaps more evident for 2-edge connectivity, where vertices on the same cycle belong to the same 2-edge connected component. We will first describe the reductions that we will make to this tree in Sect. 3.1, and adapt them to bi-connectivity in Sect. 3.2.

At times we will make use of the term "equivalent cut". By this we mean that a cut C' is equivalent to C if it has the same size and separates the vertices of W in the same way.

3.1 2-Edge Connectivity

Using depth-first search [21], we can identify all cut-edges in the graph and the 2-edge connected components that they partition the graph into. The case of edge cuts is slightly simpler conceptually, since we can combine vertices without introducing new cuts. Specifically, we show that each 2-edge connected component can be shrunk to single vertex.

Lemma 3. *Let S be a 2-edge connected component in G. Then contracting all vertices in S to a single vertex s in H^5, and endpoints of edges correspondingly, creates a 2-edge equivalent graph.*

Proof. The only cuts that we need to consider are ones that remove cut edges in G or H. Since we only contracted vertices in a component, there is a one-to-one mapping of these edges from G to H. Since S is 2-edge connected, all vertices in it will be on the same side of one of these cuts. Furthermore, removing the same edge in H leads to a cut with s instead. Therefore, all active vertices in S are mapped to s, and are therefore on the same side of the cut.

This allows us to reduce G to a tree H, but the size of this tree can be much larger than k. Therefore we need to prune the tree by removing inactive leaves and length 2 paths whose middle vertex is inactive.

---

[5] Here we slightly abuse our requirement $W \subseteq V_H$, where V_H are the vertices of H. A map of W onto V_H that preserves the cuts needed by c-edge/c-vertex equivalence suffices.

Lemma 4. *If G is a tree, the following two operations lead to 2-edge equivalent graphs H.*

– *Removing an inactive leaf.*
– *Removing an inactive vertex with degree 2 and adding an edge between its two neighbors.*

Proof. In the first case, the only cut in G that no longer exist in H is the one that removes the cut edge connecting the leaf with its unique neighbor. However, this places all active vertices in one component and thus does not separate W and need not be represented in H.

In the second case, if a cut removes either of the edges incident to the degree 2 vertex, removing the new edge creates an equivalent cut since the middle vertex is inactive. Also, for a cut that removes the new edge in H, removing either of the two original edges in G leads to an equivalent cut.

This allows us to bound the size of the tree by the number of active vertices, and therefore finish the construction.

Lemma 5. *Given a graph G with m edges and k active vertices W, a 2-edge equivalent of G of size $O(k)$, H, can be constructed in $O(m)$ time.*

Proof. We can find all the cut edges and 2-edge connected components in $O(m)$ time using depth-first search [21], and reduce the resulting structure to a tree H using Lemma 3. On H, we repeatedly apply Lemma 4 to obtain H'.

In H', all leaves are active, and any inactive internal vertex has degree at least 3. Therefore the number of such vertices can be bounded by $O(k)$, giving a total size of $O(k)$.

3.2   Bi-connectivity

All cut-vertices (articulation points) can also be identified using DFS, leading to a structure known as the block-tree. However, several modifications are needed to adapt the ideas from Sect. 3.1. The main difference is that we can no longer replace each bi-connected component with a single vertex in H, since cutting such vertices corresponds to cutting a much larger set in G. Instead, we will need to replace the bi-connected components with simpler bi-connected graphs such as cycles.

Lemma 6. *Replacing a bi-connected component with a cycle containing all its cut-vertices and active vertices gives a 2-vertex equivalent graph.*

Proof. As this mapping maintains the bi-connectivity of the component, it does not introduce any new cut-vertices. Therefore, G and H have the same set of cut vertices and the same block-tree structure. Note that the actual order the active vertices appear in does not matter, since they will never be separated. The claim follows similarly to Lemma 3.

The block-tree also needs to be shrunk in a similar manner. Note that the fact that blocks are connected by shared vertices along with Lemma 6 implies the removal of inactive leaves. Any leaf component with no active vertices aside from its cut vertex can be reduced to the cut vertex, and therefore be removed. The following is an equivalent of the degree two removal part of Lemma 4.

Lemma 7. *Two bi-connected components C_1 and C_2 with no active vertices that share cut vertex w and are only incident to one other cut vertex each, u and v respectively, can be replaced by an edge connecting u and v to create a 2-vertex equivalent graph.*

Proof. As we have removed only w, any cut vertex in H is also a cut vertex in G. As C_1 and C_2 contain no active vertices, this cut would induce the same partition of active vertices.

For the cut given by removing w in G, removing u in H gives the same cut since C_1 has no active vertices (which in turn implies that u is not active). Note that the removal of u may break the graph into more pieces, but our definition of cuts allows us to place these pieces on two sides of the cut arbitrarily.

Note that Lemma 6 may need to be applied iteratively with Lemma 7 since some of the cut vertices may no longer be cut vertices due to the removal of components attached to them.

Lemma 8. *Given a graph G with m edges and k active vertices W, a 2-vertex equivalent of G of size $O(k)$, H, can be constructed in $O(m)$ time.*

Proof. We can find all the initial block-trees using depth-first search [21]. Then we can apply Lemmas 6 and 7 repeatedly until no more reductions are possible. Several additional observations are needed to run these reduction steps in $O(m)$ time. As each cut vertex is removed at most once, we can keep a counter in each component about the number of cut vertices on it. Also, the second time we run Lemma 6 on a component, it's already a cycle, so the reductions can be done without examining the entire cycle by tracking it in a doubly linked list and removing vertices from it.

It remains to bound the size of the final block-tree. Each leaf in the block-tree has at least one active vertex that's not its cut vertex. Therefore, the block-tree contains at most $O(k)$ leaves and therefore at most $O(k)$ internal components with 3 or more cut vertices, as well as $O(k)$ components containing active vertices. If these components are connected by paths with 4 or more blocks in the block tree, then the two middle blocks on this path meet the condition of Lemma 7 and should have been removed by the above procedure. This gives a bound of $O(k)$ on the number of blocks, which in turn implies an $O(k)$ bound on the number of cut vertices. The edge count then follows from the fact that Lemma 6 replaces each component with a cycle, whose number of edges is linear in the number of vertices, and that the bi-connected components themselves are arranged in a tree.

4 3-Edge Connectivity

We now extend our algorithms to 3-edge connectivity. Our starting point is a statement similar to Lemma 3, namely that we can contract all 3-edge connected components. Though of no consequence to our algorithms, we note that unlike 2-edge or biconnected components, 3-edge connected components need not be connected.

Lemma 9. *Let S be a 3-edge connected component in G. Then contracting all vertices in S to a single vertex s in H, and endpoints of edges correspondingly, creates a 3-edge equivalent graph.*

Proof. A two-edge cut will not separate a 3-edge connected component. Therefore all active vertices in S fall on one side of the cut, to which vertex s may also fall. The proof follows analogously to Lemma 3.

Such components can also be identified in $O(m)$ time using depth-first search [35], so the preprocessing part of this algorithm is the same as with the 2-connectivity cases. However, the graph after this shrinking step is no longer a tree. Instead, it is a cactus, which in its simplest terms can be defined as:

Definition 2. *A cactus is an undirected graph where each edge belongs to at most one cycle.*

On the other hand, cactuses can also be viewed as a tree with some of the vertices turned into cycles[6]. Such a structure essentially allows us to repeat the same operations as in Sect. 3 after applying the initial contractions.

Lemma 10. *A connected undirected graph with no nontrivial 3-edge connected component is a cactus.*

Due to space restrictions, we save the proof for Appendix A.

With this structural statement, we can then repeat the reductions from the 2-edge equivalent algorithm from Sect. 3.1 to produce the 3-edge equivalent graph.

Lemma 11. *Given a graph G with m edges and k active vertices W, a 3-edge equivalent of G of size $O(k)$, H, can be constructed in $O(m)$ time.*

Proof. Lemma 10 means that we can reduce the graph to a cactus after $O(m)$ time preprocessing.

First consider the tree where the cycles are viewed as vertices. Note that in this view, a vertex that's not on any cycle is also viewed as a cycle of size 1. This can be pruned in a manner analogous to Lemma 4:

1. Cycles containing no active vertices and incident to 1 or 2 other cycles can be contracted to a single vertex.
2. Inactive single-vertex cycles incident to 1 other cycle can be removed.

---

[6] Some 'virtual' edges are needed in this construction, because a vertex can still belong to multiple cycles.

This procedure takes $O(m)$ time and produces a graph with at most $O(k)$ leaves. Correctness of the first rule follows by replacing a cut of the two edges within an inactive cycle by a cut of the single contracted vertex with one of its neighbors. The second rule does not affect any cuts separating W. It remains to reduce the length of degree 2 paths and the sizes of the cycles themselves.

As in Lemma 4, all inactive vertices of degree 2 can be replaced by an edge between its two neighbors. This bounds the length of degree 2 paths and reduces the size of each cycle to at most twice its number of incidences with other cycles. This latter number is in turn bounded by the number of leaves of the tree of cycles. Hence, this contraction procedure reduces the total size to $O(k)$.

We remark that this is not identical to iteratively removing inactive vertices of degrees at most 2. With that rule, a cycle can lead to a duplicate edge between pairs of vertices, and a chain of such cycles needs to be reduced in length.

A   Omitted Proofs

Proof (Proof of Lemma 10). We prove by contradiction. Let G be a graph with no nontrivial 3-edge connected component. Suppose there exists two simple cycles a and b in G with more than one vertex, and thus at least one edge, in common.

Call the vertices in the first simple cycle $a_1, \ldots, a_n$ and the second simple cycle $b_1, \ldots, b_m$, in order along the cycle.

Since these cycles are not the same, there must be some vertex not common to both cycles. Without loss of generality, assume (by flipping a and b) that b is not a subset of a, and (by shifting b cyclically) that b_1 is only in b and not a.

Now let b_{first} be the first vertex after b_1 in b that is common to both cycles, so

$$first \overset{\text{def}}{=} \min_i b_i \in a. \tag{1}$$

and let b_{last} be the last vertex in b common to both cycles

$$last \overset{\text{def}}{=} \max_i b_i \in a. \tag{2}$$

The assumption that these two cycles have more than 1 vertex in common means that

$$first < last. \tag{3}$$

We claim b_{first} and b_{last} are 3-edge connected.

We show this by constructing three edge-disjoint paths connecting b_{first} and b_{last}. Since both b_{first} and b_{last} occur in a, we may take the two paths formed by cycle a connecting b_{first} and b_{last}, which are clearly edge-disjoint.

By construction, vertices

$$b_{last+1}, \ldots, b_m, b_1, \ldots, b_{first-1} \tag{4}$$

are not shared with a. Thus they form a third edge-disjoint path connecting b_{first} and b_{last}, and so the claim follows. Therefore, a graph with no 3-edge connected vertices, and thus no nontrivial 3-edge connected component has the property that two simple cycles have at most one vertex in common.

References

1. Abboud, A., Dahlgaard, S.: Popular conjectures as a barrier for dynamic planar graph algorithms. In: IEEE 57th Annual Symposium on Foundations of Computer Science, pp. 477–486, Nov 2016
2. Abboud, A., Williams, V.V.: Popular conjectures imply strong lower bounds for dynamic problems. In: Proceedings of the 2014 IEEE 55th Annual Symposium on Foundations of Computer Science, FOCS 2014, pp. 434–443 (2014)
3. Abboud, A., Williams, V.V., Yu, H.: Matching triangles and basing hardness on an extremely popular conjecture. In: Proceedings of the Forty-Seventh Annual ACM on Symposium on Theory of Computing, STOC 2015, pp. 41–50 (2015)
4. Abraham, I., Durfee, D., Koutis, I., Krinninger, S., Peng, R.: On fully dynamic graph sparsifiers. In: 2016 IEEE 57th Annual Symposium on Foundations of Computer Science (FOCS), pp. 335–344, Oct 2016
5. Assadi, S., Khanna, S., Li, Y., Tannen, V.: Dynamic sketching for graph optimization problems with applications to cut-preserving sketches. In: FSTTCS (2015)
6. Bringmann, K., Kunnemann, M., Nusser, A.: Frechet distance under translation: conditional hardness and an algorithm via offline dynamic grid reachability. CoRR abs/1810.10982 (2018). https://arxiv.org/abs/1810.10982
7. Dahlgaard, S.: On the hardness of partially dynamic graph problems and connections to diameter. In: 43rd International Colloquium on Automata, Languages, and Programming (2016)
8. Durfee, D., Gao, Y., Goranci, G., Peng, R.: Fully dynamic spectral vertex sparsifiers and applications. In: Proceedings of the thirtieth annual ACM symposium on Theory of computing, STOC 2019. ACM (2019)
9. Durfee, D., Kyng, R., Peebles, J., Rao, A.B., Sachdeva, S.: Sampling random spanning trees faster than matrix multiplication. In: Proceedings of the 49th Annual ACM SIGACT Symposium on Theory of Computing, pp. 730–742 (2017)
10. Eppstein, D.: Offline algorithms for dynamic minimum spanning tree problems. J. Algorithms 17(2), 237–250 (1994)
11. Eppstein, D., Galil, Z., Italiano, G.F., Nissenzweig, A.: Sparsification-a technique for speeding up dynamic graph algorithms. J. ACM 44(5), 669–696 (1997)
12. Eppstein, D., Galil, Z., Italiano, G.F., Spencer, T.H.: Separator-based sparsification II: edge and vertex connectivity. SIAM J. Comput. 28(1), 341–381 (2006)
13. Fafianie, S., Hols, E.M.C., Kratsch, S., Quyen, V.A.: Preprocessing under uncertainty: matroid intersection. In: 41st International Symposium on Mathematical Foundations of Computer Science, MFCS 2016, vol. 58, pp. 35:1–35:14 (2016)
14. Fafianie, S., Kratsch, S., Quyen, V.A.: Preprocessing under uncertainty. In: 33rd Symposium on Theoretical Aspects of Computer Science, STACS 2016, vol. 47, pp. 33:1–33:13 (2016)
15. Goranci, G., Henzinger, M., Peng, P.: The power of vertex sparsifiers in dynamic graph algorithms. In: European Symposium on Algorithms (ESA), pp. 45:1–45:14 (2017)
16. Goranci, G., Henzinger, M., Peng, P.: Dynamic effective resistances and approximate schur complement on separable graphs. In: 26th Annual European Symposium on Algorithms, ESA 2018, vol. 112, pp. 40:1–40:15 (2018)
17. Goranci, G., Henzinger, M., Saranurak, T.: Fast incremental algorithms via local sparsifiers. CoRR (2018). https://drive.google.com/file/d/1SJrbzuz_szMwsBfeBZfGUWkDEbZKAGD5/view

18. Holm, J., de Lichtenberg, K., Thorup, M.: Poly-logarithmic deterministic fully-dynamic algorithms for connectivity, minimum spanning tree, 2-edge, and biconnectivity. In: Proceedings of the thirtieth annual ACM symposium on Theory of computing, STOC 1998, pp. 79–89. ACM, New York (1998)

19. Holm, J., Rotenberg, E., Thorup, M.: Dynamic bridge-finding in $\tilde{O}(\log^2 n)$ amortized time. In: Symposium on Discrete Algorithms (SODA) (2018)

20. Holm, J., Rotenberg, E., Wulff-Nilsen, C.: Faster fully-dynamic minimum spanning forest. In: Bansal, N., Finocchi, I. (eds.) ESA 2015. LNCS, vol. 9294, pp. 742–753. Springer, Heidelberg (2015). https://doi.org/10.1007/978-3-662-48350-3_62

21. Hopcroft, J.E., Tarjan, R.E.: Dividing a graph into triconnected components. SIAM J. Comput. **2**(3), 135–158 (1973)

22. Huang, S.E., Huang, D., Kopelowitz, T., Pettie, S.: Fully dynamic connectivity in $O(\log n (\log \log n)^2)$ amortized expected time. In: Proceedings of the Twenty-Eighth Annual ACM-SIAM Symposium on Discrete Algorithms, pp. 510–520 (2017)

23. Kapron, B., King, V., Mountjoy, B.: Dynamic graph connectivity in polylogarithmic worst case time. In: SODA (2013)

24. Karczmarz, A., Łącki, J.: Fast and simple connectivity in graph timelines. In: Dehne, F., Sack, J.-R., Stege, U. (eds.) WADS 2015. LNCS, vol. 9214, pp. 458–469. Springer, Cham (2015). https://doi.org/10.1007/978-3-319-21840-3_38

25. Karger, D.R.: Minimum cuts in near-linear time. J. ACM **47**(1), 46–76 (2000)

26. Kopeliovich, S.: Offline solution of connectivity and 2-edge-connectivity problems for fully dynamic graphs. Master's thesis, Saint Petersburg State University (2012)

27. Kratsch, S., Wahlstrom, M.: Representative sets and irrelevant vertices: New tools for kernelization. In: Proceedings of the 2012 IEEE 53rd Annual Symposium on Foundations of Computer Science, FOCS 2012, pp. 450–459 (2012)

28. Li, H., Patterson, S., Yi, Y., Zhang, Z.: Maximizing the number of spanning trees in a connected graph. CoRR abs/1804.02785 (2018). https://arxiv.org/abs/1804.02785

29. Li, H., Zhang, Z.: Kirchhoff index as a measure of edge centrality in weighted networks: nearly linear time algorithms. In: Symposium on Discrete Algorithms (SODA), pp. 2377–2396 (2018)

30. Łącki, J., Sankowski, P.: Reachability in graph timelines. In: ITCS (2013)

31. Molina, A., Sandlund, B.: Personal communication

32. Patracscu, M., Demaine, E.D.: Lower bounds for dynamic connectivity. In: Proceedings of the thirty-sixth annual ACM symposium on Theory of computing, STOC 2004, pp. 546–553. ACM, New York (2004)

33. Peng, R., Sandlund, B., Sleator, D.D.: Offline dynamic higher connectivity. CoRR abs/1708.03812 (2017). http://arxiv.org/abs/1708.03812

34. Thorup, M.: Near-optimal fully-dynamic graph connectivity. In: Proceedings of the Thirty-Second Annual ACM Symposium on Theory of omputing, STOC 2000, pp. 343–350, ACM, New York (2000)

35. Tsin, Y.H.: Yet another optimal algorithm for 3-edge-connectivity. J. Discrete Algorithms **7**(1), 130–146 (2009)

36. Wulff-Nilsen, C.: Fully-dynamic minimum spanning forest with improved worst-case update time. In: Proceedings of the 49th Annual ACM SIGACT Symposium on Theory of Computing, STOC, pp. 1130–1143 (2017)

Zip Trees

Robert E. Tarjan[1,2], Caleb C. Levy[1,2(✉)], and Stephen Timmel[3]

[1] Princeton University, Princeton, NJ, USA
ret@cs.princeton.edu, cclevy@princeton.edu
[2] Intertrust Technologies, Sunnyvale, CA, USA
[3] Virginia Polytechnic Institute and State University, Blacksburg, VA, USA
stimmel@vt.edu

Abstract. We introduce the *zip tree*, (*Zip*: "To move very fast.") a form of randomized binary search tree that integrates previous ideas into one practical, performant, and pleasant-to-implement package. A zip tree is a binary search tree in which each node has a numeric rank and the tree is (max)-heap-ordered with respect to ranks, with ties broken in favor of smaller keys. Zip trees are essentially treaps [8], except that ranks are drawn from a geometric distribution instead of a uniform distribution, and we allow rank ties. These changes enable us to use fewer random bits per node.

We perform insertions and deletions by unmerging and merging paths (*unzipping* and *zipping*) rather than by doing rotations, which avoids some pointer changes and improves efficiency. The methods of zipping and unzipping take inspiration from previous top-down approaches to insertion and deletion by Stephenson [10], Martínez and Roura [5], and Sprugnoli [9].

From a *theoretical* standpoint, this work provides two main results. First, zip trees require only $O(\log \log n)$ bits (with high probability) to represent the largest rank in an n-node binary search tree; previous data structures require $O(\log n)$ bits for the largest rank. Second, zip trees are naturally isomorphic to skip lists [7], and simplify Dean and Jones' mapping between skip lists and binary search trees [2].

1 Introducing: Zip Trees

Preliminaries. A *binary search tree* is a binary tree in which each node contains an item, each item has a key, and the items are arranged in *symmetric order*: if x is a node, all items in the left subtree of x have keys less than that of x, and all items in the right subtree of x have keys greater than that of x. Such a tree supports binary search: to find an item in the tree with a given key, proceed as follows. If the tree is empty, stop: no item in the tree has the given key. Otherwise, compare the desired key with that of the item in the root. If they are equal, stop and return the item in the root. If the given key is less than that of the item in the root, search recursively in the left subtree of the root.

Research at Princeton University partially supported by an innovation research grant from Princeton and a gift from Microsoft.

© Springer Nature Switzerland AG 2019
Z. Friggstad et al. (Eds.): WADS 2019, LNCS 11646, pp. 566–577, 2019.
https://doi.org/10.1007/978-3-030-24766-9_41

Otherwise, search recursively in the right subtree of the root. The path of nodes visited during the search is the *search path*. If the search is unsuccessful, the search path starts at the root and ends at a missing node corresponding to an empty subtree.

To keep our presentation simple, in this and the next section we do not distinguish between an item and the node containing it. (The data structure is *endogenous* [12].) We also assume that all nodes have distinct keys. It is straightforward to eliminate these assumptions. We call a node *binary*, *unary*, or a *leaf*, if it has two, one or zero children, respectively. We define the *depth* of a node recursively to be zero if it is the root, or one plus the depth of its parent if not. We define the *height* of a node recursively to be zero if it is a leaf, or one plus the maximum of the heights of its children if not. The *left* (resp. *right*) *spine* of a tree is the path from the root to the node of smallest (resp. largest) key. The left (resp. right) spine of x contains only the root and left (resp. right) children. We represent a binary search tree by storing in each node x its left child $x.left$, its right child $x.right$, and its key, $x.key$. If x has no left (resp. right) child, $x.left = null$ (resp. $x.right = null$).

Intuition. Our goal is to obtain a type of binary search tree with small depth and small update time, one that is as simple and efficient as possible. If the number of nodes n is one less than a power of two, the binary tree of minimum depth is *perfect*: each node is either binary (with two children) or a leaf (with no children), and all leaves are at the same depth. But such trees exist only for some values of n, and updating even an almost-perfect tree (say one in which all non-binary nodes are leaves and all leaves have the same depth to within one) can require rebuilding much or all of it.

We observe, though, that in a perfect binary tree the fraction of nodes of height k is about $1/2^{k+1}$ for any non-negative integer k. Our idea is to build a good tree by assigning heights to new nodes according to the distribution in a perfect tree and inserting the nodes at the corresponding heights.

We cannot do this exactly, but we can do it to within a constant factor in expectation, by assigning each node a random rank according to the desired distribution and maintaining heap order by rank. Thus we obtain zip trees.

Definition of Zip Trees. A *zip tree* is a binary search tree in which each node has a numeric *rank* and the tree is (max)-heap-ordered with respect to ranks, with ties broken in favor of smaller keys: the parent of a node has rank greater than that of its left child and no less than that of its right child. We choose the rank of a node randomly when the node is inserted into the tree. We choose node ranks independently from a geometric distribution with mean 1: the rank of a node is non-negative integer k with probability $1/2^{k+1}$. We denote by $x.rank$ the rank of node x. We can store the rank of a node in the node or compute it as a pseudo-random function of the node (or of its key) each time it is needed. The pseudo-random function method, proposed by Aragon and Seidel [8], avoids the need to store ranks but requires a stronger independence assumption for the validity of our efficiency bounds, as we discuss in Sect. 3.

To insert a new node x into a zip tree, we search for x in the tree until reaching the node y that x will replace; namely the node y such that $y.rank \leq x.rank$, with strict inequality if $y.key < x.key$. From y, we follow the rest of the search path for x, *unzipping* it by splitting it into a path P containing each node with key less than $x.key$ and a path Q containing each node with key greater than $x.key$. Along P from top to bottom, nodes are in increasing order by key and non-increasing order by rank; along Q from top to bottom, nodes are in decreasing order by both rank and key. Unzipping preserves the left subtrees of the nodes on P and the right subtrees of the nodes on Q. We make the top node of P the left child of x and the top node of Q the right child of x. Finally, if y had a parent z before the insertion, we make x the left or right child of z depending on whether its key is less than or greater than that of z, respectively (x replaces y as a child of z); if y was the root before the insertion, we make x the root.

Deletion is the inverse of insertion. To delete a node x, we do a search to find it. Let P and Q be the right spine of the left subtree of x and the left spine of the right subtree of x. *Zip* P and Q to form a single path R by merging them from top to bottom in non-decreasing rank order, breaking a tie in favor of the smaller key. Zipping preserves the left subtrees of the nodes on P and the right subtrees of the nodes on Q. Finally, if x had a parent z before the insertion, make the top node of R (or *null* if R is empty) the left or right child of z, depending on whether the key of x is less than or greater than that of z, respectively (the top node of R replaces x as a child of z); if x was the root before the insertion, make the top node of R the root. Figure 1 demonstrates both insertion and deletion in a zip tree.

An insertion or deletion requires a search plus an unzip or zip. The time for an unzip or zip is proportional to one plus the number of nodes on the unzipped path in an insertion or one plus the number of nodes on the two zipped paths in a deletion.

2   Related Work

Zip trees closely resemble two well-known data structures: the treap of Seidel and Aragon [8] and the skip list of Pugh [7]. A *treap* is a binary search tree in which each node has a real-valued random rank (called a *priority* by Seidel and Aragon) and the nodes are max-heap ordered by rank. The ranks are chosen independently for each node from a fixed, uniform distribution over a large enough set that the probability of any rank tie is small. Insertions and deletions are done using *rotations* to restore heap order. A rotation at a node x is a local transformation that makes x the parent of its old parent while preserving symmetric order. In general a rotation changes three children. To insert a new node x in a treap, we generate a rank for x, follow the search path for x until reaching a missing node, replace the missing node by x, and rotate at x until its parent has larger rank or x is the root. To delete a node x in a treap, while x is not a leaf, we rotate at whichever of its children has higher rank (or at its only child if it has only one child). Once x is a leaf, we replace it by a missing node.

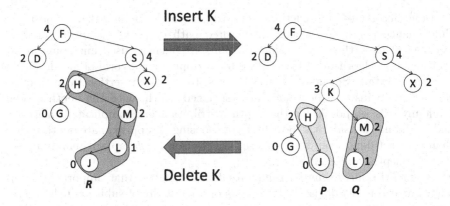

Fig. 1. Insertion and deletion of a node with key "K" assigned rank 3.

One can view a zip tree as a treap but with a different choice of ranks and with different insertion and deletion algorithms.[1] Our choice of ranks reduces the number of bits needed to represent them from $O(\log n)$ to $\lg \lg n + O(1)$ (Theorem 1), if ranks are stored rather than computed as a function of the node or its key. Treaps have the same expected depth as search trees built by uniformly random insertions, namely $2 \ln n$, about $1.39 \lg n$, as compared to $1.5 \lg n$ for zip trees. The results in Sect. 3 correspond to results for treaps. Allowing rank ties as we do thus costs about 8% in average depth (and search time) but allows much more compact representation of priorities.

A precursor of the treap is the *cartesian tree* of Jean Vuillemin [14]. This is a binary search tree built by leaf insertion (search for the item; insert it where the search leaves the bottom of the tree), with each node having a priority equal to its position in the sequence of insertions. Such a tree is min-heap ordered with respect to priorities, and its distributional properties are the same as those of a treap if items are inserted in an order corresponding to a uniformly random permutation.

Martínez and Roura [5] proposed insertion and deletion algorithms that produce trees with the same distribution as treaps. Instead of maintaining a heap order with respect to random priorities, they do insertions and deletions via random rotations that depend on subtree sizes. These sizes must be stored, at a cost of $O(\log n)$ bits per node, and they must be updated after each rotation. This suggests using their method only in an application in which subtree sizes are needed for some other purpose.

---

[1] Seidel and Aragon [8] hinted at the possibility of doing insertions and deletions by unzipping and zipping: in a footnote they say, "In practice it is preferable to approach these operations the other way around. Joins and splits of treaps can be implemented as iterative top-down procedures; insertions and deletions can then be implemented as accesses followed by splits or joins." But they provide no further details.

Doing insertions and deletions via unzipping and zipping takes at most one child change per node on the restructured path or paths, saving a constant factor of at least three over using rotations. Stephenson used unzipping in his root insertion algorithm [10]; insertion by unzipping is a hybrid of his algorithm and leaf insertion. Sprugnoli [9] was the first to propose insertion by unzipping. He used it to insert a new node at a specified depth, with the depth chosen randomly. His proposals for the depth distribution are complicated, however, and he did not consider the possibility of choosing an approximate depth rather than an exact depth. Zip trees choose the insertion *height* approximately rather than the depth, a crucial difference.

A *skip list* is an alternative randomized data structure that supports logarithmic comparison-based search. It consists of a hierarchy of sublists of the items. The level-0 list contains all the items. For $k > 0$, the level-k list is obtained by independently adding each item of the level-$(k-1)$ list with probability $1/2$ (or, more generally, some fixed probability p). Each list is in increasing order by key. A search starts in the top-level list and proceeds through the items in increasing order by key until finding the desired item, reaching an item of larger key, or reaching the end of the list. In either of the last two cases, the search backs up to the item of largest key less than the search key, descends to the copy of this item in the next lower-level list, and searches in this list in the same way. Eventually the search either finds the item or discovers that it is not in the level-0 list. To guarantee that backing up is always possible, all the lists contain a dummy item whose key is less than all others.

One can view a zip tree as a compact representation of a skip list. There is a natural isomorphism between zip trees and skip lists. (See Fig. 2.) Given a zip tree, the isomorphic skip list contains item e in the level-k sublist if and only if e has rank at least k in the zip tree. Given a skip list, the isomorphic zip tree contains item e with rank k if and only if e is in the level-k sublist but not in the level-$(k+1)$ sublist. Let e be an item in the zip tree with left and right children e' and e'', respectively. Let e, e', and e'' have ranks k, k', and k'', respectively. A search in the skip list that reaches an occurrence of e will reach it first in the level-k sublist. The next node visited during the search that is not an occurrence of e will be the occurrence of e' in the level-k' sublist or the occurrence of e'' in the level-k'' sublist, depending on whether the search key is less than or greater than the key of e. Our rule for breaking rank ties in zip trees is based on the search direction in skip lists: from smaller to larger keys.

A search in a zip tree visits the same items as the search in the isomorphic skip list, except that the latter may visit items repeatedly, at lower and lower levels. Thus a zip tree search is no slower than the isomorphic skip list search, and can be faster. The skip list has at least as many pointers as the corresponding zip tree, and its representation requires either variable-size nodes, in which each item of rank k has a node containing $k+1$ pointers; or large nodes, all of which are able to hold a number of pointers equal to the maximum rank plus one; or small nodes, one per item per level, requiring additional pointers between levels. We conclude that zip trees are at least as efficient in both time and space as skip lists.

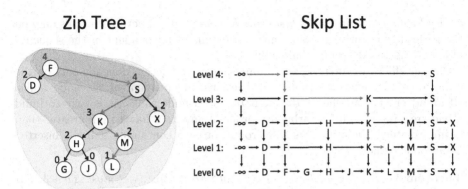

Fig. 2. Representation of the zip tree from Fig. 1 as a skip list. The level-k sublist comprises nodes of the Zip Tree of rank k or less. The search path for L is highlighted in blue in both the zip tree and the corresponding skip list.

Dean and Jones were the first to provide a mapping that converts a skip list into a binary search tree [2], but it is not the natural isomorphism given in the previous paragraph. They store ranks in the binary search tree in difference form. They map the insertion and deletion algorithms for a skip list into algorithms on the corresponding binary search tree by using rotations.

3   Properties of Zip Trees

If we ignore constant factors, the properties of zip trees are shared with treaps, skip lists, and the randomized search trees of Martínez and Roura. Because zip trees use a different rank distribution than treaps and Martínez and Roura's trees, and a different representation than skip lists, we reprove a selection of these properties. None of the proofs are difficult.

We denote by n the number of nodes in a zip tree. To simplify bounds, we assume that $n > 1$, so $\log n$ is positive. We denote by $\lg n$ the base-two logarithm. The following lemma extends a well-known result for trees symmetrically ordered by key and heap-ordered by rank [8] to allow rank ties:

Lemma 1 (From [8]). *The structure of a zip tree is uniquely determined by the keys and ranks of its nodes.*

Proof. The lemma is immediate by induction on n, since the root is the node of largest rank whose key is smallest, and the nodes in the left and right subtrees of the root are those with keys less than and greater than the key of the root, respectively. □

By Lemma 1, a zip tree is *history-independent*: its structure depends only on the nodes it currently contains (and their ranks), independent of the sequence of insertions and deletions that built it.

In our efficiency analysis we assume that each deletion depends only on the sequence of previous insertions and deletions, independent of the node ranks. (If an adversary can choose deletions based on node ranks, it is easy to build a bad tree: insert items in arbitrary order; if any item has a rank greater than 0, immediately delete it. This will produce a path containing half the inserted nodes on average.)

Theorem 1. *The expected rank of the root in a zip tree is at most* $\lg n + 3$. *For any $c > 0$, the root rank is at most $(c+1)\lg n$ with probability at most $1 - 1/n^c$.*

Proof. The root rank is the maximum of n samples of the geometric distribution with mean 1. For $c > 0$, the probability that the root rank is at least $\lg n + c$ is at most $n/2^{\lg n + c} = 1/2^c$. It follows that the expected root rank is at most $\lceil \lg n \rceil + \sum_{i=1}^{\infty} i/2^i \le \lceil \lg n \rceil + 2 \le \lg n + 3$. For $c > 0$, the probability that the root rank exceeds $(c+1)\lg n$ is at most $1/2^{c \lg n} = 1/n^c$. □

Let x be a node in a zip tree. If y is on the search path for x then y is an *ancestor* of x and x is a *descendant* of y. The *low* (respectively *high*) ancestors of x are the ancestors of x with key less than (respectively greater than) that of x.

Lemma 2. *The expected number of low ancestors of x of rank at most k is at most k. For any $\delta > 0$, this number is at most $(1 + \delta)k$ with probability at least $1 - e^{-\frac{\delta^2 k}{2+\delta}}$.*

Proof. If we order the low ancestors of x in increasing order by key, they are in non-increasing order by rank. We can think of these ancestors and their ranks as being generated by coin flips in the following way. At each successive node y less than x in decreasing key order we flip a fair coin until it comes up tails and give y a rank equal to the number of heads. Given such a y, let z be the low ancestor of smallest key greater than that of y if there is such a low ancestor; otherwise, let $z = y$. Then y is a low ancestor of x if and only if its rank is at least the rank of z. We call the first $z.rank$ coin flips at y *irrelevant* and the rest *relevant*.

Node y is a low ancestor of x if and only if at least one flip at y is relevant. The relevant flips are a sequence of Bernoulli trials in which the number of tails is the number of the number of low ancestors of x produced so far and the number of heads is at most the rank of the low ancestor of x of highest rank produced so far. Thus, the number of low ancestors of x of rank at most k is the number of tails in a sequence of flips containing at most k heads. Since the expected number of tails equals the expected number of heads, the expected number of low ancestors of x of rank at most k is at most k. The second half of the lemma follows by a Chernoff bound [1]. □

Lemma 3. *The expected number of high ancestors of x of rank at most k is at most $k/2$. For any $\delta > 0$, this number is at most $(1 + \delta)k/2$ with probability at least $1 - e^{-\frac{\delta^2 k}{2(2+\delta)}}$.*

Proof. The proof is like that of Lemma 2. We think of generating the high ancestors of x and their ranks by flipping a fair coin until it comes up tails at each node y greater than x in increasing key order and giving y a rank equal to the number of heads. Given such a y, let z be the high ancestor of x of largest key smaller than that of y, or x if there is no such high ancestor. We call the first $z.rank + 1$ flips at y irrelevant and the rest relevant.

Node y is a high ancestor of x if and only if at least one flip at y is relevant. The relevant flips are a sequence of Bernoulli trials in which the number of tails is the number of high ancestors of x produced so far and the number of flips is at most the rank of the high ancestor of x of highest rank produced so far. Thus, the number of high ancestors of x of rank at most k is the number of tails in a sequence of at most k flips. This is at most $k/2$ in expectation. The second half of the lemma follows by a Chernoff bound.   □

Theorem 2. *The expected depth of a node in a zip tree is at most $(3/2) \lg n + O(1)$. For $c \geq 1$, the depth of a zip tree is $O(c \lg n)$ with probability at least $1 - 1/n^c$, where the constant inside the big "O" is independent of n and c.*

Proof. The expected rank of the root is at most $\lg n + 3$ by Theorem 1. Adding together the bounds in Lemmas 2 and 3, the expected number of ancestors of any node x is at most $(3/2) \lg n + O(1)$. The second half of the Theorem follows from the high-probability bounds in Theorem 1 and Lemmas 2 and 3.   □

Remark 1. Rather than proceeding from scratch, one can prove Theorem 2 using results from [6].

By Theorem 2, the expected number of nodes visited during a search in a zip tree is at most $(3/2) \lg n + 2$, and the search time is $O(\log n)$ with high probability.

Theorem 3. *If x is a node of rank at most k, the expected number of nodes on the path that is unzipped during its insertion, and on the two paths that are zipped during its deletion, is at most $(3/2)k + 2$. For any $\delta >$, this number is at most $(1 + \delta)(3/2)k + 2$ with probability at least $1 - 2e^{-\frac{\delta^2 k}{2(2+\delta)}}$.*

Proof. Let x be a node of rank at most k. If x is not in the tree but is inserted, the nodes on the path unzipped during its insertion are exactly those on the two paths that would be zipped during its deletion. Thus, we need only consider deletion. Let P and Q be the two paths zipped during the deletion of x, with P containing the nodes of smaller key and Q containing the nodes of larger key. Let y and z be the predecessor and successor of x in key order, respectively. Then the nodes on P are y and the low ancestors of y of rank less than $x.rank$, and the nodes on Q are z and the high ancestors of z of rank at most $x.rank$. The theorem follows from Lemmas 2 and 3.   □

Theorem 4. *The expected number of pointer changes during an unzip or zip is* $O(1)$*. The probability that an unzip or zip changes more than* $O(1)$ *pointers is at most* $1/c^k$ *for some* $c > 1$*.*

Proof. The expected number of pointer changes is at most one plus the number of nodes on the unzipped path during an insertion or the two zipped paths during a deletion. For a given node x, these numbers are the same whether x is inserted or deleted. Thus we need only consider the case of deletion. The probability that x has rank k is $1/2^{k+1}$. Given that x has rank k, the expected number of nodes on the two zipped paths is at most $(3/2)k + 2$ by Theorem 3. Summing over all possible values of k gives the first half of the theorem.

By the second half of Theorem 3, there is a constant $a > 1$ such that if the rank of x is at most $k/(2a)$, then the probability that the insertion or deletion of x changes more than $k + O(1)$ pointers is at most $1/(2a^k)$. The probability that the rank of a node exceeds $k/(2a)$ is at most $1/2^{k/a+1}$. Choosing $c = \min\{a, 2^{1/a}\}$ gives the second half of the theorem. $\qquad\square$

By Theorem 4, the expected time to unzip or zip is $O(1)$, and the probability that an unzip or zip takes k steps is exponentially small in k.

In some applications of search trees, each node contains a secondary data structure, and making any change to a subtree may require rebuilding the entire subtree, in time linear in the number of nodes. The following result implies that zip trees are efficient in such applications.

Theorem 5. *The expected number of descendants of a node of rank k is at most* $3(2^k) - 1$*. The expected number of descendants of an arbitrary node is at most* $(3/2)\lg n + 3$*.*

Proof. Let x be a node of rank k. Consider the nodes with key less than that of x. Think of generating their ranks in decreasing order by key. The first such node that is not a descendant of x is the first one whose rank is at least k. The probability that a given node has rank at least k is $1/2^k$. The probability that the ith node is the first of rank at least k is $(1-1/2^k)^{i-1}/2^k$. The expected value of i is 2^k, which means that the expected number of descendants of x of smaller key is at most $2^k - 1$. (The expected value of i minus one is an overestimate because there are at most $n - 1$ nodes of key less than that of x and they may all have smaller rank.)

Similarly, among the nodes with key greater than that of x, the first one that is not a descendant of x is the first one with rank greater than k. A given node has rank greater than k with probability $p = 1/2^{k+1}$. The probability that the ith node is the first of rank greater than k is $(1-1/2^{k+1})^{i-1}/2^{k+1}$. The expected value of i is 2^{k+1}, so the expected number of descendants of x of larger key is at most $2^{k+1} - 1$.

We conclude that the expected number of descendants of x, including x itself, is at most $3(2^k) - 1$. The expected number of descendants of an arbitrary node is the sum over all k of the probability that the node has rank k times the expected number of descendants of the node given that its rank is k. Using the fact that the number of descendants is at most n, this sum is at most $(3/2)\lg n + 3$. $\qquad\square$

4   Comments and Extensions

Zip trees combine two independent ideas: the use of random ranks distributed geometrically and the use of unzipping and zipping to perform insertion and deletion. The former saves space as compared to treaps and makes zip trees isomorphic to skip lists but more efficient. In practice, allocating a byte (8 bits) per rank should suffice in practice. The latter makes updates faster as compared to using rotations. Either idea may be used separately.

As compared to other kinds of search trees with logarithmic search time, zip trees are simple and efficient: insertion and deletion can be done purely top-down, with $O(1)$ expected restructuring time and exponentially infrequent occurrences of expensive restructuring. Certain kinds of deterministic balanced search trees, in particular weak AVL trees and red-black trees achieve these bounds in the amortized sense [3], but at the cost of somewhat complicated update algorithms.

Zipping and unzipping make catenating and splitting zip trees simple. To catenate two zip trees T_1 and T_2 such that all items in T_1 have smaller keys than those in T_2, zip the right spine of T_1 and the left spine of T_2. The top node of the zipped path is the root of the new tree. To split a tree into two, one containing items with keys at most k and one containing items with keys greater than k, unzip the path from the root down to the node x with key k, or down to a missing node if no item has key k. The roots of the two unzipped paths are the roots of the new trees.

If the rank of a node is a pseudo-random function of its key, then search and insertion can be combined into a single top-down operation that searches until reaching the desired node or the insertion position. Similarly, search and deletion can be so combined. Furthermore ranks need not be stored in nodes, but can be computed as needed. However, for our efficiency analysis to hold this approach requires the stronger independence assumption that the sequence of insertions and deletions is independent of the function generating the ranks.[2]

One more nice feature of zip trees is that deletion does not require swapping a binary node before deleting it, as in Hibbard deletion [4].

As compared to treaps, zip trees have an average height about 8% greater. By choosing the ranks using a geometric distribution with higher mean, we can reduce this discrepancy, at the cost of increasing the number of bits needed to represent the ranks. Whether this is worthwhile is a question for experimental study.

We believe that the properties of zip trees make them a good candidate for concurrent implementation. The third author developed a preliminary, lock-based implementation of concurrent zip trees in his senior thesis [13]. We plan to develop a non-blocking implementation.

---

[2] This issue is not merely theoretical. Reuse of random seeds has led to real-world "denial-of-service" attacks for a number of programming libraries. See http://ocert. org/advisories/ocert-2011-003.html.

Implementation 1: Recursive versions of insertion and deletion.

---

insert(x, $root$):

   if $root = null$ then $\{x.left \leftarrow x.right \leftarrow null;\ x.rank \leftarrow$ RandomRank; return $x\}$
   if $x.key < root.key$ then
      | if insert(x, $root.left$) $= x$ then
      | | if $x.rank < root.rank$ then $root.left \leftarrow x$
      | | else $\{root.left \leftarrow x.right;\ x.right \leftarrow root$; return $x\}$
   else
      | if insert(x, $root.right$) $= x$ then
      | | if $x.rank \leq root.rank$ then $root.right \leftarrow x$
      | | else $\{root.right \leftarrow x.left;\ x.left \leftarrow root$; return $x\}$
   return $root$

zip(x, y):

   if $x = null$ then return y
   if $y = null$ then return x
   if $x.rank < y.rank$ then $\{y.left \leftarrow$ zip(x, $y.left$); return $y\}$
   else $\{x.right \leftarrow$ zip($x.right$, y); return $x\}$

delete(x, $root$):

   if $x.key = root.key$ then return zip($root.left$, $root.right$)
   if $x.key < root.key$ then
      | if $x.key = root.left.key$ then
      | | $root.left \leftarrow$ zip($root.left.left$, $root.left.right$)
      | else delete(x, $root.left$)
   else
      | if $x.key = root.right.key$ then
      | | $root.right \leftarrow$ zip($root.right.left$, $root.right.right$)
      | else delete(x, $root.right$)
   return $root$

---

5   Implementation

In this section we present pseudocode implementing zip tree insertion and deletion. Our implementation is recursive; we provide iterative insertion methods in an extended version of this paper [11]. Our pseudocode assumes an endogenous representation (nodes are items), with each node x having a key $x.key$, a rank $x.rank$, and pointers to the left and right children $x.left$ and $x.right$ of x respectively.

Our recursive methods for insertion and deletion appear in Implementation 1. Method insert($x, root$) inserts node x into the tree with root $root$ and returns the root of the resulting tree. It requires that x not be in the initial tree. Method delete($x, root$) deletes node x from the tree with root $root$ and returns the root of the resulting tree. It requires that x be in the initial tree. Unzipping is built into the insertion method; in deletion, zipping is done by the separate method $\texttt{zip}(x, y)$, which zips the paths with top nodes x and y and returns the top node of the resulting path. It requires that all descendants of x have smaller key than all descendants of y.

Remark 2. Once the last line of $\texttt{insert}$ ("return *root*") is reached, $\texttt{insert}$ can actually return from the outermost call: all further tests will fail, and no additional assignments will be done.

Acknowledgements. We thank Dave Long for carefully reading the manuscript and offering many useful suggestions, most importantly helping us simplify the iterative insertion and deletion algorithms that appear in the extended version of this paper [11]. We thank Sebastian Wild for correcting the bound on expected node depth in treaps in Sect. 2 and for his ideas on breaking rank ties. Finally, we are grateful to Dominik Kempa for providing us with C++ zip-tree implementations, benchmarks, and general comments.

References

1. Chernoff, H.: A measure of asymptotic efficiency for tests of a hypothesis based on the sum of observations. Ann. Math. Stat. **23**(4), 493–507 (1952)
2. Dean, B.C., Jones, Z.H.: Exploring the duality between skip lists and binary search trees. In: Proceedings of the 45th Annual Southeast Regional Conference, pp. 395–400. ACM Press (2007)
3. Haeupler, B., Sen, S., Tarjan, R.E.: Rank-balanced trees. ACM Trans. Algorithms **11**(4), 1–26 (2015)
4. Hibbard, T.N.: Some combinatorial properties of certain trees with applications to searching and sorting. J. ACM **9**(1), 13–28 (1962)
5. Martínez, C., Roura, S.: Randomized binary search trees. J. ACM **45**(2), 288–323 (1998)
6. Prodinger, H.: Combinatorics of geometrically distributed random variables: left-to-right maxima. Discrete Math. **153**(1–3), 253–270 (1996)
7. Pugh, W.: Skip lists: a probabilistic alternative to balanced trees. Commun. ACM **33**(6), 668–676 (1990)
8. Seidel, R., Aragon, C.R.: Randomized search trees. Algorithmica **16**(4–5), 464–497 (1996)
9. Sprugnoli, R.: Randomly balanced binary trees. Calcolo **17**(2), 99–117 (1980)
10. Stephenson, C.J.: A method for constructing binary search trees by making insertions at the root. Int. J. Comput. Inf. Sci. **9**(1), 15–29 (1980)
11. Tarjan, R.E., Levy, C.C., Timmel, S.: Zip trees. arXiv e-prints arXiv:1806.06726 (2018)
12. Tarjan, R.E.: Data structures and network algorithms. Society for Industrial and Applied Mathematics, Philadelphia (1983)
13. Timmel, S.: Zip trees: a new approach to concurrent binary search trees. http://arks.princeton.edu/ark:/88435/dsp01gh93h214f. Senior Thesis, Department of Mathematics, Princeton University (2017)
14. Vuillemin, J.: A unifying look at data structures. Commun. ACM **23**(4), 229–239 (1980)

Improved Algorithms for the Bichromatic Two-Center Problem for Pairs of Points

Haitao Wang[1] and Jie Xue[2($\boxtimes$)]

[1] Utah State University, Logan, UT, USA
haitao.wang@usu.edu
[2] University of Minnesota - Twin Cities, Minneapolis, MN, USA
xuexx193@umn.edu

Abstract. We consider a bichromatic two-center problem for pairs of points. Given a set S of n pairs of points in the plane, for every pair, we want to assign a red color to one point and a blue color to the other, in such a way that the value $\max\{r_1, r_2\}$ is minimized, where r_1 (resp., r_2) is the radius of the smallest enclosing disk of all red (resp., blue) points. Previously, an exact algorithm of $O(n^3 \log^2 n)$ time and a $(1+\varepsilon)$-approximate algorithm of $O(n + (1/\varepsilon)^6 \log^2(1/\varepsilon))$ time were known. In this paper, we propose a new exact algorithm of $O(n^2 \log^2 n)$ time and a new $(1 + \varepsilon)$-approximate algorithm of $O(n + (1/\varepsilon)^3 \log^2(1/\varepsilon))$ time.

1 Introduction

In this paper, we consider the following bichromatic 2-center problem for pairs of points. Given a set S of n pairs of points in the plane, for every pair, we want to assign a red color to one point and a blue color to the other, in such a way that the value $\max\{r_1, r_2\}$ is minimized, where r_1 (resp., r_2) is the radius of the smallest enclosing disk of all red (resp., blue) points.

Previously, Arkin et al. [2] proposed an $O(n^3 \log^2 n)$ time exact algorithm, as well as two $(1 + \varepsilon)$-approximate algorithms of time $O((n/\varepsilon^2) \log n \log(1/\varepsilon))$ and $O(n + (1/\varepsilon)^6 \log^2(1/\varepsilon))$, respectively. In this paper, we propose a new exact algorithm of $O(n^2 \log^2 n)$ time, which is a linear factor improvement over the exact algorithm in [2]. Also, we propose a new $(1 + \varepsilon)$-approximate algorithm of $O(n + (1/\varepsilon)^3 \log^2(1/\varepsilon))$ time, shaving off three $1/\varepsilon$ factors of the second term of the previous $O(n + (1/\varepsilon)^6 \log^2(1/\varepsilon))$ time.

1.1 Related Work

Our problem may be considered as a new type of facility location problem. Facility location problems have been studied extensively in operations research,

A full version of the paper is available at [17]. The work was partially done when Jie Xue was visiting Utah State University. The research of Jie Xue is supported, in part, by a Doctoral Dissertation Fellowship from the Graduate School of the University of Minnesota.

© Springer Nature Switzerland AG 2019
Z. Friggstad et al. (Eds.): WADS 2019, LNCS 11646, pp. 578–591, 2019.
https://doi.org/10.1007/978-3-030-24766-9_42

computational geometry, and other related areas. The classical 1-center problem for a set of points in the plane, which is also the smallest enclosing disk problem, can be solved in linear time [4,7,15]. Our problem may be more closely related to the 2-center problem for a set of n points in the plane, which has attracted much attention. Hershberger and Suri [12] first solved the decision version of the problem in $O(n^2 \log n)$ time, which was later improved to $O(n^2)$ time [11]. Using this result and with parametric search technique [14], Agarwal and Sharir [1] gave an $O(n^2 \log^3 n)$ time algorithm for the planar 2-center problem. Later, Jarom-czyk and Kowaluk [13] proposed an $O(n^2)$ time algorithm. A breakthrough was achieved by Sharir [16], who gave the first-known subquadratic algorithm for the problem, and the running time is $O(n \log^9 n)$. Afterwards, based on Sharir's algorithm scheme [16], Eppstein [8] derived a randomized algorithm with $O(n \log^2 n)$ expected time, and then Chan [3] developed an $O(n \log^2 n \log^2 \log n)$ time deterministic algorithm.

As discussed in [2], in addition to a natural variant of the planar 2-center problem, the bichromatic 2-center problem is motivated by a chromatic cluster-ing problem arising in certain applications in biology, e.g., [6], as well as in trans-portation. For example, suppose we have a set of origin/destination pairs. We want to find two centers to build airports, such that for each origin/destination pair, we can travel from the origin to the destination by first driving to the closer airport, and then flying to the other airport, and finally driving to the destination. If the goal is to minimize the maximum of the driving time, then the problem is exactly an instance of our bichromatic 2-center problem.

The distance in our bichromatic 2-center problem is measured in the Euclidean metric. Arkin et al. [2] also considered the same problem in the L_∞ metric, which is much easier and is solvable in $O(n)$ time. In addition, instead of minimizing the maximum radius of the two smallest enclosing disks for red and blue points, Arkin et al. [2] studied the problem of minimizing the sum of the radii of the two smallest enclosing disks. They gave an $O(n^4 \log^2 n)$ time exact algorithm for this min-sum problem in the Euclidean metric, along with two $(1 + \varepsilon)$-approximate algorithms, and an $O(n \log^2 n)$ time (deterministic) algo-rithm and an $O(n \log n)$ time randomized algorithm for the same problem in the L_∞ metric. Refer to [2] for some other variants of the problem.

Outline. In Sect. 2, we introduce some notation. We present our exact and approximation algorithms in Sects. 3 and 4, respectively. Due to the page limit, many proofs are omitted but can be found in the full paper [17].

2 Preliminaries

Let r^* denote the radius of the larger disk in an optimal solution for our bichro-matic 2-center problem. Note that there exists an optimal solution consisting of two congruent disks of radius equal to r^*. We use OPT to denote such an optimal solution in which the distance between the centers of the two disks is minimized. Let D_1^* and D_2^* be the two disks in OPT.

We say that two disks *bichromatically cover* S if it is possible to assign a point a red color and the other a blue color for every pair of S such that one disk covers all red points and the other covers all blue points. To solve our bichromatic 2-center problem, it is sufficient to find two congruent disks of smallest radius that bichromatically cover S.

For a subset S' of S, we denote by $P(S')$ the set of points in all pairs of S'. For a connected region B in the plane, let ∂B denote the boundary of B.

For any point c in the plane and a value r, let $D_r(c)$ denote the disk centered at c with radius r. For a set A of points in the plane, define $\mathcal{I}_r(A) = \bigcap_{c \in A} D_r(c)$, i.e., the common intersection of the disks $D_r(c)$ for all points $c \in A$. Note that $\mathcal{I}_r(A)$ is convex and can be computed in $O(|A| \log |A|)$ time [12].

For a point pair $(p, p') \in S$ and a value r, let $U_r(p, p')$ denote the union of the two disks $D_r(p)$ and $D_r(p')$. For a subset S' of pairs of S, define $\mathcal{U}_r(S') = \bigcap_{(p,p') \in S'} U_r(p, p')$. The following lemma, given by Arkin et al. [2] (specifically, in Lemma 1), will be used later in our algorithm.

Lemma 1 (Arkin et al. [2]). *Given a subset S' of pairs of S and a point c with a value r such that $D_r(c)$ covers all points of $P(S')$, $\mathcal{U}_r(S')$ can be computed in $O(|S'| \log |S'|)$ time and the combinatorial complexity of $\mathcal{U}_r(S')$ is $O(|S'| \cdot \alpha(|S'|))$, where $\alpha(\cdot)$ is the inverse Ackermann function.*

Remark. In our algorithm, we often need to solve the following subproblem. Let S', c, and r be specified as in Lemma 1. Let A be a set of $O(n)$ points in the plane. The problem is to determine whether $\mathcal{U}_r(S') \cap \mathcal{I}_r(A)$ is empty. The problem can be solved in $O(n \log n)$ time [2] (specifically, Lemma 1), as follows. We first compute $\mathcal{U}_r(S')$ and $\mathcal{I}_r(A)$ in $O(n \log n)$ time as discussed above. Then, since $\mathcal{I}_r(A)$ is convex and $\mathcal{U}_r(S')$ is star-shaped with respect to the point c, checking whether $\mathcal{U}_r(S') \cap \mathcal{I}_r(A) = \emptyset$ can be done in additional $O(n\alpha(n))$ time by an angular sweeping around the point c (see Lemma 1 in [2] for more details). Note that we can also slightly change the algorithm to check whether the interior of $\mathcal{U}_r(S')$ intersects the interior of $\mathcal{I}_r(A)$ in the same time asymptotically as above.

3   The Exact Algorithm

Before describing our algorithm in detail, we first give an overview of our approach. To obtain the $O(n^3 \log^2 n)$ time algorithm for the problem, Arkin et al. [2] first solved in $O(n^3 \log n)$ time the decision version of the problem: Given a value r, decide whether $r \geq r^*$. Then, an easy observation is that r^* is equal to the radius of the circumcircle of two or three points of S, and thus one can easily form a set of $O(n^3)$ candidate values for r^*. Consequently, r^* can be found in the set by binary search using the decision algorithm.

We take a different approach. As our problem is closely related to the planar 2-center problem for a set of points, we follow the algorithmic scheme in [3,8,16] for the planar 2-center problem. More specifically, as in [3,8], let δ^* be the distance of the centers of the two disks D_1^* and D_2^* in OPT. We consider two cases. If $\delta^* \geq r^*$, we call it the *distant case*; otherwise, it is the *nearby case*.

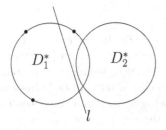

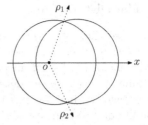

Fig. 1. Illustrating the distant case.   **Fig. 2.** Illustrating the nearby case.

In the distant case, as for the planar 2-center problem [3,8,16], we can determine a constant number of lines such that at least one line l has the following property (e.g., see Fig. 1): The subset of points of S on one side of l (say, the left side) are contained in one disk, say, D_1^*, of the optimal solution, such that the subset has a point on the boundary of D_1^* and D_1^* is the circumcircle of two or three points of S. By using this observation, we first solve the decision problem of this case in $O(n^2 \log n)$ time. Then, following a similar algorithm scheme to that in [8] and using our decision algorithm, we compute r^* in $O(n^2 \log^2 n)$ time using parametric search [5,14].

In the nearby case, as for the planar 2-center problem [3,8,16], we can determine a constant number of points such that at least one point o is contained in the intersection of D_1^* and D_2^* (e.g., see Fig. 2). In this case, we sort all points of S cyclically around o and form a matrix M of size $\Theta(n^2)$, such that r^* is the smallest element in M. The similar approach is also used in [3,8,16]. The difference, however, is that it is quite challenging to evaluate a matrix element in our problem. To this end, we first solve the decision problem in $O(n \log n)$ time and then solve the optimization problem (i.e., computing the matrix element) in $O(n \log^2 n)$ by parametric search [5,14]. Then, with help of an observation on the monotonicity properties of the matrix M, we find r^* in M in $O(n^2 \log^2 n)$ time without evaluating all elements of M (more precisely, we only need to evaluate $O(n)$ elements), by a matrix searching technique [8–10].

Given the set S, because we do not know which case happens, we will simply run our algorithms for the above two cases and then return the best solution.

Comparing with the planar 2-center problem [3,8,16], a main challenge in our problem is that we do not have an efficient data structure to dynamically compute certain values needed in the algorithm (e.g., the elements of the matrix M) in poly-logarithmic time each. Instead, in most cases we have to spend more than linear time on computing each such value. This is a main obstacle that prevents us from achieving a subquadratic time algorithm for our problem. In the next two sections, we consider the distant case and the nearby case, respectively.

3.1   The Distant Case

In this case (i.e., $\delta^* \geq r^*$), the two disks D_1^* and D_2^* in OPT are relatively far from each other, and they may intersect or not. As shown in [8], after the smallest enclosing disk of all points of $P(S)$ is obtained, which can be done in $O(n)$ time [4,7,15], we can determine in constant time a set L of $O(1)$ lines such that at least one line $l \in L$ must have the following property: The subset P_1 of the points of $P(S)$ on one particular side (e.g., the left side) of l are contained in one disk of OPT such that a point of P_1 is on the boundary of the disk and the disk is the circumcircle of two or three points of $P(S)$ (e.g., see Fig. 1).

With L, because we do not know which line of L and which side of the line has the above property, we will run the following algorithm for the subset P_1 for each side of every line of L, and finally return the best solution. In the following, we give our algorithm by assuming that we know the line l as well as the set P_1 with the property stated above.

We first consider the decision problem: Given a value r, decide whether $r \geq r^*$. The property of P_1 leads to the following observation.

Observation 1. $r \geq r^*$ *if and only if there exist two congruent disks of radius* r *bichromatically covering* S *such that one disk contains all points of* P_1 *and has one point of* P_1 *on its boundary.*

We first compute the common intersection $\mathcal{I}_r(P_1)$, which can be done in $O(n \log n)$ time as discussed in Sect. 2. Then, for each point $c \in P(S) \backslash P_1$, we compute the intersection $\partial \mathcal{I}_r(P_1) \cap \partial D_r(c)$, which consists of at most two points as argued in [12], and can be done in $O(\log n)$ time since $\mathcal{I}_r(P_1)$ is convex [12]. We sort these intersection points and the vertices of $\mathcal{I}_r(P_1)$, along $\partial \mathcal{I}_r(P_1)$, into a list I, which can be done in $O(n \log n)$ time since $|I| = O(n)$.

We run a scanning procedure to scan the list of I. For each point $c \in I$, we process it as follows. We place a disk of radius r centered at c, i.e., $D_r(c)$. We wish to answer the following question: Whether do there exist two congruent disks of radius r bichromatically covering S such that one of them is $D_r(c)$? This can be done in $O(n \log n)$ time, as follows.

First, in $O(n)$ time, we check whether $D_r(c)$ contains at least one point from each pair of S. If no, then the answer to the above question is negative and the processing of the point c is done (and we proceed to process the next point of I). Otherwise, we proceed as follows. Let $S(c)$ be the subset of pairs of S whose points are both covered by $D_r(c)$. Let $P(c)$ denote the subset of points of $P(S)$ not covered by $D_r(c)$. To answer the question, it is now sufficient to determine whether there exists a disk of radius r containing all points of $P(c)$ and at least one point from each pair of $S(c)$. To this end, we first compute $\mathcal{U}_r(S(c))$, which can be done in $O(n \log n)$ time by Lemma 1, since every point of $S(c)$ is covered by $D_r(c)$. Next, we compute $\mathcal{I}_r(P(c))$ in $O(n \log n)$ time. Finally, we determine whether $\mathcal{U}_r(S(c)) \cap \mathcal{I}_r(P(c))$ is empty, which can be done in $O(n \log n)$ time as remarked in Sect. 2. Note that the answer to our question is positive if and only if $\mathcal{U}_r(S(c)) \cap \mathcal{I}_r(P(c))$ is not empty.

If the answer to our question is positive, then we stop our decision algorithm with the assertion that $r \geq r^*$, in which case two congruent disks of radius r that bichromatically cover S are also obtained as implied by the above algorithm. Otherwise, we continue on the next point of I. If the answer to the question is negative for all points of I, then we stop with the assertion that $r < r^*$. Observation 1 guarantees the correctness of the algorithm.

Since $|I| = O(n)$ and processing each point of I takes $O(n \log n)$ time, the total time of the algorithm is $O(n^2 \log n)$.

With the decision algorithm, in Lemma 2 we solve the optimization problem, i.e., computing r^*, in $O(n^2 \log^2 n)$ time using parametric search [5,14]. The parametric search scheme is almost the same as that in [8] (i.e., in Sect. 4).

Lemma 2. *An optimal solution can be computed in $O(n^2 \log^2 n)$ time.*

3.2   The Nearby Case

In this case (i.e., $\delta^* < r^*$), the centers of the two disks D_1^* and D_2^* of OPT are relatively close and the two disks must intersect. As shown in [8,16], after the smallest enclosing disk of $P(S)$ is computed, we can determine in constant time a set of $O(1)$ points such that one point o must be in $D_1^* \cap D_2^*$. Because we do not know which point has the property, we will run the following algorithm for each such point as o, and then return the best solution. In the following, we assume that the point o has the property. We make o as the origin of the plane.

Note that ∂D_1^* and ∂D_2^* have exactly two intersections, and let ρ_1 and ρ_2 be the two rays through these intersections emanating from o (e.g., see Fig. 2). As argued in [3], one of the two coordinate axes must separate ρ_1 and ρ_2 since the angle between the two rays lies in $[\pi/2, 3\pi/2]$, and without loss of generality, we assume it is the x-axis. Again, because we do not know which axis separates the two rays, we will run the following algorithm once for the x-axis and once for the y-axis, and then return the best solution. In the following, we present the algorithm by assuming that it is the x-axis.

For ease of exposition, we make a general position assumption that no point of $P(S)$ has the same y-coordinate as o and no two points of $P(S)$ are collinear with o. The degenerate case can still be solved by our technique, but the discussion would be more tedious.

Let P^+ denote the subset of points of $P(S)$ above the x-axis, and P^- the subset below the x-axis. To simplify the discussion, let $|P^+| = |P^-| = n$. Let $p_1, p_2, \ldots, p_n$ be the sorted list of the points of P^+ counterclockwise around o, and $q_1, q_2, \ldots, q_n$ the sorted list of the points of P^- also counterclockwise around o (e.g., see Fig. 3). For each $i = 0, 1, \ldots, n$ and $j = 0, 1, \ldots, n$, define $L_{ij} = \{p_{i+1} \ldots, p_n, q_1, \ldots, q_j\}$ and $R_{ij} = \{q_{j+1}, \ldots, q_n, p_1, \ldots, p_i\}$. Note that if $i = n$, then $L_{ij} = \{q_1, \ldots, q_j\}$, and if $j = n$, then $R_{ij} = \{p_1, \ldots, p_i\}$. In other words, if we consider a ray emanating from o and between p_i and p_{i+1} and another ray emanating from o and between q_j and q_{j+1}, then L_{ij} (resp., R_{ij}) consisting of all points to the left (resp., right) of the two rays (e.g., see Fig. 3).

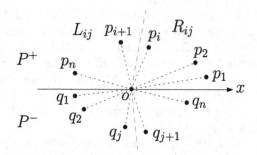

Fig. 3. Illustrating the points of P^+ and P^-.

For any pair (i, j) with $0 \leq i, j \leq n$, we consider the following *restricted bichromatic 2-center problem*. Find a pair of two congruent disks D_1 and D_2 of the smallest radius such that the following hold: (1) D_1 and D_2 bichromatically cover S; (2) D_1 covers all points of $L_{ij} \cup \{o\}$ and D_2 covers all points of $R_{ij} \cup \{o\}$. We let r_{ij}^* denote the radius of the two disks in an optimal solution. We use $RB2C(i, j)$ to refer to the problem. If a pair of disks satisfies the above two conditions, then we call them a *feasible* pair of disks for $RB2C(i, j)$.

The following lemma shows why we need to consider the problem $RB2C(i, j)$.

Lemma 3. $r^* = \min_{0 \leq i, j \leq n} r_{ij}^*$.

Define an $(n + 1) \times (n + 1)$ matrix $M[0 \dots n; 0 \dots n]$, where $M[i, j] = r_{ij}^*$ for all $0 \leq i, j \leq n$. By Lemma 3, r^* is equal to the minimum element in M. To find r^* from M, instead of computing all $(n + 1)^2$ elements of M, we will prove certain monotonicity properties of the matrix and then apply a matrix searching technique [8–10], so that it suffices to compute $O(n)$ elements of M. One of the challenges is that it is not trivial to compute even a single element of M. In the following, we first present an algorithm that can compute a single matrix element $M[i, j]$, i.e., r_{ij}^*, in $O(n \log^2 n)$ time. By using the algorithm, we describe later how to find r^* in M in $O(n^2 \log^2 n)$ time.

An Algorithm for Computing r_{ij}^*. To compute r_{ij}^* for $RB2C(i, j)$, we will resort to parametric search again. To this end, we first solve the decision problem: Given a value r, decide whether $r \geq r_{ij}^*$. We present an $O(n \log n)$ time decision algorithm for it. Let S_1 be the subset of pairs of S whose points are both in L_{ij}, and S_2 the subset of pairs of S whose points are both in R_{ij}. The following observation is self-evident.

Observation 2. $r \geq r_{ij}^*$ *if and only if there exist a pair of congruent disks of radius* r *such that one disk covers all points of* $L_{ij} \cup \{o\}$ *and at least one point from each pair of* S_2, *and the other disk covers all points of* $R_{ij} \cup \{o\}$ *and at least one point from each pair of* S_1.

Based on Observation 2, our algorithm works as follows. First, we compute the radius r_1 of the smallest enclosing disk of $L_{ij} \cup \{o\}$ and the radius r_2 of the smallest enclosing disk of $R_{ij} \cup \{o\}$. Note that Observation 2 implies that $r_1 \leq r_{ij}^*$ and $r_2 \leq r_{ij}^*$. Hence, if $r < \max\{r_1, r_2\}$, then we have $r < r_{ij}^*$, and thus we can stop the algorithm. Otherwise, we proceed as follows.

Observe that there exists a disk of radius r covering all points of $L_{ij} \cup \{o\}$ and at least one point from each pair of S_2 if and only if $\mathcal{I}_r(L_{ij} \cup \{o\}) \cap \mathcal{U}_r(S_2) \neq \emptyset$. Computing $\mathcal{I}_r(L_{ij} \cup \{o\})$ can be done in $O(n \log n)$ time. For $\mathcal{U}_r(S_2)$, notice that every point of S_2 is in the disk $D_r(c)$ because $r_2 \leq r$, where c is the center of the smallest enclosing disk of $R_{ij} \cup \{o\}$. Hence, by Lemma 1, $\mathcal{U}_r(S_2)$ can be computed in $O(n \log n)$ time. In addition, determining whether $\mathcal{I}_r(L_{ij} \cup \{o\}) \cap \mathcal{U}_r(S_2) = \emptyset$ can also be done in $O(n \log n)$ time, as remarked in Sect. 2. As such, determining whether there exists a disk of radius r covering all points of $L_{ij} \cup \{o\}$ and at least one point from each pair of S_2 can be done in $O(n \log n)$ time.

Similarly, it takes $O(n \log n)$ time to determine whether there exists a disk of radius r covering all points of $R_{ij} \cup \{o\}$ and at least one point from each pair of S_1. This solves the decision problem in $O(n \log n)$ time.

The next lemma provides a parametric search algorithm for computing r_{ij}^*.

Lemma 4. *For any pair* (i, j) *with* $0 \leq i, j \leq n$, *the value* r_{ij}^* *can be computed in* $O(n \log^2 n)$ *time.*

Searching r^* in the Matrix M. We now find r^* in M by using the algorithm in the previous subsection. The runtime of our algorithm is $O(n^2 \log^2 n)$.

To find r^* in M, a straightforward way is to compute all $(n + 1)^2$ elements of M and then return the minimum one, which would take $O(n^3 \log^2 n)$ time by Lemma 4. To reduce the time, we resort to some matrix searching techniques [8–10]. To this end, we need some sort of "stronger" solution for the problem $RB2C(i, j)$, as follows.

Consider any pair (i, j) with $0 \leq i, j \leq n$. Define S_1 to be the subset of pairs of S whose points are both in L_{ij}. Similarly, define S_2 to be the subset of pairs of S whose points are both in R_{ij}. Define D_{ij}^1 to be the smallest disk containing all points of $L_{ij} \cup \{o\}$ and at least one point of each pair of S_2, and let l_{ij} be the radius of D_{ij}^1. Similarly, define D_{ij}^2 to be the smallest disk containing all points of $R_{ij} \cup \{o\}$ and at least one point of each pair of S_1, and let r_{ij} be the radius of D_{ij}^2. We have the following observation.

Observation 3. *1.* $\max\{l_{ij}, r_{ij}\} = r_{ij}^*$.
2. If $l_{ij} < r_{ij}^*$, *then* $r_{ij} = r_{ij}^*$; *otherwise,* $l_{ij} = r_{ij}^*$.

Proof. Notice that D_{ij}^1 and D_{ij}^2 form a feasible pair of disks for the problem $RB2C(i, j)$. Therefore, $\max\{l_{ij}, r_{ij}\} \geq r_{ij}^*$ holds.

Next, we show that $\max\{l_{ij}, r_{ij}\} \leq r_{ij}^*$. Consider an optimal solution for the problem $RB2C(i, j)$, in which one disk D_1 must contain all points of $L_{ij} \cup \{o\}$ and at least one point of each pair of S_2 and the other disk D_2 must contain all points of $R_{ij} \cup \{o\}$ and at least one point of each pair of S_1, and D_1 and D_2 are congruent with radius r_{ij}^*. By the definitions of l_{ij} and r_{ij}, $l_{ij} \leq r_{ij}^*$ and $r_{ij} \leq r_{ij}^*$. Therefore, $\max\{l_{ij}, r_{ij}\} \leq r_{ij}^*$.

The above proves that $\max\{l_{ij}, r_{ij}\} = r_{ij}^*$, from which the second part of the observation easily follows. $\qquad\square$

Our algorithm for searching r^* in M needs to solve the following subproblem: decide whether $l_{ij} < r^*_{ij}$. With help of Lemma 4, we have the following result.

Lemma 5. *For any (i, j) with $0 \leq i, j \leq n$, deciding whether $l_{ij} < r^*_{ij}$ can be done in $O(n \log^2 n)$ time.*

Proof. We first compute r^*_{ij} by Lemma 4. Let $r = r^*_{ij}$, and $P_1 = L_{ij} \cup \{o\}$. We compute $\mathcal{I}_r(P_1)$ and $\mathcal{U}_r(S_2)$. $\mathcal{I}_r(P_1)$ can be computed in $O(n \log n)$ time, as discussed in Sect. 2. For $\mathcal{U}_r(S_2)$, notice that all points of S_2 are covered by the disk $D_r(c)$, where c is the center of the smallest enclosing disk of R_{ij}, since $r = r^*_{ij}$ is no smaller than the radius of the smallest enclosing disk of R_{ij}. Hence, once c is computed in $O(n)$ time [4,7,15], $\mathcal{U}_r(S_2)$ can be computed in $O(n \log n)$ time by Lemma 1. Then, observe that $l_{ij} < r$ if and only if the intersection of the interior of $\mathcal{I}_r(P_1)$ and the interior of $\mathcal{U}_r(S_2)$ is not empty. Checking whether the interior of $\mathcal{I}_r(P_1)$ intersects the interior of $\mathcal{U}_r(S_2)$ can be done in additional $O(n\alpha(n))$ time as remarked in Sect. 2. Hence, we can determine whether $l_{ij} < r^*_{ij}$ in $O(n \log^2 n)$ time, which is dominated by the algorithm for computing r^*_{ij}. □

The following lemma provides a basis for applying a matrix searching technique [8–10] to search r^* in the matrix M.

Lemma 6. *For any $0 \leq i, j \leq n$, if $l_{ij} < r^*_{ij}$, then $r^*_{ij} \leq r^*_{i'j'}$ for any $i' \in [i, n]$ and $j' \in [0, j]$; otherwise, $r^*_{ij} \leq r^*_{i'j'}$ for any $i' \in [0, i]$ and $j' \in [j, n]$.*

Proof. If $l_{ij} < r^*_{ij}$, then $r_{ij} = r^*_{ij}$ by Observation 3. Consider any pair (i', j') with $i' \in [i, n]$ and $j' \in [0, j]$. By their definitions, $R_{ij} \subseteq R_{i'j'}$ and $L_{i'j'} \subseteq L_{ij}$. Let S'_1 be the subset of pairs of S whose points are both in L'_{ij}. Then, $S'_1 \subseteq S_1$, for $L_{i'j'} \subseteq L_{ij}$. We claim that $r_{ij} \leq r_{i'j'}$. Indeed, consider a disk D of radius $r_{i'j'}$ containing all points of $R_{i'j'}$ and at least one point for each pair of S'_1. Observe that at least one point of each pair of $S_1 \backslash S'_1$ is in $R_{i'j'}$. Because $R_{ij} \subseteq R_{i'j'}$, the disk D contains all points of R_{ij} and at least one point of S_1. Therefore, by the definition of r_{ij}, since $r_{i'j'}$ is the radius of D, $r_{ij} \leq r_{i'j'}$ holds. Because $r_{ij} = r^*_{ij}$ and $r_{i'j'} \leq r^*_{i'j'}$ (by Observation 3), we obtain that $r^*_{ij} \leq r^*_{i'j'}$.

If $l_{ij} < r^*_{ij}$ does not hold, then $l_{ij} = r^*_{ij}$ by Observation 3. This is a symmetric case to the above and by a similar proof we can show that $r^*_{ij} \leq r^*_{i'j'}$ holds for any $i' \in [0, i]$ and $j' \in [j, n]$. □

Recall that r^* is equal to the smallest element of M and each matrix element $M[i, j]$ is equal to r^*_{ij}. Lemma 6 essentially tells the following: If $l_{ij} < r^*_{ij}$ for a cell $M[i, j]$ of M, then all cells of M to the southwest of $M[i, j]$ can be pruned (i.e., they are irrelevant to finding r^*); otherwise all cells of M to the northeast of $M[i, j]$ can be pruned. This is exactly the property the matrix searching algorithm in [8] (i.e., the algorithm in Lemma 5.3, which relies on the property in Lemma 5.2 that is similar to ours and follows a similar technique as in [9,10]) relies on. By using that algorithm, we can compute r^* from M with $O(n)$ matrix cell evaluations and $O(n)$ additional time, and here each matrix cell evaluation on $M[i, j]$ is to compute r^*_{ij} and determine whether $l_{ij} < r^*_{ij}$. By Lemmas 4 and

5, each matrix cell evaluation can be done in $O(n \log^2 n)$ time, which leads to an $O(n^2 \log^2 n)$ time algorithm for finding r^* from the matrix M.

Once r^* is known, we can obtain a pair of optimal disks as follows. Assume that r^* is equal to r_{ij}^* for some i and j. We apply our decision algorithm with $r = r^*$ for the problem $RB2C(i, j)$ to obtain two congruent disks of radius r^* as the optimal solution for our original bichromatic 2-center problem on S.

Theorem 1. *The bichromatic 2-center problem on a set of n pairs of points in the plane is solvable in $O(n^2 \log^2 n)$ time.*

4   The Approximation Algorithm

In this section, we give a $(1+\varepsilon)$-approximate algorithm of $O(n+(1/\varepsilon)^3 \log^2(1/\varepsilon))$ time, improving the $O(n + (1/\varepsilon)^6 \log^2(1/\varepsilon))$-time algorithm of [2]. We assume that ε is sufficiently small.

4.1   Reducing to the IB2C Problem

The first step of our algorithm is to use a grid and identify the points in the same cell. This is similar to an idea in [2] (and is in fact a standard technique used in many other geometric approximation algorithms), and here we describe it in a self-contained way. Let $\tilde{r}$ be the radius of the minimum enclosing disk of $P(S)$. Clearly, $\tilde{r} \geq r^*$. If $\tilde{r} \geq 10r^*$, the problem is actually easy.

Lemma 7. *If $\tilde{r} \geq 10r^*$, then the bichromatic 2-center problem for S can be solved exactly in $O(n)$ time.*

So it suffices to consider the case where $\tilde{r} \in [r^*, 10r^*)$. We build a grid G consisting of square cells of side-length $\delta = \varepsilon\tilde{r}/100$. For a point $x \in \mathbb{R}^2$, we denote by $\square_x$ the cell containing x. For each pair $(a_i, a_i') \in S$, we create another point-pair (b_i, b_i') where b_i is an arbitrary vertex of $\square_{a_i}$ and b_i' is an arbitrary vertex of $\square_{a_i'}$. Let S' be the set of all these pairs excluding the duplicates, and $\mathcal{D}$ be the collection of all disks whose centers are grid points of G. Consider the bichromatic 2-center problem for S' with the solution space $\mathcal{D}$ (i.e., the disks must be chosen from $\mathcal{D}$), and let $(D_r(c_1), D_r(c_2))$ be an optimal solution of this problem consisting of two congruent disks of radius r. Set $r' = (1 + \varepsilon/3)r$.

Lemma 8. *$(D_{r'}(c_1), D_{r'}(c_2))$ is a feasible solution for the bichromatic 2-center problem for S. Furthermore, $r^* \leq r' \leq (1 + \varepsilon)r^*$.*

The above lemma reduces the approximate bichromatic 2-center problem for S to the (exact) bichromatic 2-center problem for S' with the solution space $\mathcal{D}$. To solve the latter problem, we exploit its following special properties.

- All points in $P(S')$ are grid points of G.
- All points in $P(S')$ are contained in a (orthogonal) square of side-length $(2 + 2\varepsilon)\tilde{r}$. Indeed, the diameter of $P(S)$ is at most $2\tilde{r}$ and thus the diameter of $P(S')$ is at most $(2 + 2\varepsilon)\tilde{r}$.
- The two disk-centers in a solution must be grid points of G.

By scaling, we may assume that the grid points of G are the points in $\mathbb{R}^2$ with integral coordinates (or *integral points* hereafter). We say a disk is *integral* if its center is an integral point. We then pass to the following *integral* bichromatic 2-center (IB2C) problem.

The Integral Bichromatic 2-center Problem (IB2C). Given a set T of m point-pairs each consisting of two integral points in $[U] \times [U]$ where $[U] = \{1, \ldots, U\}$, find two integral disks bichromatically covering T such that the radius of the larger one is minimized.

We have $|S'| \leq n$. Before scaling, the points in $P(S')$ are contained in a square of side-length $(2 + 2\varepsilon)\tilde{r}$ and the side-length of the cells in G is $\Theta(\varepsilon\tilde{r})$. Therefore, the original problem is reduced to the IB2C problem with $m = O(n)$ and $U = O(1/\varepsilon)$. S' can be computed in $O(n)$ time from S (using the floor function, as did in [2]). Hence, if the IB2C problem can be solved in $f(m, U)$ time, then there is an $O(n + f(n, 1/\varepsilon))$-time $(1 + \varepsilon)$-approximate bichromatic 2-center algorithm.

4.2   An IB2C Algorithm

In this section, we solve the IB2C problem in $O(m + U^3 \log^2 U)$ time, and in turn establish our approximate bichromatic 2-center algorithm of $O(n + (1/\varepsilon)^3 \log^2(1/\varepsilon))$ time. The IB2C problem itself is of independent interest, as it is a natural variant of the standard bichromatic 2-center problem.

Let T be a set of m point-pairs such that $P(T) \subseteq [U] \times [U]$, which is the input of the IB2C problem. For a point $a \in [U] \times [U]$, we define $T_a \subseteq [U] \times [U]$ to be the set consisting of all points $b \in [U] \times [U]$ such that $(a, b) \in T$. Note that $\sum_{a \in [U] \times [U]} |T_a| = O(m)$. To solve the IB2C problem, we first compute a subset $T' \subseteq T$ with the property: a pair (D_1, D_2) of disks bichromatically covers T iff it bichromatically covers T'. To this end, we observe an important fact.

Lemma 9. *Let* $(a, b_1), \ldots, (a, b_k)$ *be* k *pairs of points in* $\mathbb{R}^2$ *sharing a common point* a, *and* $b \in \mathbb{R}^2$ *be a point in the convex hull of* $b_1, \ldots, b_k$. *If* $(a, b_1), \ldots, (a, b_k)$ *are all bichromatically covered by a pair* (D_1, D_2) *of disks and* $b \in D_1 \cup D_2$, *then* (a, b) *is also bichromatically covered by* (D_1, D_2).

We construct T' as follows. For a set $Z \subseteq [U] \times [U]$ and a point $z \in Z$, we say z is a *left* (resp., *right*) *extreme point* in Z if all points in Z on the same horizontal line as z are to the right (resp., left) of z, except z itself. Note that **(1)** Z is contained in the convex hull of the left and right extreme points in Z and **(2)** the number of the left/right extreme points in Z is $O(U)$. For $a \in [U] \times [U]$, let $T'_a \subseteq T_a$ be the subset consisting of all left and right extreme points in T_a. Then we define $T' = \{(a, b) : a \in [U] \times [U], b \in T'_a\}$. We have $|T'| = O(U^3)$, as $T'_a = O(U)$ for all $a \in [U] \times [U]$. Furthermore, T' can be computed from T in $O(m + U^3)$ time. The desired property of T' follows from the above lemma.

Corollary 1. *A pair* (D_1, D_2) *of disks bichromatically covers* T *iff it bichromatically covers* T'.

Now it suffices to solve the IB2C problem for T', whose size is $O(U^3)$. To this end, we consider the configuration of an optimal solution. Let r^* be the radius of the larger disk in an optimal solution.

Lemma 10. *For all $r \geq r^*$, there exists a pair (D_1, D_2) of congruent disks of radius r bichromatically covering T' such that the centers of D_1 and D_2 are both in $[U] \times [U]$. In particular, $r^* \in \{\sqrt{0}, \sqrt{1}, \ldots, \sqrt{2(U-1)^2}\}$.*

By the above lemma, we can do binary search for r^* among the $O(U^2)$ values $\sqrt{0}, \sqrt{1}, \ldots, \sqrt{2(U-1)^2}$, and pass to the decision problem, namely, deciding whether there is a feasible solution of radius r for a given number r. In addition, according to the above lemma, when solving the decision problem, we may require the centers of the two disks to be in $[U] \times [U]$. Therefore, it suffices to solve the following decision problem.

The Decision Problem. Given a set T' of $O(U^3)$ pairs of points in $[U] \times [U]$ and a value r, decide whether there exist two points $c_1, c_2 \in [U] \times [U]$ such that $(D_r(c_1), D_r(c_2))$ bichromatically covers T'.

To solve this problem, we first establish a sufficient and necessary condition for $(D_r(c_1), D_r(c_2))$ to bichromatically cover T'.

Lemma 11. *For $c_1, c_2 \in [U] \times [U]$, $(D_r(c_1), D_r(c_2))$ bichromatically covers T' iff $c_1, c_2 \in \mathcal{U}_r(T')$ and $P(T') \subseteq D_r(c_1) \cup D_r(c_2)$.*

Using Lemma 11, we solve the decision problem in two steps. In the first step, we compute the set C of all points in $[U] \times [U]$ that lie in $\mathcal{U}_r(T')$. We call the points in C *candidate centers*. In the second step, we check if there exist two candidate centers $c_1, c_2 \in C$ such that $P(T') \subseteq D_r(c_1) \cup D_r(c_2)$. By Lemma 11, the answer of the decision problem is "yes" iff such two points exist.

The difficulty of the first step is that we are not able to compute $\mathcal{U}_r(T')$ efficiently, unless the points in $P(T')$ lie in a disk of radius r. To resolve this issue, we recall the definition of T'_a for a point $a \in [U] \times [U]$. We observe that

$$\mathcal{U}_r(T') = \bigcap_{(a,b) \in T'} D_r(a) \cup D_r(b) = \bigcap_{a \in P(T')} D_r(a) \cup \mathcal{I}_r(T'_a).$$

Therefore, a point is in $\mathcal{U}_r(T')$ iff it is in $D_r(a) \cup \mathcal{I}_r(T'_a)$ for all $a \in P(T')$. Note that we can compute $\mathcal{I}_r(T'_a)$ for all $a \in P(T')$ in $O(U^3 \log U)$ time because $\sum_{a \in P(T')} |T'_a| = O(U^3)$. With $\mathcal{I}_r(T'_a)$, a direct way to compute the candidate centers is to check for every $c \in [U] \times [U]$ whether $c \in D_r(a) \cup \mathcal{I}_r(T'_a)$ for all $a \in P(T')$. However, this requires $\Omega(U^4)$ time since $|P(T')| = \Omega(U^2)$ in the worst case. In order to do it more efficiently, our idea is to compute the candidate centers in a row simultaneously. Formally, let $R_j = [U] \times \{j\}$ be the set of the points in the j-th row of $[U] \times [U]$, for $j \in [U]$. We want to find the candidate centers in R_j. For $a \in P(T')$, let I_a be the intersection of $D_r(a) \cup \mathcal{I}_r(T'_a)$ and the horizontal line $y = j$. We define the *depth* of a point in R_j as the number of I_a's containing it. A point in R_j is a candidate center iff it is contained in I_a for all $a \in P(T')$, or equivalently, its depth is $|P(T')|$. We find the candidate centers in R_j by computing the depths of these points as follows. Since $D_r(a)$ and $\mathcal{I}_r(T'_a)$ are both convex, each I_a is either an interval or a double-interval (i.e., the union of two disjoint intervals). Let E be the set of the endpoints of these intervals and double-intervals. Note that $|E| \leq 4|P(T')| = O(U^2)$, as an interval has two endpoints and a double-interval has four. We sort the points in $E \cup R_j$ from left to right in $O(U^2 \log U)$ time, and scan these points in this order. In this procedure, we maintain a number dep which is the depth of the current point. Initially, we set $dep = 0$. At every time we hit a left (resp., right) endpoint in E, we increase (resp., decrease) dep by 1. When we hit a point in R_j, its depth is just the current value of dep. In this way, we compute the depths of the points in R_j in $O(U^2 \log U)$, and find the candidate centers in R_j. After doing this for all $j \in [U]$, we obtain the set C of all candidate centers, which takes $O(U^3 \log U)$ time in total.

With C, we proceed to the second step, namely, finding two candidate centers $c_1, c_2 \in C$ such that $P(T') \subseteq D_r(c_1) \cup D_r(c_2)$, or deciding the nonexistence of them. To this end, we first build a set of data structures $\mathcal{E}_1, \ldots, \mathcal{E}_U$ as follows. For each row R_j, consider the points in $C_j = C \cap R_j$. These points lie on the horizontal line $\ell_j : y = j$. The data structure $\mathcal{E}_j$ can answer 1-dimensional range-emptiness queries on C_j: given an interval I on the line ℓ_j, $\mathcal{E}_j$ can decide whether $C_j \cap I = \emptyset$ and return a point in $C_j \cap I$ if $C_j \cap I \neq \emptyset$. Such a data structure is well-known, and can be built in $O(U \log U)$ time with $O(\log U)$ query time, as $|C_j| = O(U)$. Building all $\mathcal{E}_1, \ldots, \mathcal{E}_U$ takes $O(U^2 \log U)$ time. With the data structures in hand, we solve the problem by considering the rows $R_1, \ldots, R_U$ separately. For each row R_j, we want to check whether there exist $c_1 \in C \cap R_j$ and $c_2 \in C$ such that $P(T') \subseteq D_r(c_1) \cup D_r(c_2)$. Fix $j \in [U]$. For $i \in [U]$, define $p_i \in R_j$ as the point whose coordinate is (i, j). Assume we set $c_1 = p_i$ for some $p_i \in C \cap R_j$. Then there exists $c_2 \in C$ satisfying the desired property iff $C \cap \mathcal{I}_r(P(p_i)) \neq \emptyset$ where $P(p_i) = P(T') \backslash D_r(p_i)$. Note that $C \cap \mathcal{I}_r(P(p_i)) = \bigcup_{j' \in [U]} (C_{j'} \cap I_{i,j'})$ where $I_{i,j'} = \mathcal{I}_r(P(p_i)) \cap \ell_{j'}$ is an interval on the line $\ell_{j'}$. Therefore, if we know $I_{i,j'}$ for all $j' \in [U]$, then we can use the data structures $\mathcal{E}_1, \ldots, \mathcal{E}_U$ to determine in $O(U \log U)$ time the emptiness of $C_{j'} \cap I_{i,j'}$ for all $j' \in [U]$ and hence the emptiness of $C \cap \mathcal{I}_r(P(p_i))$; furthermore, if $C \cap \mathcal{I}_r(P(p_i)) \neq \emptyset$, a point $c_2 \in C \cap \mathcal{I}_r(P(p_i))$ can be found by one of $\mathcal{E}_1, \ldots, \mathcal{E}_U$. It follows that as long as we know $I_{i,j'}$ for all $i, j' \in [U]$, we can determine in $O(U^2 \log U)$ time if there exist $c_1 \in C \cap R_j$ and $c_2 \in C$ such that $P(T') \subseteq D_r(c_1) \cup D_r(c_2)$, by enumerating all $p_i \in C \cap R_j$. The following lemma computes $I_{i,j'}$ for all $i, j' \in [U]$.

Lemma 12. *The intervals $I_{i,j'}$ for all $i, j' \in [U]$ can be computed in $O(U^2 \log U)$ time.*

By considering all $j \in [U]$, we can complete the second step in $O(U^3 \log U)$ time.

Now we see that both steps can be done in $O(U^3 \log U)$ time, which is also the time for solving the decision version of the IB2C problem. Using the decision algorithm as a sub-routine to do binary search, we can solve the IB2C problem on T' in $O(U^3 \log^2 U)$ time. Including the time for constructing T' from T, we finally obtain an IB2C algorithm with $O(m + U^3 \log^2 U)$ running time.

Theorem 2. *There exists an $O(m + U^3 \log^2 U)$-time IB2C algorithm.*

Corollary 2. *The $(1 + \varepsilon)$-approximate bichromatic 2-center problem on a set of n pairs of points in the plane is solvable in $O(n + (1/\varepsilon)^3 \log^2(1/\varepsilon))$ time.*

References

1. Agarwal, P.K., Sharir, M.: Planar geometric location problems. Algorithmica **11**, 185–195 (1994)
2. Arkin, E.M., et al.: Bichromatic 2-center of pairs of points. Comput. Geom. **48**, 94–107 (2015)
3. Chan, T.M.: More planar two-center algorithms. Comput. Geom. **13**, 189–198 (1999)
4. Chazelle, B., Matoušek, J.: On linear-time deterministic algorithms for optimization problems in fixed dimension. J. Algorithms **21**, 579–597 (1996)
5. Cole, R.: Slowing down sorting networks to obtain faster sorting algorithms. J. ACM **34**(1), 200–208 (1987)
6. Ding, H., Xu, J.: Solving the chromatic cone clustering problem via minimum spanning sphere. In: Proceedings of the 38th International Colloquium on Automata, Languages, and Programming (ICALP), pp. 773–784 (2011)
7. Dyer, M.E.: On a multidimensional search technique and its application to the Euclidean one centre problem. SIAM J. Comput. **15**(3), 725–738 (1986)
8. Eppstein, D.: Faster construction of planar two-centers. In: Proceedings of the 8th Annual ACM-SIAM Symposium on Discrete Algorithms (SODA), pp. 131–138 (1997)
9. Frederickson, G., Johnson, D.: The complexity of selection and ranking in $X + Y$ and matrices with sorted columns. J. Comput. Syst. Sci. **24**(2), 197–208 (1982)
10. Frederickson, G., Johnson, D.: Generalized selection and ranking: sorted matrices. SIAM J. Comput. **13**(1), 14–30 (1984)
11. Hershberger, J.: A faster algorithm for the two-center decision problem. Inf. Process. Lett. **1**, 23–29 (1993)
12. Hershberger, J., Suri, S.: Finding tailored partitions. J. Algorithms **3**, 431–463 (1991)
13. Jaromczyk, J., Kowaluk, M.: An efficient algorithm for the Euclidean two-center problem. In: Proceedings of the 10th Annual Symposium on Computational Geometry (SoCG), pp. 303–311 (1994)
14. Megiddo, N.: Applying parallel computation algorithms in the design of serial algorithms. J. ACM **30**(4), 852–865 (1983)
15. Megiddo, N.: Linear-time algorithms for linear programming in R^3 and related problems. SIAM J. Comput. **12**(4), 759–776 (1983)
16. Sharir, M.: A near-linear algorithm for the planar 2-center problem. Discrete Comput. Geom. **18**, 125–134 (1997)
17. Wang, H., Xue, J.: Improved algorithms for the bichromatic two-center problem for pairs of points. arXiv:1905.00157 (2019)

Author Index

Printed in the United States
By Bookmasters

Printed in the United States
By Bookmasters